Concepts and Techniques of Geographic Information Systems

PH Series in Geographic Information Science
Keith C. Clarke, Series Editor

Arnold, *Interpretation of Airphotos and Remotely Sensed Imagery*

Avery/Berlin, *Fundamentals of Remote Sensing and Airphoto Interpretation, 5/e*

Clarke, *Analytical and Computer Cartography, 2/e*

Clarke, *Getting Started with Geographic Information Systems, 3/e*

Clarke/Parks/Crane, *Geographic Information Systems and Environmental Modeling*

Foresman, *The History of Geographic Information Systems*

Heywood/Cornelius, *Geographic Information Systems*

Jensen, *Introductory Digital Image Processing: A Remote Sensing Perspective, 2/e*

Jensen, *Remote Sensing of the Environment: An Earth Resource Perspective*

Lo/Yeung, *Concepts and Techniques of Geographic Information Systems*

Peterson, *Interactive and Animated Cartography*

Slocum, *Thematic Cartography and Visualization*

Vincent, *Fundamentals of Geological and Environmental Remote Sensing*

Concepts and Techniques of Geographic Information Systems

C.P. LO
University of Georgia
Athens, Georgia, U.S.A.

ALBERT K.W. YEUNG
Ontario Police College
Aylmer, Ontario, Canada

PH Series in Geographic Information Science
Keith C. Clarke, Series Editor

PRENTICE HALL
Upper Saddle River, New Jersey 07458

Library of Congress Cataloging-in-Publication Data

Lo, C.P. (Chor Pang)
Concepts and techniques of geographic information systems / C.P. Lo and Albert K.W. Yeung.
p. cm.
Includes bibliographical references.
ISBN 0-13-080427-4
1. Geographic information systems. I. Yeung, Albert K. W.

G70.212 .L627 2002
910'285--dc21 2001050088

Executive Editor: *Dan Kaveney*
Assistant Editor: *Amanda Griffith*
Editorial Assistant: *Margaret Ziegler*
Production Editor/Composition: *Preparé, Inc.*
Marketing Manager: *Christine Henry*
Managing Editor, Audio/Video Assets: *Grace Hazeldine*
Art Director: *Jayne Conte*
Cover Designer: *Bruce Kensalaar*
Manufacturing Manager: *Trudy Pisciotti*
Assistant Manufacturing Manager: *Michael Bell*

Printed in the United States of America
10 9 8 7 6 5 4 3

ISBN 0-13-080427-4

Pearson Education Ltd., *London*
Pearson Education Australia Pty. Ltd., *Sydney*
Pearson Education *Singapore*, Pte. Ltd.
Pearson Education North Asia Ltd., *Hong Kong*
Pearson Education Canada, Inc., *Toronto*
Pearson Educacíon de Mexico, S.A. de C.V.
Pearson Education—Japan, *Tokyo*
Pearson Education Malaysia, Pte. Ltd.
Pearson Education, *Upper Saddle River, New Jersey*

To
Christine Seen-lim Lo
and
Agnes Yuk-lan Yeung
in appreciation of their support for this book project

Brief Contents

Contents

PREFACE

This book on Geographic Information Systems (GIS) provides a rigorous but balanced treatment of concepts and techniques in a single volume. GIS has evolved from a mapping and spatial analytical tool in geography to become a full-fledged professional practice. There is now a proliferation of educational programs from universities and colleges that purport to certify or accredit the proficiency of their graduates' GIS skills in the workplace.

GIS is a multidisciplinary science. GIS practitioners may be geographers, planners, surveyors, or computer scientists. Despite the diversity in approaches, GIS has a special set of skills and knowledge needed by the professionals to use GIS in all its forms and implementations. This book is an attempt to define the set of skills and concepts for GIS professionals. One of us is a GIS practitioner who has over 20 years of work experience in the academic and public sectors as well as the private consulting industry in Hong Kong and Canada. The other is an academic with over 30 years of research and teaching experience in photogrammetry, remote sensing, and GIS in universities in the United Kingdom, Hong Kong, and the United States. We put our experience to work for this book with the hope that it will be useful to both geographic information practitioners and scientists.

We approach the study of GIS from the broader context of information technology in general, and information resource management in particular, because current GIS development is driven primarily by the needs of the business and public sectors, as evidenced by large databases, client/server computing architecture, national geospatial infrastructure, and the deployment of the Internet for disseminating geographic information by business organizations and all levels of government agencies. Such a change of perspective has led to the expansion of the scope of the study of GIS, from a narrow focus on technology-pushed applications in the past, to a broad-based approach emphasizing the total integration of data, technical, and human resources within the framework of information technology today. Our book therefore includes topics that are not always adequately covered in earlier GIS textbooks, such as the principles and practice of information resource management, information system development methodology, spatial database modeling and design, as well as the implementation of GIS in the enterprise information technology environment. This book attempts to provide a complete coverage of the concepts and techniques pertaining to every stage of the systems development life cycle of GIS and its applications in various areas of spatial problem solving and decision making.

As GIS is a subject that needs to be learned hands-on, an accompanying laboratory manual has been produced. Each laboratory exercise has been designed to relate to the major concepts covered in each chapter of the book. Students can use real-world data and appropriate GIS software to perform tasks using personal computers taking a period of about two hours. The data are chosen so that they can be easily substituted. These laboratories are not designed for teaching a specific GIS software package.

The emphasis of the book is rigor and balance. Rigor demands accuracy and detail. Balance provides a broad perspective that covers applications in both the physical and social sciences. The book has been designed for use in a 15-week semester, but it may be readily adapted to a 10-week quarter. It is assumed that students have a basic high school mathematical background in geometry, algebra, and trigonometry, as well as some knowledge of computer science. A good grounding in geography (or any science related to the study of the spatial distribution of features and phenomena) is essential.

Because GIS is such a rapidly changing technology, anything written may be out of date very quickly. The book has a Web site devoted to updating the contents of this book after its publication and to providing resources for its use in the classroom. We will try our best to ensure the accuracy and timeliness of the materials covered in the book. We appreciate comments and criticisms from readers so that we can improve the books in subsequent editions.

Acknowledgments

We greatly appreciate the encouragement and useful comments made by Professor Keith C. Clarke, Series Editor, on an earlier draft of this book. We also wish to express our thanks to Messrs. Norm Trowell and Zoran Madon, geoscientists at the Ontario Geological Survey (OGS) in Sudbury, Ontario, Canada; and Dr. Brent Hall, Associate Dean, Computing of the Faculty of Environment Studies, University of Waterloo, Waterloo, Ontario, Canada, for reviewing an earlier draft and giving very useful comments. However, any shortcomings of this book are ours.

We also would like to acknowledge the following companies, departments, institutes, and individuals for permission to use their materials as illustrations in the book: 3D Nature LLC, American Society for Photogrammetry and Remote Sensing, Aristo Graphic Systeme, Arnold, The Martin D. Adamiker LLC, Professor Michael Batty, Cambridge University Press, Walter de Gruyter, Inc., Elsevier Science, ESRI (ESRI Graphical User Interface is the intellectual property of ESRI and is used herein with permission. Copyright © ESRI. All rights reserved), Geographical Survey Institute of Japan, Georgia Department of Transportation, Professor Brent Hall, Hewlett Packard, Institute of British Geographers, Intergraph Corporation, Leica Geosystems, Inc., Longman, The McGraw-Hill Companies, NASA, MicroImages, Inc., National Center for Geographic Information and Analysis (NCGIA) University of California, Santa Barbara, National Geographic Society, Natural Resources Canada, Ontario Ministry of Transportation, Oxford University Press, POITRA Visual Communication, Prentice Hall, RADARSAT International, Sage Publications, Space Imaging, Springer-Verlag, SPOT Corporation, Taylor and Francis, Trimble Navigation, Dr. Robert Twiss, John Wiley and Sons, Inc., United States Census Bureau, United States Geological Survey, University of Chicago Library, Webscape, Wisconsin State Cartographer, Wiley and Sons, and Xerox.

Special recognition is due to those who critically reviewed the book and made valuable suggestions.

Douglas Banting, *Ryerson Polytechnic University*
Keith C. Clarke, *University of California, Santa Barbara*
Allison Feeney, *Shippensburg University*
Roberto Garza, *University of Houston*
Charles Geiger, *Millersville University*
Richard Greene, *Northern Illinois University*
Trevor Harris, *West Virginia University*
James W. Merchant, *University of Nebraska*
Morton E. O'Kelly, *Ohio State University*
Diana Sinton, *Alfred University*
Daniel Sui, *Texas A & M*

C.P. Lo
Albert K.W. Yeung

CHAPTER

1

Introduction to Geographic Information Systems (GIS)

1.1 Introduction

Geographic Information Systems (GIS) are also commonly known as *Geographical Information Systems* outside North America. When these systems were first developed in the early 1960s, they were no more than a set of innovative computer-based applications for map data processing that were used in a small number of government agencies and universities only. Today, GIS has become an important field of academic study, one of the fastest growing sectors of the computer industry, and, most important, an essential component of the information technology (IT) infrastructure of modern society.

This chapter aims to provide the context for studying GIS in both the academic and professional settings. We will first define GIS by identifying the characteristics that uniquely distinguish them from other classes of information systems. We will then explain how GIS have come to the current state of development by tracing their evolution over the years. This will be followed by an overview of the working principles of GIS from the perspectives of their key components: data, technology, application, and people. Finally, we will discuss different approaches to the study of GIS and explain how the concepts and techniques presented in this book provide a common thread for all people interested in these systems.

1.2 Definitions of GIS and Related Terminology

Attempting to define GIS is not a simple task. Some people see GIS generally as a branch of information technology; others see it more specifically as a computer-assisted mapping and cartographic application, a set of spatial-analytical tools, a type of database systems, or a field of academic study. In the following sections, we will provide a definition of GIS from first principles by examining the generic meanings of "information," "system," and "information system" and explaining why the word "geographic" makes GIS a unique class of information systems.

1.2.1 A Working Definition of GIS

In order to provide a simple working definition of GIS for the benefit of beginning students, the authors prefer the following two definitions given by Rhind (1989) and the United States Geological Survey (USGS, 1997) respectively.

1. "...a system of hardware, software, and procedures designed to support the capture, management, manipulation, analysis, modeling, and display of spatially referenced data for solving complex planning and management problems"
2. "...a computer system capable of assembling, storing, manipulating, and displaying geographically referenced information, i.e., data identified according to their locations"

The ideas expressed by these two definitions are generally adopted by authors of GIS textbooks, notably Burrough (1986), Aronoff (1989), DeMers (2000), and Clarke (2001). Simply stated, GIS is a set of computer-based systems for managing geographic data and using these data to solve spatial problems. As we will understand from the following discussion, these two characteristics distinguish GIS from other types of information systems.

1.2.2 Data, Information, Knowledge, and Intelligence

In both working definitions of GIS quoted above, the word "data" has been used. It is quite common for people to use the words "information" and "data" as if they were synonyms. However, these two words actually convey two relatively distinct concepts. When the word "data" is used, it usually refers to a collection of facts or figures that pertain to places, people, things, events, and concepts. These facts and figures can generally be represented in the following basic forms: numerical values, alphanumeric characters, symbols, and signals. When data are transformed—through processes such as structuring, formatting, conversion, and modeling—to a form that is meaningful to a user, it is referred to as "information" (Figure 1.1). Information, therefore, is processed or value-added data that have certain perceived values to a user or a community of users. The transformation of data into useful information is the core function of information systems.

The concepts of data and information are closely related to the concepts of *knowledge* and *intelligence*. The transformation of data into information is an explicit act by a user to raise his/her knowledge to a level appropriate for specific decision-making purposes. When the user deploys this knowledge to perceive relationships, formulate principles, and introduce personal values and beliefs, he/she develops what is called intelligence or wisdom. Intelligence enriches our life, and the collective intelligence of humankind drives the evolution of civilization.

The objectives of collecting geographic data and converting them into useful information by means of a GIS transcend the traditional boundary of data processing and information management. Geographic information helps us better understand the world around us. It enables us to develop spatial intelligence for logical decision making.

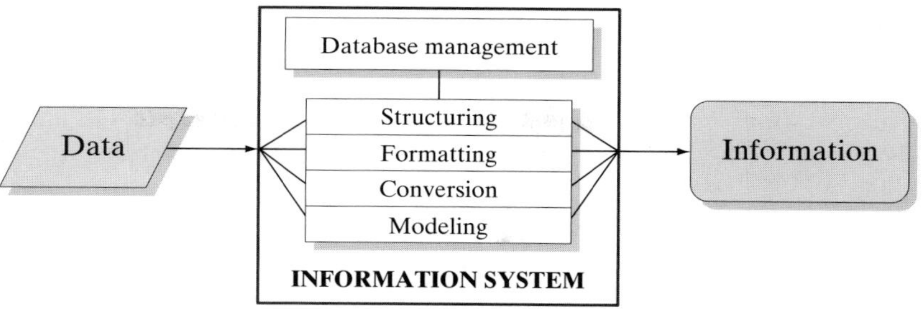

FIGURE 1.1
Transformation of data into information using an information system. Data that exist in the form in which they were collected are not usable for their intended purposes unless they are transformed into information. The primary function of an information system is to turn data into useful information.

This is the reason why any definition of GIS should include not only the data-processing functions of these systems, but also their analytical capabilities for deriving spatial knowledge and intelligence as well.

1.2.2 SYSTEMS AND INFORMATION SYSTEMS

The word "system" is used in different contexts to mean different things. It can be used to describe physical entities, such as the "solar system," the "Grand River watershed ecosystem," and the "immune system of the human body." It can also be used to describe conceptual entities, such as "democratic political system," "free market economic system," and "client/server computing system." Generally speaking, all systems, whether they are physical or conceptual, have the following characteristics:

1. They are formed or constructed to achieve certain basic objectives or functions.
2. Their continuing existence depends on the ability to satisfy the intended objectives—if this ability fails or starts to decline, the systems concerned must be upgraded or replaced.
3. An individual system is composed of many interrelated parts, which may be operational systems themselves.
4. These parts operate individually and interact with one another according to certain rules of conduct, such as procedures, laws, contractual agreements, and accepted behavior.

Information systems are a special class of systems that can be understood with respect to the above general characteristics (Figure 1.2). From the functional perspective, an information system is set up to achieve the specific objectives of collecting, storing, analyzing, and presenting information in a systematic manner. Structurally, an information system is made up of interrelated components that include a combination of data, technical, and human resources. It can also be perceived as being made up of input, processing, and output subsystems, all working according to a well-defined set of operational procedures. Finally, individual information systems can be operated independently, and at the same time linked with other information systems through standard communications protocols to form an *information system network*.

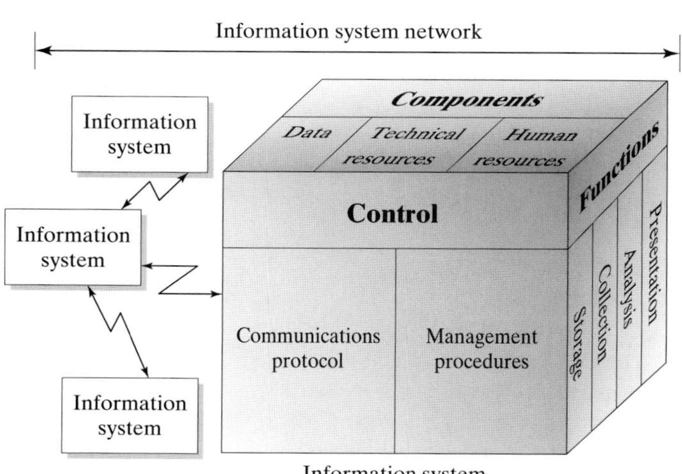

FIGURE 1.2
Different perspectives of an information system. An information system can be explained from different perspectives, such as (1) the data and technical and human resources used to form the system, (2) communications protocol and data management procedures, (3) functional subsystems, and (4) stand-alone or network mode of systems configuration.

1.2.3 GIS AS A SPECIAL CLASS OF INFORMATION SYSTEMS

GIS is a special class of information systems. By virtue of this heritage, it possesses all the characteristics of information systems noted above. The word "geographic" in GIS carries two meanings: "Earth" and "geographic space." By "Earth," it implies that all data in the system are pertinent to Earth's features and resources, including human activities based on or associated with these features and resources. By "geographic space," it means that the commonality of both the data and the problems that the systems are developed to solve is geography, i.e., location, distribution, pattern, and relationship within a specific geographical reference framework. This focus on geographic data and their applications for spatial problem solving makes GIS unique among information systems.

The relationship between GIS and different types of information systems can be identified from the typology of information systems in Figure 1.3. Depending on the nature of the data that they process, information systems in general can be classified as *spatial information systems* and *nonspatial information systems*. Nonspatial information systems are designed for processing data that are not referenced to any position in geographic space. Systems for accounting, banking, human resource management, and goods inventory are typical examples of nonspatial information systems.

Spatial information systems are those designed for processing data pertaining to real-world features or phenomena that are described in terms of locations. Contrary to the relatively common practice of equating spatial information systems to GIS, it is important to note that not every spatial information system can be regarded as a GIS. Computer-assisted drafting (CAD) and computer-assisted manufacturing (CAM) systems are typical examples of spatial information systems that are not GIS. These systems use spatial data, but they are definitely different from GIS in purpose and data-processing requirements.

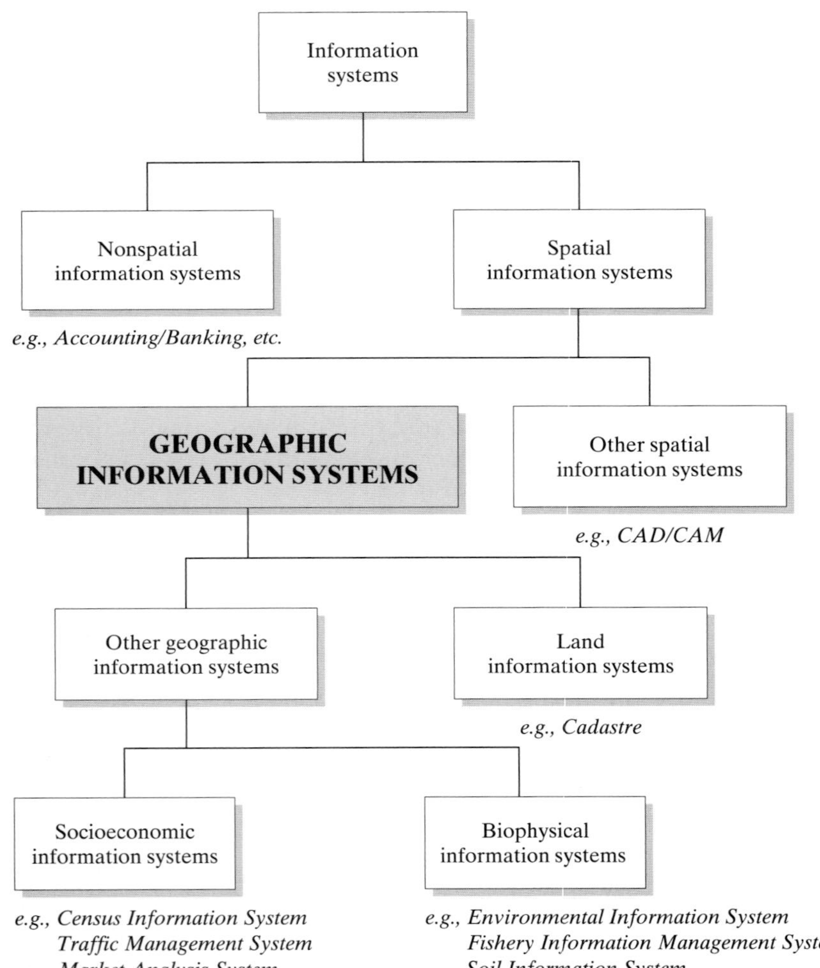

FIGURE 1.3
Typology of information systems.

Pragmatically, therefore, only those spatial information systems that are used for processing and analyzing *geographic data* (or *geographically referenced data*) can be labeled as GIS. Geographic data are a special form of spatial data that is characterized by two crucial properties.

1. the reference to *geographic space*, which means that the data are registered to an accepted geographical coordinate system of Earth's surface (see Chapter 2), so that data from different sources can be spatially cross-referenced and integrated
2. the representation at *geographic scale*, which means that the data are normally recorded at relatively small scales and, as a result, must be generalized and symbolized (see Chapter 3)

At the application level, GIS is often called by different names. For example, practitioners in land surveying, land administration, and real estate commonly use the term *land information systems* (LIS) to denote those information systems used in support of *land management* activities. Land management refers to the decision making and implementation of decisions about the land and its resources. Broadly speaking, land management includes decision making associated with property conveyance, assessment, and development; management of land and water resources, as well as the formulation of land laws and land-use policies.

Other names that are often used to denote GIS performing specific functions or covering particular geographic areas include digital topographic database, national or regional information system, natural resource information system, forest resource inventory, soil information system, environmental information system, fishery and wildlife management system, census mapping system, facility management system, business geographics, or market analysis system. Some of these systems are designed for geographically referenced biophysical data, others are used for location-based socioeconomic data. It should be noted that biophysical and socioeconomic data are not mutually exclusive to one another in GIS. By using a common geographical reference, they can be analyzed to determine the interactions between human activities and Earth's biophysical states in an integrated geographic data-processing environment.

1.2.4 GEOMATICS, GEOGRAPHIC INFORMATION SCIENCE, AND GIS

Several terms such as "geomatics," "geomatic engineering," and "geo-informatics" are now in common use pertaining to activities generally concerned with geographic information. These terms have been adopted primarily to represent the general approach that geographic information is collected, managed, and applied. Along with land surveying, photogrammetry, remote sensing, and cartography, GIS is an important component of geomatics.

When GIS became mature as a branch of information technology, the various disciplines surrounding it gradually converged to form a particular field of scientific study, leading to the proposition of the term "geographic information science" (GIScience). Goodchild (1992) defined geographic information science as the set of basic research issues raised by the handling of geographic information. These issues include, for example, the unique characteristics of geographic data, the distinct nature of research that requires geographic problem solving, the interaction between geographic information research and related academic disciplines, as well as the impacts of using geographic information on society. Geographic information science aims to provide the theoretical and organizational coherence for the scientific study of geographic information. It seeks to redefine geographic concepts and their applications in the context of GIS. Therefore, GIScience is not intended as a substitute for GIS in terminology and in practice.

The term GIS has now gained such universal acceptance that it would be difficult to replace it with other alternative names. In this book, we use GIS as the acronym of "geographic information systems" to denote both the approach and the technology pertaining to geographic information. We use the term with the understanding that GIS not only represents the skills and procedures for collecting, managing, and using geographic information, but also entails a comprehensive body of scientific knowledge from which these skills and procedures are developed. As we will demonstrate in later chapters, such a liberal view of GIS allows us to provide a balanced treatment of the subject matters between concepts and techniques throughout this book.

1.3 THE EVOLUTION OF GIS

GIS is a relatively new branch of information technology. The term GIS did not appear until the early 1960s when the Canada Geographic Information System (CGIS) was developed. During this apparently short span of history, both the technology used to construct GIS and the functions of GIS have undergone considerable changes. GIS as we understand it today is very different from its predecessors. In the following sections, we will trace the evolution of GIS from its inception in the early 1960s to the present time, as summarized in Figure 1.4, with a view to explaining how it has come to what it is today.

Stage of Development	(a) The Formative Years	(b) Maturing Technology	(c) GI Infrastructure
Time frame	1960–1980	1980–mid-1990s	Mid-1990s–present
Technical environment	• Mainframe computers • Proprietary software • Proprietary data structure • Mainly raster-based	• Mainframe and minicomputers • Geo-relational data structure • Graphical user interface • New data acquisition technologies (GPS, redefinition of datum, remote sensing)	• Workstations and PCs • Network/Internet • Open systems design • Multimedia • Data integration • Enterprise computing • Object-relational data model
Major users	• Government • Universities • Military	• Government • Universities • Utilities • Business • Military	• Government • Universities and schools • Business • Utilities • Military • The general public
Major application areas	• Land and resource management • Census • Surveying and mapping	• Land and resource management • Census • Surveying and mapping • Facilities management • Market analysis	• Land and resource management • Census • Surveying and mapping • Facilities management • Market analysis • Utilities • Geographic data browsing

FIGURE 1.4
Evolution of GIS.

1.3.1 THE FORMATIVE YEARS IN THE 1960s AND 1970s

As a computer-dependent application, the origin of GIS could be traced to research and development in electronic data processing dating back to the 1940s and 1950s. Research and development efforts in those years finally led to the successful implementation of computer-aided graphical data processing at the Massachusetts Institute of Technology (MIT) in 1955 and database management systems (DBMS) by General Electric in 1965. The idea of processing large amounts of complex data electronically was quickly picked up by many government agencies to build computer systems for geographic data handling. The CGIS, for example, was conceptualized in the early 1960s and became operational in 1971 (Tomlinson et al., 1976). The purpose of CGIS was to address the needs of land and resource information management of the federal government of Canada. This system has been generally recognized as the first GIS ever produced.

In 1973, the USGS started the development of the Geographical Information Retrieval and Analysis System (GIRAS) to handle and analyze land-use and land-cover data (Mitchell et al., 1977). There were similar systems developed in Europe. A well-known pioneering system was the Swedish Land Data Bank (SLDB) developed in the early 1970s to automate land and property registration (Andersson, 1987). In Britain, important developments included the Local Authority Management Information System (LAMIS) and the Joint Information System (JIS), both of which were used by local governments for land-use control and monitoring (Rhind, 1981).

The growing interest in computer-based map data processing also initiated several active research and development programs in universities in Europe and North America. Notable examples included, among many others (1) the Harvard Laboratory for Computer Graphics, where the arguably first contemporary vector GIS called ODESSEY was developed; (2) the Center for Urban and Regional Analysis, University of Minnesota, which housed the Minnesota Land Management Information System (MLMIS), one of the exemplary state GIS in the United States; (3) the Experimental Cartographic Unit (ECU) at the Royal College of Art in London, England, which was concerned primarily with the generation of high-quality maps using computers; and (4) the Department of Geography, University of Edinburgh, Scotland, where the widely used GIMMS (Geographic Information Mapping and Management System) package was developed.

The 1960s and 1970s thus represented the important formative years of GIS (Figure 1.4a). During these two decades, hundreds of software packages for handling and analyzing geographic information were produced (Marble, 1980). These early generations of systems were developed and used mainly by government agencies and universities, for very specific data management and research objectives. They were based on the mainframe class of computers, running in the batch mode as stand-alone software applications. In essence, all these systems were designed and implemented using proprietary software programs and data structures. Although there were several successful examples of systems using vector graphics, most early systems were constructed on the relatively simple concept of raster cells. As noted, land and resource management was the primary application area of these systems. However, there were also many other isolated attempts to apply this newly found technology to different fields of science and technology, including defense, utilities, transportation, oil exploration, urban planning, population census, as well as surveying and mapping. Map data processing was the primary focus of GIS, and spatial analysis functionality was generally rather limited. Obviously, GIS development in these formative years was application driven. This means that it was custom-built to meet the specific information needs of individual organizations. Although many of these applications were probably nowhere near what we regard as GIS today, and some were in fact never labeled as such, they laid a solid foundation for the growth of GIS into an important branch of information technology in the 1980s.

1.3.2 THE YEARS OF MATURING TECHNOLOGY FROM THE EARLY 1980S TO THE MID-1990S

The technical paper of Corbett (1979) on the concept of *topology* as applied to spatial data was a major milestone in the development of GIS concepts and techniques. In the context of geographic data structure, topology refers to the spatial relationships of *adjacency*, *connectivity*, and *containment* among topographic features. Using the concept of topology, geographic data can be stored in a simple structure that is capable of representing their attributes (i.e., what they are), locations (i.e., where they are), as well as their relationships (i.e., how they are spatially associated with one another) (see Chapter 3). The concept of topological data structure largely solved the data representation problem that hindered the development of GIS in the early years. It significantly reduced the complexity of applying geographic data for spatial analysis, thus making GIS much easier to implement and use.

In 1982, Environmental Systems Research Institute, Inc. (ESRI) released ArcInfo, based on minicomputers. This particular GIS software package was one of the first vector-based GIS to use the *georelational data model* that employed a hybrid approach to geographic data processing (Morehouse, 1989). In this approach, graphic data are stored using the topological data structure, while attribute data are stored using the relational or tabular data structure (see Section 1.4.2 and Chapter 3). Other GIS software packages developed using a similar data model in the mid-1980s included, for example, INFOMAP by Synercom in the United States and CARIS by Universal Systems Limited in Canada. The ability to port GIS applications to the microcomputer platform in the late 1980s led to the development of MapInfo, SPANS, PC-ArcInfo, and several other PC-based systems. Following its strength in the CAD area, Intergraph released its Modular GIS Environment (MGE) in 1989. This system was widely regarded as the first true CAD implementation as a graphics platform for GIS.

With the theoretical complexity of data structure largely resolved by the mid-1980s as noted above, the focus of GIS development gradually shifted to the methods of data collection, quality, and standards, as well as data analysis and database organization. This data-oriented approach of GIS development fundamentally changed the way GIS technology had been developed. It led to the integration of raster and vector geographic data, as well as the integration of geographic data with other types of business data in what is called the *enterprise computing environment*. It also nurtured the concept of open GIS that aimed to develop systems capable of sharing technical and data resources with one another.

The development of GIS was greatly accelerated by the phenomenal growth of computer technology in the 1990s. With advances in operating systems, computer graphics, database management systems, computer–human interaction, and graphical user interface (GUI) design, GIS became multiplatform applications that run on different classes of computers as stand-alone applications and as time-sharing systems. At the same time, as quantitative and analytical techniques were developed in the social and physical sciences and as more data were collected for different aspects of human activities and the environment, the needs to find suitable tools to take advantage of the new techniques and data became more pressing than ever before. The increasing access to computers and the urgent need for effective geographic data management together pushed the use of GIS to a new height. The applications of GIS were no longer limited to the traditional areas of land and resource management, but quickly extended to new areas that included facility management, vehicle navigation, market research, and decision support in business management.

While the applications of GIS were becoming increasingly diversified, they were also becoming more sophisticated. Spatial decision support was gradually replacing data management as the primary application functionality of GIS. Instead of aiming at satisfying task-specific needs and objectives, GIS applications were designed to meet the demands of corporate business goals and information requirements. GIS vendors invariably took advantage of advances in computer technology to produce software packages that allowed more user control and interactivity. Some GIS incorporated real-time data input facilities, advanced data modeling routines and scientific visualization technologies that positioned them among the most advanced and sophisticated types of computer applications.

It is generally agreed that by the mid-1990s, GIS has become relatively mature in terms of both technology and applications (Figure 1.4b). The proliferation of GIS gradually led to the formation of a specialized sector in the traditional computer industry. The GIS industry sector today includes not only companies whose core business is GIS software development. It also encompasses mainstream computer hardware and software companies, data-acquisition and supplying companies, manufacturers of data-acquisition equipment, developers of CAD systems and image-processing systems, information technology consultants and systems integrators, publishers of GIS books and journals, GIS conference organizers, as well as professional organizations to which GIS practitioners are affiliated.

1.3.3 THE AGE OF GEOGRAPHIC INFORMATION INFRASTRUCTURE STARTED IN THE MID-1990s

Since the mid-1990s, the development of GIS has entered a new era that can be aptly called the Age of Geographic Information Infrastructure (Figure 1.4c). The concept of *information infrastructure* emerged in the early 1990s when the United States government proposed the National Information Infrastructure (NII) initiative (NAE, 1994; NAPA, 1998). The objective of this initiative was to provide all U.S. citizens access to information affecting their lives that pertains to government, health care, education, and community development. As geographic information was important to well over half of all public services and economic activities in the society, it was soon recognized that this particular type of information should become a special component of the NII in its own right.

In 1994, President Clinton issued Executive Order 12906 supporting the implementation of a National Spatial Data Infrastructure (NSDI) that he defined as the "technology, policies, standards, and human resources to acquire, process, share, distribute, and improve utilization of geospatial data." Similar national geographic information infrastructure initiatives have also been established in Canada, the European Union, Australia, New Zealand, and many other countries. These initiatives have significantly raised the profile of using geographic information in government, business, industry, and academia, and by the general public.

The word "infrastructure" implies that geographic information must be treated in the same way as other economic and political infrastructures in the society, such as highways and bridges, power and communications networks, education systems, law-and-order institutions, as well as other forms of government services. The concept of geographic information infrastructure can be applied globally, where international and regional organizations are formed to deal with geographic information issues that transcend national boundaries, such as international navigation, air and maritime transportation, weather forecasting, natural hazards, and global change. At the national level, the geographic information infrastructure provides the means to reinvent government by delivering services more cost-effectively using information technology. At the same time, it allows citizens to access public information as the fundamental civil right to know. This geographic information infrastructure also enables business and industry to rejuvenate, reorganize, and reengineer to achieve a competitive advantage in the global economy. When applied at the local level, the geographic information infrastructure allows citizens to know more about the community, to participate in public affairs, to entertain, and to socialize.

Being the primary vehicle for delivering geographic information services, GIS is a critical component of geographic information infrastructure. In order to operate within the conceptual framework of geographic information infrastructure, GIS is now designed, constructed, and applied very differently from its predecessors. A typical GIS today has the following technical and operational characteristics.

1. It is part of a computer network rather than a stand-alone computer, configured using advanced processors and memory systems, and sharing input/output connectivity and data-storage facilities with different types of mainstream information technology applications.
2. It is made up of multitier, distributed software components that allow geographic information processing functions to be performed across multiple processors, computer clusters, and data-storage systems.
3. It is connected to the Internet for access to data and technical resources available on the World Wide Web (WWW), and for distributing the geographic information it contains to any interested person or organization anywhere in the world.
4. It is developed using industry standards for computer systems development methodology, software development tools, database connectivity, and communications protocols as well as user interfaces.
5. It contains a set of off-the-shelf software applications that allows it to perform common data-processing functions in input/output, database administration, spatial analysis, and generation of information products; it also provides a scripting language for users to develop task-specific applications quickly.
6. It uses data stored locally in its own hard drives and data drawn from a local server through local area networks (LAN), as well as data obtained from national, regional, and corporate data warehouses through the Internet.

7. It is capable of using raster and vector data and it can integrate geographic data with other forms of business data in a totally integrated manner.
8. It is capable of presenting the results of information retrieval and analysis using multimedia technologies that include sound, graphics, and animation.
9. It is tightly coupled with other software applications such as statistical analysis and visualization, which enhance its spatial analysis functionality, as well as spreadsheet and word processing, which are capable of using its output in other desktop applications.
10. It supports multiuser information needs ranging from sophisticated spatial problem solving and complex business decision making, to relatively simple information query and browsing.

The concept of geographic information infrastructure has brought about a dramatic philosophical and technological revolution in the development of GIS. Instead of being used simply as a set of software tools for processing and analyzing geographic data stored locally, GIS is now a gateway for accessing and integrating geographic data from different sources located locally and globally, increasingly used for interactive visualization of scenarios resulting from different business decisions, as well as for the communication of spatial knowledge and intelligence among people all over the world. GIS has popularized the use of geographic information by empowering individuals and organizations to use such information in areas that earlier generations of GIS users could never have thought of even with their wildest imagination. It is now commonplace for ordinary people to use GIS to check the weather and traffic conditions before they leave home for work, locate the nearest automated teller machine (ATM) when they need money, and find information about the country or city they are about to visit. Increasingly, business people rely on GIS to identify locations where to set up their new shops and to determine the best routes to deliver their goods and services. At the same time, GIS has become an indispensable tool for government officials to manage land and natural resources, monitor the environment, formulate economic and community development strategies, enforce law and order, and deliver social services. In view of the critical role that GIS plays in the daily life of individuals and the business operations of business and government organizations, there is definitely no better word than "infrastructure" to describe what these systems really are today.

1.4 Components of GIS

Conventionally, the construction and functions of GIS were explained by dividing them into subsystems of input, processing, analysis, and output. In this chapter, we adopt the prevailing concept that an information system is made up of four components, namely, data, technology, application, and people. We will explain how a typical GIS works from this more embracing view of information systems.

1.4.1 The Data Component of GIS

Geographic data record the locations and characteristics of natural features or human activities that occur on or near Earth's surface. Depending on their nature and use, geographic data can be categorized into three distinct types, namely, the *geodetic control network*, the *topographic base*, and the *graphical overlays*.

The geodetic control network is the foundation of all geographic data. It provides a geographical framework by which different sets of geographic data can be spatially cross-referenced with one another. Geodetic control networks are established by high-precision surveying methods and vigorous computation at the national or, as in the case of North America, continental level. The topographic base is normally created as the results of a basic mapping program by national, state/provincial, and local government mapping agencies. The contents of the topographic database can be obtained by various methods of land surveying, but generally speaking photogrammetry appears to be the method of choice. Graphical overlays are thematic data pertaining to specific GIS applications. These overlays of physical features can be derived directly from the topographic base, such as the road and drainage networks, vegetation cover, and buildings. Graphical overlays pertaining to socioeconomic activities, such as population, parcel boundaries, natural resource values, and land-use status, can be obtained only by site investigation, field surveying, remote sensing, and other forms of data collection methods.

Within a digital geographic database, all data are represented by three basic forms that include *vector*, *raster*, and *surface* (Figure 1.5). Vector data depict the real world by means of discrete points, lines, and polygons. These data are best suited for representing natural and artificial features that can be individually identified. Raster data depict the real world by means of a grid of cells with spectral or attribute values. These data are not good for representing individually identifiable features but are ideal for a variety of spatial analysis functions. Surface data depict the real world by means of a set of selected points or continuous lines of equal values. These data can be analyzed and displayed in two or three dimensions. They are most suited for natural phenomena with changing values across an extensive area.

Conventionally, geographic data have been organized as separate layers in the geographic database according to

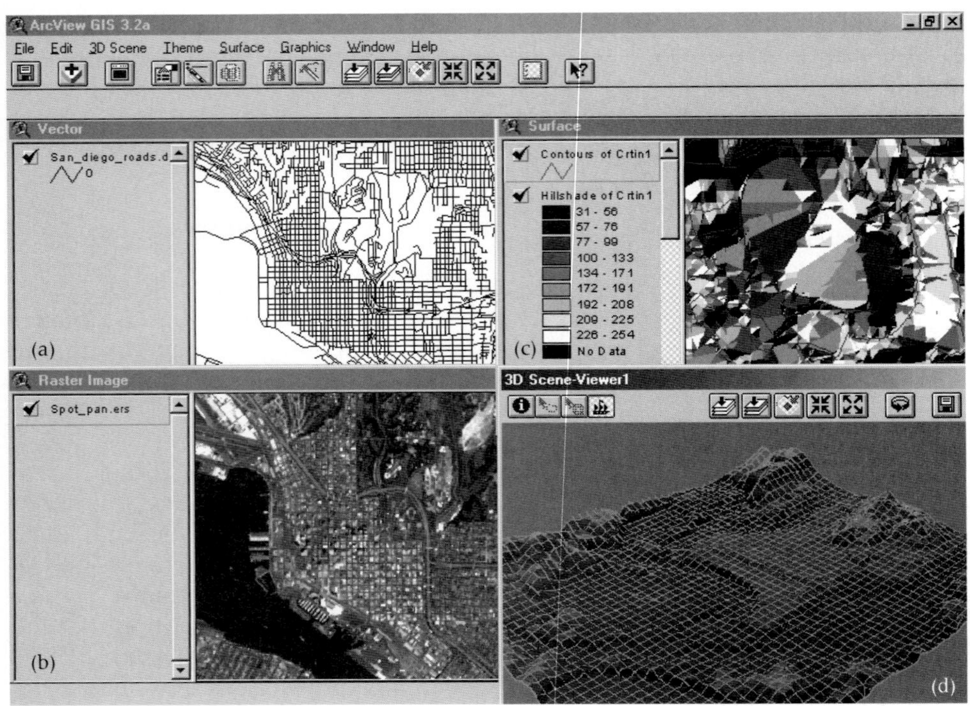

FIGURE 1.5
Basic forms of geographic data representation. Geographic data can be represented in three basic forms: (a) vector, (b) raster, and (c) surface. Surface data can be displayed in 2-D, as in (c), or in 3-D, as in (d).

the proprietary data structures of individual software vendors. The GIS software selected by an organization dictated the format and use of its geographic data. Additionally, as different methods were required to process vector, raster, and surface data, GIS in the past were developed as distinct classes of systems that could use only one of these data types. However, as the result of recent advances in computer technology and data-processing methodology, limitations associated with different data structures and segregation of vector and raster systems have been largely eliminated. In GIS implementation today, the data component is more concerned with data sharing, quality, and standards, rather than with data creation, structure, and types. As long as information about a particular geographic data set, which is called *metadata*, is available and the data conform to an accepted data collection standard, geographic data created by different systems all over the world can be as accessible as those that are created and stored locally.

1.4.2 THE TECHNOLOGY COMPONENT OF GIS

The technology component of GIS can be explained in terms of hardware and software. The hardware of GIS is made up of a configuration of core and peripheral equipment that is used for the acquisition, storage, analysis, and display of geographic information. At the heart of the GIS hardware architecture is the *central processing unit* (CPU) of the computer. The CPU performs all the data processing and analysis tasks and also controls the input/output connectivity with data acquisition, storage, and display systems. Depending on the data-processing power of the CPU, computers are classified as supercomputers, mainframes, minicomputers, workstations, and microcomputers or personal computers (PCs). All these classes of computers can be used as the hardware platforms for GIS.

Conventionally, GIS were developed as stand-alone applications that ran on one of these classes of computers. Today's GIS are mostly implemented in a network environment using the client/server model of computing (Figure 1.6).

Client/server computing is based on the concept of division of work among different machines in a local or distributed computer network. A server is the computer on which data and software are stored. A client, on the other hand, is the computer by which the users access the server. The application programs can be executed on either the server or the client computer. In the client/server environment, a client can access multiple servers, and similarly, a server can provide services to a number of clients at the same time. For GIS that are implemented on the client/server architecture, processor-intensive operations and data management are most commonly performed in the workstation class of servers, and PCs are used as the clients that provide the graphical interface to the system. Such a configuration, which combines the processing power of workstations and the economy of using PCs, has replaced the mainframes and minicomputers as the dominant hardware platforms for GIS.

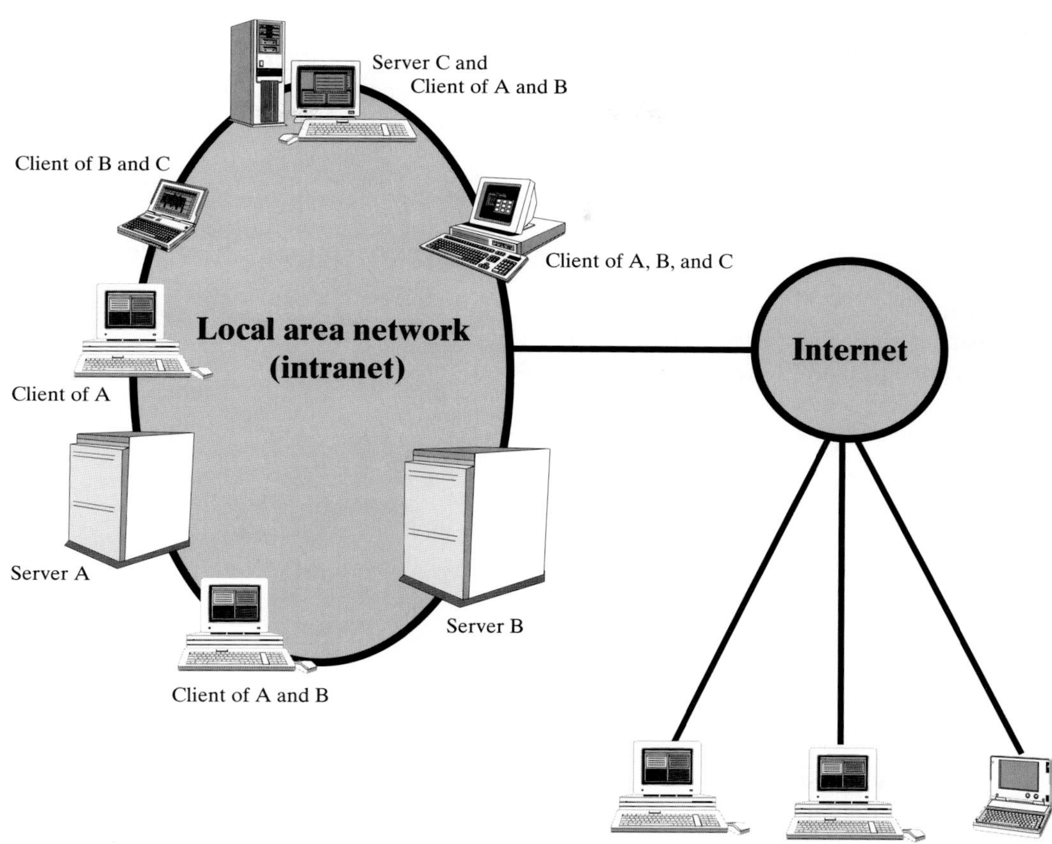

FIGURE 1.6
The client/server computing architecture. In the client/server computing architecture, a server can have many clients and a client can access multiple servers simultaneously. A computer can be used as a dedicated server or client, and it can also be used as a server and a client at the same time.

On the software side, GIS were conventionally developed using a hybrid approach that handled graphical and descriptive geographic data separately. In such an approach, the *graphical data engine*, which was usually proprietary to a GIS software vendor, handled the graphical data, while a commercial DBMS took care of the associated descriptive data. The connection between the graphical data engine and the DBMS was provided by the software vendor in the form of a proprietary interface. This form of software configuration, which is known as the geo-relational data model as noted in Section 1.3.2, was the norm for GIS implementation until the late 1990s.

Following the thrust of the computer industry toward the use of object-oriented technologies, GIS software vendors have adopted these new concepts and techniques in the design and development of their software products. The geo-relational model has been rapidly replaced by what is now known as the *object-relational* model (see Chapter 3). GIS that are constructed on this new database model allow both graphical and descriptive data to be stored in a single database. They also allow users to identify facts about the real world and transform these facts into geographic objects useful for geographic data processing and analysis. The use of object-oriented technologies has effectively transformed GIS from automated filing cabinets of maps into smart machines for geographic knowledge.

GIS software was originally developed as turnkey systems that did not allow any modification by the end users. As the concepts and techniques of systems development advanced, GIS vendors started to use the *toolbox approach* in software design and construction. Using this particular approach, GIS software packages are shipped to the user with a set of core software tools for standard geographic data processing and analysis. Users are allowed to customize their applications by using a scripting language to build software extensions to meet their specific data-processing and analysis needs. However, this approach is proprietary in nature, as the scripting language is a software product of a specific vendor and can be used only for developing applications for a particular GIS.

The current trend of GIS software development is to move away from the proprietary development environment to open industry standards. It is now possible to build application software modules with programming languages, such as Visual Basic, Visual C++, and PowerBuilder, and to integrate them with the GIS functions originally supplied by the software vendor. The concepts and techniques of using generic computer languages to build GIS applications are based on the use of *component software*. This is a software engineering methodology that has been evolving since the early 1990s. There has been considerable success in using this approach to effectively address the integration of separate computer-based applications such as document imaging, optical character recognition, database query, and fax. GIS application can obviously be benefited from this new approach to software development (Anderson et al., 1996).

1.4.3 THE APPLICATION COMPONENT OF GIS

The application components of GIS can be explained from three perspectives: areas of applications, nature of applications, and approaches of implementation. Table 1.1 summarizes the major areas of GIS application today. It is interesting to note how quickly these applications

Table 1.1
Major Application Areas of GIS

Sectors	*Application Areas*
Academic	Research in humanities, science and engineering Primary and secondary schools—school district delineation, facilities management, bus routing Spatial digital libraries
Business	Banking and insurance Real estate—development project planning and management, sales and renting services, building management Retail and market analysis Delivery of goods and services
Government	Federal government—national topographic mapping, resource and environmental management, weather services, public land management, population census, election, and voting State/provincial government—surveying and mapping, land and resource management, highway planning and management Local/municipal government—social and community development, land registration and property assessment, water and wastewater services Public safety and law enforcement—crime analysis, deployment of human resources, community policing, emergency planning and management Health care International development and humanitarian relief
Industry	Engineering—surveying and mapping, site and landscape development, pavement management Transportation—route selection for goods delivery, public transit, vehicle tracking Utilities and communications—electricity and gas distribution, pipelines, telecommunications networks Forestry—forest resource inventory, harvest planning, wildlife management and conservation Mining and mineral exploration Systems consulting and integration
Military	Training Command and control Intelligence gathering

have grown in a relatively short history of development. As noted in Section 1.3.1, when GIS was first developed, it had a relatively narrow focus on land and resource management. Today, GIS are used in all sectors of the economy and for applications pertaining to both Earth's natural environment and human activities. In fact, GIS are now in such wide use that it is hard to imagine what this world would be like without them.

As the areas of GIS application have become more diversified, the nature of GIS applications has also undergone significant changes over the years. Such changes have been particularly drastic since the mid-1990s when GIS started to be implemented within computer networks. As a result of the ability to access multiple computers simultaneously through a LAN, GIS became an integrating tool for all types of geographically referenced data housed at different locations. This prompted the development of enterprise GIS applications that aim to address the business information needs of the entire organization, rather than those of specific work groups or departments.

The advent of the Internet has fundamentally revolutionized the nature of GIS applications. The Internet began as a computer network for military communications and for the exchange of scientific information in 1969. It has now become an international computer network of networks logically consisting of millions of academic, military, government, and commercial computers in a cooperative collaboration. By using different protocols of the Internet such as the World Wide Web (WWW) and File Transfer Protocol (FTP), GIS have now become a virtual global system that offers all kinds of geographic information services via a worldwide system of computer networks (see Chapter 12 and Appendix A). At present, GIS serves not only as the means for geographic data management and spatial decision support, it also provides the mechanism for geographic information resource sharing and the communication of spatial information and knowledge as well (Harder, 1998).

1.4.4 THE PEOPLE COMPONENT OF GIS

In the past, users were seldom formally recognized as a component of GIS because the profile of the user community in the early days of GIS was extremely simple. Many users developed applications for their own use, probably with the assistance of one or more specialist computer programmers. Others used turnkey systems that were commercially obtainable from software vendors. The use of a particular GIS application was usually limited to a small number of people. There was no direct interaction between the users and the computer because data processing was done in the batch mode. The human factor in GIS was therefore largely taken for granted by all the parties concerned.

The spectacular growth of GIS that began in the early 1990s has resulted in a GIS user community that is not only very much larger in number, but also more sophisticated in requirements and more diverse in applications (Frank et al., 1996). This has in turn led to the growing awareness of and interest in the human factor in GIS applications. Because of the visual nature of GIS, most of the early studies on human factors were concerned with the cognitive characteristics of GIS and issues relating to cartographic communication, computer–human interaction, as well as graphical user interface design (Mark, 1993; Turk, 1993). More recently, increasing attention has been paid to the user aspects of the human factor in GIS development (Burrough and Frank, 1995; Nyerges, 1993). The main concern here is with the profile of the user community. Technically, this means the stratification of the users into different classes and the identification of the specific requirements of each class of users.

According to their information needs and the way they interact with the system, GIS users in general can be classified into three categories: *viewers*, *general users*, and *GIS specialists* (Figure 1.7). Viewers are the public at large whose only need is to browse a geographic database occasionally for referential information. The primary requirements for viewers are accessibility to information and ease of use of the system. Viewers in general are passive users who play no active role in the design and operation of GIS. However, since viewers probably constitute the largest class of users, their acceptance or rejection of the technology will have very significant impacts on the development of GIS as a whole.

General users are people who use GIS for conducting business, performing professional services, and making decisions. This group of users includes facility managers, resource planners, scientists, engineers, land administrators, lawyers, business entrepreneurs, and politicians. The diversity of the membership of general users means that their requirements may vary considerably among one another, ranging from relatively simple spatial queries to very complicated temporal–spatial modeling. The variation in the technical background of these users also implies that their modes of computer–human interaction are different among individual users. Unlike viewers, general users are active users because GIS is implemented to support their information needs. As a result, this group of users usually has direct and considerable influence on the successful use of GIS in an organization.

GIS specialists are people who actually make GIS work. They include GIS managers, database administrators, application specialists, systems analysts, and programmers. They are responsible for the maintenance of the geographic database and the provision of technical support to the other two classes of users. They also build applications for advanced spatial data analysis and modeling, and they produce information products according to the specifications of other users. Although GIS specialists are usually small in number, they play the most direct role in the success of GIS implementation in an organization.

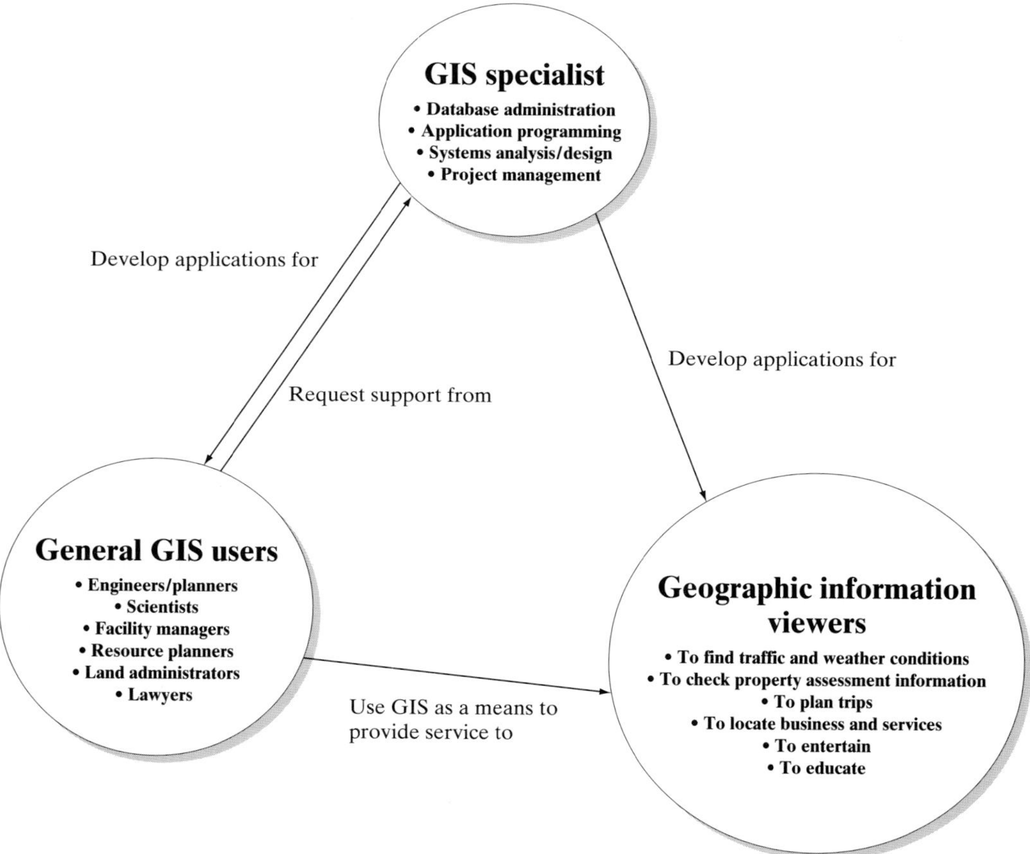

FIGURE 1.7
GIS users and their relationships.

1.5 Approaches to the Study of GIS

People now study GIS for different purposes. Many students study GIS simply as an academic pursuit; others study GIS to prepare for work as a specialist in a rapidly growing industry. There are also practicing professionals in various fields who study GIS in order to learn a new set of software tools that is increasingly used in their workplaces. In the following sections, we summarize the basic approaches to the study of GIS and explain why all people interested in these systems should have a solid understanding of the concepts and techniques presented in this book.

1.5.1 Studying GIS as a Special Field of Academic Study

Before the 1980s, GIS was a set of relatively obscure computer applications that were found only in a small number of government agencies and universities. Although articles pertaining to GIS could be found in a wide variety of learned journals in geography, cartography, computer science, and surveying, opportunities in the academic study of GIS in general were rather limited (Unwin, 1991). Today, it is hard to find a department of geography where no GIS course is offered. GIS is also widely taught in programs in Earth and environmental sciences, computer science, business administration, natural resource management, urban and regional planning, geomatics, and civil engineering.

The introduction of GIS into the academic curriculum not only means the addition of a new course of study to the degree programs, it also represents a new way of looking at geographic data. It has injected new concepts and techniques into traditional academic disciplines such as those noted here. In geography, in particular, the advent of GIS has been widely regarded as one of the most significant developments since the so-called quantitative revolution in the 1960s. Given the growing interest among geographers in GIS today, it would be hard to contest the proposition that GIS will be central to the development of geography as an academic discipline.

Until the late 1990s, GIS courses were taught mostly at the undergraduate level as a part of degree programs in geography or other disciplines. Since then, many universities started to offer specialist degree programs in GIS at both the undergraduate and graduate levels (Waters, 1999). These specialist degree courses generally have a strong academic thrust that makes them distinct from technically oriented GIS programs offered by community colleges and institutes of vocational training. Although the contents of different programs tend to vary from one another, it seems that the concepts and techniques of GIS presented in this book always constitute the fundamental component. Supplementary courses incorporated in most programs include traditional spatially oriented fields such as cartography and remote sensing, as well as those fields that GIS relies on in its development, such as computer science, information systems, statistics, mathematics, and cognitive science. These programs prepare students for an academic and research career in geographic information science as defined in Section 1.2.5 or in different disciplines in science and engineering. They also provide the basic education for students who aspire to become professional GIS practitioners in government, business, and industry.

1.5.2 Studying GIS as a Branch of Information Technology

The rapid growth of GIS as a special sector in the computer industry and the proliferation of the use of GIS in public and business sectors have created a strong demand for people with technical skills in GIS. This in turn has led to the establishment of technically oriented GIS programs in community colleges and institutes of vocational training. The focus of these GIS programs is the development of technical skills in application programming, database creation and administration, systems implementation, and user support. In other words, they aim to train people who build GIS and make sure that these systems work, i.e., GIS specialists as defined in Section 1.4.4.

A typical technical GIS program includes core courses in GIS principles and methods, as well as supporting courses in cartography, computer programming, database management, statistics and mathematics. The training on the use of GIS software packages is an essential part of these programs. It is interesting to note that many students attending these technical programs are university graduates in various disciplines who see the great opportunities and challenges as information technology (IT) workers in the information economy. Although these programs are very technical in nature, it is important that students have a solid understanding of GIS concepts and principles. This will allow them to master GIS technical skills more easily. At the same time, it will also enable them to advance to technical managerial positions, where a good understanding of GIS concepts is as important as proficiency in GIS skills.

1.5.3 Studying GIS as a Spatial Data Institution and Its Societal Implications

GIS has been developed as a technical tool for geographic information processing and analysis. Conventionally, many people perceived and used GIS simply as a special branch of information technology. As the use of GIS proliferates, its impacts on human society become very interesting and worthwhile research topics.

The history of GIS development in fact may also be examined from the perspective of the evolution of human civilization. As shown in Table 1.2, the evolution of human civilization can be roughly divided into three phases, namely, *preindustrial, industrial,* and *postindustrial.* The postindustrial society, which is also referred to as the *information society,* is the latest phase of the evolution of human civilization. In comparison with the other phases of human civilization, the information society is characterized by the domination of the service-oriented economic sectors; the emphasis of information as the basis of power and wealth; and the reliance on organized knowledge for decision making. As a result, the modern information society has created an insatiable demand for information of all kinds. At the same time, it has also generated more information than ever before in human history.

The initial serious engagements between GIS and social theory were concerned mainly with issues related to the politics of knowledge and the social impacts of its use in the modern society (Mark et al., 1997; Pickles, 1995). The early debates on issues such as the apparent lack of empiricism of GIS methodology, the acceptance of the switch from knowledge to information, and the intellectual and practical commitments of GIS have been proved to be imperative in the development of the concepts of GIS. The questions of the origin, selection, and access of data; the forms of data representation; and the economics, politics, and ethics of using geographic information have until recently been perceived largely as marginal to the more technical questions of system implementation and applications.

The interests in the role of GIS in society brought these questions to the core of GIS principles. GIS is no longer simply seen as a branch of information technology, but also as a set of institutionalized systems and practices for data management that must work within particular economic, political, cultural, and legal structures (Curry, 1995; Latour, 1993). For the first time in the history of GIS, people have come to the realization that GIS will be able to become an important technology only if it can be totally integrated into the fabric of the information society and serves its objectives well. The study of GIS, therefore, is not limited to the study of technology. It also has a very strong flavor of the humanities and social sciences as well.

Table 1.2

Comparison Between Preindustrial, Industrial and Postindustrial Societies

Characteristics \ Societies	Preindustrial (extraction-based)	Industrial (product-based)	Postindustrial (service-based)
Dominant economic sectors	Primary: agriculture, fishing, forestry, mining	Secondary: food and mineral processing, manufacturing, building, construction	Tertiary: transportation, merchandising; utilities Quaternary: finance, insurance, real estate, law Quinary: government, education, research, health, entertainment
Transforming resource	Natural power: wind, water, animals	Created energy: coal, natural gas, electricity, nuclear power	Information: computer and telecommunication systems
Basis of wealth	Land	Financial capital, labor	Information
Technology base	Craft	Machine technology	Intellectual technology
Skill base	Artisan, manual worker	Technologist, technician, semiskilled worker	Scientist, engineer, manager, other professional occupations
Decision-making methodology	Common sense, custom, tradition, trial and error	Empiricism, experimentation	Hypothesis, simulation modeling, decision theory, system analysis
Time perspective and orientation	To the past, experience, superstition	To the present, ad hoc adaptation and adoption	To the future, forecasting, planning
Organizational form	Family operation	Hierarchical corporation	Networking
Marketplace	Village square	Shopping mall	Cyberspace
Objectives of using information	Traditionalism, survival	Economic growth, economy of scale, mechanization	Codification and organization of knowledge, automation

Compiled from Bell (1973), McLaughlin (1985), and Boar (1997)

1.5.4 ORGANIZATION OF THIS BOOK

No matter what approach a student wishes to follow when studying GIS, there is a set of core concepts and techniques that he/she must fully understand and master (Figure 1.8). The primary purpose of this book is to provide students the concepts and techniques fundamental to studying GIS from different perspectives. The contents of the book have been designed with the needs of beginning students and GIS professionals in mind. Accordingly, we use the four components of GIS, data, technology, application, and people, as the framework for organizing this book. As data processing and analysis are the reason or justification for the existence of GIS, we start by giving a very detailed account of the nature, modeling, and standards of geographic data in Chapters 2, 3, and 4. Computer hardware and software as they relate to raster and vector data, which constitute the technology component, will then be discussed in Chapters 5, 6, 7, and 8. The application component is the most important, and it can be found intermingled with the technology component in Chapters 5 through 8. It is also the sole focus in Chapters 9, 10, and 11. The people component has not been given as much detailed treatment as the other three components, but some discussion can be found in Chapter 12, which

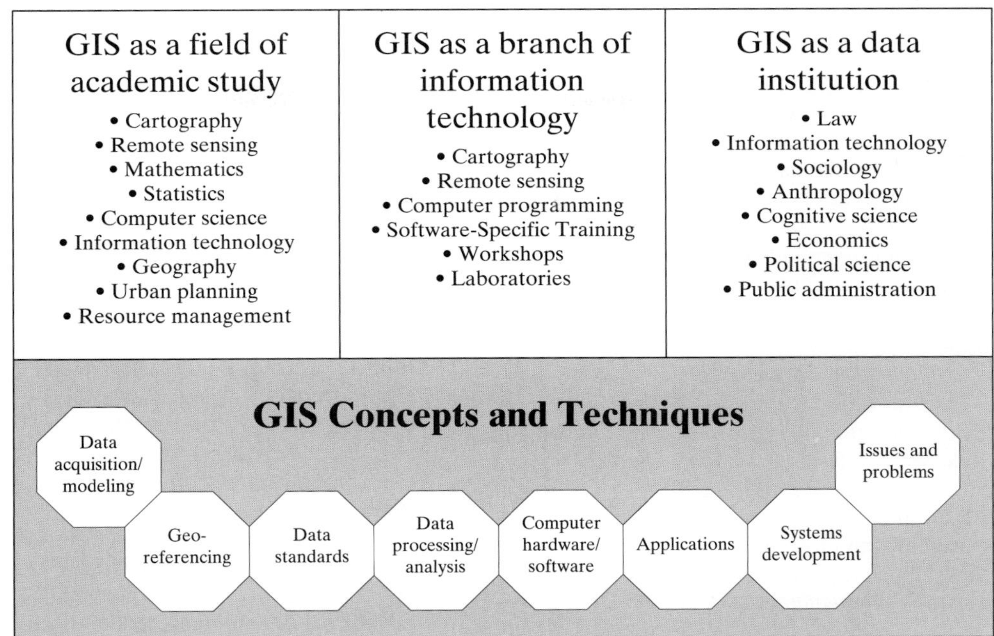

FIGURE 1.8
Core concepts and techniques as a foundation for GIS study.

also examines the problems of GIS and its trend for future development.

GIS as a special field of academic study has a strong technical flavor. Many of the concepts and techniques presented in this book can be most easily mastered by actually working on examples of using GIS. The accompanying laboratory manual has been designed and developed to reinforce what students learn from this book. It is important for students to complete the assignments associated with each of the chapters they have studied. At the same time, they should also consult this book's Web site to obtain up-to-date references and other learning resources pertaining to the themes of individual chapters.

1.6 Summary

In this chapter, we have presented a definition of GIS and described how GIS has become what it is today. We adopted a working definition of GIS as a computer-based system capable of capturing, storing, manipulating, and displaying geographically referenced data, and converting them into spatial information useful in solving complex spatial problems. The focus on geographic data and the ability to analyze the data spatially distinguished GIS from other types of information systems.

Since its inception in the early 1960s, GIS has undergone different phases of development. As a result, the GIS as we understand it today is very different from what it was before. From the perspective of technology, GIS is among the most sophisticated information systems in use in government, business, industry, and academia. From the perspective of application, GIS is now an essential software tool for delivering government services, making business decisions, analyzing and presenting spatial information in academic research, as well as helping the general public to understand the world around them.

GIS was originally developed as a technical tool that aimed to help users manage large volumes of spatial data and use them for spatial problem solving. As this objective remains one of the reasons for implementing GIS, it is still appropriate to study GIS from the perspective of technology. However, as we have demonstrated in this chapter, GIS is now perceived not merely as a branch of information technology, but also as a key component of the information technology infrastructure of society. This changing role of GIS has opened an entirely new horizon for its applications and, consequently, given its practitioners many more career opportunities. It has also moved GIS from a relatively simple and technically oriented development environment to a more complex one in which institutional and socioeconomic factors play a significant role. As a result, GIS today is concerned as much with technology as it is with the humanities and social sciences, making it much more intellectually challenging both as a field of academic study and as a profession.

REFERENCES

Anderson, D., Ledbeter, M., and Tepovich, D., (1996) "Piecing together component-based GIS application development," *GIS World*, Vol. 9, No. 6, pp. 42–44.

Andersson, S. (1987) "The Swedish Land Data Bank," *International Journal of Geographical Information Systems*, Vol. 1, No. 3, pp. 253–263.

Aronoff, S. (1989) *Geographic Information Systems: A Management Perspective*, Ottawa: WDL Publications.

Bell, D. (1973) *Coming of Post-industrial Society*, New York: Basic Publishing.

Boar, B. (1997) *Understanding Data Warehousing Strategically*, Lincroft, NJ: NCR Corporation.

Burrough, P. A. (1986) *Principles of Geographical Information Systems for Land Resources Assessment*, Oxford: Oxford University Press.

Burrough, P. A., and Frank, A. U. (1995) "Concepts and paradigms in spatial information: Are current geographical information systems truly generic?," *International Journal of Geographical Information Systems*, Vol. 9, No. 2, pp. 101–116.

Clarke, K. (2001) *Getting Started with Geographic Information Systems*, 3rd ed., Upper Saddle River, NJ: Prentice Hall.

Corbett, J. P. (1979) *Topological Principles in Cartography*, Technical Paper No. 48, US Bureau of the Census, Washington, DC: US Government Printing Office.

Curry, M. R. (1995) "Rethinking rights and responsibilities in geographic information systems: Beyond the power of imagery," *Cartography and Geographic Information Systems (Special Issues on GIS and Society)*, Vol. 22, No. 1, pp. 58–69.

DeMers, M. N. (2000) *Fundamentals of Geographic Information Systems*, 2nd ed., New York: John Wiley and Sons.

Frank, S. M., Goodchild, M. F., Onsrud, H. J., and Pinto, J. K. (1996) *Framework Data Sets for the NSDI*, Washington, DC: The Federal Geographic Data Committee.

Goodchild, M. F. (1992) "Geographical information science," *International Journal of Geographical Information Systems*, Vol. 6, No. 1, pp. 31–45.

Harder, C. (1998) *Serving Maps on the Internet: Geographic Information on the World Wide Web*, Redlands, California: Environmental Systems Research Institute, Inc.

Latour, B. (1993) *We Have Never Been So Modern*, Cambridge, MA: Harvard University Press.

Marble, D. F., ed. (1980) *Computer Software for Spatial Data Handling*, Ottawa, Canada: Commission on Geographic Data Sensing and Processing, International Geographical Union.

Mark, D. M. (1993) "Human spatial cognition," in *Human Factors in Geographical Information Systems*, by Medyckyj-Scott D., and Hearnshaw, H. M., eds., pp. 51–60, London: Belhaven Press.

Mark, D., Chrisman, N., Frank, A. U., McHaffie, P. H. and Pickles, J. (1997) *The GIS History Project*, National Center for Geographic Information and Analysis (NCGIA), Department of Geography, State University of New York at Buffalo, NY **(http://www.geog.buffalo.edu/ncgia/gishist/bar_harbor.html).**

McLaughlin, J. D. (1985) "Land information management: A Canadian Perspective," *Journal of Surveying Engineering*, Vol. 111, No. 2, pp. 93–104.

Mitchell, W. B., Guptill, S. C., Anderson, E. A., Fegeas, R. G., and Hallam, C. A. (1977) *GIRAS—A Geographic Information Retrieval and Analysis System for Handling Land Use and Land Cover Data*, Professional Paper 1059, Reston, VA: U.S. Geological Survey.

Morehouse, S. (1989) "The architecture of ARC/INFO," in *Auto-Carto 9 Proceedings*, pp. 266–277, Falls Church, VA: American Society of Photogrammetry and Remote Sensing.

NAE (National Academy of Engineering) (1994) *Revolution in the U.S. Information Infrastructure*, Washington, DC: National Academy Press.

NAPA (National Academy of Public Administration) (1998) *Geographic Information for the 21st Century: Building a Strategy for the Nation*, Washington, DC: National Academy of Public Administration.

Nyerges, T. L. (1993) "Understanding the scope of GIS: Its relationship to environmental modeling," in *Environmental Modeling with GIS*, by Goodchild, M. F., Parks, B. O., and Stayaert, L. T., eds., pp. 75–93, New York: Oxford University Press.

Pickles, J., ed. (1995) *Ground Truth: The Social Implications of Geographic Information Systems*, New York: Guilford Press.

Rhind, D. (1981) "Geographical information systems in Britain," in *Quantitative Geography: A British View* by Wrigley, N., and Bennet, R. J., eds., London: Routledge & Kegan Paul.

Rhind, D. (1989) "Why GIS?," *ARC News*, Vol. 11, No. 3 (Summer, 1989), Redlands, CA: Environmental Systems Research Institute, Inc.

Tomlinson, R. F., Calkins, H. W., and Marble, D. F. (1976) "CGIS: A mature, large geographic information system," *Computer Handling of Geographical Data*, Paris: UNESCO Press.

Turk, A. (1993) "The relevance of human factors to geographical information systems," in *Human Factors in Geographical Information Systems*, by Medyckyj-Scott, D., and Hearnshaw, H. M., eds., pp. 15–31, London: Belhaven Press.

Unwin, D. (1991) "The academic setting of GIS," in *Geographical Information Systems, Vol. 1: Principles*, by Maguire, D. J., Goodchild, M. F., and Rhind, D. W., eds., pp. 81–90, Harlow, UK: Longman Scientific and Technical.

USGS (1997) "Geographic Information Systems," an information brochure published by the United States Geological Survey, Reston, VA. **(http://info.er.usgs.gov/research/gis/title.html)**

Waters, N. (1999) "Creating a new master's degree in GIS proves rewarding," *GEOWorld*, Vol. 12, No. 9, pp. 30–31.

CHAPTER

2

MAPS AND GIS

2.1 Introduction

A fundamental characteristic of a Geographic Information System (GIS) is its ability to handle spatial data, i.e., the locations of objects in a geographic space, and the associated attributes. Normally, the geographic space is on the surface of Earth on which we live. A map is the most efficient shorthand to show locations of objects with attributes and their spatial distributions. These objects can be physical or cultural in nature. Therefore, a map is a graphical representation of the spatial structure of the physical and cultural environments. In essence, mapping is an abstraction process by which real-world objects are measured, documented, and stored on a medium (commonly paper). As the result of the abstraction, the real world is both simplified and reduced in size. This leads to the important concepts of scale, classification, symbolization, and generalization in cartography (Robinson et al., 1995). GIS therefore has its roots in the map.

Besides showing spatial locations and attributes, maps are also used as tools for spatial analysis. Such analyses can be performed on a single map sheet or multiple map sheets for the detection of spatial distribution patterns and relationships among different types of objects. By analyzing maps of the same area produced at different times, changes and trends of spatial phenomena can be identified. Because of the large number of measurements involved, spatial analysis based on maps has been a very time-consuming and tedious task. The difficulty of using paper maps for spatial analysis was one of the factors behind the development of GIS.

Conventionally, paper is the primary storage medium for maps. Although dimensional stability is one of the key considerations in the selection of the type of paper for making maps, it is unavoidable for the paper to shrink or expand over time. This will cause distortion of the location of features depicted on maps. Paper maps are cumbersome to produce and maintain. They are also difficult to handle and distribute to users. This was one of the driving forces behind computer-assisted cartography, which is sometimes also referred to as digital mapping, and the use of GIS.

Spatial data stored in maps can be described by three concepts: (1) *entity*, (2) *attribute*, and (3) *relationship*. By definition, an entity is a distinct spatial object of interest. Attribute is a description of some aspects of the entity. Relationship is the spatial association among entities. As an example, a river is an entity. The length, width, and volume of flow are examples of attributes of the river. When the river drains into a lake or the sea, it represents a relationship between the river as one entity and the lake or the sea as another entity.

These concepts of entity, attribute, and relationship are also used to describe spatial data in GIS. However, there are essential differences. In general, spatial data on maps can be described as "unstructured" representations of point, line, and area objects. Symbols are commonly used on maps to represent attributes associated with the entities, e.g., a solid line representing a perennial river, a dashed line representing an ephemeral river, or a square to represent a settlement. As such, it is not possible to generate a list of point objects within a certain range of values without extensive manual searching. In GIS, spatial data are not only properly structured according to entity types, but also by properly organized tables of attributes, which constitute the *database*. Unlike relationships among entities on maps that are only depicted visually, relationships among entities in GIS are stored either explicitly or in a form that allows them to be calculated when required. This is known as "topology" or the mathematics of spatial relationships among point, line, and area objects. As a result of these differences, maps and GIS represent two distinct approaches to the handling of spatial data, although they do share a common focus on these data.

2.2 Map Scale

Real-world objects can be represented on a map only at reduced scale. The *scale* of a map is the ratio or proportion between distances measured on the map and the corresponding distances measured on the ground (USGS, 1999a). There are three common ways to express the scale of a map.

- *Representative Fraction (RF).* This is a simple ratio in the form of 1 : 1,000,000 or 1/1,000,000. In an RF, the number on the left side is the map distance, and its value is always unity. The number on the right side is the ground distance, the value of which varies for different scales. In cartographic usage, the larger this number, the smaller the scale. There is no indication of any unit of measurement in an RF because the units on both sides of the ratio are assumed to be the same. For example, if the unit of measurement on the left is mm, the unit of measurement on the right is also mm. This means that for RF 1 : 1,000,000, one mm, cm, or inch on the map represents 1,000,000 mm, cm, or inches respectively on the ground.
- *Statement Scale.* A statement scale expresses the ratio between the map distance and the ground distance in words. For example, 1 : 1,000,000 can be stated verbally as "One millimeter to one kilometer," and 1 : 63,360 as "One inch to one mile."
- *Bar Scale.* A bar scale is a linear graphical scale drawn on the map to facilitate the estimation of ground distances from measurements made on the map (Figure 2.1).

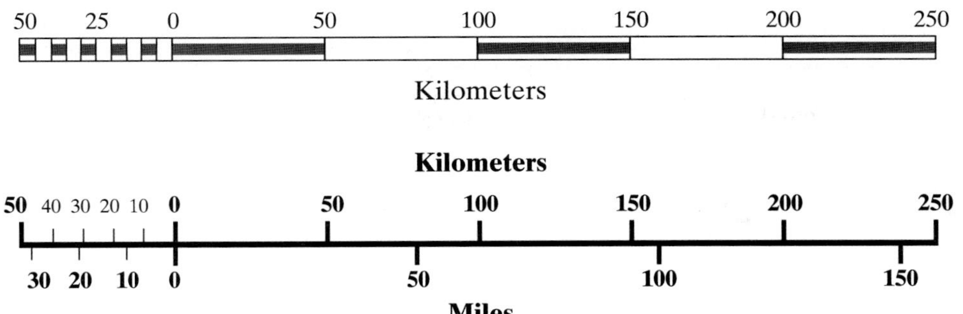

FIGURE 2.1
Examples of scale bars.

Scale is one of the principal factors that govern the design, production, and use of maps. The choice of scale for a particular map or map series is determined basically by the purpose of the map. Maps for engineering construction, land development, and land parcel registration are produced at large scales, which typically range from 1 : 1,000 to 1 : 5,000 (Figure 2.2). These maps provide a very detailed portrayal of an area by depicting small spatial features (e.g., lampposts, manholes, and fire hydrants) as individual entities and showing boundaries of land parcels, buildings, and other features in correct geometrical shapes and directions. On the other hand, atlas maps and wall maps are intended for portraying the geography of large areas. These maps are described as "small-scale," which usually refers to scales smaller than 1 : 100,000. Small-scale maps show a large area on a single map sheet, but their contents are less detailed and include only major features such as highways, railroads, airports, built-up areas, rivers, lakes, and general topography. In between these large- and small-scale maps are maps of medium scales that range from 1 : 5,000 to 1 : 100,000. These maps cover a relatively large geographic area in single map sheets with a reasonable amount of detail, including the locations of important buildings, fence lines, street layouts in urban areas, minor roads, and footbridges. As medium-scale maps represent a good compromise between geographic coverage and map detail, they are most commonly adopted for topographic or general-purpose mapping by national, state, or provincial mapping agencies.

The concept of map scale as explained here pertains to the *spatial* scale of maps. The increasing use of time-sensitive data in GIS has led to growing recognition of the need for the scale for measurements over time. Such a scale is usually called a *temporal scale* in GIS literature. The unit of measurement of a temporal scale can be in year, month, day, hour, minute, or second. The concept of temporal scale is fundamental in the development of temporal or spatiotemporal databases (Langran, 1992). These databases are important in many fields of science and technology, including social and scientific research (e.g., demographic analysis, environmental monitoring, and hydrologic modeling), public administration (e.g., census and property records, community development, crime analysis, environmental impact assessment, and resource planning), infrastructure and facility management (e.g., utilities, transportation and traffic management, and telecommunications), as well as surveying and mapping (e.g., hydrographic charts, weather charts, and spatial database maintenance). For spatial analysis in these application areas, the appropriate use of the temporal scale is important to study changes, movements, and dynamics of objects and phenomena. Temporal scale can be visualized as the number of times a certain spatial event is measured or recorded. In other words, the more frequently the spatial data are collected, the higher is the temporal resolution and the finer is the temporal scale. GIS users should make sure that the geographic data they use are compatible in both spatial and temporal scales. We will examine in detail the characteristics of temporal map scales for the representation of spatial data in Chapter 3.

2.3 Classes of Maps

There are two classes of maps based on the purpose of use: (1) general-purpose, or reference maps, and (2) special purpose, or thematic maps.

General-Purpose, or Reference Maps These are maps not designed for any specific applications. This class of maps focuses on locations and shows a variety of physical and cultural features, such as landforms, drainage, roads, railways, airports, built-up areas, forests, and cultivated areas. They serve as a geographical referencing base for developing and integrating other forms of geographic information. The 1 : 24,000 topographic maps from the USGS, commonly known as the *7.5-minute quadrangles* because each map generally covers a single geographic quadrangle bounded by 7.5 minutes of latitude and 7.5 minutes of longitude, are good examples of this class of maps. These maps are characterized by their adherence to standards of accuracy and contents, which make them good base maps for determining distances, areas, directions, and

Typical RF	1:1000 1:5,000 1:10,000	1:20,000 1:50,000 1:100,000	1:1,000,000 1:2,500,000
Description	LARGE-SCALE	MEDIUM-SCALE	SMALL-SCALE
Characteristics	• Depict small features • Show geometric shapes	• Small features disappear • Generalize geometric shapes • Good compromise between map detail and extent of map coverage	• Symbolize features, e.g., areas represented by point or line symbols • Show macro features, e.g., climatic zones

Large: 1:2,400 scale map of Athens, GA

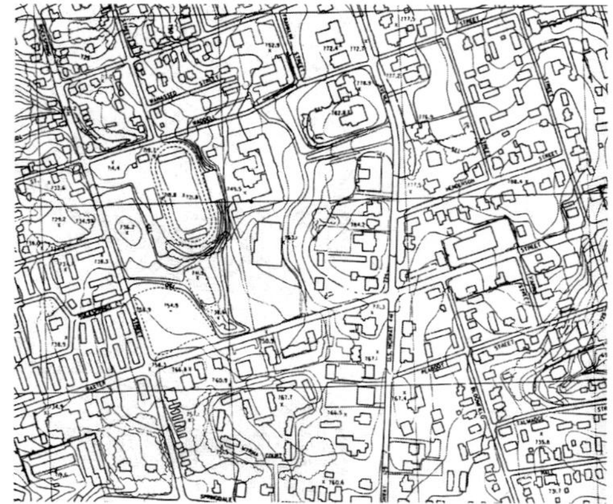

Medium: 1:24,000 scale map of Athens, GA

Small: 1:2,500,000 scale map of Athens, GA

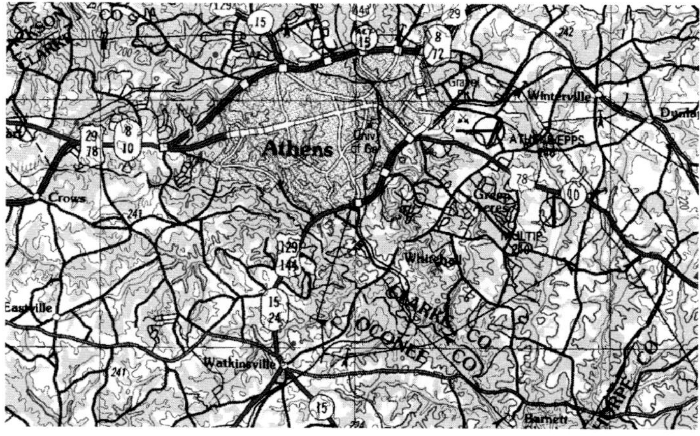

FIGURE 2.2
Continuum of large-, medium-, and small-scale maps, and their characteristics. To view this photo in color, see the color insert.

coordinates of well-defined real-world features. In GIS applications, these maps supply the locational information of the spatial database.

Special Purpose or Thematic Maps These are maps designed to depict a particular type of feature or measurement only. They are produced at different scales on the basis of a general-purpose map. Thus, maps of population distribution, rainfall distribution, rivers, landforms, and highways in the United States are good examples. In GIS applications, these maps depict geographic phenomena and processes.

2.4 The Mapping Process

Maps are produced by the combined effort of many professionals using a variety of technologies. Typically, the mapping process is made up of the following phas-

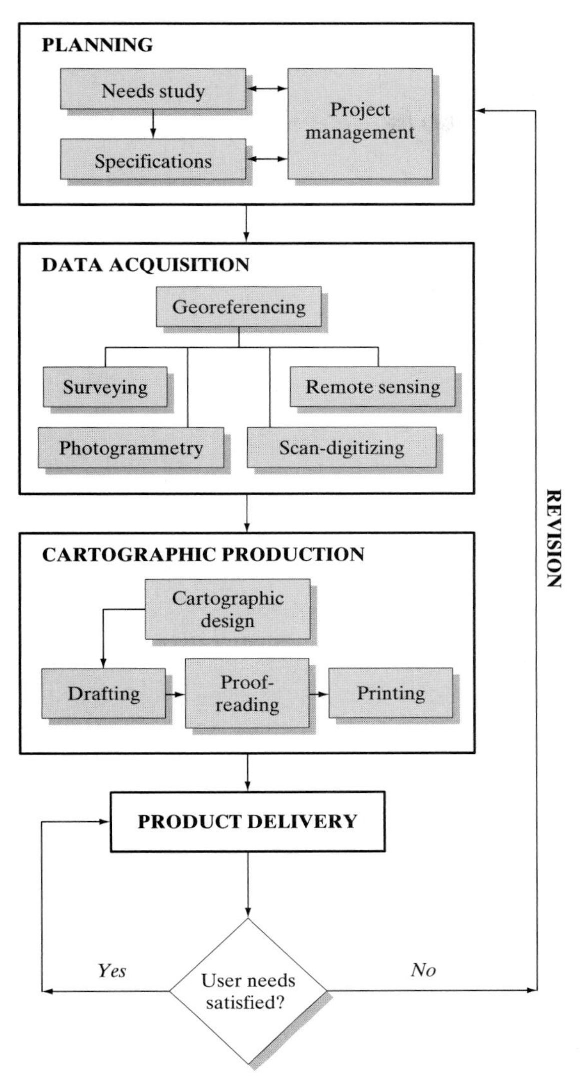

FIGURE 2.3
The mapping process.

es of work (Figure 2.3): (1) planning, (2) data acquisition, (3) production, and (4) product delivery.

The Planning Stage A mapping project is usually initiated as the result of a users' requirements study and analysis. Depending on the nature of the mapping project, the scope of the users' requirements study will be different. In general-purpose mapping at the national and regional level, for example, the agency responsible has to solicit input from the user community at large. The requirements study is always a consensus-building exercise because the agency has to balance the interests of different user groups, often with conflicting requirements. In contrast, users' requirements study for thematic mapping is relatively simpler and more straightforward because the needs of the users are largely defined by the specific purpose of the maps. After it has been proved that the users' requirements justify the cost and resources for the mapping, the agency then proceeds to organize the project office and develop the mapping specifications and standards.

Data Acquisition Phase This phase of work includes the selection and establishment of the geographical reference framework and the actual collection of spatial data by the methods of conventional land surveying, photogrammetry, and remote sensing. It also includes the computation and data processing that render data from field notes and other observation records into a form ready for use in the next phase of work. Not all mapping tasks require acquisition of new data. There are map compilation projects that aim to produce new maps from existing data sources (e.g., to produce a small-scale map series from large-scale maps). In such cases, the major work in the data-acquisition phase is to assemble and evaluate existing data sources.

Cartographic Production Phase This phase starts with cartographic design, which determines what the end product will look like. This is then followed by the actual map production processes of drafting, proofreading, and final printing of the maps. In the digital mapping environment, drafting and editing are done on the computer screen rather than on the drafting table. Proofreading is normally done both on the computer screen and by producing check plots. The objective is to ensure that the maps in both the digital and hard-copy forms have met design specifications. The greatest difference between conventional and GIS mapping at this stage is in the production of the maps. Instead of sending the compiled map data to the printing press, the resulting files from digital compilation are stored and indexed for safekeeping in a digital map repository.

Product Delivery Phase This phase involves the storage and management of printed maps and the establishment of map sales offices and other channels for distribution to the users. Advances in computer network technologies have led to the development of new concepts and methods of delivering maps to end users over the Internet (see Chapter 12). These include the geospatial data clearinghouse program of the U.S. Federal Government (FGDC, 1997; USGS, 1999e and f), digital spatial libraries (NRC, 1993 and 1997) and numerous on-line spatial data distribution sites set up by government agencies, research institutes, and commercial data suppliers. Depending on their storage format and media, maps can be delivered to end users as printed products or in CD-ROMs through the regular mail or commercial courier services. It is now increasingly common for digital map data to be directly transmitted to end users over communication networks using either the World Wide Web (WWW) or File Transfer Protocol (FTP) of the Internet.

The mapping process is always described as a "cycle" in the sense that after the completion of the four phases of work described here, the mapping agency has the obligation to maintain a map revision program. It needs to review the requirements of the users from time to time. When the requirements change, the mapping specifications have to be changed accordingly. This may lead to a full revision of the old maps or the initiation of a new mapping project. In digital mapping and GIS, map revision can be done more easily technically. However, the ability of individual map users to update maps may easily lead to loss in the integrity of the data. This means that different users may be using different versions of the same digital map or map series, making it hard for data sharing and information exchange. In this regard, map revision has become an extremely important database management issue in geographic data processing.

2.5 Plane Coordinate Systems and Transformations

A map is a record of locations of objects in a geographic space. Each location is unique, and can be represented in different ways. A pair of latitude and longitude is used to locate a place on Earth's surface. The ZIP code of your home address is another method of showing a location. The essence of all these methods of fixing locations relates to the mathematical concept of *coordinates*. A coordinate is one of a set of numbers that determines the location of a point in a space of a given dimension. In a two-dimensional space (i.e., a plane), a point can be fixed by a set of two numbers. The use of coordinates has the advantage of simplifying and standardizing the computational methods, making possible the use of computers. It also facilitates the transformation of geographic space to conform to other frameworks of entities and relationships, so often required in mapping and GIS operations (Chrisman, 1997).

There are two basic types of coordinate reference systems on a plane (two-dimensional space): (1) the plane rectangular coordinate system and (2) plane polar coordinates.

2.5.1 Plane Rectangular Coordinate System

The *rectangular coordinate system* (or *Cartesian coordinate system*) is the simplest coordinate system that we use to locate objects in space. Two straight lines intersecting one another at right angles are used to define the geographic space (Figure 2.4). These two lines are called the *axes* of the coordinate system, and they define the directions of the two families of lines (Maling, 1992). Where the two lines intersect is called the *origin* of the rectangular coordinate system (O). The horizontal axis OX is called the

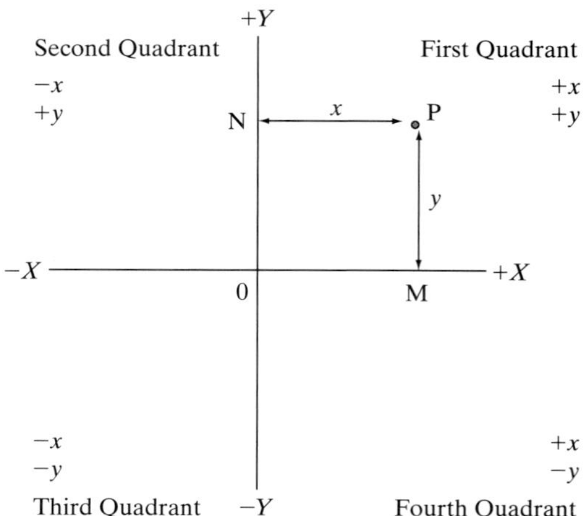

FIGURE 2.4
Plane rectangular Cartesian coordinate system.

X-axis (or *easting*) while the vertical axis OY is called the *Y-axis* (or *northing*). The position of a point (P) in this rectangular coordinate system is fixed by two distances measured perpendicularly from the point (P) to the *X*- and *Y*-axes (i.e., PM and PN respectively in Figure 2.4).

The rectangular coordinate system partitions the geographic space into four *quadrants*. The quadrants are numbered 1 to 4 in a counterclockwise direction starting from the top right quadrant (Figure 2.4). A sign convention has to be observed to distinguish locations of objects with the same x and y coordinate measurements but in different quadrants. The *X*-axis is positive toward the right and the *Y*-axis is positive toward the top of the page. Thus, the sign convention for each quadrant for x and y coordinates as shown in Figure 2.4 is obtained. Therefore, the point P located in the first quadrant has coordinates of $+x$ and $+y$.

2.5.2 Plane Polar Coordinate System

The plane polar coordinate system makes use of an angular measurement and a linear measurement to fix the position of a point (Figure 2.5). The system consists of a single line passing through the origin, called the *pole*. This line is the *polar axis*. In Figure 2.5, the position

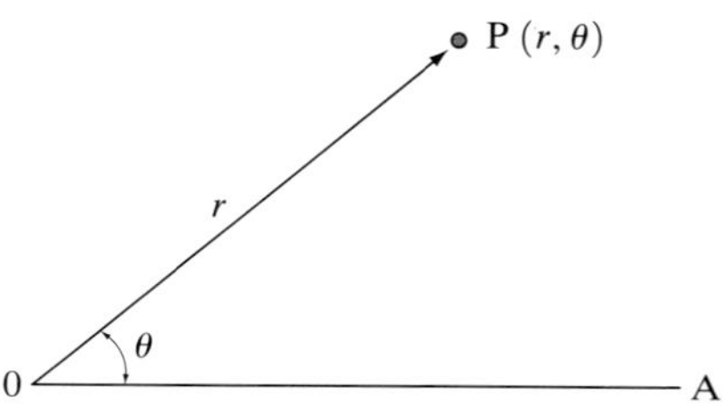

FIGURE 2.5
Plane polar coordinate system.

of point P is determined by the distance OP (r) and the angle it forms at the pole with the polar axis, i.e., the angle POA (θ). The *radius vector* OP should be stated first to be followed by the *vectorial angle* (θ) in defining the point. A positive angle designates the counterclockwise angle measured from the initial line as used in vector algebra. In surveying and cartography, however, a positive angle should be measured in the clockwise direction, as this is the convention of measuring azimuths and bearings clockwise from the north.

Polar coordinates can be readily converted to rectangular coordinates and vice versa. From Figure 2.6, which shows the relationship between the rectangular and polar coordinates, it can be seen that polar coordinates can be converted to rectangular coordinates by using the following formulas:

$$x = r \cdot \sin(\theta) \quad \textbf{(2.1)}$$

$$y = r \cdot \cos(\theta) \quad \textbf{(2.2)}$$

and the conversion from rectangular to polar coordinates by using the following formulas:

$$\tan(\theta) = x/y \quad \textbf{(2.3)}$$

$$r = y \cdot \sec(\theta) \quad \textbf{(2.4)}$$

$$r = x \cdot \mathrm{cosec}(\theta) \quad \textbf{(2.5)}$$

$$r^2 = x^2 + y^2 \quad \textbf{(2.6)}$$

Formula 2.6 is known as the Pythagorean theorem, which is particularly useful in finding a distance from the origin of a rectangular system to a point in a right-angled triangle. Note that in these calculations using formulas 2.1 to 2.5, the angle (θ) is assumed to be measured clockwise from the grid north.

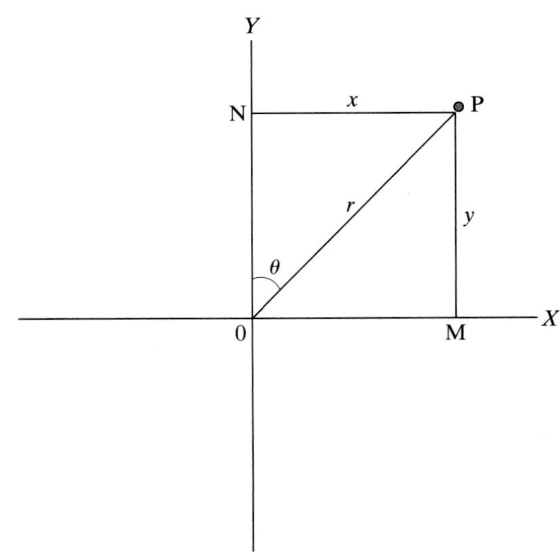

FIGURE 2.6
The relationship between plane rectangular and plane polar coordinate systems.

2.5.3 LINEAR TRANSFORMATIONS

It is always necessary to unify into one general coordinate system objects recorded in different coordinate systems. This can be achieved through coordinate transformations. Transformation is the derivation of one set of coordinates for a point whose coordinates in another coordinate system are known. The conversion from polar coordinates to rectangular coordinates explained in Section 2.5.2 is an example of simple transformation that does not involve any change in scale or shape of the geographic space. In map projections (see Section 2.7), positions of points on the curved surface of Earth are transformed to their corresponding positions on a flat piece of paper. There are two useful types of simple coordinate transformations employed in surveying, cartography, and photogrammetry: the *similarity transformation* and the *affine transformation*. They are linear transformations because the terms in the mathematical equations are of a single power.

Similarity Transformation

In this transformation (which is also known as the linear conformal, or Helmert transformation), a subsidiary coordinate system is brought into coincidence with the principal coordinate system by means of a translation (horizontal shifts), a rotation through an angle, and a change of scale by a factor (Richardus, 1966). The following equations for similarity transformation assume the use of a rectangular coordinate system:

$$X = ax - by + c \quad \textbf{(2.7a)}$$

$$Y = bx + ay + d \quad \textbf{(2.7b)}$$

where x and y are coordinates in the subsidiary system and X and Y are coordinates in the principal system. Since there are four unknowns to solve in these simultaneous equations, a minimum of two points must be given in both coordinate systems. These two points must refer to the same objects in the two coordinate systems, and they are generally known as *control points*. In geographic space transformation, these two control points should be well spread out. If there are more than the minimum number of control points, the unknowns can be more accurately determined using least-squares adjustment. After the similarity transformation, the shape of the original figure will not change, but the size and orientation toward the axes may change. Because this transformation keeps the shape unchanged, it is also known as the *orthogonal transformation*.

Affine Transformation

The affine transformation makes use of the following equations in a rectangular coordinate system.

$$X = ax + by + c \quad \textbf{(2.8a)}$$

$$Y = dx + ey + f \quad \textbf{(2.8b)}$$

where the definitions for *X*, *Y*, *x*, and *y* are the same as those given for the similarity transformation. Because six unknowns are involved, a minimum of three points in both systems is required to solve the simultaneous equations.

In this transformation, straight lines remain unchanged and parallel lines remain parallel, but in order to eliminate discrepancies between the two coordinate systems, the angles will undergo slight changes. As a result, the shape of the original figure in the subsidiary system will change. As an example, a circle could be transformed into an ellipse.

In applying to geographic space, the affine transformation is used more often than the similarity transformation because transformation of geographic space invariably involves changes in shape. A common application of the affine transformation in GIS is to transform the coordinates of a digitized map in digitizer units (most often in inches) to those in the real-world coordinate system of the map (such as the Universal Transverse Mercator [UTM] Coordinate System measured in meters [see Section 2.8.3]). Another application is to convert a digital image from its image-based coordinate system into the ground-based terrestrial coordinate system in remote sensing, a process known as *rectification* (see Chapter 8). These two types of transformations are easily programmed for use in the computer, and have been incorporated in most digital image processing packages.

The similarity and affine transformations may also be extended to multidimensional space. More commonly in analytical and digital photogrammetry, affine transformation is used to determine the elements of exterior orientation of a camera, which fix the position of the camera in the sky at the time of exposure (Moffitt and Mikhail, 1980; Wolf and Dewitt, 2000).

2.6 Geographic Coordinate System of Earth

In order to locate places on Earth, a three-dimensional coordinate reference system has to be developed that takes into account its shape. The spherical coordinate system in use for over 2000 years is known as the Geographic Coordinate System, which makes use of a network of latitude and longitude (known as a *graticule*) to fix the positions of points on Earth. The concept of latitude and longitude can be traced back to the work of the Greek astronomer Hipparchus in the middle of the second century B.C., formalized by Claudius Ptolemy 300 years later (Berthon and Robinson, 1991). The two primary reference points on Earth are the North and South Geographic Poles, the two points on the surface of Earth intersected by its axis of rotation. Halfway between the two poles is an imaginary line called the Equator. The polar axis and the circle containing the Equator intersect at right angles at the center of Earth, which is regarded as the origin of this geographic coordinate system. The position of a point on the spherical surface of Earth is determined by two angles: two orthogonal planes that intersect at the origin or the center of Earth. One of these planes is the plane of Equator, which is used as the reference plane for measurement of the first angle (a vertical angle) called *latitude* (ϕ). The other plane is that of the meridian chosen as *zero longitude* (known as the *prime meridian*), which is used as the origin to measure the second angle (a horizontal angle) along the Equator to another meridian passing through the point on the surface of Earth to be fixed. This angle is known as *longitude* (λ) (Figure 2.7).

The latitude of a point is therefore a vertical angle measured at the center of Earth between the plane of the Equator and the radius drawn to the point (Figure 2.7). It is measured in angular units north or south of the Equator, so that the Equator's latitude is 0°, the North Pole's latitude is 90° N and the South Pole's latitude is 90° S. The sign convention is that north latitude is positive ($+\phi$) and south latitude is negative ($-\phi$). The longitude of a point is a horizontal angle measured in the plane of the Equator between the plane of the meridian through the point and the plane of a datum meridian (i.e., the prime meridian as noted

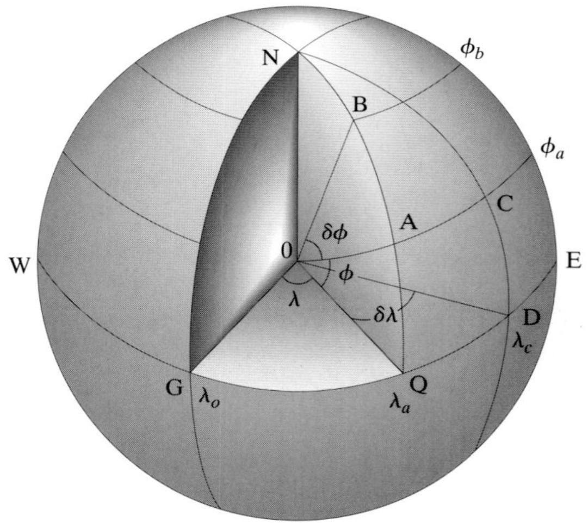

FIGURE 2.7
The geographic coordinate system of Earth, showing the horizontal angles of longitude measured along the equatorial plane and vertical angles of latitude measured from the equatorial plane. The center of Earth is the origin of the system. The plane of the Greenwich Meridian (NGO) is highlighted. *(Source: Maling, 1992)*

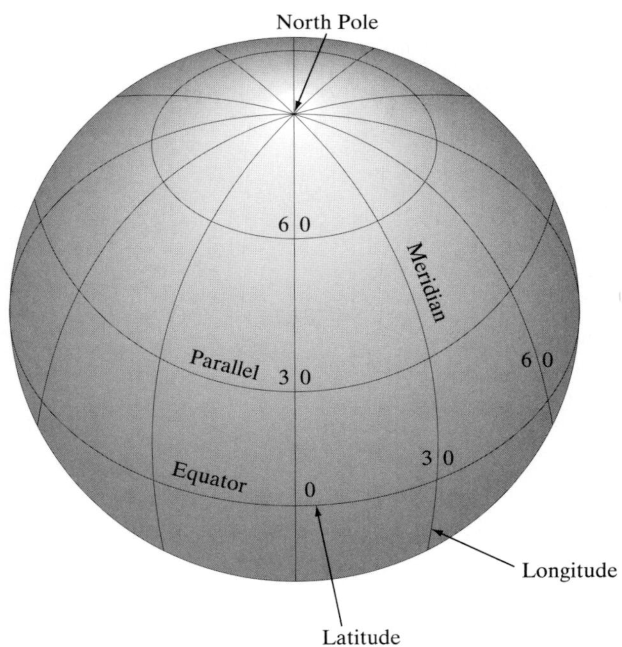

FIGURE 2.8
The graticule of Earth: the parallels of latitude and the meridians of longitude. *(Source: Snyder, 1987)*

above). Any meridian can be selected as the prime meridian. In 1884, the world mapping community agreed to use the Greenwich meridian in London, England, as the prime meridian (i.e., with a longitude of 0°). Longitude is measured from the prime meridian plane east and west of Greenwich up to 180 degrees. The sign convention is that east longitude is positive ($+\lambda$) and west longitude is negative ($-\lambda$).

On the surface of Earth, a series of parallels of latitude can be drawn parallel to the Equator, and a series of meridians of longitude can be drawn from pole to pole, crossing each parallel of latitude at right angles, but striking Earth at various points. These series of imaginary lines form a network of parallels and meridians that comprise the system of geographic coordinates explained earlier. This is known as the graticule (Figure 2.8).

2.7 MAP PROJECTION

2.7.1 BASICS OF MAP PROJECTIONS

Because we need to produce a map of Earth on a flat surface, some methods of transforming Earth's graticule of geographical coordinates, as described in Section 2.6, from three dimensions to two dimensions are required. A map projection is a systematic representation of all or part of the surface of a round body, especially Earth, on a plane (Snyder, 1987). It can be *perspective* or *nonperspective*. A perspective projection is strictly geometric in nature, characterized by the use of a point of origin (or viewpoint) and a surface of projection. The viewpoint is chosen to suit a particular requirement and can be in the center of the globe, at an infinite distance, or on the surface of the globe (see Figure 2.13 and Section 2.7.5). The surface of projection can be a plane, a cylinder, or a cone, each a developable surface, i.e., one that can be unfolded or unrolled to a plane without distortion. A nonperspective projection is obtained by modifying the perspective projection so that certain desired properties can be maintained, such as equal area, equal distance, or correct shape. A good example is the Mercator Projection.

Mathematically, map projection can be expressed in the following generalized functional relationships between the geographical coordinates [latitude (ϕ) and longitude (λ)] of a point on Earth's surface and the coordinates on the plane.

for the rectangular coordinate system

$$x = f_1(\phi, \lambda) \quad \textbf{(2.9a)}$$
$$y = f_2(\phi, \lambda) \quad \textbf{(2.9b)}$$

for the polar coordinate system

$$r = f_3(\phi, \lambda) \quad \textbf{(2.10a)}$$
$$\theta = f_4(\phi, \lambda) \quad \textbf{(2.10b)}$$

where f_1, f_2, f_3, and f_4 are functions, implying different forms of mathematical transformation. Thus, map projection is a transformation of geographic space, a concept extended by Tobler (1961) to the analysis of distance decay and time–distance distortion in economic geography.

2.7.2 PROPERTIES OF MAP PROJECTIONS

If you squash an orange on a flat surface, the skin of the orange will split and spread out in all directions, and the original properties of the orange, such as its shape, will change. In map projection, Earth can be regarded as a perfect sphere. To represent it on a flat plane is like squashing an orange on a flat surface; some of the properties of the spherical Earth will be lost. There are four properties to consider: (1) area, (2) shape, (3) distance, and (4) direction. For a spherical Earth, *all* these four properties are correct. However, once the earth is transformed into a plane, only *some* of these properties can be maintained. Thus, all the different map projections have been designed to produce a network of meridians and parallels that can achieve one or two of these properties in the final map for specific purposes of representation (Figure 2.9).

Projected grid of graticule	Scale		Properties			
	Parallel	Meridian	Area	Shape	Distance	Direction
(a) 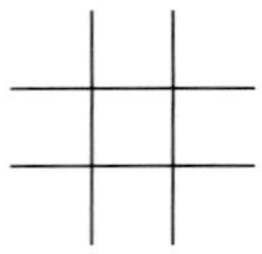	Correct	Correct	Preserved	Preserved	Preserved	Preserved
(b) 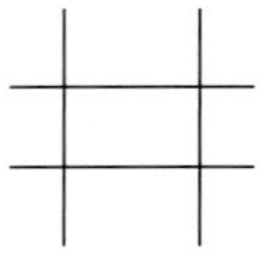	Doubled	Correct	Distorted	Distorted	Preserved N–S only	Preserved only along parallels and meridians
(c) 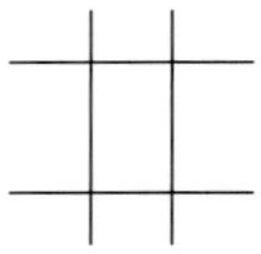	Correct	Doubled	Distorted	Distorted	Preserved E–W only	Preserved only along parallels and meridians
(d) 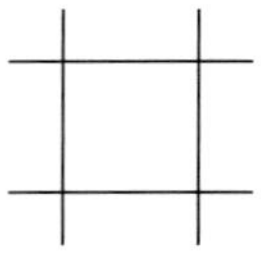	Doubled	Doubled	Distorted	Preserved	Distorted	Preserved
(e) 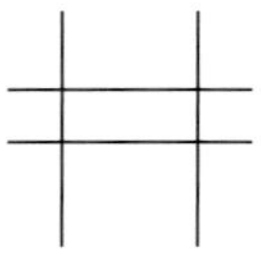	Doubled	Halved	Preserved	Distorted	Distorted	Preserved only along parallels and meridians
(f) 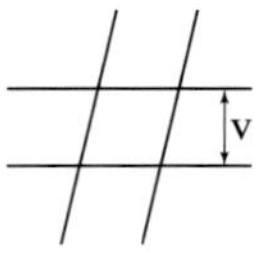 v	Correct; correct vertical interval (v)	Distorted	Preserved	Distorted	Preserved E–W; slightly distorted N–S	Preserved only along parallels

FIGURE 2.9
Preservation of projection properties by manipulating scales of projected grids of graticule.

Area A map projection may be designed to be equal area, so that any area measured on the map is the same as that measured on Earth. The map projection is called *equal-area* or an *equivalent* projection. This property can be accomplished by distorting the shape of the graticule. Equal-area projections are best employed to show spatial distributions and relative sizes of spatial features, such as political units, population, land use and land cover, soils, wetlands, wildlife habitats, and natural resource inventories. The equal-area property allows the sizes of real-world features to be visually compared on the same areal basis. However, the trade-off of preserving true area is that spatial features on the maps will inevitably be distorted in shapes, distance, and, occasionally, directions (Figure 2.9e and f). Examples are the Albers Equal-Area Conic Projection, the Lambert Azimuthal Equal-Area Projection, and the Sinusoidal Equal-Area Projection.

Shape A map projection can maintain the correct shape of the spatial features represented. This is possible only by making the scale along the meridian and the parallel the same in both directions. As a result, the relative local angles about every point on the map are shown correctly. The meridians intersect parallels at right an-

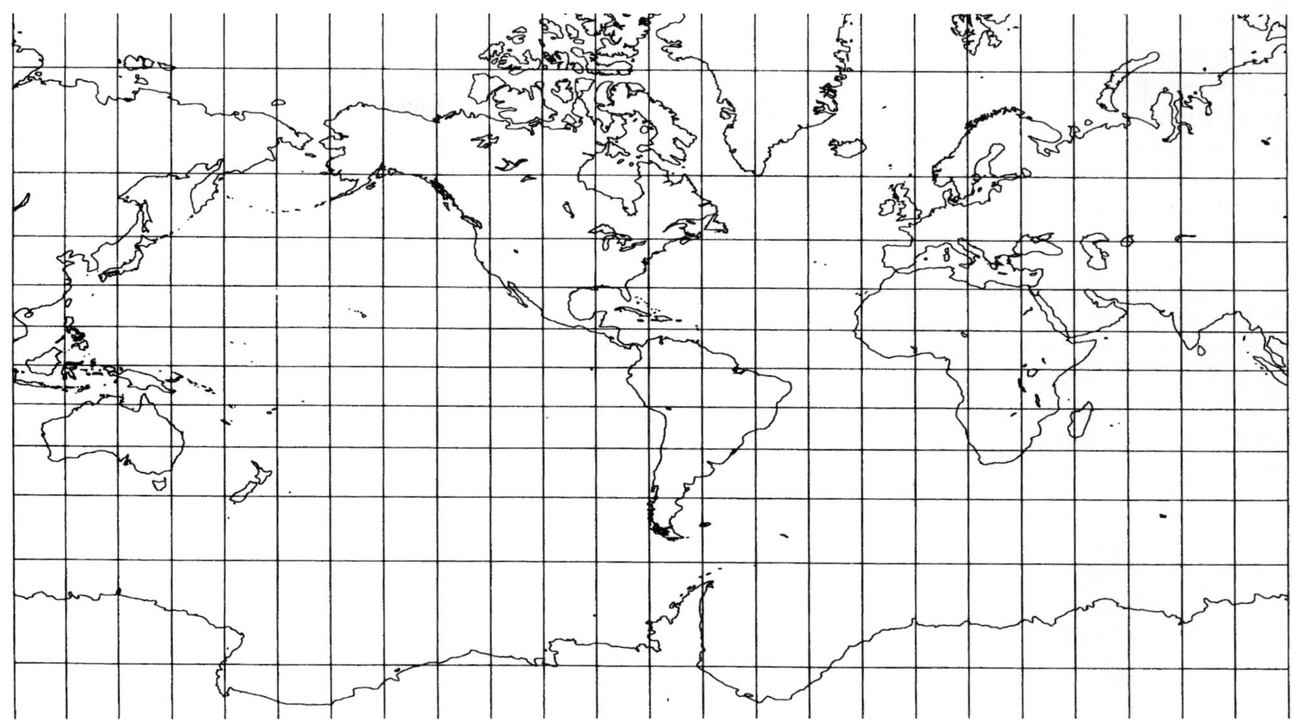

FIGURE 2.10
Mercator projection. *(Source: Snyder, 1987)*

gles, just as they do on the globe. This type of projection is known as *conformal*, or *orthomorphic* ("straight shape" in Greek), and is very important for topographic mapping and navigation purposes. The need to retain shape inevitably distorts both area and distance (Figure 2.9d). A good example is the Mercator Projection in which the area of Greenland is shown to be larger than South America, although in reality Greenland is only one-eighth the size of South America (Figure 2.10). There are a few variants of the Mercator Projection: Transverse Mercator, Oblique Mercator, and Space Oblique Mercator. These all belong to the cylindrical class of map projections. A conformal map projection in the conic type is the Lambert Conformal Conic Projection.

Distance The distance between two points measured on the map is equal to that between the same two points measured on Earth's surface and scaled. Obviously, this property is not possible at all points throughout the map. It can be achieved only by selecting certain lines along which the scale remains true. These lines can be along every meridian. In other words, distances can be measured correctly in only one direction. These lines of true scale include the *central meridians* of the cylindrical class of projections, as well as the *standard parallels* of the conic class of projections. This type of map projection is called *equidistant*. The property of equidistance is very sensitive to scale change (Figure 2.9b, c, d, e, and f). All measurements made away from the lines of true scale are subject to distance distortion due to changing scales. It is a useful compromise between the conformal and equal-area projections because the area scale of an equidistant map projection increases more slowly than that of a conformal map projection. As a result, the equidistant map projection is used more often in atlas maps. Examples are Azimuthal Equidistant Projection and Equidistant Conic Projection.

Direction Direction measurements made on the map are the same as those made on the ground. This is the property of a true direction map projection. True direction is an inherent property of the azimuthal class of map projections because all meridians pass through the pole. This particular property can also be retained simultaneously with one or two of the other three properties described above. Since the conformal map projections preserve shapes, they naturally preserve true direction. However, true direction in many conformal map projections is not real true direction in the sense that accurate direction measurements are obtainable in only one or two directions (Figure 2.9b, c, d, e, and f). The most notable exception is the Mercator Projection that preserves true direction in all directions and in all parts of the map. True direction is a useful property for air and sea navigation charts. Examples of true direction map projections include the different variants of azimuthal and Mercator projections.

2.7.3 CLASSIFICATION OF MAP PROJECTIONS

There are different ways to classify map projections (Tobler, 1962). One simple scheme is to classify map projections according to the type of developable surface onto which the network of meridians and parallels is projected. A developable surface is a surface that can be laid out flat without distortion. There are three types of developable surfaces: (1) cylindrical, (2) conical, and (3) planar.

Cylindrical Projection A cylinder is assumed to circumscribe a transparent globe (marked with meridians and parallels) so that the cylinder touches the Equator throughout its circumference (Figure 2.11a). Assuming that a lightbulb is placed at the center of the globe, the graticule of the globe is projected onto the cylinder. By cutting open the cylinder along a meridian and unfolding it, a rectangle-shaped cylindrical projection is obtained. The meridians are vertical and parallel straight lines, intersecting the Equator at right angles and dividing it into 360 equal parts. The parallels will be horizontal straight lines at some selected distance from the Equator and from each other.

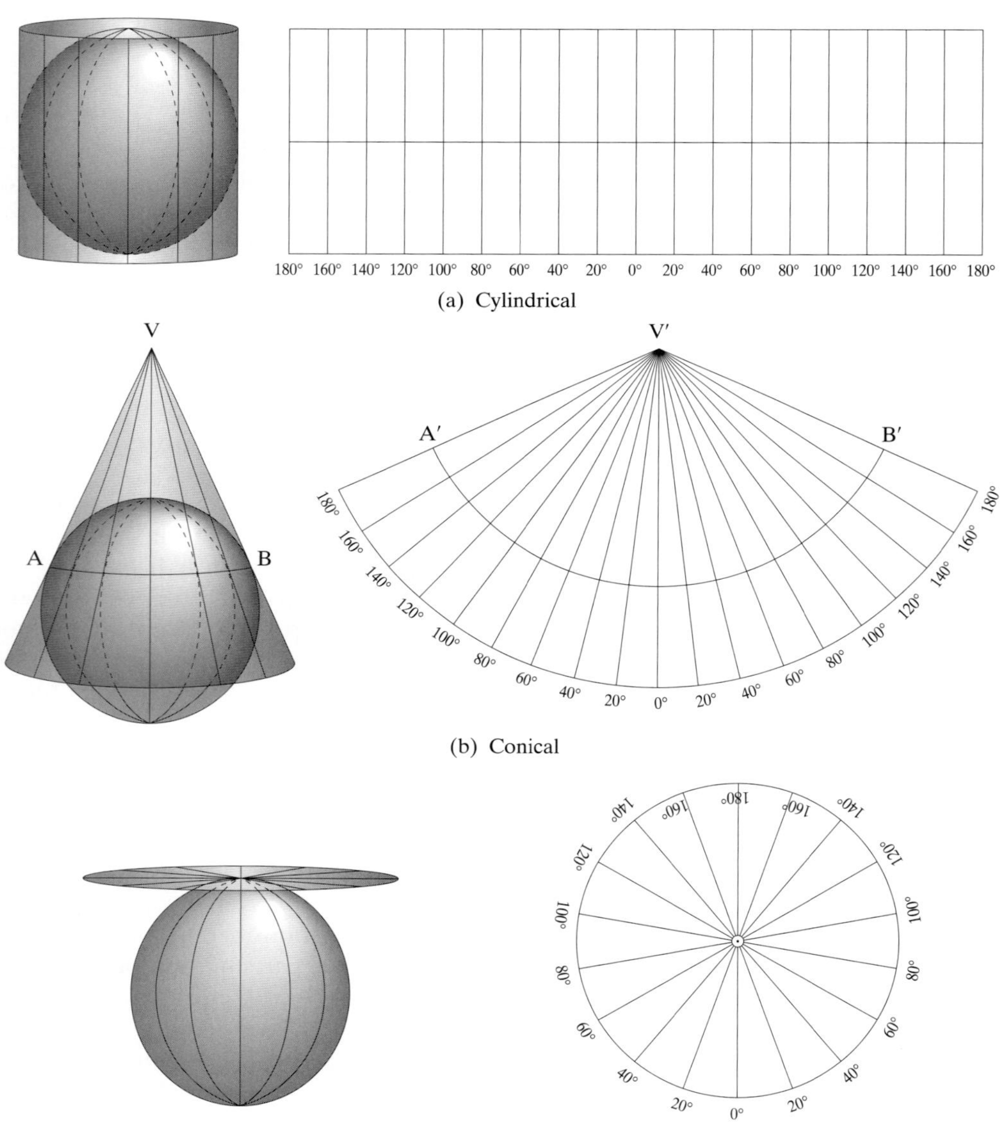

FIGURE 2.11
Three basic types of map projections. *(Source: Mainwaring, 1960)*

Conical Projection A cone is placed over the globe in such a way that the apex of the cone is exactly over the polar axis (Figure 2.11b). A cone must touch the globe along a parallel of latitude, known as the standard parallel, which can be selected by the cartographer. Along this standard parallel, scale is correct and distortion is the least. When the cone is cut open along a meridian and laid flat, a fan-shaped map is produced, with meridians as straight lines radiating from the vertex at equal angles, while parallels are arcs of circles, all drawn using the vertex as the center.

Planar or Azimuthal Projection A plane is placed so that it touches the globe at the North or South Pole (Figure 2.11c). This can be conceived as the cone becoming increasingly flattened until its vertex reaches the limit of 180°. The projection resulting is better known as the Polar Azimuthal Projection. It is circular in shape with meridians projected as straight lines radiating from the center of the circle (the tangent point of the plane), which is the pole. These meridians are spaced at their true angles. The parallels are complete circles centered at the pole. The radii of these circles may be computed with reference to some mathematical formula, such as the Azimuthal Equidistant Projection, or geometrically projected from some viewpoint situated somewhere on the polar axis or its extension, such as the Stereographic Projection (viewpoint on the opposite pole of the globe) or the Gnomonic Projection (viewpoint at the center of the globe).

In these three projection types, the developable surface is assumed to touch the surface of the globe to form a tangent cylinder, a tangent cone, and a tangent plane. Mathematically, it is possible to make the developable surface cut through the globe as a secant cylinder, a secant cone, and a secant plane. The purpose of using a secant developable surface rather than a tangent one is to minimize the amount of distortion that occurs away from the standard parallel or the pole. In the case of the secant cylinder and secant cone, two standard parallels are produced, where the scales will be more correct than in other parts of the map. In a planar projection, using a secant plane means that true scale is obtainable along the selected standard parallel rather than at one single point. Although this will introduce distortion for measurements around the pole, it will also minimize the magnitudes of the errors that occur for measurements made away from the pole.

2.7.4 ASPECTS OF MAP PROJECTIONS

There is still one more consideration known as the *aspect* of a map projection. The developable surface may be placed in three different ways relative to the globe: (1) normal, (2) transverse, or (3) oblique. This will affect the appearance of the graticule.

The Normal Aspect This is directly related in form to the graticule of the globe. The axis of the cylinder or the cone is coincident or parallel to the polar axis of Earth, and the plane is tangent to the pole (Figure 2.12a). Our previous discussions on map projection are based on the normal aspect of the map projection.

The Transverse Aspect The axis of the cylinder or cone may be placed at a different direction from the polar axis of Earth. If the axis of the cylinder is placed at 90° to the polar axis, this will make the true "north pole" of Earth lie on the equator of the map projection, and the poles of the map projection lie on the Equator of Earth (Figure 2.12b). This produces the *transverse* aspect of the map projection and can be regarded as a 90° rotation. The appearance of the graticule will be totally different from that of the conventional. This transverse aspect may also be obtained with a plane in the

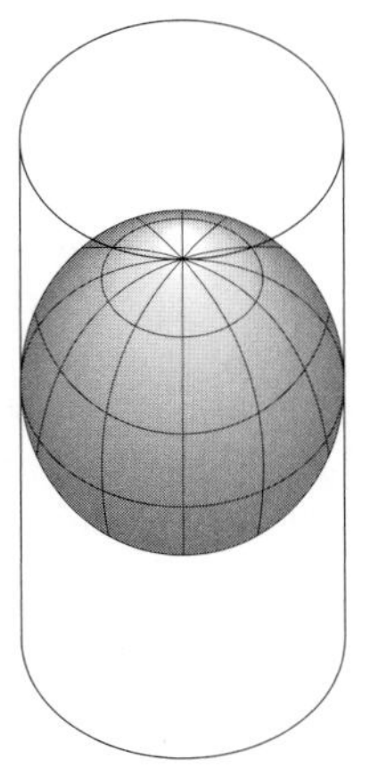

(a) Normal cylindrical

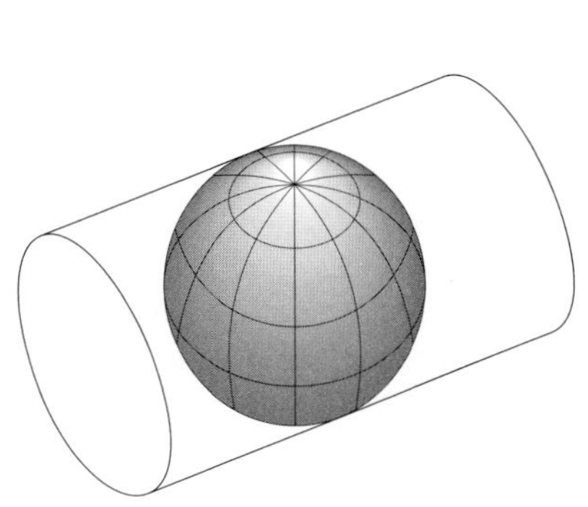

(b) Transverse cylindrical

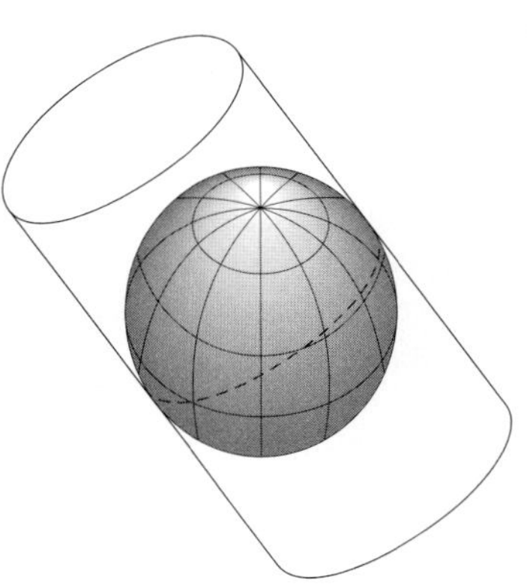

(c) Oblique cylindrical

FIGURE 2.12
Three different aspects of map projections. *(Source: Snyder, 1987)*

case of the azimuthal projection in which the plane is placed at a tangent at the Equator of Earth and is more commonly called the *equatorial aspect*.

The Oblique Aspect If the axis of the cylinder or the cone or the center of the plane is located somewhere between the poles and the Equator of Earth, an *oblique* aspect is produced (Figure 2.12c). The oblique projections are usually applied to the mapping of areas that lie at an angle to the latitudes or longitudes.

The transverse and oblique aspects are transformations of the normal aspect of the map projection so that certain desired properties may be preserved for a particular area to be mapped. A notable example is the Transverse Mercator Projection in which the equator of the projection is rotated 90° to coincide with the desired central meridian, thus making the central meridian true to scale no matter how far north and south the map extends (thus solving the excessive scale change problem of the regular Mercator Projection as the parallels and meridians move north and south away from the Equator) while retaining conformality. This explains why the Transverse Mercator Projection is used in the State Plane Coordinate System (SPCS) in the United States for states with predominantly north– south extent (Snyder, 1987).

2.7.5 VIEWPOINTS OF MAP PROJECTIONS

As has been explained in Section 2.7.1, a map projection may be produced from three viewpoints: at the center, from infinity, and on the surface of the globe opposite the developable surface. This gives rise to three basic variants of map projections.

Gnomonic projections are projections with the viewpoint (or projection point) at the center of the globe. Great circles are straight lines (Figure 2.13a). Since a great circle is the shortest distance between two points on a sphere, gnomonic projections are used for navigation charts.

Orthographic projections are projections obtained by placing the viewpoint at infinity. When applied to the planar equatorial case, Earth will appear to be similar to a view from space (Figure 2.13b). When applied to the cylindrical normal case, the longitudes will remain parallel and are at equal distances apart while the latitudes will remain parallel but are at successfully decreasing distances apart toward the pole, thus preserving the areas locally.

Stereographic projections are projections obtained by placing the viewpoint on the surface at the far side of the globe. When applied to the planar equatorial case, both latitudes and longitudes appear to be curved lines (Figure 2.13c). However, since the projected latitudes and longitudes intersect at almost 90°, the shape property is preserved for small areas around the intersections. Therefore, stereographic planar projections are conformal projections. When applied to the cylindrical normal case, the longitudes will remain parallel and are at equal distances apart while the latitudes will remain parallel but are at successfully decreasing distances apart toward the Equator. Using two standard parallels minimizes distortions around the Equator. The result will be a projection that will distort every property a little rather than preserving one special property exactly, thus giving rise to a *compromise projection*.

2.7.6 SOME USEFUL MAP PROJECTIONS FOR GIS

In the preceding sections, several map projections have been mentioned as examples that illustrate specific properties and projection types. These map projections represent the ones that GIS users in general are most likely to encounter. For ease of reference and comparison, the characteristics and major uses of these map projections are summarized in Table 2.1.

2.8 ESTABLISHING A SPATIAL FRAMEWORK FOR MAPPING LOCATIONS ON EARTH: GEOREFERENCING

Handling spatial information requires the establishment of a *spatial reference system* to which all spatial measurements must relate. The primary function of the map is to portray accurately real-world features that occur on the curved surface of Earth. Geographic referencing, which is sometimes simply called *georeferencing*, is defined as the representation of the location of real-world features within the spatial framework of a particular coordinate system. The objective of georeferencing is to provide a rigid spatial framework by which the positions of real-world features are measured, computed, recorded, and analyzed. In practice, georeferencing can be seen as a series of concepts and techniques that progressively transform measurements carried out on the irregular surface of Earth to a flat surface of a map, and make it easily and readily measurable on this flat surface by means of a coordinate system (Figure 2.14). Map data are different from all other forms of data by this characteristic of georeferencing and, as noted in Chapter 1, the ability to manipulate and analyze georeferenced spatial data is what distinguishes GIS from CAD and other types of computer graphics systems. The concept of representing the physical shape of Earth by means of a mathematical surface and the realization of this concept by the definitions of the *geoid* and the *ellipsoid* are fundamental to georeferencing.

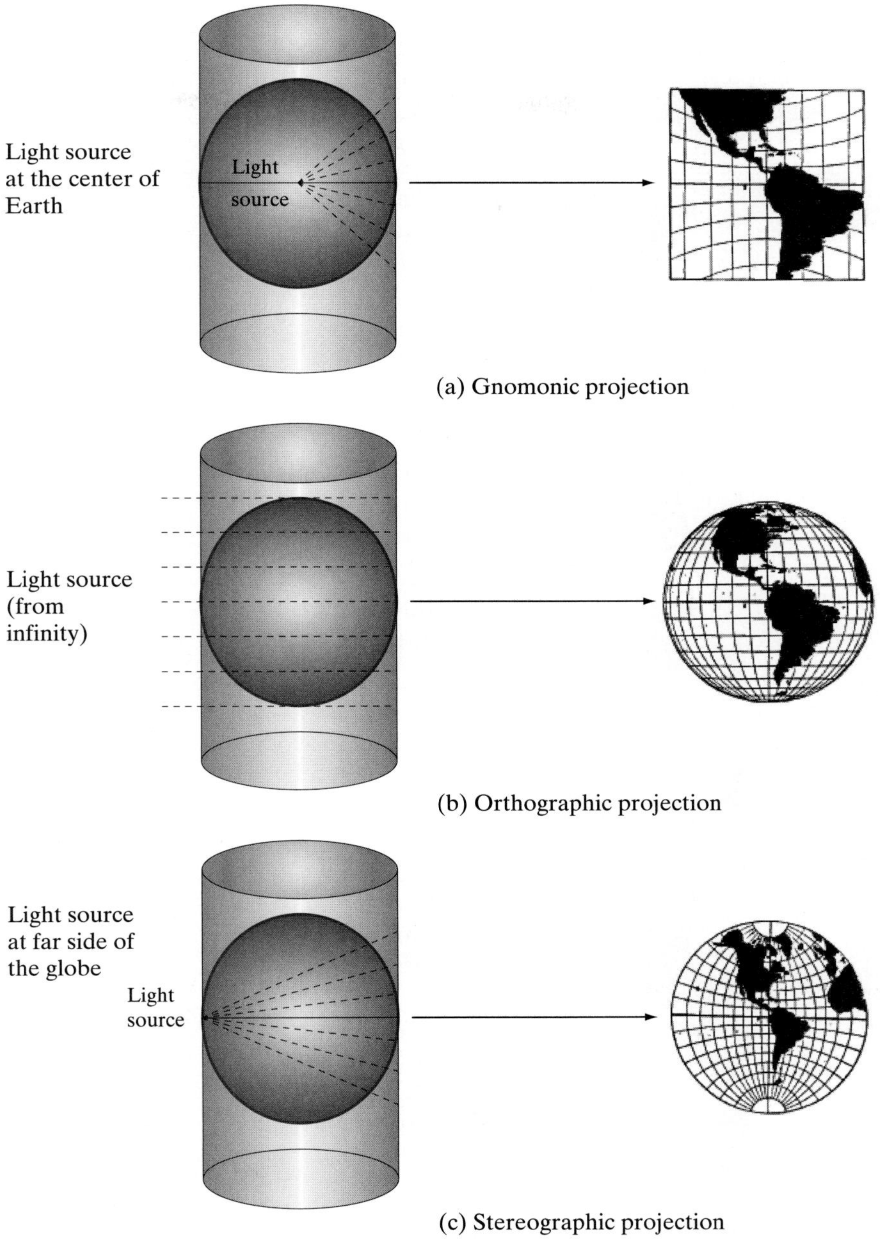

Light source at the center of Earth

Light source

(a) Gnomonic projection

Light source (from infinity)

(b) Orthographic projection

Light source at far side of the globe

Light source

(c) Stereographic projection

FIGURE 2.13
Map projections obtained by varying locations of light source (viewing point).

2.8.1 THE SHAPE OF EARTH: THE ELLIPSOID AND GEOID MODELS

In the map projections discussed previously, we assumed that the shape of Earth was a perfect sphere. From a geophysical perspective, the shape of Earth is highly irregular. The irregular shape of Earth makes it impossible to transform systematically the geometric relations from the three-dimensional surface of Earth to the two-dimensional surface of a map without some assumptions. The most fundamental assumption in this regard is to conceptualize Earth as a simple, solid shape that can be represented mathematically.

The *ellipsoid-geoid* model is the commonly used mathematical surface that represents the shape of Earth. In the context of georeferencing, the geoid and the ellipsoid are two distinct surfaces for different purposes: The ellipsoid is the reference surface for horizontal positions and the geoid is the reference surface for elevations. The relationships between Earth's irregular surface, the geoid, and the ellipsoid are shown in Figure 2.15.

Table 2.1

Common Map Projections, Their Properties, and Major Uses

Projection/Construction	*Appearance*	*Properties*	*Major Uses*
Albers Equal-Area/Conical	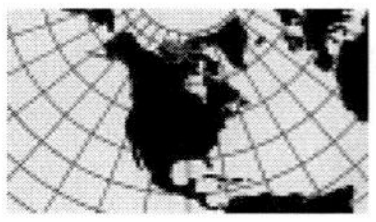 (a)	Equal area; conformal along standard parallels	Small regional and national maps
Azimuth Equidistant/Planar	 (b)	Equidistant; true directions from map center	Air and sea navigation charts; equatorial and polar area large-scale maps
Equidistant Conic/Conical	 (c)	Equidistant along standard parallel and central meridian	Region mapping of midlatitude areas with east–west extent; atlas maps for small countries
Lambert Conformal Conic/Conical	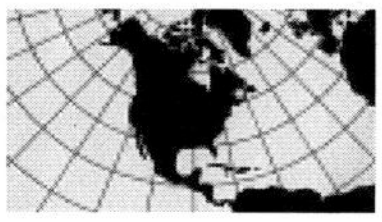 (d)	Conformal; true local directions	Navigation charts; U.S. State Plane Coordinate System (SPCS) for all east–west State Plane Zones; continental U.S. maps; Canadian maps
Mercator/Cylindrical	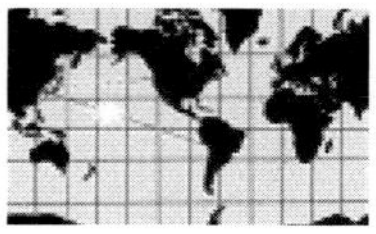 (e)	Conformal; true direction	Navigation charts; conformal world maps
Polyconic /Conical	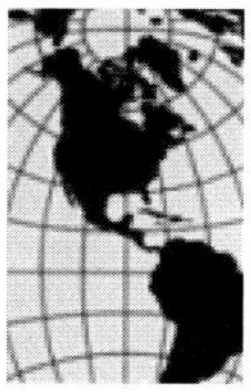 (f)	Equidistant along each standard parallel and central meridian	Topographic maps; USGS 7.5- and 15-minute quadrangles
Robinson/ Pseudo-Cylindrical	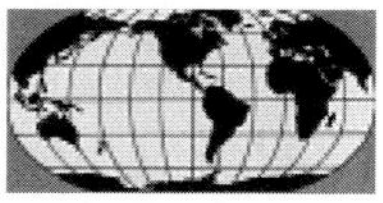 (g)	Compromise between properties	Thematic world maps
Sinusoidal/ Pseudo-Cylindrical	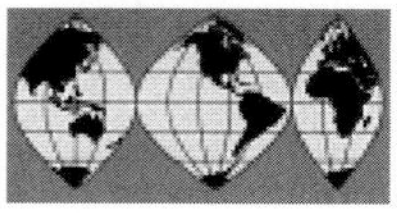 (h)	Equal area; local directions correct along central meridian and equator	World maps and continental maps
Stereographic/Planar	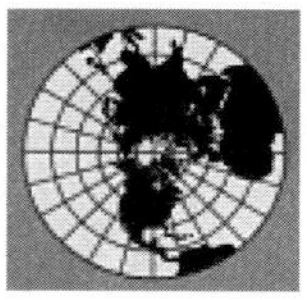 (i)	Conformal; true directions from map center	Navigation charts; polar region maps
Transverse Mercator/Cylindrical	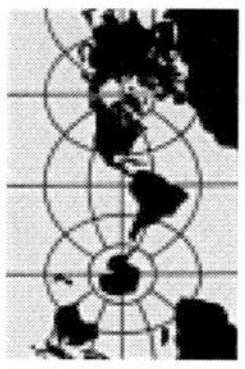 (j)	Conformal; true local directions	Topographic mapping for areas with north–south extent; U.S. State Plane Coordinate System (SPCS) for all north–south State Plane Zones

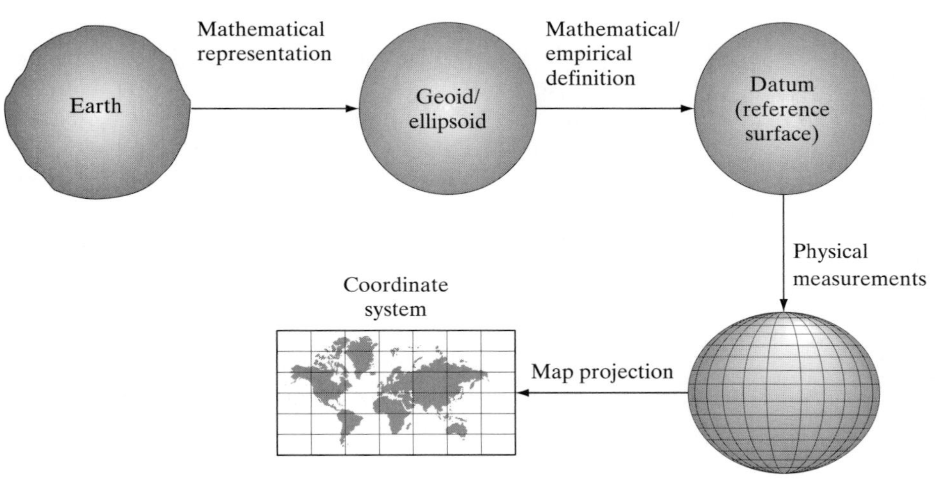

FIGURE 2.14
The concept of georeferencing. Georeferencing involves a series of transformations that progressively flattens the irregular surface of Earth so that measurements on the curved surface on Earth can be represented on a flat surface of a map.

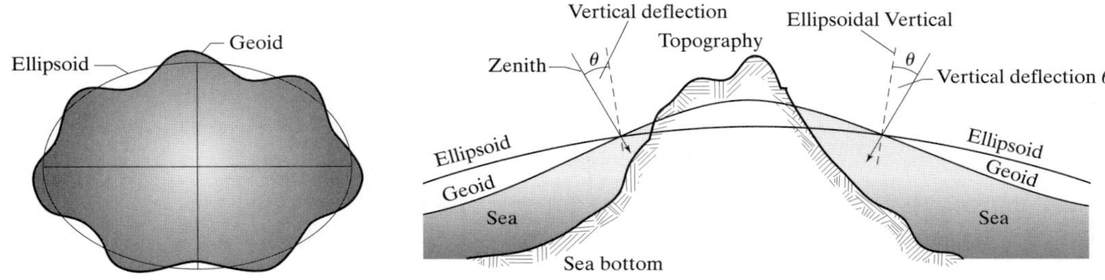

FIGURE 2.15
The relationships between Earth's irregular surface, ellipsoid, and geoid. *(Source: Seeber, 1993)*

The Ellipsoid Earth has been found to be slightly flattened at the poles, and the physical shape of the real Earth is closely approximated by the mathematical surface of the *rotational ellipsoid* or the solid obtained by rotating an ellipse on its minor axis (Seeber, 1993). The ellipsoid is widely used as the *reference surface for horizontal coordinates* (latitude and longitude) in geodetic networks. Because the flattening occurs at the poles, the figure may be further defined as an *oblate spheroid.* The amount of polar flattening can be determined by the following formula:

$$f = (a - b)/a \qquad \textbf{(2.11)}$$

where a and b are the lengths of the major and minor semi-axes of the ellipse, which correspond to the equatorial and polar radii of Earth respectively (Figure 2.16). Alternatively, the *eccentricity* (e) is used.

$$e^2 = (a^2 - b^2)/a^2 \qquad \textbf{(2.12)}$$

Also,

$$e^2 = 2f - f^2 \qquad \textbf{(2.13)}$$

and

$$1 - e^2 = (1 - f)^2 \qquad \textbf{(2.14)}$$

The flattening (f) is very close to 1/300. The difference in length between the equatorial and polar radii is about 11.5 km, and the polar axis is about 23 km shorter than the equatorial axis.

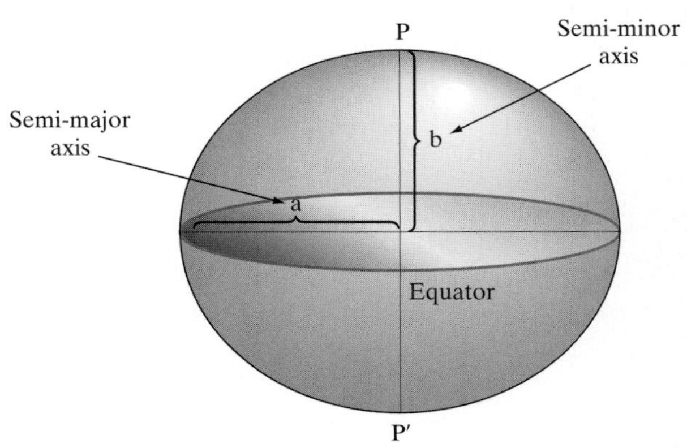

a = Major semi-axis of ellipsoid = Equatorial radius of the Earth
b = Minor semi-axis of ellipsoid = Polar radius of the Earth

PP′ = Axis of revolution of Earth's ellipsoid

FIGURE 2.16
The major and minor semi-axes for the computation of polar flattening of Earth.

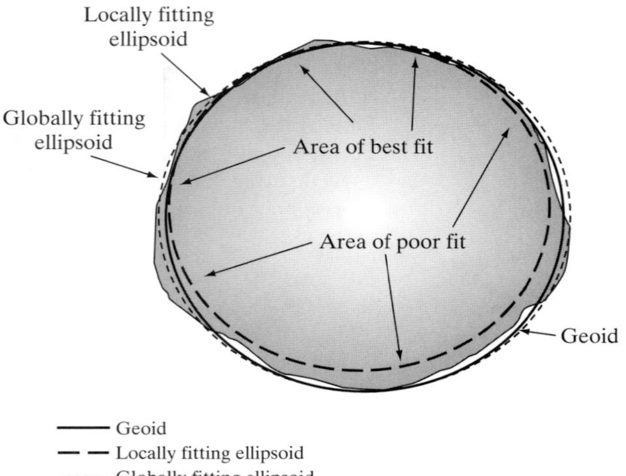

FIGURE 2.17
Different ellipsoids were developed to fit Earth's surface in different parts of the world. Locally fitting ellipsoids match the shape of Earth's surface in or around the area of interest; globally fitting ellipsoids are defined to match the shape of the entire planet Earth.

Since the 18th century, geodesists have been attempting to determine the polar flattening value that would produce the best Earth-fitting ellipsoid. They came up with tens of ellipsoids because many different values of equatorial and polar radii were used in the computations. Practically all these early ellipsoids were defined using constants determined by measurements in a particular area of interest, for example, a country or a continent. Ellipsoids defined in this way could naturally fit well that part of Earth's surface in or around the area of interest but not necessarily in other parts of the world (Figure 2.17). Depending on how well they can fit the local surface of Earth, different ellipsoids have been adopted by different countries. Some 30 ellipsoids are in common use today. These include, for example, Clarke 1866, International 1924, and the Geodetic Reference System of 1980 (GRS 80) (Table 2.2).

Since the 1960s, new values of the equatorial and polar radii of Earth have been obtained by satellite-based observations. There has also been greater international cooperation in the scientific study of the shape of Earth, through organizations such as the International Union of Geodesy and Geophysics (IUGG). These efforts have allowed new ellipsoids to be defined. Unlike ground-determined ellipsoids defined by using a physical origin on Earth's surface, satellite-determined ellipsoids are defined by using the *center of mass* of the whole planet Earth as the origin. These ellipsoids are called *geocentric* or *Earth-centered* ellipsoids. As these newly proposed ellipsoids are able to represent the entire Earth more precisely, many countries have adopted them in their georeferencing systems. In the United States, for example, the USGS has now adopted the Geodetic Reference System of 1980 (GRS 80), originally proposed by the International IUGG, instead of the Clarke 1866 ellipsoid that has been in use for over a century.

The Geoid The term "geoid" means earthlike. It is the shape of Earth that would be formed if the oceans were allowed to flow freely under the continents to create a single undisrupted global sea level covering the entire planet Earth and adjusted to gravity. Geophysically, the geoid is defined as an *equipotential* surface (i.e., a surface on which the *gravity potential* is everywhere constant) to which the direction of gravity is everywhere perpendicular. Therefore, the method of determining the geoid, known as *geoidal modeling*, is based on precise measurements of gravity across the continents and around the world. As the densities of Earth's rock constituents vary from one another and as the rock constituents are irregularly distributed over the world, readings of gravity measurements tend to vary

Table 2.2
Examples of Ellipsoids Used in National and Regional Mapping

Ellipsoid	*Semi-major Axis (a)*	*Inverse Flattening (1/f)*
Airy 1830	6,377,563.396	299.3249646
Australian National	6,377,340.189	298.25
Clarke 1866	6,378,206.4	294.9786982
Clarke 1880	6,378,249.145	293.465
Everest 1956	6,377,301.243	300.8017
International 1924	6,378,388	297
GRS 80	6,378,137	298.257222101
WGS 84	6,378,137	298.257223563

from place to place on Earth's surface. This means that Earth's equipotential surface and, by extension, the geoid are both irregular surfaces. Since gravitational forces are greater over the continents, where Earth's crust is thicker, than those over the ocean floors, where Earth's crust is thinner, the geoid generally rises over the continents and is depressed over the oceans. More locally, the geoid also shows various bumps and hollows that can depart from the average smooth surface by as much as 60 m in some instances. In general, the geoid coincides very well with the *mean sea level* (MSL) in the open oceans that make up most of Earth's surface. Therefore, all elevations in surveying and mapping are computed relative to the geoid as represented by the MSL. In other words, geoid is a *reference surface for vertical coordinates*.

Since the ellipsoid is defined entirely by the mathematical method, it is a smooth surface that is different from the geoid obtained by gravity measurement. The separation between these two surfaces at a particular point on Earth's surface is known as *geoid undulation, geoid separation*, or *geoid height*. The angle between the perpendicular to the ellipsoid and the perpendicular to the geoid (i.e., the plumb line) is called the *vertical deflection* (θ) (Figure 2.15).

2.8.2 GEODETIC AND VERTICAL DATUMS

The geoid and the ellipsoid, in the context of vertical and horizontal positions in georeferencing, are called *datums*. By definition, a "datum" is a model that describes the position, direction, and scale relationships of a reference surface to positions on the surface of Earth. There are two datums in georeferencing: *geodetic datums* and *vertical datums*. Although these two datums are spatially and conceptually related, conventionally they are defined separately because they describe the relationships between two different reference surfaces to Earth. Since various ellipsoids and different points on the geoid have been used to define datums, there are now tens of geodetic and vertical datums in use around the world. Some of these are used locally, others are used globally.

Geodetic Datums Geodetic datums are established to provide positional control that supports surveying and mapping projects covering large geographic areas such as a country, a continent, or the whole world. Classic geodetic datums are defined by five elements that include the position of the *origin* (in terms of latitude and longitude or other systems), the orientation of the geodetic network (i.e., the azimuth to another point), the parameters (i.e., major semi-axis and flattening) of the reference ellipsoid selected for the computation, and the geoid undulation at the origin. Starting with these geodetic parameters of the origin as the absolute minimum information, a national, continental, or global network of geodetic control points is then established, using standard geodetic surveying methods of *triangulation* and *trilateration*. The general technical name of the resulting network of known control points is a *horizontal geodetic reference system* but is commonly called the datum, such as the North American Datum (NAD), Australian Geodetic Datum (AGD), European Geodetic Datum (EGD), and Tokyo Datum (TD).

By using a geocentric reference ellipsoid, geodetic datums provide a better fit to the geoid globally than conventional datums and, consequently, are used in today's satellite-based navigation systems such as the Global Positioning System (GPS). Geocentric datums are referenced by three-dimensional Cartesian coordinates (i.e., x, y and z) with the origin coincident with the center of the ellipsoid. This allows the positions of geodetic control points to be expressed in terms of geocentric coordinates (i.e., x, y, z) rather than the conventional ellipsoidal coordinates (i.e., latitude, longitude, ellipsoidal height). There are well-developed algorithms for the conversion of geocentric coordinates to and from ellipsoidal coordinates. As a result, horizontal positions can be measured, computed, and documented by using either of these two coordinate systems.

Geodetic datums can be classified according to the geographical area that they cover into global and local datums. *Global geodetic datums* are those that have been developed for georeferencing based on a single point at the center of Earth. The World Geodetic Systems of 1984 (WGS 84) is such a datum (NIMA, 1997). WGS 84, a geocentric datum, was established by the World Geodetic Society using measurements of the National Imagery and Mapping Agency (NIMA) (formerly U.S. Defense Mapping Agency (DMA)). The initial definition of WGS 84 relied on Transit Satellite Doppler observations, which permitted its positions to be computed to accuracy at only the 1- to 2-meter level (NIMA, 1997). WGS 84 is now the standard ellipsoid model of Earth for GPS (Kaplan, 1996).

Local geodetic datums are based on ellipsoids that best fit Earth's surface in a particular area of interest, also known as *area of validity*. They are usually established to serve the georeferencing needs of a particular country or a group of adjacent countries. The NAD is a local datum (Doyle, 1997). This datum has its origin in the New England Datum, which was established by early surveys in the eastern and northeastern states referenced to the Clarke ellipsoid of 1866. The New England Datum was later extended through triangulation to the south and west of the United States without major readjustment of the original surveys in the east. In 1901, this expanded network was designated the United States Standard Datum. A triangulation station at Meades Ranch, Kansas, was selected as the origin because this is approximately the geographic center of the 48 conterminous states of the United States. In 1913, Canada and Mexico agreed to base their triangulation networks on the United States network, and the New England Datum was renamed the North American Datum (NAD). In the period 1927–1932, adjustment of all survey data into the system called North American Datum of 1927 (NAD 27) was completed (NAS, 1971).

Rapid advances in geodetic surveying, especially the GPS, soon made the weaknesses of the NAD 27 control network apparent. Geodetic measurements obtained using new technologies were hard to link to the existing survey control. In the 1970s and 1980s, the NGS undertook a readjustment of the horizontal datum by redefining the ellipsoid and incorporating many new measurements (NGS, 1994; NOAA, 1989). This resulted in the establishment of North American Datum of 1983 (NAD 83) (Table 2.3). The readjustment from NAD 27 to NAD 83 resulted in the locations (i.e., latitude/longitude) of all previous control points to shift, sometimes as much as over 200 m (Welch and Homsey, 1997) (Figure 2.18). NAD 83 is more robust than NAD 27 because it is a geocentric datum defined by satellite- and ground-based measurements, as well as using the GRS 80 ellipsoid as the reference surface (Doyle, 1997).

Vertical Datums A vertical datum is the *zero surface* from which all elevations or heights are measured. The MSL was used as a vertical datum for mapping because the sea surface is available worldwide. For the purpose of establishing the MSL for use as a vertical datum, it is necessary to continuously measure the rise and fall of the ocean level at one or more tidal gauge stations for a period of ten to twenty years. The purpose is to average out the highs and lows of the tides caused by the changing effects of the gravitational forces of the sun and the moon. It should be noted that MSLs determined at tidal gauge stations at different locations, even along the same seacoast, will vary from one another. Therefore, the MSL selected by a particular country or region as the vertical datum is by no means the true zero surface for measuring elevations. It is simply an arbitrarily selected surface that can be more conveniently determined than the true zero point on the surface of the geoid obtainable only by sophisticated gravity measurements. Since the geoid is not a physical or tangible surface, it is not possible to actually measure elevations above or below it. For all practical purposes in georeferencing, it is assumed that the geoid and the MSL are coincident along coastlines.

Vertical datums may also be classified into global and local vertical datums. Advances in surveying technologies such as GPS have enabled the National Imagery and Mapping Agency (NIMA) in the United States to develop a geoid model to an accuracy of about one meter (NIMA, 1996). An international initiative, the Global Sea Level Observing System (GSLOS) established by the Intergovernmental Oceanographic Commission in 1985, will help make the definition of a global vertical datum a reality. GSLOS includes some 300 sea level observation stations maintained by 80 countries around the world (Tolkatchev, 1996).

Table 2.3
Comparison of Datum Elements between NAD 27 and NAD 83

Datum Elements	*NAD 27*	*NAD 83*
Reference ellipsoid	Clarke 1866 a = 6,378,206.4 m f = 1/294.9786982	GRS 80 a = 6,378,137.0 m f = 1/298.2572221
Datum origin	Trig. Station Meades Ranch, Kansas	Center of mass of Earth
Longitude origin	Greenwich meridian	Greenwich meridian
Azimuth orientation	From south	From north
Adjustment	• 25,000 geodetic control points • Several hundred baselines • Several hundred azimuths	• 250,000 geodetic control points • Approximately 30,000 EDM baselines • 5,000 astro-azimuths • Numerous Doppler points
Best fitting	North America	Worldwide

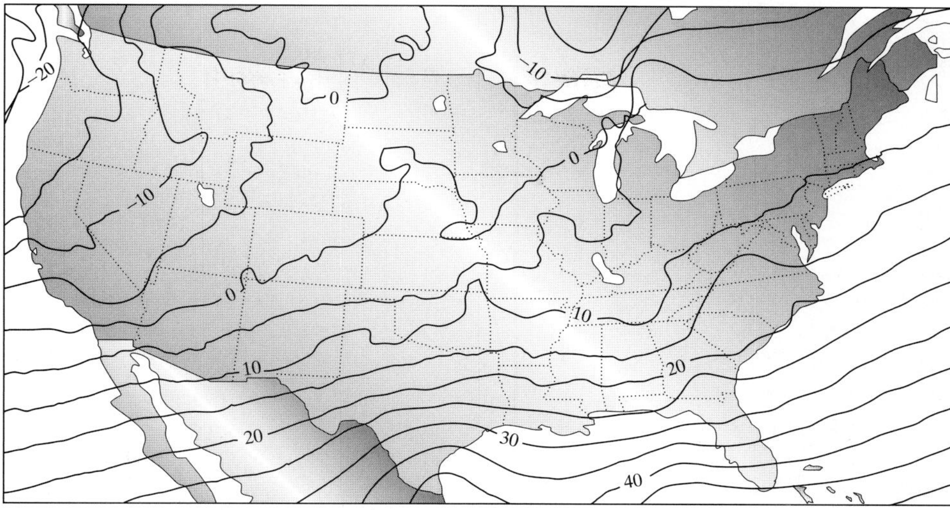

(a) Latitude datum shift in meters (NAD 83 minus NAD 27)

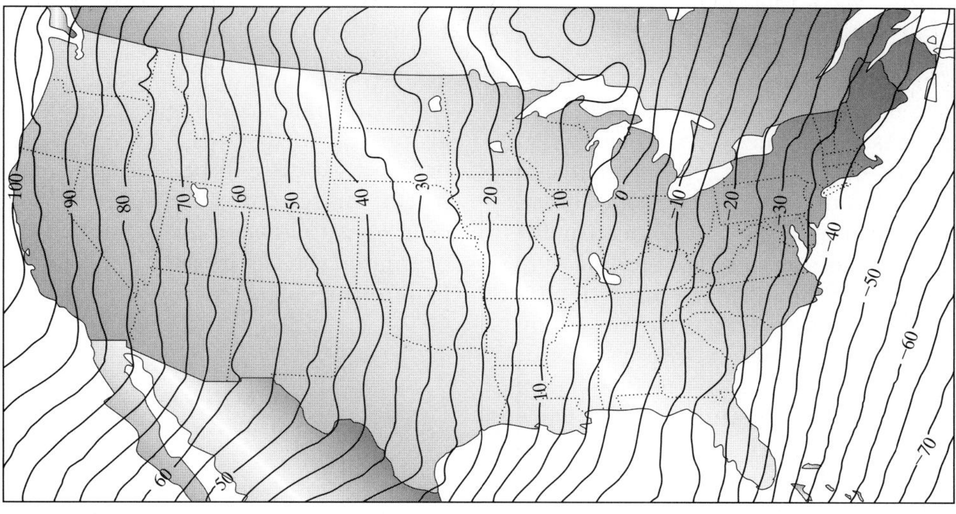

(b) Longitude datum shift in meters (NAD 83 minus NAD 27)

FIGURE 2.18
Datum shifts between NAD 27 and NAD 83. *(Source: Wisconsin SCO, 1992)*

Local vertical datums are defined to serve the georeferencing needs of a country or a group of adjacent countries. In the United States, the National Geodetic Vertical Datum of 1929 (NGVD 29) is a local vertical datum. This datum was based on the MSL readings of 21 tidal gauge stations in the United States and 5 in Canada. As with NAD 27, errors and distortions were detected over the years and it became increasingly difficult to fit new control stations established by modern geodetic surveying methods to the old control network. Also, it has been found that thousands of the original NGVD 29 control stations were lost due to railway abandonment and highway construction. This prompted the NGS to start an extensive releveling and densification program in the 1970s. The data obtained in these surveys, along with the International Great Lakes Datum of 1985 (IGLD 85), were readjusted as the North American Vertical Datum of 1988 (NAVD 88). This vertical datum, implemented by the U.S. Federal Government in 1993, was redesigned using the latest geoid definition to provide elevation information compatible with GPS. The redefinition of the NAVD has caused changes to all elevations previously referenced to the old NAVD, in the same way the redefinition of the North American geodetic datum has had on horizontal positions. The magnitudes of elevation changes are as great as 15 meters in some instances (Figure 2.19).

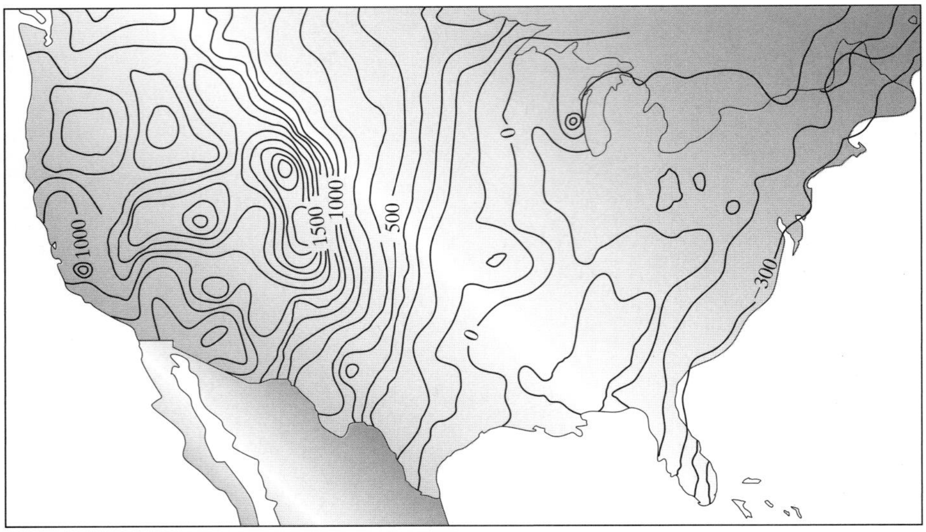

FIGURE 2.19
Magnitude of vertical datum shift in millimeters (NAVD 88 minus NAVD 29). *(Source: Wisconsin SCO, 1992)*

A knowledge of the horizontal and vertical datums employed by a map or a georeferenced image is important because it is often necessary to carry out transformation of the map or image so that it can conform with other maps or images. As noted above, the change from one datum to another can cause differences in rectangular coordinates. Since old map data based on NAD 27 will inevitably be used in GIS, they have to be transformed to match the new map data based on NAD 83. Fortunately, today's major image processing software (such as the Projection Chooser in ERDAS IMAGINE) and GIS software (such as the PROJECT command and its SPHEROID subcommand in ArcInfo) have stored all the parameters for different ellipsoids that can be selected for use in converting from one type of reference surface to another for different map projections, ellipsoids, and datums.

2.8.3 RELATIONSHIP BETWEEN COORDINATE SYSTEMS AND MAP PROJECTIONS

The relationship between coordinate systems and map projections is confusing because coordinate systems are constructed on the basis of map projections, but they are not map projections themselves. Map projections and coordinate systems are distinct concepts that serve two different purposes in georeferencing. The function of map projections is to define how positions on the Earth's curved surface are transformed onto a flat map surface. A coordinate system is then superimposed on the surface to provide the referencing framework by which positions are measured and computed (Figure 2.20). There are well-established computer programs that allow rectangular coordinates to be transformed to and from geographic coordinates. Many of these computer programs for use in personal computers are available free of charge from the Internet Web site of the NGS *(http://www.ngs.noaa.gov)*. The following two major coordinate systems have been used for the topographic maps produced by the USGS.

The Universal Transverse Mercator (UTM) Coordinate System

The Universal Transverse Mercator Coordinate System is a coordinate system based on the Transverse Mercator Projection (known in Europe as the Gauss–Kruger Projection) invented by Johann Heinrich Lambert (1728–1777). Lambert modified the Mercator Projection by changing the aspect of the projection to transverse, placing the true poles of Earth to lie on the equator of the basic projection. In this way, the distortion from pole to pole (for a North–South strip) is minimized. Distances are true only along the central meridian selected (where the "cylinder," the developable surface, touches the surface of Earth). All distances, directions, shapes, and areas are reasonably accurate close to the central meridian. Because this is a conformal projection, shapes and angles within any small area are essentially true. However, the meridians and parallels of the Transverse Mercator Projection are no longer straight lines as in the case of the regular Mercator Projection (Figure 2.21).

In order to minimize the distortions that occur in the Transverse Mercator Projection so that it can be used for georeferencing, several further modifications have been made. The resulting projection is called the Universal Transverse Mercator (UTM) Projection. The

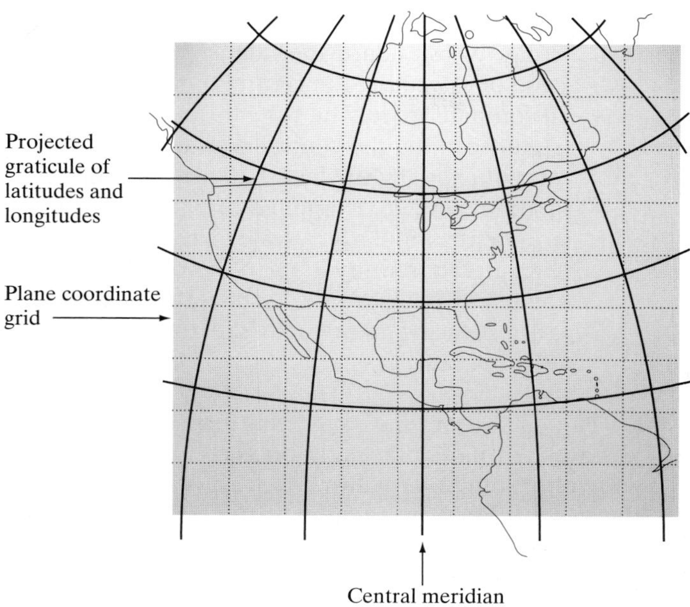

FIGURE 2.20
Relationship between map projection and plane coordinate system. The graticule represents positions on the three-dimensional surface of Earth that have been projected or transformed onto the two-dimensional surface of the map. The rectangular coordinate grid is superimposed on the graticule so that positions of mapped features can be georeferenced by means of linear measurements rather than by angular measurements.

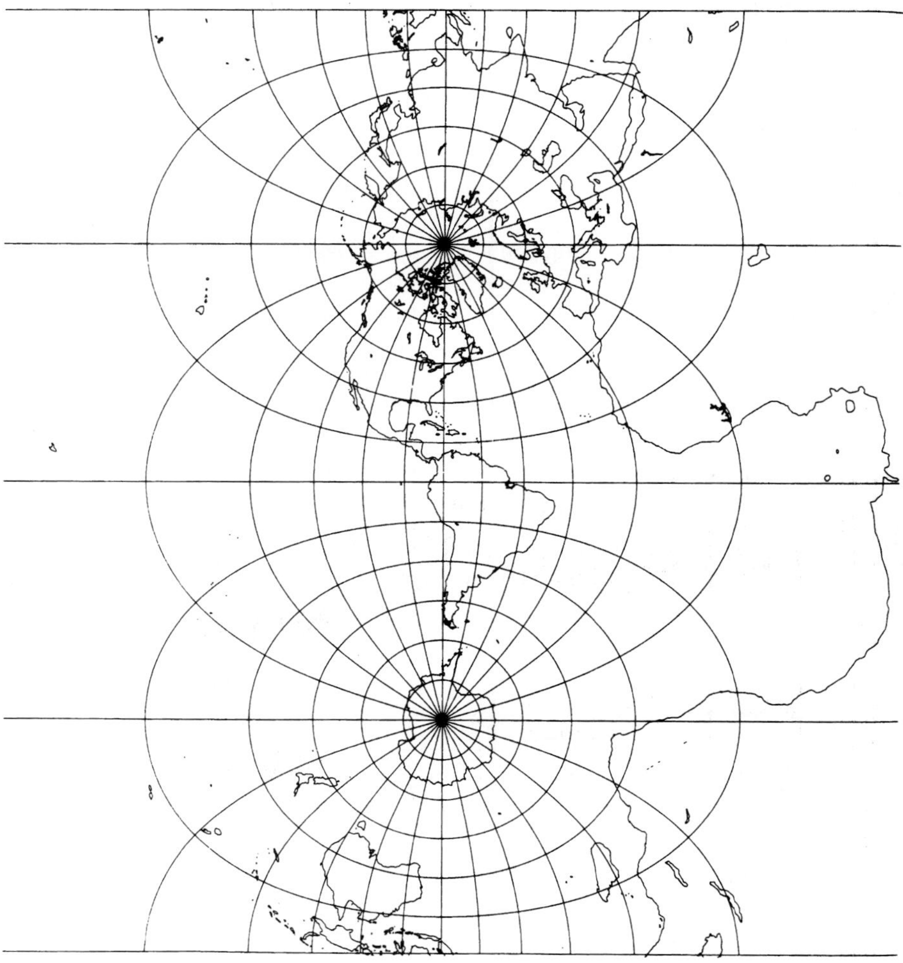

FIGURE 2.21
The Transverse Mercator Projection. *(Source: Snyder, 1987)*

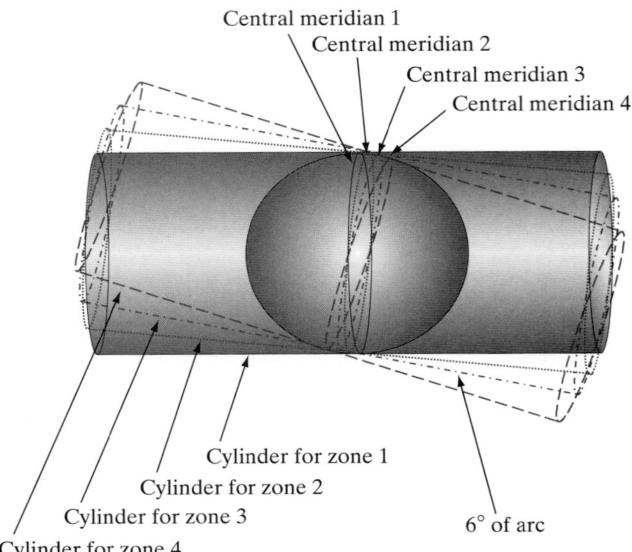

FIGURE 2.22
Creation of the Universal Transverse Mercator (UTM) Projection. The UTM projection is created by using multiple cylinders that touch the globe at 6° longitude intervals, creating 60 projection zones. (See Figure 2.24.)

UTM projection is different from the regular Transverse Mercator Projection in the following ways:

- The projection is applied repeatedly by using multiple cylinders that touch the globe at 6° intervals, resulting in 60 projection zones each 6° of longitude wide (Figure 2.22).
- To avoid the extreme distortions that occur in the polar areas, the projection zones are limited at 84° N and 80° S.
- To improve the overall accuracy of measurements within a projection zone, the cylinder is made to intersect the globe at two standard meridians that are 180 km east and west of the central meridian (i.e., a secant cylinder). This results in true scale along two standard meridians of longitude instead of one along the central meridian.
- To compensate for the scale distortion that is introduced along the central meridian, a scale factor, which is slightly less than unity in the order of about 0.9996, is applied to all distance measurements. Similarly, a scale factor that is slightly greater than unity in the order of 1.0004 is also applied to compensate for distortions in all distance measurements near the zone boundaries (Figure 2.23).

The UTM coordinate system is formed by superimposing a regular square grid on each UTM projection zone of 6° longitude width. The grid is aligned so that the vertical lines are parallel to the central meridian. The UTM zones are numbered from 1 through 60, starting at the International Date Line (also known as the *antimeridian*, $\lambda = 180°$) and proceeding east. Thus, Zone 1 extends from 180° W to 174° W, with the central meridian at 177° W. North America lies between Zone 3 and Zone 21.

Each UTM zone is divided into horizontal bands spanning 8° of latitude. These bands are identified by letters, south to north, beginning at 80° S with the letter C and ending with the letter X at 84° N (Figure 2.24). The letters I and O are skipped to avoid confusion with the numbers 1 (one) and 0 (zero). The band X, which is the northernmost band, spans 12° of latitude.

UTM coordinates are expressed as a distance in meters to the east, referred to as the "easting," and a distance in meters to the north, referred to as the "northing." UTM eastings are referenced to the central meridian, which is assigned a value of 500,000. This gives the zone a false origin that is 500,000 m west of the central meridian, thus eliminating the use of negative values for the western half of the zone. UTM northings are referenced to the Equator. For position fixing in the northern hemisphere, the Equator is assigned the northing value of 0 (zero) m N. To avoid

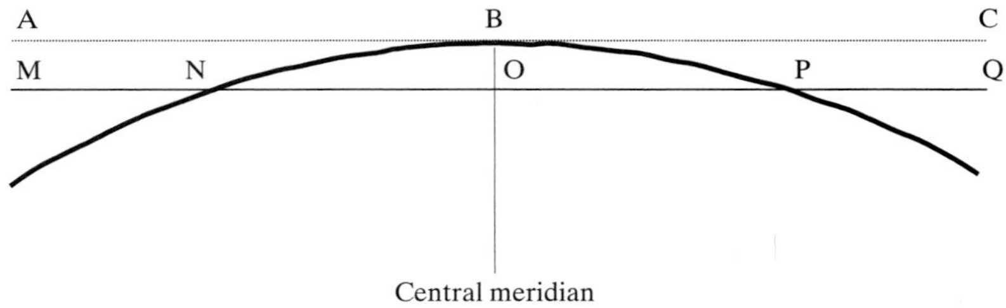

FIGURE 2.23
Balancing the scale variation in Universal Transverse Mercator (UTM) Projection by having the cylinder touching the globe at two standards rather than a single central meridian and applying a scale factor to measurements.

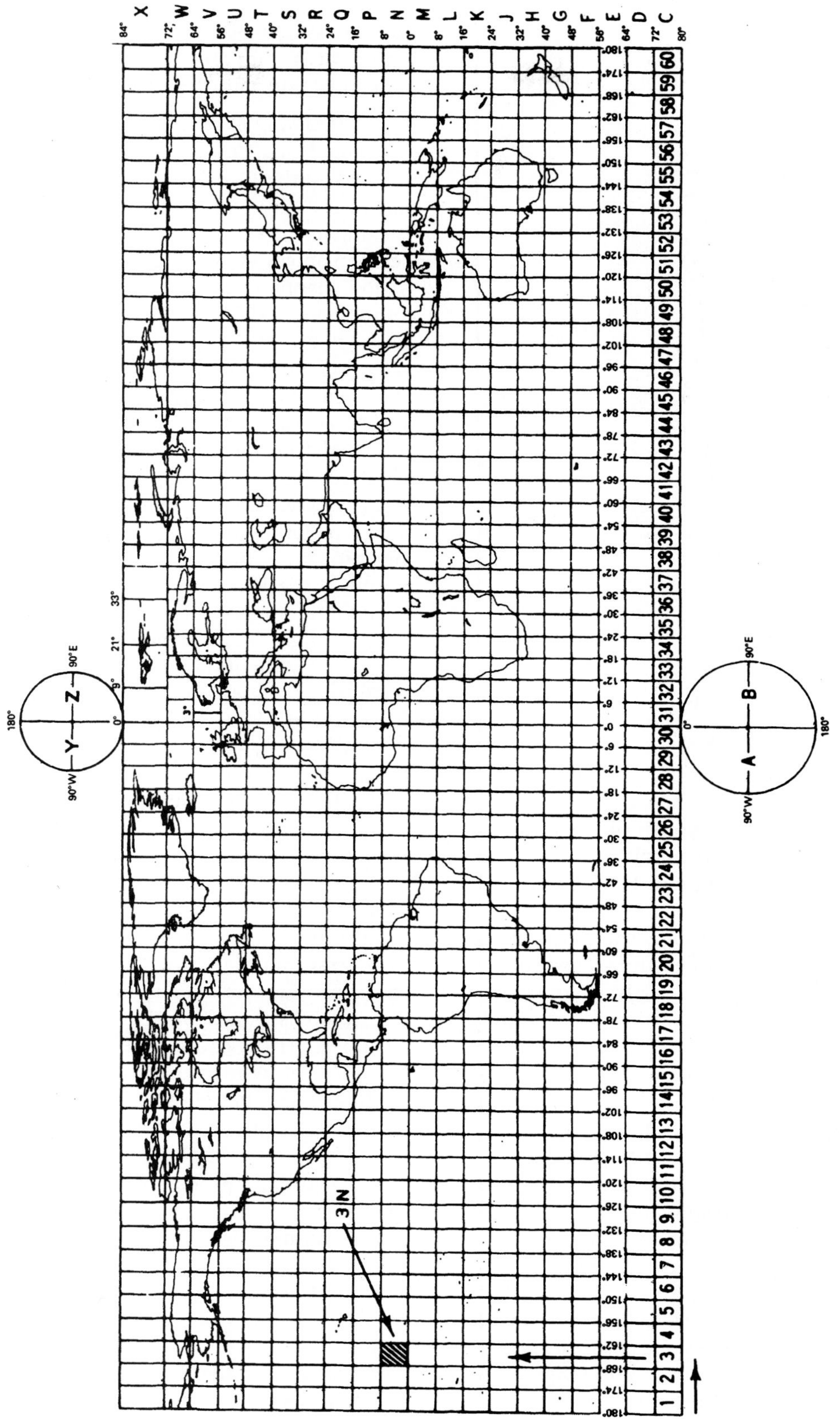

FIGURE 2.24
Universal Transverse Mercator (UTM) grid zone designations for the world shown on Equidistant Cylindrical Projection. *(Source: Snyder, 1987)*

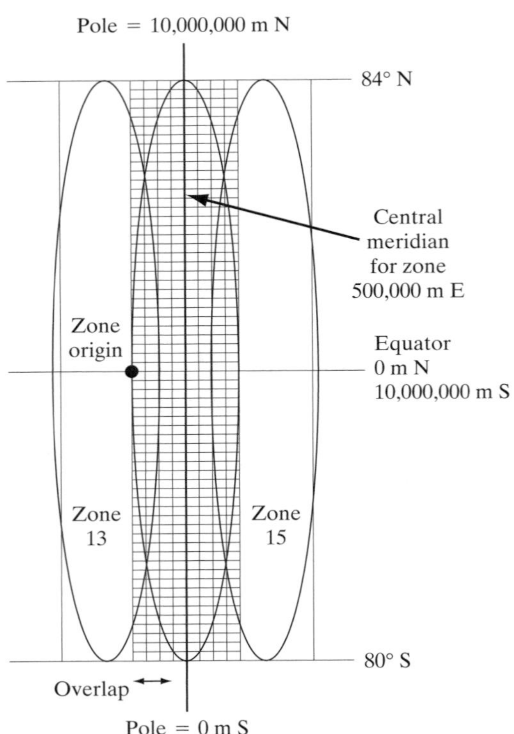

FIGURE 2.25
The Universal Transverse Mercator (UTM) coordinate system. *(Source: Clarke, 1999)*

negative numbers for locations south of the Equator, position fixing in the southern hemisphere is made with the Equator assigned a value of 10,000,000 m S. (Figure 2.25).

By combining the zone number, the hemisphere (which indicates whether the zone is north or south of the Equator), and the easting and northing coordinate values, any point on Earth's surface can be uniquely located.

Since its adoption by the U.S. Army in 1947 for designating rectangular coordinates on large-scale maps for the entire world, the UTM coordinate system has also been widely used by national mapping agencies for local applications. The UTM is probably the most popularly used coordinate system in medium-scale mapping today. Both the United States and Canada use this coordinate system for their national mapping programs (Colvocoresses, 1997; Moore, 1997). All USGS topographic maps published in the last 30 years or so include UTM either in full grid lines or as blue tick marks on the margin of the maps. On 1:24,000 maps, UTM grid lines are spaced 1000 m apart. On 1:250,000 and 1:100,000 maps, UTM grid lines are 10,000 m apart. Realizing that the UTM is the most popular coordinate system among map users, the USGS decided to produce most of its digital products on the UTM (Moore, 1997). Many manufacturers of GPS receivers include this coordinate system as an option in their products, making it a de facto standard coordinate system in the spatial data collection industry.

The State Plane Coordinate System (SPCS)

The State Plane Coordinate System (SPCS) was first developed by the U.S. Coast and Geodetic Survey (now a part of the National Ocean Survey) in the early 1930s. The original objective of this system was to provide a conformal mapping system that would accommodate the surveying, mapping, and engineering needs at the state and county levels. SPCS divides the fifty states of the United States, Puerto Rico, and the U.S. Virgin Islands into over 120 numbered sections, referred to as *zones* (Figure 2.26). These zones were derived from three conformal projections: Transverse Mercator, Oblique Transverse Mercator, and Lambert Conformal Conic, with two standard parallels. All states use either the Transverse Mercator or Lambert projections, with the exception of Alaska, Florida, and New York, which use both. The Oblique Transverse Mercator Projection is used only for the Aleutian panhandle of Alaska (zone 5010) because it is neither predominantly north–south nor east–west.

By using a different map projection according to the shape of a state and by dividing the states into two or more zones that are projected independently, the scale distortion of SPCS is kept within the accuracy of one part in 10,000. This was considered the limit of surveying accuracy when the system was set up in the 1930s. In order that the coordinates are positive in the rectangular system, the origin of the coordinate system is arbitrarily chosen at a point in the southwesternmost corner of the map. For example, the State of Georgia, which uses Transverse Mercator for its SPCS, is divided into two zones (East and West). The East Zone has its central meridian at 82° 10′ W, and the West Zone has its central meridian at 84° 10′ W. The origin of the coordinate system for the two zones is at latitude 30° 00′ N. The *x* coordinate (easting) at the origin is 500,000 feet, while the *y* coordinate (northing) at the origin is 0 feet, based on NAD 27 (Figure 2.27).

In general, the zone boundaries within each state follow county lines to facilitate local surveying and mapping needs (Figure 2.26). Each zone has a unique code designated by the USGS, and the standard USGS 7.5-minute and 15-minute quadrangle sheets carry tick marks showing the locations of SPCS grids. In order to be consistent with local use and legal precedent in the definition of property boundaries, the unit of measurement of SPCS is the *U.S. Survey Foot*, which is slightly different from the *International Foot*. SPCS as it was originally developed is based on NAD 27 using the Clarke 1866 ellipsoid. This original set of systems is now commonly referred to as SPCS 27 in order to distinguish it from the new SPCS that has been redefined using NAD 83. This new set of coordinate systems is called SPCS 83.

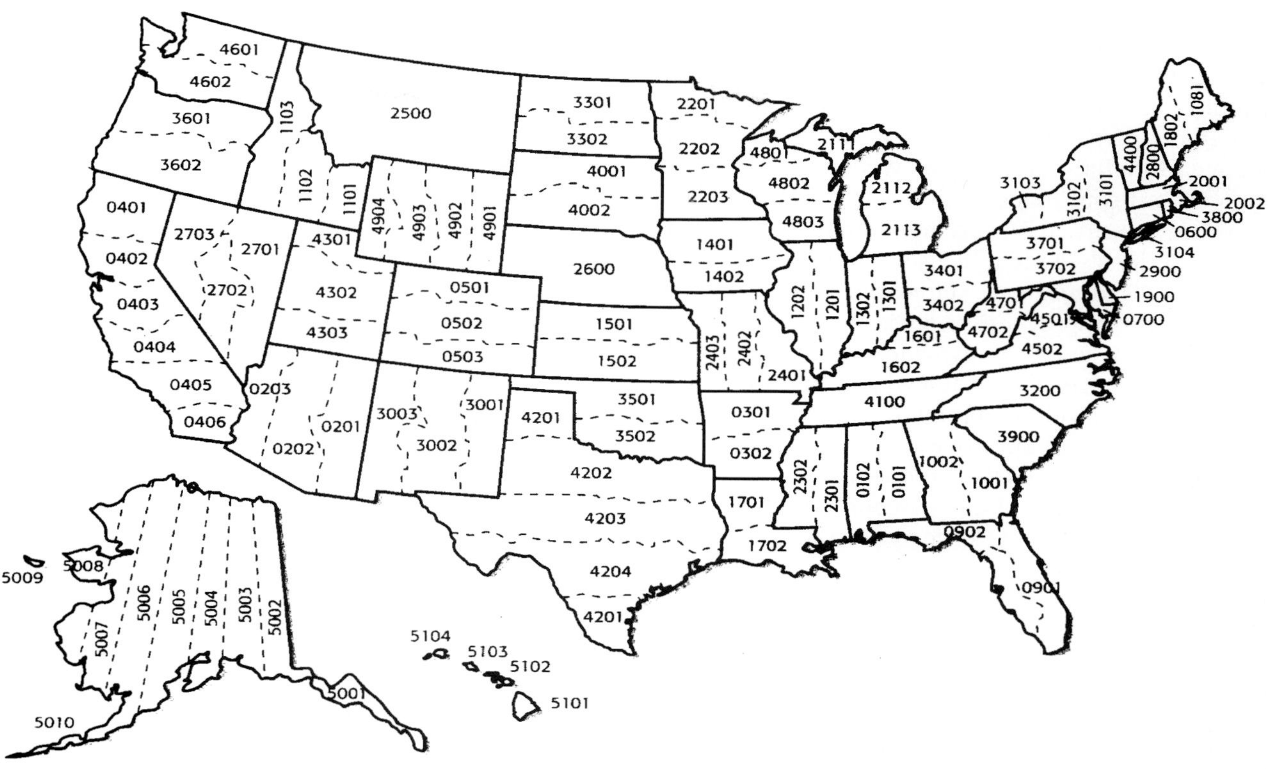

FIGURE 2.26
Zones of the State Plane Coordinate System (SPCS).

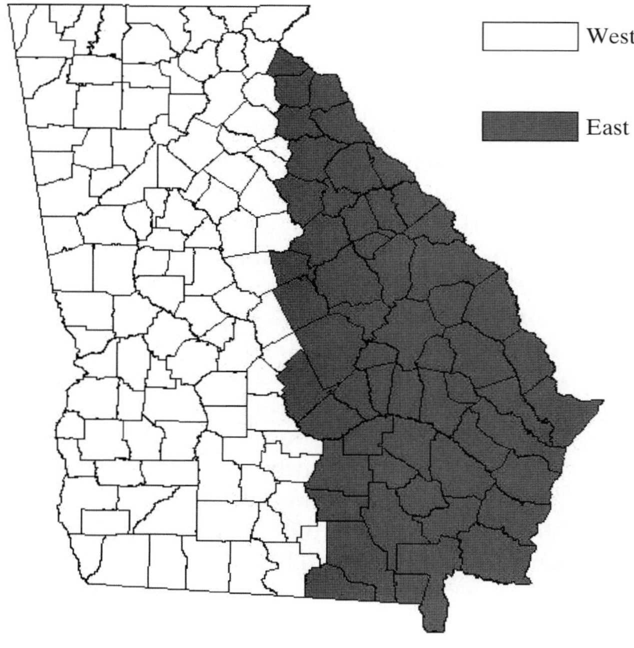

FIGURE 2.27
The State Plane Coordinate System of Georgia. Georgia is divided into an East Zone and a West Zone in the SPCS. Note how the two zones are divided along county boundaries.

In addition to the use of a different datum, SPCS 83 is different from its predecessor in several important ways. Each SPCS zone has a new FIPSZONE number defined from the Federal Information Processing Standards (FIPS), which is used concurrently with its identification number in the old system. SPCS 83 utilizes the meter as the standard unit of measurement, although Wisconsin continues to use the U.S. Survey Foot and Arizona replaced the U.S. Survey Foot by the International Foot. For the State of Georgia using SPCS 83, for example, the *x* coordinate of origin for the East Zone (located at longitude 82° 10′ W and latitude 30° 00′ N) is 200,000 m and the *y* coordinate is 0 m, while the *x* coordinate of origin for the West Zone (located at longitude 84° 10′ W and latitude 30° 00′ N) is 700,000 m and the *y* coordinate is 0 m (Snyder, 1987).

There have also been changes to the number of zones in several states (e.g., California uses six instead of the original seven, Montana uses one zone rather than three, and Nebraska, South Carolina, and Puerto Rico/Virgin Islands are now each covered by one zone instead of two). These changes in the number of zones necessitated other changes, such as the false origin and central meridians in those zones using the Transverse Mercator Projection and the standard parallels in those zones using the Lambert Conformal Conic Projection. As a result, the characteristics of SPCS 83 are not as consistent as those of the original SPCS 27.

For several decades, SPCS has been recognized by all states as the standard coordinate system for cadastral surveying and the compilation of engineering maps. It is useful for the compilation of regional topographic and engineering maps and for the input, storage, and exchange of digital map data, as well as for the production of hard-copy maps. However, the need to maintain the one part in 10,000 accuracy standard as noted above has seriously limited its use in other application areas. The division of many states into multiple zones means that, as mapping must not extend beyond these zones, it is not possible to provide an uninterrupted coverage of a particular state as a whole. In response to the need for statewide mapping and spatial database creation in GIS, some state governments have now developed their own projection and coordinate systems to supplement SPCS. In SPCS, Wisconsin is covered by three Lambert Conformal Conic projection zones. The need to avoid splitting the state into different mapping zones in many applications has prompted the Wisconsin Department of Natural Resources to develop the Wisconsin Transverse Mercator (WTM) projection and coordinate system (Wisconsin SCO, 1998). This system centers a UTM-like zone on the 90° W meridian, thereby covering the entire state with one single zone. This coordinate system was originally based on NAD 27 but has been redefined for NAD 83. WTM illustrates how a state coordinate system can be designed and created to satisfy the needs of particular applications and to avoid the inherent problems of SPCS for wide-area mapping and GIS purposes.

2.8.4 COORDINATE TRANSFORMATION AND THE PROBLEM OF UTM BOUNDARIES

In many situations, a GIS project may require the use of data from different sources. It is inevitable that the data are in different scales and projections. Therefore, coordinate transformation is always an essential function of GIS. The similarity and affine transformations explained in Section 2.5.3 can be used for direct coordinate transformation. If more accurate results are required, higher order polynomials have to be used for the transformation, such as in the case of transforming the geodetic datum for the North American continent from NAD 27 to NAD 83. Maling (1992) suggested the use of second-order polynomial equations for transforming from a local network to UTM as follows:

$$X = a_0 + a_1x - b_1y + a_2(x^2 - y^2) - 2b_2xy \qquad \textbf{(2.15a)}$$

$$Y = b_0 + b_1x + a_1y + b_2(x^2 - y^2) + 2a_2xy \qquad \textbf{(2.15b)}$$

This is really a second-order conformal transformation equation (cf Equation 2.7) derived from a complex number formulation by Lucas (1977), which takes the following form:

$$(X + iY) = \sum_{k=0}^{n-1}(a_k + ib_k)(x + iy)^k \qquad \textbf{(2.16)}$$

Where i is an imaginary number and is equal to $\sqrt{(-1)}$. Equation 2.16 was expanded with K=2, to cope with the second order scale change for UTM. The result of the expansion was split into two equations by equating real and imaginary parts as 2.15a and 2.15b. With the availability of computer power today, this type of transformation can be easily programmed and executed.

Coordinate transformation in GIS is used not only for data represented in different coordinate systems, but also for data represented in the same coordinate system, as in the case of the UTM system explained here. The use of the UTM grid system in mapping presents a zone boundary problem. Because each zone is only 6° in east–west extent, it is very common to find an area covering two or more zones in a single GIS project. The representation of geographical features becomes a problem at the zone boundaries. When a particular feature is located in two adjacent UTM zones, the portions of the feature on the two sides of the zone boundary are separately referenced to different UTM grid systems. How can we handle this problem of cross-boundary geographical features if a digital GIS database is to be created? This will cause problems in data retrieval, display, calculation of area and perimeter, and many other types of spatial analysis for GIS applications. Different approaches have been developed to deal with these zone boundary problems.

- Develop a wider UTM zone, say, of 10° so that it will cover a larger east–west extent, an approach adopted by the Province of Alberta in Canada to make the whole province in one zone.
- Store data in the UTM zone with the larger portion of the project area, and use the coordinate system of that zone to locate features in the smaller project area.
- Replace UTM Projection with Lambert Conformal Conic Projection as in the case of the SPCS.
- Geographical features in a map that covers two UTM zones can be brought to a unique georeferencing system for the purpose of data processing by reprojecting the portion of one zone to the grid system of the other zone or by projecting both portions to a third grid system. This is a very popular approach because it is quite easy to implement, as many software packages are equipped with this function.
- Extend the map coverage at the border of one grid system, say, by 40 km, into an adjacent zone to avoid the use of two grid systems.
- Use geographical coordinates (i.e., latitude and longitude).

Map projection and coordinate transformation is an essential function of raster- and vector-based geographic data processing. This function is becoming more important today than ever before because of the growing trend toward the use of spatial data from different sources, as well as the increasing use of GIS in the multiuser environment where different users have different requirements in the analysis and display of geographic information. Most GIS software vendors have responded to the need for map projection and coordinate transformation by providing built-in facilities in their software products. Also, several stand-alone software packages have been developed and commercially marketed for such a purpose.

Although the mathematics behind map projection and coordinate transformation is relatively advanced, GIS software vendors have simplified the use of transformation function. In the Windows environment, this is usually done with the aid of wizards that will guide the user through the different steps by means of a sequences of graphical user interfaces (Figure 2.28). All the user needs to do is to provide the system with the projection or coordinate system of the source and output data and properties of the input and output coordinate systems such as datum, origin, standard parallel, and central meridian. The system then displays all the input parameters and prompts the user to confirm before proceeding to perform the transformation.

Map projection and coordinate transformation is most often used and perceived as an input function by which spatial data in different map projection and coordinate systems are converted into a standard form.

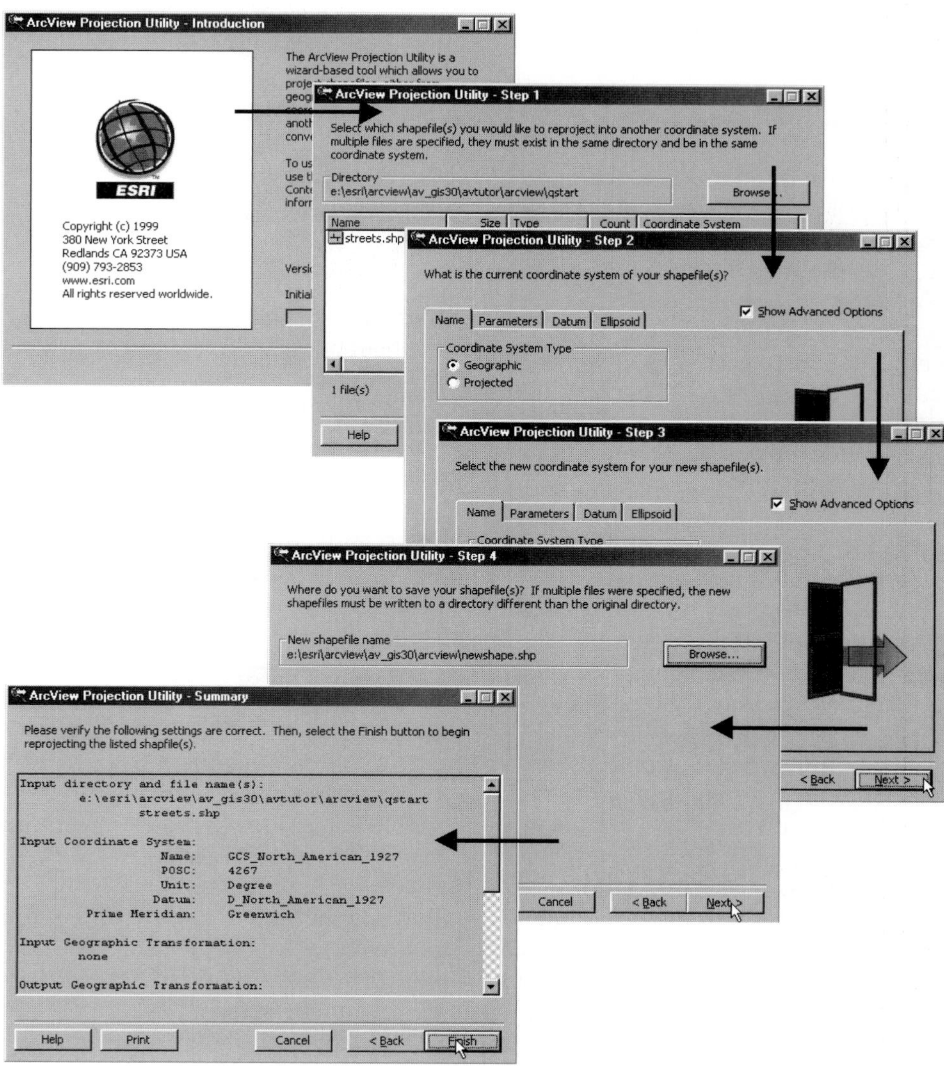

FIGURE 2.28
Coordinate transformation using the Projection Utility wizard of ArcView GIS.

for organizations to store the data using geographical coordinates and transform them into UTM coordinates when the data are displayed and measured. The processing power of today's computers has the flexibility of storing and using spatial data in different map projections and coordinate systems, thus making transformation a more frequently used data processing function than ever before.

2.9 Acquisition of Spatial Data for the Terrain: Topographic Mapping

Spatial data for topographic maps, the general purpose and reference maps for GIS, are collected using the methods of land surveying, aerial and satellite photogrammetry, and global positioning system (GPS).

2.9.1 Land Surveying

Land or field surveying is used to determine the positions (plane coordinates) of terrain features in the field. The positions of features can be fixed by the following methods (Figure 2.29): (1) *triangulation* using two angles (Shafer, 1995); (2) *traversing* using one distance and an angle (Evett, 1995); (3) *off-set* using two distances and a right angle, and (4) *trilateration* using two distances (Sturgess and Carey, 1995). Elevation data are mainly obtained by the method of differential leveling using an instrument called a "level" (Figure 2.30) (Moffitt and Bossler, 1998). Linear measurements in land surveying have been made by using either a steel measuring tape or an optical instrument called a tacheometer or a laser range finder. Angles are measured using an optical measurement device known as the theodolite. It should be noted that no matter what surveying instruments and techniques are used, field measurements represent only a small portion of the work involved in field mapping. All types of surveying involve a considerable amount of computation that is required to turn field measurements into coordinates and graphical representations on maps.

The most accurate form of land surveying is called *geodetic surveying*, with the objective of establishing the control network in support of the definition of geodetic datums as explained in Section 2.8.2. Geodetic surveying makes use of angular and distance measurements between points to establish a spatial framework necessary to control all surveys. Geodetic measurements are used for computing the positions of the control points relative to one another on Earth's surface. In order to determine the *absolute* positions of these control points, the geographic coordinates (i.e., latitude and longitude) of the origin of the datum as well as the orientation (i.e., azimuth or bearing with respect to true north) of a line of sight from the origin to another control point must be accurately determined by the method of geodetic astronomy (Vanicek and Krakiwsky, 1986).

Method	Diagram
Measure AC, BC	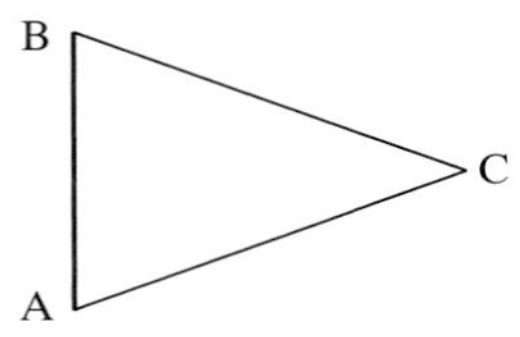 B, C, A (triangle)
Measure AD or BD and DC at right angle to AB	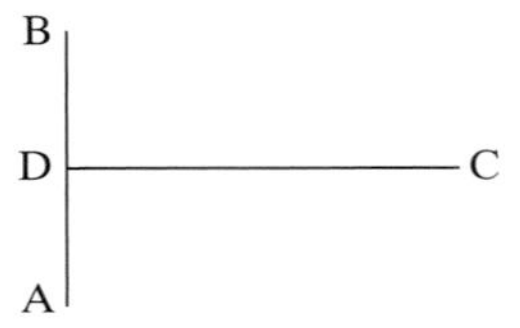 B, D, C, A
Measure AC and angle BAC	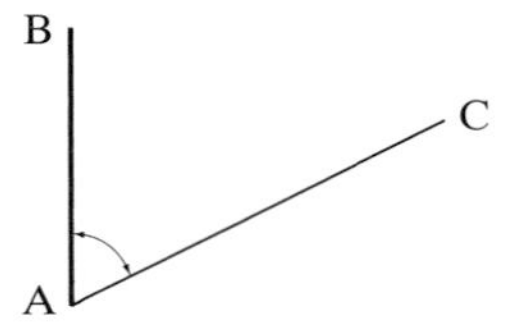 B, C, A
Measure angles CBA and BAC	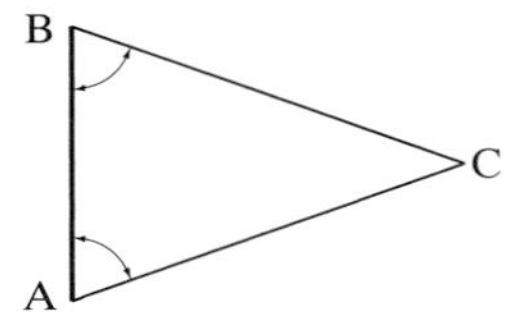 B, C, A

FIGURE 2.29
Methods of detail survey. Detail survey makes use of different measurement methods to determine the position of C relative to A and B with known locations.

Because a large number of measurements are required to make a map, conventional land surveying is a very tedious and time-consuming process (Moffitt and Bossler, 1998). The development of instruments based on electronic distance measurement (EDM) utilizing infrared or visible lasers, such as the electronic theodolite, has made it possible to obtain distance and angle measurements at the same time (Burnside, 1971). These measurements may also be stored digitally in the instrument instead of manually in survey field books, thus allowing detailed surveys to be carried out much more expeditiously. This type of instrument is known as a Total Station (Plate 2.1).

The ability to obtain survey measurements directly in digital form has led to the development of digital mapping and has drastically changed the ways by which topographic surveying is done. The USGS (1999d), for example, now depends entirely on digital mapping methods

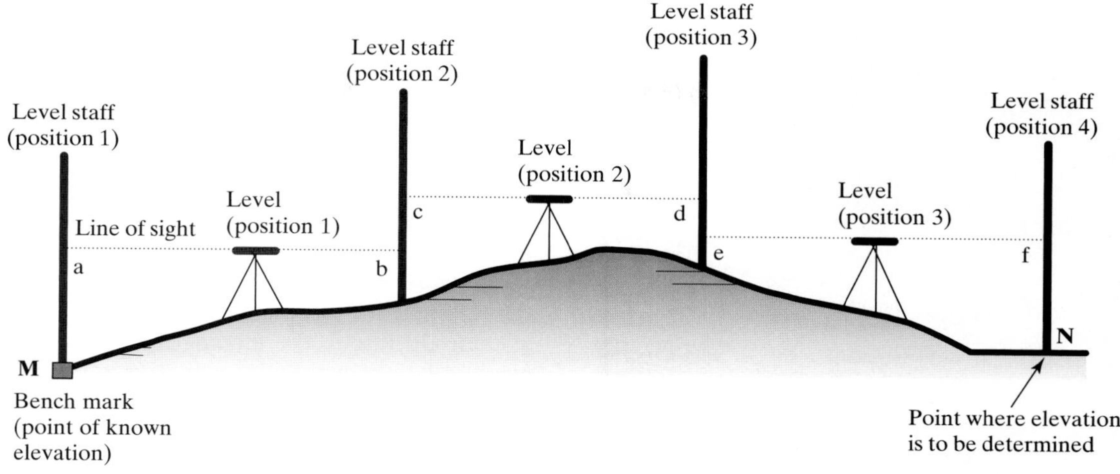

FIGURE 2.30
Determination of elevation by differential leveling.

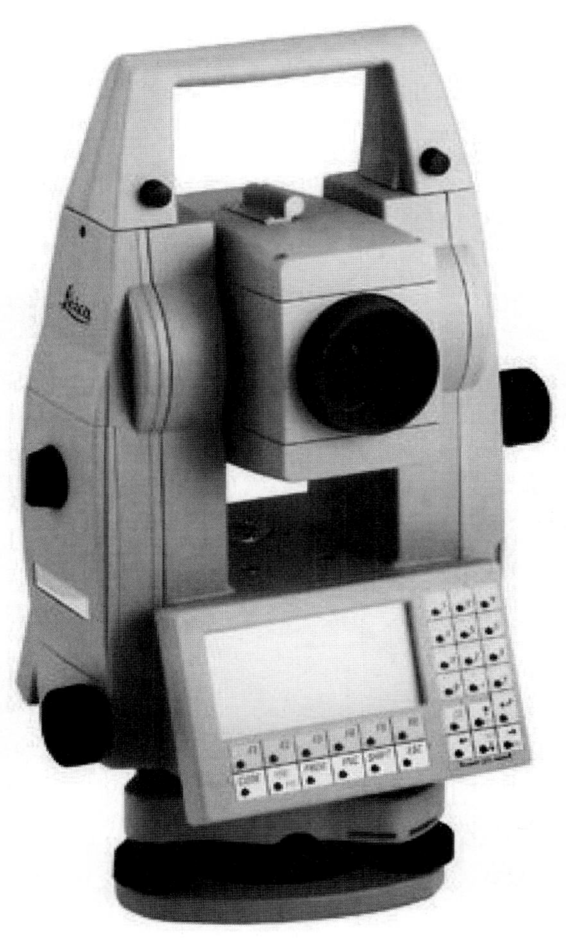

PLATE 2.1
A Total Station. *(Courtesy: Leica Geosystems)*

in its national topographic mapping and map revision programs (Leman, 1999). Since the maps are already in digital form, they can be directly used in GIS applications.

2.9.2 AERIAL AND SATELLITE PHOTOGRAMMETRY

The invention of photography and the airplane stimulated the application of aerial photography and photogrammetry for topographic mapping. Aerial photography provides a maplike view of the terrain and covers an extensive area. Using a stereo-pair of aerial photographs, a three-dimensional view of the terrain is also obtainable, from which terrain heights and contours can be extracted (Plate 2.2). Photogrammetry provides the technology for accurate planimetric and height data of the terrain to be extracted from aerial photographs. This has been the preferred method of topographic mapping for the past 50 years. Analytically, photogrammetric mapping involves the concept of coordinate transformations explained in Section 2.5. Because an aerial photograph is a central perspective projection of the terrain, its scale varies, and image positions are displaced as a function of relief. Tilts of the airplane also cause further displacement of the image positions. The coordinates of points as measured on the photograph are transformed to the corresponding ground coordinates. These transformations can be computed in two-dimensional space if only planimetric details are mapped. However, if terrain heights are required, three-dimensional transformation is necessary. The transformation equations can be solved only with the use of a number of horizontal and vertical controls, which can be obtained from an existing control network or by field

PLATE 2.2
An aerial photograph of the city of Athens, Georgia.

surveying. Alternatively, the method of spatial aerial triangulation, based on the concept of triangulation in land surveying, can be used to extend ground-surveyed horizontal and vertical controls to a much larger area covered by the block of aerial photographs, thus minimizing the amount of field surveying required (Wolf and Dewitt, 2000).

Photogrammetric coordinate transformation can be carried out by analog, analytical, or digital means. *Digital* or *softcopy photogrammetry* is the most popular approach in recent years, as a result of the availability of aerial photographs in digital form (acquired by means of image scanning or directly from digital cameras). Digital photogrammetry facilitates the production of orthophotographs (or orthoimages), in which the displacements caused by tilts and relief are removed, so that the orthophotograph exhibits a constant scale as does a map. Since orthophotographs are stored in digital form, they can be readily integrated with other forms of spatial data in GIS applications. The Digital Orthophoto Quadrangles (DOQ) of the USGS are based on the use of digital photogrammetric technology (USGS, 1999b) and are very popular among GIS users.

A new development in surveying technology to extract terrain height is the use of LIDAR (Light Detection and Ranging) carried on board an aircraft. This is similar in principle to EDM mentioned earlier. In LIDAR, a laser with a frequency of 15 kHz is used and covers a swath of 50 degrees. The laser is beamed down to the ground from the air, and the returned signal to the aircraft is timed. Knowing the speed of the laser through the atmosphere, it is possible to compute the aircraft altitude to the terrain point by point. A GPS is also used on board the aircraft. The returned signals are digitized and stored on magnetic media. When combined with GPS and inertial navigation data, a digital three-dimensional representation of the land surface can be generated. In this way, a digital elevation model (DEM) can be easily produced for an area of interest (Watkins and Conner, 2000). Using the terrain height data, orthophotographs can be easily produced.

The development of digital photogrammetry has been stimulated by the availability of satellite images in digital form. Ever since the successful launching of the Earth Resources Technology Satellite-1 (ERTS-1), later renamed Landsat-1 by the United States in July 1972, interest in the use of satellite images for topographic mapping has been intense. These satellite images have the advantage of global coverage (e.g., Landsat-1 covering an area of 185 km $\times$ 185 km in one scene) and are digital in form, which suits processing by computers (Plate 2.3).

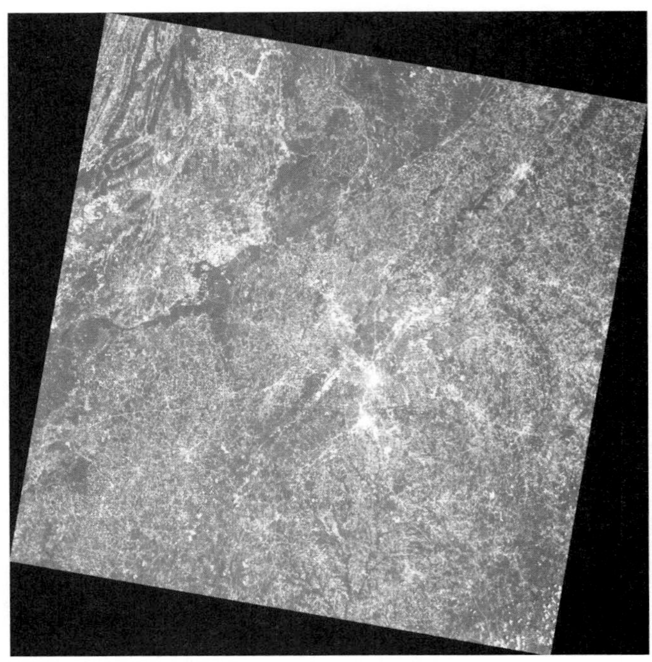

PLATE 2.3
Landsat Multispectral Scanner (MSS) band 2 (0.6–0.7 μm) image of the Atlanta, Georgia, area acquired on May 14, 1988.

sources Technology Satellite-1 (ERTS-1), later renamed Landsat-1 by the United States in July 1972, interest in the use of satellite images for topographic mapping has been intense. These satellite images have the advantage of global coverage (e.g., Landsat-1 covering an area of 185 km × 185 km in one scene) and are digital in form, which suits processing by computers (Plate 2.3).

Early research has focused on the Landsat Multispectral Scanner (MSS), which has four spectral bands and a spatial resolution of only 79 meters. The side lap of MSS images between two adjacent orbits provides some stereo coverage, which increases from the Equator to the pole. Wong (1975) showed that by using four or more control points and general polynomial equations to model the geometric distortions of the MSS images, topographic maps at the scale of 1 : 500,000 meeting United States National Map Accuracy Standards (NMAS) could be produced. Dowman and Mohamed (1981) demonstrated that the MSS images in analog form could be used in a stereoplotter to produce a topographic map. In July, 1982, when Landsat-4 was launched, a new multispectral scanner with seven bands known as the Thematic Mapper (TM) provided high resolution images (30 meter pixel), which made it possible to produce topographic maps meeting NMAS at a scale of 1 : 100,000 using 20 control points (Doyle, 1984) (Plate 2.4).

In February, 1986, France launched a satellite known as SPOT-1 (Systeme Probatoire d'Observation de la Terre), which was specifically designed for topographic mapping, among other applications. SPOT-1 carried two advanced linear-array multispectral scanners known as HRV (high resolution visible) in three spectral bands with a resolution of 20 m, using the method of *pushbroom scanning*, which eliminates the use of rotating mirrors in scanning (known as *whiskbroom scanning*) employed by the Landsat MSS and TM systems. As a result, the SPOT images are geometrically more accurate than those from Landsat. The HRV scanners also collect data in panchromatic mode, which produces broadband images with a resolution of 10 m (Plate 2.5). The two HRV scanners can be pointed so as to cover adjacent fields in one orbital pass, giving a total swath width of 117 km and an overlap of 3 km from an altitude of 832 km. This allows stereoscopic pairs of images of a given area to be acquired in the same overpass. As a result of improved spatial resolution and stereo coverage, the topographic mapping capability provided by SPOT images is much superior to that of Landsat TM images (Rodriguez et al., 1988). Gugan and Dowman (1988a) found that by using an analytical and digital approach, SPOT images could be used to produce topographic maps at 1 : 50,000 scale and smaller with 25-m contours. At these scales, a planimetric accuracy of 20 m, a height accuracy of 8 m, and 82.3% completeness in planimetric detail are achievable (Gugan and Dowman, 1988b). With the advent in computer technology and GPS to provide controls (see Section 2.9.3), the stereoscopic SPOT images can be used to produce DEMs for orthoimage production and terrain analysis in GIS applications.

New and higher spatial resolution stereoscopic satellite imaging systems have been developed, and

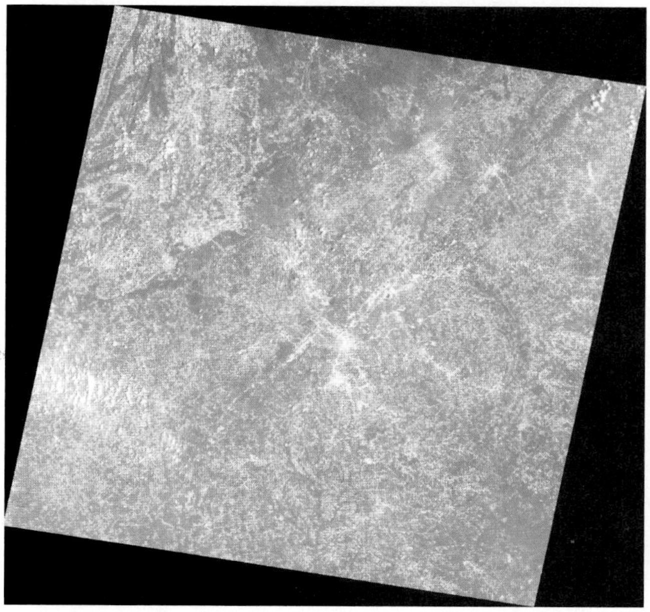

PLATE 2.4
Landsat Thematic Mapper (TM) band 2 (0.52–0.60 μm) image of the Atlanta, Georgia, area acquired on July 10, 1997.

PLATE 2.5
SPOT multispectral band 1 (0.50–0.59 μm) image of Hong Kong acquired on December 16, 1993 *(© CNES 2002/Courtesy: SPOT Image)*.

the potential of satellite images for topographic mapping at small to medium scales has greatly improved, e.g., the Modular Opto-electronic Multispectral Scanner (MOMS-02) acquires along-track scanned stereoscopic images at the panchromatic mode with a ground resolution of 4.5 m, making it possible to produce topographic maps at the scales of 1:25,000 and 1:50,000 (Konecny and Schiewe, 1996). In April 1999, Landsat 7 Enhanced Thematic Mapper Plus (ETM+) with seven multispectral bands of 30-m spatial resolution and a broad panchromatic band of 15-m resolution was successfully launched by NASA. This was followed by the successful launch of the 1-m resolution IKONOS sensor by Space Imaging-EOSAT, a private company, in January 2000. The potential of highly accurate topographic mapping using a digital satellite photogrammetric approach has now been realized.

Finally, satellite imaging at the microwave region of the electromagnetic spectrum was made possible by RADARSAT, launched by a Canadian company in 1995. RADARSAT, which can image Earth using spatial resolutions ranging from 10 m to 100 m, is an active remote sensing system because it transmits its own energy using a single microwave frequency known as C-band (5.3 GHz frequency or 5.6-cm wavelength) toward Earth and then receives the energy reflected back. In this way, RADARSAT can image Earth all the time (day and night) regardless of the weather or cloud cover condition (Plate 2.6). All the satellite data are particularly suited for use with grid-based raster GIS.

2.9.3 GLOBAL POSITIONING SYSTEM (GPS)

Fixing of positions on the surface of the Earth was made easier by the advent of the GPS, which was originally designed by the U.S. Department of Defense (DoD) as a worldwide radio-navigation system but has been made available for civilian positioning applications. GPS is based on a constellation of 24 high-altitude satellites (called Navigation System with Time and Ranging, or NAVSTAR) (Kaplan, 1996). The satellites are positioned in six Earth-centered orbital planes with four satellites in each plane. The nominal orbital period of a GPS satellite (i.e., the time for the satellite to complete one revolution around Earth) is 11 hr 58 min. The orbits are nearly circular and equally spaced above the Equator at a 60° separation, with an inclination relative to the Equator of 55°. The altitude of the satellite from the center of Earth is about 26,600 km. The satellite constellation is designed to allow 24-hr global user navigation and time determination capability.

GPS makes use of *time of arrival* (TOA) of the GPS signal to determine positions on Earth's surface. A GPS satellite, which has a known position in space, sends out a signal to a receiver on Earth's surface. The time interval, known as the *signal propagation time*, recorded at

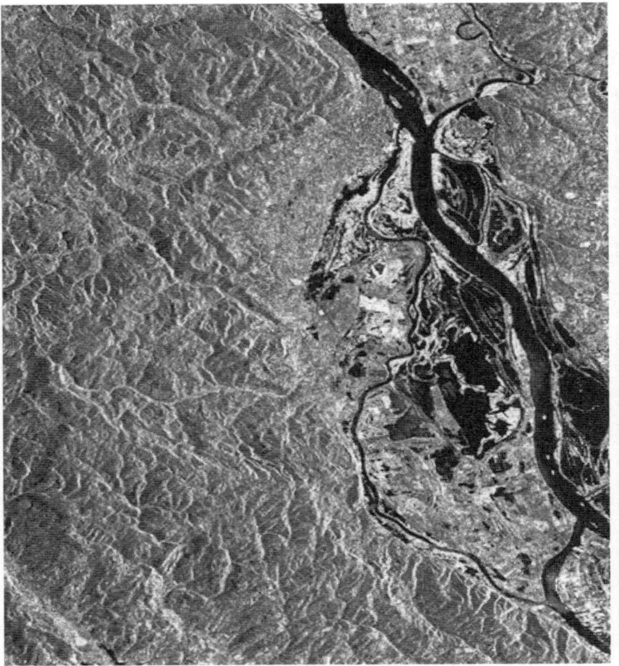

PLATE 2.6
RADARSAT images of Portland, Oregon, showing the peak of the flood of the Columbia River on February 13, 1996 (left image), and the area damaged by flood waters on February 14, 1996 (right image). The spatial resolution of the two images is 25 m. *(RADARSAT-1 Imagery© Canadian Space Agency 1996. Received by the Canada Centre for Remote Sensing. Processed and distributed by RADARSAT International).*

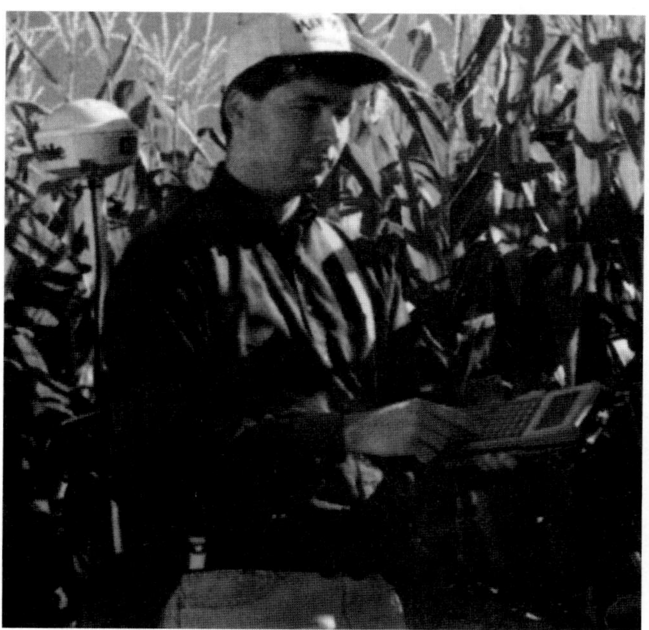

PLATE 2.7
GPS receiver and antenna are being used to collect coordinates in the field. *(Courtesy: Trimble Navigation Limited)*

the receiver is then multiplied by the speed of the signal to give the emitter-to-receiver distance (Plate 2.7). By measuring the propagation time of signals broadcast from multiple satellites at known locations, the receiver can determine its position by means of the method of *space resection* (Figure 2.31). The concept of GPS, therefore, is relatively simple. However, the actual implementation of the system is much more complicated because it involves extremely accurate time measurement and synchronization, very sophisticated geodetic computations, and precise tracking of the positions of the entire constellation of satellites. There is also the need to compute and manipulate the various types of errors that occur during the measurement process such as clock error, orbit errors, receiver noise, and the effects of the atmosphere (Trimble, 1997).

A GPS receiver can be employed as a stationary or mobile unit in point positioning, known respectively as *static* and *kinematic* GPS surveys (Figure 2.32a). A single receiver in a static point positioning can give moderate accuracy (5 to 10 m). Kinematic point positioning, which is used to determine the position of moving objects such as a vehicle's trajectory in space, is capable of achieving an accuracy of 10 to 100 m. For accurate distance determination, *static relative positioning* using two stationary receivers is the most commonly employed method by surveyors (Figure 2.32b). The accuracy achievable is 1 ppm (part per million) to 0.1 ppm. The method of *differential GPS* (DGPS) has been developed for accurate point determination. In this method, at least two GPS receivers are required. One receiver is placed at a station whose precise position is already known (hence known as the *base*, or *reference station*), and the other receiver is moving from one station to another (Figure 2.32c). For the two receivers, code pseudo-ranges (a term used to denote true ranges that contain an error caused by the lack of synchronization

Satellite A
Satellite B
Radio signal
Radio signal
P_1
P_2

(a) Spatial resection by two GPS satellites

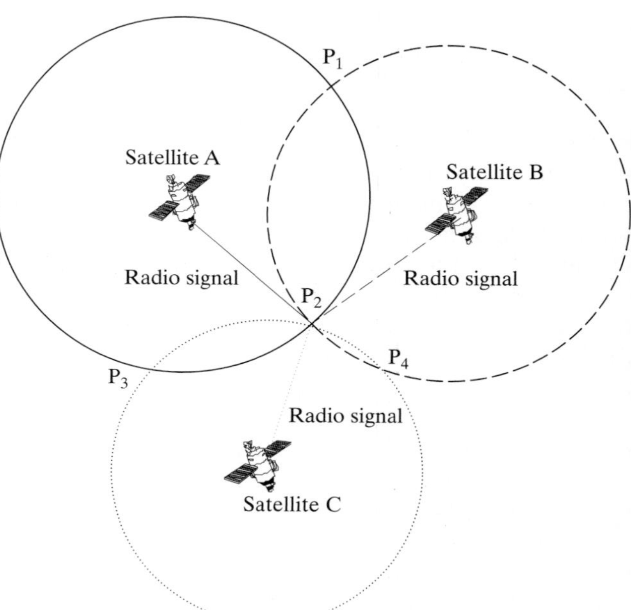

(b) Spatial resection by three GPS satellites

FIGURE 2.31
Spatial resection by satellite ranging. Two satellites cannot fix a position in space because the circles representing the satellite ranges meet at two locations, P_1 and P_2, as shown in (a). To fix a position, a minimum of three satellites is needed, as shown in (b). To find three-dimensional elevation, four satellites are required.

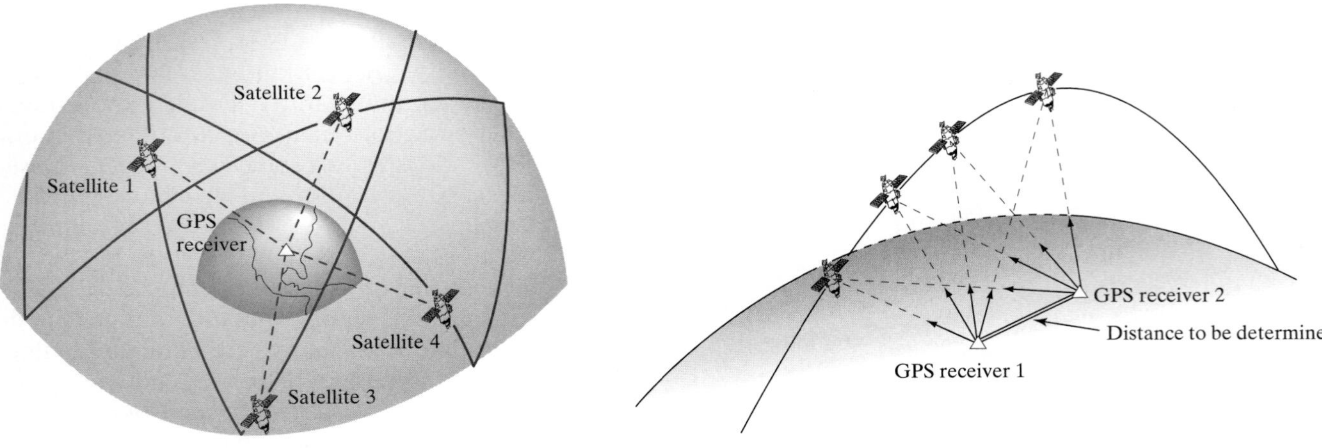

(a) Single point static and kinematic GPS measurement

(b) Relative static GPS measurement of distance

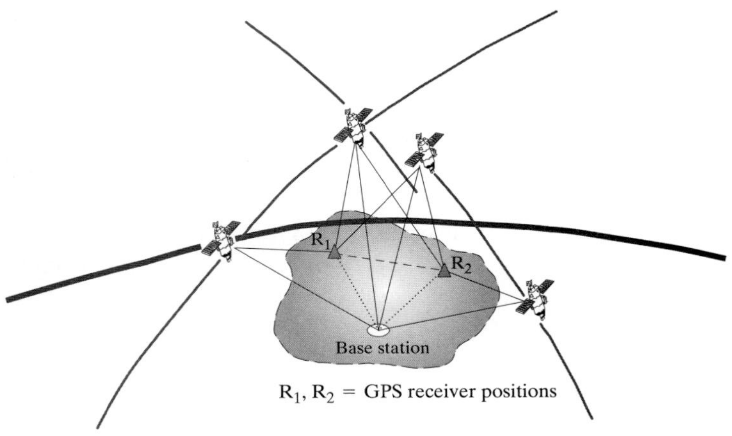

(c) Differential GPS measurements

FIGURE 2.32
Methods of GPS positioning.

between the satellite clock and the receiver clock) are simultaneously observed to four or more satellites. The base station calculates the actual corrections to the observed code pseudo-ranges. With the aid of communication links, these corrections are transmitted to the station where the rover receiver is located (or stored for later correction). In this way, the positions fixed by the rover receiver are greatly improved, often with submeter accuracies.

In recent years, differential GPS has been greatly facilitated by the presence of base stations set up by the Canadian and the U.S. Federal Governments especially around harbors, waterways, and airports (NGS, 1999). In the United States, these stations are established under the Continuously Operating Reference Stations (CORS) program. Each CORS station consists of one permanent GPS receiver, a computer, and a telecommunications support system. Anyone working within a 200-km range of these stations can receive the corrections transmitted on the radio beacons to improve their GPS measurements accurate to 3 cm horizontally and 5 cm vertically. The correction data are also archived by NGS and made available to users who are doing postprocessing of field observations. GPS has now been integrated with GIS (Kennedy, 1996). GPS measurements are available in a format that can be directly input to a GIS database (e.g., the shapefile format for ArcView GIS) (see Chapter 9). GPS has revolutionized surveying and mapping.

2.10 Attribute Data for Thematic Mapping

A map does not show just locations, but also the attributes of features found at these locations. A map shows the spatial distribution of a particular attribute or theme very well. Many thematic maps are constructed using the topographic maps as the map base. It is particularly true for thematic maps for soil surveys, land cover, environmental indices, and population distribution, among many others in both the social and physical sciences, where the ability to measure the map is desirable if not absolutely necessary. These maps are geographically referenced like topographic maps, using coordinates to denote the locations of mapped attributes.

2.10.1 GEOGRAPHICAL REFERENCING FOR ATTRIBUTE DATA

Attributes in thematic mapping can be geographically referenced by a *hierarchical referencing system,* which is sometimes referred to as the *discrete* system. Hierarchical systems are commonly used to identify locations in the delivery of public services, in the collection of socioeconomic statistics, and in defining data collection units in natural resource inventories.

As its name implies, a hierarchical referencing system makes use of a hierarchy of labels or codes to identify individual entities and relate them to one another in a particular data set. An address, which contains the names of a country, a state, a city, a street, and a house number, is a typical example of a hierarchical referencing system. By using a hierarchy of place names and numbers (i.e., country, state, city, street, and house number), it is possible to "describe" where a particular house is and to use these names and numbers as the key to look up the location of the house on a map. A hierarchical referencing system is an indirect referencing system that is used for spatial indexing rather than spatial positioning.

Hierarchical referencing systems are mostly parcel-based, i.e., they use the *land parcel* as the basic unit of measurement and documentation. In GIS literature, these land parcels are often referred to as *polygons*. Land parcels in hierarchical referencing systems can be natural or artificial. Many natural resource and environmental data are measured and documented in terms of irregularly shaped areas that characterize the land cover or physical conditions of a particular geographic space. These are examples of natural land parcels. There are many land parcels that are created particularly for the purpose of government administration (e.g., dividing a state/province into counties), land and community development (e.g., residential subdivision), business or service delivery (e.g., sales territories and postal delivery), and the collection of socioeconomic statistics (e.g., census tracts). The boundaries of artificial land parcels are usually delineated arbitrarily, which means that they do not necessarily follow the natural characteristics of the land.

Hierarchical referencing systems make use of unique identifiers in the forms of names or numerical codes to distinguish individual parcels from one another. These identifiers can be assigned either in an arbitrary manner or in a particular numerical sequence. By making the identifiers hierarchical, land parcels at one level of a hierarchical system can be aggregated to form larger parcels at a higher level for data management and processing purposes. This relationship between land parcels at different levels sometimes provides an implicit location reference of the geographic features that the data represent. For example, all counties that form a particular state are geographically located within the boundaries of that state, and all states that form a country are geographically located within the boundaries of that country.

Data in hierarchical referencing systems can be made useful for GIS applications by giving them the proper spatial references in geographical or plane coordinates. This can be done by one of two ways, depending on the form of the data. Some hierarchical data, such as land parcels in the township system, are recorded on index maps. Such data can be digitized to pinpoint their positions in geographic space. Others, such as municipal addresses, are not related to any index map. Instead of digitizing these addresses one by one, which can be a very tedious and time-consuming task, an automated method known as *geocoding* can be used (see Chapter 6). The objective of geocoding is to match street addresses to locations in a digital street map that contains the range of street numbers in individual street segments. This process allows GIS users to take advantage of the vast amount of address-based geographic data now in use in business and government organizations.

2.10.2 SOME IMPORTANT HIERARCHICAL REFERENCING SYSTEMS

There are many types of hierarchical referencing systems in use today. The following four systems are particularly important for GIS users, as they contain a wealth of socioeconomic information associated with some aspects of geography. These include (1) land administration and title systems, (2) municipal addresses, (3) postal codes, and (4) census enumeration areas.

Land Administration and Title Systems These systems are designed to provide the legal definition of land parcels in land administration or property registration. The major functions of these systems are to portray the position of land parcels relative to each other and to index other forms of data associated with the parcels, such as ownership, address, tax roll number, and property assessment. Examples of such systems include the Public Land Rectangular Surveys in the U.S. and the Dominion Lands System in Canada, both of which are used for land administration outside urban areas. As the principal land parcel in these systems is the township, these systems are sometimes referred to as the *township system*. Typically, the township system is made up of the following hierarchy of land parcels (Figure 2.33).

- township—36 square miles, approximately equal to 10 × 10 km

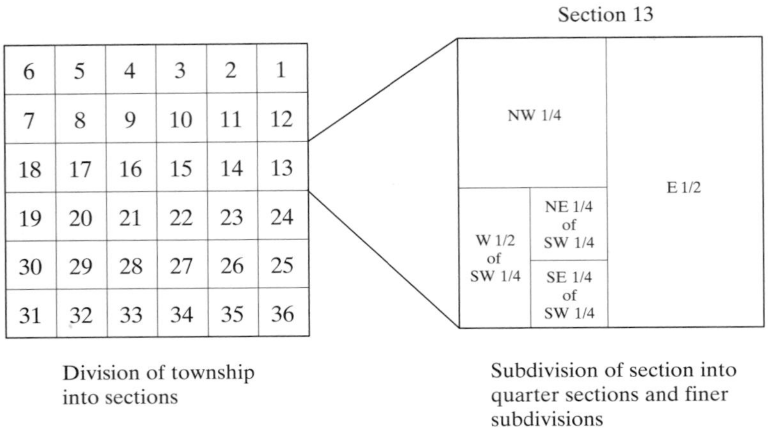

Section 13

6	5	4	3	2	1
7	8	9	10	11	12
18	17	16	15	14	13
19	20	21	22	23	24
30	29	28	27	26	25
31	32	33	34	35	36

Division of township into sections

NW 1/4

E 1/2

W 1/2 of SW 1/4

NE 1/4 of SW 1/4

SE 1/4 of SW 1/4

Subdivision of section into quarter sections and finer subdivisions

FIGURE 2.33
Township system of land subdivision.

- section—640 acres, approximately equal to 1.6 × 1.6 km
- quarter section—160 acres, approximately equal to 800 m × 800 m

In urban areas, a lot/block/plan system is used instead of the township system. The function of this system is essentially the same as the township system. It provides an index to the location of individual ownership parcels within a given jurisdiction such as a town or a city. Conventionally, there is no obvious spatial relationship between registered plans in this system. The growing recognition of the importance of accurate land information in the 1980s led to the development of *integrated surveys* and modern *land record systems* (Hamilton and McLaughlin, 1985). In these systems, all land parcels are tied to survey control points and encoded in the digital database, thus making possible the integration of parcel data with other forms of spatial data used in GIS.

Municipal Addresses Municipal addresses describe the location of a particular land parcel by means of a hierarchy of geographic names and identifiers such as country, state, city and street, and house number. They probably represent the most extensively used hierarchical referencing system by government agencies and commercial organizations. Addresses are descriptive data and cannot be used directly to pinpoint the positions that they describe in space. The most common way of making address-based data usable by GIS is to match them to locations in a digital street map by the method of geocoding as noted earlier. However, this method can be used only in urban areas where there are well-defined street networks with address ranges properly encoded between street intersections (see Chapter 6). Another way of making address-based data useful is to take advantage of their associated postal codes to relate them to geographic areas represented by these codes as explained here.

Postal Codes Postal codes are designed primarily for sorting of mail for postal delivery. Each postal code defines a geographic area that can vary in size, depending on the distribution of the sorting stations and the areas that they serve (Figure 2.34). Postal codes are made up of a combination of digits, such as 30602-2502 in the U.S. ZIP code system, or alphanumerical characters, such as P3E 5N3 in the Canadian postal code system, that are directly related to a hierarchy of geographic areas. In the 5-digit ZIP code, for example, the first digit designates a broad geographic area of the United States, ranging from 0 (zero) for the northeast to 9 for the far west (USPS, 1999). This is followed by the next two digits that more closely pinpoint population concentrations and those sectional centers accessible to common transportation networks. The final two digits designate small post offices or postal zones. The ZIP + 4 code, introduced in 1983, adds further hierarchies to the system. The first two numbers of the 4-digit code identify a delivery section, which may embrace several street blocks, a group of streets, a group of post office boxes, several office buildings, a single high-rise office building, a large apartment building, or a small geographic area. The last two digits denote a delivery segment, which may be one floor of an office building, one side of a street, specific departments in a business organization, or a group of post office boxes.

By using postal codes and the associated index maps, geographic data can be represented at a relatively fine spatial resolution. Many organizations in the public and business sectors have taken advantage of postal codes as the spatial reference to store, manage, and analyze data. Postal code–based data are a particularly important source of data for GIS applications in business geographics, such as market analysis, sales territory delineation, and the identification of new outlets or franchise locations.

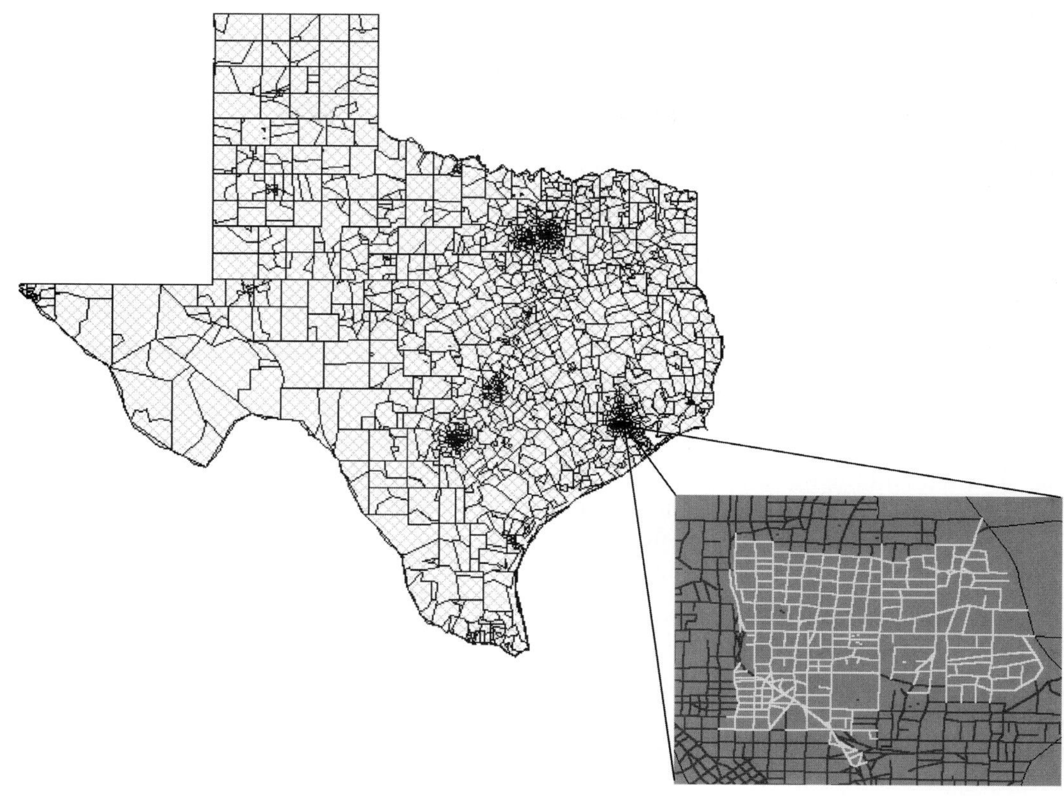

FIGURE 2.34
Five-digit ZIP code zones of Texas. Originally developed for sorting mail for postal delivery, postal zones provide a relatively fine spatial resolution, particularly in urban areas (see insert) for the referencing of geographic data.

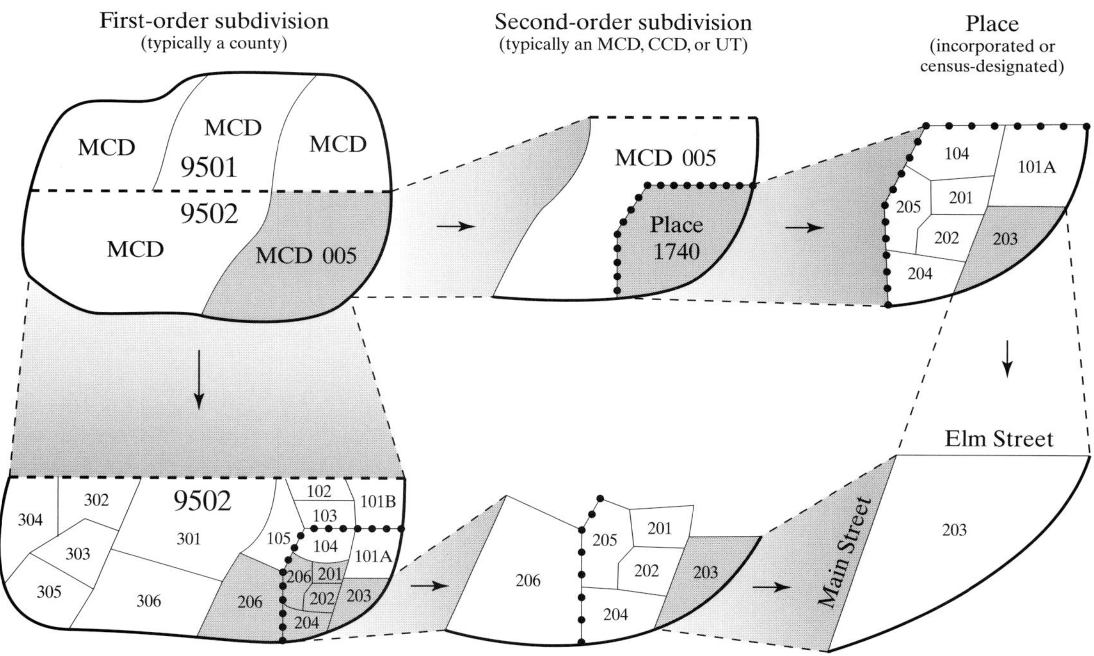

FIGURE 2.35
Small-area geography in the 1990 census. *(Source: U.S. Census Bureau, 1999a)*

Census Enumeration Areas The census is designed primarily to record national or regional population and other social data. For data collection purposes, the country is divided into different classes of geographic areas that are hierarchically related. In the United States, the building block of the census geographic area system is the *census tract* (Figure 2.35). By definition, census tracts are small, relatively permanent geographic subdivisions of a county or equivalent entity. The primary purpose of census tracts is to provide a nationwide set of geographic units that have stable boundaries. In general, census tracts follow visible and identifiable features such as roads, rivers, canals, railroads, and aboveground high-tension power lines. In cases where such features are not available, nonvisible administrative boundaries such as town and township boundaries are also used. Census tracts in a county are identified uniquely by 4-digit numbers ranging from 0001 to 9989. For some census tracts, a 2-digit decimal suffix may also be used.

Within each census tract, the basic unit for data collection is the *census block*. In general, a census block is a street block bounded by streets and other features such as railroads and rivers. For statistical reporting purposes, contiguous clusters of census blocks are grouped into *census block groups*. Within a census tract, a block group consists of all census blocks whose identification numbers begin with the same digit.

As illustrated in Figure 2.36, census tract data can be aggregated into progressively higher hierarchies at the state, regional, and national levels. They can also be aggregated by *metropolitan areas* (MAs). MAs are defined by the U.S. Office of Management and Budget (OMB) according to published standards that are applied to Census Bureau data. The general concept of an MA is that of a core area containing a large population nucleus, together with adjacent communities having a high degree of economic and social integration with the core. Since the first introduction of the definition of metropolitan areas under the designation of "standard metropolitan area" (SMA) in 1949, the term has been changed several times. It became known as "standard metropolitan statistical area" (SMSA) in 1959, and "metropolitan statistical area" (MSA) in 1983. The current definition of "metropolitan area," which became effective in June 1999, is based on one of the two following provisions: one city with at least 50,000 or more inhabitants or a Census Bureau–defined urbanized area of at least 50,000 inhabi-

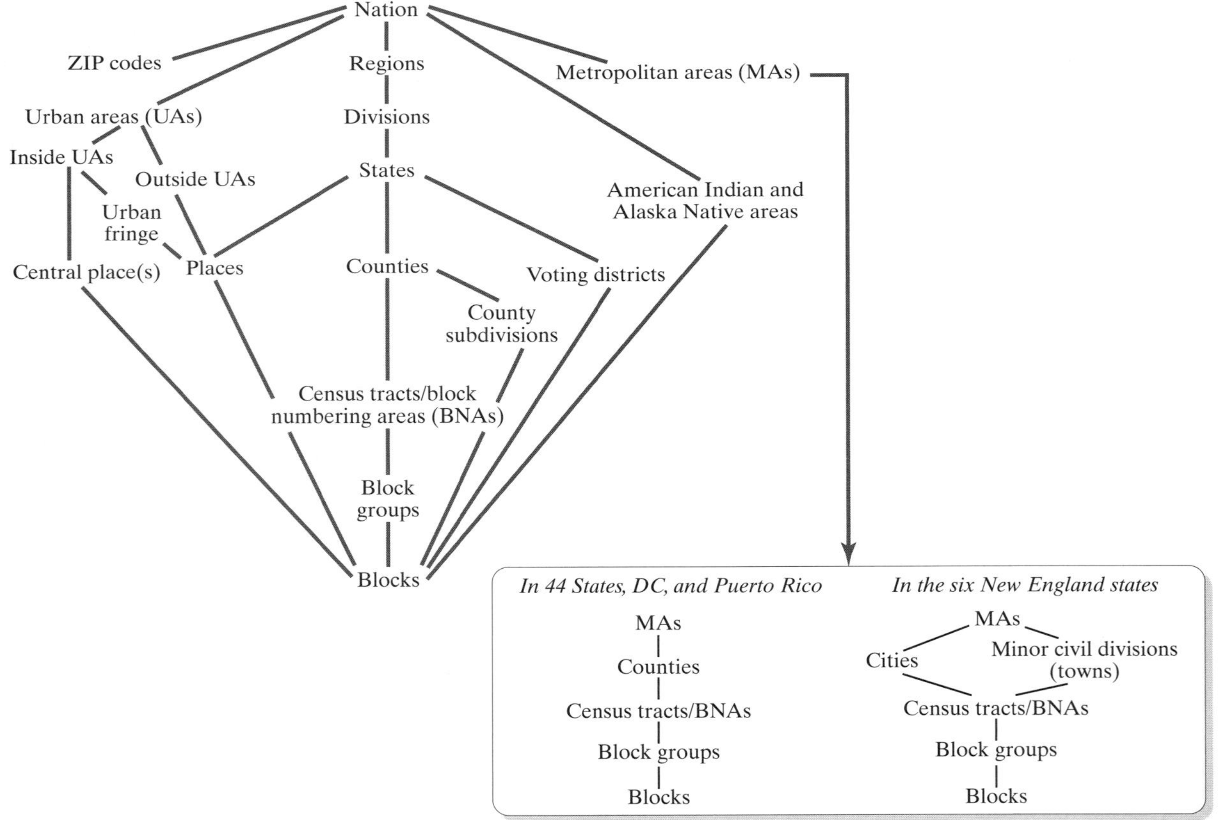

FIGURE 2.36
Geographic hierarchy for the 1990 census. *(Source: U.S. Census Bureau, 1999a)*

tants and a total metropolitan population of at least 100,000 (75,000 in New England).

2.10.3 COLLECTION OF ATTRIBUTE DATA

Attribute data of spatial objects are collected by a variety of means, including field investigation, site visits, remote sensing image processing and interpretation, administrative procedures, and extraction from existing data files. Just like topographic data, the collection of thematic data is also a costly and time-consuming undertaking. As it is not possible or practical to take a large number of observations in the data collection process, thematic data as a rule are collected by sampling. As a result, the sampling design to ensure that the data collected are statistically representative of the "population" is a very important consideration (Berry and Baker, 1968).

By virtue of the synoptic view provided by remote sensing from aerospace platforms, notably aerial photography and satellite imaging, our ability to collect thematic data has been enhanced considerably. It does not mean, however, that the process is much less costly and involved. Satellite images and aerial photographs are expensive. Image processing equipment also incurs considerable cost. In addition, the process of ground-truthing, or field checking to verify the results of photo-interpretation or image analysis, can also be prohibitively expensive and time-consuming.

2.10.4 MEASUREMENT SCALES

The attributes pertaining to spatial objects shown in a thematic map can be recorded by four different levels of measurement according to Stevens (1946), a Harvard psychologist, which have become the framework for cartography and GIS (Chrisman, 1997). In the order of increasing complexity, these four levels of measurement (also known as scales of measurement) are (1) nominal, (2) ordinal, (3) interval, and (4) ratio.

Nominal Scale Nominal scale is the simplest of all because it just uses names as labels. Thus, a map of agricultural regions of the United States will show such descriptive labels such as "wheat region," "corn region," "soybean region," or "cotton region." In this example, the nominal scale is applied over polygons. Nominal scale may also be applied to lines, such as in the names and numbers of the highways, and to points, such as the number codes of power poles or the names of settlements in a world map.

Ordinal Scale This particular scale shows ordering or ranks. A map can show the rank of cities in the United States by size, or in a study of residential desirability, the rank of each city in the United States according to its quality of life (e.g., the annual survey done by *Money* magazine). Similarly, a classification of roads into first, second, and third class is very common on road maps. Also, countries in the world can be classified according to their population size. Soils can be classified according to their drainage condition. Thus, ordinal scale can be expressed in different ways and is applicable to points, lines, and polygons.

Interval Scale Interval scale does not have a natural zero and uses an arbitrary one instead. Temperatures in °F or °C are good examples, because temperatures in °F and °C use the freezing point of water at 32 °F and 0 °C respectively as the arbitrary origin. In other words, a temperature of 40 °F is 8 °F above 32 °F, and a temperature of 20 °F is 12 °F below 32 °F, giving a difference in temperature of 20 °F (8 °F + 12 °F) between 40 °F and 20 °F because they are all measured relative to 32 °F. Therefore, a temperature of 40 °F cannot be twice as hot as 20 °F. In other words, subtraction makes sense on an interval scale, but not multiplication because the value is *relative* from an arbitrary origin. In topographic maps, we deal with terrain heights. As we have learned from this chapter, terrain height is sometimes measured relative to the MSL (see Vertical Datums in Section 2.8.2). The MSL is not an absolute origin. As a result, terrain elevations are measured on an the interval scale.

Ratio Scale Unlike the interval scale, a ratio scale makes use of an absolute zero (or a true origin). Weight is an example of a ratio scale because the true origin is zero. If your friend weighs 200 lbs and you weigh 100 lbs, your friend is two times heavier than you are. Speed is a ratio scale because it is quite clear what the true zero is. If you drive your car at 60 miles per hour, you are driving twice as fast as someone driving at 30 miles per hour. In other words, division makes sense in ratio scale. Ratio scale supports the arithmetic operations of addition, subtraction, multiplication, and division.

Unfortunately, the distinction of the four measurement scales is not always very clear. As Chrisman (1997) observed, angular measurements are in ratio scale because it has a natural origin as 0°, but that origin also becomes 360° when the angle of a circle is measured. The angle 359° is the same as 1° from 0°, the natural origin. Even in the simplest nominal scale, the boundaries of the agricultural regions given in the example are only arbitrarily delineated, so that the "wheat region" may contain a part of the "corn region" and vice versa. This gives rise to the concept of membership grades now increasingly being used in soft classification based on fuzzy set theory (see Chapter 8, Section 8.6.2).

The impact of scales of measurement will be on cartographic representation and statistical data analysis and spatial modeling in GIS. The general rule is to ensure that different measurement scales are not mixed up. However, it is quite common that after, say, an overlay of two maps in ratio scale, the resultant new map will be in ordinal scale (such as in determining store location or evaluating the erosion potential of the terrain). In cartographic representation of attributes, the scale of measurement can be combined with the concept of continuous (fractions) and discrete (integers) data to determine whether shading, dot, or line should be used to display the attribute data in thematic mapping (Robinson et al., 1995).

2.11 SUMMARY

Maps are the most efficient *analog* means to display spatial and nonspatial data (attributes). GIS, which represents a *digital* approach to handling spatial data, has its roots in cartography. A good understanding of cartography is therefore necessary for any GIS user. In this chapter, some fundamental concepts of maps, such as map scale, the distinction between topographic and thematic maps, as well as the mapping process are explained. The main focus of this chapter is the mathematical methods of representing positions of spatial objects on a map, and how these positions can be transformed to relate to Earth's frames of reference.

For GIS, the most important coordinate system is the three-dimensional geographic coordinate system of Earth, expressed in the form of latitude and longitude (in degree-minute-second of arc). The representation of the three-dimensional coordinates of points on Earth to their corresponding positions on a two-dimensional plane gives rise to map projections. Inevitably, each type of map projection developed preserves only one or two but not all of the following properties of spatial objects to be mapped: shape, area, distance, and direction.

The most important operation in GIS is georeferencing, or the establishment of a spatial reference system to which all spatial measurements must relate. This requires a mathematical model of the shape of Earth or the choice of an ellipsoid to represent Earth. For example, the Clarke 1866 ellipsoid model was adopted by the United States as the basis for North American Datum 1927 (NAD 27), until its replacement by the geocentric Geodetic Reference System of 1980 (GRS 80) as the North American Datum of 1983 (NAD 83). The concepts of horizontal and vertical datums are introduced. NAD 27 and NAD 83 are horizontal datums, whereas mean sea level (MSL) as determined by National Geodetic Survey (NGS) in 1929 (NGVD 29) and in 1988 (NAVD 88) is a vertical datum.

Two map coordinate systems, Universal Transverse Mercator (UTM) and State Plane (SPCS), have been explained because they are among the most commonly used coordinate systems in North America. Some practical advice on the transformation of UTM coordinates straddling two or more zones is also given.

This chapter also briefly examines how spatial data of the terrain are acquired. The methods of land surveying, photogrammetry, satellite remote sensing, and Global Positioning System (GPS) are briefly explained. All these surveying methods produce an accurate geographic framework within which attribute (or thematic) data can be added. Attribute data may also make use of a different approach to geographical referencing, known as a hierarchical georeferencing system, to which we should be quite familiar because they are land and title systems, municipal addresses, postal codes, and census enumeration area systems. Collection of attribute data has increasingly made use of aerial photographs and satellite images in recent years. This chapter concludes with a discussion of the four levels (or scales) of measurements relating to attribute data: nominal, ordinal, interval, and ratio, and how they have affected the choice of the method of cartographic representation and GIS operations.

REFERENCES

Berry, B. J. L., and Baker, A. M. (1968) "Geographic sampling," in Berry, B. J. L., and Marble, D. F., eds. *Spatial Analysis: A Reader in Statistical Geography*. Englewood Cliffs, NJ: Prentice-Hall, pp. 91–100.

Berthon, S., and Robinson, A. (1991) *The Shape of the World: The Mapping and Discovery of the Earth*. Chicago: Rand McNally.

Burnside, C. D. (1971) *Electromagnetic Distance Measurement*. London: Crosby Lockwood and Son Ltd.

Chrisman, N. (1997) *Exploring Geographic Information Systems*, New York: John Wiley and Sons.

Clarke, K. C. (1999) *Getting Started with Geographic Information Systems*. Upper Saddle River, New Jersey: Prentice Hall.

Colvocoresses, A. P. (1997) "The gridded map," *Photogrammetric Engineering and Remote Sensing*, Vol. 63, pp. 377–380.

Dowman, I. J., and Mohamed, M. A. (1981) "Photogrammetric applications of Landsat MSS imagery," *International Journal of Remote Sensing*, Vol. 2, No. 2, pp. 105–113.

Doyle, F. J. (1984) "Surveying and mapping with space data," *ITC Journal*, No. 4, pp. 314–321.

Doyle, F. J. (1997) "Map conversion and the UTM grid," *Photogrammetric Engineering and Remote Sensing*, Vol. 63, No. 4, pp. 367–370.

Evett, J. B. (1995) Traversing, in Brinker, R. C., and Minnick, R., eds. *The Surveying Handbook*, 2nd Edition, New York: Chapman and Hall, pp. 156–179.

FGDC (1997) *A Strategy for the National Spatial Data Infrastructure*. Reston, VA: Federal Geographic Data Committee

Gugan, D. J., and Dowman, I. J. (1988a) "Topographic mapping from SPOT imagery," *Photogrammetric Engineering and Remote Sensing*, Vol. 54, No. 10, pp. 1409–1414.

Gugan, D. J., and Dowman, I. J. (1988b) "Accuracy and completeness of topographic mapping from SPOT imagery," *Photogrammetric Record*, Vol. 12, No. 72, pp. 787–796.

Hamilton, A. C., and McLaughlin, J. D. (eds.) (1985) *The Decision Maker and Land Information System* (Papers and Proceedings from the FIG International Symposium, Edmonton, Alberta, October 1984), Ottawa, ON: The Canadian Institute of Surveying.

Kaplan, E. D., ed. (1996) *Understanding GPS Principles and Applications*. Boston: Artech House.

Kennedy, M. (1996) *The Global Positioning System and GIS: An Introduction*. Chelsea, MI: Ann Arbor Press, Inc.

Konecny, G., and Schiewe, J. (1996) "Mapping from digital satellite image data with special reference to MOMS-02," *ISPRS Journal of Photogrammetry and Remote Sensing*, Vol. 51, pp. 173–181.

Langran, G. (1992) *Time in Geographic Information Systems*. London: Taylor & Francis.

Leman, R. M. (1999) "The evolution of topographic mapping in the U.S. Geological Survey's Nation Mapping Program," in *Digital Mapping Techniques '99 Workshop Proceedings* (USGS Open File Report 99–386), pp. 27–30. Reston, VA: United States Geological Survey, Mapping Applications Center.

Lucas, E.F.(1977) "Transformation from local to UTM coorinates," *Survey Review*, Vol. 24, No.183, pp. 42–48.

Mainwaring, J. (1960) *An Introduction to the Study of Map Projection*. London: Macmillan.

Maling, D. H. (1992) *Coordinate Systems and Map Projections*. Oxford: Pergamon Press.

Moffitt, F. H., and Bossler, J. D. (1998) *Surveying*, Menlo-Park, Calif: Addison-Wesley.

Moffitt, F. H., and Mikhail, E. M. (1980) *Photogrammetry*. New York: Harper and Row.

Moore, L. (1997) "Transverse Mercator Projections and U.S. Geological Survey Digital Products," Mid-Continent Mapping Center, United States Geological Survey. http://mcmcweb.er.usgs.gov/drg/mercproj/index.html

NAS (National Academy of Sciences) (1971) *North American Datum*. National Ocean Survey Contract Report E-53–69(N). Washington, DC: National Academy of Sciences.

NGS (1994) *National Geodetic Survey: Its Mission, Vision and Strategic Goals*. Silver Spring, MD: National Geodetic Survey.

NGS (1999) *National CORS* (CORS World Wide Web Home Page). Silver Spring, MD: National Geodetic Survey.

NIMA (1996) "Vertical datums, elevations, and heights." Fairfax, VA: National Imagery and Mapping Agency. **http://164.214.2.59/publications/guides/VerticalDatums**

NIMA (1997) *Department of Defense World Geodetic System 1984: Its Definition and Relationship with Local Geodetic Systems* (NIMA TR 8350.2), 3rd ed. Bethesda, MD: National Imagery and Mapping Agency.

NOAA (1989) *North American Datum of 1983* (NOAA Professional Paper No. 2). Rockville, MD: National Oceanic and Atmospheric Administration.

NRC (1993) *Toward a Coordinated Spatial Data Infrastructure for the Nation*. Washington, DC: National Research Council, Mapping Science Committee.

NRC (1997) *Distributed Geolibrary Workshop White Papers*. Washington, DC: National Research Council, Mapping Science Committee.

Richardus (1966) *Project Surveying: General Adjustment and Optimization Techniques with Applications to Engineering Surveying*, Amsterdam: North-Holland Publishing Company.

Robinson, A. H., Morrison, J. L., Muehrcke, P. C., Kimberling, A. J., and Guptill, S. C. (1995) *Elements of Cartography*, 6th ed. New York: John Wiley & Sons.

Rodriguez, V., Gigord, P., de Gaujac, A. C., and Munier, P. (1988) "Evaluation of the stereoscopic accuracy of the SPOT satellite," *Photogrammetric Engineering and Remote Sensing*, Vol. 54, No. 2, pp. 217–221.

Seeber, G. (1993) *Satellite Geodesy Foundations, Methods, and Applications*. Berlin: Walter de Gruyter.

Snyder, J. P. (1987) *Map Projections—A Working Manual*. U.S. Geological Survey Professional Paper 1395, Washington, DC: United States Government Printing Office.

Shafer, M. L. (1995) Triangulation, in Brinker, R. C., and Minnick, R., eds. *The Surveying Handbook*, 2nd ed. New York: Chapman and Hall, pp. 204–233.

Stevens, S. S. (1946) On the theory of scales of measurement. *Science* 103, pp. 677–680.

Sturgess, B. N., and Carey, F. T. (1995) Trilateration, in Brinker, R. C., and Minnick, R., eds. *The Surveying Handbook*, 2nd Edition, New York: Chapman and Hall, pp. 234–270.

Tobler, W. R. (1961) *Map Transformations of Geographic Space*. Ph.D. Dissertation, Department of Geography, University of Washington. Ann Arbor, MI: University Microfilms, Inc.

Tobler, W. R. (1962) "A classification of map projections," *Annals of the Association of American Geographers*, Vol. 52, pp. 167–175.

Tolkatchev, A. (1996) "Global Sea Level Observing System (GLOSS)," *Marine Geodesy*, Vol. 19, pp. 21–62.

Trimble (1997) *GPS Tutorial*. Trimble Navigation Limited. **http://www.trimble.com/gps/aa_abt.htm**

U.S. Census Bureau (1999a) *Geographic Areas Reference Manual*. Washington, DC: United States Census Bureau. **http://www.census.gov/geo/www/garm.html**

U.S. Census Bureau (1999b) "U.S. Census Bureau Resources for locating 1990 census tracks, matching addresses to census tracks and listing of qualifying poverty tracts." Washington, DC: United States Census Bureau. **http://www.census.gov/geo/www/tractez.html**

USGS (1999a) *Map Scale (Fact Sheet 056–98)*. Reston, VA: United States Geological Survey, Mapping Applications Center. **(http://mapping.usgs.gov/mac/isb/pubs/factsheets/fs03800.html)**

USGS (1999b) *Digital Orthophoto Quadrangles* (Fact Sheet 129–95). Reston, VA: United States Geological Survey, Mapping Applications Center. **(http://mapping.usgs.gov/mac/isb/pubs/factsheets/fs12995.html)**

USGS (1999d) *Topographic Mapping*, On-line ed. Reston,

VA: United States Geological Survey, Mapping Applications Center. **(http://mapping.usgs.gov/mac/isb/pubs/booklets/topo/topo.html)**

USGS (1999e) *United States Map Indexes* (Fact Sheet 190–95). Reston, VA: United States Geological Survey, Mapping Applications Center. **(http://mapping.usgs.gov/mac/isb/pubs/factsheets/fs19095.html)**

USGS (1999f) *Global Land Information System* (Fact Sheet 069–99). Reston, VA: United States Geological Survey, Mapping Applications Center. **http://mapping.usgs.gov/mac/isb/pubs/factsheets/fs06999.html**

USPS (1999) *History of the U.S. Postal Service, 1775–1993.* Washington, DC: United States Postal Service. **http://www.usps.gov/history/history/his3_5.htm**

Vanicek, P., and Krakiwsky, E. (1986) *Geodesy*, 2nd ed. New York: North-Holland Publishing Company.

Watkins, R., and Conner, J. (2000) "Mapping America's oldest city," *Point of Beginning* (*POB*), Vol. 25, No. 5, pp. 14–16.

Welch, R., and Homsey, A. (1997) "Datum shifts for UTM coordinates," *Photogrammetric Engineering and Remote Sensing*, Vol. 63, No. 4, pp. 371–375.

Wisconsin SCO (1992) "Wisconsin Geodetic Control." Madison, WI: Wisconsin State Cartographer's Office, University of Wisconsin-Madison.

Wisconsin SCO (1998) "Wisconsin Coordinate System." Madison, WI: Wisconsin State Cartographer's Office, University of Wisconsin-Madison.

Wolf, P. R., and Dewitt, B. A. (2000) *Elements of Photogrammetry with Applications in GIS,* 3rd ed. New York: McGraw-Hill.

Wong, K. W. (1975) "Geometric and cartographic accuracy of ERTS-1 imagery," *Photogrammetric Engineering and Remote Sensing*, Vol. 41, No. 5, pp. 621–635.

CHAPTER

3

Digital Representation of Geographic Data

3.1 Introduction

Digital geographic data are numerical representations that describe real-world features and phenomena, coded in specific ways in support of GIS and mapping applications using the computer. For geographic data to be useful to GIS, they must be encoded in digital form and organized as a geographic database. This digital database then creates our perception of the real world, very much in the same way as the conventional paper map does.

There is a basic difference between these two methods of representation. Whereas the conventional paper map represents a general-purpose snapshot of the real world at a given time only, the geographic database allows a range of functions for storing, processing, analyzing, and visualizing spatial data. In other words, the geographic database is a dynamic, rather than static, view of specific aspects of geographic space, together with the necessary tools that allow users to interact with the data in support of their objectives.

In database terminology, the ways of representing data are known as *data models*. In this chapter, we will explain the concepts and techniques of representing geographic data in digital form using different data models. We start by examining the general characteristics of digital geographic data with special reference to their use in GIS.

3.1.1 THE GEOGRAPHIC MATRIX

The quest for a logical way to represent the real world has for centuries been one of the central themes of the science of geography. The geographic matrix proposed by Berry (1964) was a major milestone in the development of data organization methods in geographic studies. In the geographic matrix (Figure 3.1), a row represents the variations of a specific aspect of the natural or socioeconomic characteristics across some geographic space (that is, a spatial pattern of the variables that can be identified and mapped). A column, on the other hand, denotes a specific location in geographic space. Therefore, each cell formed by the rows and columns of the geographic matrix contains a specific item of geographic fact that can be found at a particular location. By comparing the rows in the geographic matrix, we study the spatial variation of geographic characteristics. By comparing the columns, we study the differences among different places (known as *areal differentiation* in geography). This implies that geographic data possess two distinct components: locations and attributes.

When geographic matrices pertaining to different points in time are assembled according to their temporal sequence, the time dimension of geographic data is introduced. Such an array of geographic matrices provides the conceptual framework for identifying the changes that occur at specific locations over a particular period of time. Therefore, the geographic matrix can be perceived as a representation of both the organization of geographic data and how they can be studied from a spatiotemporal perspective.

The geographic matrix by itself had very little to do directly with the development of geographic databases. The original purpose of the geographic matrix was to provide a coherent framework for bringing the different approaches of geographic studies together. However, it has been proposed that by representing geographic characteristics according to specific themes on the basis of a common locational reference, the geographic matrix provided an elegant theoretical foundation for the initial development of modern GIS (Chrisman et al., 1989; Sui, 1995). The geographic matrix was indeed the blueprint for many of the concepts for representing geographic data to be discussed in this chapter.

3.1.2 REPRESENTING GEOGRAPHIC SPACE

Real-world features exist in two basic forms: *objects* and *phenomena*. Objects are discrete and definite, such as buildings, highways, cities, and national parks. Phenomena are distributed continuously over a large area, such as terrain, temperature, rainfall, noise levels, and other environmental indices. Geographic data depict the real world in these two basic forms, leading to two distinct approaches of representing the real world in geographic databases: the *object-based* model and the *field-based* model (Figure 3.2) (Goodchild, 1992; Wang and Howarth, 1994).

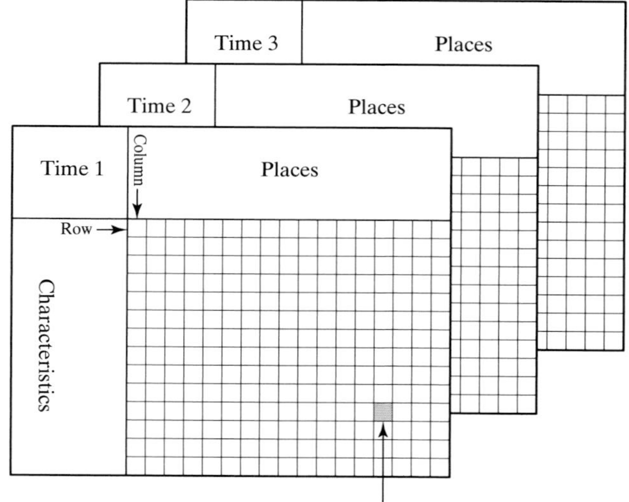

FIGURE 3.1
The geographic matrix.

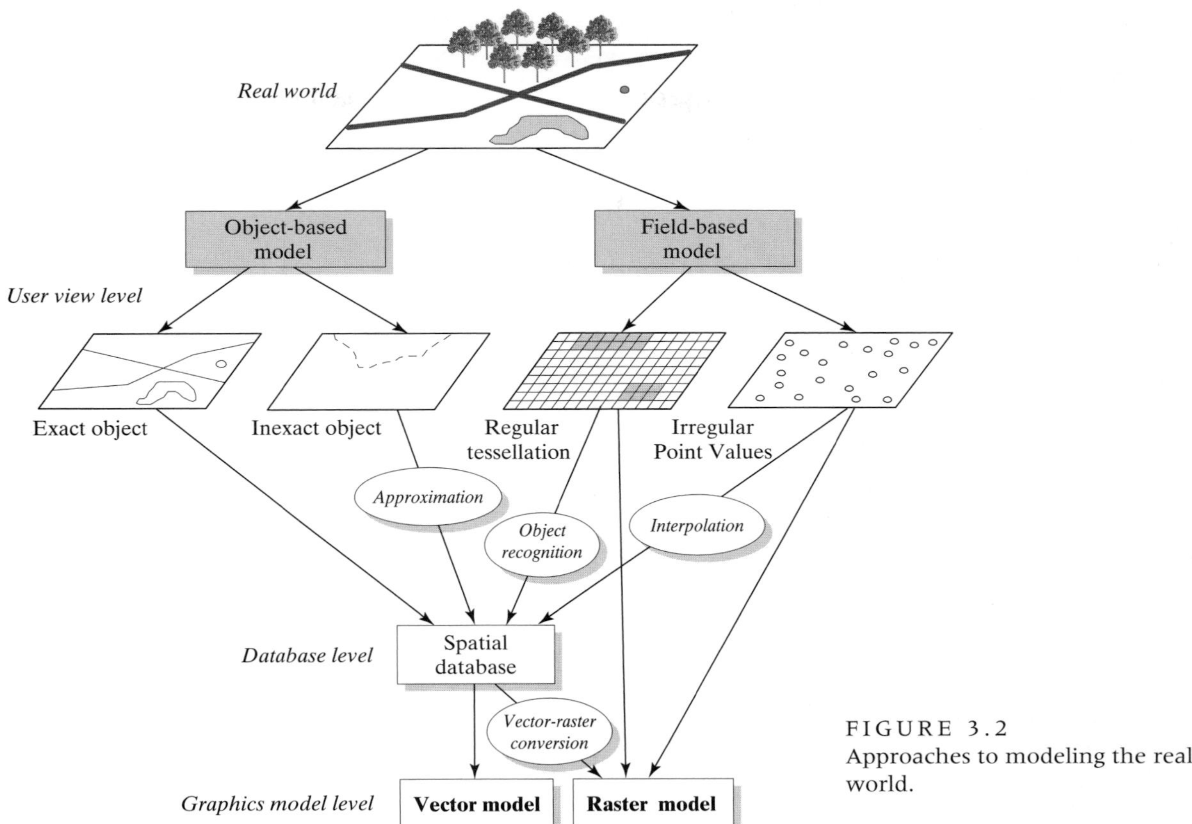

FIGURE 3.2
Approaches to modeling the real world.

The object-based model treats geographic space as populated by discrete and identifiable objects. By definition, an object is a spatial feature that satisfies the following conditions: (1) It has identifiable boundaries or spatial extent; (2) it is relevant to some intended application; and (3) it is describable by one or more characteristics commonly referred to as *attributes*. There are actually two classes of spatial objects. Spatial objects are said to be *exact objects* if they represent discrete features with well-defined boundaries. Man-made features such as buildings and land parcels are examples of exact objects. In many instances, the boundaries between spatial objects are identifiable but not well defined. The characteristics of these objects are transitional (i.e., changing gradually) across the assumed boundaries between neighboring spatial objects. These spatial objects are called *inexact objects* or *fuzzy entities*. Landform features and natural resource features are mostly inexact objects such as soil types, wildlife habitats, and forest stands. Data in the object-based model are obtained by field surveying methods (e.g., land and engineering surveying, site investigations) or laboratory methods (e.g., photogrammetric mapping, map and air photo interpretation, remote sensing image analysis, and map digitizing). Depending on the nature of the objects and the geographical scales at which they are recorded, spatial objects are represented as graphical elements of *points*, *lines*, and *polygons* (Figure 3.3).

The field-based model, in contrast, treats geographic space as populated by one or more spatial phenomena. Spatial phenomena are real-world features that vary continuously over space with no obvious or specific extent. Data for spatial phenomena structured as fields can be obtained either directly or indirectly. Direct data acquisition includes aerial photography, remotely sensed imagery, map scanning, and field measurements made at selected or sampled locations, for example, topographic data for *triangulated irregular networks* (TIN) (see Chapter 9). Indirect data acquisition methods generate data by applying mathematical functions, such as interpolation, reclassification, or resampling, to measurements made at selected or sampled locations. Topographic data such as contours and *digital elevation models* (DEM) are examples of data usually obtained by indirect measurements (see Chapter 9). Spatial phenomena are represented as *surfaces*, which can be conceptualized as being made up of spatial data units in the forms of either *regular tessellations* or *irregular tessellations* (Figure 3.4).

At the database level, a spatial database can be structured using either the object-based model or the field-based model. In object-based databases, the basic spatial units are discrete objects. Inexact objects are abstracted as discrete objects by approximation. Discrete objects may also be obtained from field-based data by means of object recognition, (e.g., image understanding and

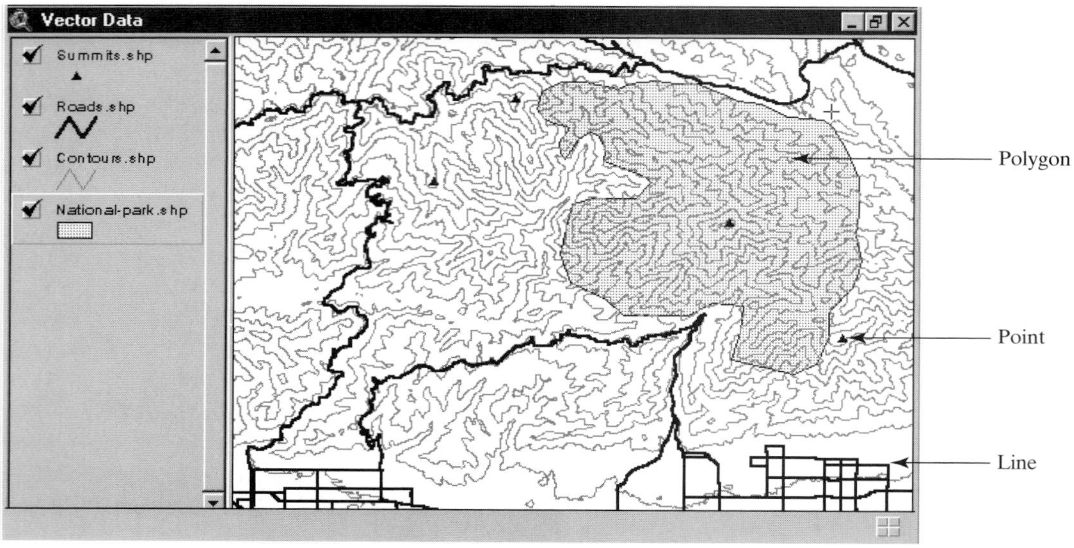

FIGURE 3.3
Representing spatial objects by points, lines, and polygons.

classification in remote sensing, digital photogrammetry, and air photo interpretation) and mathematical interpolation (e.g., contouring from TIN and DEM). As data in object-based spatial databases are mostly represented in the form of coordinate lists (i.e., vector lines), this type of spatial database is generally labeled as the *vector data model*. When a spatial database is structured on the field-based model, the basic spatial units are the different forms of tessellation by which spatial phenomena are depicted. As the most commonly used type of tessellation is a fi-

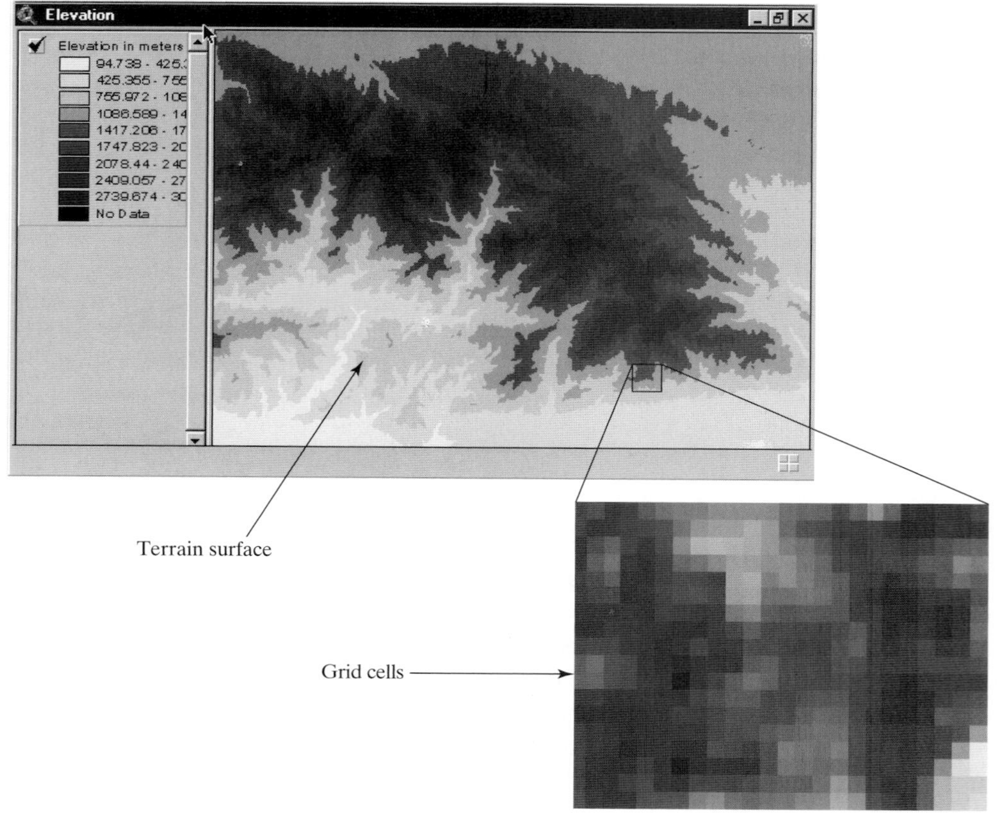

FIGURE 3.4
Representing surface by regular tessellations.

nite grid of square or rectangular cells, field-based databases are generally labeled as the *raster data model*.

In Chapter 1, we talked about three types of geographic data: vector, raster, and surface. Surface data are most commonly represented by the raster model, but there are exceptions. Topography, for example, can be conveniently represented either by the vector model using contour or TIN, or by the raster model using DEM. Similarly, temperature and rainfall maps are surfaces created by using vector lines known as isotherms and isohyets.

3.1.3 THE REPRESENTATION OF SPATIAL RELATIONSHIPS

Spatial relationships are the associations or connections between different real-world features. Such relationships can be *geometric* when adjacent features share the same boundary. They can also be *proximal* when one feature is said to be close to another feature. Spatial relationships serve several important purposes in geographic data processing and analysis. Many GIS applications rely on spatial relationships rather than coordinates to solve spatial problems (see Section 3.5.3). A notable example is network analysis, which makes use of the spatial relationships between line segments in the vector data to determine the shortest route between two points (Figure 3.5a). When raster data are used, GIS applications make use of the spatial relationships between adjacent grid cells to determine, for example, the direction of flow in hydrologic modeling and the areas affected by a point source in environmental pollution investigations (Figure 3.5b).

FIGURE 3.5
Examples of using spatial relationships in geographic data processing and analysis.

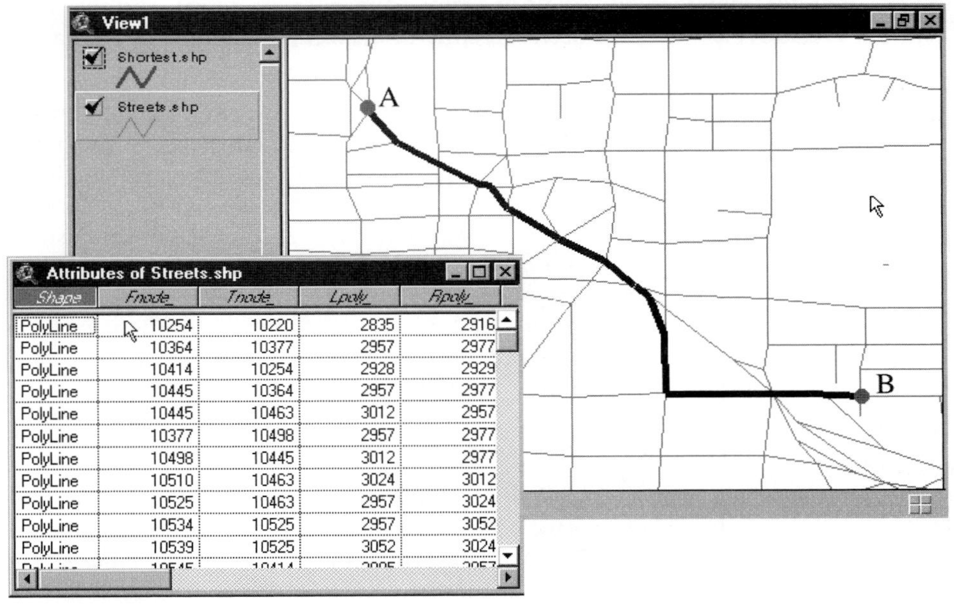

(a) The spatial relationships stored in the database: Fnode_ (from_node) and Tnode_ (to_node) are used to determine the shortest route between two points A and B.

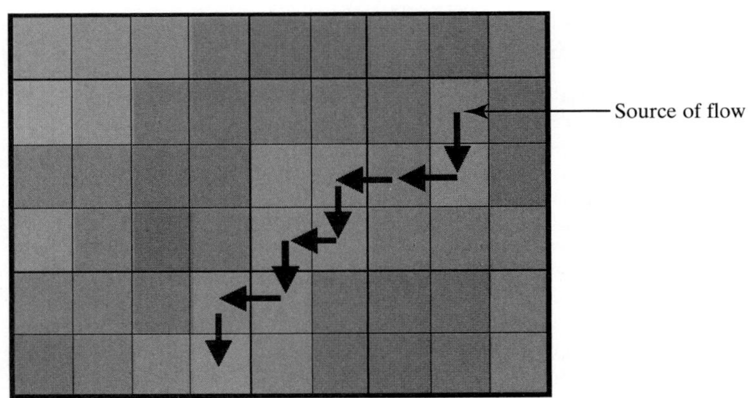

(b) Spatial relationships are used to determine direction of flow in hydrologic modeling using raster data.

Spatial relationships in geographic databases are normally represented by storing them explicitly in the data or by computing them when they are required. The "from_nodes" and "to_nodes" relationships in Figure 3.5a are typical examples of stored spatial relationships in the database. However, spatial relationships may also be readily computed when the data are used, for example, to determine the intersection of two lines, the containment of one polygon inside another, and the position of a point relative to a polygon.

The representation of spatial relationships looks relatively simple, but in practice it is no trivial problem. One of the reasons is the numerous possible relationships that exist among different types of spatial objects (Figure 3.6). The storage of spatial relationships demands considerable storage space in the computer, but "on-the-fly" computation of spatial relationships will slow down the speed of processing, particularly if the relationships are required frequently. What is perhaps more problematic is the fuzzy nature of human languages that describe these relationships. For example, while it is relatively straightforward to determine geometric spatial relationships such as "end at," "is within," "intersect," and "overlap," it is difficult to quantify semantic spatial relationships such as "near to" or "not far away from" because different people tend to have different ideas about what is meant by "near" and "not far away." Generally speaking, today's GIS can handle geometric spatial relationships very well, but special software tools are necessary in order to handle semantic relationships in geographic data processing and analysis.

3.1.4 THE REPRESENTATION OF TEMPORAL RELATIONSHIP

As the geographic matrix in Section 3.1.1 shows, time as an intrinsic property of geographic space has long been recognized by geographers. The concept of geography being made up of an integrated geometric-temporal space has also been widely accepted by practitioners of GIS. Until recently, however, the use of temporal data has been limited by the inability of conventional database management systems (DBMSs) to model and represent the time dimension efficiently and effectively. The state of the database at any point in time represents only a snapshot of the real world, but not the continuum that it actually is. This inability to handle the dynamic nature of environmental features has been recognized as one of the major shortcomings of conventional GIS technology (Burrough, 1986).

In recent years, there has been an obvious growth of interest in temporal data in GIS. Advances in computing technologies in the last few years have made it possible to implement many of the concepts developed to address temporal problems, such as new database structures, extended relational algebra, and augmented query languages (Kemp and Kowalczyk, 1994). The representation of time in digital geographic data can be explained in terms of its measurement, the relationship between time and space, and the temporal attribution of geographic processes.

The Measurement of Time Time is the yardstick for recording the existence and occurrences of spatial objects. The measurement of time for these objects can be made at an *instance* (from a point of reference), which is a measurement of their *existence*. It can also be made for a certain *duration*, which is a measurement of their *evolution*, *occurrence*, or *permanence*. For application in computerized databases, the measurement of time can be recorded in terms of *world time* or *database time*. World time refers to the time when an event occurs, whereas database time is the time when data pertaining to the event is entered into the database.

The Relationship Between Time and Space Spatial objects may change over time in terms of space and content. Spatial changes refer to the *geometrical transformation* of the objects that may include a change in the location, size, orientation, and form. As a consequence of these

	Point	Line	Area
Point	Is nearest to Is neighbor of	Ends at Is nearest to Lies on	Is within Is outside of Can be seen from
Line		Crosses Joins Flows into Comes within Is parallel to	Crosses Borders Intersects Is contained in
Area			Overlaps Is nearest to Is adjacent to Is contained in

FIGURE 3.6
Matrix of point-line-area spatial relationships.

changes, the spatial relationships among different objects may also change. Occasionally, changes such as those in object size may result in changes of the representation of spatial objects. An object represented as a point, at a certain scale, at a particular point in time may grow to become an object that has to be represented as a line or an area at the same scale at another point in time. When changes in the locations of a group of objects are considered together, changes in the pattern of spatial distribution of these objects can be revealed. The combined effects of all of the above relationships between time and space can be potentially very complex when changes in location, topology, and attribute are considered simultaneously in a temporal database, as illustrated in Figure 3.7.

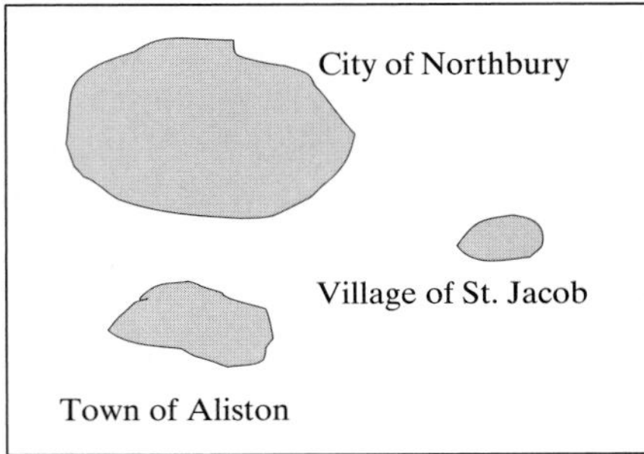

(a) Original settlements at time A.

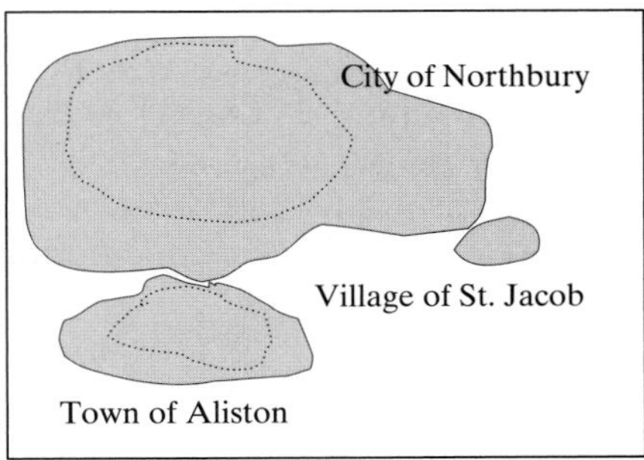

(b) Expansion of settlements led to change of sizes and spatial relationships at time B.

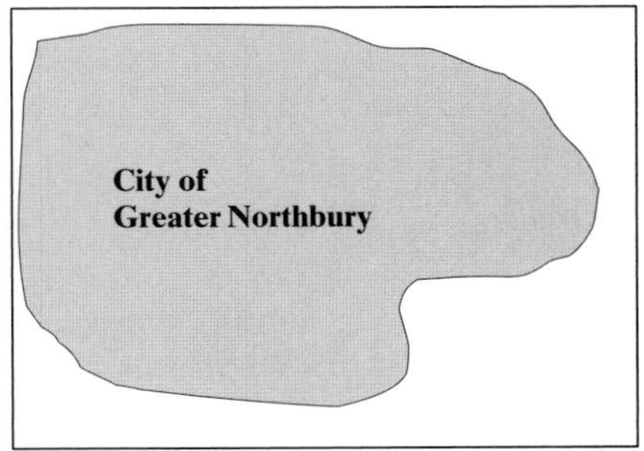

(c) Amalgamation of settlements led to change of sizes, attributes, and spatial relationships at time C.

FIGURE 3.7
Changing spatiotemporal relationships over time.

Temporal Attributes of Geographical Processes In general, there are four temporal attributes.

1. *Generation time* is the time at which an object is created. Since it is not always possible to know beforehand when something will happen, the determination of generation time is usually a retrospective process.
2. *Duration time,* also known as the *degree of permanence,* is the time during which an object is in existence or is observed.
3. *Temporal significance* refers to the importance of a particular event. It is usually the primary factor used to determine the inclusion or exclusion of an object in a database, in much the same way that spatial significance is used to include or exclude an object in a database.
4. *Temporal scale* is similar to map scale. Map scale is the ratio between ground and map measurements. Temporal scale is the ratio between actual and map time. As an example, an animated sequence that takes six seconds to display images recorded over a 12-hour observation interval, represents a temporal scale of 1:7200.

3.2 Technical Issues Pertaining to Digital Representation of Geographic Data

For digital data to be usable by the computer, they must be properly encoded as well as properly organized. Proper encoding only enables the data to be used, and proper organization is necessary for the data to be used efficiently. In the computer, digital data are organized logically and physically according to some rules and protocols. *Logical organization* is concerned with the ways data are classified and feature-coded so that the interrelationships between individual data items can be readily identified. *Physical organization,* on the other hand, is the way by which data items are actually stored in the computer's memory. In the following discussion, we will examine some of the key issues pertaining to the organizational aspects of digital representation of geographic data.

3.2.1 Data Classification and Classification Scheme

Digital geographic data are often collected and stored as categorical data (on a nominal scale), which means that they are represented as members of a class of spatial objects rather than as individual spatial objects themselves. In essence, collection of this particular type of geographic data is a data *discriminant analysis* process whereby a score or a class identifier is assigned to spatial objects or phenomena

Main Class	Subclass
1 Urban or built-up land	11 Residential
	12 Commercial and services
	13 Industrial
	14 Transport and communication
	....
	....
2 Agricultural land	21 Cropland and pasture
	22 Orchards, groves, and vineyards
	....
3 Rangeland	31 Herbaceous rangeland
	32 Shrub and brush rangeland
	......
4 Forest land	41 Deciduous forest
	42 Coniferous forest
	......
5 Water areas	51 Streams and canals
	......
6 Wetland	61 Forested wetland
	......
7 Barren land	71 Dry salt flats
	....

FIGURE 3.8
Partial listing of the USGS Land Use/Land Cover Classification Scheme. *(Source: Anderson et al., 1976)*

according to a predefined classification scheme. The purpose of a classification scheme, such as the USGS land-use/land cover classification system for remote sensing data (Anderson et al., 1976) (Figure 3.8), is to provide an a priori standard with which individual observations can be compared and recorded during the data collection process.

A classification scheme is made up of two components: *descriptive names* of the principal classes and their subclasses (there may be more than one level of subclasses), and *definitions* of each of the principal classes and the subclasses. The descriptive names can be based on either the *form* or the *function* of the spatial objects being observed. A 20-floor office tower in a land-use classification scheme, for example, will be recorded as a "high-rise structure" according to its form, yet it will be recorded as "commercial" in terms of its function. Whether the classification scheme is form-based or function-based is dependent on the application of the geographic data. The definition component of the classification scheme is usually made up of verbal descriptions of the physical or functional characteristics of the individual classes and subclasses that uniquely identify them from one another. In some cases, such as data collection by aerial photographic interpretation, the verbal descriptions are also supplemented by pictorial illustrations, similar in form to a photo interpretation key.

The development of a classification scheme is a rather complex process that requires considerable research and planning. In large-scale GIS implementation projects, unless an existing classification can be used, the classification scheme will be developed only after an extensive user requirements study in data modeling (see Chapter 11). An ideal classification scheme is one that is capable of exactly meeting the requirements of the applications for which the data are intended. A scheme that is too detailed (has too many classes) or too coarse (contains too few classes) may lead to potential problems when the data are used. It should also be noted that once the data collection process has started, it will be very difficult to make even a minor change in the scheme. This will involve a very laborious process of reclassification of all the data already collected. Also, whether or not it is possible to integrate and share data with other organizations depends very much on how the classification scheme is constructed. The classification scheme therefore has profound impacts on both the quality and the value of the geographic data collected.

Rhind and Hudson (1980) listed the following factors that must be considered when a classification scheme is designed.

1. The classification must be mutually exclusive, i.e., any spatial object can fall into one category only.
2. It must meet the detailed needs of the primary user and as many of the needs of the secondary users of the data as possible.
3. It has to be easily understood and applied.
4. It must be able to produce repeatable results when used by different observers using the same measurement technology.
5. It has to be hierarchical in order to cater to the need to make measurements at differing levels of resolution in different areas.
6. It has to be sufficiently stable for measurements carried out at different points in time to be compared.
7. It has to be sufficiently flexible for new interests and tasks to be met from a modified, rather than a completely new, classification.
8. It must incorporate some recognition of seasonal or other cyclic changes.
9. Wherever possible, it must be based on quantitative criteria.

Obviously, it is very unlikely that a classification scheme will meet every one of these criteria. In fact, some of the criteria conflict with one another, such as the need for stability (factor 6) and flexibility (factor 7). There may also be conflicting requirements between the primary user and the secondary users (factor 2) in terms of, for example, the number of classes of data. The design of a classification scheme, therefore, is always based on the compromise between the anticipated conflicting needs of different applications in GIS implementation.

3.2.2 FEATURE CODES AND FEATURE CODING

The classification scheme helps the data collector identify spatial objects in the geographic data-collection process. In database terminology, each spatial object identified and recorded is referred to as an *entity*. An entity is a spatial object that has specific properties that categorically separate it from other entities. These properties are known as descriptive *attributes*. Entities that share common properties are collectively called an *entity class, entity type*, or *feature class*. For example, an interstate highway is an entity in a transportation information system. All the interstate highways in the system collectively constitute an entity class that shares common attributes such as highway number, number of lanes, starting point, ending point, year of construction, and the latest date of maintenance.

In a geographic database, entities and attributes are represented in the form of *feature codes*. The process of encoding the values of the entities and attributes to graphical elements during the data-collection processes (by the methods of direct field data collection and digitizing) is referred to as *feature coding*. In its simplest form, a feature code is made up of two components: a *major code* that identifies the entity type to which a particular spatial entity belongs and a *minor code* that identifies the attributes that an entity has. For this reason, feature codes are also referred to as *attribute codes*. Feature codes may be alphabetic, numeric, or alphanumeric. Using the major and minor codes, feature codes may be organized into classes and subclasses. For example, the feature code "RLD" can be used to identify a class of residential structures (R for residential; LD for low-density); and the feature code "RLD0102PD" to identify a subclass of low-density structure that is of the single level type (01), with double garage (02) and paved driveway (PD).

Feature codes form a mini-database within a geographic database that can be used for various purposes. The principal application of feature codes is to provide a means for systematically representing spatial entities and their attributes in geographic databases as illustrated in the preceding example. For such an application, the feature code is always constructed to provide a hierarchical ordering of the entity classes and subclasses in the geographic database. One practical example is the Canadian Council on Surveys and Mapping (CCSM) classification scheme. This scheme uses a ten-character code with four levels that denote the class, category, feature code, and attribute code associated with an entity (Figure 3.9) (BCMELP, 1996).

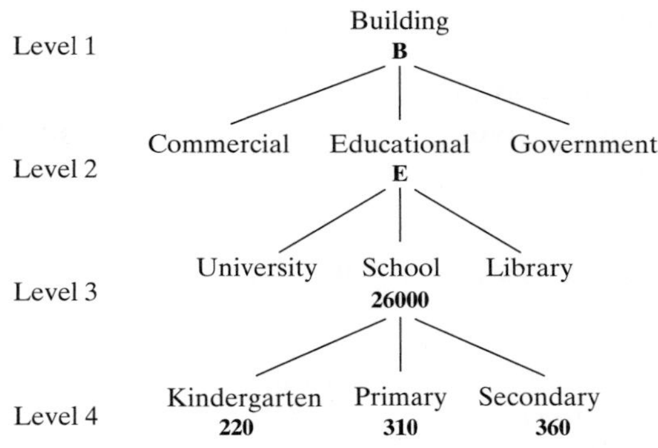

FIGURE 3.9
The Canadian Council on Surveys and Mapping (CCSM) Feature Coding Hierarchy.

In addition to their common use as the digital representation of geographic entities, feature codes may also be used to control the geographic data collection process as well as the cartographic display of the data, as exemplified by the feature coding scheme of the Digital Line Graphs (DLGs) of USGS (Figure 3.10). In this particular scheme, the feature code is made up of a major code and a minor code. The major code has three digits, the first two of which uniquely identify the data category to which the described entity belongs, and the third digit is used to designate the interpretation of the minor code (USGS, 1997). The minor code is the attribute code, which describes the subclasses within a major data category and the cartographic elements (i.e., a single point, a node, a line, or a polygon) that represent the data. When using feature-coded digital geographic data,

Major Code	*Category*
020	Hypsography
050	Hydrography
070	Vegetative surface cover
080	Nonvegetative feature
090	Boundaries
150	Survey control and markers
170	Roads and trails
180	Railroads
190	Pipelines, transmission lines, etc.
200	Man-made features
300	U.S. Public Land Survey system

Minor Code (for 070 vegetative surface cover)	
Area attribute code	
070 0101	Woods or brushwood
070 0102	Scrub
070 0103	Orchard or plantation
070 0104	Vineyard
070 0105	Scattered trees
070 0106	Void area

FIGURE 3.10
The USGS Digital Line Graphics (DLG) Attribute Codes.

the codes may be conveniently used to extract selected spatial objects from the database. The codes may also be used in conjunction with a look-up table of color palette or font type to control the appearance of cartographic displays.

3.2.3 GEOGRAPHIC DATA PRECISION

The precision of digital representation of geographic data is dependent on the number of *bits* used by the computer to store spatial (coordinate) and descriptive (attribute) values. A single bit, having a value of 0 or 1, is the most basic unit of data in the computer. To store data, bits are grouped into *words*. The number of bits that make up a word is known as the *word length*, which typically ranges from 16 to 64 bits. However, the smallest *addressable* unit in the computer is a *byte* rather than a word. By definition, a byte is an 8-bit data item. If the word length of a particular computer is 32 bits, for example, four bytes can be placed in a word, each individually addressable (Hamacher et al., 1996).

The number of bytes (and therefore the number of bits) that a computer uses to store a number varies, depending on the class of the computer, the type of the number, and the format of storage. In general, attribute data are stored as *integers* (whole numbers with no decimal points), and coordinates may be represented as either integers or *floating-point numbers* (numbers with decimal points, also known as *real numbers*). Each of these two types of numbers is stored with a fixed number of digits. Integers can be stored by using as many as 32 bits, whereas floating-point numbers can be stored by using 128 bits on most computers. A 32-bit integer has a range of values from −2,147,483,648 to +2,147,483,648. Floating-point numbers of various sizes have very large ranges and provide different numbers of decimal digits (7 places for 32 bits; 15 places for 64 bits, and 33 places for 128 bits).

Computer numbers, both integers and floating-point numbers, are *discrete* (that is, they "jump" from one value to another) and cannot cover all the possible measurement values (which by nature are *continuous*). As a result, not all coordinates can be precisely represented in the computer. When the original data contain more precise measurements than those supported by the computer, rounding occurs and the precision of the data is reduced. In practice, coordinates in GIS are stored by floating-point numbers, usually in *double-precision* mode, in order to minimize the impacts of coordinate rounding during data processing (Figure 3.11). The tradeoff of using this mode of data representation, however, is the increase in the volume of the data because the number of bits required to represent a coordinate value is increased.

As GIS applications are always data-intensive, the volume of data is a critical consideration in GIS implementation. The volume of data does not only have an impact on the storage requirement of a particular geographic database, but also the transmission of the data across computer networks (see Chapter 12). Hence, the issue of the precision of coordinate representation is definitely one of the most important considerations because of the large number of coordinates stored in a typical geographic database. Other factors that affect the volume of data in digital geographic databases include: the spacing between points along a line, the number of entities to be stored, the amount of attribute data to be stored, and the amount of redundant data required by specific data models. The aim of data modeling is to identify the best possible solutions to address these issues, with a view to optimizing the usability for as many applications as possible as well as the performance of the system by minimizing the data-storage and transmission requirement.

Original Length of Line Subject to Scaling	Long-integer **801.4454**	Double-precision **127780.25508**
Multiplied by	Results	
100	80,106.191	12778025.5084148
0.01	801.44026	127780.2550841
1000	80,105.41	127780255.084148
0.001	801.43613	127780.2550841
10000	8,010,563.7	1277802550.8415
0.0001	801.43342	127780.255084
1000	801,053.36	127780255.084147
0.0001	81.928767	12778.0255084
100000	8,009,658.7	1277802550.8414
0.0001	**801.34934**	**127780.2550841**

FIGURE 3.11
Comparison between the values of a long integer and a double-precision floating-point number during data processing. This table shows how the values of the lengths of two spatial objects change during data processing. Note how the values of the longer integer length change in an unpredictable manner after each scaling operation. In the case of the double-precision length, its values practically remain unchanged throughout the whole process. *(Source: Autodesk, 1996)*

3.2.4 DATA ORGANIZATION IN THE COMPUTER

The basic building block of data organization in the computer is a *data item*. By definition, a data item is an *occurrence* or *instance* of a particular characteristic pertaining to an entity, which can be a person, thing, event, or phenomenon (Figure 3.12a). In database terminology, a data item is called an *attribute* or a *stored field*. The value of an attribute can be in the form of a number, a character string, a date, or a logical expression (e.g., T for "true" and F for "false").

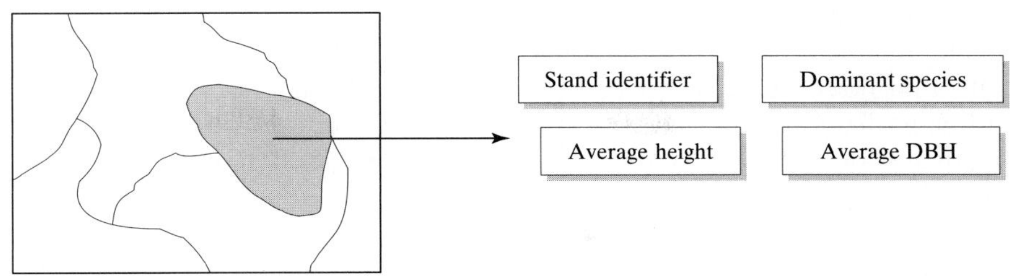

Forest resource inventory map

(a) Data items pertaining to a forest stand

Stand identifier	Dominant species	Average height	Average DBH

(b) A record of data items pertaining to a forest stand

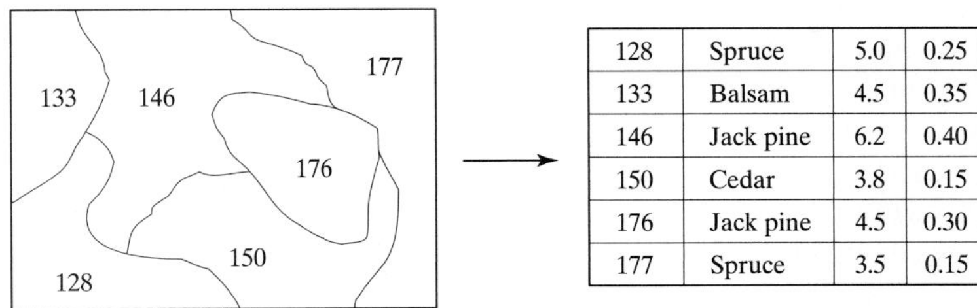

128	Spruce	5.0	0.25
133	Balsam	4.5	0.35
146	Jack pine	6.2	0.40
150	Cedar	3.8	0.15
176	Jack pine	4.5	0.30
177	Spruce	3.5	0.15

(c) A table or flat file of related records

Census District	Population				Number of Households		Medium Income ($/year)	
	1990		2000		1990	2000	1990	2000
	M	F	M	F				
C13	210	200	235	220	45	50	35,000	37,500
C14	354	288	390	356	80	92	40,000	42,900
C15	400	427	380	350	84	75	50,000	57,500
C16	260	279	300	320	45	52	37,000	42,000
C17	320	298	350	320	76	82	45,000	52,000

(d) A hierarchical data file of related records

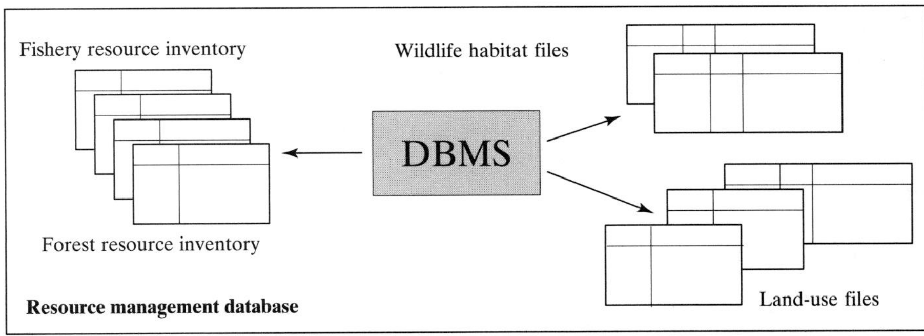

(e) A database of centrally controlled but physically distributed data files

FIGURE 3.12
Organization of digital data in the computer.

When related data items are grouped together, a *record* is formed (Figure 3.12b). "Related data items" means that the data items are occurrences of different characteristics pertaining to the same entity. For example, in a forest resource inventory, a record may contain related data items such as stand identifier, dominant tree species, average height, and average diameter at breast height (DBH). Data items in a record can have different types of values. In a forest resource inventory, for example, a record may have two character strings representing the stand identifier and dominant tree species, an integer representing the average tree height rounded to the nearest meter, and a floating-point number representing the average breast height diameter. In database terminology, a record is always formally referred to as a *stored record* or a *tuple* (see Section 3.3.3).

The next level of data organization is a *data file* that is formed by grouping related records together. "Related records" means that the records represent different occurrences of the same type of class of entities. A data file that is made up of a single record type with single-valued data items is called a *flat file* (Figure 3.12c). A data file that is made up of a single record type with nested repeating groups of data items forming a multilevel structure is called a *hierarchical file* (Figure 3.12d). In the computer, each data file is identified by a unique file name. A data file may contain records having different types of values. A file that contains records made up of character strings is called a *text* or *ASCII file*, whereas one that contains records made up of numerical values in binary form is called a *binary file*.

Computer-based data processing is based on *databases* rather than data files. A database is an automated, formally defined, and centrally controlled collection of persistent data, used and shared by different users in an organization (Figure 3.12e) (Date, 1995; Elmasri and Navathe, 1994; also see Section 3.3.2). This definition excludes the informal, private, and manual collection of data. A database is set up to serve the information needs of an organization and, as such, data sharing is key to the concept of databases. In the above definition, "centrally controlled" does not mean "physically centralized." Databases today tend to be physically distributed on different computer systems, either at the same or different locations. Data in a database are described as "persistent" in the sense that they are different from "transient" data such as input to and output from an information system. This means that data in a database often remain there for a considerable length of time, although the actual values of the data can change frequently.

The use of databases for data organization does not mean the demise of data files. Data in a database are still organized and stored as data files. In *relational databases*, for example, data are organized in files with data items logically arranged in rows and columns. Such data files are formally called *tables* or *relations* (see Section 3.3.3). Using databases rather than data files represents a change in the perception of data, the mode of data processing, and the purpose of using the data. In other words, using a database is simply an approach to data processing and management. It does not necessarily involve any physical change in the storage of the data (Table 3.1).

The concept of using data items, records, data files, and databases to organize digital attribute data also applies to graphical data. In this case, coordinate values are the equivalents of data items that serve as the most elementary unit of data organization. Coordinates that form the three basic graphical elements (i.e., points, lines, and areas) constitute individual records. Graphical data files are collections of related graphical elements. These files are commonly referred to as *layers*, *themes*, or *coverages*. A collection of layers associated with a particular application forms the spatial database. Attribute data files and their associated graphical layers that cover a particular geographic space of interest constitute a geographic database.

Table 3.1

Comparison Between a Data File and a Database

Characteristics of a Data File	*Characteristics of a Database*
Data are collected for a specific purpose	Data are generic for the business of an organization
A collection of records usually of the same data type and format	A collection of interrelated records, organized in one or more data files that may have different data types and formats
Data file processing is usually associated with computer programming that aims at solving a particular problem, i.e., it stops when an answer is obtained	Database processing is always associated with DBMS that aims at solving the operation or production needs of an organization, i.e., it involves routine, largely repetitive applications executed over and over again
Mainly used in support of the information needs of an ad hoc application	Used mainly in support of the day-to-day operation of business transaction processing but increasingly used in support of decision making

3.3 Database and Database Management Systems

The database approach provides a robust method to data organization and processing in the computer. However, it also requires more effort and resources to implement. The design and building of geographic databases, therefore, is an elaborate process by necessity. This process starts with the relatively general observations of the real world and progresses through the abstract, user-oriented data organization, to the concrete, system-dependent data-storage structure in the computer. Generally, the progression of data abstraction falls into three relatively distinct modeling levels: *conceptual, logical*, and *physical*. The characteristics of these different levels of data abstraction can be explained in terms of the concept of *data models* in the following discussion.

3.3.1 Levels of Data Abstraction and Data Models

Simply defined, a data model is a *description* or *view* of the real world and data modeling is the process that formalizes the description or view at different levels of data abstraction. Since the real world is made up of complex spatial objects and phenomena, it is practically impossible for a single data model to represent everything that is present. A data model therefore always tends to be tailored to a specific application or problem context. This means that different users may have different data models when they attempt to collect data in the same location. For example, a municipal engineer, a land developer, and an urban geographer all wish to obtain information about the same urban area by interpreting a set of aerial photographs. It is obvious that the interest of these three people will be very different from one another. The engineer wants to examine the physical conditions of the roads and buildings; the developer has the objective of seeking locations for potential development and redevelopment; whereas the geographer is more interested in the spatial pattern of the land use.

In data modeling terminology, the different *views* of the same urban area obtained by the engineer, the developer, and the geographer are called *conceptual models*. Conceptual models represent the user's perception of the real world. Data abstraction at the conceptual model level is strictly limited to the description of the *information contents* of the user's view of the real world, without any concern for computer implementation. This implies that although conceptual data modeling is part of the database design process, conceptual models are basically database-independent. The function of conceptual data models is to provide the necessary language to describe how we naturally conceptualize data organization.

The *entity-relationship (E-R) data model* is one of the more commonly used conceptual data models for GIS data modeling (Chen, 1976) (see Chapter 11). This particular conceptual data model uses the concepts of *entities, attributes*, and *relationships* to represent real-world features, their properties, and the relationships between the entities. Figure 3.13a is an example of a small portion of the E-R data model for a land title information system. It depicts diagrammatically how the three entities (land title, parcel, and owner) are related, what entities they have, and how these entities are related to one another. The development of the conceptual model is concerned only with data and how they represent the real world. There is no consideration for hardware and software requirements and implementation at this stage. The only purpose of the E-R diagram is to show our conceptualization of the entities, their attributes, and relationships.

The next level of data abstraction is called logical data modeling. A *logical data model* represents an implementation-oriented view of the database. As such, a logical data model represents the real world by means of diagrams, lists, and tables designed to reflect the recording of the data in terms of some formal language. This implies that, unlike conceptual data models, logical models are software-dependent (that is, they must be expressed in terms of the language of a specific DBMS). There are three classic logical data models: the *relational data model*, the *network data model*, and the *hierarchical data model*. The *object-oriented data model*, which is increasingly prominent in the database world, is a relatively new development (see Section 3.5.2). As these data models have actually been implemented in DBMS, logical data models are sometimes also referred to as *database models* or *implementation models*. Figure 3.13b illustrates the logical model that is created from the conceptual model in Figure 3.13a. This particular logical model is developed on the assumption that the database will be a relational database. If a network or hierarchical database is used, the logical model will be different.

Physical data modeling, which represents the hardware implementation-oriented view of the database, is the third level of data abstraction. A *physical data model* describes the physical storage (on file format) of the data in the computer by record formats, record ordering, and access paths. It is therefore hardware-dependent and is concerned primarily with the implementation details of a database. Physical data models are generally intended for the system programmer and database administrator, not for the general end users. In practice, the complete physical model is a very long and technically complex document. Figure 3.13c illustrates a small portion of the physical model that can be developed from the logical model in Figure 3.13b. It shows how the data elements of a particular record in the database are physically stored in computer memory.

(a) A conceptual data model

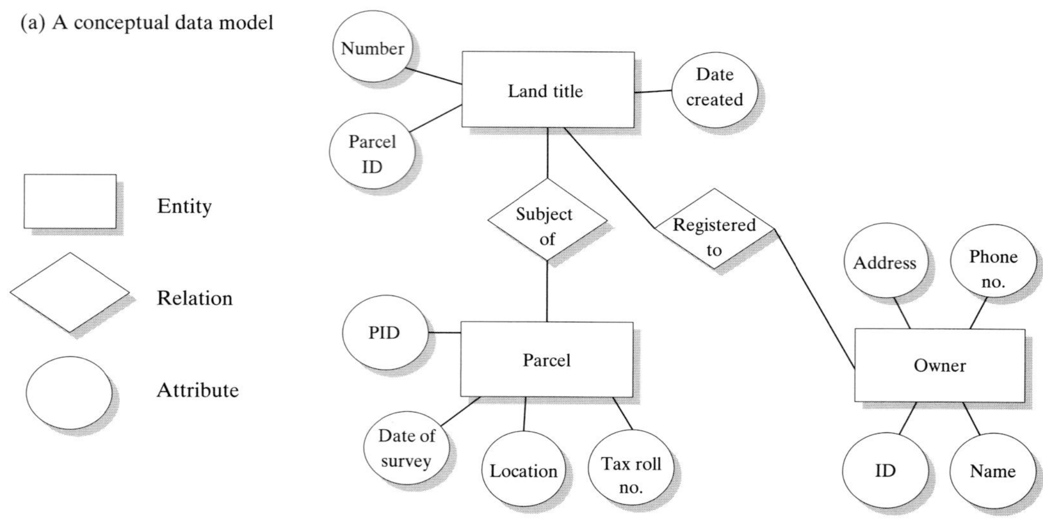

(b) A logical data model

Landtitle

Num	PID	
0001	0101	
0002	0234	
0003	0901	
0004	1201	
0007	8765	
0008	1225561	
0010	2314561	1998

Parcel Table

PID	Date of Survey	Location	Tax Roll Num
0101000	1998 12 01	123 John St. Whitby	1345-0103-002
0023451	1997 09 23	111 St George Oshawa	02231-99879
0901231			
1201234			
8765431			
1225561			
2314561			

Owner Table

ID	Name	Address	Phone Num.
0101000	J Doe	123 John St. Whitby	668-0103
0023451	K Jones	111 St George, Oshawa	679-02231
0901231	K Smith	23 Green Ave., Toronto	413-9222
1201234	L Law	45 DupontSt. Whitby	678-9988
8765431	P Pope	334 Kingsway, Whitby	668-0104
1225561	W Bush	88 High St., Kingston	780-6600
2314561	C Clayton	331Leeside St.,Whitby	668-2347

(c) A physical data model

```
STORE_LANDTITLE          LENGTH=20
  PREFIX                 TYPE=BYTE(6),OFFSET=0
  OWNER ID #             TYPE=BYTE(6),OFFSET=6,
                         INDEX=HBTX
  DISTCODE               TYPE=BYTE(4),OFFSET=12
  AREA                   TYPE=FULLWORD,
                         OFFSET=16
```

FIGURE 3.13
Levels of data abstraction and data models. The conceptual, logical, and physical data models represent three levels of data abstraction. The conceptual data model is our understanding of real-world objects and their relationships. It is independent of hardware and software considerations. The logical data model is the organization of data within the framework of a DBMS. The physical data model is the actual storage of the data in the computer. It is therefore both hardware- and software-dependent.

The conceptual, logical, and physical data models form a sequence of data modeling processes used in database design and development. The database design process starts with the conceptual data model. The conceptual model is mapped onto a logical data model when a decision is made to implement it in a DBMS. When it comes to the actual implementation of the database, the logical data model is mapped onto the physical data model (Date, 1995; Elmasri and Navathe, 1994). Database design and mapping require special training and skills and are usually the responsibility of the systems specialist. However, it is essential for GIS users to have a basic understanding of the concepts and processes involved. We will study the process of GIS data modeling in greater detail in Chapter 11.

3.3.2 CHARACTERISTICS AND FUNCTIONALITY OF DATABASE SYSTEMS

The implementation of a physical data model results in the creation of a database. Depending on its intended purposes, a database may be of different sizes and varying complexity. It may be created and managed manually or by the computer. A computer-based database is maintained by a DBMS. Functionally, a DBMS is a collection of software programs that facilitates the processes of defining, constructing, and manipulating the database for various applications. The database, including the system catalog where the data dictionaries are stored, and the DBMS together are usually referred to as a *database system* (Figure 3.14).

Computer-based data processing was traditionally done by the method of file processing. In file processing, individual users define and implement the data files for specific applications. This means that data files are stored as disparate files by their respective custodians. Modern data processing by the database approach stores all data in a single repository that is created for use by a multitude of users. Such an approach has been made possible by applying the following general paradigms (Elmasri and Navathe, 1994).

1. the use of a system catalog that stores the metadata (i.e., data about data, or data dictionaries) about the structure of the data in addition to the data themselves
2. the separation of applications and data by using information of the data structure stored in the system catalog for access to the database rather than by embedding the data structure in the application programs (thus making possible the change of applications without changing the structure of the database)
3. the support of multiple views of the data by employing the layered approaches of data modeling and definitions (as explained in Section 3.4.1)
4. the support of multiuser transaction processing by employing concurrency control to database access

The database approach is different from traditional file processing in the centralized control of the data. There are many advantages of using the database approach for data processing, including (Date, 1995)

1. reduction of redundancy in stored data by sharing them among multiple users
2. minimizing inconsistency of data by a single data-collection and input process
3. sharing data as a corporate resource by allowing concurrent access to the data
4. ability to enforce standards in corporate data processing (standards may include any or all of the following: international, national, industry, corporate, departmental, hardware, software, and installation)
5. provision of an interface for nonprogrammers (i.e., users without programming background can use the database by following predefined procedures)
6. application of security restrictions by preventing unauthorized access to the data
7. maintenance of data integrity (i.e., accuracy, compatibility, completeness, consistency, currency, database backup, and recovery)
8. balancing the conflicting needs of different users

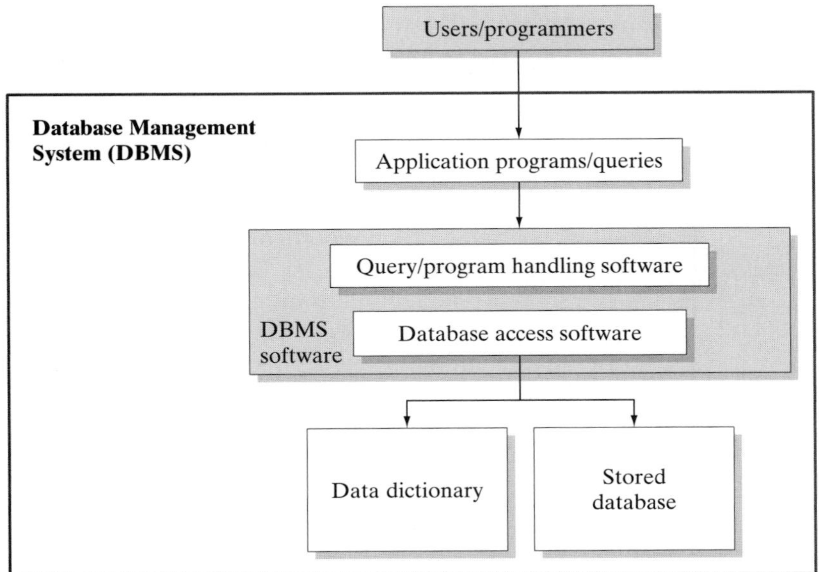

FIGURE 3.14
Components of a database management system (DBMS).

Conventionally, DBMSs have been designed for alphanumeric data processing in business transactions and recordkeeping. They are not suitable for geographic data processing, which requires the use of both graphical and descriptive data. Among the many deficiencies of conventional DBMSs for geographic data processing, the more critical ones include

1. *Transaction management*: In database terminology, a *transaction* is the process that accesses or changes the contents of the database. Transactions in commercial database applications, such as updating a banking account, are usually of very short duration, consist of simple procedures, require immediate attention, and are executed almost in real-time. In contrast, transactions in GIS applications such as updating a map often take days or even weeks to complete. As a result, the usual method of *record locking* that controls concurrent usage in conventional DBMSs is not applicable to GIS transactions.
2. *Spatial data indexing*: DBMSs are capable of accessing very large databases quickly with the aid of data indexing. Data indexing is a relatively complex process in commercial DBMSs. Spatial data indexing is even more complex because spatial data are two-dimensional.
3. *Graphical/descriptive data integration*: Graphical data, in the form of coordinates, represent spatial objects and phenomena. Descriptive data, on the other hand, represent the properties associated with spatial objects and phenomena. These two types of data are different in nature and consequently have different data access and storage requirements.
4. *Handling of spatial relationships and spatial operations*: Conventional DBMSs are designed for processing alphanumeric data that requires relatively simple operators such as addition, subtraction, multiplication, and division. These operators are not adequate for spatial operations, such as overlay, intersecting, and buffering that are essential for spatial data processing and analysis

Different GIS software vendors have tackled these problems in different ways. To address the problem of *long transactions*, for example, the common approach is to use a transaction management method based on the concept of *map librarian*. In this approach, the geographic database is partitioned into discrete storage units that are usually referred to as *tiles*. When a certain part of the data is retrieved (for example, for map revision), it is marked as having been checked out. Other users can still access the data for viewing but are not allowed to make any modifications to them (that is, that particular part of the database becomes *read-only*). On completion of the work on the data, the map librarian function will use the data to update the database and perform the necessary integrity checks before unlocking the data. The concept of the map librarian is simple and easy to understand. However, implementation is relatively complicated in practice because of the complexity of indexing geographic data as noted above. In addition, the volume of data involved in the transaction process is usually very large because it is always necessary to retrieve the whole tile, no matter how minor the changes may be. In other words, even though it is possible to overcome the problems of conventional DBMSs in one way or another, the inherent functionality of these systems has remained the major limiting factor for using them for GIS.

The growing market demands and opportunities have prompted GIS and DBMS vendors to develop new database tools that can handle the special requirements of geographic data processing. By taking advantage of new object-oriented technology, many software tools and design paradigms have now been successfully implemented that allow both attribute and graphical data to be manipulated by the same DBMS. Examples include Oracle Spatial by Oracle Corp. (1999), DB2 Spatial Extender by IBM (Davis, 1998) and ArcSDE by ESRI (1998 and 1999). The advent of this new breed of DBMS tool allows geographic data to be stored directly in large corporate databases, thus enabling the integration of geographic data and other forms of business data in an organization. Further explanations of these software tools is given in Section 3.6.4.

3.3.3 CONCEPTS AND TERMINOLOGY OF RELATIONAL DBMS

The *relational database model* is probably the most widely used database model in the computer industry. The concept of this particular database model, which was developed by IBM's San Jose Research Laboratory in the early 1970s (Codd, 1970), represents the database as a collection of *tables* or simple files. Each table in the database represents an entity type identified in the data modeling process (Figure 3.15). In a particular table, each column represents an attribute that describes the entity type and each row represents a collection of data values pertaining to the occurrence of an entity. The table name (i.e., entity name) and the column names (i.e., attribute names) are used to assist in the interpretation of the meaning of the values in the rows of the table. In relational database terminology, the table is called a relation. A row of the table is called a tuple and the column name is called an attribute or a field. The type of values that can appear in each column is called a *domain*. The properties of these terms are explained in detail here.

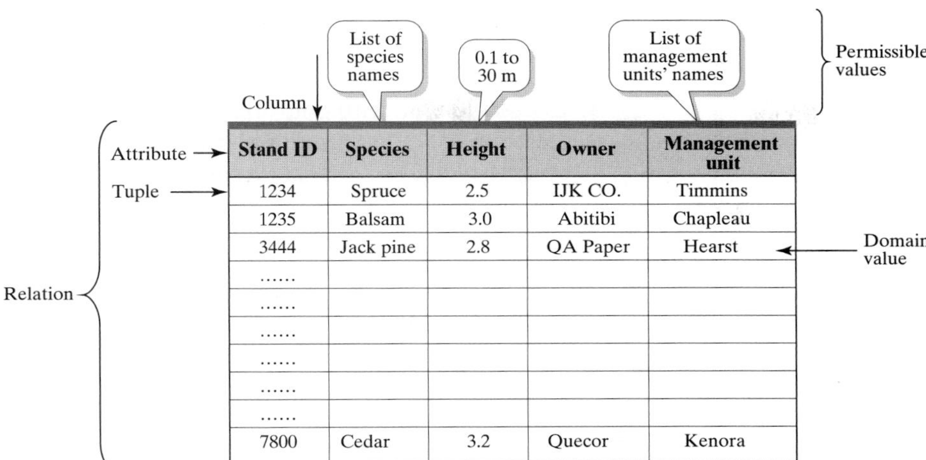

FIGURE 3.15
Terminology pertaining to the relational data model.

Domain The domain is the smallest unit of data representation in the relational database. A domain is a set of *atomic* attribute values (similar to the "geographic fact" in each cell of the Geographic Matrix). Atomic means that each value is *indivisible* into smaller components within the framework of the relational model. It follows that composite and multivalued attributes are not allowed. Much of the concept of the relational data model has been developed with this assumption, which is called the *first normal form* assumption (see Chapter 11). A domain is identified by an attribute name or identifier (e.g., City_name, Landuse_type, Highway_number) that is used to interpret its values. The values of a domain can be in one of several *data types* (which include *character, numeric, logical,* and *date*). The values of some domains are limited to a specified list of names (such as category names in a land-use classification scheme) or a specific range of numbers (such as the age of people and the height of trees). Such a list of names or range of numbers are usually referred to as the *permissible values* of a domain (Figure 3.15). There are occasions when the values of some attributes within a particular tuple are unknown or missing. A special value, called *null* (not zero), is then assigned.

Tuple A tuple is a set of attribute values that forms a row of a relation. There can be any number of attributes in a tuple, as long as the attributes conform to the first normal form assumption. Each tuple represents an occurrence of a particular type of spatial objects identified in the conceptual data model.

Relation A relation is made up of a set of tuples. There can be any number of tuples in a relation. Within a relation, the logical order of the tuples is not important. This means that the tuples can be ordered according to the values of any of the attributes of the relation. There is no preference for one tuple ordering over another. Logically, a relation can be simply thought of as a table of data. Functionally in the relational model, however, a relation must satisfy a number of constraints, including (Elmasri and Navathe, 1994)

1. *domain constraints,* which specify that the value of each attribute must be an atomic value from the domain of that attribute
2. *key constraints,* which require that no two tuples have the same combination of values for all of their attributes (i.e., no two tuples are identical in all their values) so that individual tuples can be identified by one or more attributes as the key (one of the attributes is assigned as a *primary key*)
3. *data schema integrity constraints,* which require the same real-world object, when represented by an attribute, to have identical attribute names in all relations (thus allowing different relations to be logically joined in data processing)
4. *entity integrity constraints,* which state that no primary key value can be null

Additionally, relations must also satisfy the condition of *data dependencies.* This means that they must satisfy the conditions of various *normal forms.* The concepts of data dependencies and *normalization* are used mainly for database design, which will be explained in detail in Chapter 11.

In order to query the relational database, a front-end query language called *Structured Query Language* (SQL) is used. This is an English-like language that consists of a set of powerful and flexible *commands* for the manipulation of the data in the relational tables (Figure 3.16) (Elmasri and Navathe, 1994). The use of SQL allows relational database systems to clearly separate the physical storage of data from their logical representation. Data are made to appear as simple tables that mask the complexity of the storage and access mechanisms. This effectively frees the user from the need to

```
SELECT c.name, c.address
    FROM customers
    WHERE state = 'GA' AND
              c.income > 50000 AND
              city = 'ATHENS'
```

FIGURE 3.16
Example of SQL query in data processing. The purpose of this SQL query is to find the names (c.name) and addresses (c.address) of all Georgia (state = 'GA') customers whose incomes (c.income) are greater than $50,000 per year and who live in Athens (city = 'ATHENS'), from a data table called 'customers.'

know the technical details of how the data are stored, so that they can access the data logically. By using SQL statements, the user needs only to specify the tables, columns, and row qualifiers to retrieve any data item in the entire database. The DBMS will take care of the complex sequence of procedures for the retrieval of the data and their display in the desired reporting format.

3.4 Raster Geographic Data Representation

The raster model is one of the variants of the field-based model of geographic data representation introduced in Section 3.1.2. It is best employed to represent geographic phenomena that are continuous over a large area. The field-based model is also commonly referred to as a *tessellation model*. Tessellations are geometric arrangements of figures that completely cover a flat surface. In this particular model, the basic spatial data unit is a predefined area of space for which attribute data are explicitly recorded. The spatial data unit can be of regular shapes (*regular tessellation*) or irregular shapes (*irregular tessellation*). There are three basic forms of regular tessellation: triangular, square, and hexagonal (Peuquet, 1984). All of these tessellations have been used as the basis for spatial data models. However, the regular square or rectangular tessellation, which is called the raster data model, has been the most widely used for the following practical reasons:

1. It is compatible with different types of hardware devices for data capture and output (image scanners, computer monitors, and electrostatic plotters).
2. It is compatible with the concepts and methods of bit-mapped images in computer graphics (and is therefore able to take advantage of the theories and software tools developed in this particular field of computer science).
3. It is compatible with grid-oriented coordinate systems, such as the plane rectangular coordinate system (Section 2.5.1 in Chapter 2).

3.4.1 Nature and Characteristics of Raster Data

The raster data model is characterized by subdividing a geographic space into grid cells (Figure 3.17). The linear dimensions of each cell define the spatial resolution of the data, which is determined by the size of the smallest object in the geographic space to be represented. The size is also known as the *minimum mapping unit* (MMU). A general rule is that the grid size should be less than half of the size of the MMU. Each grid cell must contain a value. This value can be an integer, a floating-point number, or a character (a code value). The value indicates the quantity or characteristic of the spatial object or phenomenon that is found at the location of the cell. These grid cell values can be used directly for computation (for example, interpolation of contours, isotherms, and isohyets) or indirectly as code numbers referenced to an associated table (a *look-up table* for classification data or a *color palette* for cartographic display). In a raster database, values pertaining to different characteristics at the same cell location are stored in separate files (*map layers*). For example, the soil type and forest cover for the same area are stored as separate soil

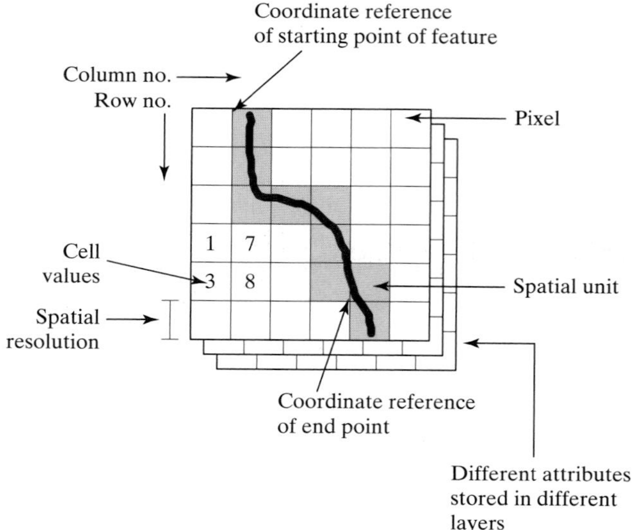

FIGURE 3.17
Characteristics of raster data.

and forest data layers. When the data are used for processing, the appropriate layers are retrieved. This means that raster data processing always involves the use of multiple raster files, in the same way that different layers are used in vector data processing.

Because the spatial data entities are the cells, the positions of spatial objects or phenomena in the raster model are recorded only to the nearest cell. Also, the representation does not always correspond to the spatial objects or phenomena in the real world. For example, a watershed does not exist as a distinct spatial entity in the database. The individual cells that make up the watershed are the entities. In other words, the identities of individual spatial objects are completely lost in the raster data model.

When a specific raster layer is displayed, it is shown as a two-dimensional matrix of grid cells. In computer storage, the raster data are stored as a linear array of attribute values. Since the dimension of the data (the number of rows and columns) is known, the location of each cell is implicitly defined by its row and column numbers. There is no need to store the coordinates of the cells in the data file. The locations of the cells can be computed when the data are used for display and analysis.

In order to translate a linear array storage to a two-dimensional display, enough information must be stored in the header section of the data file as well. In general, the file header contains information about the number of bits used to represent the value in each cell, the number of rows and columns, the type of image (e.g., whether the cell values should be interpreted directly as RGB color or indirectly as indices to a color palette or a look-up table), the legend, the name of the color palette if the file uses one, and the name of the look-up table if the file uses one. Some file headers also contain parameters for coordinate transformation so that the raster data in the files can be georeferenced. This is, however, a system-dependent feature. Many systems use a separate file for the storage of the information for coordinate transformation (see Chapter 5).

Raster data files are stored in different file formats. The differences between these file formats are due mainly to the different algorithms used to compress the raster data files (see Section 3.4.2). In order to minimize the data-storage requirements, raster data are often stored in a compressed form. The data are decompressed "on-the-fly" when they are used by an application program. There are now more than ten industry standard raster data formats in common use. Many GIS software products are now equipped with translators that allow them to accept a variety of raster formats. Some commonly used raster file formats are shown in Table 3.2.

Many government survey agencies have now adopted the raster data model for the provision of geographic data services to the public. A good example is the *Digital Raster Graphics (DRG)* of the USGS (1996a). A DRG file is a scanned image of a USGS topographic map. The purpose of DRG is to provide a backdrop onto which other digital data can be overlayed. At the USGS, for example, the DRG is used for validating Digital Line Graphs (DLGs) in the data collection and revision processes of these files. It is also widely used by other U.S. government agencies to assess the completeness of their digital geographic data and to create new map products (for example, by combining DRG with digital elevation models for creating shaded relief maps).

3.4.2 THE PRINCIPLES OF RASTER DATA COMPRESSION

It is common for a single raster data file to contain several million grid cells. For example, a black and white line map measuring 50 cm × 50 cm will produce 400 million pixels (picture elements) when it is scanned at a resolution of 25 micrometers (approximately 1000 dpi [dots per inch]). This will create a very large data file. The actual size of the file is dependent on the bit depth—the number of bits used to represent the value of the pixel. For example, 8-bit data will give values from 0 to 255 (2^8), 16-bit data from 0 to 65535 (2^{16}), and 32-bit data from 0 to 4294967295 (2^{32}). A few raster files can easily consume all the space on the hard drive of a small computer. The large file size also poses severe systems performance problems when the data are transmitted across computer networks. Data compression is therefore an important feature of digital representation of raster data. Most of the raster data formats listed in Table 3.2 support some form of data compression.

All approaches to raster data file compression are based on the realization that cells representing areas of the same entity type have identical values and that the patterns of the values tend to be spatially clumped. A variety of algorithms have been developed to handle adjacent cells of identical values. One such algorithm is called *run length encoding*. In this method, adjacent cells along a row with the same value are treated as a group called a "run." Instead of repeatedly storing the same value for each cell, the value is stored once together with the number of the cells that make the run. In Figure 3.18, for example, the 6 × 6 raster data matrix has 36 ASCII codes. In run length code, the matrix becomes a string of "10A2B3A8B1C4B2C3B3C." This string is made up of 18 ASCII codes, which represent a 50% reduction in storage requirement.

Table 3.2
Some Commonly Used Raster File Formats

Format Name	*Description*
BMP	The format used by bit map graphics in Microsoft Windows applications
PCX	A proprietary bit map format of ZSoft that is supported by many image scanners
TIFF (Tagged Image File Format)	A nonproprietary system-independent format designed as a nonproprietary format for the storage of scanned images and the exchange of data between graphics packages. TIFF/JPEG, commonly referred to as TIFF 6.0, is an extension of the TIFF format.
GeoTIFF	An extension of the TIFF format that contains georeferencing information in its file header. GeoTIFF is developed and maintained as the result of the concerted effort of GIS software developers, commercial data suppliers, and data users to provide a publicly available and platform-interoperable standard for the support of raster geographic data in TIFF.
GIF (Graphic Interchange Format)	A cross-platform format, proprietary to Compuserve, that is widely used for the transmission of images on the World Wide Web
JPEG (Joint Photographic Experts Group)	Another cross-platform format developed primarily for storage of photographic images and is also widely used for graphics on the World Wide Web
PNG (Portable Network Graphics)	A patent-free raster format that is intended to replace the proprietary GIF format
MrSID (Multi-resolution Seamless Image Database)	A raster format using compressing technology originally developed by the Computer Research and Applications Group at Los Alamos National Laboratory. A Seattle-based company known as LizardTech, Inc. has acquired the right to commercialize this technology using the U.S. Federal Technology Transfer Acts.
GRID	The proprietary raster format used by ESRI (Environmental Systems Research Institute, Inc.) in its software products such as ArcInfo and ArcView GIS

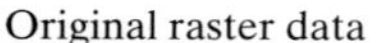

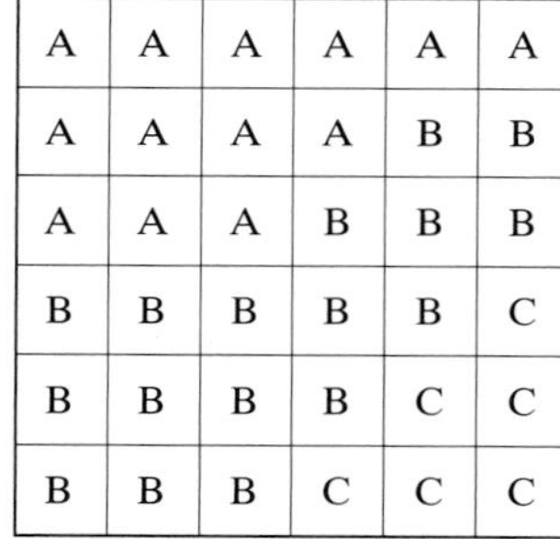

Total no. of values = 36

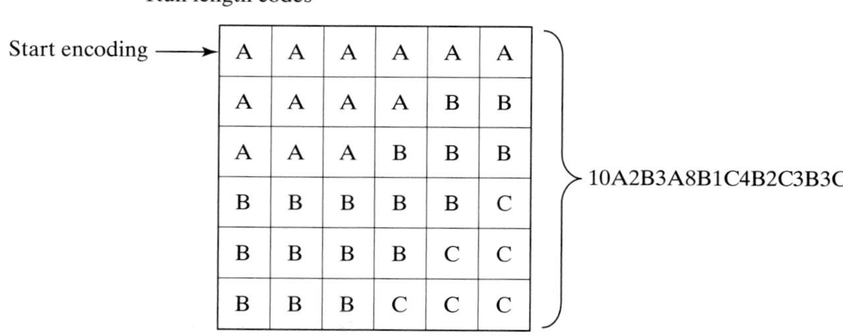

Total no. of values = 18

FIGURE 3.18
The method of run length encoding.

Run length encoding is conceptually simple to understand and technically easy to implement. However, its usefulness in practice is rather limited because it does not lead to a good compression ratio, i.e., the ratio between the original file size and the compressed file size. For example, using a more efficient compression technology known as wavelet compression, the format known as Multi-resolution Seamless Image Database (MrSID), (Table 3.2) is capable of achieving a compression ratio between 15 : 1 to 20 : 1 for 8-bit gray scale images and between 30:1 to 40:1 for 24-bit color images (see Section 5.1.2). Compression technology like this has greatly enhanced the use of raster images in GIS applications.

3.4.3 QUADTREE DATA REPRESENTATION

The *quadtree data model* is a hierarchical tessellation model that uses grid cells of variable sizes. In this model, the geographic space is divided by the process of *recursive decomposition*. Instead of dividing the entire geographic space into grid cells of the same size as in the case of the raster data model, the quadtree model uses finer subdivisions in areas where finer detail occurs. For example, if it happens that the entire map contains only a single land-use type, the whole map will be represented by one single cell. If more than one land-use type is present, the map will be subdivided into four equal-sized quadrants. The same test is repeated for each of the four quadrants. Any quadrant that contains more than one land-use type will again be subdivided into four equal parts; and any quadrant that contains a homogeneous land-use type will not be subject to further subdivision (Figure 3.19). The decomposition process will continue until it reaches a predefined maximum number of iterations (which in effect determines the minimum cell size that can be represented). For a typical topographic map that measures 50 cm × 50 cm, it takes 11 iterations to reach a resolution of 0.5 mm and 13 iterations to reach a resolution of 0.12 mm.

Just like the raster data model, no explicit storage of coordinates is necessary for the quadtree model. The positions of individual cells (the basic spatial data units) in the resulting quadtree can be determined by the cell identification number. In Figure 3.19, for example, cell 121 is contained in cell 12 at the second level of decomposition, which is in turn contained in cell 1 at the first level of decomposition. Many numbering schemes

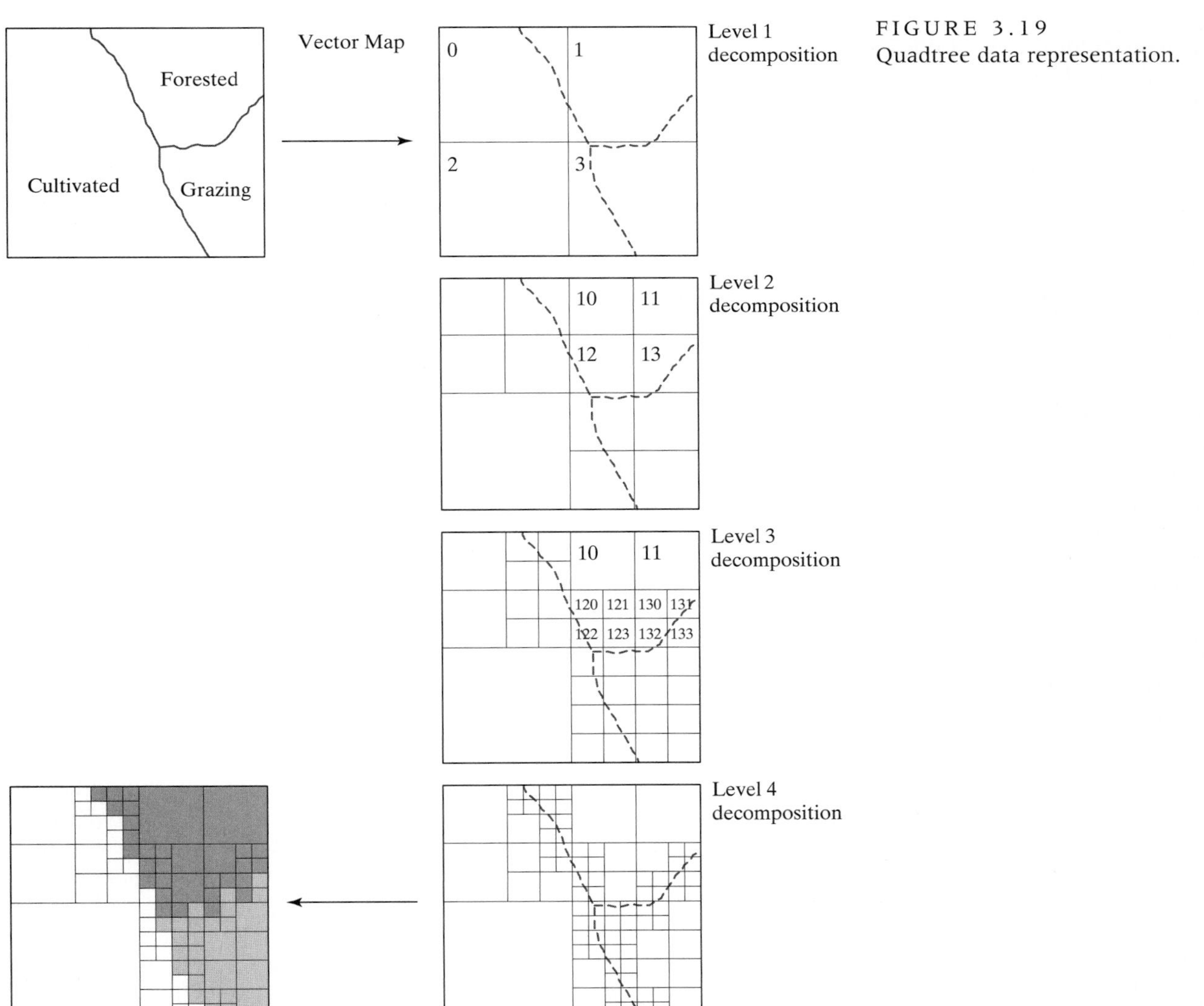

FIGURE 3.19
Quadtree data representation.

have been developed to identify the cells. The example in Figure 3.19 is only one of these schemes. The hierarchical numbering system of cell identification allows the location of each cell to be readily computed relative to the map origin. It also facilitates the determination of neighborhood and distance information. The descriptive attributes associated with the cells are stored as feature codes of individual cells. They can be accessed and displayed by cell locations or attribute values as required by specific applications.

The quadtree data model offers a number of advantages of representing geographic data, including

1. Data-storage and search techniques by quadtree are well researched and understood in computer science, with well-documented techniques for implementation.
2. It is compatible with the Cartesian coordinate system for cartographic applications.
3. The recursive subdivision of the geographic space facilitates physically distributed storage, allows an easy mechanism for the economic use of central memory, and greatly expedites browsing operations.
4. It allows variable spatial resolution to be represented in accordance with the degree of complexity of the geographic surface (Peuquet, 1984).

There are, however, a couple of trade-offs in using the quadtree model. It is more difficult, compared with a simple raster model, to create quadtree indices and tables. It is also a relatively complex process to modify quadtree indices and tables. The quadtree model is therefore best suited for areas where the data are relatively homogeneous and for applications that require frequent high-performance spatial search of the database. In recent years, there has been considerable interest in the use of the quadtree data model for GIS (Samet, 1990). This data model has also been actually implemented in one GIS software package, PCI SPANS (Ebdon, 1992).

3.5 Vector Data Representation

The vector data model is an object-based approach to the representation of real-world features and is best used to represent discrete objects. In this object-based approach, spatial objects are identified individually and represented mathematically (by coordinates). As a result, the vector data model is conceptually more complex than the raster data model. It is also more technically difficult to implement. Different GIS software vendors have used different approaches to tackle the conceptual and technical problems of using vector data. This has resulted in the development of a multitude of variants of the vector data model. However, no matter what methods and algorithms have been used, all vector data models are built on two common and interrelated concepts: (1) the decomposition of spatial objects into basic graphical elements (i.e., points, lines and polygons), and (2) the use of *topology* (spatial relationships) to represent spatial objects in addition to the use of geometry (coordinates). We will explain these concepts in the following four sections. Our discussion will focus on the representation of spatial objects and their relationships largely at the conceptual level of data representation and organization, without any detail of their implementation in GIS.

3.5.1 NATURE AND CHARACTERISTICS OF VECTOR DATA

Vector data are represented by means of coordinates. By using one or more pairs of coordinates, spatial objects are identified and represented by one of the three forms of *basic graphical elements*: (1) *points*, represented as single pairs of coordinates, are the simplest type of vector data; (2) *lines* or *arcs*, represented as a string of coordinates, begin and end with a *node*; and (3) *polygons* or *areas*, are represented as a closed loop of coordinates. When graphical elements representing an individually identifiable real-world feature are logically grouped together, a *graphical entity* is formed. For example, the different line segments that represent a highway are graphical elements. When these line segments are logically joined together as an identifiable highway in the database, the highway is a graphical entity. Similarly, polygons representing individual islands of, say, Hawaii, are graphical elements. When these island are grouped and identified as a single geographical unit, all the islands form a graphical entity called "Hawaii." Simply put, graphical elements are "dumb" graphics, whereas graphical entities are "intelligent" graphics.

Vector data that have been collected but not structured are said to be in the *spaghetti data model*. Vector data obtained by map digitizing are said to be in this particular data model because they are not yet structured (see Chapter 6). Data for computer-assisted design (CAD) systems are often stored in this data model. The spaghetti data model stores graphical elements, not graphical entities, defined by strings of coordinates (Figure 3.20). There is considerable redundancy with this data model, as the boundaries between adjacent polygons are stored twice, one for each polygon. Vector data in the spaghetti data model are usually not directly usable by GIS. For use in GIS, they must be properly structured according to either the specific format of a particular GIS, or a transfer format acceptable to the GIS (see Chapter 4).

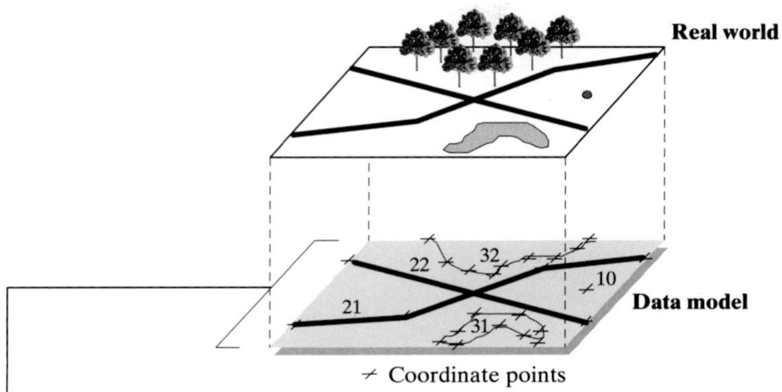

Stored data

Feature ID	Feature Type	Location
10	Point	xy
21	Line	$x_1y_1, x_2y_2, x_3y_3 \ldots x_ny_n$ (string)
22	Line	$x_1y_1, x_2y_2, x_3y_3 \ldots x_ny_n$ (string)
31	Polygon	$x_1y_1, x_2y_2, x_3y_3 \ldots x_1y_1$ (closed loop)
32	Polygon	$x_1y_1, x_2y_2, x_3y_3 \ldots x_1y_1$ (closed loop)

FIGURE 3.20
The Spaghetti Data Model. The Spaghetti Data Model stores geographic data in the form of graphical elements (point, line, and polygon), but not as individually identifiable graphical entities such as a lake, a wooded area, or a highway.

Structured vector data may be depicted at different levels of abstraction and by different data models (conceptual, logical, and physical) that represent a progression of increasing information content and corresponding complexity. Since these data models are built on the concept of topology, they are sometimes collectively referred to as *topological data models*. Of the many variants of topological data models, the most commonly used one is the *arc-node data model*. "Arc" is a line segment, and "node" refers to the end points of the line segment. Just like the spaghetti data model, the arc-node data model also stores graphical elements rather than graphical entities. However, this model explicitly stores spatial relations between the graphical elements, as well as the relationships between the arcs and their respective nodes (Figure 3.21). The stored topological relationships allow graphical entities to be constructed from the basic graphical elements.

Vector data stored in the forms of basic graphical elements can be used to represent simple spatial objects as well as complex objects. The construction of complex objects from simple graphical elements is based primarily on the use of *topological relationships* among spatial objects to be explained in Section 3.5.3. However, while this concept itself is quite generic, the actual methods used by individual GIS software vendors are mostly proprietary, i.e., they are system-specific and unavailable. This accounts for the different implementation details of different GIS based on the vector data model.

Vector data files can be stored in different file formats. Examples of commonly used vector data formats are shown in Table 3.3.

In addition to the formatting of the basic graphical elements, it is also necessary for the vector data to be properly linked to the descriptive data in geographic databases. This is usually achieved by the use of a unique *feature identifier* (FID) that is assigned to individual spatial objects. By using common FIDs, the graphical and descriptive elements of vector data can be correctly cross-referenced during database creation and spatial data processing (Figure 3.22). Usually, the assignment of the feature identifier is an automated procedure during the topology building process (see Section 3.5.2), but the linkage to descriptive data is normally a manual process that can be only partially automated.

The representation of vector data is governed by the scale of the input data. For example, a building that is represented as a polygon on a large-scale map will become a point on a medium-scale map; and it will not be represented at all as an individual entity on a small-scale map (unless it is a very important landmark). The possibility of representing vector data differently at different scales is associated with two important concepts: (1) *cartographic generalization*, whereby line and area objects are represented by more coordinates at a larger scale than at a smaller scale, and (2) *cartographic symbolization*, whereby vector data are represented by different symbols that serve to visually distinguish them from one another

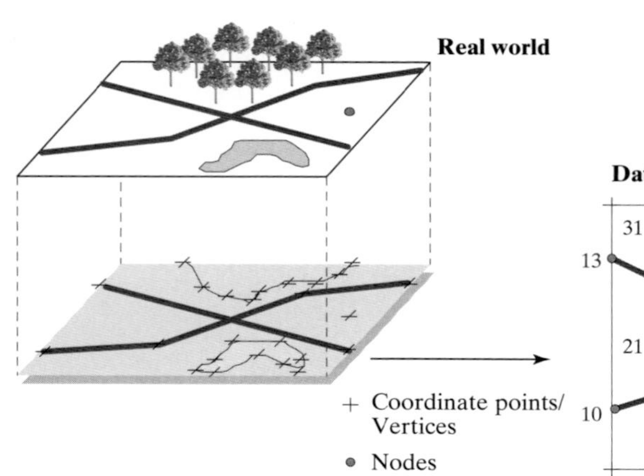

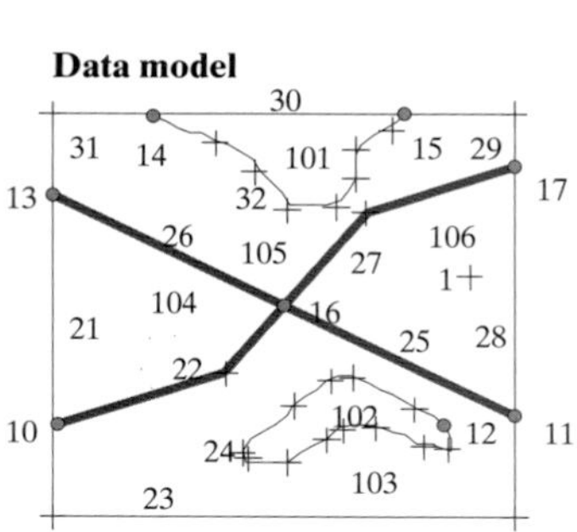

Point Topology

Pt ID	F. Code
1	10001

Pt ID	Coord.
1	*xy*

Arc Topology

Arc ID	F. Code
21	20001
22	20300
23	20001
24	20400
...	

Arc ID	FNode	TNode
21	13	10
22	10	16
23	10	11
24	12	12
...		

Arc ID	Start	Intermediate	End
21	*xy*	*xy, xy, xy* ...	*xy*
22	*xy*	*xy, xy, xy* ...	*xy*
23	*xy*	*xy, xy, xy* ...	*xy*
...			
...			

Polygon Topology

Poly ID	F. Code
101	30200
102	40500
...	
...	

Poly ID	Arcs List
101	30,32
102	24

Arc ID	FNode	TNode	L-Poly	R-Poly
24	12	12	103	102
30	14	15	World	101
32	14	15	101	105

Arc ID	Start	Intermediate	End
24	*xy*	*xy, xy, xy* ...	*xy*
30	*xy*	*xy, xy, xy* ...	*xy*
32	*xy*	*xy, xy, xy* ...	*xy*

FIGURE 3.21
The arc-node topological data model.

when the data are displayed. Further details on these two concepts can be found in Chapter 7.

In the computer, vector data can be stored as integers or floating-point numbers. In order to avoid the problem of rounding errors that occur during data processing, most if not all GIS software products store vector data by using double-precision floating-point numbers (see Section 3.2.3). This creates the impression that vector data are accurate and precise representations of spatial objects in the real world. In reality, however, this is not necessarily true. This is because the precision of data storage does not always mean accurate description of the data. Also, the boundaries of many spatial objects are fuzzy rather than exact entities. Storing vector data by double-precision floating-point numbers does not improve the quality of the data. It simply serves to avoid degradation of data quality due to rounding errors during data processing.

3.5.2 THE CONCEPT OF A TOPOLOGICAL MAP

Topology is a branch of mathematics that studies those properties of geometric figures that are unchanged when the shape of a figure is twisted, stretched, shrunk, or otherwise distorted without breaking (West et al., 1982). In other words, topology is geometry on a rubber sheet. It is the particular field of geometry that is

Table 3.3
Some Common Vector Data Formats

Name	*Description*
GBF/DIME (Geographic Base File/Dual Independent Map Encoding)	The data format developed by the U.S. Bureau of the Census for digitally storing street maps to assist in the gathering and tabulation of data for the 1970 census, which was used again for the 1980 census (U.S. Bureau of the Census, 1970)
TIGER (Topologically Integrated Geographic Encoding and Referencing System)	Created by the U.S. Bureau of the Census as an improvement to the GBF/DIME and was used in the 1990 census (Marx, 1986). This format is known as TIGER/SDTS for the 2000 census
DLG (Digital Line Graphs)	The format used for USGS topographic maps (USGS, 1996b)
AutoCAD DXF (Data Exchange Format)	One of the file formats of AutoCAD and AutoCAD Map that has become a de facto industry standard and is widely used as an export format in many GIS
IGDS (Intergraph Design System) DGN file	A proprietary file format for Intergraph CAD systems, which is also widely used in the mapping industry
ArcInfo Coverage	The proprietary data format of ArcInfo; stores vector graphical data using a topological structure explicitly defining spatial relationships
ArcInfo E00	The export format of ArcInfo
Shapefiles	The data format for ArcView GIS that defines the geometry and attribute of geographically referenced objects by three specific files (i.e., a main file, an index file, and a database table)
CGM (Computer Graphics Metafile)	An ISO (International Organization of Standards) standard for vector data format that is widely used in PC-based computer graphics applications

concerned with spatial relationships rather than with rigid coordinates.

The use of topological relationships to represent geographic information is not new. Our cognition or perception of geographic space, usually referred to as *mental maps,* is based on conceptual topological relationships rather than actual coordinates. The concept of topological relationships as applied to geographic data was formalized in a definitive paper by Corbett (1979) that described the structure of the GBF/DIME files of the U.S. Bureau of the Census. The concept of topological relationships includes three basic elements (Figure 3.23): (1) *adjacency,* which is information about neighborhood among spatial objects; (2) *containment,* which is information about inclusion of one spatial object within another spatial object; and (3) *connectivity,* which is information about linkages among spatial objects.

The concept of topological relationships can be best explained by means of a *topological map*. In essence, a topological map is a map that contains explicit topological information on top of the geometric information expressed in coordinates (Masry and Lee, 1988). In a topological map, all spatial entities are decomposed and represented in the forms of the three basic graphical elements (Figure 3.24). Each point entity is identified by a unique feature identifier. Each line entity (or *arc*) is identified by the line itself as well as its *nodes*. The topological relationships among the line entities are computed and stored by using the identifier of the nodes (nodes of different line entities having identical identifiers meet at a common node). Polygon entities are formed by using the line entities and their respective nodes. Once formed, polygon entities are individually identified by a unique identification number. The topological relationship among the polygon entities are computed and stored by using the adjacency information stored with the line entities. The adjacency information includes the nodes of the line (which are assigned as a *from_node* and a *to_node*), as well as the identifiers of the polygons to the left and the right of the line. The process of computing the topological relationships is called *topology building* (see Chapter 6).

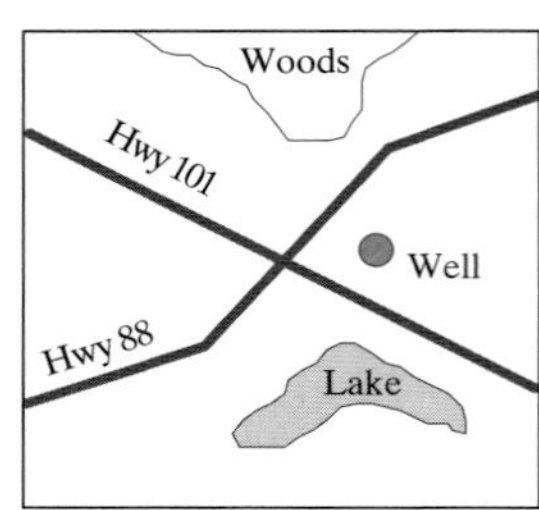

(a) Original Map

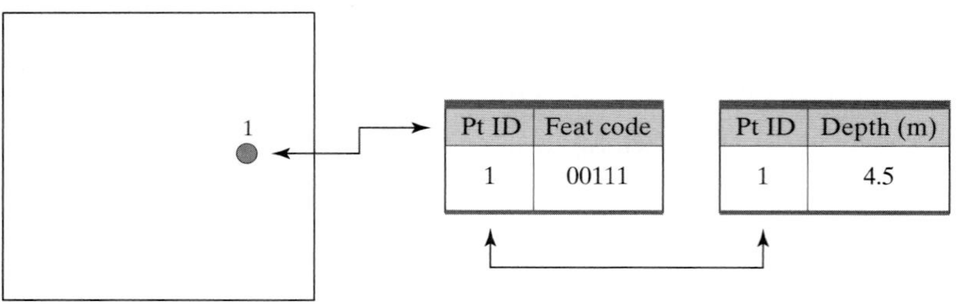

Pt ID	Feat code
1	00111

Pt ID	Depth (m)
1	4.5

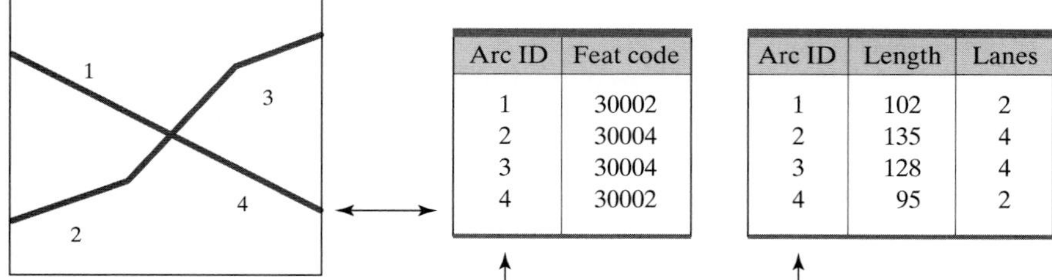

Arc ID	Feat code
1	30002
2	30004
3	30004
4	30002

Arc ID	Length	Lanes
1	102	2
2	135	4
3	128	4
4	95	2

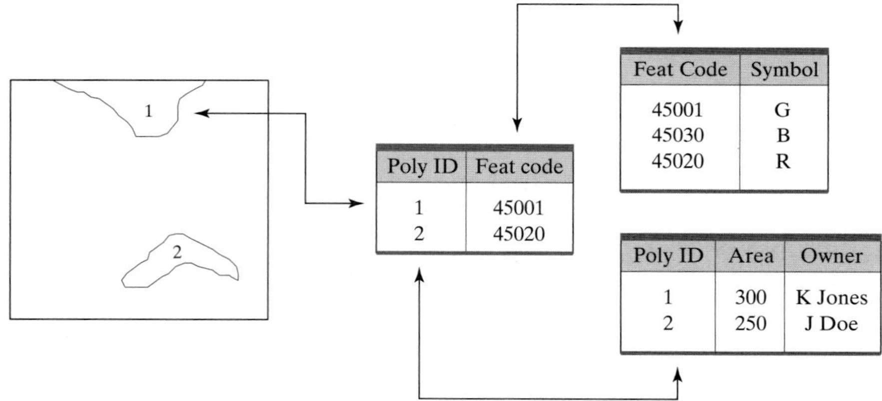

Feat Code	Symbol
45001	G
45030	B
45020	R

Poly ID	Feat code
1	45001
2	45020

Poly ID	Area	Owner
1	300	K Jones
2	250	J Doe

(b) Linking graphical elements to their associated attributes by means of common feature identification numbers (Pt ID, Arc ID, and Poly ID)

FIGURE 3.22
Linking vector data to attributes in relational tables.

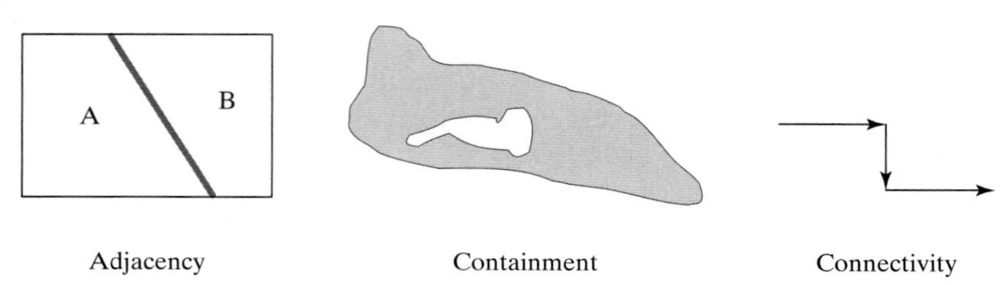

FIGURE 3.23
Three basic elements of topological relationships.

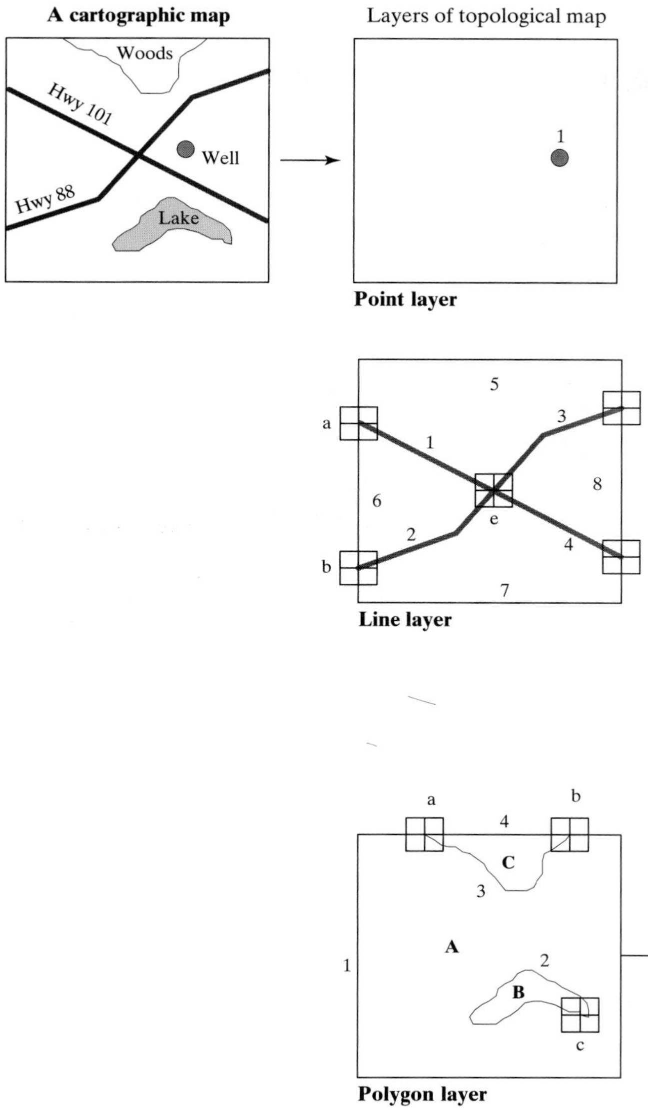

A cartographic map | Layers of topological map | Associated tables

Point layer

Pt ID	X, Y
1	15000,13450

Line layer

Arc ID	From	To
1	a	e
4	e	d
2	b	e
7	b	d
...	...	...

Polygon layer

Poly ID	Arcs
A	1, 2, 3
B	2
C	3, 4

Arc ID	From	To
1	a	b
2	c	c
3	a	b
4	a	b

Arc ID	Left Poly	Right Poly
1	A	World
2	B	A
3	C	A
4	World	C

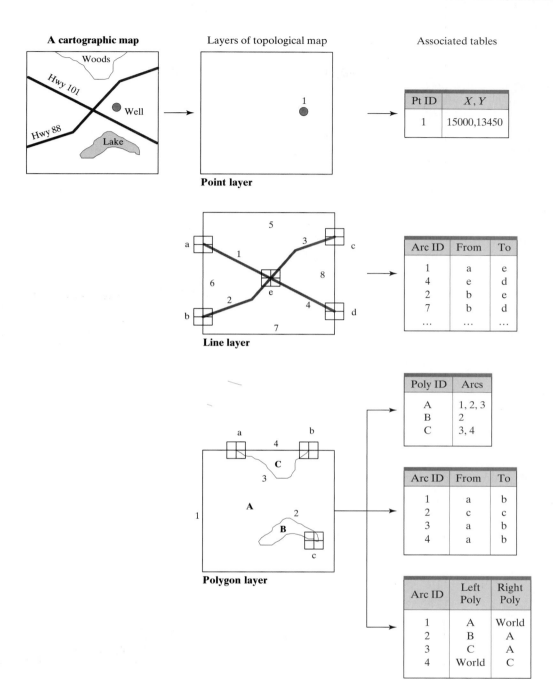

FIGURE 3.24
The concept of a topological map.

3.5.3 THE USE OF TOPOLOGICAL RELATIONSHIPS IN GIS

In GIS, many of the vector-based geographic data-processing algorithms are based on topological relationship rather than coordinates. In general, topological relationships are applied in the following GIS functional areas: (1) data input and representation, (2) spatial search, (3) construction of complex spatial objects from basic graphical elements, and (4) integrity checks in database creation.

Data Input and Representation To create a topologically structured geographic database manually is an extremely time-consuming and error-prone process. This requires all points, lines, and polygons to be digitized and numbered sequentially. For many maps used in natural resource management (such as a forest resource inventory map), manual creation of a topological map is practically impossible because of the complexity of the map features. By using the principle of topological relationships, it is possible to adopt the method of *spaghetti digitizing* (so-called because the data obtained are in the spaghetti data model) for graphical data input. This is a very simple arc-based digitizing procedure. There is no need to follow any particular sequence in digitizing the lines or to number the nodes and polygons individually. As a result, the manual part of the data input process is kept very simple and straightforward, leaving the tedious processes of structuring the spatial database to the computer (Figure 3.25).

Storing the adjacency information in the form of identifiers of polygons to the left and to the right of a line removes the need to duplicate the storage of the line data. This means that boundaries shared by adjacent polygons are digitized and stored only once. Adjacency information allows the computer to identify the lines shared by different polygons. When the data are plotted to show all the polygons, for example, it is the lines that are plotted, not the polygons themselves. This means that the boundaries between neighboring polygons are drawn only once. For applications in which a large number of polygons are involved such as a forest resource inventory, the savings in digitizing, storage, and plotting requirements are significant.

Spatial Search by Topological Relationships The use of topological relationships is the mechanism by which a GIS performs spatial searches efficiently. It is a very te-

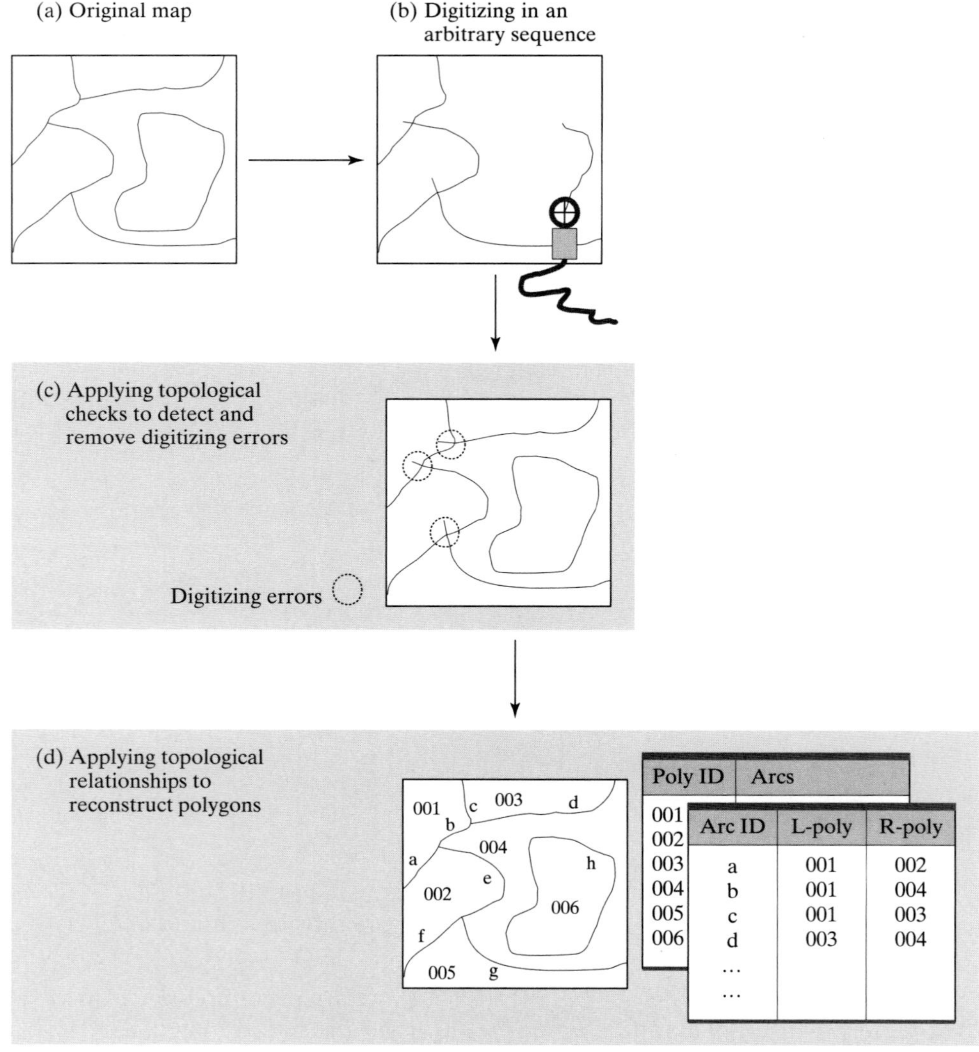

FIGURE 3.25
Using topology in map digitizing. By making use of topological relationships, it is possible to expedite the process of map digitizing by keeping the work of the digitizing operator to the minimum and by leaving the tedious task of database creation to the computer.

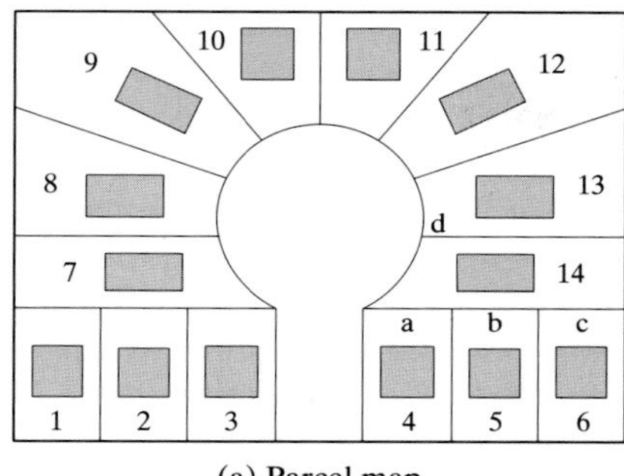

(a) Parcel map

Arc ID	From_node	To_node	Left_Poly	Right_Poly
...				
...				
a			14	4
b			14	5
c			14	6
d			13	14
...				

(b) Arc attribute table

FIGURE 3.26
Spatial search using topological relationships. In order to spatially search for parcels located next to Parcel #14 in (a), the computer makes use of the Left_Poly and Right_Poly topological relationships of the arcs stored in the attribute table in (b) to find all those parcels sharing a common boundary with this particular parcel of interest, thus identifying them as adjacent parcels.

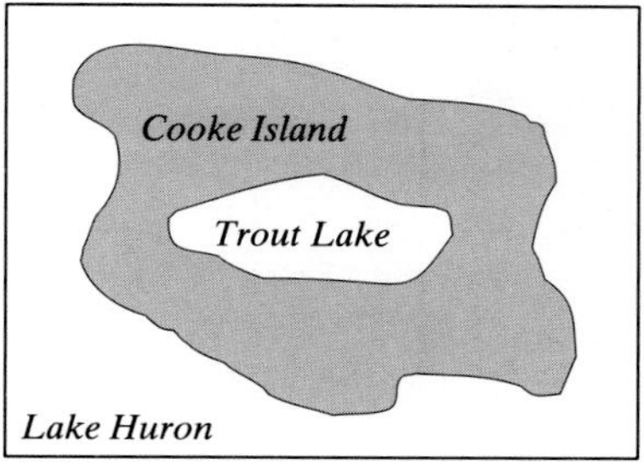

(a) Cartographic map

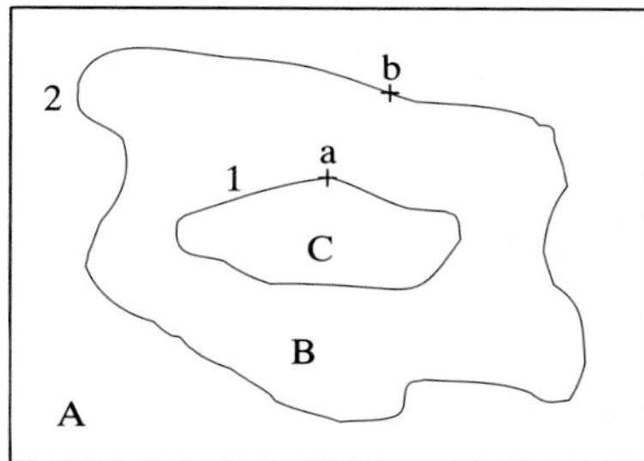

(b) Topologic map

Arc ID	From_node	To_node	Left_Poly	Right_Poly
1	a	a	B	C
2	b	b	A	B

(c) Arc attribute table

FIGURE 3.27
Constructing complex spatial objects by topology. Trout Lake in the cartographic map (a) is identified as an enclave of Cooke Island by using the Left_Poly and Right_Poly topological relationships of Arc 1 stored in the attribute table (c). Similarly, Cooke Island is identified as an enclave of Lake Huron by using the Left_Poly and Right_Poly topological relationship of Arc 2 stored in the attribute table.

dious process, for example, to find manually the data for parcels surrounding a particular parcel in a land title information system. The process requires first going to an index map to obtain the *parcel identification numbers* (PID) of all the neighboring parcels, and then retrieving the data from the appropriate records, which may be stored in different places. In a GIS, the computer makes use of the adjacency information to identify immediately all polygons that share a common boundary with the parcel of interest and applies their respective PIDs as a key to retrieve the descriptive data from the geographic database (Figure 3.26). Similarly, connectivity information can be used for spatial search using line data. The commonly used GIS function of finding the optimum path (shortest or least-cost route) between two points, for example, is based on the use of the arc–node relationships stored in the geographic database. The methods and algorithms of spatial search by topological relationships will be explained in detail in Chapter 6.

Construction of Complex Spatial Objects Complex spatial objects are represented as *complex polygons* in geographic databases. There are two types of complex polygons: those containing one or more *holes,* also referred to as *islands* or *enclaves,* and those made up of two or more polygons that are not physically connected. Complex polygons can be easily visualized when displayed on the screen or printed on a map. In the computer, however, these polygons must be constructed from vector lines using topological information (for the first type of complex polygons) (Figure 3.27) or a common identifier (for the second type of complex polygons).

Integrity Checks in Database Creation Graphical data must be topologically cleaned before they can be used for GIS applications. This means that they must not contain any of the topological errors depicted in Figure 3.28. During the topology building process, the computer will detect topological errors and flag them automatically. This allows the data input operator to

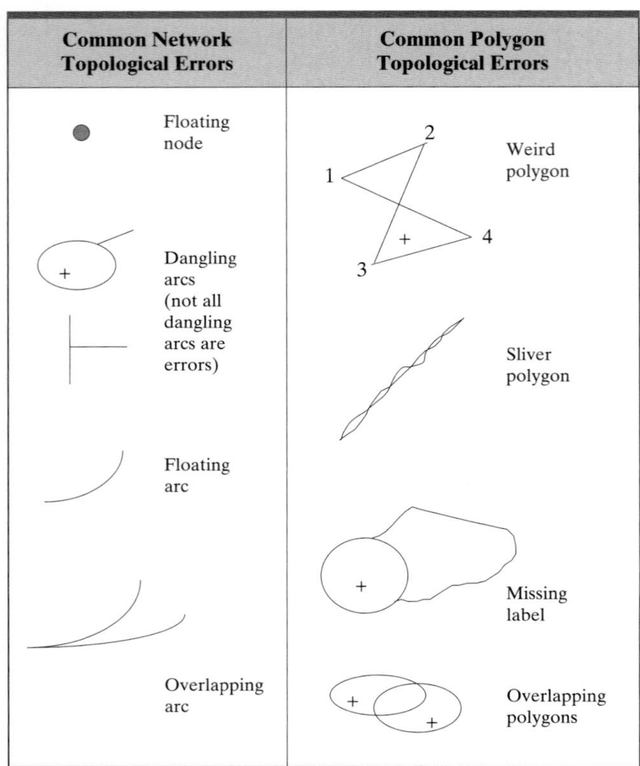

FIGURE 3.28
Common topological errors.

check the errors and decide how they must be corrected. Not all errors need to be corrected, for example, cul-de-sacs in a street network are permissible *dangling arcs,* defined as arcs with one of their end nodes not connected to the nodes of other arcs. Corrections of some errors may be done automatically, for example, using the concept of tolerance as in ArcInfo, or by interactive manual editing (see Chapter 6).

3.5.4 THE GEORELATIONAL DATA MODEL

The georelational data model is a *hybrid data model* that has been implemented in several vector-based GIS (Morehouse, 1985 and 1989). The data model is described as hybrid because it is a combination of a topological data mode (which represents spatial data) and a relational DBMS (which represents attribute data). Since the georelational data model describes geographic data by following the rules and conceptual constructs of the relational data model (i.e., it is DBMS-specific), it is a logical data model in terms of the three levels of data abstraction as noted in Section 3.3.1.

In the georelational data model, geographic data are abstracted into a series of independently defined *layers.* Each of these layers represents a selected set of closely associated spatial objects such as roads, streams, and land cover types (Figure 3.29). This means that spatial objects in a particular geographic space are classified and stored separately according to the forms of the basic graphical elements that represent them. Spatial objects in point form are separately stored from those represented by lines, which are in turn separately stored from those represented by polygons. Spatial objects of the same basic graphical element type, such as lines, will be classified and separately stored again according to their respective entity types. For example, roads and streams, which are both represented by lines, will be stored in separate layers because they belong to different entity types. The process of identifying layers is called *layer design* in geographic database design (see Chapter 11).

The attribute data for the spatial objects in each of the layers are stored in separate relational tables (also known as *attribute tables*). The descriptive and spatial data are logically linked by means of the unique feature identifiers common to both. By using the common attribute data in the different relational tables as search indices or keys, data in these tables can be logically joined with one another when required during geographic data processing and analysis.

3.6 OBJECT-ORIENTED GEOGRAPHIC DATA REPRESENTATION

Object orientation was first introduced in the early 1960s. However, it was not widely accepted until the late 1980s. Since then, there has been considerable research on its applicability to GIS (Egenhofer and Frank, 1989). Today, many of the methods of object orientation have been adopted in GIS application development. Object orientation is built on the fundamental principle of reusing program components. Reuse leads to a shorter *systems development life cycle* (SDLE) and higher quality programs (see Chapter 11). Object-oriented software is easier to maintain and easily scalable (that is, large systems can be constructed by assembling small reusable subsystems).

The concept of object orientation includes four basic elements.

1. *Object-oriented user interface* uses point-and-click icons and dialog boxes to build the *graphical user interface* (GUI) of today's software applications.
2. *Object-oriented programming languages* use widgets in the development of computer codes by flowcharting the logic of the programs (Visual Basic and Visual C are examples of object-oriented programming languages).

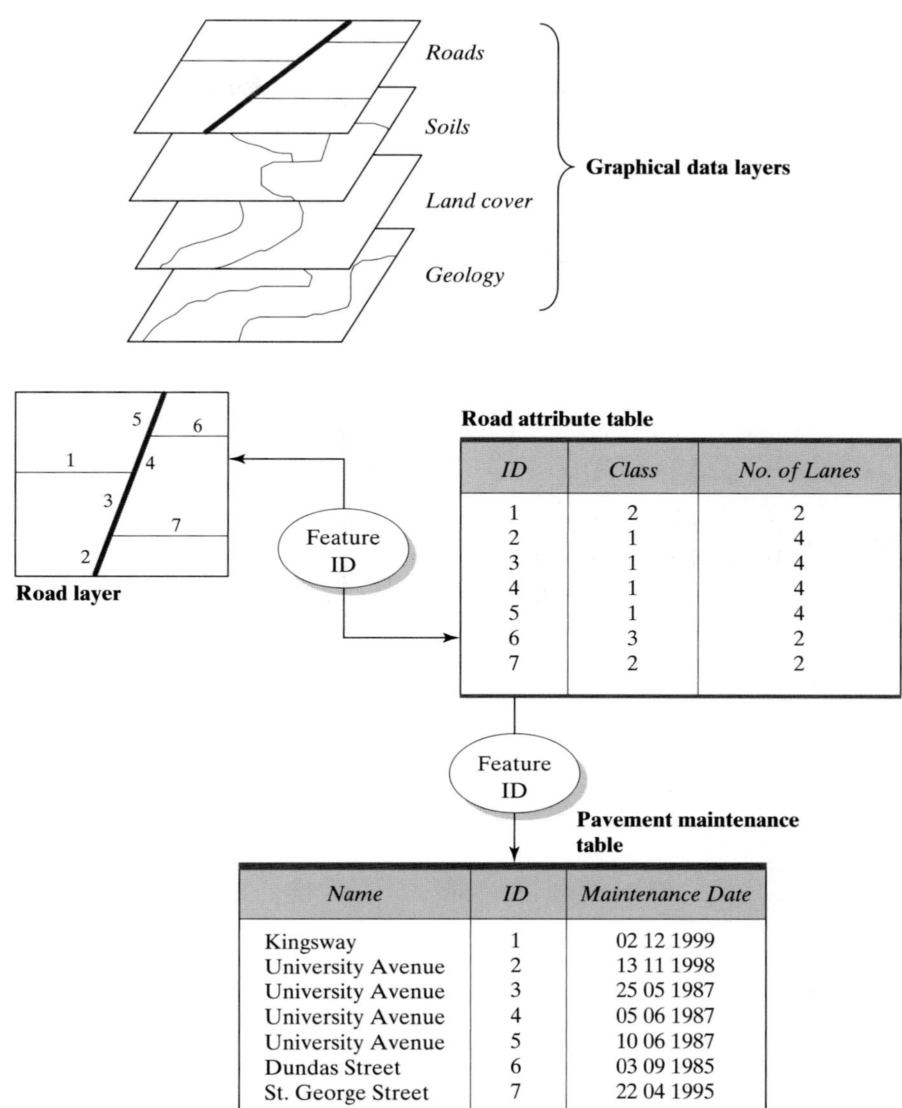

Road attribute table

ID	Class	No. of Lanes
1	2	2
2	1	4
3	1	4
4	1	4
5	1	4
6	3	2
7	2	2

Pavement maintenance table

Name	ID	Maintenance Date
Kingsway	1	02 12 1999
University Avenue	2	13 11 1998
University Avenue	3	25 05 1987
University Avenue	4	05 06 1987
University Avenue	5	10 06 1987
Dundas Street	6	03 09 1985
St. George Street	7	22 04 1995

FIGURE 3.29
The georelational data model. The georelational data model stores graphical and attribute data separately. It makes use of common feature identifiers to link the data in different tables in the database.

3. *Object-oriented analysis* and *design methodology* is the software engineering approach that identifies objects (more generally referred to as *data objects*) and their attributes, as well as the modeling of the representation of these objects in databases.
4. *Object-oriented database management* provides procedures (through the use of an object-oriented DBMS) for establishing and relating data objects in databases (by using a *unique object identifier*).

Unlike the relational data model we learned earlier, object orientation is more an abstract concept rather than a formalized data model with well-defined rules and procedures (although the term object-oriented data model has been frequently used in DBMS literature). The study of object orientation is a relatively advanced and broad subject in its own right. In this chapter, our discussion will focus only on the nature and characteristics of data objects in the context of modeling and representing geographic data.

3.6.1 DATA OBJECTS

The *data object* is central to object technology. A data object represents a real-world feature (place, person, thing, event, and concept), similar to an entity in the relational model we learned above. In object-orientation literature, the term "entity" is usually used to refer to features in the real world, and the term "object" is used to refer to properties of features as they are represented in the database. Just like the entity in the relational model, a data object must be atomic (that is, it cannot be decomposed into data objects of the same class) and can be described by attributes. What makes a data object different from a relational entity are its definition and behavior. These two properties can be conveniently grouped into the following categories: *classification, inheritance, operations, encapsulation*, and *polymorphism*.

Classification and Inheritance A *class* is a collection of data objects with the same attributes. A *super class* is a collection of classes, and a *subclass* is an instance (member) of a class. To understand the concept of the data object, consider Georgia as an example of an object in the geographic database of the United States. Georgia is an instance (member) of a much larger class of objects that is called "states." There is a set of generic attributes associated with every object in the class, such as location (coordinates), name of capital city, population, unemployment rate, and other socioeconomic characteristics. These attributes apply whether we are talking about Alabama, Colorado, Maryland, or any other state.

Once the class has been defined, the attributes can be reused when new instances of the class are created. If, for example, we expand the above database to include Canadian provinces and territories, the same attributes can be reused to describe these new instances. The addition of Canadian provinces and territories helps explain *class hierarchy*. The states, provinces, and territories together represent a class of data objects (state/provincial level of political units) that belongs to a *super class* (the national level of political units). This super class has two instances: the United States and Canada. As instances of a super class, the two countries can be described by the same attributes that are used to describe the states and provinces. Reusing the same attributes for instances in the same class and instances in the super class is called *inheritance*.

Operations, Encapsulation, and Polymorphism In the object-oriented approach, every object in the class can be manipulated in a number of ways. The states and provinces in this example can be plotted, classified (according to a specific socioeconomic index such as per capita annual income), and modified (e.g., cartographic generalization of boundaries). It is also possible to use the attribute values to derive new data (e.g., coordinates to calculate area, and area and population to calculate population density). These processes are called *operations* (sometimes also referred to as *services* or *methods*) and are inherited by all instances of the class (Figure 3.30).

The concept of bundling data (the attributes that describe the object) and operations (the actions that are applied to the attributes) for specific functions in the object-oriented approach is known as *encapsulation*. This concept is the cornerstone of object-oriented data processing. In the above example, the definition of each data object includes a set of attributes (coordinates, population, unemployment rates, and other socioeconomic indices) and a set of operations (plot, classify, calculate

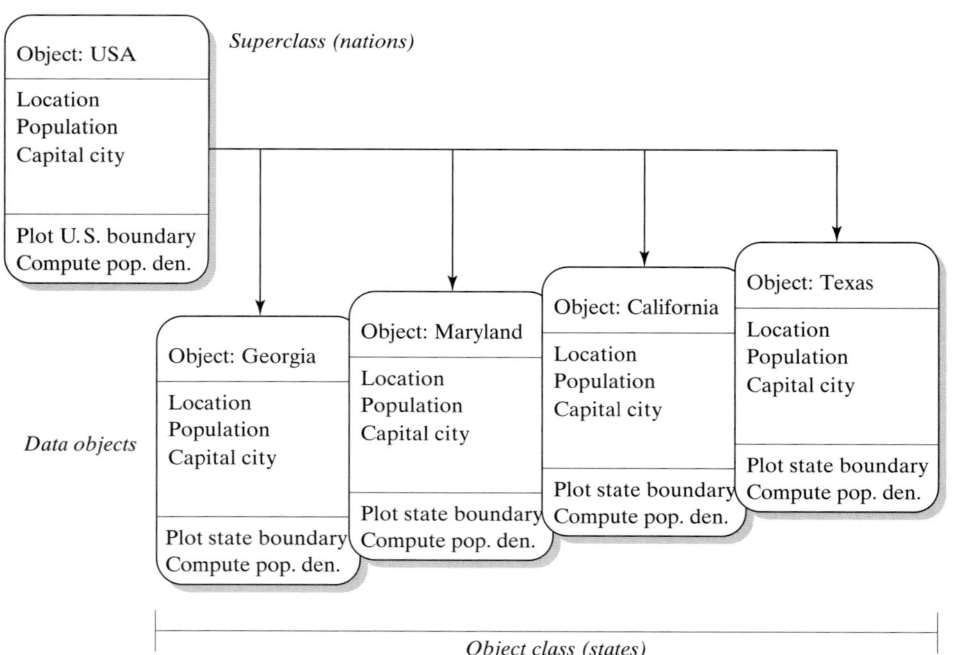

FIGURE 3.30
Inheritance of attributes and operations in data objects.

dimension, and calculate population density). When the data/operation bundle is accessed, the operation will be executed on the data. For example, when the "calculate area" bundle is executed, the operation (procedure for calculating area from coordinates) will act on the data (the coordinates) to return the size of the states concerned.

Polymorphism is related to encapsulation and refers to the fact that the same data/operation bundle can have different results depending on the operation executed. In the "calculate area" example, the operation may contain separate procedures for calculating regularly shaped and irregularly shaped objects. When the bundle is used, it first checks the type of the object and then automatically calls the appropriate procedure to calculate the result. The use of such an operation is transparent to the user. This means that the user needs only to supply the data object, and the system will sort out the details and the procedure to be applied.

This discussion explains the characteristics of data objects in their most basic and generic form. In practice, object technology allows for complex objects. Complex objects lead to the development of complex classification/inheritance patterns. These patterns embed more information and commonality in data objects that can be translated into higher levels of database performance. The trade-off, however, is that such databases are extremely difficult to design and implement.

3.6.2 OBJECT-ORIENTED DATA REPRESENTATION IN GIS

The identification of data objects is the first step in the object-oriented approach to GIS application and database development. This is a process that is somewhat equivalent to the process of conceptual data modeling in conventional database design noted above. It is carried out in three sequential steps: (1) identifying objects and classes, (2) specifying attributes, and (3) defining operations (Pressman, 1997).

Identification of Objects and Classes This process is usually done by examining the documents pertaining to user requirements for the system to be built. Objects are identified by underlying each noun or noun clause and entering it in a simple table. Objects manifest themselves by showing one or more of the following characteristics or functions.

1. external entities that produce or consume information in the system, for example, other systems, hardware devices, people
2. things that are part of the information domain of the problem to be solved, for example, reports, displays, signals
3. events that occur within the context of system operation, for example, exchange or transfer of data, plotting, generation of summary statistics
4. roles played by people associated with an application, for example, manager, engineer, technician
5. organizational units that are relevant to an application, for example, department, agency, group, team
6. places that establish the context of the problem, for example, laboratory, field survey sites
7. structures and contents that define a class of object, for example, high-rise or low-rise buildings, land cover types.

The decision to include or exclude an object in the data analysis model is somewhat subjective, and later evaluation may cause an object to be discarded or reinstated.

Specification of Attributes Attributes are properties or descriptions of the selected data objects. In essence, it is the attributes that define the object. To develop a meaningful set of attributes for a particular object, it is necessary to examine the user requirements and select those things that define the object in the context of the problem at hand.

Definition of Operations Operations define the behavior of objects and change their attributes. In general, operations can be divided into three broad categories: operations that *manipulate* data (e.g., inserting, deleting, reformatting, and selecting); operations that *process* the data (e.g., plot, display, summarize, and perform other forms of computation); and operations that *monitor* an object for the occurrence of a controlling event (e.g., topological checks in spatial data editing).

The process of representing geographic data in the object-oriented approach is an iterative process. The generic life cycle of a data object includes creation, modification, manipulation (i.e., examining it from other perspectives), and possibly deletion. The completion of the definition of data objects signifies the beginning of a object-oriented data analysis model.

3.6.3 OBJECT-ORIENTED DATA MODELING IN GIS

The intent of object-oriented data modeling is to define all data objects (and the classes, relationships, and behavior associated with them) that are relevant to the problem to be solved. The popularity of object orientation has led to the development of dozens of object-oriented modeling methods (Pressman, 1997). These methods vary considerably in terms of notation and procedures, but they are all characterized by the combination of the specifications of conceptual/logical data types and data-processing procedures into a single modeling process. In this modeling process, the data objects used to depict the real-world entity types will also list their attributes as part of the objects. In addition, the data objects list the operations to be performed on a particular attribute. The operations include all the different actions to be performed on the individual attributes, as well as the actions to be performed on collections of these attributes (Figure 3.31).

Object-oriented data modeling is therefore very similar to the relational data modeling concepts discussed in Section 3.3.1. Although the terminology and procedures are somewhat different from the conventional methodology, object-oriented modeling addresses essentially the same underlying objectives. There are two major problems with this particular method of data modeling. One of these is the lack of a standard set of modeling tools. It is not yet possible, for example, to present an object-oriented analysis model with the same level of universal acceptability (in terms of standard notation and symbology) as the E-R modeling. The other problem is the slow acceptance of object-oriented databases for large-scale applications. At present, the DBMS world is still largely dominated by relational systems. It is very likely that after the completion of the data modeling exercise, the object-oriented data model may have to be mapped onto a relational database implementation rather than an object-oriented database. In essence, this is one of the driving forces behind the proposition for the rapprochement of object-orientation and relational technologies (Date, 1995) and for the emergence of the object-relational concept in GIS development as exemplified in the next section.

3.6.4 OBJECT-RELATIONAL DATABASE AND SOFTWARE TOOLS

Advances in object-oriented technology have resulted in the development of several new approaches to digital geographic data representation that enable geographic data to be stored in large corporate databases and managed by commercial DBMS. As these approaches are based on the concepts and techniques of object-oriented technologies and conventional relational DBMS, the resulting databases are generally referred to as *object-relational databases*. The advent of these new breeds of databases and associated DBMS software tools has changed the conventional way of representing digital geographic data and expedited the integration between GIS and conventional database management in the enterprise computing environment.

Different object-relational software products, such as those noted in Section 3.3.2, have different implementation details with respect to representing and storing spatial objects, referencing them spatially, and operating on them during data processing and analysis. However, these products share some common characteristics that can be summarized in the following topics.

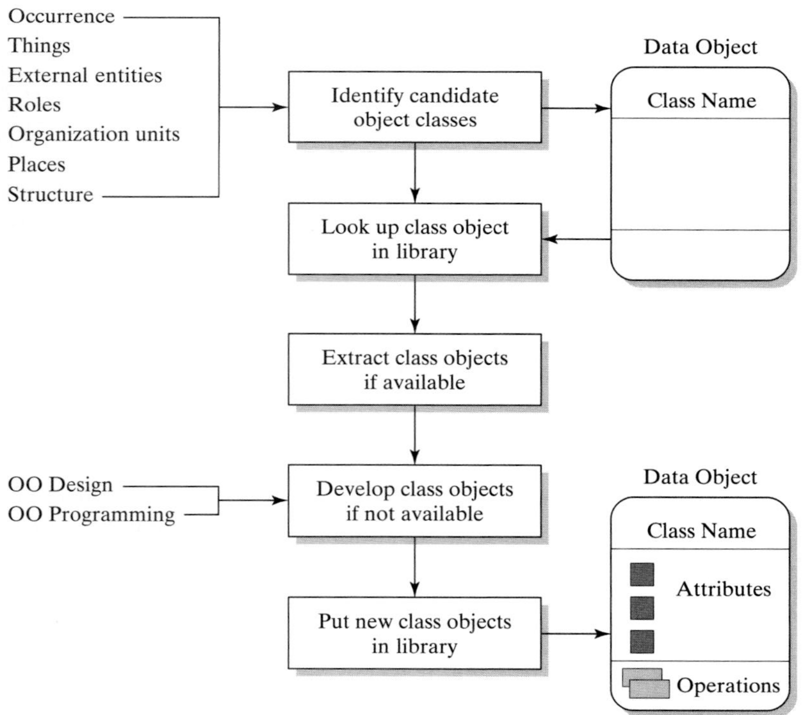

FIGURE 3.31
Stages of object-oriented analysis.

Representation of Spatial Objects Spatial objects in object-relational databases are represented as discrete data objects, as defined in Section 3.6.1, using graphical elements in the same way as the georelational database. The major difference is that the object-relational model often uses an enriched set of graphical element types on top of the three basic types of points, lines, and polygons. Oracle Spatial, for example, has nine graphical elements called *geometric primitive type*, including points and point clusters, line strings, polygons, arc line strings, arc polygons, compound polygons, compound line strings, circles, and optimized rectangles (Figure 3.32a). IBM's DB2 Spatial Extender uses the term *geometry type* to describe its graphical elements. The graphical elements are organized in a hierarchy of subtypes of geometry, including points, multipoints, line strings, multiline strings, polygons, multipolygons, and ellipses (Figure 3.32b). In ESRI's ArcSDE, spatial objects are represented as shapes

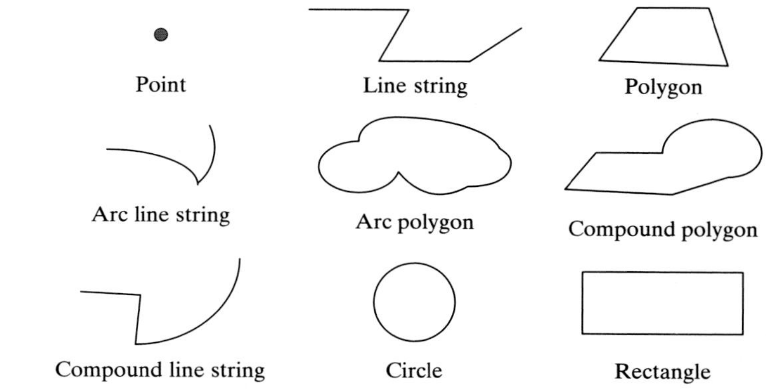

(a) Geometry types used in the object-relational model of Oracle Spatial

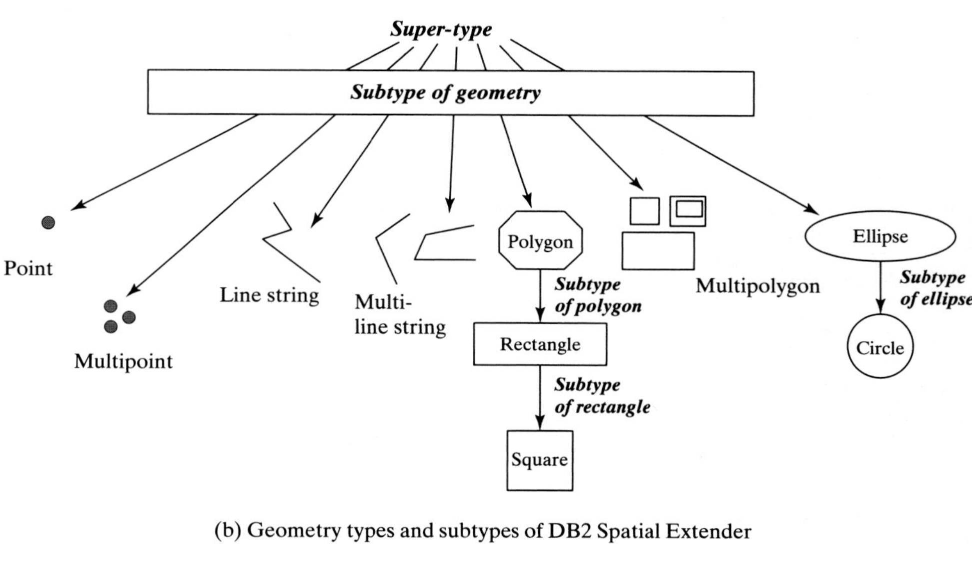

(b) Geometry types and subtypes of DB2 Spatial Extender

Attribute

Point Line Area Nil

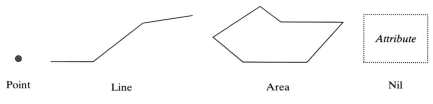

(c) Types of shapes of ArcSDE

FIGURE 3.32
Graphical elements used by Oracle Spatial, IBM DB2 Spatial Extender and ESRI ArcSDE. *(Sources: Oracle, 1999; Davis, 1998; and ESRI, 1999)*

that belong to one of the following four types: points, lines, polygons, and nil. Points, lines, and polygons are defined in the usual way, and nil is a particular shape that has no geometry but has attribute (Figure 3.32c). All these shapes have single part or multipart geometric properties, which allow complex spatial features such as a group of islands to be represented as a single geometric object in the database.

Organizing and Storing Spatial Data The object-relational DBMS model is remarkably similar to the georelational data model with respect to the logical organization of spatial data. This means that the data models are often based on the same paradigm of organizing the data in a hierarchical structure consisting of graphical elements, geographic objects, and layers. The graphical elements, just like their counterparts in the georelational model, are the basic building blocks of a geometry. They are formed by one or more pairs of coordinates, depending on their respective graphical types. Individual geographic objects, called *geometries* or *shapes*, are the actual spatial objects formed by the graphical elements. They are individually identified by a unique identifier, and they are collectively organized as different layers according to the attribute characteristics of the spatial objects that they represent.

The greatest difference between the georelational and object-relational data models is in the physical storage of the data. In the object-relational database, the data objects are stored in conventional relational tables. Obviously, different implementations have different table structures. ArcSDE, for example, stores and organizes geographic data by adding a *spatial data type* to relational databases. By doing so, there is no need to change the existing database or affect existing applications. ArcSDE simply adds a shape column to existing relational tables and uses software tools to manipulate and access the shape data referenced by the shape column (Figure 3.33). In this way, both the graphical and attribute components of geographic data can be stored and managed in a single DBMS environment.

Spatial Referencing Spatial Objects In order to provide high-performance spatial search and retrieval of spatial objects from very large databases, object-relational DBMS software products are developed with powerful spatial indexing methods. The function of spatial indexing is to provide the mechanism to limit spatial search within database tables, thus speeding up database access and retrieval. Again, different object-relational DBMS software products have different implementation details. For example, ArcSDE builds its spatial index by dividing the data layers into grid cells, identifying the spatial objects that exist in each cell, and writing the information to an index table (Figure 3.34). Oracle Spatial, similarly, builds its spatial index by dividing the data layers into cells, but it uses the method of quadtree decomposition instead (see Section 3.4.3). The creation and maintenance of spatial indices are often done by means of generic DBMS data definition language (DDL) and data manipulation language (DML) commands, such as CREATE, ALTER, INSERT, UPDATE, and DELETE.

Working on Spatial Objects One of the major characteristics of object-orientation technology, as noted, is the integration between data and processes. Therefore, object-relational DBMSs are usually constructed with a rich set of spatial operators for manipulating and analyzing spatial data (Figure 3.35). Since these commands are built using industry standards and

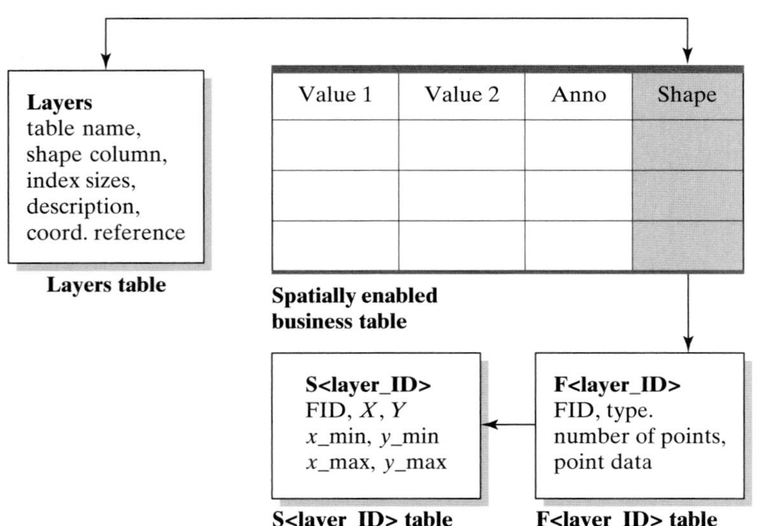

FIGURE 3.33
Storing graphical and spatial data in an object-relational database: the example of ArcSDE. Geographic data are stored in a spatially enabled business table that is created by adding a shape column to it. ArcSDE manages spatially enabled tables by storing information about the shape column in the LAYERS table that helps maintain the links between the business table and spatial data. The actual spatial data are stored in the related F<layer_ID> table, which in turn is related to the S<layer_ID> table where coordinates are stored.

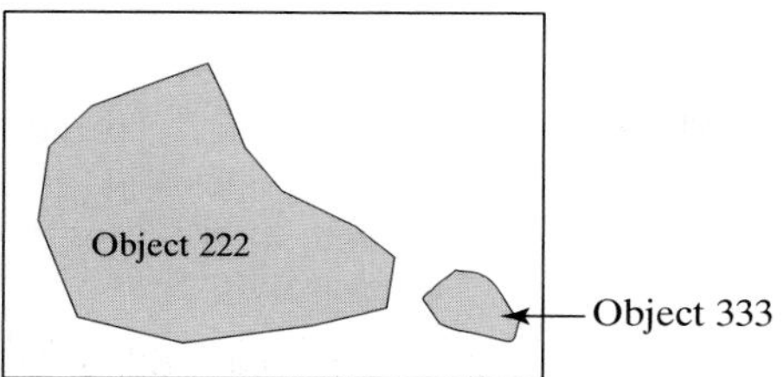

(a) Data layer containing geographic objects having the same attribute values

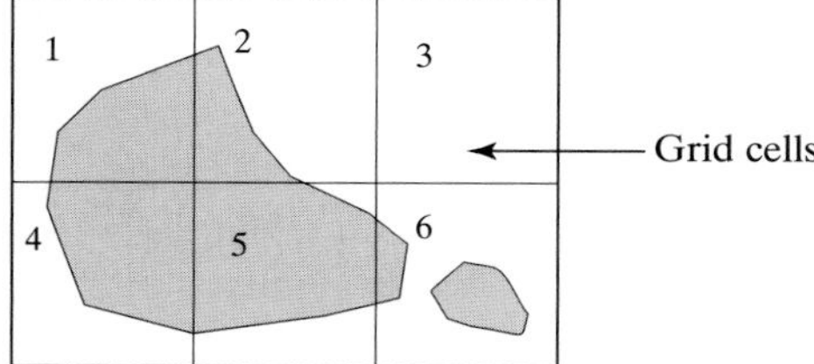

(b) Geographic objects identified in each grid cell

Grid Cell No.	Object Identifier
1	222
2	222
3	
4	222
5	222
6	222, 333

(c) Storing indexing information in an index table

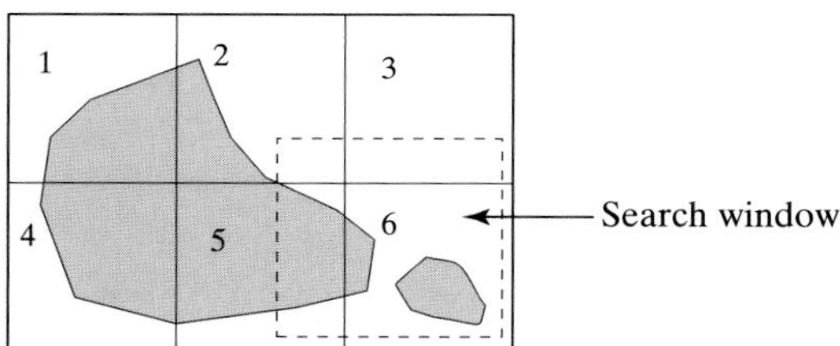

(d) Spatial search using a window

FIGURE 3.34
Spatial indexing in object-relational geographic data processing. The purpose of spatial indexing is to expedite the search of spatial objects in the database. When performing a spatial search using a user-defined window as illustrated in (d), the system starts by determining the grid cells that the search window covers. It then looks up the index table to find the geographic objects that are present in these cells, and executes a join with the table that contains data pertaining to these geographic objects.

```
CREATE type SDO_GEOMETRY as object (
  SDO_GTYPE NUMBER,
  SDO_SRID NUMBER,
  SDO_POINT SDO_POINT_TYPE,
  SDO_ELEM_INFO MDSYS.SDO_ELEM_INFO_ARRAY,
  SDO_ORDINATES MDSYS.SDO_ORDINATE_ARRAY);
```

(a) Syntax of spatial operator to create a geometry in Oracle Spatial

```
SDO_GEOMETRY Column = (
  SDO_GTYPE = 3
  SDO_SRID = NULL
  SDO_POINT  = NULL
  SDO_ELEM_INFO = (1.3.1.19.3.1)
  SDO_ORDINATES = (6.15.10.10.20.10.
                   25.15.25.35.19.40.
                   11.40.6.25.6.15.
                   12.15.15.15.15.24.
                   12.24.12.15) );
```

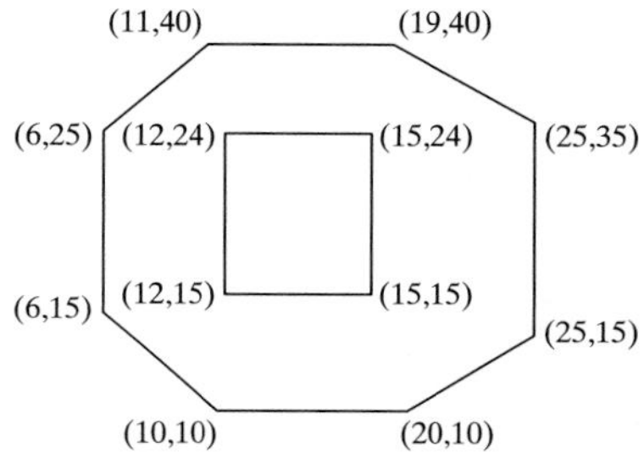

(b) Using the spatial operator SDO_GEOMETRY in (a) to generate a spatial object. Note how the coordinates are stored and used, starting at (6,15) and (12,15) for the two graphical elements, and ending at the same points.

FIGURE 3.35
Example of using a spatial operator in object-relational geographic data processing. *(Source: Oracle, 1999)*

protocols, the application development environments of object-relational DBMSs are open, efficient, and highly customizable. This means that the developer can construct database queries and interfaces to meet very specific needs of particular applications by using a variety of application development methods and languages. At the same time, this also allows GIS applications to be embedded within the enterprise DBMS operating environment of business organizations and government agencies.

3.7 The Relationship Between Data Representation and Data Analysis in GIS

The raster and vector data models for geographic data represent two distinct views of the real world. They also represent two distinct approaches to geographic data processing and analysis. These two data models are essentially designed for different types of geographic problem solving. This means that the representation of geographic data is essentially determined by the applications for which the data are intended. In order to understand the relationships between data representation and data analysis in GIS, it is necessary to understand the relative merits and limitations of the raster and vector data models as summarized in Table 3.4.

It is obvious that the raster data model is best suited for GIS applications aiming at the analysis of the *spatial variability* of geographic phenomena. Advances in computer graphics technology in recent years have made it possible to effectively display variations in spatial phenomena in both two and three dimensions. Examples of the applications of raster data for visual display of spatial variability include temperature, rainfall, topography, and environmental indices. Raster data are also useful for the analysis of the spatial relationships between data pertaining to different aspects of the environment, particularly at the regional and national levels. Examples include environmental impact analysis, wildlife habit studies, and biodiversity studies. The concepts and techniques of geographic data processing using raster data will be explained in detail in Chapter 5.

Where the information of interest to geographic data analysis is the distribution and location of individual spatial objects, the vector data model is obviously the only option. Vector data are needed for applications that require frequent queries and cross-referencing with descriptive data, as well as those that are used to show movement of spatial objects. The vector data model is therefore best suited for applications in land title information, computer-assisted cartography, facility management, transportation management and planning, as well as environmental and natural resource data management at the local (as opposed to regional) level. The concepts and techniques of geographic data processing using vector data will be explained in detail in Chapter 6.

As noted in Chapter 1, GIS in the past tended to be designed either on the raster or the vector data models. Advances in computer technology have now made it possible for GIS to use raster and vector data simultaneously. Raster and vector data covering the same geographic area can now be processed and displayed within a single system. This has greatly enhanced the usefulness of GIS as a geographic data management and analysis tool. The use of vector data as a backdrop to raster-based applications provides the latter with more precise location information. On the other hand, the use of raster data as a backdrop to vector-based applications enriches the information content of the vector data. The concepts and techniques of integrating raster and vector data, with special reference to remote sensing applications, will be explained in detail in Chapter 8.

3.8 Summary

In this chapter, we introduced the concepts of representing and organizing digital geographic data. We built our understanding of geographic data around the concept of the geographic matrix that describes geographic facts by three components: location, attribute, and time. We described the object-based and field-based

Table 3.4
Relative Merits and Limitations of Raster and Vector Data

Raster Model	*Vector Model*
Merits	**Merits**
• Simple data model	• More compact data structure
• Use of cheap technology	• Topological processing
• Ease of data collection	• Cartographic quality
• Ease of data processing	• Sophisticated attribute data handling
Limitations	**Limitations**
• No topological processing	• Complex data model
• Limited attribute data handling	• Difficult overlay processing
• Less compact data structure	• Difficult presentation of spatial variability
• Low cartographic output quality	• Expensive data collection
	• Use of expensive technology

views of the real world and explained how they are related to vector and raster data models in digital geographic databases. Although digital and conventional analog geographic data show little difference when displayed on the computer screen and printed on paper, they are distinct approaches to depicting spatial objects. This implies that we have to understand digital geographic data from different perspectives with respect to the ways data are coded, classified, and stored in the database.

As relational DBMSs now represent the dominant technology for GIS, this chapter explained in detail the concepts pertaining to this particular type of DBMS, with special reference to applications in GIS. This chapter also covered topological relationships in geographic data and explained how these concepts are used to develop the georelational data model of GIS software products. In the last few years, significant changes have occurred in the design and architecture of GIS software products as a result of advances and growing acceptance of object-oriented technology. GIS software products are now gradually moving away from the conventional georelational model to what is commonly called the object-relational model. As basic knowledge of how object-relational software products work is essential for students and practitioners in order that they can master the skills of using the new generation of GIS software products, this chapter explained the emerging concepts of object orientation and their potential applications in GIS.

In this chapter, we have answered three basic questions pertaining to geographic data: (1) What are digital geographic data? (2) How are they are represented? and (3) What relationships exist between geographic data representation and the ways they are organized in computer databases? Materials in this chapter provide the prerequisite knowledge for studying the two important topics about geographic data to be discussed in the next chapter: the quality of geographic data and data standards with special reference to geographic data.

REFERENCES

Anderson, J. R., Hardy, E. E., Roach, J. T., and Witmer, R. E. (1976) *A Land Use and Land Cover Classification System for Use with Remote Sensor Data* (USGS Professional Paper 964). Washington, DC: United States Government Printing Office.

Autodesk (1996) *Coordinate Precision in Mapping and GIS: Why You Should Care*. San Rafael, CA: Autodesk, Inc.

BCMELP (1996) *CCSM Classification System*. Victoria, BC: British Columbia Ministry of Environment, Lands & Parks. **http://www.env.gov.bc.ca/gis/ccsmclassification.html**

Berry, B. J. L. (1964) "Approaches to regional analysis: A synthesis," *Annals of the Association of American Geographers*, Vol. 54, No. 2, pp. 2–11.

Burrough, P. A. (1986) "Five reasons why geographical information systems are not being used efficiently for land resource management," *Proceedings*, pp. 139–148, AutoCarto London, London.

Chen, P. (1976) "The Entity-Relationship Model—towards a unified view of data," *Association of Computing Machinery Transactions on Database Systems*, Vol. 1, No. 1, pp. 9–36.

Chrisman, N. R., Cowen, D. J., Fisher, P. F., Goodchild, M. F., and Mark, D. M. (1989) "Geographic information systems," in *Geography in America* by Gaile, G. L., and Willmott, C. J. eds. Columbus, OH: Merrill Publishing.

Codd, E. F. (1970) "A relational model of data for large shared data banks," Communications of the *Association of Computing Machinery*, Vol. 13, No. 6, pp. 377–387.

Corbett, J. P. (1979) *Topological Principles in Cartography* (Technical Paper No. 48), US Bureau of the Census. Washington, DC: US Government Printing Office.

Date, C. J. (1995) *An Introduction to Database Systems*, 6th ed. Reading, MA: Addison-Wesley.

Davis, J. R. (1998) *IBM's DB2 Spatial Extender: Managing Geospatial Information within the DBMS*, an IBM White Paper. IBM Corporation. **http://www.software.ibm.com/data/pubs/papers**

Ebdon, D. (1992) "SPANS—A quadtree-based GIS," *Computers and Geosciences*, Vol. 18, No. 4, pp. 471–475.

Elmasri, R., and Navathe, S. B. (1994) *Fundamentals of Database Systems*, 2th ed. Reading, MA: Addison-Wesley.

Egenhofer, M. J., and Frank, A. U. (1989) "Object-oriented modeling in GIS: Inheritance and propagation," *Proceedings*, pp. 588–598, AutoCarto 9, Baltimore, MD.

ESRI (1998) *Spatial Database Engine*, an ESRI White Paper. Redlands, CA: Environmental System Research Institute, Inc.

ESRI (1999) *Getting Started with SDE*, an ESRI White Paper. Redlands, CA: Environmental System Research Institute, Inc.

Goodchild, M. F. (1992) "Geographic data modeling," *Computers & Geosciences*, Vol. 18, No. 4, pp. 401–408.

Hamacher, V. C., Vranesic, Z. G., and Zaky, S. G. (1996) *Computer Organization*, 4th ed. New York: McGraw-Hill.

Kemp, Z., and Kowalczyk, A. (1994) "Incorporating the temporal dimension in a GIS," in *Innovations in GIS 1*, by Worboys, M. F. ed. pp. 89–104. London: Francis & Taylor.

Marx, R. W. (1986) "The TIGER system: Automating the geographic structure of the United States Census," *Government Publications Review*, No. 13, pp. 181–120.

Masry, S. E., and Lee, Y. C. (1988) *An Introduction to Digital Mapping* (Lecture Notes No. 56). Fredericton, NB: Department of Surveying Engineering, University of New Brunswick.

Morehouse, S. (1985) "ARC/INFO: A georelational model spatial information," *Proceedings*, pp. 388–397, AutoCarto 7, Washington, DC.

Morehouse, S. (1989) "The architecture of ARC/INFO," *Proceedings*, pp. 266–277, AutoCarto 9, Baltimore, MD.

Oracle (1999) *Oracle 8i Spatial User's Guide and Reference*, Release 8.1.5. Redwoodshores, CA: Oracle Corporation.

Peuquet, D. J. (1984) "A conceptual framework and comparison of spatial data models," *Cartographica*, Vol. 21, No. 4, pp. 66–113.

Pressman, R. S. (1997) *Software Engineering: A Practitioner's Approach*, 4th ed. New York: McGraw-Hill.

Rhind, D., and Hudson, R. (1980) *Land Use*. London and New York: Methuen & Co.

Samet, H. (1990) *Applications of Spatial Data Structures*. New York: Addison-Wesley.

Sui, D. Z. (1995) "Geographic matrix and GIS education: Toward a new pedagogic framework for teaching GIS," in *Proceedings*, Vol. 2, pp. 940–949,GIS/LIS '95, Nashville, TN.

US Bureau of the Census (1970) "The DIME geocoding system," *Census Use Study (Report No. 4)*. Washington, DC: U.S. Department of Commerce.

USGS (1996a) *Digital Raster Graphics* (Fact Sheet FS-122-95). Reston, VA: United States Geological Survey. **http://mapping.usgs.gov/did/drg/fs960305.html**

USGS (1996b) *USGS DLG Data*. Reston, VA: United States Geological Survey. **http://bard.wr.usgs.gov/htmldir/dlginfo.html**

USGS (1997) *Standards (Section 3.0—General Principles)*. Reston, VA: United States Geological Survey. **http://mcmcweb.er.usgs.gov/standards/part3_std/gp.html**

Wang, M., and Howarth, P. J. (1994) "Multi-source spatial data integration: Problems and some solutions," *Canadian Journal of Remote Sensing*, Vol. 20, No. 4, pp. 360–367.

West, B. H., Griesback, E. H., Taylor, J. D., and Taylor, L. T. (1982) *The Prentice Hall Encyclopedia of Mathematics*. Englewood Cliffs, NJ: Prentice Hall.

CHAPTER

4

DATA QUALITY AND DATA STANDARDS

4.1 Introduction

Data are fundamental to the implementation of GIS. Without data, a GIS simply serves no practical purpose. In the early years of the development of GIS, the focus was mainly on the acquisition and structure of geographic data. Since the 1990s, when GIS began to mature into a practical information management tool, the problem of data quality emerged as a new dimension of the data component of GIS. Users of GIS have come to realize that merely having data is not enough to use the technology beneficially. Using geographic data of dubious quality often creates, rather than solves, spatial problems. In order for geographic data to be useful, therefore, they must be of a known quality compatible with their intended applications.

Data quality is a relatively abstract construct that is sometimes difficult to interpret. Data standards, which are documented as a set of specifications, rules, and procedures, provide the necessary context for measuring and controlling data quality. The adherence to accepted standards allows geographic data to be shared among GIS users. This has effectively broken the conventional boundaries between proprietary systems and led to the development of interoperable, or open, GIS. The objective of this chapter is to provide an overview of the concepts and techniques pertaining to data quality and data standards, and explain their importance to the development and implementation of GIS.

4.2 Concepts and Definitions of Data Quality

In the broadest sense, *data quality* refers to the "fitness for use" of data for intended applications (Chrisman, 1983). Whether a particular data set is fit for use can be a rather abstract and subjective judgment. Different people tend to use different terms to describe similar concepts pertaining to data quality. Therefore, it is important to define the basic concepts and terminology commonly used for the description and evaluation of data quality.

4.2.1 The Scope of Geographic Data Quality

Several qualitative criteria are commonly used to describe data quality. For example, data must be *reliable* and *accurate* in order that they can be considered as usable. In other words, they should be in agreement with the real world they represent. Data must be sufficiently *current* and *up to date* for the applications for which they are intended, as well as *relevant* (in terms of both contents and the level of detail) and *timely* to the application. They must be *complete* and *precise*; otherwise the degree of *uncertainty* must be indicated. Data must be represented in a data model that is *concise* and *intelligible* (that is, comprehensible by the user). They must be stored in a format that can be *conveniently handled* (maintained, transmitted, distributed, classified, resampled, retrieved, and updated) and *adequately protected* (with controlled access to maintain data integrity). It is also essential for data to be documented as to how they were derived (metadata) to enable users to determine their suitability for a certain application.

In addition to these general characteristics of high-quality data, there are several considerations specifically associated with geographic data. For example, geographic data must be stored in the *map projection* that best meets the requirements of the application with respect to the preservation of area, shape, distance, and direction (see Chapter 2). They must also be captured at a *scale* using a *classification scheme* that is commensurate with the applications to which they will be used (see Chapter 3). For geographic data on maps, *cartographic properties* play a significant part in determining the quality of the data. These properties include the type and physical condition of mapping media, quality of line work, use of color and symbology, classification of features, as well as map maintenance and revision cycle. For geographic data in digital form, they must be stored in a format that can be directly used by a particular GIS software product or in a standard *transfer format* that is accepted by different systems.

The growing awareness of the importance of high-quality data for the functioning of GIS has led to considerable interest in data quality among researchers and practitioners alike (Chrisman, 1983). Such a surge of interest in data quality has been basically driven by three motivations (Goodchild and Gopal, 1989).

1. the imminence and reality of the problems of geographic data quality in practice and the need to find better solutions to address them
2. the general lack of understanding of the nature of errors in geographic data and the need to improve user awareness of their characteristics and impacts on geographic data analysis
3. the technical need to develop methods for tracking and reporting data errors that may occur in GIS applications, as well as algorithms and data structures for directly identifying errors in geographic data processing

The scope of geographic data quality is, therefore, much broader than the qualitative assessment and description of the fitness for use as noted above. It embraces the theoretical and empirical studies of the origin, characteristics, as well as behavior of errors and uncertainties in spatial data processing and modeling (see Section 4.3). It also includes the various methods and models developed to assist GIS users in identifying and quantifying data quality (see Section 4.4). In addition, the scope of data quality covers the concepts and techniques for error management in GIS, as well as the reporting of data quality by means of metadata so that users can find, access, and apply the data with confidence (see Section 4.5). Therefore, a good understanding of data quality is essential for GIS users.

4.2.2 ACCURACY, PRECISION, ERROR, AND UNCERTAINTY

Data quality is largely determined by four generic measures of quality, namely, *accuracy*, *precision*, *error*, and the *uncertainty* that is associated with using data of unknown quality. The best way to understand data quality is to explain how these four terms are used to describe different aspects of data quality in practice.

Accuracy By definition, accuracy is the degree to which data agree with the values or descriptions of the real-world features that they represent. In other words, accuracy is a measure of how "close" data match the true values or descriptions (Figure 4.1a). In practice, some true values can never be exactly determined or known, for example, an angle and a distance in survey measurement. This is due to the limitations of instruments and the human inability to perform perfect observations. In such cases, an assumed value, such as the mean value of repeated measurements, can be accepted as the true value (Figure 4.1b).

Accuracy is usually application-specific. Data that are considered to be accurate for one application may not necessarily be adequately accurate for another. Accuracy is one of the most important factors governing the cost of data collection. High-accuracy data are time-consuming and costly to collect. In GIS projects, it is imperative to make sure that data are always collected at a level of accuracy commensurate with the objectives of the intended applications. Data collected at a lower level of accuracy than is required may not be able to generate information useful to the user. On the contrary, data collected at a higher level of accuracy than necessary may be a waste of valuable financial, human, and technical resources. In other words, geographic data of excessive accuracy are a burden rather than an asset to the GIS user.

Precision Whereas accuracy is a measure of how "close" data are to true or accepted values, precision is a measure of how "exact" data are measured and stored (Figure 4.2a). In mathematics, the exactness of representation is the number of significant digits used to record the data. For digital geographic data, this is the number of bits and the form (i.e., long integer, single-precision floating points, and double-precision floating

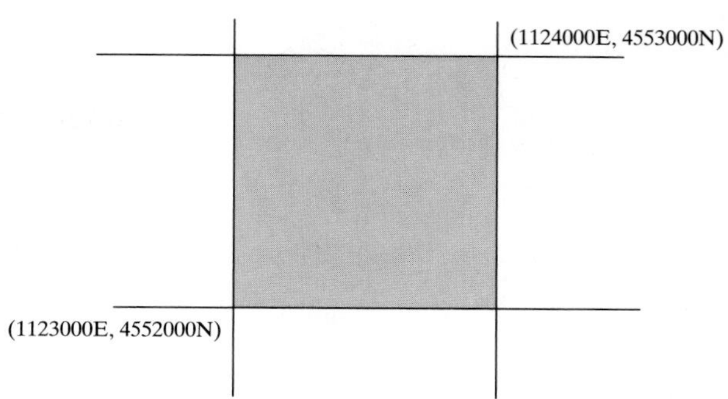

calculated area = 1000 m^2; measured area 1 = 1012.5 m^2; measured area 2 = 985.7 m^2

(a) The area of a UTM grid cell can be accurately calculated by using the coordinate values of the grid lines. If this area is actually measured, for example by using a planimeter, and the value is close to the computed value, the measured area is said to be "accurate." In the diagram, measurement 1 (1012.5 m^2) is more accurate than measurement 2 (985.7 m^2) because its value is closer to the computed value.

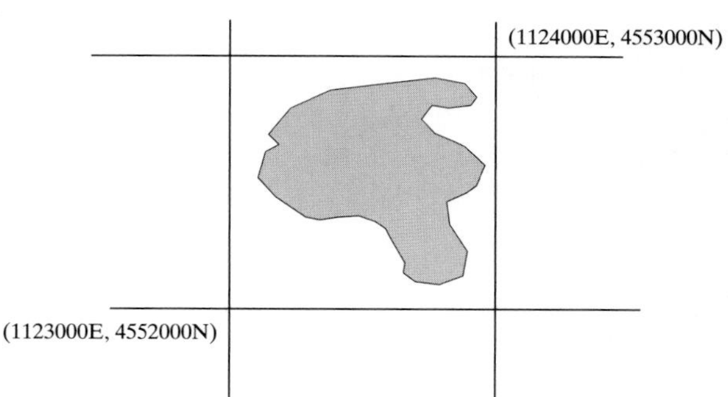

mean area = 584.5 m^2; measured area 1 = 592.5 m^2; measured area 2 = 580.7 m^2

(b) If the area of an irregularly shaped feature, such as a lake, is to be determined, it is necessary to measure the area several times in order to obtain an average value that is assumed to be the "true" value of the lake's area. Any subsequent measurement that yields a value close to this assumed true value is said to be "accurate." In the diagram, measurement 2 is more accurate than measurement 1 because its value is closer to the mean value.

FIGURE 4.1
Accuracy of data.

points) used for data capture and storage in the computer (Figure 4.2b) (see Chapter 3).

Generally speaking, high precision does not necessarily mean high accuracy. For example, a wrong reading in an angle measurement made to the tenth of a second of arc is precise, but definitely not accurate. On the other hand, high accuracy does not necessarily always require high-precision data representation. In a global change information system, for example, coordinates may be accurately plotted without the need to

(a) Different arms of the clock measure time to different degrees of precision: the hour, the minute, and the second. If a precision of 1/100 of a second is measured, a stopwatch is required. If 1/1,000,000 of a second is measured, an atomic clock must be used.

Format	Bits of Storage	Significant Digits of Precision	True Floating-Point Decimals?
Long integer	16	9	No
Single-precision floating points	16	7	Yes
Double-precision floating points	32	13	Yes

(b) Comparison of the precision of storing data by the three storage formats in computers

FIGURE 4.2
Precision of data.

store them to the precision of millimeters. Using 32-bit integers, it is possible to create a global database at a precision of approximately 2 cm. Such a precision far exceeds the accuracy requirement for any typical database at the global scale. From these examples, it is obvious that accuracy and precision refer to different aspects of data quality, and their meaning must be interpreted in the context of the scale of the data and the objective of the application.

The measure of precision when applied to categorical data carries a different meaning. Categorical data are nonnumerical descriptive data (nominal or ordinal) such as land-use types (e.g., residential, commercial, and industrial), income groups (e.g., high, medium, and low), classes of slopes (e.g., steep, gentle, and flat) and environmental indices (e.g., good, average, and poor). For this type of data, the measure of precision is based on the *level of detail* to which the data are observed and recorded, rather than on the number of significant digits by which the data are represented in the computer. This level of detail is a function of the number of categories in a classification or feature coding scheme (see Chapter 3). The more categories there are in the scheme, the more precisely the phenomena can be identified and recorded. Obviously, by using a classification scheme that contains five land-use classes (e.g., residential, industrial, commercial, institutional, and recreational), it is possible to record the land use in an urban area more precisely than using a classification scheme that has only two classes (e.g., built-up areas and forests).

Error The measure of error is relative to the measure of accuracy in that high-accuracy data are supposed to be free of errors. In practice, however, the two words are usually used in different contexts. Generally speaking, accuracy is used to imply "closeness" between the measured value and the true value of a real-world feature, but error is used to describe the "deviation" between these two values (Muller, 1987; Taylor, 1997). In a broader sense, error also includes the concept of "variation" or "discrepancy" in statistics (Burrough, 1986). The idea of deviation, variation, and discrepancy from the true or accepted value implies the lack of accuracy and precision in the data. This is probably the reason why data quality is often expressed in terms of error, rather than accuracy and precision of data.

There are three types of errors that may occur in measurements and observations: *gross errors*, *systematic errors*, and *random errors* (Thapa and Bossler, 1992). Gross errors are blunders or mistakes. These errors can be detected relatively easily by independent checks. Gross errors can be avoided by giving data-collecting staff adequate training and by enforcing standard rules and procedures in the data-collection processes. Systematic errors occur as the result of a variety of factors, such as human bias in measurement, mechanical defects in the instruments used, and, in the case of high-precision measurements in geodesy and land surveying, changing environmental conditions, such as temperature and humidity, at the time of field observation. The effect of systematic errors tends to be cumulative. If these errors are neglected, the impact can be very significant. Systematic errors cannot be detected by repeated measurements. They are usually eliminated by instrumental calibration and applying corrections in computation.

Random errors are those discrepancies in the measurements that remain after gross and systematic errors have been eliminated. These errors are caused by the limitation of instruments and the human operator to make perfect measurements. If a sufficiently large number of measurements are made, random errors will exhibit the following characteristics: The magnitudes are always very small; very large errors rarely occur; and the errors are equally likely to be positive and negative, which means that they tend to cancel the effects of one another. These characteristics are typical of errors of measurements that are *normally distributed* in the statistical sense. Consequently, random errors in geographic data can be treated by mathematical methods based on the theory of normal distribution such as least-square adjustment in survey computation.

Uncertainty When data of an unknown quality are used, there is always a certain degree of doubt or *uncertainty* about the validity of the information derived from the data. The words "uncertainty" and "error" are commonly used as synonyms in GIS literature, but there is a basic difference between the concepts conveyed by these two words (Davis and Keller, 1997). Whereas "error" refers to the lack of accuracy and precision in the data, "uncertainty" implies the lack of confidence in the use of the data that is due to the incomplete knowledge of the data. In other words, uncertainty is a measure of what we do not know. Obviously, the more that is known about the source and characteristics of the data, the more the user can use the data with certainty. Since errors are not avoidable in data collection, it is not possible to eliminate them by being careful. The objective of maintaining high-quality geographic data is not to eliminate errors totally, but to make sure that errors are as small as reasonably possible and to have a reliable estimate of their magnitudes (Taylor, 1997). This allows data to be used with the minimum amount of uncertainty. In this regard, error and uncertainty have a cause-and-effect relationship.

4.2.3 SOURCES OF ERROR IN GEOGRAPHIC DATA

From the previous discussion, it is obvious that both absolute accuracy and absolute precision of geographic data are practically impossible to attain. The presence of errors in geographic data is a fact of life. As noted in Chapter 2, geographic data collection is a relatively complex process that is carried out in different stages, by different people, using different technologies, and represented by different data models. As a result, geographic data errors have multiple sources because they may be introduced at any of the stages, by any of the people involved, with any of the technologies, and consequential to the choice of a particular data model (Thapa and Bossler, 1992). In general, errors in geographic data come from three principal sources: original source documents, data automation and compilation, and data processing and analysis. Based on the characteristics of these errors, Vitek et al. (1984) grouped them into two categories: *inherent errors* and *operational errors* (Table 4.1).

Table 4.1
Sources and Types of Possible Errors in Geographic Data

Sources	*Inherent*	*Operational*
1. Original source maps		
Map projection	X	
Map scale	X	
Cartographic generalization	X	X
Cartographic revision	X	X
Feature classification/coding	X	X
Field survey measurements	X	X
Photogrammetric measurements	X	X
Image analysis	X	X
Sampling design	X	X
Aging of maps	X	
2. Data automation and compilation		
Digitizing	X	X
Attribute data input	X	X
Format translation	X	
Map projection transformation	X	
Vectorization of raster data	X	
3. Data processing and analysis		
Numerical rounding in computing	X	
Overlay analysis	X	
Classification and reclassification	X	
Generalization and aggregation	X	
Interpolation	X	
Inappropriate use of algorithm		X

Inherent Errors in Geographic Data Inherent errors occur as a result of the special nature of geographic data. Geographic data, as representations of the real world in a certain data model, are necessarily incomplete and generalized. The real world is far too complex for faithful representation by a data model at the true scale (i.e., 1 : 1 scale). Subject to the rules of scale, geographic data collection is essentially a process of selection, generalization, and symbolization. It can be perceived as a filtering process that aims at removing detailed geographical variations in the real world. As a rule, therefore, geographic data inherently contain errors of an unknown magnitude.

Further errors occur as a result of the limitations of the instruments and techniques for obtaining measurements with absolute accuracy, as well as the inability of the computer to represent coordinates with absolute precision. Errors also occur as the result of the limitations of map projections to represent Earth without compromising one geometrical requirement in favor of another (see Chapter 2). In addition, errors can occur when data are translated from one storage format to another or from one coordinate system to another. Errors tend to develop gradually when data become out of date. Procedures such as map revision and database update may be able to slow down the process, but cannot stop it completely in the long run (Blakemore, 1990). For categorical descriptive data, errors always occur as the result of the fuzzy nature of human languages and other factors that impose constraints on the design of classification schemes for data collection (see Chapter 3).

The various sources of errors noted here are called *inherent* because they occur naturally in geographic data no matter what instruments and procedures are used for data collection and representation. These errors are sometimes referred to as *source errors* because they basically describe the differences that exist between the data model and the geographical truth that the model represents (Goodchild, 1989). In practice, inherent data errors in geographic data may occur as gross, systematic, or random errors as explained in Section 4.2.2.

Operational Errors in Geographic Data Operational errors occur mostly during the operation of the procedures for collecting, managing, and using geographic data. Since these errors occur mainly during the processing of geographic data, they are also sometimes referred to as *user* or *processing errors*. Operational errors occur as the result of the imperfection (both mechanical and procedural) of the instruments and methods used for geographic data collection, management, and application. Operational errors also include human bias and mistakes in field surveying measurements, site visits and investigation, keyboard data input, air photo interpretation, remote sensing image analysis, table digitizing, as well as the inappropriate application of algorithms in spatial data analysis. Just like inherent errors, operational errors in geographic data may also occur as gross, systematic, or random errors.

In practice, both inherent and operational errors occur essentially at the same time. In addition, errors that have occurred at one stage in data collection or processing may be carried over to subsequent stages. A typical example is the map digitizing process. The source maps contain both inherent and operational errors that occurred when they were compiled. When the maps are digitized, further inherent errors (e.g., limitation of instrumental precision) and operational errors (e.g., operator biases) will be added. The resulting digital data therefore contain an accumulation of both types of errors from two separate data-collection stages. The amount of errors will continue to accumulate when the data are combined with data from other sources or when they are translated into another storage format or map projection.

In conventional cartography and geographic data processing using manual methods, many of the problems associated with errors are visible. When a problem shows up, the data user will probably be able to make the necessary adjustments and, as a result, intimately know how far the information can be relied upon. In the GIS environment, data processing is a more "transparent" operation to the user. The user is not always aware of the problems and limitations of the data. The ease with which digital data from different sources, in different formats, of different scales, and with different levels of accuracy can be integrated and manipulated in a single system environment often tends to hide the potential problems from the user. Consequently, applications that use data from different sources and at different scales will potentially generate more uncertain results (Openshaw, 1989). In contrast to popular belief, the use of digital data does not automatically lead to better spatial decision making.

4.3 Components of Geographic Data Quality

The U.S. National Committee for Digital Cartographic Data Standards (NCDCDS, 1988; Morrison, 1995) identified five dimensions for geographic data quality: (1) *lineage*, (2) *positional accuracy*, (3) *attribute accuracy*, (4) *logical consistency*, and (5) *completeness*. In addition, the International Cartographic Association (ICA) proposed two more dimensions of geographic data quality: *temporal accuracy* and *semantic accuracy* (Morrison, 1995). The NCDCDS indices were included in the USGS quality reporting standards for data transfer

(USGS, 1988). These indices have also been adopted by Statistics Canada for its information products (Lundin et al., 1989) and are used by other jurisdictions as the blueprint for the development of standards for reporting digital geographic data quality, notably in Australia and New Zealand (Clarke, 1991; Henderson, 1989).

4.3.1 LINEAGE OF GEOGRAPHIC DATA

Lineage is a documentation of the source materials from which a specific set of geographic data was derived (Clarke and Clark, 1995). It also describes the method of derivation, including all transformations involved, in producing the final data files. The purpose of lineage is to provide an account of the data collection processes at a level of detail sufficient for the user to understand the quality of the source data. Typically, lineage is designed to answer the following questions:

1. Who collected the data?
2. When were the data collected?
3. How were the data collected (e.g., by field surveys, photogrammetry, remote sensing image analysis)?
4. Why were the data collected (e.g., topographic mapping or special project)?
5. How were the data converted (e.g., by table digitizing, image scanning and automated vectorization, image understanding and recognition)?
6. What algorithms have been used to process the data (e.g., coordinate transformation for georeferencing image data)?
7. What was the precision of computation?

Additional documentation containing greater detail of the source data, such as specifications for collection and compilation, may be included in a reference list of the lineage. The lineage of source data may not be able to answer all questions about the development process. However, it will allow the user to formulate pertinent questions about the data.

4.3.2 POSITIONAL ACCURACY OF GEOGRAPHIC DATA

Positional accuracy is defined as the "closeness" of coordinate values in the geographic database to the "true" positions of the real-world features that they represent (Drummond, 1995). Conventionally, maps are accurate to roughly one line width, or 0.5 mm. This is also roughly the accuracy with which an average digitizer operator can position the cross hairs of a cursor over a point or line feature. As noted in Chapter 2, such a value that represents the smallest depictable or plottable object is known as the *minimum mapping unit*. A 0.5-mm resolution is equivalent to 5 m on 1:10,000 scale maps or 125 m on 1:250,000 scale maps. Potential accuracy of geographic data therefore increases proportionally with the scales of the data (Table 4.2).

Positional accuracy may be specified by standards established by government agencies by legislation, such as the *National Map Accuracy Standard* (NMAS) of the U.S. Bureau of the Budget (1947) (Table 4.3). It may also be established by professional bodies, such as the *Accuracy Standards for Large-scale Maps* of the American Society for Photogrammetry and Remote Sensing

Table 4.2

Map Scale and Effective Resolution

Scale	*Effective Resolution (m)*
1:2500	1.25
1:10,000	5
1:24,000	12
1:50,000	25
1:100,000	50
1:250,000	125
1:500,000	250
1:1,000,000	500
1:10,000,000	5000

Table 4.3

Summary of the National Map Accuracy Standard

- On maps smaller than 1:20,000, no more than 10% of points tested should be more than 1/50 inch in horizontal error, where points refer only to points that can be well defined on the ground
- On maps with scales larger than 1:20,000, the corresponding error term is 1/30 inch
- No more than 10% of the elevation points tested on contour lines can be in error by more than one-half the contour interval
- Accuracy tests should be carried out by comparison of actual map data with survey data of higher accuracy
- If maps have been tested and met these standards, a statement should be made to that effect in the legend
- Maps that have been tested but failed to meet the standards should omit all mention of the standards on the legend

(Source: Bureau of the Budget, 1947)

Table 4.4
Planimetric Coordinate Accuracy Requirements

Planimetric (X or Y) Accuracy on Ground (limiting RMSE in meters)	*Typical Map Scales*
0.0125	1:50
0.025	1:100
0.05	1:200
0.125	1:500
0.25	1:1000
0.50	1:2000
1.00	1:4000
1.25	1:5000
2.50	1:10,000
5.00	1:20,000

(Source: ASPRS, 1989)

(ASPRS, 1989) (Table 4.4). The test of positional accuracy may be carried out as a quality assurance and quality control (QA/QC) process during data collection. Alternatively, it may be carried out as a stand-alone data quality reporting process (see Section 4.4.1).

Although these standards for positional accuracy have been widely accepted, few tests have actually been carried out (Chrisman, 1991; Fisher, 1991). Many suppliers of geographic data who claim to be in compliance with accepted accuracy standards might not have properly tested their data. This is probably due to the time and cost constraints in many mapping programs. The USGS, for example, did not start systematic testing of its maps until 1958. At present, accuracy testing is performed on only 10% of the mapping projects at each contour interval as a method of controlling overall quality (see Section 4.4.2). It is understood that maps rarely fail the test, but this happens on occasion (USGS, 1997a). We will examine in detail the methods of testing positional accuracy in Section 4.4.1.

4.3.3 ATTRIBUTE ACCURACY OF GEOGRAPHIC DATA

Attribute accuracy is defined as the "closeness" of the descriptive data in the geographic database to the true or assumed values of the real-world features that they represent (Goodchild, 1995). Attribute accuracy may be determined in different ways depending on the nature of the data. For *metric* attributes such as those in a digital elevation model (DEM) or triangulated irregular network (TIN), accuracy may always be simply expressed as measurement error (elevation accurate to 1 m). For *categorical* attributes, such as classified polygons in a soil survey, tests of accuracy are much more difficult because there is no such thing as a *close value* or a *metric of deviations* (i.e., accuracy to a certain unit of measurement). In such cases, attribute accuracy is usually evaluated in terms of other factors, including

1. the classification scheme (e.g., Is the scheme appropriate to the application, sufficiently detailed and adequately defined?)
2. the amount of gross errors (e.g., misidentification during data collection)
3. the degree of heterogeneity of the polygons that are assumed to be homogeneous (e.g., a polygon in a soil map classified as sandy soil may in fact be only 70% sand, and 25% clay and 5% gravel)

Of these three factors, the magnitude of gross error is usually considered the most significant index of the quality of attribute data. Gross errors are described by an *error matrix* (see Section 4.4.2). Quantitative indices of the quality obtained from the error matrix are used to determine the fitness for use of the data by comparing them with an accepted standard. Just like positional accuracy standards, attribute accuracy standards can be established by either government agencies or professional bodies for particular mapping programs. An example is the *Land Use and Land Cover* mapping program of the USGS (Table 4.5) (Anderson et al., 1976). However, as accuracy testing is a time-consuming and costly undertaking, actual testing of attribute accuracy has seldom been carried out in practice (Fisher, 1991).

4.3.4 LOGICAL CONSISTENCY OF GEOGRAPHIC DATA

Logical consistency is a description of the fidelity of the relationships between the real world and encoded geographic data (Kainz, 1995). In essence, logical consistency in geographic data includes the following elements:

Table 4.5
Summary of the Accuracy Specifications for USGS Land-Use and Land Cover Maps

The minimum level of accuracy in identifying land-use and land cover categories from remote sensor data should be at least 85 percent
The several categories shown should have about the same level of accuracy
Accuracy should be maintained between different interpreters and times of data collection

(Source: Anderson et al., 1976)

1. consistency of the data model to the real world, which means how well the data model can adequately represent spatial features in the real world
2. consistency of the positional and attribute data to the real world, which means how close they agree with one another
3. consistency within the data model, which implies the agreement between the data and the rules of the data model
4. consistency within the system, which implies the ability to construct complex features from simple data-storage formats
5. consistency between different parts of a data set, which implies the matching of data across the boundaries of maps
6. consistency between data files, which means that similar data in two or more layers (e.g., rivers and highways used as background information) agree with one another

Testing logical consistency by conventional manual methods is a very time-consuming and laborious process. In GIS, the use of the topological model provides a convenient means for automated testing of logical consistency (see Chapter 3). It allows a variety of errors in the data such as missing boundaries and missing labels of polygons to be detected. The ability to detect logical inconsistency helps eliminate errors that would affect positional or attribute accuracy. In this regard, logical consistency may be perceived as a natural extension of positional and attribute accuracy in geographic data.

4.3.5 COMPLETENESS OF GEOGRAPHIC DATA

Completeness of geographic data refers to the degree to which the data exhaust the universe of all possible items (Brassel et al., 1995). This particular quality is critically important to GIS applications because using information obtained from incomplete data can lead to potentially serious consequences in decision making. Completeness of geographic data has two elements: *spatial completeness*, which means that the data cover the entire area of interest; and *thematic completeness*, which means that data contain all the layers required for the intended applications.

For geographic data to be considered high quality, they must be both spatially and thematically complete. Generally speaking, it is easy to determine the *spatial completeness* of geographic data. This is done by displaying or plotting the data and visually inspecting them for omissions in any part of the area of interest. *Thematic completeness*, on the other hand, is relatively more difficult to assure. Whether or not a particular geographic data set has all the necessary layers is often application-specific. This means that the layers that satisfy the requirements of one application do not necessarily satisfy those of another. In order to ensure that geographic data have all the thematic layers to satisfy the requirements of intended applications, a carefully designed user-needs study is always necessary before the data are collected (see Chapter 11).

4.3.6 TEMPORAL ACCURACY

Temporal accuracy is the measure of data quality with respect to the representation of time in geographic databases (Guptill, 1995). As noted in Chapter 3, time can be represented in geographic databases in terms of *world time* or *database time*. World time refers to when events actually occur in the real world, and database time is when data pertaining to these events are collected. When temporal accuracy is used in terms of world time, it is associated with the observation or sampling of the events at regular time intervals. Obviously, the shorter the time period between the observations, the more accurately the state of the event can be recorded. For example, when two events occurring at different times on the same day are observed at daily intervals, they will be recorded as if they occurred simultaneously. However, if these two events are observed at hourly intervals, they will be accurately recorded as events that occur at different times.

When temporal accuracy is used in terms of database time, it is a measure of how up to date the data are. In practice, this is the difference between the time when the data are entered into the database and the time when the data are used. Temporal accuracy with respect to database time is application-specific and is dependent on the *occurrence rate* of real-world events. For example, data in an air-traffic control system where the positions of aircraft are constantly changing have an extremely high occurrence rate. Such data become outdated by the second, and their temporal accuracy must be reckoned by the second. On the other hand, data in a land-use monitoring system have a much lower occurrence rate because land-use change seldom occurs overnight. In this particular case, a temporal accuracy of a month or even a year may be able to satisfy the needs of most users.

4.3.7. SEMANTIC ACCURACY

Semantic accuracy is how correctly spatial objects are labeled or named. Specifically, semantic accuracy describes the number of features, relationships, or attributes that have been correctly encoded in accordance with a set of feature representation rules (Salge, 1995). Semantic accuracy is closely related to the measure of precision of categorical data (see Section 4.2.2) and the attribute accuracy of data (see Section 4.3.3). There is, however, a basic distinction between the concepts behind these terms. Whereas the precision of categorical data is concerned with the number of classes by which data can be

Database Name	Layer Name	Attributes
Gas company database	STREET SEGMENT POINT PIPE NODE	#street, street_name, width #segment, #point1, #point2 #point, x, y #node1, #node2 #node, x, y, z, type
Water supply database	STREET SEGMENT POINT PIPE NODE	#street, (#right_segment, order), (#left_segment, order) #segment, #from_point, #to_node #point, x, y #edge, #from_node, #to_node #node, x, y, z, #edge, category
Public work (street maintenance) database	STREET SEGMENT POINT GAS-PIPE GAS-NODE WTR-PIPE WTR-NODE	#street, street_name, #parcel_segment #segment, #point1, #point2, begin_address, end_address #point, x, y #edge, #node1, #node2 #node, x, y, depth, type #edge, #node1, #node2 #node, x, y, depth, type

FIGURE 4.3
Examples of semantic discrepancies in distributed geographic databases. Different organizations in a GIS network may use different names to describe the same spatial objects. All the names are semantically accurate to the organizations themselves, but discrepancies among different databases will cause problems in the design and operation of the distributed system.

sampled and the attribute accuracy of data is concerned with the degree to which data can be identified correctly, semantic accuracy is concerned with the correct labeling of spatial objects and their attribute characteristics. Obviously, a higher degree of semantic accuracy helps remove uncertainty in data collection and will potentially improve the attribute accuracy of data.

Just like temporal accuracy, semantic accuracy is also application-specific. Whether a label is semantically accurate is often dependent on the practice of a particular profession or organization. It is not uncommon for different professions or organizations to use different labels to describe the same spatial features (Figure 4.3). The semantic discrepancies between different professions or organizations often make it difficult to assess and standardize semantic accuracy in distributed geographic databases (Laurini, 1993).

4.4 Assessment of Data Quality

Of the seven data quality indices noted in Section 4.3, positional and attribute accuracies are usually considered the most critical factors in determining the fitness for use of geographic data. The level of positional and attribute accuracies of a particular data set can be quantified by checking a portion of the data, usually called the *sample data*, against *reference data* of a higher degree of accuracy. These reference data can be obtained from different independent sources, such as data of a larger scale, field verification, and ground-truthing, field checks by Global Positioning System (GPS) for positional accuracy, reference to raw survey materials, and using internal evidence such as logical consistency testing for unclosed polygons, undershoots, and overshoots. The following sections explain the concepts and techniques for the quantitative assessment of positional and attribute accuracies commonly used in practice. These concepts and methods are applicable to both raster and vector data.

4.4.1 EVALUATION OF POSITIONAL ACCURACY

Positional accuracy is made up of two elements: *planimetric accuracy* and *height accuracy*. Most GIS applications require only the evaluation of planimetric accuracy. This is done by comparing the coordinates (x and y) of sample points on maps (either a hard-copy map or a digital map) to the coordinates (X and Y) of corresponding reference points as determined by one of the methods noted earlier. Evaluation of height accuracy involves comparison of elevation values of sample and reference data points. To be used as a sample point, the point must be "well defined," which means that it can be unambiguously identified both on the map and on the ground. If the reference data are from an existing source, the points must be well defined in the source document. Examples of well-defined points include survey monuments and bench marks, road intersections, corners of buildings, and point features such as lampposts and fire hydrants. It is important for both the sample and reference data to be in the same map projection and based on the same datum (see Chapter 2).

The number and spatial distribution of sample points for each map are determined at the discretion of the user. The *Accuracy Standards for Large-scale Maps* of ASPRS noted earlier, however, specifies that a minimum of 20 check points must be established throughout the area covered by the map. It also recommends that these sample points should be spatially distributed in such a way that at least 20% of the points be located in each quadrant of the map, with individual points spaced at intervals equal to at least 10% of the diagonal of the map sheet.

The discrepancies between the coordinate values of the sample points and their corresponding reference coordinate values are used to compute the overall accuracy of the map as represented by the *root mean-square error* (RMSE). The RMSE is defined as the square root of the average of the squared discrepancies. The RMSE for discrepancies in the X coordinate direction (rms_x), Y coordinate direction (rms_y) and elevation (rms_e) are computed from

$$rms_x = \sqrt{\frac{\Sigma X^2}{n}} \qquad \textbf{(4.1)}$$

$$rms_y = \sqrt{\frac{\Sigma Y^2}{n}} \qquad \textbf{(4.2)}$$

$$rms_e = \sqrt{\frac{\Sigma E^2}{n}} \qquad \textbf{(4.3)}$$

where

$\Sigma X^2 = dx_1^2 + dx_2^2 + dx_3^2 + \ldots dx_n^2$

$dx =$ discrepancies in X coordinate direction

$= X_{\text{reference}} - X_{\text{sample}}$

$\Sigma Y^2 = dy_1^2 + dy_2^2 + dy_3^2 + \ldots dy_n^2$

$dy =$ discrepancies in Y coordinate direction

$= Y_{\text{reference}} - Y_{\text{sample}}$

$\Sigma E^2 = e_1^2 + e_2^2 + e_3^2 + \ldots e_n^2$

$e =$ discrepancies in elevation $= E_{\text{reference}} - E_{\text{sample}}$

$n =$ total number of points checked (sampled)

From rms_x (4.1) and rms_y (4.2), a single RMSE of planimetry (rms_p) can be computed as follows.

$$rms_p = \sqrt{(rms_x)^2 + (rms_y)^2} \qquad \textbf{(4.4)}$$

The RMSEs of planimetry and elevation have now been generally accepted as the overall accuracy of the map. RMSE is used as the index to check against specific standards to determine the fitness for use of the map. The major drawback of the RMSE is that it provides information of only the overall accuracy. It does not give any indication of the spatial variation of the errors. For users who require such information, a map showing the positional discrepancies at the sample points can be generated. Separate maps can be generated for discrepancies in easting and northing. Alternatively a map showing the vectors of discrepancies at each point can be plotted (Figure 4.4). As the discrepancies are normally very small numerical values, they have to be plotted with an exaggerated scale. The

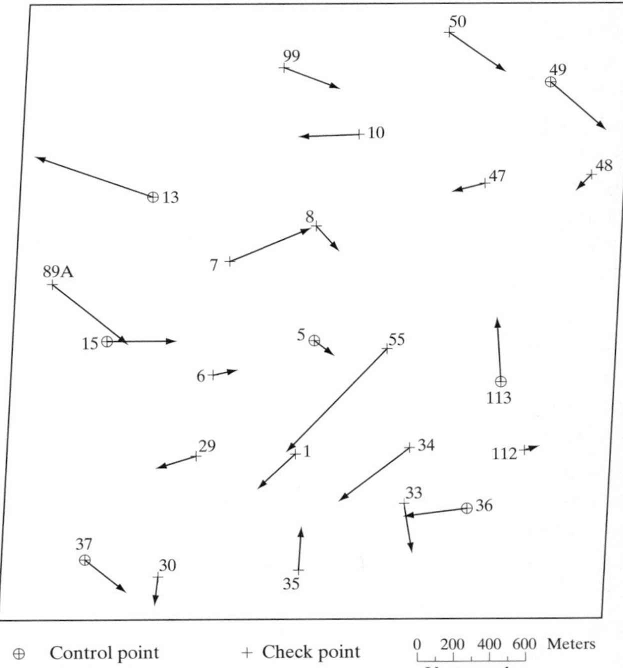

FIGURE 4.4
An error vector map showing the magnitude and displacement directions of planimetric errors at different locations of check points on the map. Look whether the pattern is random or systematic and where the errors tend to concentrate. The error pattern as exhibited here is quite random. *(Source: Lo, 1981)*

RMSE and the error maps, if available, will form part of the data quality report of the data sets concerned (see Section 4.5.4).

4.4.2 EVALUATION OF ATTRIBUTE ACCURACY

The assessment of attribute accuracy is similar to the assessment of positional accuracy. Attribute accuracy is obtained by comparing values of sample spatial data units with reference data obtained either by field checks or from sources of data with a higher degree of accuracy. These sample spatial units can be raster cells; raster image pixels; or sample points, lines, and polygons. There are different methods for presenting the resulting discrepancies, depending on the measurement scale by which the attribute data are recorded (see Chapter 2). For interval and ratio data, errors can be measured by the mean deviation between actual and observed values at sample locations. This approach provides an error index similar to the RMSE for positional accuracy as explained in Section 4.4.1.

Ordinal and nominal data are ranked and categorical data, respectively. Error as a deviation from a true value does not apply in these cases. Consequently, no RMSE can be computed. Instead, an *error matrix* is constructed to show the frequency of discrepancies between encoded values (i.e., data values on a map or in a database) and their corresponding actual or reference values for a sample of locations. The error matrix has been widely used as a method for assessing classification accuracy of remotely sensed images (Story and Congalton, 1986; Congalton, 1991; Congalton and Green, 1993). It applies equally well for accuracy tests of both vector and raster geographic data.

An error matrix, also known as *classification error matrix* or *confusion matrix*, is a square array of values, denoted as C, which cross-tabulates the number of sample spatial data units assigned to a particular category relative to the actual category as verified by the reference data (Figure 4.5). Conventionally, the rows of the error matrix represent the categories of the classification of the database, while the columns indicate the classification of the reference data. In the error matrix, the element c_{ij} represents the frequency of spatial data units assigned to category i that actually belong to category j. The numbers along the diagonal of the error matrix (i.e., when $i = j$) indicate the frequencies of correctly classified spatial data units in each category; and the off-diagonal numbers (when $i \neq j$) represent the frequencies of misclassification in the various categories.

The error matrix is an effective way to describe attribute accuracy of geographic data. If in a particular error matrix, all the nonzero entries lie on the diagonal, it indicates that no misclassification at the sample locations has occurred and an overall accuracy of 100% is obtained.

Sample Data	Reference Data						Total
	A	B	C	D	E	F	
A	**1**	2	0	0	0	0	3
B	0	**5**	0	2	3	0	10
C	0	3	**5**	1	0	0	9
D	0	0	4	**4**	0	0	8
E	0	0	0	0	**4**	0	4
F	0	0	0	0	0	**1**	1
Total	1	10	9	7	7	1	35

A = exposed soil
B = cropland
C = range
D = sparse woodland
E = forest
F = water body

PCC = (1 + 5 + 5 + 4 + 4 + 1)*100/35 = 57.1%

Producer's accuracy
A = 1/1 = 100% D = 4/7 = 57.1%
B = 5/10 = 50% E = 4/7 = 57.1%
C = 5/9 = 55.6% F = 1/1 = 100%

User's accuracy
A = 1/3 = 33.3% D = 4/8 = 50%
B = 5/10 = 50% E = 4/4 = 100%
C = 5/9 = 55.6% F = 1/1 = 100%

FIGURE 4.5
An error matrix, showing percent correctly classified (PCC), producer's and user's accuracies.

When misclassification occurs, it can be identified either as an *error of commission* or an *error of omission*. Any misclassification is simultaneously an error of commission and an error of omission. Errors of commission, also known as *errors of inclusion*, are defined as wrongful inclusion of a sample location in a particular category due to misclassification. When this happens, it means that the same sample location is omitted from another category in the reference data, which is an error of omission.

Errors of commission are identified by off-diagonal values across the rows. In Figure 4.5, for example, there are two errors of commission for row 1. The value of 2 at $c_{1,2}$ means that two spatial data units of cropland were misclassified as exposed soil. Similarly, the values of 2 and 3 in row 2 (at $c_{2,4}$ and $c_{2,5}$, respectively) indicate that two spatial data units of sparse woodland and three spatial data units of forest were misclassified as cropland. The errors of commission in rows 3 and 4 (i.e., range and sparse woodland) may be interpreted in the same way. There is no error of commission for rows 5 and 6, which means that all the sample spatial data units of classes forest and water (not the whole map) were correctly classified.

Errors of omission, also known as *errors of exclusion*, are identified by those off-diagonal values down the columns. For example, in Figure 4.5 there are five errors of omission in column 2. This means that five spatial data units of cropland were missing from that particular category in the reference data. The numerical values of 2 at $c_{1,2}$ and 3 at $c_{3,2}$ indicate that two of these spatial data units were misclassified as exposed soil and three as range. The errors of omission in the rest of the columns may be interpreted in the same way. Since there is no off-diagonal nonzero value for columns 1 and 6, all the sample spatial data units of exposed soil and water body were correctly classified.

In addition to the interpretation of errors of commission and omission, the error matrix may also be used to compute a series of descriptive indices to quantify the attribute accuracy of the data. These include the *overall accuracy*, *producer's accuracy*, and *user's accuracy* that are explained as follows.

Overall Accuracy The PCC (Percent Correctly Classified) index represents the overall accuracy of the data. In the case of simple random sampling, the PCC is defined as the trace of the error matrix (i.e., the sum of the diagonal values) divided by n, the total number of sample locations.

$$PCC = (S_d/n)*100\% \quad \textbf{(4.5)}$$

where

S_d = sum of values along diagonal

n = total number of sample locations

An example of the computation of PCC is shown in Figure 4.5. The maximum value of the PCC index is 100 when there is perfect agreement between the database and the reference data. The minimum value is 0, which indicates no agreement.

Although the PCC index provides a relatively simple measure for attribute accuracy, it has a number of deficiencies that must be understood (Veregin, 1995). In the first place, since the sample points are randomly selected, the index is sensitive to the structure of the error matrix. This means that if one category of data dominates the sample (this occurs when the category covers a much larger area than others), the PCC index can be quite high even if the other classes are poorly classified. Second, the computation of the PCC index does not take into account the chance agreements that might occur between sample and reference data. The index therefore always tends to overestimate the accuracy of the data. Third, the PCC index does not differentiate between errors of omission and commission. Indices of these two types of errors are provided by the producer's accuracy and the user's accuracy.

Producer's Accuracy This is the probability of a sample spatial data unit being correctly classified and is a measure of the error of omission for the particular category to which the sample data belong. The producer's accuracy is so-called because it indicates how accurate the classification is at the time when the data are produced. The producer's accuracy is computed by

$$\text{producer's accuracy} = (C_i/C_t)*100\% \quad \textbf{(4.6)}$$

where

C_i = correctly classified sample locations in column

C_t = total number of sample locations in column

error of omission = 100 − producer's accuracy

An example of the computation is shown in Figure 4.5.

User's Accuracy This is the probability that a spatial data unit classified on the map or image actually represents that particular category on the ground. This index of attribute accuracy, which is actually a measure of the error of commission, is of more interest to the user than the producer of the data. The user's accuracy is computed by

$$\text{user's accuracy} = (R_i/R_t)*100 \quad \textbf{(4.7)}$$

where

R_i = correctly classified sample locations in row

R_t = total number of sample locations in row

error of commission = 100 − user's accuracy

An example of the computation is shown in Figure 4.5.

More advanced analytical techniques can be applied to the error matrix in order to improve the information about attribute accuracy, as indicated by these descriptive indices. One of these techniques is the normalization of the error matrix. An error matrix can be normalized by using a proportional fitting procedure that iteratively balances all the values in the rows and columns until the sums of the values in each row and column are equal to unity. Congalton (1991) suggested that the normalized error matrix is more representative of the level of accuracy of the data for two reasons. First, as part of the iterative process, the rows and columns are totaled, resulting in an error matrix that is more indicative of the off-diagonal values. Second, as the overall accuracy of a normalized error matrix is computed by summing the major diagonal and dividing by the total of the entire matrix, it contains information about the off-diagonal values and is a better representation than the overall accuracy computed from the original non-normalized error matrix.

In addition to computing a normalized overall accuracy, the normalized matrix may be used to directly compare cell values between different matrices. Intermatrix evaluations are sometimes required in order to compare the results obtained by using different classification methods, such as supervised and unsupervised classification in remote sensing image analysis, and by different data collectors (Hardin and Shumway, 1997). Such comparisons are possible only with a normalized error matrix because it removes the problem of using unequal number of samples in deriving individual matrices.

Sample Data	Reference Data						Total ($i+$)
	A	B	C	D	E	F	
A	**1**	2	0	0	0	0	3
B	0	**5**	0	2	3	0	10
C	0	3	**5**	1	0	0	9
D	0	0	4	**4**	0	0	8
E	0	0	0	0	**4**	0	4
F	0	0	0	0	0	**1**	1
Total ($+i$)	1	10	9	7	7	1	35

Kappa coefficient
Pc = (row sum/N)* (column sum/N)

A = 3*1 = 3 ***D = 8*7 = 56***
B = 10*10 = 100 ***E = 4*7 = 28***
C = 9*9 = 81 ***F = 1*1 = 1***

Sum = A + B + C + D + E + F = 269
Pc = 269/(35*35) = 0.2196
Po = 20/35 = 0.5714

Kappa = (Po – Pc)/(1 – Pc)
= (0.5714 – 0.2196)/(1 – 0.2196)
= 0.3518/0.7804 = 0.451

Tau coefficient
(Equal a priori probabilities)
Pr = 1/m
For m = 6, Pr = 0.1667
Tau = (Po – Pr)/(1 – Pr)
= (0.5714 – 0.1667)/(1 – 0.1667)
= 0.4047/0.8333 = 0.486

Tau coefficient
(Unequal a priori probabilities)
Pr = (column sum)**2/(N*N)
A = 1*1 = 1 ***D = 7*7 = 49***
B = 10*10 = 100 ***E = 7*7 = 49***
C = 9*9 = 81 ***F = 1*1 = 1***
Sum = A + B + C + D + E + F = 281
Pr = 281/(35*35) = 0.2294
Tau = (Po – Pr)/(1 – Pr)
= (0.5714 – 0.2294)/(1 – 0.2294)
= 0.3420/0.7706 = 0.444

FIGURE 4.6
An error matrix, showing computation of kappa and tau coefficients.

Another useful analytical technique is the computation of the kappa coefficient or *Kappa Index of Agreement* (KIA) (Rosenfield and Fitzpatrick-Lins, 1986; Congalton, 1991; Foote and Huebner, 1996). The kappa coefficient has become widely used because it is capable of controlling the tendency of the PCC index to overestimate by incorporating all the off-diagonal values in its computation (i.e., it takes into account the chance agreements). The use of the off-diagonal values in the computation of the kappa coefficients also makes them useful for testing the statistical significance of the differences in different error matrices (Congalton, 1991). The coefficient (κ), first developed by Cohen (1960) for nominal scale data, takes the following mathematical form (Ma and Redmond, 1995).

$$\kappa = \frac{P_o - P_c}{1 - P_c} \tag{4.8}$$

where

$$P_o = \sum_{i=1}^{m} P_{ii} = \frac{1}{N}\sum_{i=1}^{m} n_{ii}$$

$$P_c = \sum_{i=1}^{m} P_{i+}P_{+i} = \frac{1}{N^2}\sum_{i=1}^{m} n_{i+}n_{+i}$$

In this equation, P_o is the proportion of agreement between the reference and sample data (PCC), n_{ii} is the total number of correctly classified points by class along the diagonal of the error matrix, N is the total number of points checked (sampled), P_{ii} is the proportion of correctly classified sample points by class at the diagonal of the error matrix (i.e., n_{ii}/N), P_{i+} is the marginal distribution of the sample data (n_{i+}/N where n_{i+} is the row sum by class), P_{+i} is the marginal distribution of the reference data (n_{+i}/N where n_{+i} is the column sum by class), and m is the total number of classes. Kappa coefficient varies from a minimum of 1 to a maximum of 0. An example of computation of kappa coefficient is shown in Figure 4.6.

Research by Foody (1992) indicated that the kappa coefficient tends to overestimate the agreement between data sets that is due to chance and to underestimate the overall classification accuracy. Foody (1992) described a modified kappa coefficient based on equal probability of group membership that resembles and is derived more properly from the tau coefficient. Expanding on Foody's findings, Ma and Redmond (1995) developed an alternative method for assessing classification accuracy from an error matrix using the tau coefficient (τ) as follows.

$$\tau = \frac{P_o - P_r}{1 - P_r} \tag{4.9}$$

where

$$P_r = \sum_{i=1}^{m} P_{i+}P_i = \frac{1}{N^2}\sum_{i=1}^{m} n_{i+}x_i$$

It was demonstrated that the tau coefficient, which is based on the a priori probabilities of group membership, provides an intuitive and relatively more precise quantitative measure of classification accuracy than the kappa coefficient, which is based on the a posteriori probabilities (Ma and Redmond, 1995). The a priori probabilities of group membership are given by P_i, which are computed from x_i/N where x_i is the number of sampled points belonging to class i and N is the total number of sample points checked. If the a priori probabilities of group membership are equal (as in unsupervised image classification), $P_r = \frac{1}{m}$ where m is the total number of classes. If the a priori probabilities of group membership are unequal, the marginal distribution of the reference data is used to determine the a priori probabilities, or in other words, $x_i = n_{+i}$. An example of computation for both cases is shown in Figure 4.6. A conditional tau coefficient was developed by Naesset (1996) as an improvement of the tau coefficient in assessing producer's accuracy.

4.4.3 SPECIAL CONSIDERATIONS IN THE EVALUATION OF ATTRIBUTE ACCURACY

The use of the error matrix is based on the assumption that all discrepancies between the sample data and the reference data are due to classification errors only. There are in fact other sources of errors that contribute to the confusion. These errors include digitizing error, data input error, interpretation error, seasonal or temporal variation, and errors in obtaining reference data (Congalton and Green, 1993). In order to ascertain the statistical validity of the attribute accuracy test, it is necessary to pay attention to the factors noted below when generating the error matrix.

Collection of Reference Data Depending on the nature and characteristics of the original data, the collection of reference data can be a very expensive and time-consuming task. Field checks as a rule are always much more expensive than obtaining reference data from an existing source of a higher degree of accuracy. It is also not possible to get rid of inherent errors entirely in the reference data themselves, nor is it possible to know exactly the level of accuracy of the reference data. Conventionally, it is assumed that the reference data are of an acceptable accuracy, no matter whether they are from field checks or from existing data. It is important for the user to ensure that reference data of the highest possible degree of accuracy are used, subject to the constraints of the cost and time available for the data collection.

Classification Scheme The classification scheme is one of the most important factors governing attribute accuracy of geographic data. As a rule, the larger the number of categories in a classification scheme (such as the level II land-use/land cover classes in Anderson's (1976) scheme), the more precisely the data can be represented. However, more classes also lead to a low degree of accuracy because of the fuzzy nature of both real-world features and the human language used to describe them (see discussion on semantic accuracy discussed in Section 4.3.7). In tests for attribute accuracy, it is important for the collector of reference data to be knowledgeable about the classification scheme, so that discrepancies due to human bias can be minimized.

Spatial Autocorrelation The presence, absence, or characteristics of some spatial objects may sometimes have significant impacts on the presence, absence, or characteristics of the neighboring objects (Moran, 1948). The relationship between spatial objects and their neighbors is called *spatial autocorrelation* (Cliff and Ord, 1973). Depending on the relative distribution of the spatial objects of interests and their neighbors, three basic classes of spatial autocorrelation can be identified (Figure 4.7).

1. positive autocorrelation, when spatial objects with a particular property group vary together
2. random autocorrelation, when spatial objects with a particular property show no patterns of clustering
3. negative autocorrelation, when spatial objects with a particular property are distributed evenly over a large geographic space

The various quantitative measures of spatial autocorrelation are fully explained in Section 10.3 of Chapter 10. Spatial autocorrelation can shed light on the spatial structure of uncertainty or the spatial patterns of error (Goodchild, 1995). Work by Congalton (1988) on Landsat Multi-Spectral Scanner (MSS) data from areas of varying spatial diversity showed that the impact of spatial autocorrelation could be felt as much as 30 pixels away. This represents more than 1.6 km in ground distance. Since spatial autocorrelation affects the assumption of sample independence, its magnitude should always be kept in mind when determining the sample size and sampling scheme in attribute accuracy tests.

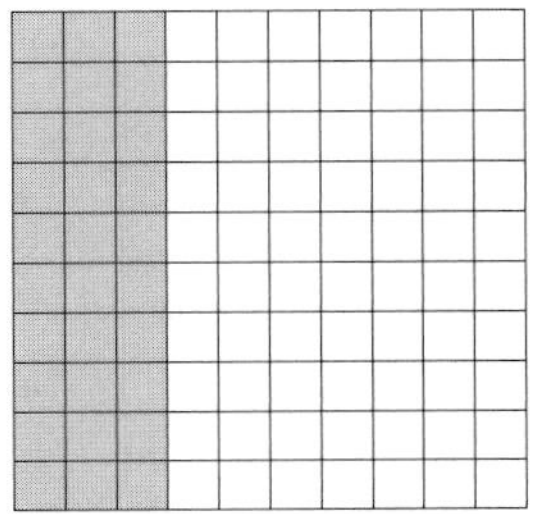

(a) Positive autocorrelation

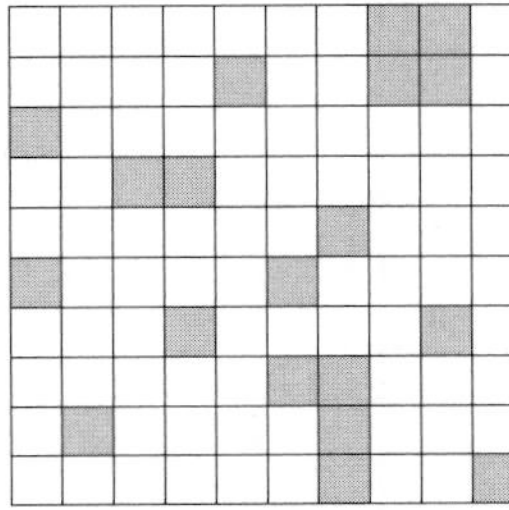

(b) Random autocorrelation

(c) Negative autocorrelation

FIGURE 4.7
The concept of spatial autocorrelation. When spatial objects have the same property group together as in (a), they are said to be positively autocorrelated; when the distribution of these spatial objects shows no sign of clustering as in (b), they are said to be randomly autocorrelated; and when these spatial objects are distributed evenly over space as in (c), they are said to be negatively autocorrelated.

Sample Size Sample size is a critical consideration in attribute accuracy tests. Due to the cost of collecting reference data, there is a tendency for sample size to be kept to the minimum. However, to ensure the statistical validity of the accuracy test, the sample must be of a sufficiently large size. The issue of sample size has been widely researched and debated, but there does not seem to be any agreement on the definition of an adequate sample size. As a general guideline for Landsat MSS images, Congalton (1991) suggested that a minimum of 50 samples be collected for each vegetation or land-use category in the error matrix. This number may be adjusted based on the relative importance of the different categories of the data or by the inherent variability within each of the categories.

Geographic Sampling Scheme While sample size is concerned with the number of sample points for an accuracy test, the geographic sampling scheme is concerned with the spatial distribution of the sample points. The objective of designing the geographic sampling scheme is to avoid spatial bias that may be introduced into the error matrix. Just like sample size, sampling scheme has also been widely researched and debated. Opinions vary greatly among researchers and include everything from simple random sampling to relatively elaborated stratified systematic unaligned sampling (Figure 4.8) (Congalton, 1991; Haggett, 1970).

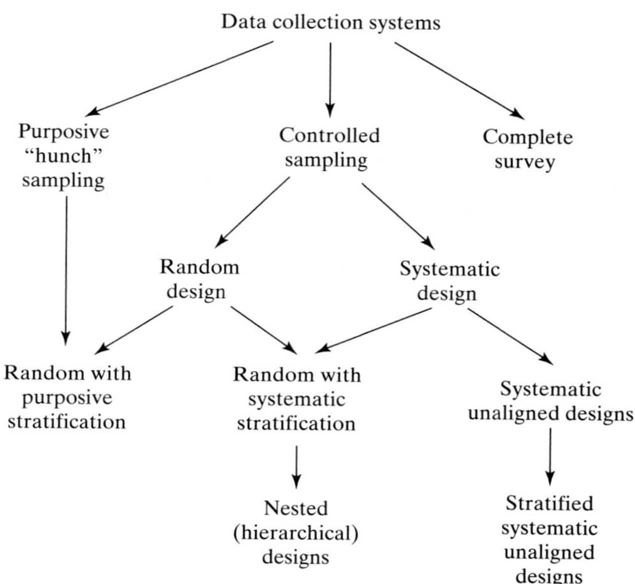

FIGURE 4.8
Classification of sampling methods. *(Source: Haggett, 1970)*

Of the many sampling schemes that have been proposed, four have been particularly common. A *simple random sample* is one in which each point is chosen randomly, each point having an equal chance of being chosen (Figure 4.9a). A *systematic sample* randomly selects an initial point first and then uses a fixed interval to determine the other points (Figure 4.9b). A *stratified sample* subdivides the study area into strata, and within each stratum, sample points are usually selected randomly (Figure 4.9c). Berry and Baker (1968) introduced a *stratified systematic unaligned sample* in which point *A* is first selected at random, and the *x* coordinate of *A* is used with a new random *y* coordinate to locate *B*. Using

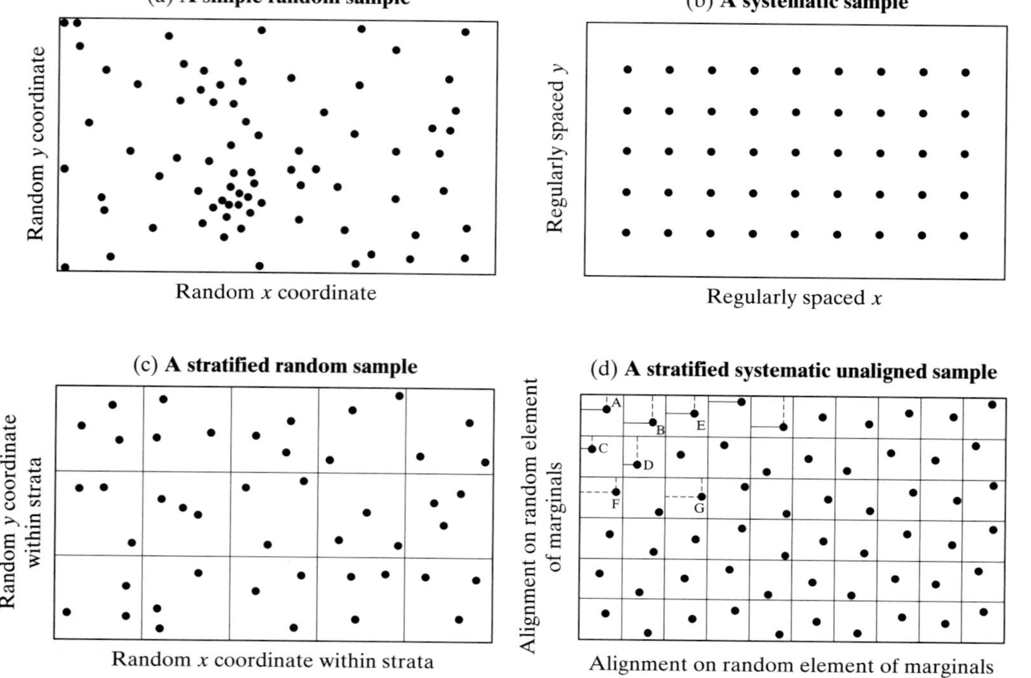

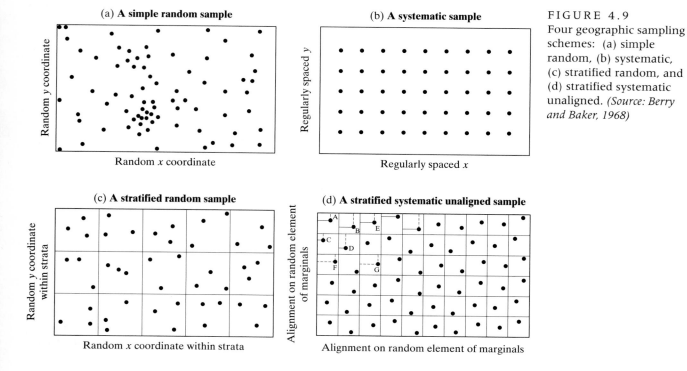

FIGURE 4.9
Four geographic sampling schemes: (a) simple random, (b) systematic, (c) stratified random, and (d) stratified systematic unaligned. *(Source: Berry and Baker, 1968)*

a second random y coordinate, E is located, and so on across the top row of strata (Figure 4.9d). Similarly, the y coordinate of A is used in combination with the random x coordinates to locate point C and all successive points in the first column of strata. The random x coordinate of C and y coordinate of B are then used to locate D, of E, and F to locate G, and so on until all strata have sample elements. It has been claimed that the resulting sample possesses the advantages of random, systematic, and stratified sampling.

Each of these sampling schemes has its merits and drawbacks. Simple random sampling is statistically acceptable but sometimes difficult to apply in practice. Random sampling tends to underrepresent small but important areas unless the sample size is large. Sampling points may fall in locations that are difficult to access. Systematic sampling is easy to execute but suffers from possible periodic changes or variations of data, also known as the *periodicity* of data. Stratified random sampling appears to be most suitable where a minimum number of samples are selected from each category. However, this particular method also suffers from the general limitation of random sampling as noted earlier. This explains why the stratified systematic unaligned sampling has been a popular geographic sampling scheme employed to evaluate the thematic accuracy of maps.

Research undertaken by Congalton (1988b) revealed the importance of spatial complexity in affecting the choice of geographic sampling scheme based on an evaluation of agriculture (spatially least complex), rangeland, and forest (spatially most complex) areas. The conclusion was that simple random sampling and stratified random sampling performed well in all situations, while systematic sampling and stratified systematic unaligned sampling were affected by spatial complexity and should be used with care. Spatially complex patterns were found to exhibit a strong positive spatial autocorrelation between errors (Congalton, 1988a), suggesting periodicity in the geographic data. Stehman (1992) confirmed that systematic sampling and stratified systematic unaligned sampling gave more precise estimates of kappa coefficient than simple random sampling if periodicity of data was absent.

4.5 Managing Spatial Data Errors

From the previous discussion, it is clear that the presence of errors is a norm rather than an exception in geographic data. GIS users have generally come to the realization that the objective of handling errors is not to eliminate them, but rather to manage them so that the data can be used with the minimum amount of uncertainty. The management of spatial data errors can be effected from three perspectives: *data production*, by developing methods to assure and control the quality of data during the data acquisition process; *data use*, by understanding the behavior of errors when the data are used; and *communication between data producer and data user*, by evaluating and documenting the quality of the data so that users are fully aware of the level of uncertainty in the data.

4.5.1 Quality Assurance and Quality Control of Geographic Data

Quality assurance and *quality control*, commonly referred to as QA/QC, refer to the aggregate of activities that aims at ensuring that the end product of a production process meets anticipated quality requirements. Conventionally, very stringent QA/QC procedures are applied to primary geographic data collection by geodesy, land surveying, and photogrammetry. All the methods of field or laboratory observations and computation are designed in such a way that gross errors may be easily detected, usually by compulsory independent checks. These methods also force systematic errors to be corrected procedurally (e.g., instrumental calibration) or mathematically (e.g., applying corrections for temperature change), leaving random errors to be treated by stochastic models such as least-squares adjustment. Since such a preventive approach to error management is always strictly applied in surveying and detailed mapping, it is generally safe to assume that geographic data obtained by the primary data collection methods are of a quality that is compliant with accepted mapping specifications and standards.

Most of the geographic data used today are converted from existing maps by different methods of digitizing. Experience has shown that these methods of secondary data collection are by far the greatest source of errors in digital geographic data. Since the best way to minimize the occurrence of errors is to contain them at the source, trying to understand the map digitizing process as a source of error is therefore the most logical first step in error management. Burrough (1986) presented an assessment of errors associated with table digitizing. He showed that the accuracy of digitizing is influenced by two factors: the width of the line and the scale of the source map (this affects the accuracy of positioning the cursor of the digitizer, which in turn affects the accuracy of the coordinates collected); and the number of points collected while digitizing curves (this is usually referred to as *line following inaccuracy*). More detailed analysis of digitizing errors by Walsby (1995) indicated that errors generated during manual digitizing are acceptable according to conventional mapping standards. Such a finding was confirmed by another study by Bolstad et al. (1990).

Walsby also found that digitizing errors are influenced by line characteristics and intolerable errors are always operator-specific. On the basis of these findings, Walsby suggested that the most practical way to rectify the problem is to provide operators with proper training.

Advances in image scanning technology have led to increasing use of what has now come to be known as "heads-up" digitizing. Instead of using a table digitizer, this approach uses scanned images of the source maps (see Chapter 6). The images are vectorized either manually or, more commonly, with the aid of computer-assisted or fully automated techniques. Experience has shown that this approach has not always been able to produce high-quality results (Flanagan et al., 1994). However, Mayo (1994) showed that by using a combination of operator-based and computer-assisted techniques, it was possible to improve map digitizing efficiency, both in terms of increased capture speed and reduced error rate. Relatively simple techniques may increase effective point placement rates by around 50% compared to conventional methods.

Another important source of error during secondary data collection comes from coordinate transformation. Coordinate transformation is required because data from different sources may be georeferenced to coordinate systems based on different datums and map projections (see Chapters 2 and 6). Errors resulting from mathematical transformation between different map projections and coordinate systems are normally tolerable according to conventional mapping standards. More problematic is the transformation between data on different datums, which in North America often means the transformation between NAD 27 and NAD 83 (see Chapter 6). Empirical investigations have shown that positions in UTM coordinates can "shift" up to 300 meters when transformed from NAD 27 to NAD 83 (Snyder, 1987; Welch and Homsey, 1997). If ignored, the errors introduced by datum transformation can seriously affect the usefulness of the converted data.

The USGS (1989) has made available the *North American Datum of 1983 Map Data Conversion Table* to assist data users with their conversion computation. Special computer programs are also available to handle the necessary conversion computation (Welch and Homsey, 1997). Most GIS and image analysis software packages have built-in programs in their projection transformation functions that take into account the shift of coordinates due to the change of datum. However, for users who are not fully aware of this particular problem or those who do not have access to computer programs or GIS software products capable of handling the computation properly, errors due to the change of datum can be a serious problem.

4.5.2 ERROR PROPAGATION AND ERROR MANAGEMENT

Geographic analysis with GIS always involves the use of multiple data sets from different sources. If these data sets contain errors, the end product will contain errors that represent the cumulative effect of all the errors combined. The accumulation of the effect of errors during the process of geographic analysis is generally referred to as *error propagation*. Error propagation is easy to understand and to prove empirically (Newcomer and Szajgin, 1984; Walsh et al., 1987; see also Section 4.2.3). In Figure 4.10, for example, if the user's accuracy of the soil type (clay) in polygon A in Map 1 is 0.8, and that of the land cover type (scrub) in polygon B in Map 2 is 0.6, the probability of polygon C in Map 3 having clay soil and a scrubland cover is $0.8 \times 0.6 = 0.48$. This means that if we use Map 3 for decision making (for example, to determine the suitability of an area for cultivation), we will be working with an uncertainty

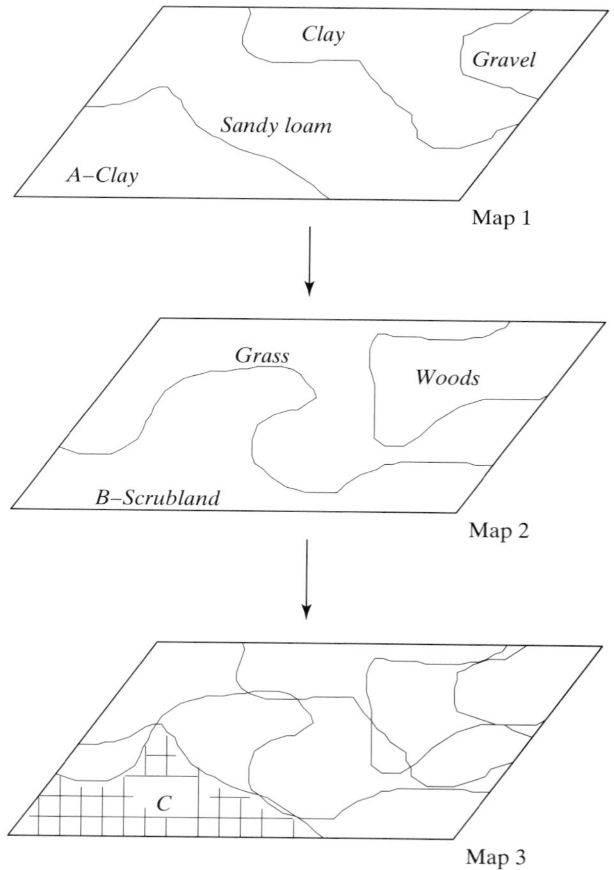

FIGURE 4.10
Error propagation in GIS. If the user's accuracy of polygon A in Map 1 is 0.8, and that of polygon B in Map 2 is 0.6, the probability of polygon C in Map 3 having clay soil and a scrubland cover is $0.8 \times 0.6 = 0.48$.

of about 50% at polygon C. If Map 3 is combined with another source of data for further analysis, the level of uncertainty will change again.

From the perspective of error management in GIS, it is more important to know *how* error propagation behaves during the process of geographic analysis rather than *why* error propagation occurs. The behavior of error propagation is a very intriguing problem in geographic data processing because it involves all three components of data, namely, position, attribute, and time. Therefore, error propagation is not a simple additive model of summing up the errors in individual data sets as in the example in Figure 4.10. Instead, it is a relatively complex stochastic model because of the following reasons.

1. The characteristics of the errors in different data sets are different, both in the levels of uncertainty and the spatial variability of occurrence.
2. Geographic analysis usually involves overlaying two or more layers of data, using operations such as intersect and union (Section 6.6.2), each of which has different influences on the ways that errors propagate (e.g., intersect increases uncertainty, while union reduces uncertainty).
3. The contributions of different data sets to the solution of the spatial problem may be different, which means that their uncertainties carry different weights.

Error propagation in geographic analysis is usually treated in terms of the spatial or attribute components in practice (Veregin, 1995). For some applications, it is possible to treat these two components as completely independent error sources. The attribute accuracy of the population estimates in the census, for example, is independent of the positional accuracy of census tract boundaries. For other applications such as soil maps and land-use maps, the assumption of independence does not hold because the spatial component (polygon boundaries) is derived directly from the attribute component (class values). Error propagation in these types of data therefore impacts on both the thematic and the spatial components of the end product of spatial analysis.

Propagation of positional errors results in the generation of spurious or sliver polygons that may probably degrade further analysis and interpretation of the data. The problem of the propagation of positional errors has appeared to be relatively well understood and well handled in GIS practically (Pullar, 1991; Zhang and Tulip, 1990). Many GIS software products are now capable of automatically eliminating sliver polygons smaller than a given threshold size. In contrast, the functionality for handling attribute error propagation is obviously lacking in commercial GIS, although there have been considerable efforts in developing models for propagation of error in attribute data (Burrough, 1986; Heuvelink and Burrough, 1993; Lanter and Veregin, 1992; Veregin, 1995). In Chapter 5, we will demonstrate how the concept of error propagation can be applied to cartographic modeling in practice.

4.5.3 SENSITIVITY ANALYSIS AND ITS APPLICATION

The complex and stochastic nature of modeling attribute error propagation is probably one of the reasons for the lack of its acceptance as a practical tool for improving spatial problem solving. A more pragmatic approach to managing errors in attribute data is to use *sensitivity analysis* to calibrate the results of geographic analysis (Foote and Huebner, 1995 and 1996). Since spatial data analysis is not an exact science, it is allowable for the user to perturb the input data in order to test how variation in data and modeling procedure influence a specific solution (Figure 4.11).

Sensitivity analysis is, therefore, a *dynamic modeling technique* for dealing with subjectivity and variability in the parameters of spatial problem-solving models.

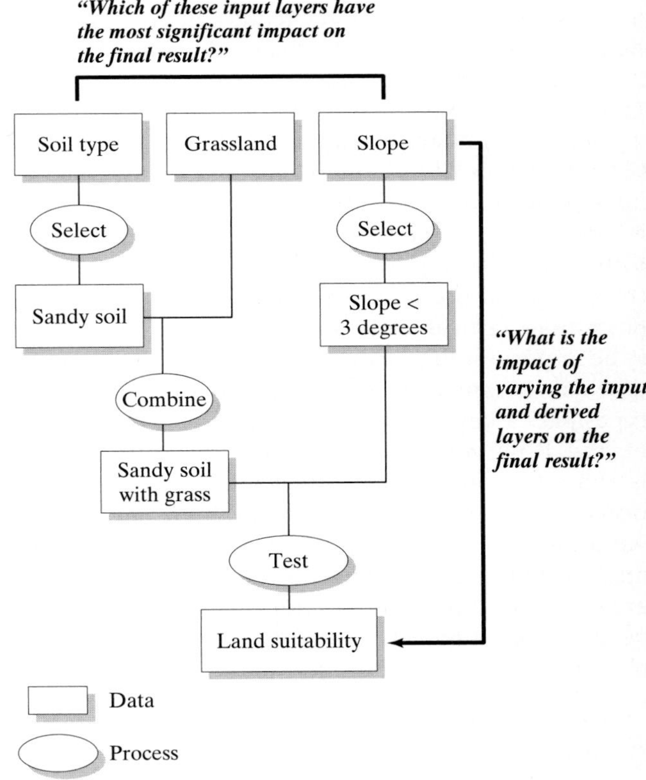

FIGURE 4.11
Sensitivity analysis in map overlay.

A sensitivity analysis assesses the variability of the modeling results in response to changes in parameter values. The range of parameter values used in a sensitivity analysis should represent a range of logical alternatives. In other words, the purpose of the sensitivity analysis is to test the model for output over a range of legitimate uncertainties. By changing the parametric input of the model, different yet equally valid scenarios are created. The capability of GIS to repetitively perform a task makes it an effective tool for this type of testing.

Sensitivity analysis has been proven to be a practical tool for managing errors in GIS. It allows useful applications of geographic data under uncertainty. The usefulness of sensitivity analysis can be understood from the following two perspectives: (1) systems design and development and (2) data analysis for spatial problem solving.

Using Sensitivity Analysis in Systems Design and Development During the systems design and development stage of a GIS application, the user can experiment with test data of differing levels of uncertainty to see how the system performs. By trial and error, the user will be able to establish the attribute accuracy requirements of the particular GIS application. Such a test is particularly useful for large projects in which an extensive quantity of data will be required. As noted in Section 4.2.2, high accuracy and high precision data are costly to collect. A good estimate of the accuracy requirements of data can help GIS users avoid the possibility of wasting valuable resources on acquiring data at a level of accuracy and precision that is not warranted for the application.

Sensitivity analysis may also be applied at the systems design and development stage to test the relative importance of the different parameters of a proposed GIS solution. In many applications, it is essential to distinguish the importance of a particular data set in theory and in practice. A particular data set that appears to be critical in the solution may not necessarily be important in practice. A good example is the soil type data in land capability modeling. In principle, soil types play a key role in determining the productivity of land. However, if the entire area under investigation is of a relatively uniform soil type, soil type data will not be an important factor in modeling the spatial variation of land capability in the study. A sensitivity analysis will enable the user to identify data sets that are of this nature. In the extreme case, the user may be able to eliminate an entire data set from the modeling process.

Using Sensitivity Analysis in Spatial Analysis and Problem Solving Sensitivity analysis has been successfully used as a practical spatial problem-solving tool. A typical example is its application in *composite mapping analysis* (CMA) (Lowry et al., 1995). Composite mapping analysis is a technique commonly used in environmental applications such as land-use suitability analysis and environmental impact analysis. A composite mapping analysis characterizes locations based on the spatial coincidence of relevant variables that affect an existing or proposed activity. Composite mapping analysis is based on spatial coincidence, so it is a particularly good candidate for GIS application. A recurring question associated with composite mapping analysis is how to weight the various factors that impact the activity under investigation. A typical composite mapping analysis has to make use of a myriad of hazard-related and human-related variables, which include sources of hazards, potential contaminant pathways, population density, and locations of sensitive sites such as schools and hospitals.

Conventionally, the relative weights of the factors are determined through consensus building in an expert panel. However, the availability of expertise may be limited in some cases, and consensus is often difficult to attain because of the large number of factors at stake. The use of sensitivity analysis in GIS is an obvious alternative. Using the ability of GIS to quickly generate different scenarios by varying weightings for different factors, composite mapping analysis using the sensitivity analysis approach is capable of providing the user with the necessary information for logical decision making. Success in the use of sensitivity analysis technique has been reported in different application areas, for example, by Stoms et al. (1992) and Dean et al. (1997) in wildlife habitat modeling and by Emmi and Horton (1997) in the assessment of seismic risk.

4.5.4 REPORTING DATA QUALITY

So far we have learned the methods of evaluating the quality of geographic data and explained how the indices of data accuracy can be used to enable users to perform spatial data analysis using uncertain data. However, all the efforts in data evaluation will be meaningless unless the information of data quality can be effectively communicated to all potential users. Therefore, data quality reporting is a very important component of error management in GIS.

Conventional methods of reporting data quality in the form of accuracy indices are useful for presenting the general uncertainty of individual data sets to the user. Such information is not able to satisfy users who have to rely on their own knowledge of the spatial variability of uncertainty for decision-making purposes. This has led to a growing interest in the visualization of uncertainty among GIS users, particularly those in the academic sector (Beard and Buttenfield, 1991; Davis and Keller, 1997). One approach is to generate an error map (Figure 4.4) as explained in Section 4.4.1. This approach is useful for the presentation of spatial variation of positional uncertainty. Another approach,

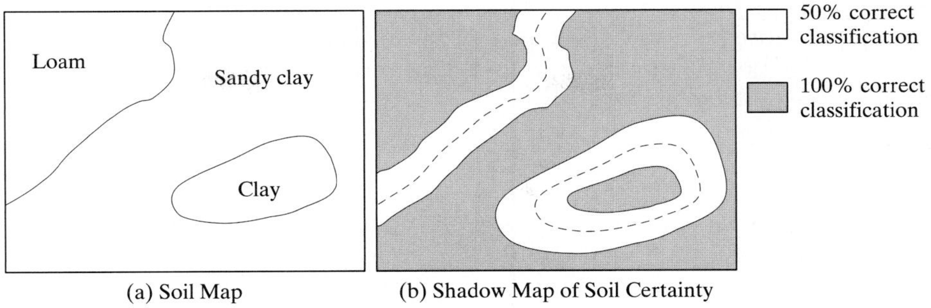

(a) Soil Map (b) Shadow Map of Soil Certainty

FIGURE 4.12
Using a shadow map to show certainty of data. When the soil map in (a) is displayed together with its shadow map of certainty in (b), the user will have an immediate impression of how accurate the data are. Areas around the boundaries of the soil types are transitional zones and thus have a lower degree of certainty than those that are away from the boundaries.

which is more appropriate for attribute uncertainty, is to produce a *shadow map of certainty* that displays the estimated degrees of uncertainties associated with individual features (Figure 4.12) (Berry, 1995). Such a map is difficult to produce and use manually. Hansen (1999) showed that a digital map of certainty can be easily generated in GIS by varying the symbols according to accuracy values assigned to the data. Research in visualization of uncertainty has gradually extended this static map approach to the use of *dynamic cartography*, which allows the user to interactively toggle between the actual data and uncertainty information (Davis and Keller, 1997; Wittenbrink et al., 1996). The concepts and techniques for the visualization of uncertainty will be explained in detail in Chapter 7.

4.6 Geographic Data Standards

As noted, data quality is a rather abstract construct that is sometimes hard to comprehend. It is also subjective because different people tend to use different criteria to determine whether or not a particular data set is of an acceptable quality. Standards provide the yardstick against which quality can be evaluated, quantified, and documented. In the discussion that follows, we introduce the definition and methods of data standards, with special reference to development and application of standards in geographic data processing.

4.6.1 The Definition and Role of Standards

A "standard" is defined by the *International Standards Organization* (ISO) as "a document, established by consensus and approved by a recognized body, that provides, for common and repeated use, rules, guidelines, or characteristics for activities or their results, aimed at the achievement of the optimum degree of order in a given context" (OMNR, 1994). It is also defined by the *American Society for Testing and Materials* (ASTM) as "a physical reference used as a basis for comparison or calibration, and a concept that has been established by authority, custom, or agreement to serve as a model or rule in the measurement of quality or the establishment of a practice or procedure." Simply put, a standard is a physical object or a document that is established by government or authorized standard development organizations, through popular use or by general agreement as a model or example for practice in a specific profession or for procedures to complete a certain task.

Standards play a very important role in our daily lives: the clothes that we wear, the food that we eat, the cars that we drive, the equipment that we use, and the places where we live. In fact, standards are so commonplace in almost every aspect of our daily lives that people simply take them for granted. Usually, people are not aware of the existence and importance of standards unless they experience the inconvenience resulting from the absence or deficiency of standards. This implies that the benefits of standards do not come directly from the standards *per se*, but from what they support. Many activities and tasks in our daily lives would be simply impossible or very difficult to carry out without standards.

Standardization is an indispensable component of the information system development strategy of any organization. Without standards, users would tend to go their own way in systems operation, data management, and application development. Sooner or later this would result in the proliferation of many small systems that might be very different from and incompatible with one another. Such a scenario will make the use of computer systems extremely difficult and expensive. The variations in hardware architecture, for example, will not allow cost and resource sharing in systems maintenance and upgrade. It will also lead to duplication of efforts in application development because software developed on one computer may not be portable to other computers. What is perhaps more problematic is the inability to integrate data within the organization and to share data among different applications residing on different computers. This would defeat the whole purpose of using information technology in the enterprise computing environment. Therefore, the compliance with standards must be perceived as one of the most important considerations in the use of information technology, including GIS.

4.6.2 THE CLASSIFICATION OF STANDARDS IN GIS

Standards may be classified according to the ways by which they are developed, and the legal weight that they carry, into three categories: *de facto standards*, also always referred to as *industry standards*, which are standards that gain wide acceptance through popular use rather than through a formal standard development process (see Section 4.7.2); *de jure standards*, which are developed by an organization empowered to create standards, such as the American National Standards Institute (ANSI), the Canadian General Standards Board (CGSB), and the British Standards Institute (BSI); and *regulatory standards*, which are usually, but not always, de jure standards that have been established by government legislation and, therefore, have certain legal power behind their enforcement.

Within a particular discipline or branch of technology, standards may also be classified according to their functions. In GIS, for example, at least four categories of standards may be functionally identified (Ventura, 1993): (1) *application standards*, (2) *data standards*, (3) *technology standards*, and (4) *professional practice standards*. These standards may have status as de facto, de jure, or regulatory standards, as exemplified in the following explanation and in Table 4.6.

Table 4.6
Components and Types of GIS Standards

Standards	*De Facto*	*De Jure*	*Regulatory*
Application	URISA ACSM/ALTA ASPRS Cartographic symbols	Engineering and topographic surveying Natural resource classification codes	Environmental impact assessment Land-use planning Data protected under freedom of information legislation
Data	Vector data structure Raster data compression Attribute data storage Data transfer Socioeconomic data collection and analysis	Geodetic control Data transfer Spatial data sets Census data collection and management	Cadastre Land title and registration
Technology	Hardware Software including operating systems, graphical interface and application development tools Communication and networks	Hardware DBMS protocols Programming languages Communication and networks FIPS	
Professional practices	Application development methodology including software test	ISO standards Certification of surveyors, planners, engineers, and data processing professionals	Licensing of cadastral surveyors and professional engineers

Application Standards These are standards that provide guidelines for the use of GIS. Application standards may come from specific legislative requirements (e.g., environmental impact assessment, land-use planning, and freedom of information legislation), or from government agencies that have a substantial stake in an application area (e.g., *National Map Accuracy Standard* of the USGS noted in Section 4.3.2). Sometimes, application standards are established by professional bodies, such as the American Society of Civil Engineering (ASCE, 1983), the American Congress on Surveying and Mapping, the American Land Title Association (ACSM/ALTA, 1988), American Society of Photogrammetry and Remote Sensing (ASPRS, 1989), the International Association of Assessing Officers, the Urban and Regional Information Association (IAAO/URISA, 1993), and the Canadian Council on Geomatics (CCSM, 1984). In addition, map projections, cartographic symbology, species codes for natural resource inventories, and specifications for engineering and topographic surveying are also examples of application standards. Application standards may be established as de facto, de jure, or regulatory standards as the case may be in different jurisdictions.

Data Standards These are standards that describe the organization and exchange format of geographic data, as well as the documentation of metadata pertaining to them. Data standards may come from government agencies such as cadastral surveys, land title and land registration, maintenance and accuracy for geodetic control, data transfer formats, collection and processing of census and other forms of socioeconomic data. Very often data standards are sponsored by international and national standards development organizations (e.g., ISO data transfer standards) or are established by popular use in the industry (e.g., many of the vector and raster data file formats discussed in Chapter 3). Of the four standards in GIS, data standards are probably the most important and the most diverse. Data standards will be the focus of our discussion in the later sections of this chapter.

Technology Standards Technology standards include standards and protocols for the various aspects of the use of computer technology in GIS, including computer hardware components, software (e.g., operating systems, programming languages, graphical interfaces, and DBMS), as well as communication and telecommunication protocols. The common goal of all technology standards is to ensure compatibility among computer systems. Many technology standards are industry standards, or de jure standards that had their origins as industry standards, for example, operating systems such as DOS and UNIX, Microsoft Windows graphical interface, as well as printing and plotting formats. There are also technology standards established by government agencies, such as the U.S. *Federal Information Processing Standards* (*FIPS*), and by standard development organizations, such as ANSI, ISO, BSI, and CGSB noted previously.

In GIS, the software component has conventionally been largely proprietary, but the software and application tools used are always based on industry or de jure technology standards, for example, programming languages such as FORTRAN, C++, and Visual Basic for application development, and Structured Query Language (SQL) for database query. There is now a movement toward standard-compliant GIS software, based on the concept of *interoperability* or cross-platform application of GIS (Buehler and McKee, 1996) (see Section 4.7.1).

Professional Practice Standards These standards cover the certification of professional competency and licensing of GIS practitioners, as well as the accreditation of academic programs in GIS, as explained in Chapter 1. Professional practice standards also include standards for application development methodology and quality management in business and industry, such as *ISO 9000* developed by ISO. ISO 9000 is not a product standard; it does not certify the quality of goods directly, but a management standard that is implemented to help corporations ensure consistency in the quality of their services. In many jurisdictions, GIS software vendors and consultants are required to register as an ISO 9000–compliant company in order to tender for government contracts.

4.6.3 COMPONENTS OF GEOGRAPHIC DATA STANDARDS

Typically a geographic data standard is made up of one or more of the following four components: *standard data products, data transfer standard, data quality standard,* and *metadata standards*. These components are developed for different purposes and have different degrees of complexity. However, they all share the common function of providing a means of communication between suppliers and users of geographic data.

Standard Data Products Standard data products basically include data products resulting from the basic mapping programs at the national and state/provincial levels, usually at small or medium scales. The primary function of these data sets is to provide the georeferencing framework for developing GIS applications. Examples of standard data products in the United States include the following digital products of the USGS: *Digital Line Graphs* (*DLG*), *Digital Elevation Models* (*DEM*), *Digital Orthophoto Quadrangles* (*DOQ*), *Geographic Names Information System (GNIS)*, and *Digital Raster Graphics* (*DRG*). In Canada, the *National Topographic Database* (*NTDB*) of Geomatics Canada is the standard data product. Of particular interest to GIS users worldwide is *the Digital Chart of the World* (*DCW*) (Danko, 1990). This data product provides 1 : 1 million scale coverage of the world using a topologically structured vector representation. It was originally developed by the U.S. Defense Mapping Agency (DMA) but is now available for sale to the general public.

In the United States and Canada, digital map files produced for census purposes are increasingly used as the basis for GIS applications in many social science disciplines and business geographics. Notable examples are the *Topologically Integrated Geographic Encoding and Referencing* (*TIGER*) files of the U.S. Bureau of the Census and *Area Master Files* (*AMF*) of Statistics Canada. Many third-party data suppliers, such as ETAK, MapInfo, and Strategic Mapping Inc., have refined and packaged TIGER files for sale commercially. Because of the increasing popularity of these data in the market, these refined data products themselves may one day become de facto standards (Coleman and McLaughlin, 1992).

Data Transfer Standards Data transfer standards, which are also known as *data exchange* or *interchange* standards, were originally conceptualized as a way to assist in the translation of geographic data from one GIS to another. Geographic data used by GIS are always structured in proprietary data formats. Sharing of data between individual pairs of GIS is normally not a big problem because translators can be programmed to facilitate the exchange of data between the two systems. However, as the need to share data among many different systems arises, the magnitude of the problem multiplies very quickly. To share data among eight systems, for example, it is necessary for two translators to be programmed for each system to change data to and from another system. A total of 112 different data translators will be required (Figure 4.13a). If, however, a common transfer format is available to serve as a hub, the required number of translators will be reduced to 16. It is now necessary for only two translators to be programmed for each system to communicate with the transfer format. Once the data are in the transfer format, they can be effectively used by all the systems by means of their respective translators (Figure 4.13b).

Geographic data transfer standards may be developed as de facto standards as the result of the popular use of a particular software package. The most notable example is ArcView's shapefiles format. Many geographic data transfer standards used today are de jure standards developed by standards development organizations with the support of government agencies. Examples of geographic data standards developed in this way include *Spatial Data Transfer Standard* (*SDTS*) of the United States, the *Canadian Geomatics Interchange Standard* (*CGIS*) of Canada, and the *National Transfer Standard* (*NTS*) of the United Kingdom.

Although the concept of geographic data transfer standards is relatively simple, the development and implementation of the standards may be a very complicated and lengthy process. Geographic data are represented by a wide variety of data models. There are different, and often conflicting, application requirements and system architecture among different GIS. This means that data transfer standards are by necessity complex and voluminous. In general, there are two approaches to geographic data transfer standards: the *product-oriented approach* and the *modeling approach*. The product-oriented approach leads to a "defined" or "restrictive" data transfer standard that specifies an explicit format for the exchange of geographic data (i.e., the product). Such a standard is labeled as "defined" because it has a tightly defined data structure, feature and attribute coding catalog, and content specifications that ensure ease of direct use of the data. It is labeled as "restrictive" because it can be used only for data that are structured in the way specified by the standard. This type of transfer standard is relatively easy to develop and implement but not flexible to use. The shapefiles format noted previously is an example of a product-oriented standard.

In contrast, a modeling standard is a "generic" or "permissive" standard that allows any data set to be interchanged without altering the basic organization of the data involved. It is more flexible than a product-oriented standard because, instead of describing a specific data structure, it relies on a *profile* to describe a set of data structures that will receive the data. Geographic data transfer standards of this type are much more flexible than the product-oriented standards. However, they are also much more complex to develop and to implement. Most of the international and national geographic data transfer standards to be discussed in Sections 4.6.4, 4.6.5 and 4.6.6 belong to this class of transfer standards.

Data Quality Standards A data quality standard is a document that lists the requirements of data quality for specific applications or for specific spatial scales. Like data transfer standards, data quality standards have been developed as de facto standards and as de jure standards. De facto data quality standards are usually related to a particular profession and developed by the professional bodies concerned. A typical example is the *ASPRS Accuracy Standards for Large-scale Maps* noted in Section 4.3.2. De jure data quality standards are developed by government agencies or standards development organizations as a guideline for geographic data. The *National Map Accuracy Standard* is a typical example of de jure data quality standards. Data quality standards

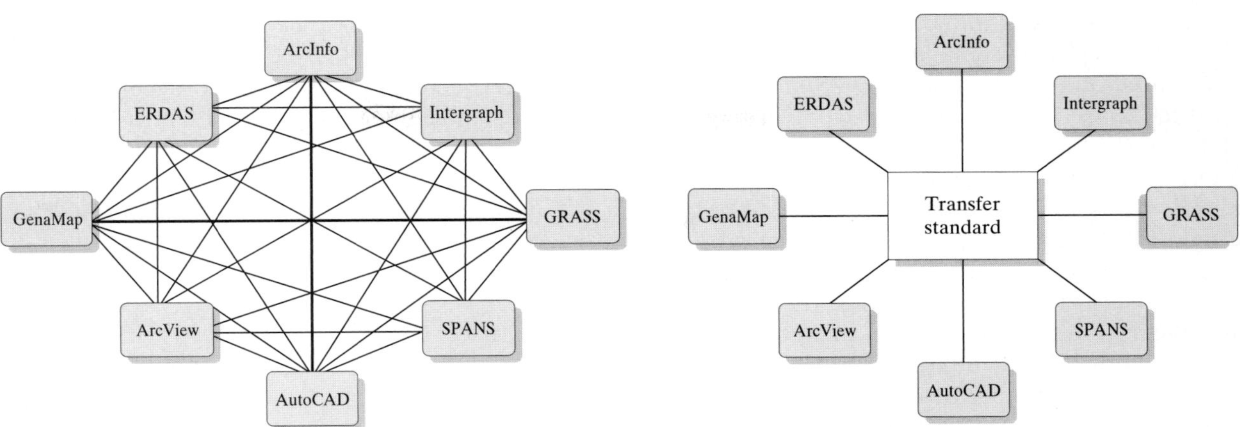

(a) Data transfer without a common transfer format (112 conversion programs required)

(b) Data transfer with a common transfer format as a hub (16 conversion programs required)

FIGURE 4.13
The concept of a data transfer standard. Using a data transfer standard, it is necessary for each software system to write translators to and from the transfer standard only, thus reducing the number of format translators between different systems.

can be developed as stand-alone documents. In many cases, however, data quality standards have been incorporated as a part of data transfer standards. The SDTS, for example, describes several elements as part of its data quality report components, including spatial accuracy, attribute accuracy, logical consistency, completeness, and lineage.

Metadata Standards Metadata standards are also sometimes referred to as *data documentation standards*. Until relatively recently, there was no standard in the United States for describing metadata. This has greatly limited the function of metadata because the lack of standardization makes it very difficult to access geographic data in different organizations in a systematic and unified manner. As a result, many organizations developed standards for metadata. Standardization of metadata collection and storage is particularly important for the establishment of geographic information distribution networks such as the *National Geospatial Data Clearinghouse* of the U.S. Federal Geographic Data Committee (FGDC). The National Geospatial Data Clearinghouse is a collection of over 50 spatial data servers in federal and state government agencies, nonprofit organizations, and commercial corporations that have digital geographic data available for use by the public. These data collections can be searched through several interfaces based on their metadata. The successful implementation of a metadata standard is obviously a prerequisite for the success of projects of this nature.

Of all the four components of geographic data standards, metadata standards are probably the most difficult to develop and implement. The problem is that the standard must be able to describe all possible properties of geographic data, which includes the *who, what, when, where, why*, and *how* of every facet of the data that are being documented. The FGDC metadata standard, for example, has 344 different elements, 119 of which exist only to contain other elements. These compound elements are important because they describe the relationships among other elements. Because of the technical complexity of establishing metadata standards, together with the difficulty and cost in collecting and maintaining metadata, few jurisdictions now have a metadata standard in place.

4.6.4 INTERNATIONAL GEOGRAPHIC DATA STANDARDS

The International Standards Organization (ISO) is the internationally recognized organization for developing and sponsoring standards. The growing importance of geographic data and the increasing need to harmonize geographic data standards developed by different countries prompted ISO to establish a Technical Committee on Geographic Information and Geomatics, called TC211. This committee brought together the work of the *Digital Geographic Information Working Group* (*DGIWG*) and *International Hydrographic Organization* (*IHO*), and the *Comite Europeen de Normalization* (*CEN*). As a result, ISO plays a much more important role in the development of standards pertaining to geographic data than ever before.

The geographic data transfer standards sponsored by ISO include *ISO 6709: Standard Representation of Latitude, Longitude and Altitude for Geographic Point Locations*, which is used as standard for scientific notation for global positioning; *ISO 8211: Specification for a Data Descriptive File for Information Interchange*, which is the standard used to transfer data over connections conforming to ISO's Open Systems Interconnection (OSI) specifications; and *ISO 15046: Geographic Information*. Of these three standards, ISO 15046 is the most comprehensive and important geographic data standard. The document for this standard contains 20 parts that specify the environment within which the standardization of geographic information should take place, the fundamental principles that should apply, and the architectural framework required for standardization.

The work of the DGIWG noted earlier played a significant role in the development of internationally used geographic data standards. DGIWG was formed by countries of the North Atlantic Treaty Organization (NATO) (but it was not an entity of NATO). The purpose of this group was to encourage the development of geographic data standards for both military and civilian use. In 1991, DGIWG published the *Digital Geographic Information Exchange Standard* (*DIGEST*). This is a family of international standards that provides a uniform method for exchanging digital geographic information at medium to small scale. The data format of DIGEST allows for the interchange of cartographic text, and raster and vector data, including spatially intelligent spaghetti vector and topological vector data formats.

The DIGEST format resembles, at the conceptual level, the US SDTS. It uses ISO 8211 for the transmission layer and ANSI standards for the nomenclature of geographic objects. DIGEST provides a standard method for coding features and attributes, through the use of the Feature and Attribute Coding Catalogue (FACC). This is a dynamic document that facilitates data transfer without the use of a comprehensive data dictionary each time the data transfer occurs. DIGEST has now been adopted as the geographic data transfer standard for NATO. Canada is the custodian country responsible for maintaining this standard.

The move toward political and economic union in Europe has prompted the formation of a couple of standardization organizations relating to geographic data standards in the European Norms for Geographic Information. *CEN TC287* is the European Standardization Organization for Geographic Information. It has produced a suite of interrelated standards that cover the reference model, spatial schema, quality, metadata, transfer, position, query, and update, as well as geographic identifiers. *CEN TC278* is the European Standardization Organization for Road Transport. The geographic data standard sponsored by CEN TC278 is known as Geographic Data File (GDF). GDF provides a general data model compatible with the TC287 geometric model, a feature catalog for road-related features, an attribute catalog, a relationship catalog, a feature representation scheme, a quality description specification, a global catalog scheme, logical data structures, and media record specifications.

4.6.5 GEOGRAPHIC DATA STANDARDS IN THE UNITED STATES

Work on the development of a national geographic data transfer standard for the United States started in the 1980s with the National Committee for Digital Cartographic Data Standards (NCDCDS). In the late 1980s, the work of NCDCDS was merged with that of the Federal Interagency Coordinating Committee on Digital Cartography (the predecessor of the present Federal Geographic Data Committee [FGDC]) by a task force drawn from members of both groups. This resulted in the *Spatial Data Transfer Standard* (*SDTS*). In 1992, the National Institute of Standards and Technology (NIST) approved SDTS as a FIPS standard and designated the USGS as the maintenance agency. Since then, SDTS has become the backbone of geographic data transfer in the United States (Morrison and Wortman, 1992).

The specifications of SDTS are organized into the basic specification (Parts 1 to 3) and multiple profiles (Parts 4 to 6) (Table 4.7). The basic specification describes the underlying conceptual model and the detailed specifications for the content, structure, and format for exchange of spatial data. The multiple profiles, on the other hand, define specific rules and formats for applying SDTS for the exchange of a particular type of data.

The USGS has taken the lead in using the SDTS by converting its digital cartographic holdings to SDTS format. Mass conversion started with DLG data. The raster profile is complete, and all USGS data holdings that conform to this profile are being converted, including DEM and DOQ. Compliance with SDTS is mandatory for federal agencies. The U.S. Army Corps of Engineers, for example, has listed SDTS as a mandatory standard in its Engineer Circular Policies, Guidance and Requirements for Geospatial Data and Systems. The Bureau of the Census has also provided a prototype version of the 2000 TIGER data in SDTS format. SDTS is available for use by state and local governments, the private sector, and the research and academic community. GIS vendors and third-party suppliers have started to provide translators for data transfer with SDTS. For example, translators are now available for ArcInfo, Intergraph MGE, and ERDAS Imagine.

Table 4.7
Summary of Specifications of SDTS

Part	*Specification*	*Functions*
1	Logical specifications	To describe the underlying conceptual model and the detailed specifications for the content, structure, and format for exchange of geographic data
2	Spatial features	To explain the SDTS conceptual model and SDTS spatial object types, components of a data quality report, and the layout of SDTS modules that contain all necessary information for a spatial data transfer compliant with SDTS
3	ISO 8211 encoding	To explain the use of a general-purpose file exchange standard, ISO 8211, to create SDTS transfers
4	Topological vector profile	This is the first of a potential series of SDTS profiles, each of which defines how the STDS base specification (in parts 1 to 3) must be implemented for a particular type of data.
5	Raster profile with BIIF extension	This profile is for two-dimensional image and gridded raster data. It permits alternate image file format using the ISO Basic Image Interchange Format (BIIF)
6	Point profile	This profile contains specifications for use with geographic point data only, with the option to carry high-precision coordinates such as those required for geodetic network control points.

Source: USGS, 1997b

4.6.6 GEOGRAPHIC DATA STANDARDS IN CANADA

Canada has made significant progress and contributions to the development of a geographic data transfer standard. Internationally, Canada is the custodian country of the DIGEST as noted earlier. In this capacity, the Canadian Department of National Defense (DND) is responsible for the continued development of this particular standard. The Canadian DND realized that the acceptance of DIGEST will not be dependent only on the standard, but on other factors such as the stability of the standard, the availability of a vast amount of data encoded in the format, and a cheap and easy-to-use toolbox to exploit the standardized data (Coleman, 1994). As a result, DND started two initiatives on DIGEST: the development of the DIGEST Multidimensional Extensions, which will enable DIGEST to support multidimensional data using the HHCode of Oracle, as well as to provide new mechanisms to handle topology and develop a more versatile visualization tool for the new data sets; and DIGESTView software, which represents an initial effort to develop a software tool kit that makes use of standardized data sets across multiple hardware platforms in a consistent application development.

Another Canadian standard that has considerable impact internationally is the *Spatial Archive and Interchange Format* (*SAIF*). This standard was originally developed by the British Columbia Ministry of Environment, Lands and Parks, but is now widely used throughout Canada and elsewhere in North America. SAIF is designed to facilitate interoperability, particularly in the context of data exchange. It provides an efficient means of archiving data in a vendor-neutral format by using a file format called SAIF/ZIP. Data can be translated into or out of a number of commercial GIS using SAIF/ZIP. This vendor-neutral format can also be used as the hub for interproduct translation. SAIF translation is carried out using commercial software called the Feature Manipulation Engine (FME). A noncommercial version of FME, known as FMEBC and owned by the British Columbia government, is available without charge.

The Canadian standards for geographic data as accepted by the Canadian General Standards Board are made up of a suite of four standards: *Geomatics Data Set Cataloguing Rules*, developed jointly by the Canadian General Standards Board Committee on Geomatics (CGSB-COG) and the Canadian Library Association; *Standard on Metadata for Geomatics Data; Canadian Geomatics Interchange Standard-DIGEST* (*CGIS-DIGEST*); and *Canadian Geomatics Interchange Standard-SAIF* (*CGIS-SAIF*). The first two standards are metadata standards, and the last two are interchange standards. CGIS-DIGEST is a product-oriented or defined standard (see Section 4.7.1) and is equivalent to the international DIGEST standard noted previously. In contrast, CGIS-SAIF is a modeling or generic standard (see Section 4.6.3) that is based on the SAIF standard noted here. The two data transfer standards are made complementary by incorporating a SAIF profile in DIGEST. The two standards will also be served by a common approach toward the development of a Canadian Feature Classification Catalogue based on the DIGEST Feature and Access Coding Catalogue (FACC).

4.7 Geographic Data Standards and GIS Development

One of the major impediments to the use of GIS in the past was the proprietary nature of the systems. The development and acceptance of data standards has eventually broken the barriers that stopped individual systems from communicating with one another. The ability to share data created the condition for the development of what is known as interoperable, or open, GIS. Interoperability is the fundamental concept of enterprise computing. It allows a user or an application to access a variety of data and systems resources residing in different systems by means of a single operational interface. In this regard, the establishment of data standards not only makes geographic data more accessible and available to users, but also has significant impact on the development and use of GIS as well.

4.7.1 Interoperability of GIS

Interoperability, which is one of the major cornerstones of mainstream information technology, has fundamentally changed the course of GIS development in recent years. Conventionally, GIS software packages were monolithic applications characterized by a highly structured and architecturally closed operating environment. They tightly coupled graphical display with proprietary spatial database design and did not allow exchange of data except by means of tape import or batch transfer over the network. In order to take advantage of the distributed computing architecture that allows enterprise-wide and worldwide resource sharing, many GIS software vendors are quick to adopt the principle of interoperability in the design and operation of GIS software products.

Although the advantages of the interoperable approach are clear and the objectives of interoperability are well defined, the implementation of interoperability is by no means an easy task. Interoperability is dependent on the universal acceptance and adoption of standards. However, standardization as a rule is a long and tedious process that is often at odds with the needs of individual users. The adherence to a standard always implies some compromise on the part of many users, which may include

1. lower performance because of the need to accommodate technologies and data models that may not necessarily be the most optimal or best performance in many cases
2. more limited functionality as the result of using generalized rather than specialized technology and methodology
3. reduced security that may occur as the result of using open technology
4. inconvenience and cost that may occur at the initial stage of adopting standards

The formation of the Open GIS Consortium (OGIS) represents the concerted effort of the GIS industry to address and overcome the problems pertaining to interoperability. This consortium is a membership organization, currently based in the United States, that is made up of vendors in different sectors of the GIS industry, with support from government and the academic sectors. OGIS promotes interoperability by publishing product specifications and by certifying GIS software products that conform to these specifications (Buehler and McKee, 1996). The first Open GIS Specification addresses "simple features," which refer to the most common vector-based software expressions of geographic features, including simple geometry (e.g., points, lines, and polygons), spatial referencing (e.g., map projections and coordinate systems) and attributes (e.g., zoning category and road types). Other specifications will include Open GIS Coverages and Open GIS Catalogs, among others (McKee, 1998).

The Open GIS specifications are not just data transfer standards, but have a much broader scope. Their goals are (1) to provide a single universal spatiotemporal data and process model that will cover all existing and potential spatiotemporal applications, (2) to provide a specification for each of the major database languages to implement the OGIS data model, and (3) to provide a specification for each of the major distributed computing environments to complement the OGIS process model. The objective of the Open GIS Consortium, therefore, is to promote interoperability through the use of data and application standards. This implies that data standardization is the foundation of, as well as the most important force behind, the movement toward interoperable GIS.

4.7.2 GEOGRAPHIC DATA STANDARDS AS A COMPONENT OF GEOSPATIAL DATA INFRASTRUCTURE

As a result of the proliferation of GIS in all levels of business operation in the public and private sectors, geographic data standards are now developed as an essential component of the geospatial data infrastructure in many jurisdictions (see Chapter 1). Geospatial data infrastructure has its origin in the term "information infrastructure" that refers collectively to the various media and physical infrastructures used for information delivery (Branscombe, 1982). The term is now used in a much broader context. The notion of infrastructure as an "enabling agent" has been widely adopted by the larger information processing community. When applied to geographic data, the concept of data infrastructure emphasizes the role of this particular type of data as an enabling agent for the essential activities and services affecting our lives in much the same way that highways, utilities, and other community facilities do in our daily lives. Obviously, geographic data standardization is no longer an issue that is of interest to only conventional GIS users. It is now high on the agenda of the economic development strategies of international organizations as well as national governments.

Geospatial data infrastructure at the international level encompasses the policies, technologies, standards, and human resources necessary for the effective collection, management, access, delivery, and utilization of geographic data in a global community (Coleman and McLaughlin, 1977). Such an infrastructure is required to support the requirements of geographic data for military mapping (for combat applications as well as for emergency relief assignments and peacekeeping duties), scientific and environmental research (e.g., global change, earth observation satellites) and navigation (e.g., hydrographic charts). At the national (and subnational) level, the same geospatial data infrastructure (i.e., policies, technologies, standards, and human resources) are required in support of the use of geographic data for applications in government administration, resource management, and environmental protection.

The establishment of the geospatial data infrastructure will make possible the incorporation of national, state/provincial, and local geographic databases into an integrated information highway that allows both the horizontal (institutional, economic, and environmental) and vertical (local, regional, and national) integration of geographic data (Coleman and McLaughlin, 1992). Such an infrastructure not only includes the geographic databases, but also the communication and telecommunication networks, and the institutional arrangement necessary for the effective flow and transfer of information. All these components of the infrastructure depend on the development and deployment of geographic data standards. In this regard, the importance of geographic data standards is felt far beyond the realm of data themselves. These standards represent one of the key components that form the foundation of today's information society.

4.8 SUMMARY

In this chapter, we explained the importance of data quality and data standards in the context of GIS. The issue of geographic data quality is more critical in GIS than in the conventional ways of using geographic data. The ease with which geographic data from different sources can be combined and analyzed digitally is very inducible to the indiscriminate use of the data without knowing and considering their quality. This will probably lead to abuse or misuse of geographic data for spatial problem solving or decision making. However, GIS users have now generally recognized that the acquisition of digital geographic data is only the first step in spatial data analysis and modeling. In order to use geographic data with the minimum amount of uncertainty, it is necessary to ascertain the quality of the data first.

As a result of the growing awareness of the importance of data quality in GIS, government agencies and professional bodies have established standards aiming to safeguard the appropriate use of geographic data in different application areas. The academic GIS community has also been directing considerable research effort to issues associated with geographic data quality and standards. This chapter explained the relationship between data quality and data standards. We also introduced the methods for quantitatively assessing the positional and attribute accuracy of geographic data in practice. It was noted that the cost and time incurred in data quality assessment have prevented many data suppliers from actually performing the necessary tests. As a result, it is usually up to GIS users to carry out their own tests using these techniques to ensure that the data are of a quality commensurate to intended applications.

As GIS becomes increasingly integrated with mainstream enterprise computing, geographic data quality and data standards are no longer the concern of the

GIS community alone. Government agencies and standard development organizations, both at the international and national levels, are now actively engaged in the sponsorship and development of geographic data standards. GIS users are therefore expected to be conversant with standards development and aware of the status of the standards in force in their respective jurisdictions and application areas. This allows them to comply with the appropriate standards in both the use and generation of geographic data. In this regard, data quality and data standards are an indispensable part of the study and implementation of GIS.

REFERENCES

Anderson, J. R., Hardy, E. E., Roach, J. T., and Witmer, R. E. (1976) *A Land Use and Land Cover Classification System for Use with Remote Sensor Data* (Professional Paper 964). Reston, VA: USGS.

ACSM/ALTA (American Congress on Surveying and Mapping and American Land Title Association) (1988) *Minimum Standard Detail Requirements for ALTA/ACSM Land Title Surveys*, ACSM.

ASCE (American Society of Civil Engineering) (1983) *Map Uses, Scales and Accuracies for Engineering and Associated Purposes*. New York: ASCE.

ASPRS (American Society of Photogrammetry and Remote Sensing) (1989) "ASPRS Accuracy Standards for Large-scale Maps," *Photogrammetric Engineering and Remote Sensing*, Vol. 55, No. 10, pp. 1068–1070.

Beard, M. K., and Buttenfield, B. P. eds. (1991) *NCGIA Research Initiative 7: Visualization of Spatial Data Quality* (NCGIA Technical Paper 91–26). Santa Barbara, CA: National Center for Geographic Information and Analysis.

Berry, B. J. L., and Baker, A. M. (1968) "Geographic Sampling," *in Spatial Analysis: A Reader in Statistical Geography*, by Berry, B. J. L., and Marble, D. F. eds. pp. 91–100. Englewood Cliffs, NJ: Prentice-Hall.

Berry, J. K. (1995) "The this, that, there rule: Creating a shadow map of certainty," Chapter 10, *Spatial Reasoning for Effective GIS*. Fort Collins, CO: GIS World.

Blakemore, M. (1990) "Progress Report: Cartography," *Progress in Human Geography*, Vol. 14, No. 1, pp. 101–111.

Bolstad, P. V., Gessler, P., and Lillesand, T. M. (1990) "Positional uncertainty in manually digitized map data," *International Journal of Geographical Information Systems*, Vol. 4, No. 4, pp. 399–421.

Branscombe, A. (1982) "Beyond deregulation: Designing the information infrastructure," *The Information Society Journal*, Vol. 1, No. 3, pp. 167–190.

Brassel, K., Bucher, F., Stephan, E-M., and Vckovski, A. (1995) "Completeness," in *Elements of Spatial Data Quality*, by Guptill, S. C., and Morrison, J. L. eds. pp. 81–108. Oxford: Elsevier Science Ltd.

Buehler, K., and McKee, L. eds. (1996) *The Open GIS Guide: Introduction to Interoperable Geoprocessing*. Wayland, MA: Open GIS Consortium.

Bureau of the Budget (1947) *National Map Accuracy Standards*. Washington, DC: U.S. Government Printing Office.

Burrough, P. A. (1986) *Principles of Geographical Information Systems for Land Resources Assessment*. Oxford: Oxford University Press.

CCSM (Canadian Council on Surveying and Mapping) (1984) "EPD standards applied to digital topographic data," *National Standards for Exchange of Digital Topographic Data*. Ottawa, ON: Technical Committee No. 3, CCSM.

Chrisman, N. R. (1983) "The role of quality information in the long-term functioning of a geographic information system," *Proceedings*, AutoCarto 6, Ottawa, ON, pp. 303–312.

Chrisman, N. R. (1991) "The error component in spatial data," in *Geographical Information Systems, Vol. 1: Principles*, by Maguire, D. J., Goodchild, M. F., and Rhind, D. W. eds. pp. 165–174. Harlow, UK: Longman Scientific and Technical.

Clarke, A. L. (1991) "Data quality reporting in the proposed Australian spatial transfer standard," in *Proceedings*, Symposium of Spatial Data Accuracy, pp. 252–255. Melbourne, Australia: University of Melbourne.

Clarke, D. G., and Clark, D. M. (1995) "Lineage," in *Elements of Spatial Data Quality*, by Guptill, S. C., and Morrison, J. L. eds. pp. 13–30. Oxford: Elsevier Science Ltd.

Cliff, A. D., and Ord, J. K. (1973) *Spatial Autocorrelation*. London: Pion.

Cohen, J. (1960) "A coefficient of agreement of nominal scales," *Educational and Psychological Measurement*, Vol. 20, No. 1, pp. 37–46.

Coleman, D. J. (1994) "The DIGEST spatial data standard: Getting on with the job," *GIS World Report/Canada*, Vol. 1, No. 4, pp. 12–13.

Coleman, D. J., and McLaughlin, J. D. (1992) "Standards for spatial information interchange: A management perspective," *CISM Journal*, Vol. 46, No. 2, pp. 133–142.

Coleman, D. J., and McLaughlin, J. D. (1997) "Defining global geospatial data infrastructure (GGDI): Components, stakeholders and interfaces," Theme Paper #1, International Seminar on Global Geospatial Data Infrastructure, Chapel Hill, NC.

Congalton, R. G. (1988a) "Using spatial autocorrelation analysis to explore errors in maps generated from remotely sensed data," *Photogrammetric Engineering and Remote Sensing*, Vol. 54, No. 5, pp. 587–592.

Congalton, R. G. (1988b) "A comparison of sampling schemes used in generated error matrices for assessing the accuracy of maps generated from remotely sensed data," *Photogrammetric Engineering and Remote Sensing*, Vol. 54, No. 5, pp. 593–600.

Congalton, R. G. (1991) "A review of assessing the accuracy of classifications of remotely sensed data," *Remote Sensing of Environment*, Vol. 37, pp. 35–46.

Congalton, R. G., and Green, K. (1993) "A practical look at the sources of confusion in error matrix generation," *Photogrammetric Engineering and Remote Sensing*, Vol. 59, No. 5, pp. 641–644.

Danko, D. M. (1990) "The Digital Chart of the World project," *Proceedings*, Vol. 1, pp. 392–401, GIS/LIS 90, Anaheim, CA.

Davis, T. J., and Keller, C. P. (1997) "Modelling and visualization multiple spatial uncertainties," *Computers and Geosciences*, Vol. 23, No. 4, pp. 397–408.

Dean, D. J., Wilson, K. R., and Flather, C. H. (1997) "Spatial error analysis of species richness for a Gap Analysis map," *Photogrammetric Engineering and Remote Sensing*, Vol. 63, No. 10, pp. 1211–1217.

Drummond, J. (1995) "Positional Accuracy," in *Elements of Spatial Data Quality*, by Guptill, S. C., and Morrison, J. L. eds. pp. 31–58, Oxford: Elsevier Science Ltd.

Emmi, P. C., and Horton, C. A. (1997) "A Monte Carlo simulation of error propagation in a GIS-based assessment of seismic risk," *International Journal of Geographical Information Systems*, Vol. 9, No. 4, pp. 447–461.

Fisher, P. F. (1991) "Spatial data sources and data problems," in *Geographical Information Systems, Vol. 1: Principles*, by Maguire, D. J., Goodchild, M. F., and Rhind, D. W. eds. pp. 175–189. Harlow, UK: Longman Scientific and Technical.

Flanagan, C., Jennings, C., and Flanagan, N. (1994) "Automated GIS data capture and conversion," in *Innovations in GIS 1* by Worboys, M. F. ed. pp. 25–38. London: Taylor & Francis.

Foody, G. M. (1992) "On the compensation for chance agreement in image classification accuracy assessment," *Photogrammetric Engineering and Remote Sensing*, Vol. 58, No. 10, pp. 1459–1460.

Foote, K. E., and Huebner, D. J. (1995) *Error, Accuracy and Precision*, The Geographer's Craft Project, Department of Geography, University of Texas at Austin. **http://www.utexas.edu/ftp/depts/grg/gcraft/notes/error/error.html**

Foote, K. E., and Huebner, D. J. (1996) *Managing Error*, The Geographer's Craft Project, Department of Geography, University of Texas at Austin. **http://www.utexas.edu/ftp/depts/grg/gcraft/notes/manerror/manerror.html**

Goodchild, M. F. (1978) "Statistical aspects of the polygon overlay problem," in *Harvard Papers on Geographic Information Systems*, No. 6. Reading, MA: Addison-Wesley.

Goodchild, M. F. (1979) "Effects of generalization in geographical data encoding," in *Map Data Processing*, by Freeman, H., and Pieroni, G. eds. pp. 191–206. New York: Academic Press.

Goodchild, M. F. (1989) "Modeling errors in objects and fields," in *Accuracy of Spatial Databases* by Goodchild, M. F., and Gopal, S. eds. pp. 107–113. London: Taylor & Francis.

Goodchild, M. F. (1995) "Attribute Accuracy," in *Elements of Spatial Data Quality*, by Guptill, S. C., and Morrison, J. L. eds. pp. 59–79. Oxford: Elsevier Science Ltd.

Goodchild, M. F., and Gopal, S. eds. (1989) *Accuracy of Spatial Databases*. London: Taylor & Francis.

Guptill, S. C. (1995) "Temporal Information," in *Elements of Spatial Data Quality*, by Guptill, S. C., and Morrison, J. L. eds. pp. 153–165. Oxford: Elsevier Science Ltd.

Haggett, P. (1970) "Scale components in geographical problems," in *Frontiers in Geographical Teaching*, by Chorley, R. J., and Haggett, P. eds. pp. 164–185. London: Methuen.

Hansen, D. T. (1999) "Using accuracy or uncertainty in the spatial characteristics of themes in ArcView GIS," *Proceedings*, ESRI Annual Users Conference, Paper No. 233. Redlands, CA: Environmental Systems Research Institute.

Hardin, P. J., and Shumway, J. M. (1997) "Statistical significance and normalized confusion matrices," *Photogrammetric Engineering and Remote Sensing*, Vol. 63, No. 6, pp. 735–740.

Henderson, J. P. (1989) "Digital geographic data quality: Two case studies," *New Zealand Surveyor*, Vol. 32, No. 274, pp. 375–386.

Heuvelink, G. B. M., and Burrough, P. A. (1993) "Error propagation in cartographic modelling using Boolean logic and continuous classification," *International Journal of Geographical Information Systems*, Vol. 7, No. 3, pp. 231–246.

IAAO/URISA (International Association of Assessing Officers/Urban and Regional Information Systems Association) (1993) *GIS Guidelines for Assessors*. Washington, DC: URISA.

Kainz, W. (1995) "Logical Consistency," in *Elements of Spatial Data Quality*, by Guptill, S. C., and Morrison, J. L. eds. pp. 109–137. Oxford: Elsevier Science Ltd.

Lanter, D. P., and Veregin, H. (1992) "A research paradigm for propagating error in layer-based GIS," *Photogrammetric Engineering and Remote Sensing*, Vol. 58, No. 6, pp. 825–833.

Laurini, R. (1993) "Sharing geographic information in distributed databases," *Proceedings*, 16th Urban Data Management Symposium, Vienna, Austria, pp. 26–41.

Lo, C. P. (1981) "Land use mapping of Hong Kong from Landsat images: an evaluation," *International Journal of Remote Sensing*, Vol. 2, No. 3, pp. 231–252.

Lowry, H. Jr., Miller, H. J., and Hepner, G. F. (1995) "A GIS-based sensitivity analysis of community vulnerability to hazardous contaminants on the Mexico/US border," *Photogrammetric Engineering and Remote Sensing*, Vol. 61, No. 11, pp. 1347–1359.

Lundin, B., Yan, J., and Parker, J.-P. (1989) "Data quality reporting methods for digital geographical products at Statistics Canada," *Proceedings*, pp. 236–251, National GIS Conference, Ottawa, ON.

Ma, Z., and Redmond, R. L. (1995) "Tau coefficients for accuracy assessment of classification of remote sensing data," *Photogrammetric Engineering and Remote Sensing*, Vol. 61, No. 4, pp. 435–439.

Mayo, T. (1994) "Computer-assisted tools for cartographic data capture," in *Innovations in GIS 1* by Worboys, M. F. ed. pp. 39–52. London: Taylor & Francis.

McKee, L. (1998) "What does Open GIS Specification Conformance mean?," *GIS World*, Vol. 11, No. 8, p. 38.

Moran, P. A. P. (1948) "The interpretation of statistical maps," *Journal of the Royal Statistical Society, B*, Vol. 10, pp. 243–251.

Morrison, J. L. (1995) "Spatial Data Quality," in *Elements of Spatial Data Quality*, by Guptill, S. C., and Morrison, J. L. eds. pp. 1–12. Oxford: Elsevier Science Ltd.

Morrison, J. L., and Wortman, K. (1992) *Special Issue: Implementing the Spatial Data Transfer Standard, Cartography and Geographic Information Systems*, Vol. 19, No. 5, December 1992.

Muller, J.-C. (1987) "The concept of errors in cartography," *Cartographica*, Vol. 24, No. 2, pp. 1–15.

Naesset, E. (1996) "Conditional tau coefficient for assessment of producer's accuracy of classified remotely sensed data," *ISPRS Journal of Photogrammetry and Remote Sensing*, Vol. 51, pp. 91–98.

NCDCDS (National Committee for Digital Cartographic Data Standards) (1988) "The proposed standards for digital cartographic data," *The American Cartographer (Special Issue)*, Vol. 15, No. 1, pp. 9–140.

Newcomer, J. A., and Szasgin, J. (1984) "Accumulation of thematic map error in digital overlay analysis,"*The American Cartographer*, Vol. 11, No. 1, pp. 58–62.

OMNR (Ontario Ministry of Natural Resources) (1994) *Data Standards in the Ontario Ministry of Natural Resources: Background and Context*. Toronto, ON: OMNR.

Openshaw, S. (1989) "Learning to live with errors in spatial databases," in *Accuracy of Spatial Databases* by Goodchild, M. F., and Gopal, S. eds. pp. 263–276. London: Taylor & Francis.

Pullar, D. (1991) "Spatial overlay with inexact numerical data," *Proceedings*, pp. 313–329, AutoCarto 10, Baltimore, MD.

Rosenfield, G. H., and Fitzpatrick-Lins, K. (1986) "A coefficient of agreement as a measure of thematic classification accuracy," *Photogrammetric Engineering and Remote Sensing*, Vol. 52, No. 2, pp. 223–227.

Salge, F. (1995) "Semantic Accuracy," in *Elements of Spatial Data Quality*, by Guptill, S. C., and Morrison, J. L. eds. pp. 139–151. Oxford: Elsevier Science Ltd.

Snyder, J. P. (1987) *Map Projections—A Working Manual* (U.S. Geological Survey Professional Paper 1395). Washington, DC: U.S. Government Printing Office.

Stehman, S. V. (1992) "Comparison of systematic and random sampling for estimating accuracy of maps generated from remotely sensed data," *Photogrammetric Engineering and Remote Sensing*, Vol. 58, No. 9, pp. 1343–1350.

Stoms, D. M., Davis, F. W., and Cogan, C. B. (1992) "Sensitivity of wildlife habitat models to uncertainties in GIS data," *Photogrammetric Engineering and Remote Sensing*, Vol. 58, No. 6, pp. 843–850.

Story, M., and Congalton, R. G. (1986) "Accuracy assessment: A user's perspective," *Photogrammetric Engineering and Remote Sensing*, Vol. 52, pp. 397–399.

Taylor, J. (1997) *An Introduction to Error Analysis: The Study of Uncertainties in Physical Measurements*. Sausalito, CA: University Science Books.

Thapa, K., and Bossler, J. (1992) "Accuracy of spatial data used in geographic information systems," *Photogrammetric Engineering and Remote Sensing*, Vol. 58, No. 6, pp. 835–841.

USGS (1988) *Digital Cartographic Data Standards, Procedure Manual* (Draft/RDR/9-88). Reston, VA: USGS.

USGS (1989) *North American Datum of 1983, Map Data Conversion Tables* (USGS Bulletin 1875 (3 volumes)). Washington, DC: U.S. Government Printing Office.

USGS (1997a) *Map Accuracy Standards* (Fact Sheet FS-078-96). Reston, VA: USGS.

USGS (1997b) " What is SDTS?" Reston, VA: USGS. **http://mcmcweb.er.usgs.gov/stdts/whatstds.html**

Ventura, S. (1993) "Standards offer LIS/GIS support," *GIS World*, Vol. 6, No. 8, pp. 48–50.

Veregin, H. (1995) "Developing and testing of an error propagation model for GIS overlay operations," *International Journal of Geographical Information Systems*, Vol. 9, No. 6, pp. 595–619.

Vitek, J. D., Walsh, S. J., and Gregory, M. S. (1984) "Accuracy in geographic information systems: An assessment of inherent and operational errors," *Proceedings*, pp. 296–302, PECORA IX Symposium.

Walsby, J. (1995) "The causes and effects of manual digitizing on error creation in data input to GIS," in *Innovations in GIS 2* by Fisher, P. ed. pp. 113–122. London: Taylor & Francis.

Walsh, S. J., Lightfoot, D. R., and Butler, D. R. (1987) "Recognition and assessment of error in geographic information systems," *Photogrammetric Engineering and Remote Sensing*, Vol. 53, No. 10, pp. 1423–1430.

Welch, R., and Homsey, A. (1997) "Datum shifts for UTM coordinates," *Photogrammetric Engineering and Remote Sensing*, Vol. 63, No. 4, pp. 371–375.

Wittenbrink, C. M., Pang, A. T., and Lodha, S. K. (1996) "Glyphs for visualizing uncertainty in vector fields," *IEEE Transactions on Visualization and Computer Graphics*, Vol. 2, No. 3, pp. 266–279.

Zhang, G., and Tulip, J. (1990) "An algorithm for the avoidance of sliver polygons and clusters of points in spatial overlay," *Proceedings*, pp. 141–150, 4th International Symposium on Spatial Data Handling, Zurich, Switzerland.

CHAPTER

5

Raster-Based GIS Data Processing

5.1 Introduction

As we learned in Chapter 3, raster and vector data are two different approaches to representing the real world. Raster- and vector-based GIS are distinct not only in the types of data that they use, but also in the areas to which they are applied. While vector-based methods are used mainly for digital mapping and resource inventories, raster-based methods are more concerned with spatio-temporal modeling in environmental applications. Advances in computer technology have largely eliminated the boundaries between raster and vector data for GIS applications. The ability to use both types of data in a single integrated working environment appears to be the rule rather than the exception in GIS software packages today.

In many organizations, geographic data are mostly acquired in raster form. Raster data have become the primary source of spatial data in geographic databases and are used increasingly in a wide variety of GIS applications. These include topographic and thematic mapping, terrain and hydrologic analysis, wildlife habitat analysis, predictive and dispersion modeling, as well as scientific visualization. This chapter aims to provide an overview of raster-based techniques in geographic data processing. The focus of discussion will be the generic raster-based processes that will enable users to take advantage of gridded data in spatial problem solving. We start by revisiting the major characteristics of raster data as they pertain to applications in GIS.

5.1.1 Characteristics of Raster Geographic Data

The term raster data when applied to GIS and mapping includes scanned monochrome and color printing separates, scanned black-and-white and colored aerial photographs, remote sensing images, digital elevation models (DEM), as well as thematic spatial data created by manual and computer-based methods.

In these types of raster data, the geographic space of interest is divided into regular cells of specific dimensions, and the measurement or attribute of each cell is represented by a digital code. Unlike vector data, which are explicitly recorded by means of coordinates, raster data are recorded implicitly as the attributes that are found at different points in space. The locations of raster cells are not explicitly recorded but are inferred from their positions in the image. If the ground coordinates of two or more cells are known, the locations of all the cells in the entire raster image can be computed by using their positions in the image as defined by row and column numbers (see Section 5.1.2). This raster model of data presentation allows nonspatial attribute data to be spatially indexed by location and is particularly suitable for answering spatial questions that aim to find the characteristics of specific points in geographic space.

Logically, raster data are organized into *layers*, which are also variously called *bands*, *themes*, or *overlays* (Figure 5.1). Depending on the application objectives, a raster geographic database often contains multiple layers of data. Each of the layers depicts a specific characteristic of Earth's surface, such as topography, soil type, hydrology, vegetation cover, or land use. In each cell of a particular layer, therefore, there is one and only one thematic attribute or value. For applications in GIS, these raster layers must be spatially registered with one another, and georeferenced to a particular map coordinate system.

Raster layers are best used to represent continuous spatial phenomena such as elevation, temperature, and pollution concentration, but discrete spatial features such as roads, rivers and lakes, land parcels, as well as land-use and land cover types may be represented too. However, discrete spatial features in raster form no longer exist as distinct individually identifiable entities. This means that although we are able to visually recognize, for example, a lake on a raster map by differentiating cells that form the lake from surrounding cells, the database does not store or identify the lake as a single entity. Instead, the "lake" is simply stored as a set of contiguous grid cells having the same attribute value. In the raster database, the stored entities are the cells, not the individual spatial features that they represent. This characteristic makes raster data particularly useful for spatial queries and modeling that focus on determining the characteristics of geographic fields, rather than the properties of individual features.

Raster cells are sometimes referred to as *picture elements*, or *pixels*, a term that has its origin in digital image processing. The minimum linear dimension of the cells, expressed in terms of actual ground distance in meters or kilometers, is known as *spatial resolution*. Spatial resolution determines the precision of spatial representation by raster data. Obviously, the smaller the size of the cells, the higher the resolution and the better the precision of spatial representation. Monochrome scanning of printing separates for cartographic reproduction and digital photogrammetric mapping are normally carried out at a resolution of 2000 dots per inch (dpi). Color-coded and gray-scale scanning of maps usually requires a lower resolution that is typically between 300 to 600 dpi. Resolutions of remote sensing images and DEM data are commonly expressed in terms of actual ground measurements, which range from a fraction of a meter for high-resolution remote sensing images to hundreds of kilometers for national and global DEM data sets.

The amount of computer memory space, expressed in terms of *bits*, that is used to store the value of a raster cell is called *image depth*. The minimum depth is one bit,

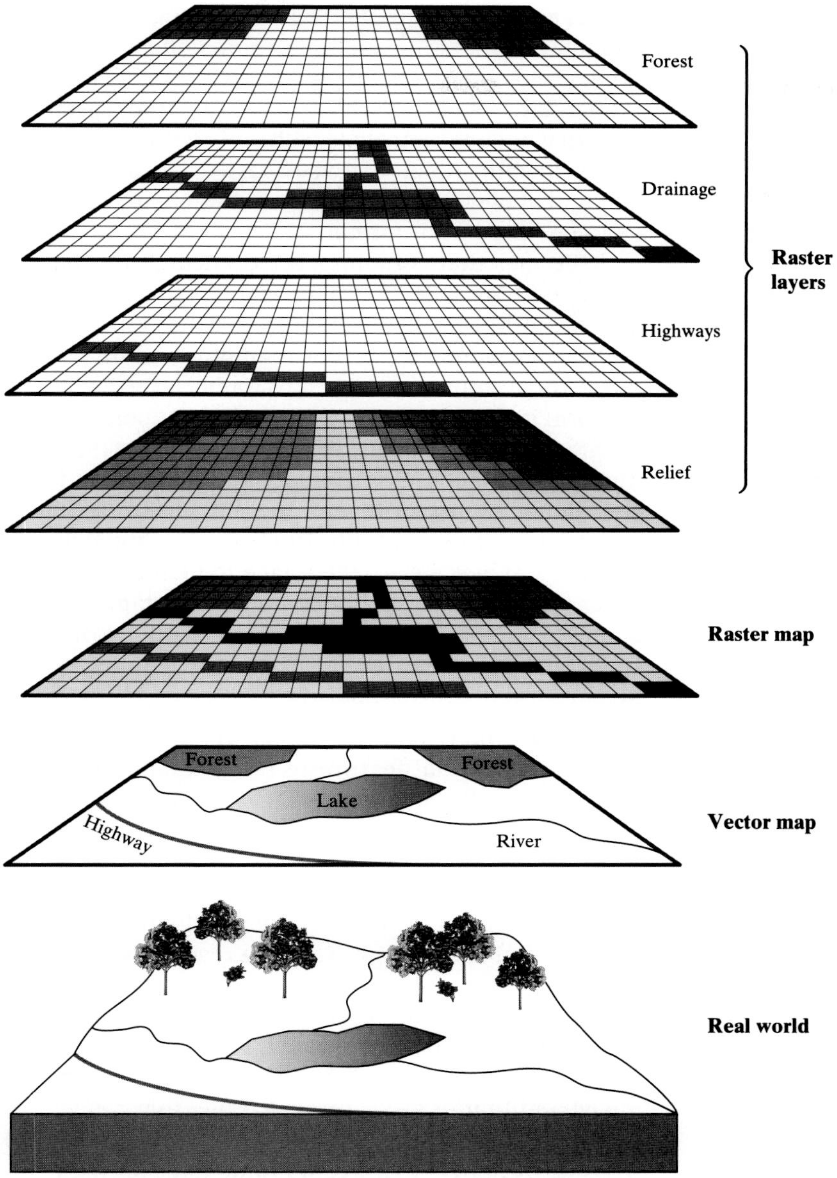

FIGURE 5.1
The raster method of representing real-world features.

which allows a raster cell to be stored using only two integer values: ones and zeros. These two values denote the presence or absence of a particular spatial feature or phenomenon at the locations represented by the raster cells. Binary scanning of black-and-white map separates generates raster data files containing only ones and zeros. This type of raster data is displayed or printed as black-and-white images. Gray-scale scanning of maps and black-and-white aerial photographs is usually carried out using eight bits of computer memory per cell to store the output data. The resulting raster image, commonly referred to as an *8-bit image*, has 256 levels of gray. An 8-bit image may be displayed or printed in color by cross-referencing its cell values to a color look-up table. Such colored images are called *pseudo-color images* in order to distinguish them from natural or true-color images. In color scanning and aerial photography, each cell contains a value between 0 and 255 for each of the red, green, and blue primary colors. These colors can be combined in different proportions to form black, white, 254 levels of gray, and over 16 million intermediate colors. Therefore, scanned colored maps and colored aerial photographs are 24-bit true-color images.

The conceptual representation of raster data in the form of layered two-dimensional arrays closely resembles the architecture of data storage and display in digital computers. It is relatively simple to display a raster layer on the computer screen with colors being controlled by the attribute values of its cells. The simple storage structure of raster data also means that it can be effectively used to represent both continuous spatial surfaces and discrete features of points, lines, and polygons. This allows different types of spatial data stored in multiple layers to be mathematically combined in geographic data analysis. Within individual layers, it is possible to use raster data for very sophisticated spatial data analysis and modeling applications because of the tight integration between locations and their associated attributes in the raster model of data representation. All these characteristics of raster data make them invaluable for GIS applications.

5.1.2 FILE FORMATS OF RASTER DATA

Representing raster data by two-dimensional arrays as explained here is relatively simple. However, when this data structure is implemented to store the data physically in the computer, it becomes a complex and sometimes confusing technical issue. In the computer, raster data can be stored in several ways that are formally referred to as *file formats*. Different file formats have different characteristics, depending on such factors as the sources of the data, the methods of compression, and the design specifications of different raster-based GIS software packages. Generally, raster file formats can be classified into the following categories: (1) generic raster file formats, (2) raster data interchange formats, (3) raster data compression formats, (4) remote sensing image formats, and (5) proprietary formats of GIS software products.

Generic Raster File Formats A generic raster file format is a simple format that is closest to the conceptual raster model of data representation. There are two generic raster file formats: the ASCII file format, which stores cell values by means of ASCII characters; and the binary file format, which stores cell values using 32-bit signed IEEE floating-point numbers. ASCII and binary raster files are device-independent, which means that they are usable by different applications regardless of hardware and software platforms. Generic raster file formats are, therefore, most suitable for cross-platform exchange of raster data. However, since the cell values are stored generically as they are collected, the size of the files are huge for large raster images. Generic raster file formats are commonly used to store satellite image data (such as Landsat Multi-Spectral Scanner and Thematic Mapper data). Their function as a data exchange format has been largely overtaken by the interchange formats described below.

Raster Data Interchange Format The function of an interchange format is to facilitate the sharing of raster data between different organizations and different GIS applications. Over the years, the *Tagged Image File Format*, or TIFF, has emerged as the most popular raster file format for exchanging raster images between application programs. TIFF is nonproprietary and device-independent and, as a result, is supported by a wide range of scanners and image-processing applications, making it a de facto raster standard for desktop publishing. However, the TIFF format has several variants based on different specifications that are not compatible with one another. This often causes incompatibility between raster files in TIFF format and application programs in GIS that are supposed to accept files in this format.

TIFF files that contain georeferencing information are called "GeoTIFF." GeoTIFF is the result of the concerted effort of over 160 GIS software developers, commercial data suppliers, and cartographic and surveying-related organizations to develop a publicly available and platform interoperable standard for the support of raster geographic data in TIFF. Since 1995, the GeoTIFF specification has been extensively revised and made an openly available exchange format for georeferenced raster data (Ritter and Ruth, 1995). In the United States, the SDTS allows a GeoTIFF image to be included with other spatial data files that contain cartographic metadata. The USGS is implementing GeoTIFF by offering Digital Orthoquadrangle image data (DOQ) in this particular format (USGS, 2000).

Raster Data Compression Formats Data compression is an essential feature of raster data storage because raster data files are generally very large. The objective of data compression is to store the same amount of data using a smaller amount of computer memory space. In the computer industry, considerable efforts have been devoted to the development of data compression technology. This has resulted in the creation of a variety of data compression formats, such as RLC, GIF, JPEG, and others that were noted in Chapter 3. These compression formats normally have a *compression ratio*, or the ratio between the size of the original image and that of the compressed file, that ranges between 5 : 1 and 10 : 1. When compressed at these ratios, the size of many raster geographic data files still remains too large to be simply manageable.

Using *wavelet algorithms* that decompose a raster image recursively into layers at different levels of resolution, the new generation of compression methods such as MrSID (Multi-resolution Seamless Image Database) and ECW (ER Mapper Compressed Wavelet) are capable of achieving compression ratios of 20 : 1 to 50 : 1 without noticeable loss of the original image quality (ER Mapper, 1998; Triglav, 1999). These properties of high compression ratio and retention of original image quality represent a tremendous improvement in data compression technology. Most GIS vendors have responded to the advent of the new breed of compression formats by supporting them with their software products. As a result, these raster file formats are getting increasingly popular not only for data compression purposes, but also as an interchange format for transferring raster images between different applications.

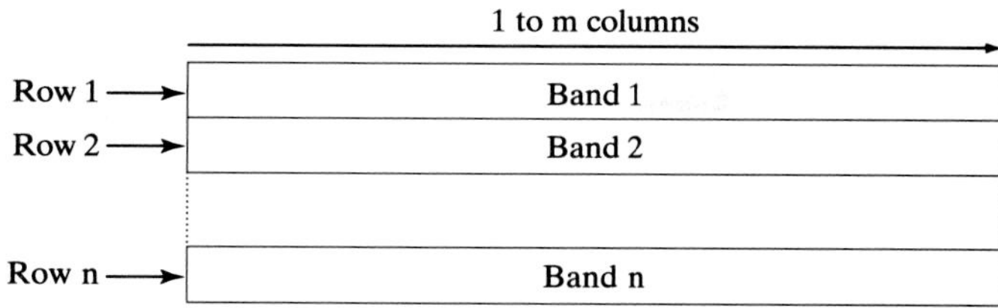

(a) Band sequential (BSQ) format

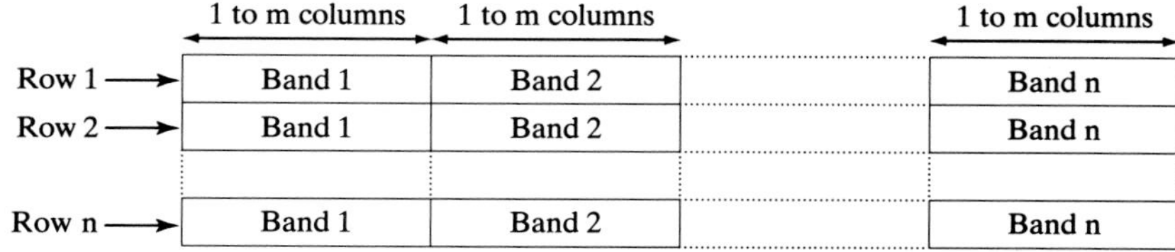

(b) Band interleaved by line (BIL) format

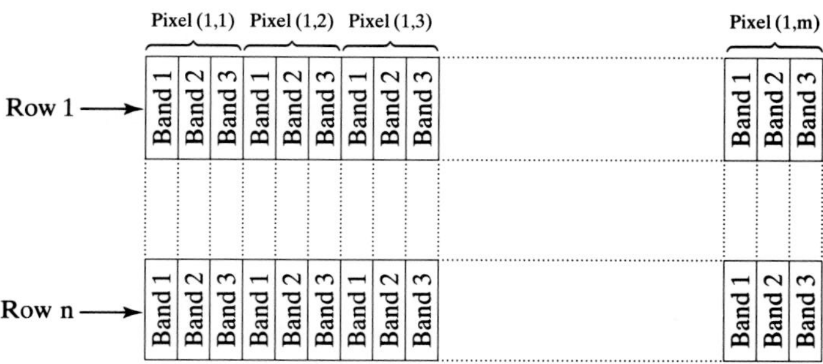

(c) Band interleaved by pixel (BIP) format

FIGURE 5.2
Methods of storing multiband remote sensing images.

Remote Sensing Image Formats Remote sensing data such as those acquired by Landsat MSS, Landsat TM, and SPOT are normally stored in the binary generic format. Because they are multiband data, the storing of each image band can be in one of the following three formats (Figure 5.2): *band sequential* (BSQ) by which each image band is kept as a separate file, *band interleaved by pixel* (BIP) that places all the different bands from a single pixel together, or *band interleaved by line* (BIL) in which pixels of each band are recorded band by band for each line or row of the image. The file header of these remote sensing image formats contains georeferencing information. This allows the images to be displayed with respect to a map coordinate system and integrated with other sources of raster and vector data.

Images stored by the interleaved options are seldom used by GIS software directly for geographic data-processing operations other than simple screen displays. This is because interleaved cells are difficult to manipulate when cells in different layers are added or subtracted, or when cells of only one layer are operated upon. Also, as the principles of data compression are not applicable to interleaved layers, the sizes of these files are often too big to be transferred between different applications or systems. For interleaved raster data to be used, it is necessary to convert them to the conventional layered structure in a proprietary or exchange format first.

Proprietary Formats of a Particular GIS Software Product All raster-based GIS software packages have their own data file formats. These file formats are sometimes proprietary in nature and store data differently from one another. Some proprietary formats, such as the grid format of ArcInfo GRID, have an associated value attribute table (VAT). This allows the use of a database management system (DBMS) to manipulate the cell values and makes it possible to relate the cells to other attributes pertaining to the cell values in the original raster (Figure 5.3).

In the past, the difficulty of exchanging raster data in different proprietary formats was one of the greatest impediments in using raster-based GIS. Many GIS software packages now have built-in data conversion capabilities that allow them to use raster data in different proprietary formats directly. It is now commonplace for GIS and digital image-processing software products to directly exchange raster data with one another, thus greatly enhancing the interoperability between different applications and systems (Limp, 1999).

1	1	2	2	2	3
1	1	1	2	2	3
1	1	4	4	3	3
5	5	4	4	4	4
5	3	3	5	1	1
5	3	3	5	1	1

Grid cells

Value	Count	Soil Type	Suitability
1	11	Clay	1
2	5	Loam	3
3	8	Sand	2
4	6	Gravel	0
5	6	Rock	0

Associated value attribute table (VAT)

FIGURE 5.3
The grid format of ArcInfo GRID module and its associated value attribute table (VAT).

5.1.3 RASTER-BASED DATA PROCESSING AND DIGITAL IMAGE ANALYSIS

The raster data model in GIS is similar to digital images in remote sensing and other computer-based applications such as robotics and pattern recognition. Because of this, it is possible to make use of many of the concepts and techniques originally developed for digital image processing in raster-based geographic data processing. These include such commonly used techniques as image classification (see Chapter 8, Section 8.6.2), resampling, edge enhancement, geometric rectification, and mosaicking.

However, although it is not always easy to draw the boundary between raster-based geographic data processing and digital image analysis, the objectives of these two classes of processes are in fact quite distinct from one another. As we will demonstrate in this chapter, the emphasis of raster-based GIS processing is on the processing of geographic data, rather than on the interpretation of the images. This means that the scope of raster geographic data processing is much broader than that of digital image analysis. The objectives of raster-based geographic data processing include a multitude of purposes that are sometimes considered to be beyond the realm of digital image analysis, for example, to determine spatial relationships, detect spatial patterns, generate new data layers, and, more important, develop models for spatial problem solving and decision making. Obviously, it is the focus on spatial characteristics of the real world as represented by the data, rather than the inherent image properties of the data themselves, that distinguishes raster geographic data processing from digital image analysis.

Raster-based geographic data processing is also different from digital image analysis with respect to working procedures. A typical task in raster-based GIS data processing is made up of a sequence of processes carried out in a particular logical order that aims to generate information for scientific research or spatial problem solving (see Section 5.5.1). This is distinctly different from tasks in digital image analysis, which are often carried out as independent and discrete procedures. In the context of the discussion in this chapter, therefore, digital image analysis is simply treated as an intermediate step that prepares raster data for spatial analysis and modeling. In other words, digital image processing is assumed to be a subset of techniques of raster-based geographic data processing.

5.1.4 RASTER-BASED AND VECTOR-BASED GIS DATA PROCESSING

The layered structure of raster data in GIS makes this model extremely effective at displaying individual characteristics of Earth's surface, either selectively or collectively, with colors being controlled by the numeric values of the grid cells. It has also led to the development of very powerful analytical methods that process raster layers graphically and mathematically (Berry, 1995; Gao et al., 1996; also see Sections 5.3.1 through 5.3.5). Therefore, raster-based GIS data processing is based on the use of one or more map layers, rather than on the use of individual spatial features in vector-based GIS data processing (see Chapter 6).

In the past, raster- and vector-based methods were often treated as distinct techniques for different application domains. Modern computer technologies and new application requirements have now largely eliminated the conventional boundaries between these two approaches of geographic data processing. Most GIS software products today are equipped to convert raster data to vector data and vice versa, and are therefore capable of using raster and vector data simultaneously. Screen digitizing using scanned raster images, for example, has superceded traditional table digitizing as the major way of converting map data into digital form (see Section 6.3.2 for more detailed explanations). Raster images have also been commonly used in cartographic production to enhance the presentation of both topographic and thematic maps by giving them a pictorial background.

The integration of raster and vector data in geographic data processing has greatly improved the interpretability of raster images. For example, spatial information resulting from raster-based analysis and modeling is not always easy to comprehend because of the absence of spatial references on the raster surface (Figure 5.4a). The superimposition of vector-based

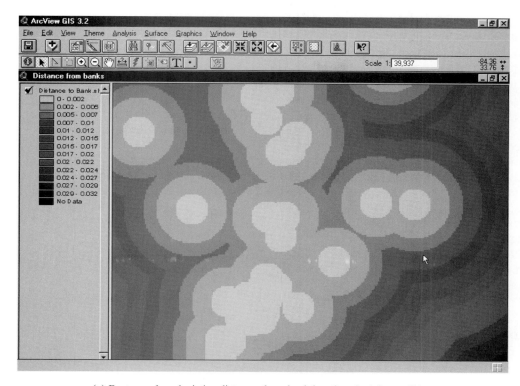

(a) Raster surface depicting distances from bank locations in Atlanta, GA—the user has little information about where the banks are actually located.

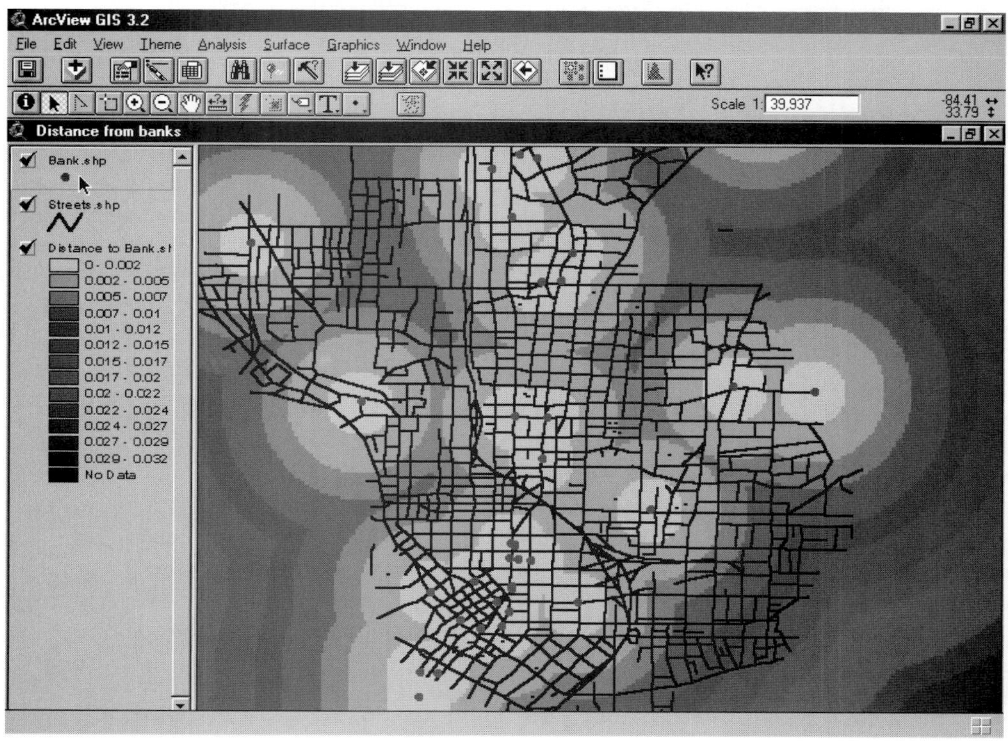

(b) The street fabric provides spatial referencing to raster surface data—users who are familiar with Atlanta, GA, know where the banks are and will be able to identify the neighborhoods that are not adequately served because of their distances from the banks.

FIGURE 5.4
Using vector data to give raster data a geographic reference framework.

spatial features, such as street networks, on top of the raster data enables the user to easily recognize the location and extent of the geographic space represented by the raster surface (Figure 5.4b).

The ability to combine raster and vector data in GIS applications has eliminated one of the major inherent limitations of raster-based methods of spatial analysis and modeling. The opportunities of integrating these two types of geographic data to enhance the usefulness of GIS are growing rapidly. Geographic data today are collected mostly in raster form and converted to vector form by automated, semiautomated, or manual means. Raster topographic maps and orthophotographs have also been produced for georeferencing of vector data, and raster images are increasingly used as backdrops in vector maps. Raster- and vector-based methods should be perceived more as complementary rather than mutually exclusive technologies.

5.1.5 ADVANTAGES AND LIMITATIONS OF RASTER-BASED GIS

Many spatial analysis and modeling functions have been developed and are now in practical use in raster-based GIS. Raster-based methods of geographic data processing have many advantages. These include

- *The Ability of Raster Data to Represent Different Types of Continuous Surfaces.* The raster data model represents spatial phenomena such as topography, land use and land cover, and air quality as categorical or continuous surfaces. This makes raster-based methods particularly suitable for spatial modeling that involves multiple surface data sets. By having all the surfaces in raster format, it is conceptually simple to overlay raster cells representing different attributes at the same location to generate a new layer showing their combined effects.
- *Fast Computer Processing.* The process of overlaying different layers of spatial data in raster form is not only conceptually simple, but also computationally less complex when compared with overlaying operations using vector data. Since raster layers are all registered to a common geographical framework, they can be mathematically combined easily without any need for geometric intersection, topological building, and error checking that are necessary in vector-based geographic data processing (see Chapter 6).
- *Fast Display of Surface Data.* Raster surfaces may be drawn more quickly than their vector counterparts. By using the values of the grid cells to directly control the color of the displayed data, for example, there is no need for a computationally intensive process to perform shading that is required for vector surfaces. The display and plotting of raster data can be further optimized by image resampling methods to produce what are known as *pyramid raster images.* These are raster images that have been progressively down-sampled at a reduced resolution. When the user zooms out and the raster cells grow smaller than the screen display pixels, the system will choose one of the generalized rasters to display. The method of generating image pyramids works faster than the vector equivalent of cartographic generalization and allows raster data to be displayed and plotted much faster.
- *Ability to Handle Very Large Databases.* Large raster data sets can be handled effectively by GIS using two techniques: *tiling,* which optimizes both random and sequential access to the grid cells during data processing by dividing large raster grids into smaller blocks; and *adaptive raster compression,* which reduces raster data-storage requirements and greatly expedites access to cell values for spatial display and analysis of very large raster files.
- *Applications That Are Difficult or Impossible to Perform Using Vector Data.* There are many spatial analysis and modeling applications that can be done only in raster format because of the limitations of the vector approach. In hydrologic modeling, for example, the flow of water across the terrain would be easier to compute and display by using the continuous surface represented by raster grids than by using the discrete surface represented by contour lines. Similarly, it is much easier to analyze and portray spatial processes such as the spread of wildfire, movement of pollutants from a point source, and the growth of settlements using spatial data in raster form than in vector form. Also, the wealth of spatial information contained in high-resolution images and the pictorial nature of these images make them superior to vector data for visualization and interpretation of the spatial phenomena.

On the other hand, raster-based processing also has some inherent limitations for GIS applications. Since the raster data structure is based on grid cells, raster geographic data processing is not suitable for applications that rely on individual spatial features represented by points, lines, and polygons. Network analysis, for example, is difficult if not impossible with raster data because linear real-world features such as streets are not identified as discrete features in raster databases. Similarly, land parcel–based applications such as land title registration and forest resource inventories must be done in vector mode because they rely heavily on the use of linear boundaries. The applicability of raster data processing is also restricted by the resolution of the source data. Although some raster data have very high resolution, the 300–500 dpi resolutions of most raster data in general use are a limiting factor for many GIS applications. For mapping applications, for example, linear features such as roads, streams, coastlines, building outlines, and parcel boundaries are clearly defined in vector form using coordinates. In raster form, these features are typically generalized and do not always appear as cartographically pleasing as vector lines on the computer screen or paper.

5.2 Acquiring and Handling Raster Geographic Data

In their proposed taxonomy of GIS functions, Giordano et al. (1994) classified procedures in geographic data processing into three categories.

1. *Input functions* include functions that prepare and structure raster data for use in GIS, such as restructuring, compilation, and editing.
2. *Analysis domain* includes all functions that will derive spatial relationships implicit in the source data, such as logical operations, arithmetic operations, overlay operations, and geometric operations.
3. *Output functions* are those that present the results of the analysis functions in a suitable form for communication and spatial problem solving, such as maps, graphs, and statistical reports.

This provides a convenient way to describe data-processing operations in a typical GIS. In the following discussion, we will first examine the concepts and techniques that constitute the input functions of a typical GIS based on the raster data model. Methods pertaining to the GIS analysis domain and output functions will be explained in Sections 5.3 and 5.4 respectively.

5.2.1 Acquisition of Raster Geographic Data

Conventionally, GIS users used to create their own data, but the current trend is toward data sharing. Different types of raster geographic data are now obtained from government mapping agencies and commercial data suppliers. In the United States, for example, the USGS produces and maintains three important raster geographic data sets that cover the entire country. These raster data sets are

- *A digital orthoquadrangle (DOQ)* is a computer-generated image of an aerial photograph in which image displacement caused by camera tilt and terrain relief has been removed (USGS, 1999a).
- *A digital raster graphic (DRG)* is a scanned image of a USGS topographic map that may be in one of the following scales: 1:24,000, 1:25,000, 1:63,360 (Alaska), 1:100,000, and 1:250,000 (USGS, 1999b).
- *A digital elevation model (DEM)* is a digital file consisting of terrain elevations for ground positions at regularly spaced horizontal intervals (USGS, 1997a and 1977b).

Thematic raster data pertaining to various aspects of the environment and human activities are collected and maintained by government agencies, commercial organizations, and academic institutions as the result of ongoing business operations and scientific research. In the past, access to such data was restricted by the lack of information about the existence and quality of the data, the difficulty of delivering data from the source to the end users, and the incompatibility between the data file formats adopted by different organizations. Today, these problems have been largely overcome. The Internet, for example, has provided end users with a very effective means to search for usable data and to acquire them through electronic delivery (see Chapter 12). At the same time, software technology has made it possible for raster-based GIS software products to directly read the proprietary raster formats of different systems. ArcInfo, for example, is capable of using raster data stored in numerous formats such as ADRG (ARC Digital Raster Graphics), BMP (Windows Bitmap), BSQ, BIL, BIP, ERDAS, IMAGINE, GRASS, JPEG, NIFF (National Image Transfer Format), RLC (run-length code), and SUN Raster files. In ArcView GIS, it is possible to display raster data in practically any file format with the aid of programmable extensions. The ability to use raster data in different file formats appears to be a norm rather than an exception among other GIS software packages today. This has greatly facilitated the use of raster data by eliminating the problem of incompatibility of data formats among different systems, thus allowing GIS users to take advantage of the huge amount of raster geographic databases in existence in different organizations.

Remote sensing is another major source of raster geographic data. The most noticeable examples are the multiband images obtained by Landsat MSS, Landsat TM, and SPOT, among many other air- and space-borne imaging systems (see Chapter 2). Remote sensing images in BSQ, BIL, and BIP formats are georeferenced raster data. Their file headers contain the necessary geometric transformation information for spatially registering the images to the ground coordinate systems (such as the UTM). This allows the images to be displayed and measured correctly. However, although remote sensing images are supported by most GIS software products, their direct use is largely limited to displaying operations as explained in Section 5.1.3. For other geographic data-processing operations, these images must always be first converted to the proprietary format of the particular GIS software product. They are also often subject to further image manipulation and interpretation operations in order to extract information relevant to the intended applications.

Users who have to create their own raster data can do so in different ways. One way is to manually enter the data as ASCII characters from a keyboard using a text editing program. This will create a generic raster file that can be manipulated and displayed by any GIS software product as noted in Section 5.1.2. This manual method is very handy for a simple and small map, but it becomes tedious and error-prone when the map contents are even marginally complex. A more practical approach is to use a digitizer to convert the map into a digital vector map, which is then changed into raster format with the aid of a computer program. This process, which is known as *rasterization*, is a standard feature of many raster-based GIS (Eastman, 1997). It can be used for point, line, or polygon vector data.

Advances in imaging technologies and reduced hardware costs have made scanning readily accessible and affordable to computer users. Scanning is now the most widely used method for converting hardcopy maps and aerial photographs into digital images (Flanagan et al., 1994; Mayo, 1994). An image scanner is equipped with light-sensitive sensors, usually an array of charge-coupled device (CCD) elements. During scanning, these CCD elements measure the brightness of the light either transmitted through film or reflected off opaque surfaces, and convert it into electric signals. The electronic signals from each CCD element are then passed down a serial register to an output amplifier where they are converted to digital numbers according to their strength. These digital numbers are most simply represented in the form of 8-bit encoding, which allows the images to have 256 gray levels, with 0 indicating black or no signal and 255 indicating white or a strong signal (Photometrics, 1990; Kolbl et al., 1996). In scanning, the resolution and format of the resulting digital image may be specified by the user. Some scanner manufacturers also supply image-processing software packages that provide the user a full range of image editing capabilities to despeckle, deskew, rotate, clip, and clean the image before it is used in GIS operations (see Section 5.2.3). As a result, scanning can be used to generate raster data of any desired resolution and image quality that will meet the specific requirements of various types of GIS applications.

5.2.2 GEOREFERENCING RASTER DATA

Raster data for GIS applications must be georeferenced to a particular map coordinate system (see Section 2.8). Fundamental to the georeferencing of raster data are the concepts of *raster space* and *raster-to-world* relationship. A raster space is defined as the digital image of a particular area as represented by a grid of x (horizontal) by y (vertical) cells. The values of the cells, as they are read from a data file, are organized by computer software into a two-dimensional array. The computer makes use of the information stored in the header of the data file to determine the dimension of the digital image and to control the display of its contents.

The position of a particular cell in raster space is referenced by (row, column) = (i, j), where i = 1, 2, 3, . . . m and j = 1, 2, 3 . . . n. This is called the *raster coordinate system*. In this coordinate system, the origin is normally defined at the upper left corner because the computer monitor displays an image from left to right and from top to bottom. The unit of measurement is the cell size or resolution. The location of a particular cell is identified according to its row number and column number, starting from the origin and with the positive direction pointing downward and to the right respectively (Figure 5.5a). This way of defining positions in raster space is obviously different from the conventional way of defining positions on maps. In map coordinate systems, locations are expressed in terms of (X, Y) or (eastings, northings) in the form of a plane rectangular coordinate system (see Section 2.5.1). The origin of the coordinates is defined at the lower left corner. Positive X or eastings values are in the same direction as columns in the raster coordinate system but positive Y or northings values are in the opposite direction as rows (Figure 5.5b).

Raster data are often recorded with respect to the raster coordinate system. In order to display and analyze such raster data spatially, it is necessary to transform them into a map coordinate system (such as the UTM). This is achieved by performing a transformation that converts the raster coordinates (row, column) into the corresponding ground coordinates (E, N). This type of coordinate transformation, known as georeferencing, is an important first step in raster-based geographic data processing for GIS.

In raster-based geographic data processing, there are two approaches to georeferencing: (1) *image-to-map rectification* and (2) *image-to-image registration*.

Image-to-Map Rectification The procedure for georeferencing by image-to-map rectification involves the measurement of the image coordinates of the reference cells and their corresponding ground coordi-

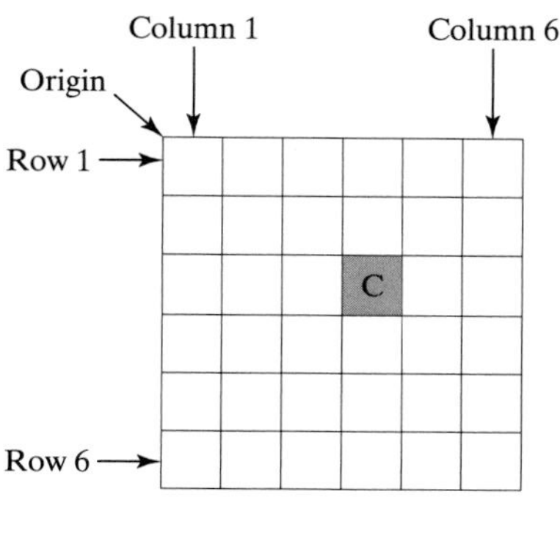

Position of C = (row, column) = (3,4)

(a) A raster coordinate system

Northing

P

E_P

N_P

Origin (0, 0)

Easting

Position of P = (E_P, N_P)

(b) A map coordinate system

FIGURE 5.5
The raster and map coordinate systems.

nates. The process of image rectification is made up of two interrelated interpolation processes.

- *Spatial Interpolation by Coordinate Transformation.* The image coordinates (rows and columns) of a number of control points and their corresponding ground coordinates (E, N), which relate the image with the real world, are used to solve a set of polynomial equations. The order of the polynomial equations depends on the size of the area covered and the amount of geometric distortion in the raster image. Higher-order polynomial equations require a greater number of ground control points to solve. Each term in the polynomial represents one type of distortion (such as translation in x and y, scale changes in x and y, skew, and rotation). In general, the affine transformation (Equation 2.8) given in Chapter 2 is quite adequate for this purpose.

$$X = ax + by + c$$
$$Y = dx + ey + f$$

where X, Y are the ground coordinates corresponding to E, N respectively in Earth's mapping coordinate system, and x and y are the image coordinates corresponding to columns and rows respectively. A minimum of three ground control points are sufficient to solve these equations, but with more ground control points, solution by least-squares adjustment is possible (Wolf and Dewitt, 2000). These ground control points should be evenly distributed to cover the whole image. A large number of ground control points is desirable for more accurate rectification. The typical number of ground control points varies from 20 to 30 but depends on the size and relief condition of the image to be transformed. The terrestrial coordinates of these ground control points are obtained either from field surveys using GPS receivers or from large-scale topographic maps. After the polynomial equations are solved, the six coefficients (a, b, c, d, e, and f) will be known. Any image coordinates can be substituted into the equations to calculate the corresponding ground coordinates in Earth's mapping coordinate system (such as the UTM). The position of the original grid cells will have to be interpolated in the new mapping coordinate system. One should note that the transformed data are still raster data, but each grid cell has now acquired the values of the ground coordinates.
- *Attribute Interpolation by Resampling.* After the coordinate transformation, the positions of the original grid cells in the raster layer may have been reoriented. New attribute values will have to be interpolated for cells oriented to the coordinate system. This process is known as *resampling*. There are three commonly used resampling methods (Figure 5.6): (a) *nearest-neighbor* by which the attribute value equal to that of the nearest original cell is assigned to the corresponding cell in the output raster layer; (b) *bilinear interpolation* by which the attribute value of a particular cell in the output raster layer is determined by taking a proximity-weighted average of the attribute values of four original grid cells nearest to it; and (c) *cubic convolution resampling* by which the attribute value of a particular cell in the output raster layer is computed on the basis of a weighted average of 16 original grid cells surrounding it (Jensen, 1996). One should note that only the nearest-neighbor resampling method will not change the attribute values of the original

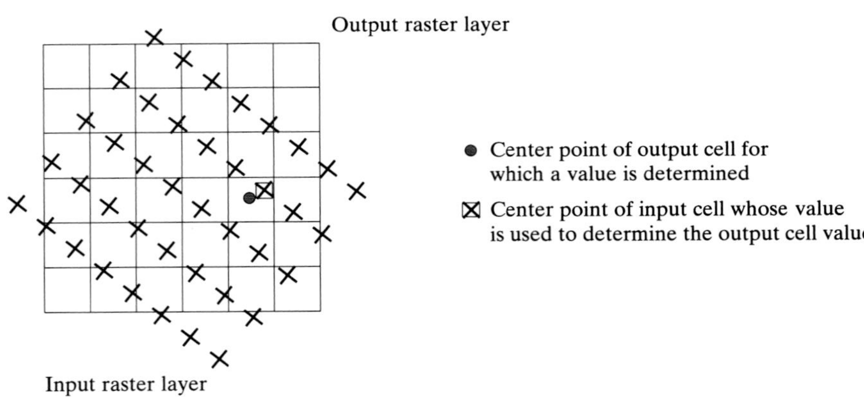

(a) Nearest neighborhood assignment

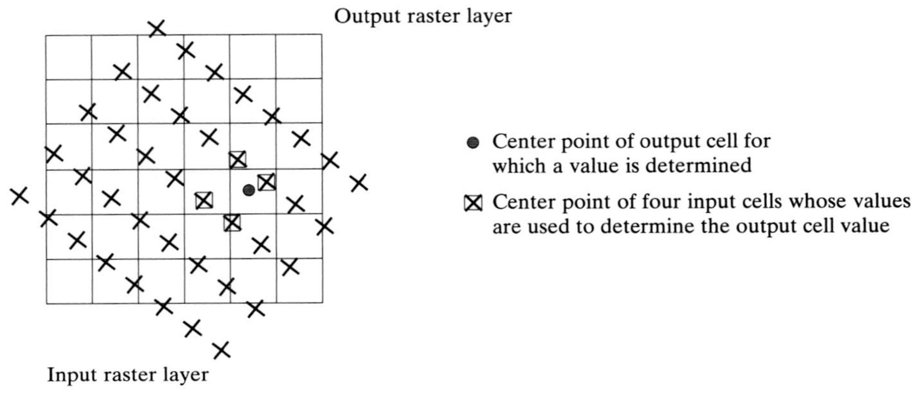

(b) Bilinear interpolation

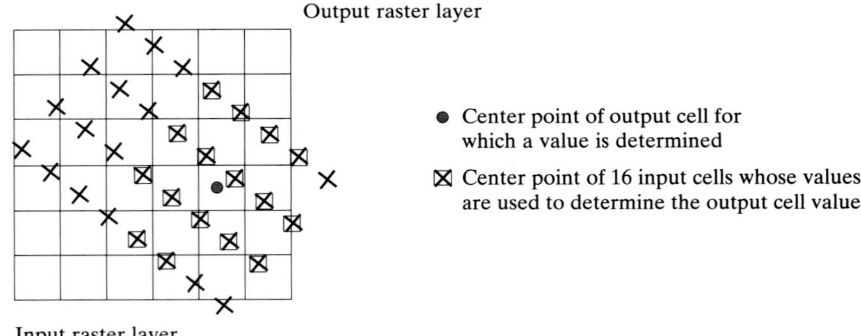

(c) Cubic convolution

FIGURE 5.6
Methods of resampling.

grid cells assigned to the reoriented grid cells. Therefore, this is the preferred method to use when applied to map layers in GIS (as opposed to image layers). The nearest-neighbor resampling method is also computationally simple, but it tends to produce blocky images and jagged edges for linear features. On the other hand, the cubic convolution resampling is more computationally intensive, but it generates the most visually appealing image of all three resampling methods. It is the preferred method to use for digital images from remote sensing systems, and it is often the default method of the image-processing software packages on the market.

The result of image-to-map rectification is a geometrically correct, map coordinate-based matrix or grid of raster cells. A georeferenced raster file obtained in

20.175410003212	(Dimension of a pixel in map units in the x-direction, i.e., x-scale)
0.0000000000000	(Rotation term for row)
0.0000000000000	(Rotation term for column)
−20.175410003212	(Dimension of a pixel in map units in the y-direction, i.e., y-scale)
3569981.2345699878	(The x coordinate of the center of the upper left pixel in map units)
50009879.009988712	(The y coordinate of the center of the upper left pixel in map units)

FIGURE 5.7
Contents of the world file for ArcInfo and ArcView GIS.

this way may be stored in the original file format by which it was originally stored, or it may be converted to an exchange format such as TIFF. There are software applications that allow users to encode the metadata such as the datum and coordinate system to which the raster image is georeferenced in the file header, thus making the raster file a GeoTIFF file.

Image-to-Image Registration This is used to register one raster layer with another that has already been georeferenced by the method of image-to-map rectification. The rectified raster layer is now the reference for another unrectified raster layer. Control points (cells) are selected from these two raster layers, and the same coordinate transformation and resampling processes as explained under "Georeferencing by Image-to-Map Rectification" are used. This method is commonly used to spatially match several different raster layers to a single reference raster layer.

Image-to-map rectification and image-to-image registration can be carried out interactively within a GIS or remote sensing program from the image window on the computer screen. For image-to-image registration, the reference raster layer and the raster layer to be registered can be displayed side by side, and the coordinates of control points can be directly extracted from their corresponding positions. For image-to-map rectification, the coordinates of the corresponding ground control points can be obtained interactively from the map window on the computer screen if digital maps are available. If ground coordinates have to be measured on paper maps, the user has to enter the values manually. Using these two sets of coordinates, the computer calculates the transformation parameters based on the affine transformation equations given earlier and stores the results in a data file specified by the user. During this process, the raster cells remain intact. The computer simply makes use of the measured cell and ground coordinates to determine the transformation parameters. It does not geometrically transform the raster image itself.

The data file by which computed georeferencing parameters are stored is known by different names in different systems, for example, "world file" in ArcInfo and ArcView GIS, "table file" in MapInfo, and "header file" in some other software packages. The world file, which is supported by many other software products other than ArcInfo and ArcView GIS, stores the six parameters of the affine transformation equation that are required to compute the image-to-world relationship for georeferencing a raster image (Figure 5.7).

Whenever a raster image in a supported file format is loaded in ArcInfo or ArcView GIS, the software will look for its accompanying world file and, if one is found, use it to draw the image in a map coordinate system. The image can then be displayed simultaneously with vector geographic data, and all measurements on the image are made in map coordinates rather than raster coordinates.

5.2.3 PREPROCESSING AND EDITING RASTER DATA

Raster data acquired either from in-house data collection projects or from government agencies and commercial data suppliers often contain errors and deficiencies that degrade their usability. Different classes of raster data tend to have different types of errors and deficiencies caused by different factors (Table 5.1). Both satellite images and digital aerial photographs, for example, suffer from geometric distortions and loss of image quality that result from the combined effects of imperfection in sensor design and construction, variations in sensor altitude, movement of the sensor platforms, Earth's curvature, as well as atmospheric refraction (Lillesand and Kiefer, 2000). On the other hand, deficiencies of scanned map images are more likely caused by the quality of the original map documents rather than by the scanner. In the case of digital elevation models, most of the errors can be traced to the sampling process and data input errors.

The capability to preprocess and edit raster data is consequently an essential feature of the data input functions of raster-based geographic data processing. The objective of preprocessing raster data is to minimize the amount and magnitude of errors or deficiencies in the original data. Generally, this capability is made up of two relatively distinct suites of processes: (1) image rectification and restoration and (2) raster data editing.

Table 5.1
Common Errors and Degradations of Raster Data

Classes of Raster Data	*Common Errors and Degradations*	*Major Causes*
Satellite Images	Cross- and along-scan distortions	Forward movements of sensor, variation of platform velocity, Earth's rotation
	Scale changes	Variations in platform altitude, Earth's curvature
	Geometric distortions	Imperfections in sensor design and construction, tilts of sensor
	Radiometric inconsistencies	Atmospheric refraction, scene illumination, viewing geometry and instrumental response characteristics
Scanned Aerial Photographs	Relief displacements	Variations in terrain heights
	Scale changes	Variations in camera altitude, variations in terrain heights, camera lens geometry
	Geometric distortion	Imperfections in camera lens design and construction, tilts of camera, movement of platform
	Radiometric inconsistencies	Atmospheric refraction, scene illumination, viewing geometry, instrumental response characteristics
Scanned Maps	Skew	Imperfections in scanner head
	Scale change	Stretching and shrinking of original documents
	Geometric distortion	Imperfections in scanner head, stretching and shrinking of original documents
	Speckles	Dirty scanner head, dirty original documents
Digital Elevation Models	Erroneous cell values	Sampling errors, data input mistakes

Image Rectification and Restoration Image rectification and restoration are the processes that correct the distortions and errors introduced during raster data acquisition. The aim is to produce an image that is as close as possible to the geometrical and pictorial characteristics of an idealized image free of geometric distortions and errors. Since these processes are normally applied before the data are actually used in any applications, they are commonly called *image preprocessing.*

The geometric characteristics of an image, which refer to its ability to match the geometric integrity of a map, can be restored by the process of image-to-map rectification as described in the previous section. Image quality is restored by *radiometric corrections.* These corrections are applied particularly to remote sensing images to improve their interpretability by modifying the brightness values of the pixels. Full explanations of the method of radiometric correction are found in Chapter 8 (Section 8.6.2), where the preprocessing of remotely sensed digital image data is discussed.

Raster Data Editing Raster data editing refers to the processes that are more concerned with correcting the specific contents of raster images than their general geometric and pictorial characteristics. The objective of raster data editing is to produce a clean raster image that will meet specific data-processing requirements. Many raster GIS software packages are equipped with some basic capability for raster editing. However, digital image-processing packages generally have much more robust tools and should be used whenever possible (Figure 5.8). The types of data edit-

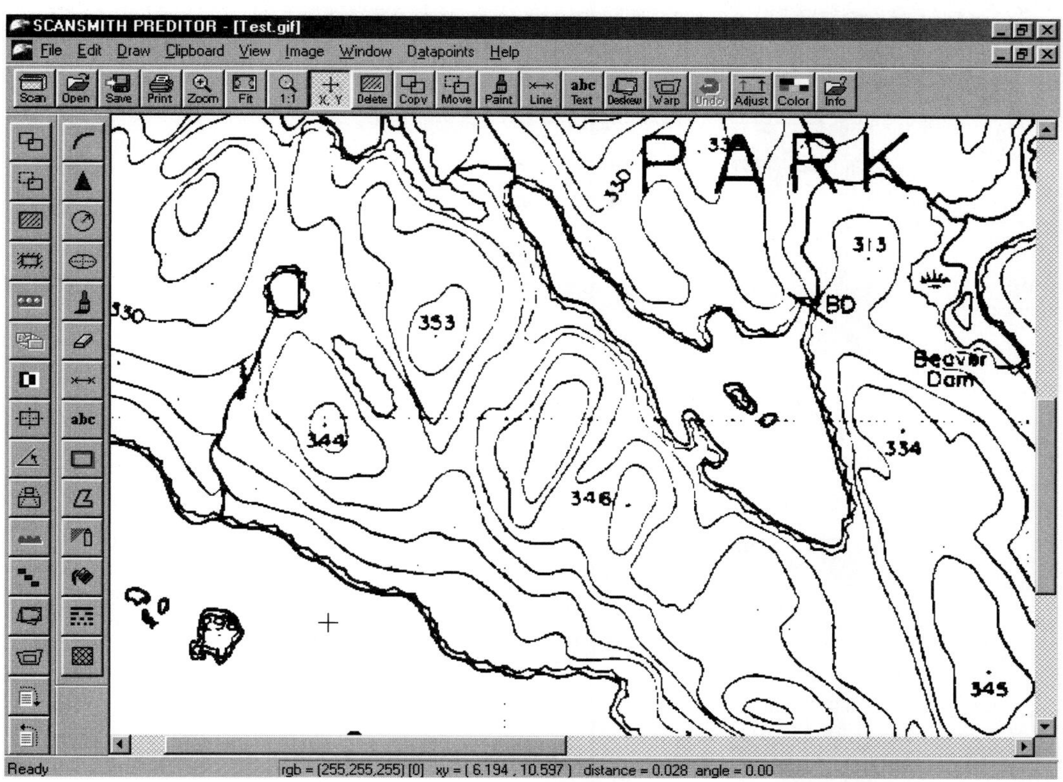

FIGURE 5.8
Editing a raster image by means of a digital image-processing software package. Note the large number of editing tools that are available for modifying the image geometrically and pictorially.

ing available in different packages tend to vary considerably, but commonly used raster data editing functions include (Figure 5.9):

1. *Filling Holes and Gaps.* To fill holes and gaps that appear in the raster image
2. *Edge Smoothing* or *Boundary Simplification.* To remove or fill single-pixel irregularities in the foreground pixels and background pixels along line edges
3. *Deskewing.* To rotate the image by a small angle so that it is aligned orthogonally to the *x* and *y* axes of the computer screen
4. *Speckle Removal* or *Filtering.* To remove speckles or random high- and low-valued pixels in the image, usually done automatically according to a particular threshold size specified by the operator
5. *Erase and Delete.* To remove speckles or unwanted pixels, usually done manually
6. *Thinning.* To reduce the representation of linear features to single cell width, usually done automatically with the options to preserve sharp corners (for street or utility networks) or round corners (for elevation contours or streams)
7. *Clipping.* To cut and remove a specific portion of the raster image and to make it a new image
8. *Drawing and Rasterization.* To add vector graphics or text to a raster image and convert the vector graphics to raster form in a new image

5.2.4 MOSAICKING RASTER IMAGES

A mosaic is a composite picture that is made by piecing together two or more aerial photographs or images to provide a continuous view of a large geographical area. The production of mosaics has long been a standard practice in photogrammetry. Mosaics portray the relative planimetric positions of spatial features in pictorial form. They can be produced in less time and at lower cost than topographic maps and have been used in a wide range of applications in resource management, urban planning, and civil engineering.

Conventionally, mosaics have been made by using hard-copy aerial photographs or satellite images that were mounted together by cut-and-paste, with or without reference to a spatial framework of control points (Mullen, 1980; Vickers, 1993). When scanned aerial

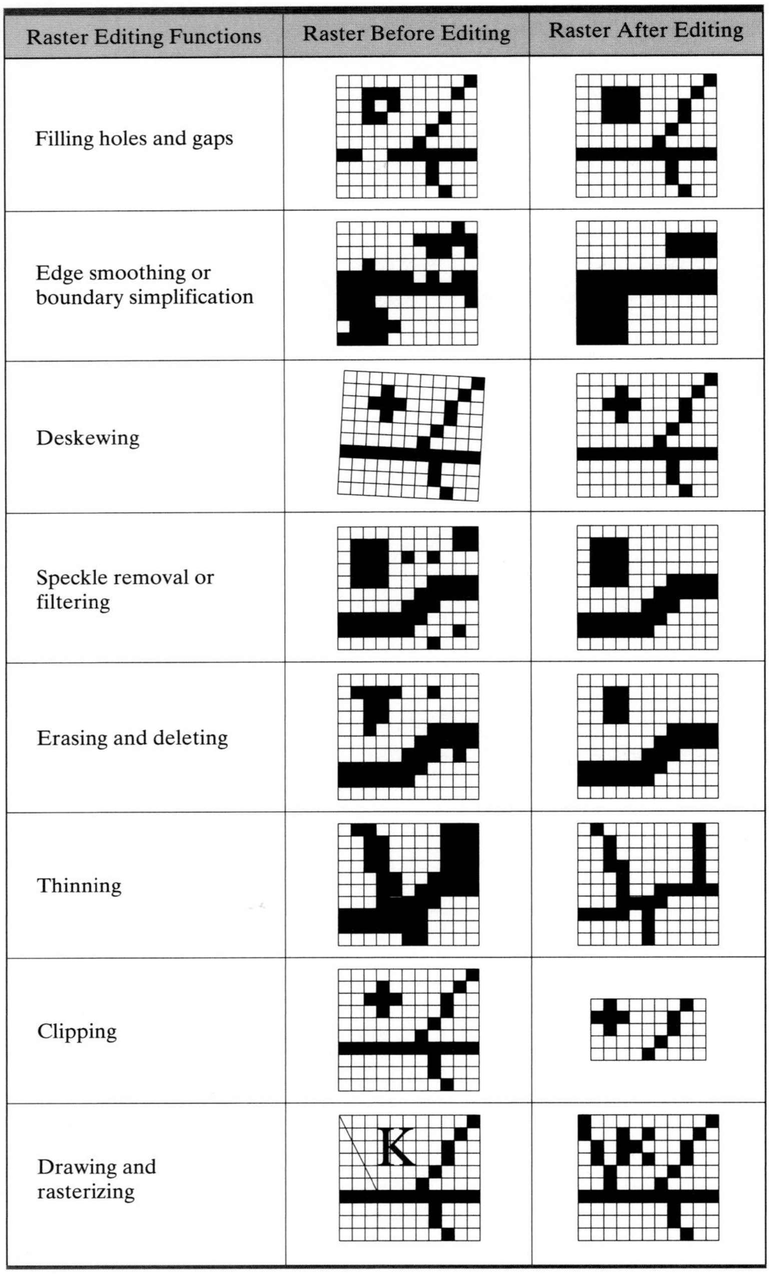

Raster Editing Functions	Raster Before Editing	Raster After Editing
Filling holes and gaps		
Edge smoothing or boundary simplification		
Deskewing		
Speckle removal or filtering		
Erasing and deleting		
Thinning		
Clipping		
Drawing and rasterizing		

FIGURE 5.9
Common raster editing functions in GIS.

photographs and digital satellite images are available, they can be used to create a digital mosaic that will serve the same purpose as a conventional mosaic (Afek and Brand, 1998). Digital mosaics have the same advantages as conventional mosaics and may be produced more quickly and economically. Many digital image-processing and raster-based GIS software packages include a mosaicking function to meet the growing demand for this particular type of digital information product.

Mosaics generally fall into three classes: controlled mosaics, semicontrolled mosaics, and uncontrolled mosaics. A controlled mosaic, which is the most accurate of the three classes, is prepared using rectified aerial photographs and horizontal ground control points. An uncontrolled mosaic is prepared by simply visually matching the image details of adjacent aerial photographs. Since no ground control points are used and the photographs are not rectified, an uncontrolled mosaic is not as accurate as a controlled mosaic. A semicontrolled mosaic is created using a combination of the techniques for controlled and uncontrolled mosaics. It can be prepared, for example, by using ground control points but unrectified photographs.

Digital mosaics may be classified in the same way. An uncontrolled digital mosaic is created by simply putting contiguous raster images together. This may be most easily done by using the clip, cut, paste, and other image editing tools that can be found in a general-purpose digital image-processing package such as Adobe PhotoShop, Corel Photo Paint, PaintShop Pro, and Scansmith PREDITOR (ANA Tech, 1997). Creating a digital mosaic in this way is based solely on the visual matching of image details across adjacent images. It does not use any ground control points, nor does it attempt to balance the pictorial qualities between adjacent images. The end product is a composite picture that covers a specific geographic space of interest. The quality of an uncontrolled digital mosaic depends on the quality of the input images and the technical skills of the operator who assembles the images. Although it is not possible to obtain accurate measurements on an uncontrolled digital mosaic, it is useful for reconnaissance and planning purposes in environmental and engineering projects.

A controlled digital mosaic is produced using georeferenced images. There are various approaches to producing a controlled digital mosaic, depending on how the georeferencing is done. If the input raster images are georeferenced by image-to-image registration using a world file, the simplest way is to modify the world file of one image to make it align neatly with adjacent images, assuming that no rotation or warping is required. Since the world file essentially specifies the coordinates of the center of the upper left pixel of the image and contains the pixel size, it is relatively easy to figure out where the upper left corner of one image is relative to another. Once the necessary changes are done, the images can be displayed on the computer screen simultaneously as a composite view. If they do not fit well, it is necessary to adjust the image slightly until no obvious gaps are apparent between adjacent images. Since the input images are not physically rectified, the resulting digital composite image is a semicontrolled mosaic.

If the raster images are georeferenced by image-to-map rectification, it is possible to construct a fully controlled digital mosaic. This can be done by using the raster management function of a GIS. In ArcInfo, for example, there is an image catalog function that allows the user to organize a collection of spatially referenced images so that they can be accessed as one single logical image. With the aid of this function, the user first creates the template for the image catalog and then adds and removes individual images to or from the catalog as necessary. The user also defines the extent of the composite image to be displayed so that different portions of the image can be viewed selectively. A digital mosaic created in this way is a *virtual mosaic* in the sense that although the composite image itself appears as a single view, the original images remain physically separate files. If a single file is required for the composite image, the user has to deploy other image manipulation commands, such as MERGE or GRIDINSERT in ArcInfo, to create the necessary output file. This file is then used to distribute the digital mosaic or to generate a hard copy of the mosaic on a large-format electrostatic printer.

Alternatively, a composite image may also be created using the mosaic-making function of a digital image-processing package, such as ENVI, ER MAPPER, ERDAS IMAGINE, and PCI OrthoEngine. In the Windows environment, these packages make use of a wizard to guide the user through the steps of mosaic construction in an interactive manner. Since the images are in digital form, the user can easily preprocess the images to enhance and manipulate their geometric and pictorial qualities. For example, it is much quicker and less costly to rectify an image digitally than to produce a rectified photographic print in the conventional way (Mullen, 1980). At the same time, with the aid of digital image analysis tools, the user can interactively balance the pictorial quality of adjacent images to ensure a gradual transition of the gray scale from one image to another in the digital mosaic. Two digital image analysis techniques are particularly useful for this purpose.

- *Histogram Matching*. This technique aims to balance the gray scales of all the input images (Figure 5.10). The user first uses the contrast stretching function of an image analysis package to generate the histograms of all the input images. Choosing a particular image as the base image, the user then changes the gray scales of other images by matching their histograms to that of the base image. This will result in all the images having the same gray-scale characteristics, thus giving them a uniform appearance when displayed on the computer screen or plotted to generate a hard copy.
- *Feathering*. The purpose of feathering is to blend or blur the seams between mosaicked images. This is achieved by changing the gray-scale values of the cells according to the weighted average of their distances from the edge of the image. The weighted average for each cell is zero at the boundary and rises toward one at the feathering distance. This means that, for example, if the specified distance is 20 pixels, 0% of the top image is used in the blending at the edge and 100% of the bottom image is used to make the output image. At the blending distance (i.e., 20 pixels from the edge), 100% of the top image is used to make the output image and 0% of the bottom image is used. 50% of each image is used to make the output image at 10 pixels from the edge. In practice, there are two approaches to feathering: *edge feathering*, which blends the image

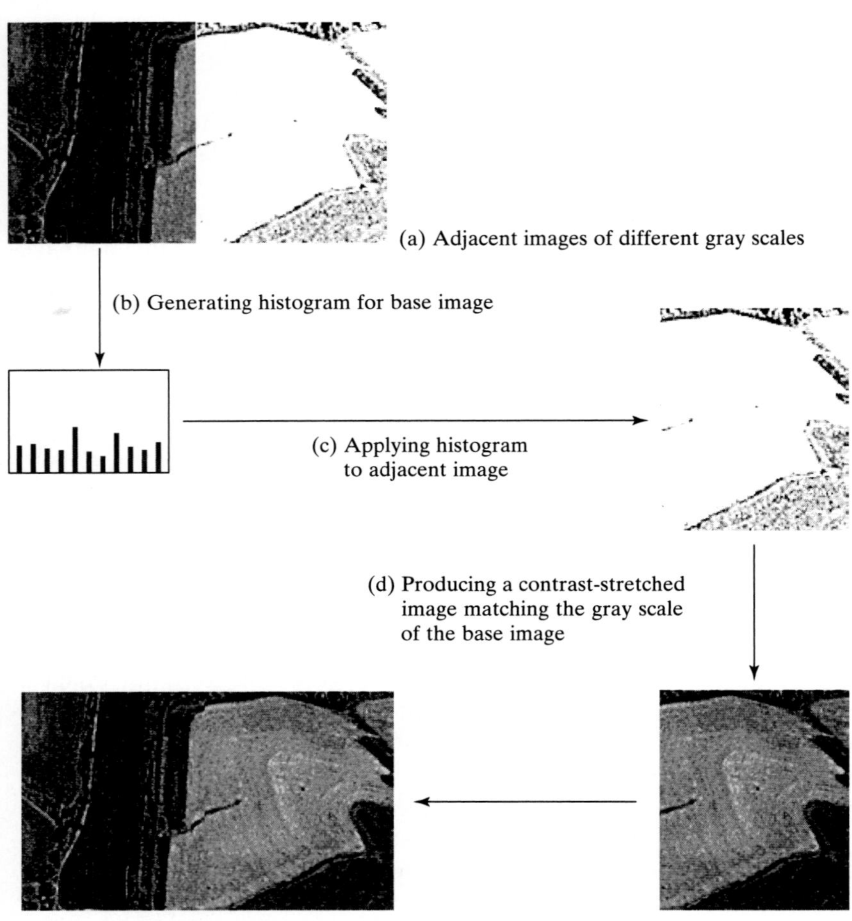
(a) Adjacent images of different gray scales
(b) Generating histogram for base image
(c) Applying histogram
to adjacent image
(d) Producing a contrast-stretched
image matching the gray scale
of the base image
(e) Resultant mosaic of uniform gray scale

FIGURE 5.10
The principle and process of histogram matching.

seams along the edge of the input images (Figure 5.11a); and *cutline feathering*, which blends the image seams along a selected "cutline" drawn from edge to edge (Figure 5.11b). Computationally, the cutline approach is slightly more complex but it is the obvious choice if linear features such as hedges, rivers, and roads are present near the edge of the images. These features form excellent cutlines because their distinct curvilinear characteristics can easily hide the image seams on the resulting digital mosaics.

Producing a mosaic digitally is apparently easier than conventional manual methods using photographic paper prints. The end product may also be of a higher quality because of the user's ability to enhance and balance the quality of the input images before they are assembled. This ability to manipulate the input images also gives digital mosaicking the additional advantage of using images from different sources and at different spatial resolutions, such as Landsat and SPOT images. The methods of digital mosaicking described here apply equally well to 8-bit gray scale as well as 8-bit palette color and 24-bit true-color images. In the past, the prohibitively large file size was the major impediment to creating and using digital mosaics. With the new raster data compression techniques noted in Section 5.1.2, digital mosaics covering large geographic areas have now become possible.

5.3 Raster-based GIS Data Analysis

The data analysis domain of a GIS, no matter whether it is raster- or vector-based, includes a variety of data-processing functions that aim to derive spatial relationships, patterns, and trends that are implicit in the source data. These functions accept spatial data represented by the raster or vector data model, and return the results in the same data model. The results of data analysis may be used immediately for spatial problem solving and decision making or as input for further spatial analysis and modeling.

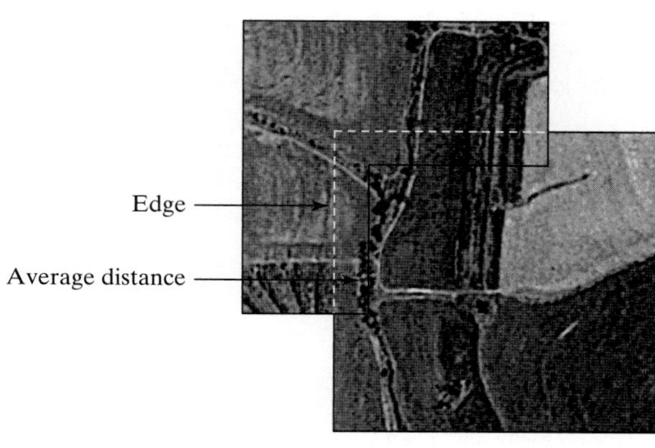

(a) Edge feathering

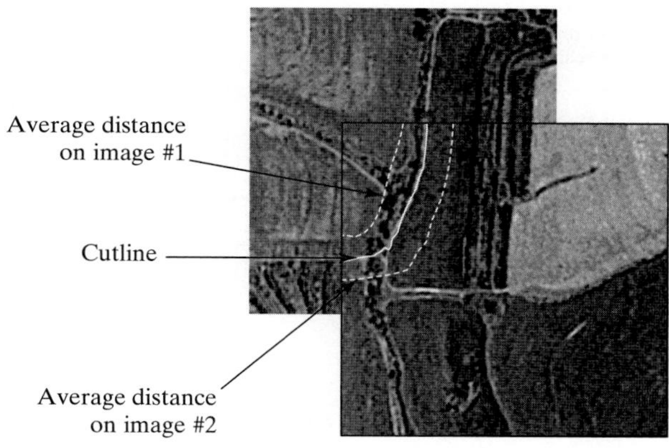

(b) Cutline feathering

FIGURE 5.11
The two approaches to feathering.

5. *Geometric transformation operations* are processes that modify the spatial properties of features on a layer by applying linear transformations, such as scale change and translation, and nonlinear transformations, such as warping in image rectification.
6. *Geometric derivation operations* are processes that create new features from existing features on a layer, using techniques such as generalization, triangulation, filtering, and surface interpolation.

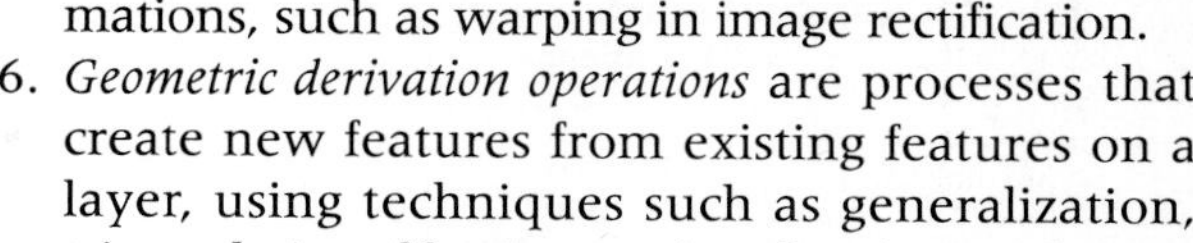

Another common method of classifying raster data-processing operations is based on the ways of using raster cells in the operations. On a raster layer, the smallest addressable unit is a *point*, which is represented by a cell. Each point can be addressed as a part of a *neighborhood* of surrounding values. When all neighboring points having the same attribute value are grouped together, a *region* is identified (Figure 5.12). In raster-based data processing, some processes use the values of individual cells only, and others rely on neighborhood relationships or regional associations. This provides a logical basis for classifying raster data analysis operations into the following four categories:

1. *Local operations* are processes that create an output layer on which the value of each cell is a function of the cell at the same location on the input layer.
2. *Neighborhood operations* are processes that create an output layer on which the value of each cell is a function of the cells neighboring the cell at the same location on the input layer.

5.3.1 A MATRIX OF RASTER-BASED GIS ANALYSIS TECHNIQUES

Raster-based GIS analysis techniques can be classified in different ways. Giordano et al. (1994), for example, identified six categories of geographic data analysis operations.

1. *Logical operations* are processes that use logical operators (i.e., "AND," "OR," and "XOR") to create a new layer from two input layers.
2. *Arithmetic operations* are processes that use arithmetic operators (i.e., addition, subtraction, multiplication, division, and assignment) to create a new layer from two input layers or by transforming the cell values in an input layer.
3. *Overlay operations* are processes that merge attribute values from two or more layers, using logical operators or arithmetic operators.
4. *Geometric property operations* are processes that compute indices that describe the geometric properties pertaining to spatial features on a layer, such as shape, size, angle, and topological relationships.

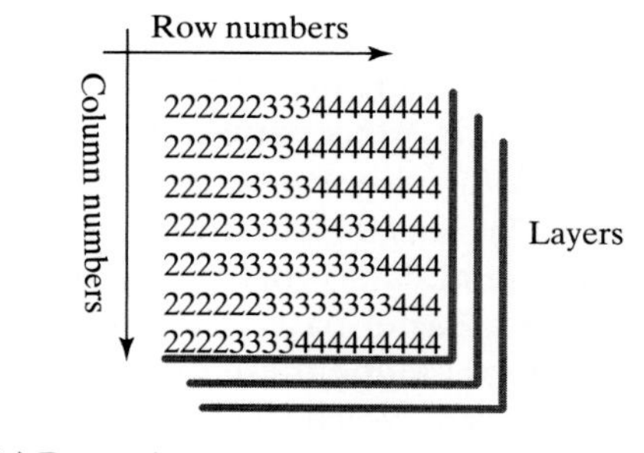

(a) Raster data structure

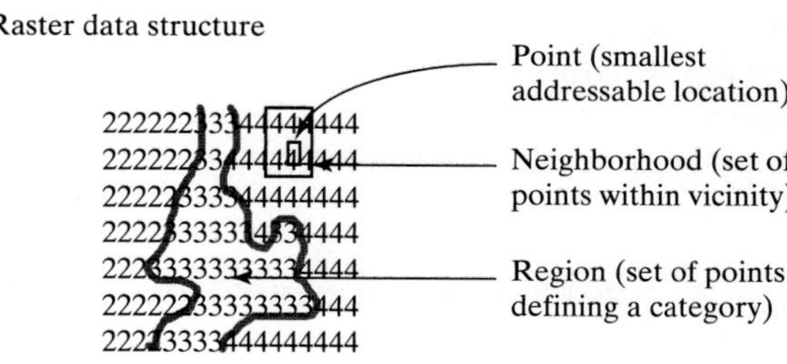

(b) Point/neighborhood/region

FIGURE 5.12
(a) Raster data structure referencing system and multiple map layers (b) The concept of point, neighborhood, and region.

	Local Operations	Neighborhood Operations	Extended Neighborhood Operations	Regional Operations
Logical Operations	• Reclassification			
Arithmetic Operations	• Reclassification	• Aggregation • Filtering	• Statistical analysis	
Overlay Operations	• Logical • Arithmetic			• Category-wide overlay
Geometric Property Operations		• Slope and aspects	• Distance, proximity, and connectivity	• Area • Perimeter • Shape
Geometric Transformation Operations			• Rotation • Translation • Scaling	
Geometric Derivation Operations			• Buffering • Viewshed analysis	• Identification and reclassification

FIGURE 5.13
A matrix of raster-based geographic data-processing techniques.

3. *Extended neighborhood operations* are processes that create an output layer on which the value of each cell is a function of the cells neighboring and beyond the immediate neighborhood of the cell at the same location on the input layer.
4. *Regional operations* are processes that create an output layer by identifying cells that intersect with or fall within each region on the input layer.

By combining the characteristics of these two approaches of classifying data-processing operations, a matrix of raster-based data analysis techniques can be constructed (Figure 5.13). This matrix summarizes the characteristics and nature of data-processing functions that are found in a typical raster-based GIS. It also depicts the interrelationships among these functions and the ways raster cells are used, which GIS users involved in raster-based geographic data processing should understand. The functions identified in the matrix and examples of their applications, where appropriate, are explained in the following sections.

5.3.2 LOCAL OPERATIONS

Local operations perform raster-based data analysis on the point-by-point or cell-by-cell basis. There are two important groups of local operations: (1) *reclassification,* also commonly referred to as *recoding,* and (2) *overlay analysis.*

Reclassification The purpose of reclassification is to create a new raster layer by changing the attribute values of the cells of the input layer. This usually takes one of the following forms that use either logical or arithmetic operators.

- assigning a new value to each value on the input map layer with the purpose of developing a binary (0 and 1) mask for use in subsequent GIS analysis, a process known as *binary masking* (Figure 5.14a)
- assigning new values to classes or ranges of old values with the purpose of reducing the number of classes in the original input layer or to group values into categories in a new classification scheme (Figure 5.14b)
- assigning ranks to unique values or categories of values found on the input layer (Figure 5.14c)
- assigning ranks or weighting to a qualitative (nominal scale) map layer to generate a quantitative map layer (ordinal, interval, or ratio scales) (Figure 5.14d)

Reclassification operations are useful for a wide variety of purposes in raster-based data analysis. These include, for example, simplification of raster images for land use and land-use change studies (Lo and Shipman, 1990), and conversion of cell values from one measurement scale to another for terrain and landform analysis (Gao and Lo, 1995; Mitchell, 1973). Reclassi-

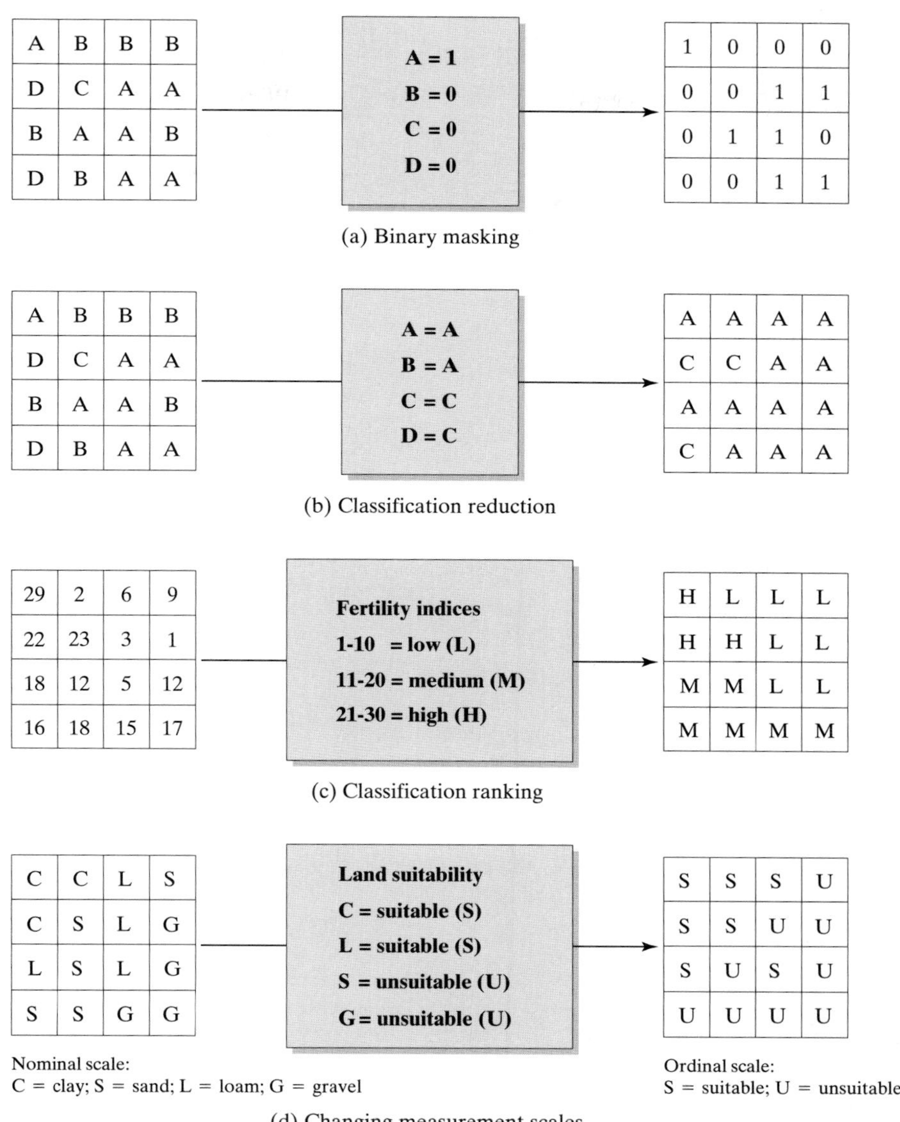

FIGURE 5.14
Methods of reclassification.

fication is also often required to generate a binary layer from results of overlay analysis as explained here.

Overlay Analysis Overlay has been widely regarded as one of the oldest map analysis methods. Map users have a long history of putting one map on top of another in order to detect the association between their contents (Goodchild, 1992; McHarg, 1969). The manual method of overlay analysis was time-consuming and error-prone because it was often necessary to redraft the maps before they could be overlaid. There was also a limit to the number of maps that could be analyzed by overlaying. To find ways that would automate overlay analysis was one of the major objectives of developing GIS (see Chapter 1).

In raster-based geographic data processing, overlay analysis uses either *logical* or *arithmetic* operators. Logical overlay methods use logical AND, OR, and XOR (exclusive OR) to generate a new layer from two input layers. Mathematically, the logical AND operation multiplies individual values of corresponding cells in the input layers, and the logical OR and XOR operations add individual values of corresponding cells in the input layers. The difference between logical OR and XOR is dependent on the way of producing the resulting binary layer. Whereas logical OR reclassifies output cell values that are equal to 2 from 2 to 1, logical XOR reclassifies these cell values from 2 to 0 (Figure 5.15).

Arithmetic overlay methods use ADDITION, SUBTRACTION, MULTIPLICATION, DIVISION, ASSIGNMENT,

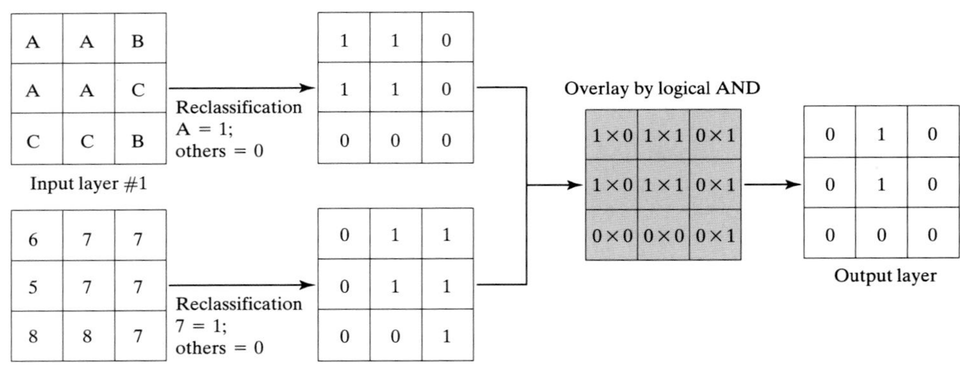

(a) Overlay by logical AND to find "A" and "7" in input raster layer

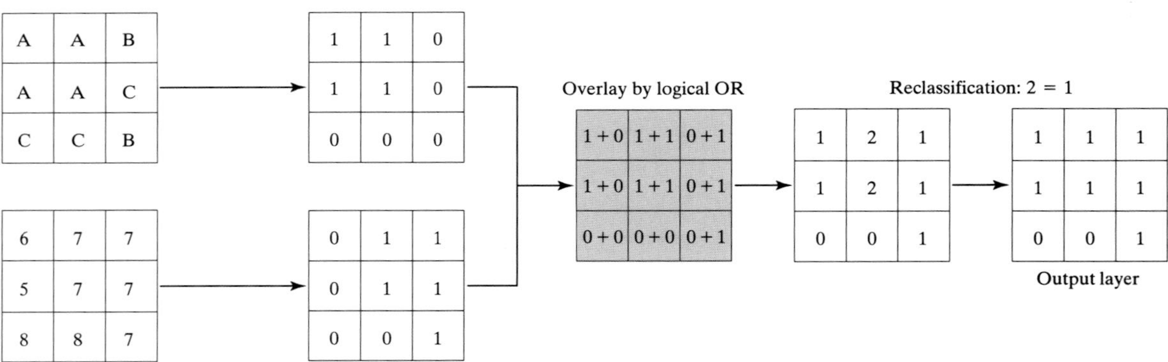

(b) Overlay by logical OR to find either "A" or "7" in input raster layer

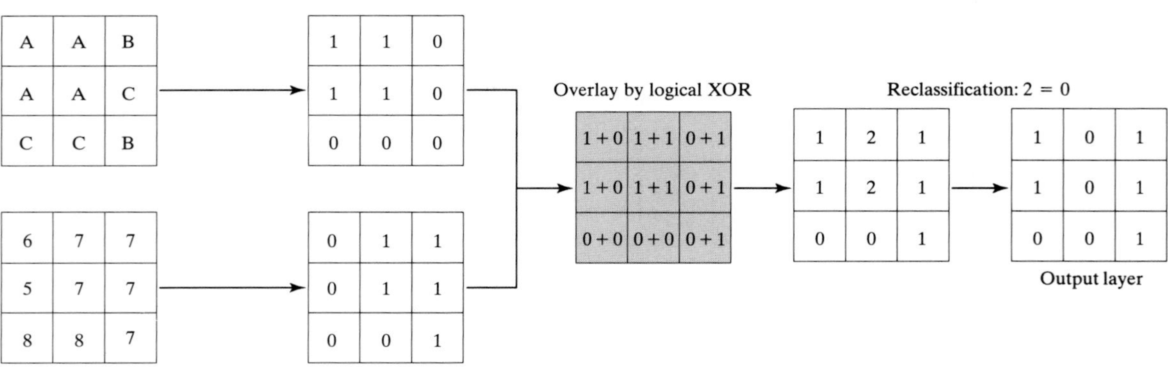

(c) Overlay by logical XOR to find only "A" or only "7" in input raster layer

FIGURE 5.15
Logical overlaying operations. Overlaying operations often require reclassification before and after the operation itself.

and other arithmetic operations to combine two or more map layers (Figure 5.16). If [A] is input layer 1, [B] is input layer 2, and [O] is the output layer created by overlaying, the following arithmetic operations are possible.

$$[O_a] = [A] + [B] \quad \textbf{(5.1a)}$$

$$[O_b] = ([A] + [B])/2 \quad \textbf{(5.1b)}$$

$$[O_c] = [A] - [B] \quad \textbf{(5.1c)}$$

$$[O_d] = [A]*[B] \quad \textbf{(5.1d)}$$

$$[O_e] = [A]/[B] \quad \textbf{(5.1e)}$$

$$[O_f] = [A] \text{ if } [A] > [B] \quad \textbf{(5.1f)}$$

$$[O_g] = [A] \text{ if } [A] < [B] \quad \textbf{(5.1g)}$$

$$[O_h] = [B] \text{ if } [B] > 0; \text{ otherwise}$$

$$[O_h] = [A] \quad \textbf{(5.1h)}$$

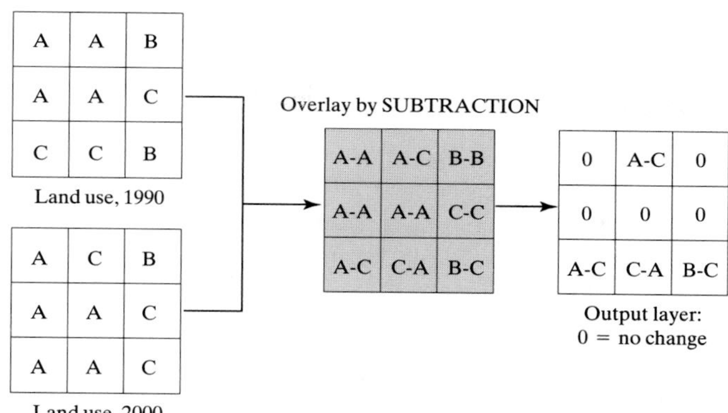

(a) Overlay by arithmetic SUBTRACTION to detect land-use change

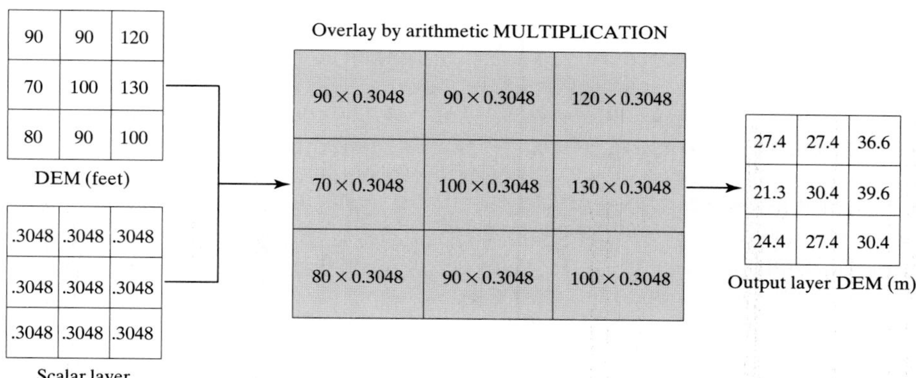

(b) Overlay by arithmetic MULTIPLICATION to convert DEM data from feet to meters

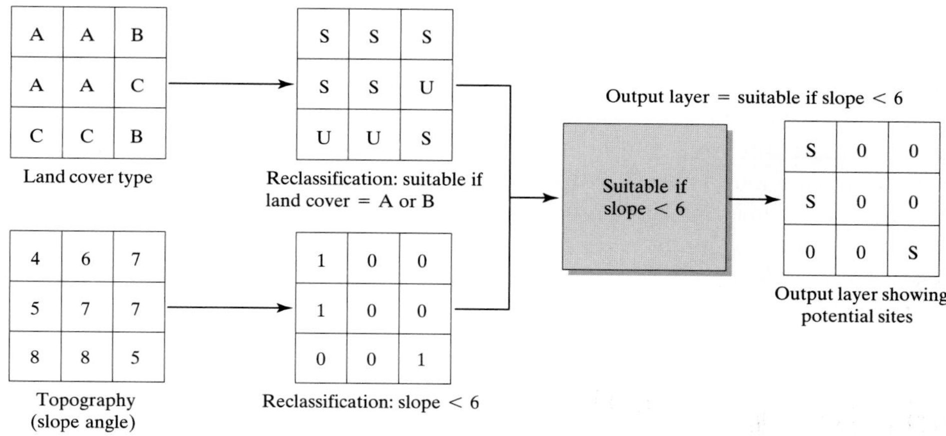

(c) Overlay by arithmetic ASSIGNMENT to select suitable sites

FIGURE 5.16
Examples of arithmetic overlaying operations.

In the logical and arithmetic overlaying operations noted here, the new layer is created by assigning to each of its cells a value that is a function of the independent values associated with the cells at the same location on two or more existing layers. These operations are generally called *location-specific overlaying* in order to distinguish them from another approach of overlay analysis called *category-wide overlaying*, where the new layer is created by assigning values to the entire thematic regions as a function of the values on the input layers that are associated with the categories (see Section 5.3.5).

In location-specific overlay operations, the output layer from one operation can be used as input layer for further overlay analysis operations. For example, the output layer $[O_a]$ from the overlay operation represented by Eq. 5.1a, can be combined with a raster layer $[C]$ to create a new layer called $[O_i]$. Using $[O_i]$ and another output $[O_e]$ as input, a new raster $[O_j]$ can be generated. This ability of location-specific overlay analysis to successively combine multiple raster layers in a variety of ways is the conceptual basis of what is called *cartographic modeling* (see Section 5.5). Overlay analysis is now the core operation in a wide range of raster-based GIS applications. These include database update, change detection, site selection, suitability studies, multivariate classification, and accuracy assessment (Morain and Baros, 1996).

5.3.3 OPERATIONS ON LOCAL NEIGHBORHOOD

Neighborhood operations, also known as *context* or *focal operations*, make use of the topological relationship of adjacency between cells in the input raster layer to create a new raster layer. The neighborhood is defined by using a "window" that is moved over each cell on the input layer to extract new cells values for the output raster layer (Figure 5.17). Local neighborhood operations are developed on the assumption that the value of a particular cell in a raster layer is often affected by the values of its neighboring cells. A commonly used window size is 3 × 3 cells, but there is no theoretical limit to the window size that can be used. Obviously, the larger the window size, the more input cells are included in the calculation of the value of each output cell. However, this also necessitates more intensive computation. Neighborhood operations that are found in a raster-based GIS can be classified according to their objectives into the following three groups: (1) *spatial aggregation*, (2) *filtering*, and (3) *computation of slope and aspects*.

Spatial Aggregation Spatial aggregation of raster cells is the process of progressively down-sampling a large raster layer to reduce it to a smaller number of cells. Aggregation is not a data compression process because it aims to represent the same geographic space using a smaller number of cells by means of a coarser spatial resolution, rather than by compressing the file size mathematically as noted in Section 5.1.2. It is basically a cartographic generalization process.

As a neighborhood operation, spatial aggregation can be carried out using a window of any size to compute the values of the cells on the output layer. A larger window size will result in a higher aggregation level. A high aggregation level means a smaller aggregated layer, but greater loss of detail as well. In addition to the choice of the window size, the user also has the option to use one of three aggregating methods (Bian and Butler, 1999) (Figure 5.18):

- *The averaging method* computes the average value of all the cells over the window and uses it as the value of the aggregated cell.
- *The central-cell method* assumes the value of the cell at the center of the window to be the value of the aggregated cell.
- *The median method* computes the median value of all the cells over the window and uses it as the value of the aggregate cell.

Spatial data aggregation is an important process in environmental modeling and scientific research today. Conventionally, spatial data are often collected at scales suitable for local analysis and modeling. These spatial data are not useful for analysis and modeling at regional and global levels because of the incompatibility between scales and application objectives (Bian and Walsh, 1993). Many spatial patterns and trends implicit in raster images will not show up unless the data are analyzed synoptically at regional, national, or even continental levels. Spatial aggregation is necessary to turn high-resolution images into low-resolution ones so that

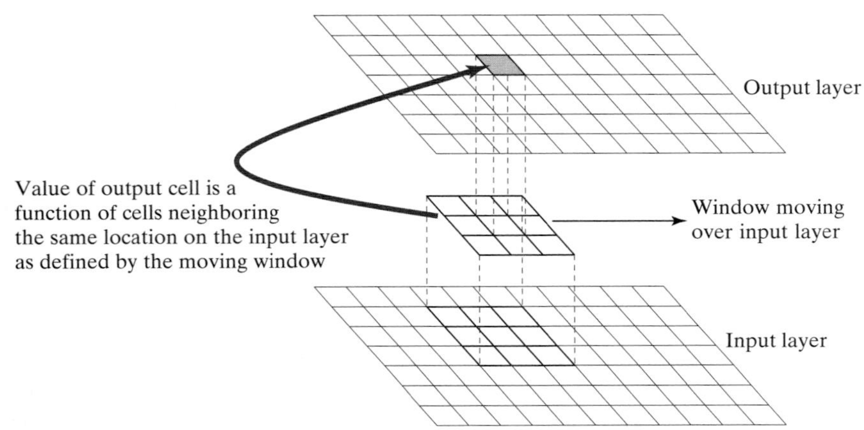

FIGURE 5.17
The concept of using a moving window to extract cells values from an input raster image for the computation of the value of an output raster cell in neighborhood operations.

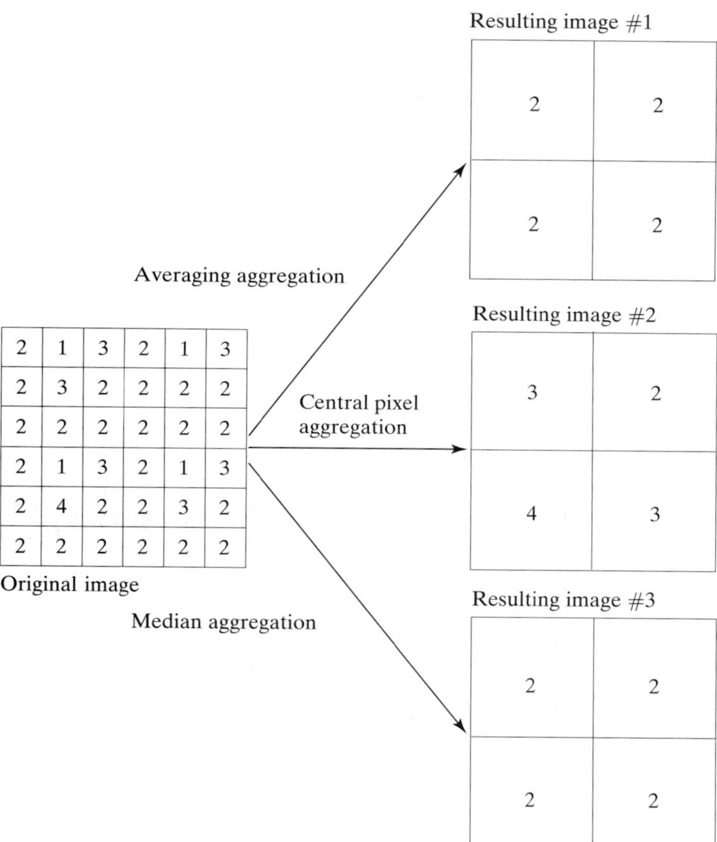

FIGURE 5.18
Methods of spatial aggregation using a 3 × 3 window. Spatial aggregation leads to loss of local details but allows regional patterns and trends to show up when the raster images are displayed and analyzed synoptically.

they can be used according to the needs of particular GIS applications.

Filtering Filtering is a digital image-processing function for image enhancement (Jensen, 1996). From the remote sensing point of view, the value in each cell on the raster image indicates the degree of brightness at that point. The change in brightness value per unit distance for any part of the image gives rise to an image characteristic called *spatial frequency*. If the change in brightness value over an area in the image is small, this is called *low-frequency area*. If the brightness values change rapidly over short distances, this is an *area of high frequency*. The spatial frequency of the raster image can be subdued or enhanced using the method of filtering.

The most commonly used method of filtering is called *spatial convolution filtering* (Jensen, 1996). This involves passing a *filter window*, also called a *kernel*, across the entire image. This filter window is commonly a 3 × 3 matrix, but larger sizes are possible. It contains user-defined coefficients that are used to either subdue or enhance the spatial frequency of the image (Figure 5.19). When the filter window is passed over the image, the coefficients in the filter window are multiplied by the brightness values of the corresponding cells on the input raster image, and the average of the results of multiplication is the new brightness value assigned to the image cell at the center of the filter window. The filter window is then shifted to the next cell, the computation repeated, and a new average value is assigned to the image cell at the center of the window. This operation is repeated for every cell in the input image.

Filtering is useful for improving image quality through removal of noise and other defects inherent in the original data. It is particularly helpful in bringing out linear features on raster images, and therefore commonly used in the analysis of drainage networks. Other applications include the enhancement of road and rail alignments, vegetation and soil type boundaries, land-use interfaces, water body limits, urban area infrastructure, as well as settlement patterns.

Determining Slopes and Aspects These neighborhood operations are used for digital elevation models. Local slope angles and aspects can be computed to create two new layers. The angle of slope (a) is determined by the following formula (Figure 5.20).

$$\tan a = vd/hd \qquad \textbf{(5.2)}$$

$$a = \tan^{-1} vd/hd \qquad \textbf{(5.3)}$$

where *vd* is the difference in height between two points, and *hd* is the horizontal distance between the two points for which the slope angle is to be determined. Aspect is

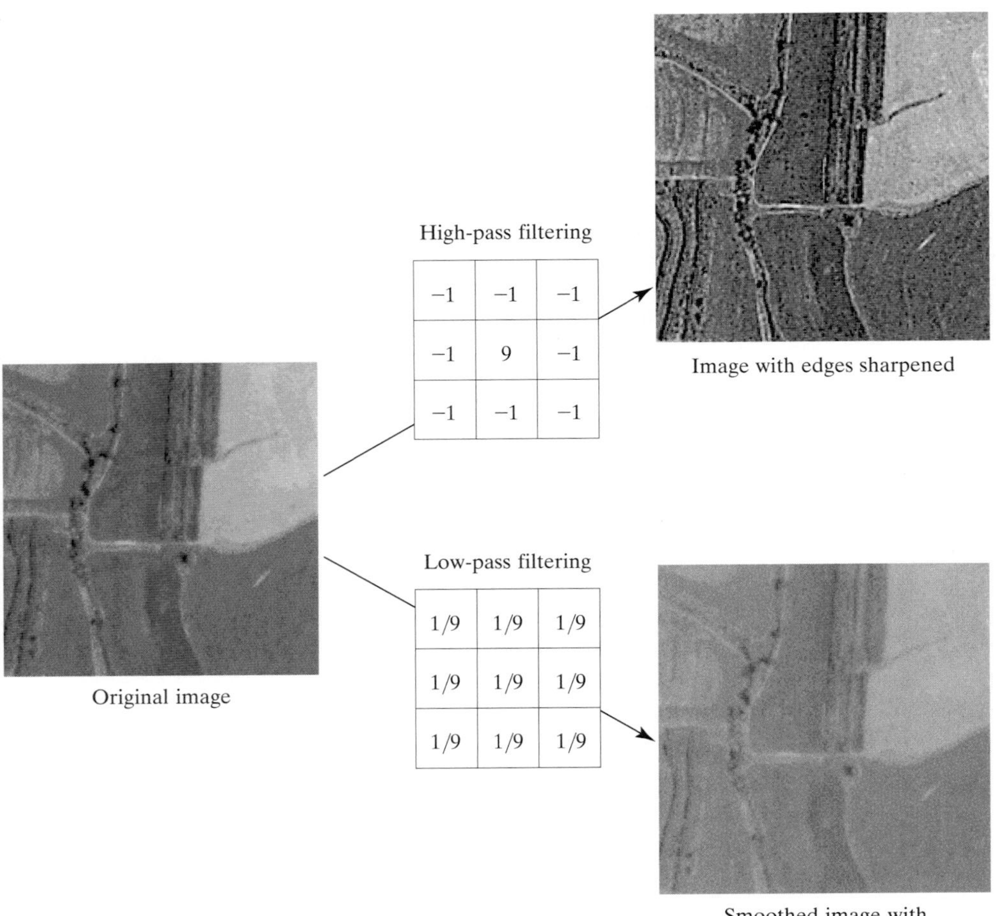

FIGURE 5.19
High-pass and low-pass image filtering.

normally defined as the direction that the maximum slope faces.

Figure 5.21 illustrates some elevation data in raster format. For a 3 × 3 window in the raster layer, the center cell is surrounded by 8 cells, and each neighboring cell can be designated a cardinal direction as N, NE, E, SE, S, SW, W, and NW. It is possible to compute a slope angle of the center cell to each of these 8 directions. To compute the slope angles to N, E, S, and W directions, the difference in height between the center cell and the cell in each of these four directions can be computed to give the vertical distance (*vd*) in Equation 5.3, and the horizontal distance (*hd*) between the center cell and each of the four cells is just the cell size or resolution of the raster data (assuming that square cells are used).

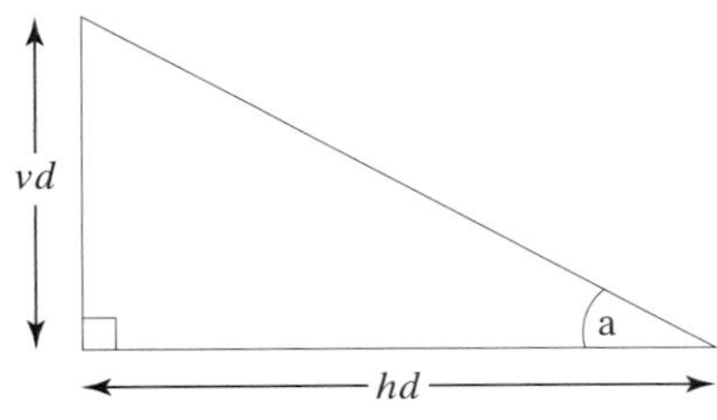

FIGURE 5.20
Computation of the slope angle.

If slope angles to NE, SE, SW, and NW are required, the vertical distance (*vd*) is the difference in height between the center cell and each of these diagonal cells. The horizontal distance (*hd*), however, has to be computed using the Pythagorean theorem. Assuming that we have to compute the slope angle between the center cell and the SE cell, the horizontal distance (*hd*) between the center cell and the SE cell can be found from the following formula.

$$hd = \sqrt{\{(s - c)^2 + (s - se)^2\}} \quad \textbf{(5.4)}$$

The values of *vd* and *hd* can be substituted in Equation 5.3 to determine the slope angle. By comparing the slope angles computed for all the eight directions, the maximum slope angle can be determined. The direction

NW	N	NE
W		E
SW	S	SE

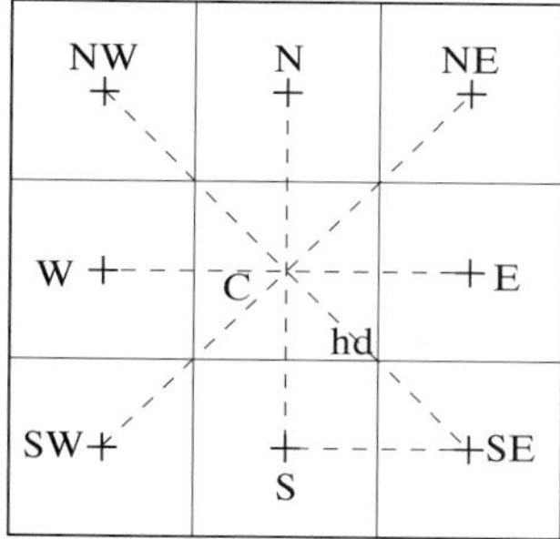

FIGURE 5.21
Eight cardinal directions surrounding a central cell in a 3 × 3 window.

where the maximum slope faces defines the *aspect* of the slope. In other words, the aspect of the slope is the direction of maximum descent of the slope. The aspect is normally measured as the angle clockwise from north.

The computation of slopes and aspects is an essential step in terrain analysis using DEM data. This particular neighborhood operation is used in hydrologic modeling to delineate watershed boundaries and river channels, as well as to compute direction and volume of flow (see Chapter 9). Conventionally, slope and aspect data were used mainly for calculating the intensity of light in terrain visualization. However, such data are now increasingly integrated with other sources of topographic and environmental data for studies in microclimates, wildlife habitat analysis, biodiversity, and land-use suitability mapping.

5.3.4 OPERATIONS ON EXTENDED NEIGHBORHOODS

Operations on extended neighborhoods process a set of points that extend over a much larger area than the local neighborhoods described in the previous section, and hence involve a greater number of cells. A new raster layer is produced as a result of an extended neighborhood operation. There are five important groups of operations under this category: (1) statistical analysis of the raster; (2) determination of distance, proximity, and connectivity; (3) geometric transformation of raster layers; (4) buffering; and (5) viewshed analysis. Since the spatial extent of an extended neighborhood can include the entire raster layer, these operations are sometimes referred to as *global operations*.

Statistical Analysis of Raster Layers Geographic analysis and modeling often start with statistics that describe the characteristics of the entire raster layer or selected parts of its contents. Therefore, statistical analysis of the raster image is an important function in raster-based geographic data processing. Statistical analysis is based on the use of attribute values rather than the location of the cells on a raster layer. The resulting descriptive statistics pertaining to a single raster layer include the mean, the median and the most common value, frequency, spatial autocorrelation, among many other statistical indices, calculated for all the raster cells on the entire layer (Bonham-Carter, 1994). When two raster layers are compared, a variety of descriptive statistics can be generated to establish the correlation between the themes of data that they represent. These include, for example, area cross-tabulation, coefficient of agreement (kappa), chi-square test, regression, and analysis of variance (Bonham-Carter, 1994).

Descriptive statistics can be presented as text-based reports, statistical charts, and graphs (Figure 5.22). These information products turn spatial information from the pictorial form into a textual or graphical form, making it more comprehensible to the user. They are particularly useful when new layers are created by raster-based GIS operations because they provide the user with a quick summary of characteristics of the new layer. Descriptive statistics can be used directly for spatial problem solving and decision making or for input to further spatial analysis and modeling using more advanced techniques. We will study in detail the concepts and techniques of spatial analysis using GIS in Chapter 10.

Determination of Distance, Proximity, and Connectivity In raster-based geographic data processing, distances are normally measured as straight-line or Euclidean distances between two cells. If the two cells have the same row number, their distance is computed by using the difference between their column numbers times the spatial resolution. If they have the same column number, the distance is computed by using the difference between their row numbers times the spatial resolution. If they have different row and column numbers, the distance is computed by using Pythagorean theorem.

When the distances between a particular cell and all the cells on the raster layer are computed, the resulting raster layer will be made up of cells representing concentric distance values from that cell. From these distance values, concentric equidistant zones can be developed by reclassification, leading to the creation of a *proximity map* (Figure 5.23). If distance-modifying effects and traveling constraints, which include *absolute*

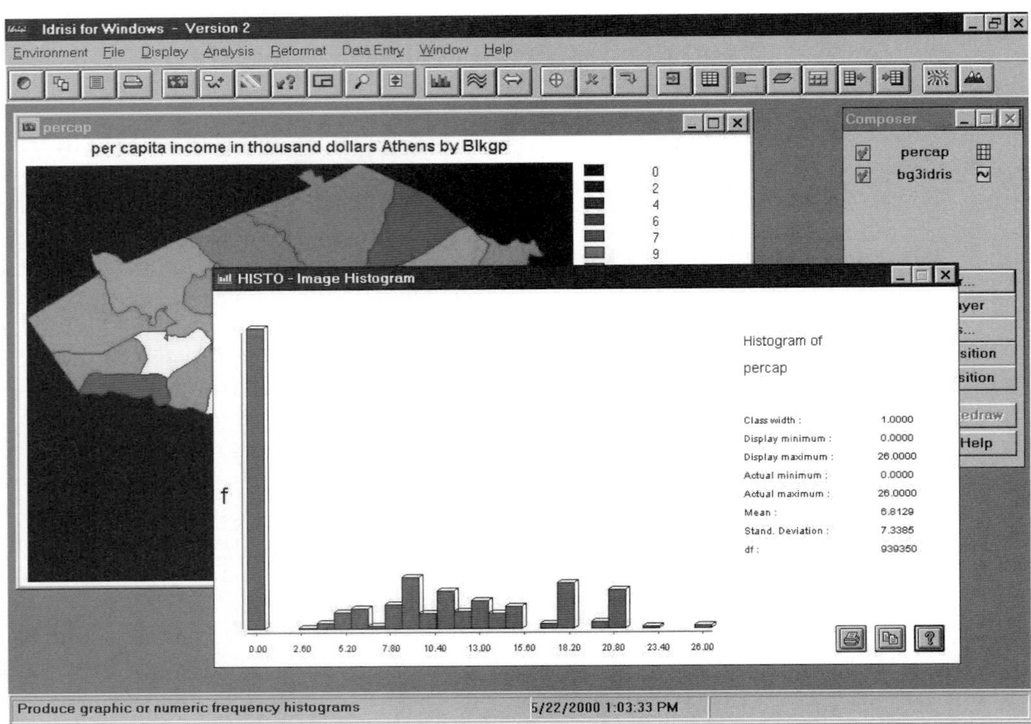

FIGURE 5.22
A histogram showing the frequency of occurrence of each class of attributes and other statistics of the displayed map layer.

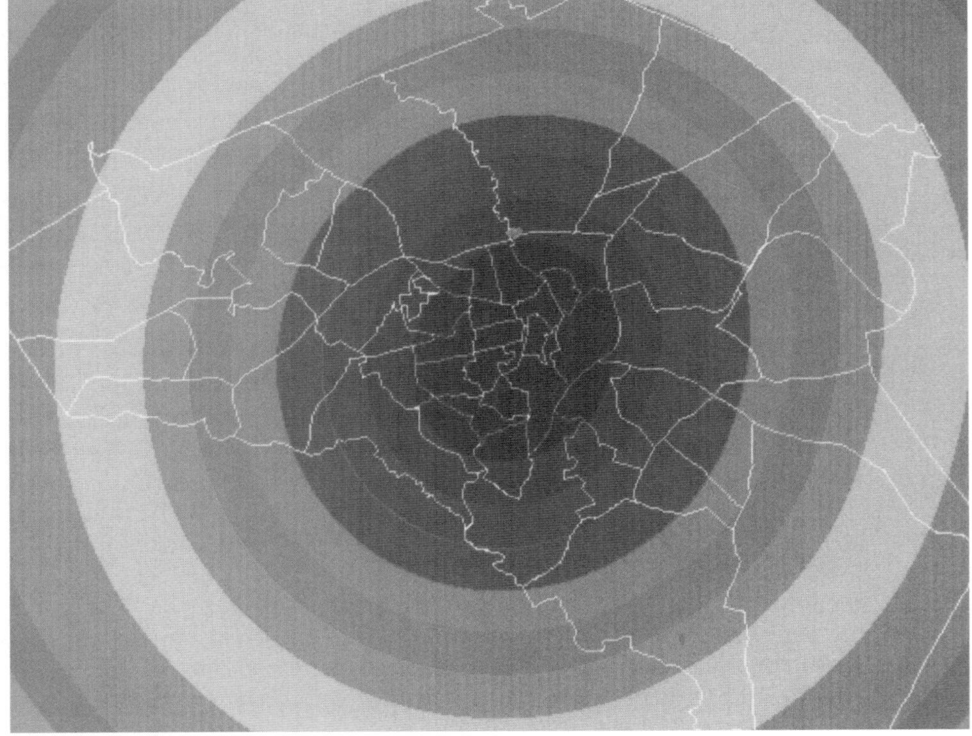

FIGURE 5.23
Concentric rings of distance from downtown Athens, Georgia. The census block group boundaries of the Athens-Clarke County are indicated in white.

barriers such as mountains and rivers and *relative barriers* such as traveling costs, are known, it is possible to establish the *connectivity* between two cells on the raster layer by means of the shortest-path or best-distance algorithms (Berry, 1993a).

Distance is the most fundamental property in geography. It has been used to determine the degree of interaction between two spatial entities. According to Newton's law of universal gravitation (1687), two bodies attract each other in proportion to the product of their masses and inversely as the square of their distances apart (Haggett et al., 1977). This law has been applied to the study of population migration and retailing, which represent two of the most important application areas of raster-based GIS in human geography today. Techniques of determining distance, proximity, and connectivity are also important building blocks for the construction of spatial models in natural resource and environmental studies such as the spread of wildfire and insect infestation, timber harvest planning, wildlife habitat analysis, and the dispersion of pollutants from a point source.

Rotation, Translation, and Scaling In essence, these are data-preprocessing functions, described in Section 5.2.2, that will ensure the compatibility between the spatial and geometric qualities of a raster layer and the objectives of a particular task in raster-based geographic data processing. The aim of these functions is to spatially position a raster layer, using the method of image-to-image registration, or to produce a new raster layer, using the method of image-to-map rectification, so that reliable measurements can be made in raster-based GIS analysis.

Buffering In the context of geographic data processing, a *buffer* is an area that is created around a particular spatial feature. This feature can be a point, a line, or an area. In the raster data model, the concept of spatial feature does not apply because individual features are not represented as independently identifiable entities but as a collection of contiguous raster cells having the same attribute values (see Section 5.1.1). In raster-based geographic data processing, therefore, a buffer is defined as the raster cells that are at a specific distance from a particular cell, a linear array of cells, or a cluster of cells (Figure 5.24).

In raster-based geographic data processing, buffers are created in one of two ways. The first approach makes use of the topological relationship of adjacency to spread the value of the identified cells to neighboring cells up to a specific buffer width from these cells. The second, more general, approach is carried out in two steps. It starts by first using the distance operator to compute the distances of all the cells in a raster layer from those cells to be buffered, and then reclassifying the cell distances into two categories: Those less than or equal to the buffer width are set to one (1), and those greater than the buffer width are set to zero (0). The results of these extended

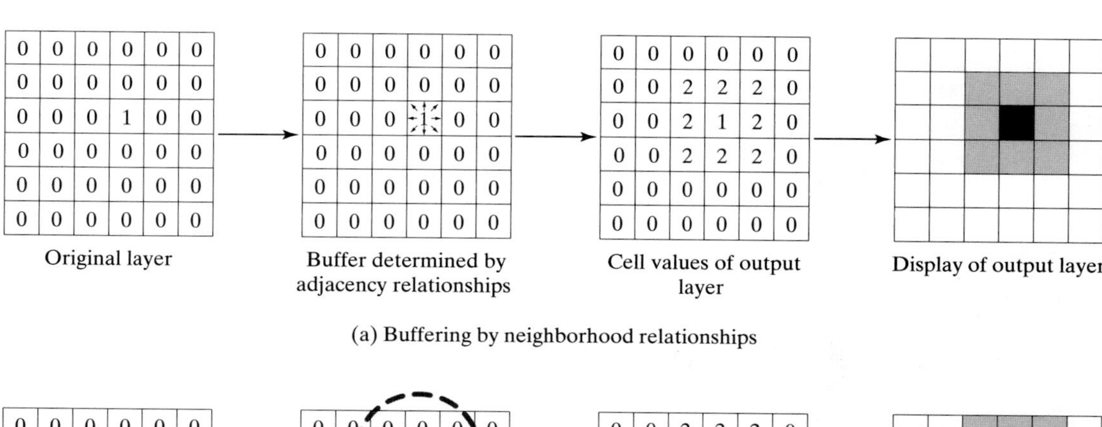

(a) Buffering by neighborhood relationships

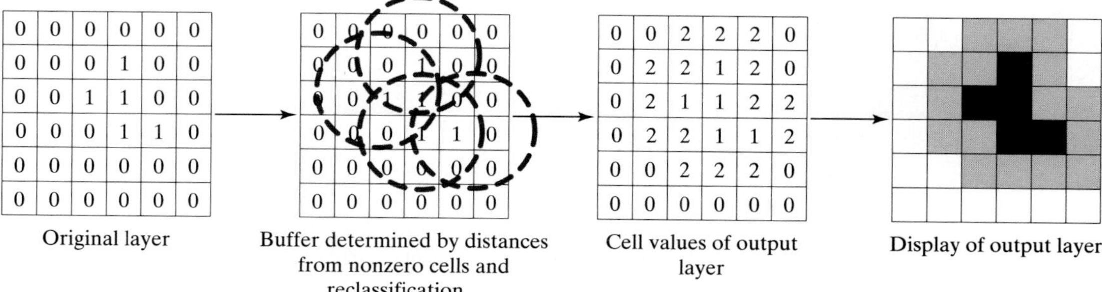

(b) Buffering by defined distance and reclassification

FIGURE 5.24
Methods of buffering on raster layers.

neighborhood operations are new raster layers containing three classes of cells: those representing the original cells, those forming the buffer, and those outside it.

The process of creating buffers, or *buffering* as it is usually called, is a very important function in GIS applications. In raster-based geographic data processing, the primary function of buffering is to identify all those cells that are adjacent to, or affected by, selected cells having a particular value on a raster layer. There are many occasions where buffering is required, for example, to provide greenways or easements around lakes and water courses, to show areas along highways where the traffic noise is above a certain level, or to delineate a safety zone around hazardous facilities.

Viewshed Analysis Viewshed analysis is the raster-based operation that makes use of terrain elevation data to determine all areas visible to an observer located at a specified point in space. It also determines all locations on the surface where the observation point can be seen. In other words, viewshed analysis is capable of determining the intervisibility between any two points on a raster layer of elevation values.

Viewshed analysis requires two input layers: a raster layer showing the locations of one or more viewpoint cells, which can be points, lines, or polygons; and a DEM layer. To determine the viewshed of a particular viewpoint cell, its elevation value is first obtained from the DEM by an overlay operation. Then, this elevation value is compared with the values of all the cells on the DEM, one cell at a time along a particular direction. These cells will be classified as visible or invisible according to three factors: their elevation values relative to the value of the viewpoint cell, their positions along the line of sight, and whether they are higher or lower than the highest point between them and the viewpoint cell (Figure 5.25a). As the viewshed is affected by the viewer's height above the elevation of the viewpoint cell, as well as the heights of vegetation cover or man-made structures along the line of sight, it is common practice to apply an average height for the viewer and an average height for trees or buildings in the computation. This can result in a more accurate viewshed (Figure 5.25b).

Intervisibility information is required in many applications using GIS. In terrain analysis, for example,

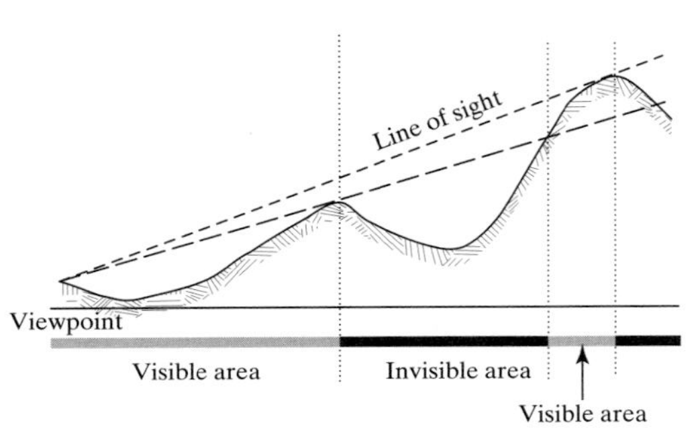

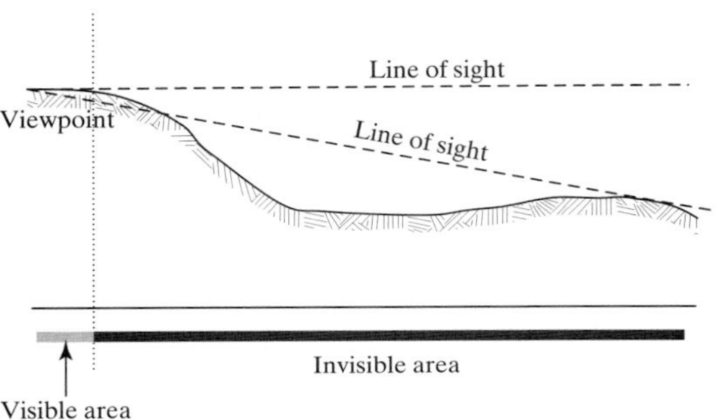

(a) Viewshed analysis using the elevation of the viewpoint

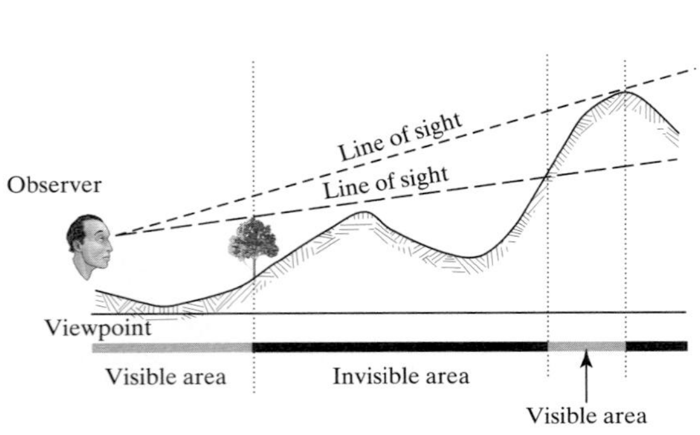

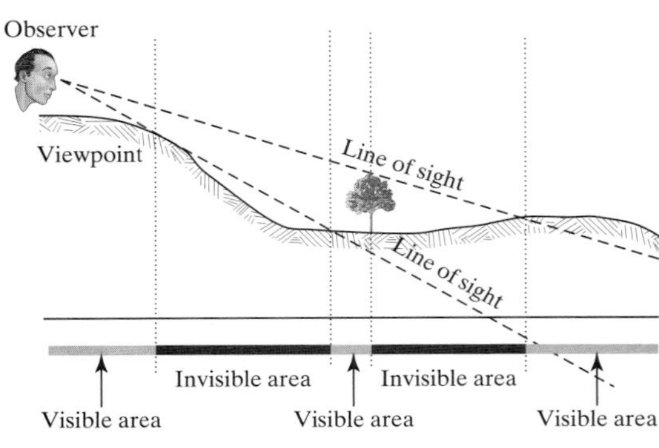

(b) Viewshed analysis using the height of the observer above the viewpoint, and taking into account the height of spatial features along line of sight

FIGURE 5.25
Viewshed analysis. The area visible from a certain viewpoint is affected by the location of the viewpoint relative to the terrain. A more realistic viewshed is obtained by adding the heights of the observer and spatial features found along the line of sight. Note how the amount of visible area changes after these heights have been taken into account in viewshed analysis in (b).

intervisibility helps the user determine what people can see from specific vantage points. This will in turn enable the user to make logical decisions pertaining to such applications as the visual impact of proposed land development projects, evaluation of real estate properties, as well as site selection for setting up telecommunication transmission networks and wildfire observation towers. It also finds important use by the military. Viewshed analysis by manual methods is a very tedious task. In raster-based GIS, viewshed analysis allows the user to obtain intervisibility information quickly and easily. The growing demand for intervisibility information in GIS applications has prompted GIS software vendors to make viewshed analysis a standard function in their products.

Users of viewshed analysis should be aware of the problems of using this particular operation in GIS applications. One of the problems is that DEM data are not always collected with a very high degree of accuracy. The USGS, for example, has reported up to 15 m root mean-square error (RMSE) in elevation. This implies that there is often a level of uncertainty in the intervisibility information obtained by viewshed analysis (Fisher, 1994). Additionally, Fisher (1993) has demonstrated that different algorithms and different GIS software packages yield vastly different results. When using viewshed analysis for spatial problem solving and decision making, these limitations must be borne in mind.

5.3.5 OPERATIONS ON REGIONS

On a raster layer, a region is a collection of cells that exhibits the same attribute characteristic. It is sometimes referred to as a *zone*, and operations on regions are called *zonal operations*. In geography, a region is an extended area with homogeneous characteristics. A region can be of different shapes, such as compact, elongated, prorupt, fragmented, or perforated (Figure 5.26). It is important to note that a region does not have to be connected on a raster layer. For example, all the wheat-growing areas in the United States, which are distributed all over the country as disparate individual clumps, can be defined as one single region since they are all characterized by the

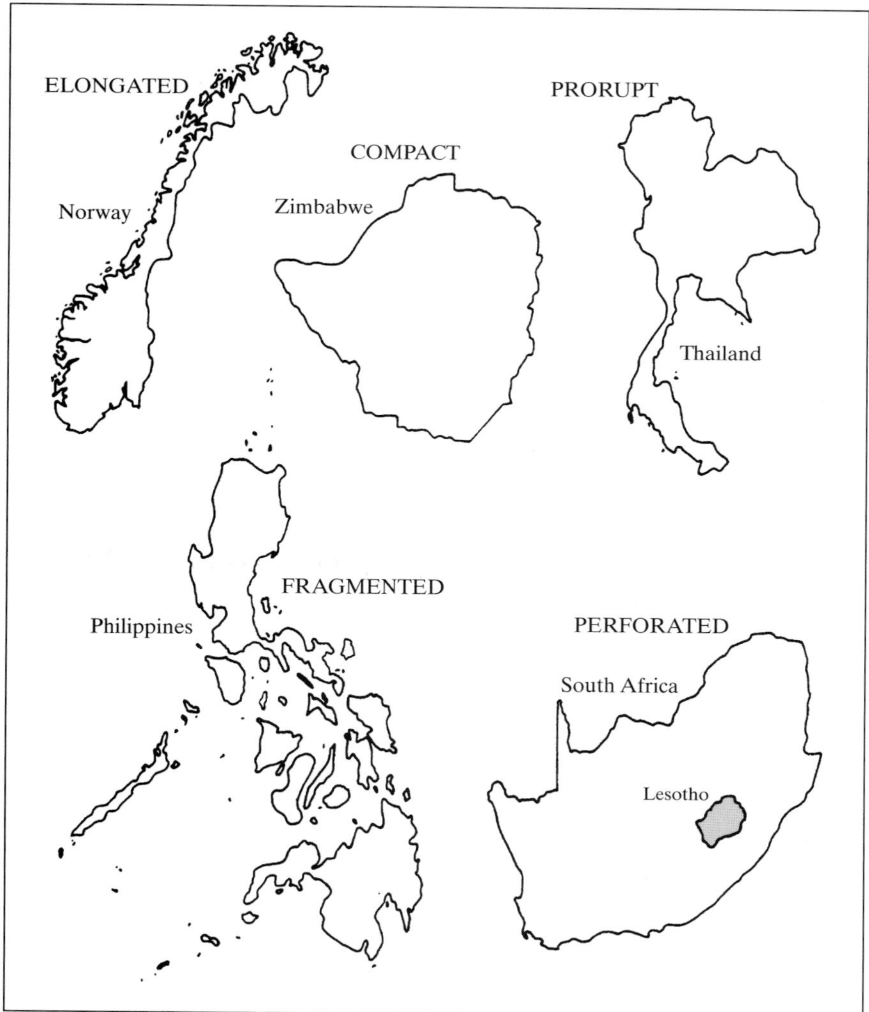

FIGURE 5.26
Compact, elongated, prorupt, fragmented, and perforated shapes as represented by some countries of the world. Note that the perimeter/area ratio will be large for complex shapes such as the Philippines and Norway. *(Source: Fellmann, Getis, and Getis, 1995)*

same attribute value. In raster-based geographic data processing, there are three major categories of zonal operations: (1) identification of regions and reclassification, (2) category-wide overlay, and (3) calculation of area, perimeter, and shape.

Identification of Regions and Reclassification On a raster layer, a region is identifiable visually by differentiating its cell values, as represented by a particular color, from values of other cells that are represented by another color. In the simplest case, a region is made up of only one single cluster or clump of cells, but, as noted, it is common for a region to include many individual clusters that are scattered all over the layer. In many applications, it is necessary to identify each cluster of a region individually. The zonal operation that does this function in raster-based data processing is known as *parceling*. This operation identifies clusters of spatially contiguous cells that have the same values and reclassifies them by giving each identified cluster a unique value (Figure 5.27).

By applying this operation, for example, all wheat-growing areas in the United States can be individually identified from a generalized raster layer representing all wheat-growing areas as one single category. The output from this particular zonal operation is a new raster layer on which individual clusters of cells, which originally have identical values, will be uniquely identified by different values. Raster layers on which individual members or clusters of a region are independently identified are useful for many applications, for example, as input for category-wide overlay analysis described next.

Category-Wide Overlay Category-wide overlay operations make use of two input raster layers to create a new raster layer. However, unlike the location-specific analysis operations that combine the input layers on a cell-by-cell basis as described in Section 5.3.2, category-wide overlay operations summarize the spatial coincidence or intersection of the entire category of values represented on one layer with the values contained on another layer (Figure 5.28). Overlaying raster layers in this way is different from location-specific overlaying operations in both concepts and objectives. Whereas location-specific overlay operations generate a new layer by combining existing layers, category-wide overlaying operations simply use the boundaries represented on one raster layer as a cookie cutter to extract cell values from other layers. The aim of category-specific overlay operations is not to find the effect of combining two or more input raster layers, but to obtain relevant data from an existing layer necessary for further analysis.

Category-wide overlay is used when category-wide statistical summaries of values of raster layers are required. In market analysis, for example, it is standard practice to superimpose a raster layer containing individually identified regions defined by ZIP code boundaries over raster layers of demographic data to determine the size, income, age, and education of the households within each ZIP code area. Similar overlay operations are also required in many natural resource and environmental applications. Summary statistics that can be generated include total, average, maximum, minimum, median, mode, and majority or minority values; standard deviation and variance; and the correlation, deviation, and other indices pertaining to particular combinations of cell values from different layers (Berry, 1993a).

Calculation of Area, Perimeter, and Shape Obtaining information about the area, perimeter, and shape of a region often constitutes the first step in many GIS ap-

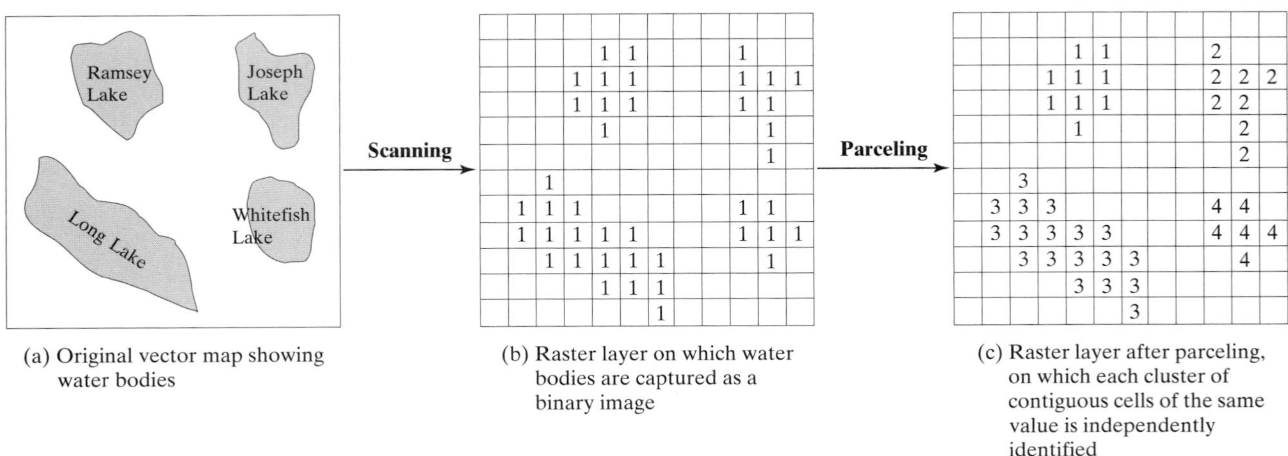

(a) Original vector map showing water bodies

(b) Raster layer on which water bodies are captured as a binary image

(c) Raster layer after parceling, on which each cluster of contiguous cells of the same value is independently identified

FIGURE 5.27
The method of parceling.

D	D	K	K	K	K
D	C	C	C	K	K
D	C	C	C	C	G
P	P	C	C	G	G
P	P	P	P	G	G
P	P	P	P	P	G

D = Dube County
K = King County
C = Clarington County
G = Green County
P = Princess Patricia County

Reclassification:
C = 1; others = "NODATA"

Input layer #1

	1	1	1		
	1	1	1	1	
		1	1		

Cells that have no value are "NODATA" cells

Input layer #2

1	0	0	1	1	0
1	1	0	0	1	0
0	1	1	0	0	0
0	1	1	1	0	0
0	0	1	0	1	0
0	0	0	0	0	0

Category-wide overlaying

Output layer (cells that have no value are "NODATA" cells)

	1	0	0		
	1	1	0	0	
		1	1		

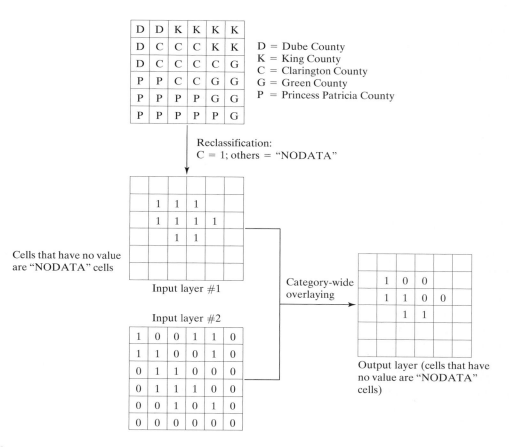

FIGURE 5.28
Category-wide overlay analysis. The aim of category-wide overlaying is to use a binary category-wide layer as a cookie cutter to extract data from an existing layer. In a raster layer, cells that have no value are commonly known as "NODATA" cells. When a NODATA cell is overlaid on another layer, the resulting output cell is always a NODATA cell.

plications. On a raster layer, area is determined by counting the number of grid cells within the boundary of a region, while perimeter is determined by adding the number of exterior cell edges of the region. The accuracy of the area and perimeter of a region is affected by the orientation of the region in relation to the grid cell. Only when the region's orientation coincides with that of the grid cells will accurate area and perimeter measurements be obtained (Figure 5.29).

Using area and perimeter measurements, it is possible to calculate the *perimeter/area ratio* that is commonly used to describe the shape of a region. The perimeter/area ratio of a circular region, which is a compact shape, is the smallest, and the ratio increases for elongated and prorupt regions. In this way, the perimeter/area ratio is a good indicator of the complexity of the form of a region. The perimeter/area ratio actually measures how much the region deviates from a circle. As such, it is also known as a *convexity index (C.I.)*.

$$\text{C.I.} = k[P/A] \quad \textbf{(5.5)}$$

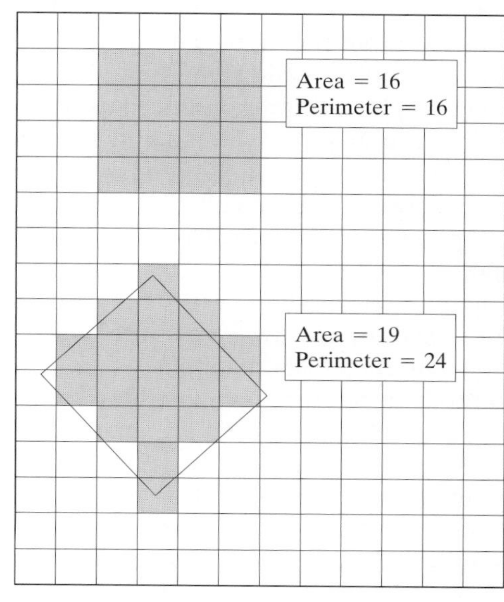

FIGURE 5.29
Errors in areas and perimeters of polygons in a raster data layer caused by orientation relative to the grid cells.

where P is perimeter, A is area, and k is a constant related to the size of the circle that would inscribe the region (Berry, 1993b). A C.I. of 100 means that the region is 100% similar to a circle.

5.4 Output Functions of Raster Data Processing

The output functions of raster data processing can be generally divided into two categories: *presentation graphics*, which makes use of the results of the analysis stage to generate graphics and statistical reports that are suitable for spatial decision making and publication; and *raster data conversion*, which is concerned with the restructuring of data extracted from the raster database for export to other applications or software systems. The emphasis of the first category of output functions is the effective communication and presentation of spatial information to end users. The second category of output functions, on the other hand, entails the conversion of raster data from one format to another. The conversion may be required between variants of the raster data model (for example, from the BIL format to a particular proprietary layered format) or between the raster and vector data models (for example, vectorization of scanned images).

5.4.1 Producing Color in Computer Graphics

Good color quality makes it easier to understand and interpret raster images. Therefore, the effective deployment of raster output functions requires a good understanding of the concepts and techniques of using color. In computer graphics, color is produced using two basic color models: the *RGB* (red, green, and blue) and the *CMY* (cyan, magenta, and yellow) models. Both models make use of three primary colors that can be mixed in varying proportions to produce all other colors. The RGB model applies when light of different colors is mixed or added (Figure 5.30a). When two of the three primary colors are mixed, their wavelengths are added to produce immediate colors. For example, red and green are mixed to produce yellow; green and blue are mixed to produce cyan, and blue and red are mixed to produce magenta. When all three primary colors are mixed together, white light is produced. This way of producing color is known as *additive mixing*, and the three primary colors are known as *additive primary colors*.

Based on the additive RGB color model, common computer monitors can produce 256 gray levels for each of the three primary colors by varying their intensities (see Chapter 7). This means that over 16 million colors

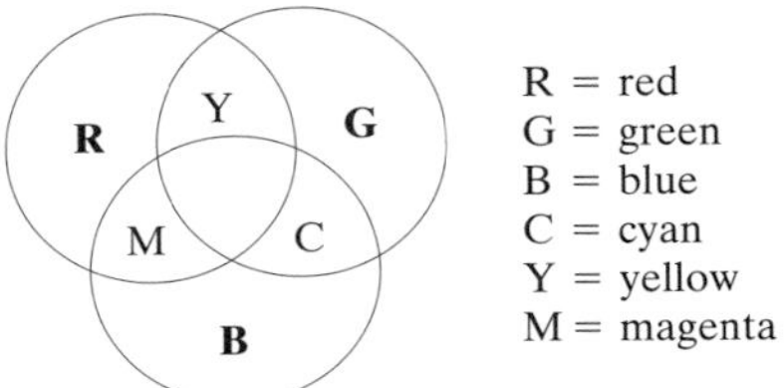

(a) The RGB color model

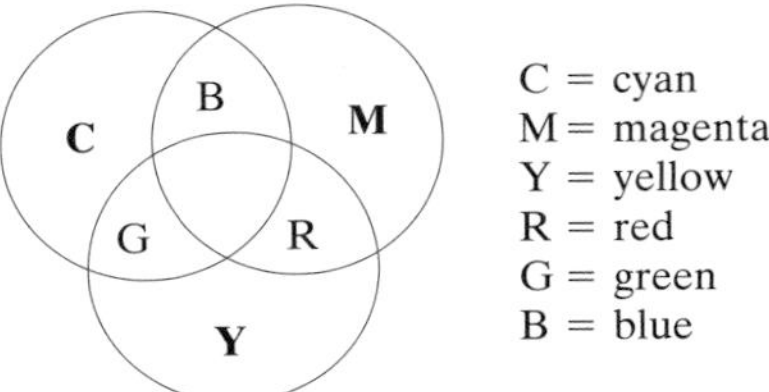

(b) The CMY color model

FIGURE 5.30
The RGB and CMY color models.

can be produced by varying the combination of these intensity levels. As noted in Section 5.1.1, images that have this full range of colors are described as "true-color," and require 24 bits of computer memory per pixel to store. In practice, raster geographic data are usually stored using display modes that require less computer memory. Common display standards include the 16-bit mode, which provides about 32,000 colors, and the 8-bit mode, which displays colors selected from a table of 256 colors supplied by the operating system or by a particular application program. This table is called a *color palette*. Since colors produced with respect to a palette do not always represent the natural color of the spatial phenomena that the raster cells are supposed to represent, images displayed in this way are said to have "pseudo-color."

The CMY color model applies to color produced by mixing or overlaying translucent inks, dyes, or glass filters. When one of these materials is illuminated by white light, certain wavelengths are absorbed or subtracted from it. This leaves the remaining reflected wavelengths to determine the color. For example, by applying translucent cyan ink (which absorbs red light and reflects green and blue light) and magenta ink (which absorbs green but reflects red and blue) in ink-jet printing, a blue dot is formed. The dot appears to be blue because it is the only light that is reflected after red and green lights are absorbed by the cyan and magenta ink respectively. Other pairs of these three primary colors can be mixed to produce red and green colors (Figure 5.30b). When all three primary colors are mixed, the entire spectrum of visible light is absorbed, which makes the dot black. On a white surface where no ink is applied, the surface remains white because no visible light is absorbed. This way of producing color by absorbing different light using different primary colors is known as *subtractive mixing*, and the three primary colors are known as *subtractive primary colors*.

In the publication and printing industry, a color model known as the CMYK model is used in favor of the CMY model. The CMYK model is formed by adding black to the CMY model. This allows black color to be produced directly rather than mixing the three subtractive primary colors. Direct black color production is not only simpler in hard-copy printing, it also results in cleaner and sharper images. CMYK is the industry standard color model for electronic publishing, including offset printing with color plates. This particular color model is also widely supported in GIS software products to specify color characteristics when hard-copy output documents are generated.

Using the subtractive primary colors in the CMY and CMYK color models, printers and plotters can produce individual dots in one of eight basic colors. This is obviously not sufficient to reproduce the hundreds or thousands of colors that are used to display raster images on the computer screen. Instead of producing intermediate colors by varying the proportion of primary colors as it is done on the computer screen, a technique known as *dithering* is used in printing. Dithering is based on the way in which the human eye perceives color on a printed image. On such an image, the printing dots are so tiny and are placed so close together, that it is impossible for the human eye to differentiate individual dots and identify their colors. This allows the printer to create an image that visually simulates the color characteristics of an image as it appears on the computer screen, by blending groups of nearby dots printed in the eight basic colors. In this way, a wide spectrum of colors is produced by false perception of the human eye, but not by mixing the subtractive primary colors physically.

5.4.2 SPECIFYING AND MANIPULATING COLOR

The RGB, CMY, and CMYK color models described here explain the concept of how color is formed in computer graphics. The actual production of color is implemented by specifying which colors are to be used in the output processes. For graphical objects to be produced in color, the user must specify their color characteristics in a way that is understood by the color manager of the output device. Therefore, color specification and manipulation are crucial elements of the output function of GIS. In order to master the use of color for the production of output graphics, GIS users have to be familiar with the ways by which color is specified and manipulated on the computer and by various output devices.

In GIS applications, color generally can be specified in terms of either *hardware colors* or *software colors*. Hardware colors are specified using the index numbers associated with the color cells in the color palette of the output device. The number of color cells, and hence the range of the index numbers, in a color palette is device-dependent. Most workstations and plotters have eight default index numbers associated with the eight basic colors (Table 5.2). In pen plotting, colors are usually specified by hardware colors because plotter pens can be conveniently referenced by the index numbers representing their respective colors.

Unlike hardware colors, software colors are not specified using the index numbers in the color palette of the output device directly. Instead, they are specified by varying the values of the gray levels of the basic colors of the RGB, CMY, and CMYK color models. Some software packages also support two additional color models known as HLS and HSV. The HLS model represents color by means of hue (the "color" of color), lightness (the percentage of gray), and saturation (the intensity of color); whereas the HSV model represents color by means of hue, saturation, and value (another term for the percentage of gray) (Table 5.3). Colors specified by software colors are not tied to the color palette of any hardware output device. When colors are displayed or printed, the output device will match the specified color to the closest color cell in its color palette. As a result, the actual colors produced by software color specifications may vary between one device and another, depending on the size of the color palette and the color allocation of the color cells.

The method of software color specification is obviously far more flexible than the method of hardware color specification. It is possible to specify far more colors too. In the past, software color specification was a very time-consuming task because the user had to figure out manually the combination of gray levels that

Table 5.2
Default Hardware Colors on Most Workstations and Plotters

Index Number	*Color*
0	Background (white)
1	Foreground (black)
2	Red
3	Green
4	Blue
5	Cyan
6	Magenta
7	Yellow

Table 5.3
Color Models and Their Parameters

Color Model	*Parameters*	*Valid Range of Values*	*Examples of Color-Component Codes and the Colors They Represent*
RGB	Red	0–255	255 0 0 = red
	Green	0–255	125 125 125 = gray
	Blue	0–255	255 255 0 = yellow
CMY	Cyan	0–100	0 0 0 = white
	Magenta	0–100	0 100 0 = magenta
	Yellow	0–100	50 50 50 = gray
CMYK	Cyan	0–100	0 0 0 0 = white
	Magenta	0–100	0 100 0 0 = magenta
	Yellow	0–100	50 50 50 0 = gray
	Black	0–100	0 0 0 100 = black
HLS	Hue	0–360*	0 100 100 = a fully saturated shade of red
	Lightness	0–100	240 25 0 = a light gray shade of blue
	Saturation	0–100	0 0 100 = black
HSV	Hue	0–360*	0 100 100 = a fully saturated tint of red
	Saturation	0–100	120 100 50 = a fully saturated shade of geen
	Value	0–100	0 0 100 = white

*Color is specified as an integer of 0 to 360 based on the Tektronic color standard; some systems use other ranges of values such as 0 to 100 and 0 to 255.

would produce a desired color. Most GIS software products are now equipped with a color editor that enables the user to interactively specify the color to be used for displaying a particular type of spatial feature or phenomenon (Figure 5.31). This has greatly improved the user's ability to master the use of color in generating output graphics in GIS.

Some GIS software packages also have a palette editor that allows the user to exercise even greater control over the use of color by creating and using an application-specific color palette. In the context of raster geographic data presentation and communication, there are many occasions when the use of an application-specific palette is necessary.

- To ensure the device-independence of a particular application, i.e., by using a specific palette for the application, the color characteristics of output products will remain the same, regardless of any changes that may happen to the default palette of the computer. Also, when the application is used

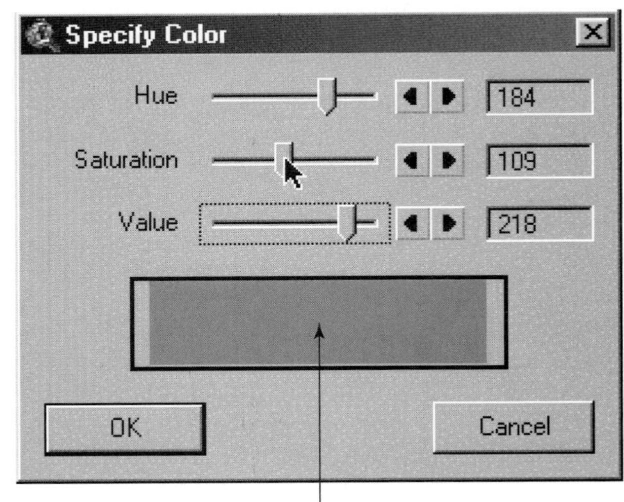

FIGURE 5.31
The color editor of ArcView GIS.

on other computer systems of the same configuration, there is no need to modify the color palette of those systems.

- Color-code a gray-scale image without the need to physically change its original cell values.
- Modify the color characteristics of a colored image without the need to physically change its original cell values.
- Portray an image using a color scheme that better reflects the color characteristics of the spatial phenomenon that it represents, for example, using a brown-dominated palette for depicting arid and semiarid terrain surfaces and a green-dominated palette for depicting vegetation cover.
- Improve the pictorial appearance of output products by portraying gradual changes of spatial phenomena using smaller and smoother color gradations in the palette.

An application-specific color palette is created by changing the values of an existing one using a process called *color spread*. This process may be applied to all or part of the palette. The user first assigns a color for the low and high ends of the selected value range and then applies the spread that will create a gradual color variation across the range of the intervening values (Figure 5.32). The creation of a color palette using the palette editor is a relatively simple and straightforward task. GIS users should always take advantage of this particular functionality to improve the visual effect of output graphics in raster-based geographic data processing.

When generating a hard-copy map from a screen display, it is not always possible to produce a color on paper that will look exactly the same as the corresponding color on the computer screen. Because of the different ways by which color is produced, there is usually a significant shift in colors between the image displayed on the computer screen and the image generated by the printer. Colors produced by ink-jet printers are often darker and less saturated than the corresponding screen colors. In order to help the user control the translation of colors from the screen to the printer, some GIS software packages have a color panel that allows the user to adjust the output colors to the printer interactively (Figure 5.33). With the aid of this color panel, the user can manipulate the colors produced in the hard-copy output without the need to alter the color palette that provides screen colors.

5.4.3 RASTER DATA CONVERSION

Data conversion is an important process of the output functions of raster geographic data processing. There are many occasions when raster data conversion is required, for example: to extract part of a particular raster database for populating the database of another system, to export the results of analysis to another system or application for further processing, and to publish the results of analysis using a word-processing or presentation graphics software package. The objective of raster data conversion is to generate a raster or vector file in a format that is usable by the target application or system.

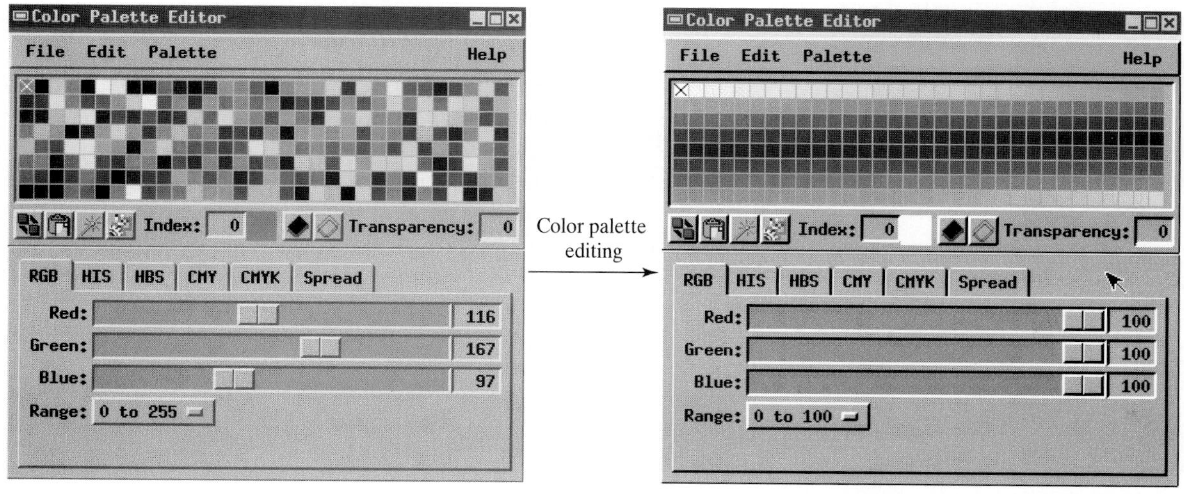

FIGURE 5.32
The color palette editor of TNTmips. A raster image displayed by the application-specific palette will show smooth transition from one class of cell values to another because the change of the colors in the color cells is gradual, rather than random as shown in the default color palette. To view this photo in color, see the color insert. *(Source: Microimages, 1998)*

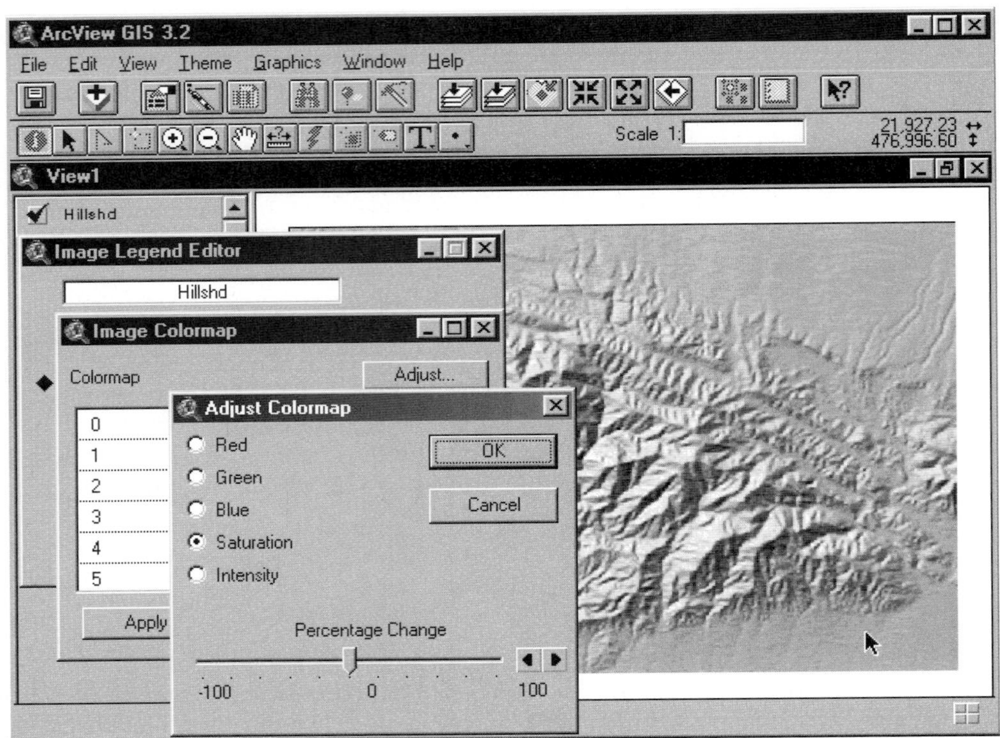

FIGURE 5.33
The color panel of ArcView GIS. The color of the display changes as the user moves the slide bar on the "Adjust Colormap" window.

Generally speaking, the data conversion capability of raster-based GIS software packages falls into four categories, depending on the objective of the conversion and the formats of the data that the target system will accept. These categories are (1) conversion to the format of a particular GIS or digital image-processing software package, (2) conversion to a standard transfer or interchange format, (3) conversion to a generic raster file format, and (4) conversion to a high-ratio compression format.

Conversion to the Format of a Particular GIS or Digital Image-Processing Software Package This is an efficient way of exporting raster data from one system to another since it is done by direct data transfer between the source and target systems without the need to use an interchange format. Many software vendors have jointly developed direct data import/export facilities in their software products so that data created in one system can be used transparently by another system. A good example is the ArcView Image Analyst Extension (ERDAS, 1999). This software product, developed as a collaborative effort between ERDAS and ESRI, provides basic image-processing functions for ArcView GIS and at the same time allows raster data used in ArcView GIS to be processed by the broad range of image-processing capability in ERDAS IMAGINE.

Conversion to a Standard Transfer or Interchange Format This method of raster data conversion is used when the target application is unable to directly use the raster data in the format of the source system. One of the most commonly used transfer formats today is TIFF, which is supported by practically all raster-based GIS software products. As noted in Section 5.1.2, the major problem with TIFF is that it has several specifications that are not compatible with one another. A particular GIS software package that is designed to accept data files in one variant of TIFF is not able to accept data files in another variant of TIFF. Therefore, the user must always take into account the variant of TIFF acceptable by the target application and convert the output data to that particular variant accordingly.

A good alternative to the TIFF format is JPEG, which is designed primarily for the transfer of large, true-color raster images. However, it is also possible to export either gray-scale or pseudo-color raster images to this format. Unlike the TIFF format, JPEG has no variant formats and, therefore, has no potential incompatibility problem between data and applications. This has made it a popular choice for outputting raster images to GIS and digital imaging systems as well as to desktop mapping and publication applications. Also, since JPEG has a much better compression ratio than TIFF, it is widely used for transferring raster images over the Internet.

Conversion to a Generic Raster File Format Many raster-based GIS software products are able to generate an output file in the generic ASCII or binary file formats. Since these formats are device-independent, they provide the safest way to transfer data between applications residing on different hardware platforms, for example, UNIX workstations and PCs (Section 5.1.2).

Conversion to a High-Ratio Compression Format The excellent compression ratios achievable by the new generation of compression formats, such as MrSID and ECW, make them practical choices for distributing large raster images and digital mosaics over communications networks. Users who convert and distribute data in these formats have to purchase the compression software licenses in order to generate image files in these formats. However, users at the receiving end can obtain from the developers of these formats, free of charge, software extensions or plug-ins that allow them to display the compressed images in various GIS or digital image-processing systems.

Raster–Vector Data Conversion This category of data conversion is different from the previous four categories of functions in that the output file is in a vector format. Raster–vector conversion is a built-in function that can be found in many raster-based GIS software products. In practice, however, a special raster–vector conversion software product is commonly used because such a special software package often provides a more robust suite of data editing tools than that available in an ordinary GIS. Raster–vector conversion is now more commonly used than the conventional method of table digitizing in acquiring vector geographic data. This process is commonly called *heads-up* or *on-screen* digitizing.

Raster–vector data conversion can be carried out manually, where the user picks the features to capture, or automatically, where the computer converts the entire raster image without any user intervention. Experience has shown that the best result is obtained by semiautomated methods that combine manual point picking with computer assistance in line tracing (see Chapter 6). In most raster–vector operations, the user is normally provided with a limited number of options for the output vector format. This is because many GIS software vendors do not allow other parties to export output data directly into their proprietary formats. At present, the most commonly used output format appears to be the AutoCAD DXF format. Many software products also generate vector graphics in the shapefiles format because of the popularity of ArcView GIS and the willingness of its vendor to open it to third-party software developers.

In raster–vector conversion, the actual process of converting a raster image to its equivalent vector graphics is only a small portion of the work involved. Considerable effort is still required to clean the resulting vector data, organize them into the desired layers, and link them to their associate attribute tables. Since these tasks are concerned primarily with vector rather than raster data, we will explain them in detail in Chapter 6.

5.4.4 INFORMATION GRAPHICS AND MULTIMEDIA PRESENTATIONS

The generation of information products (maps, charts, graphs, and tables) is a fundamental process of the output functions of all GIS. Conventionally, results of raster-based GIS data processing are plotted by means of electrostatic plotters which, using raster-based technology, offer a good combination of efficiency and economy. If high-quality products are required, such as in the production of topographic maps and hydrographic charts, the output data can be exported to a high-precision raster photo-plotter. This class of output devices produces a reproduction-quality image on stable base films, which can then be used as separates in the mechanical printing process.

It is now commonplace for the results of raster geographic data analysis to be transferred to a page description language such as PostScript, so that they can be printed economically using relatively low-cost printers and plotters. Raster images in TIFF and JPEG formats have also been widely used in word processing, desktop publication, and presentation graphics. There are special software products, such as ERDAS MapSheets, that are designed specifically to facilitate the use of raster images in the desktop computing environment. These software products have very powerful map composition and presentation tools that allow the user to create maps quickly for inclusion in reports and documents (Figure 5.34).

The increasing integration of raster and vector in GIS has led to the development of methods that generate hybrid maps. These are maps that are produced with vector data drawn on top of raster data as a backdrop. The use of raster data is particularly effective in presenting relief information such as hill shading. At the same time, advances in electronic data-storage technology have made it possible for raster data output to be generated digitally. This means that instead of paper maps, the products are stored and distributed in CD-ROMs or other forms of digital storage media. Such products are accompanied by comprehensive header metadata, including transformation coefficients that relate

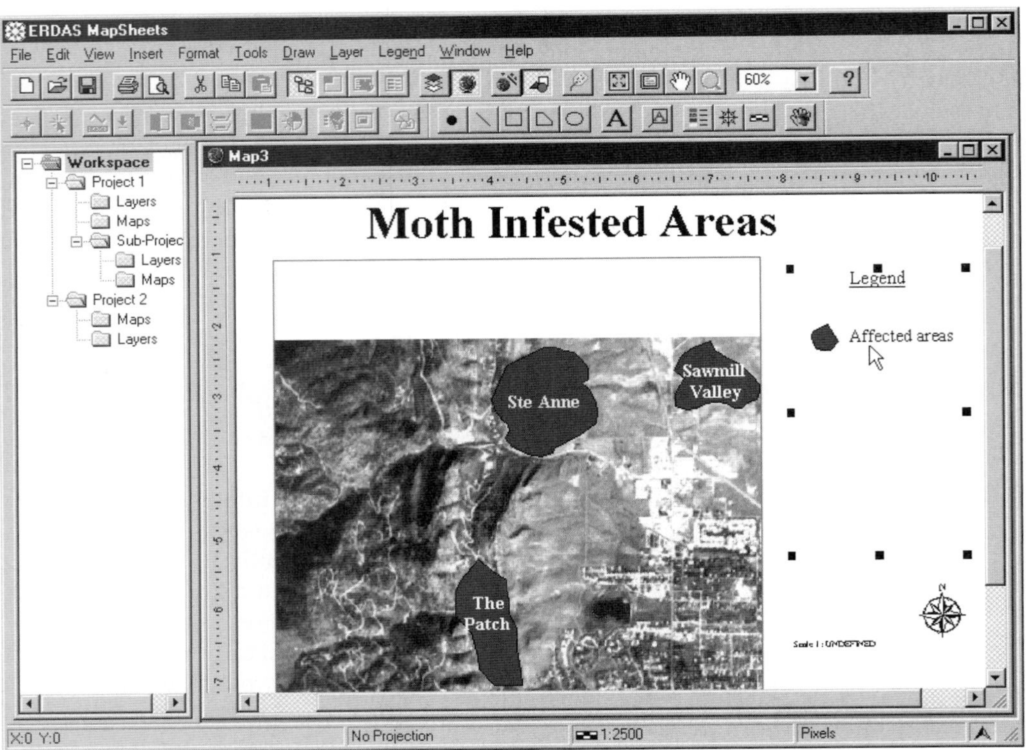

FIGURE 5.34
Using raster images in desktop computing. There are desktop computing software packages that allow the user to make use of raster images in applications such as word processing and desktop publishing. The user can add text and other forms of graphics on top of the raster images to compose a map for inclusion in reports and other documents.

the raster cells to ground positions. This allows the digital data to be used on any system for further application or analysis. Experience has shown that digital raster data are very useful for applications where a large number of maps are required, such as hydrographic charting, marine navigation, and regional parcel mapping. The use of raster-based digital data can considerably shorten the time for the introduction of digital mapping technologies in an organization. It also allows the adoption of an incremental approach to map data conversion that will eventually enable the user to take advantage of both raster- and vector-based GIS applications.

The growing interest in animated cartography, multimedia GIS, and spatial data visualization since the mid-1990s has resulted in the development of some very innovative ways of using raster images. The method of time-series animation, for example, is a visualization technique ideally suited for the display and analysis of temporal geographic data (Acevedo and Masuoka, 1997). It is based on the use of frame-based raster animation created by showing a series of raster images sequentially on the computer screen (Peterson, 1995). More advanced techniques such as fly-by or fly-through sequences make use of raster images draped over a digital elevation model to produce a series of perspective images (Figure 5.35).

Cartographic animation and multimedia visualization are now high on the agenda of the cartographic research community (MacEachren and Kraak, 1997). These new applications combine spatial data from different sources and present them in a much more intuitive manner than conventional static maps. They are rapidly gaining popularity in a wide range of fields including urban development, environmental management, population studies, regional planning, and cartographic design. Up to now, the role of raster-based GIS in cartographic animation and multimedia visualization has been as a passive data supplier. Raster-based GIS software products today have no native function that will generate animated sequences. The only exception is probably the very rudimentary slide show function available in the ArcPlot module of ArcInfo. However, attempts in the last several years to develop integrated cartographic systems using GIS, image processing, and animation techniques have generated very encouraging results (Cook et al., 1997; Rhyne, 1997). Since raster spatial data constitute the key ingredient of these systems, it is anticipated that the development of concepts and techniques to generate raster images for

FIGURE 5.35
Example of a fly-by sequence. A fly-by, also known as a fly-through, sequence is a movielike presentation that shows the user an aerial view along a predefined path. More advanced packages allow the user to define the path on the terrain *(Courtesy of Martin Adamiker).*

dynamic spatial data visualization will become one of the most challenging tasks of raster-based geographic data processing in the years to come.

5.5 Cartographic Modeling

The spatial analytical techniques using raster data described in Section 5.3 are *primitive operations* in the sense that they are developed to address basic functions relating to the encoding, storage, processing, and display of geographic data in a raster-based GIS. By organizing the different primary operations used for a particular data analysis application in a logical manner, a generalized approach to *cartographic modeling* may be developed (Tomlin, 1983). A cartographic model may be visualized as a collection of raster layers registered to a common cartographic frame of reference. Cartographic modeling is the process of linking the primitive operations that work on different raster layers in a logical sequence to solve spatial problems. In the modeling process, the sequencing of the execution of the operations is similar to the algebraic solution of equations to determine the unknowns, and this approach has been aptly called *map algebra* (Berry, 1987; Tomlin, 1990).

5.5.1 Map Algebra

Map algebra as originally proposed by Tomlin (1991) can be most easily understood by explaining it from the perspectives of its three key concepts: (1) data processing, (2) data-processing control, and (3) characterizing locations.

Data Processing In the context of map algebra, data processing refers to the *operations* and *procedures* by which geographic data are prepared, managed, interpreted, and presented. Operations are data-processing activities that perform these tasks using a well-defined suite of operators. Operations accept input from one or more existing raster layers and generate one or more layers as output. When a finite sequence of two or more operations is applied to solve a spatial problem, a procedure is created. By organizing operations into logical procedures, map algebra is used for spatial problem solving in exactly the same way that high-level computational languages are used for numerical problem solving.

Depending on the nature of particular applications, cartographic modeling may be an extremely complex process involving numerous data-processing operations and raster data sets. In order to help the user organize a cartographic modeling task and communicate

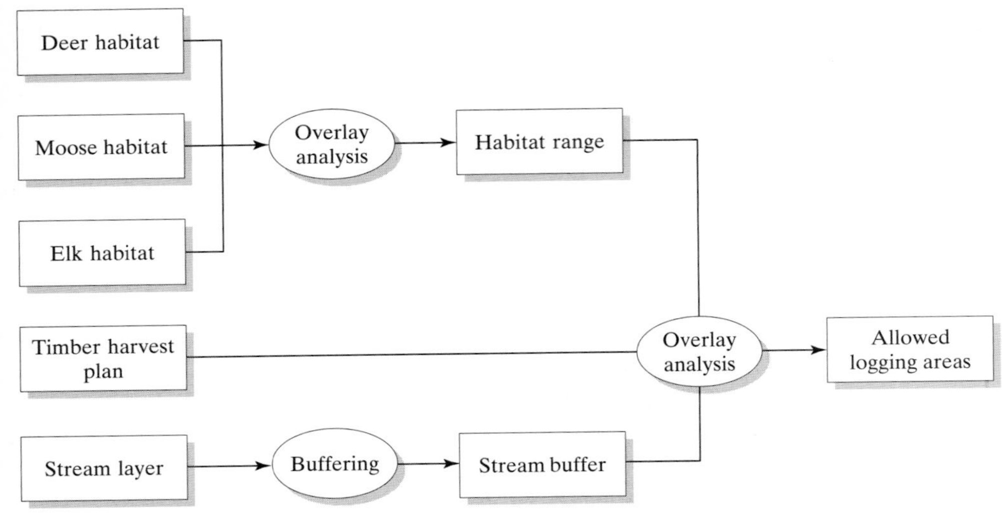

FIGURE 5.36
Example of a flowchart.

with other users in a clear and effective manner, it is usually necessary to construct a flowchart showing the logical sequence of operations and procedures necessary to obtain the final spatial solution. The flowchart is a diagrammatic representation showing the generic relationships between data and processes in cartographic modeling (Figure 5.36). It is system-independent and plays an essential role in the planning of cartographic modeling tasks. When the user is satisfied with flow of data and procedures on the flowchart, the operations are then translated into the statements or command modules used by a particular GIS software package.

Data-Processing Control Data processing in cartographic modeling is controlled by using *statements* and *programs*. In map algebra, a *statement* is the most elementary expression that represents a primitive data-processing operation and the data to which it is applied. It is an ordered sequence of letters, numerals, symbols, and/or blank spaces (Tomlin, 1990), as exemplified by the following statement that uses a primitive operation called *LocalMean*:

NEWLAYER = *LocalMean of* FIRSTLAYER *and* SECONDLAYER *and* THIRDLAYER *and* FOURTH LAYER

In this statement, NEWLAYER is the subject or the title of a new raster layer to be created. Using three modifiers "=", "*of*," and "*and*," this statement computes the mean of each cell location on four input layers, and hence the operation is named *LocalMean*.

When two or more statements are required to complete a certain task, they have to be performed in a particular order. In map algebra, a *program* controls the sequence by which statements are applied. In other words, a program is the notational representation of procedures in data processing.

Characterizing Locations In cartographic modeling, the characteristic of a certain location (i.e., the value of a particular cell), on an output raster layer is determined or affected by the characteristics of the input raster layers. Some operations compute the value of a particular cell on an output raster layer by using only the values of the cells at the same location on the input layers. Others compute the value of this cell by using the values of cells neighboring its equivalent locations on the input raster layers. There are also operations that compute the value of output cells using the values in a defined zone on the input layers. In algebraic terminology, the value of the output cells is said to be a *function* of the input cells. This function can be the mean, the sum, the difference, the standard deviation, and other arithmetical or statistical indices. Depending on the nature of the function, primitive data-processing operations can be classified into the four categories as identified and described in Sections 5.3.2 through 5.3.5. These categories are *local operations*, *local neighborhood* or *focal operations*, *extended neighborhood* or *global operations*, and *regional* or *zonal operations*.

Map algebra can be perceived as map overlay metaphors designed for manipulating geographic data according to specific processes in a way that mimics the conventional manual map overlay methodology. Map algebra provides the conceptual framework and tools (i.e., the internal formal language) that enable GIS users to formalize the conventional overlay methods. With the aid of map algebra, very sophisticated applications can be constructed from relatively simple operators (Chan and White, 1987; Hodgson and Gaile, 1998). Map algebra has become the de facto standard for geographic data analysis using raster data. Different versions of map algebra have been implemented in raster-based GIS, for example, the GRID module of ArcInfo (ESRI, 1999; Menon et al., 1991), ERDAS (ERDAS, 1992), IDRISI (Eastman, 1997), and GRASS (Shapiro, 1993). Unlike the map algebra originally proposed by Tomlin, which used a natural language form, these new implementations use function-based modeling languages designed to support the flow of data from one processing operation to the next (Gao et al., 1996). This new approach to map algebra has greatly enhanced the spatial analytical capabilities of the GIS packages concerned.

Using the map algebra capabilities explained here, two types of cartographic modeling can be identified: descriptive and prescriptive (Tomlin, 1990). *Descriptive cartographic modeling* attempts to answer the problem "what is." *Prescriptive cartographic modeling* aims to solve problems like "what should be done" (prescriptive) or "what would happen if (condition) is true" (predictive). The boundary between descriptive and prescriptive modeling is often vague, and it is not uncommon for these two approaches of cartographic modeling to be used to complement one another in spatial problem solving and decision making.

5.5.2 AN EXAMPLE OF DESCRIPTIVE CARTOGRAPHIC MODELING

This example is taken from research conducted by one of the authors to evaluate and map the quality of life (QOL) in Athens-Clarke County, Georgia (Lo and Faber, 1997). The data used were Landsat TM data and 1990 census data from the U.S. Bureau of the Census. The analysis was carried out using census block groups as the areal unit. A raster-based GIS software package IDRISI was used.

QOL was evaluated based on demographic, economic, education, housing, and environmental factors. From the Landsat TM images, a land-use/cover map, with land-use/cover classes at level II of the Anderson scheme (Anderson et al., 1976), was extracted using a supervised digital image classification. From this land-use/cover map, the "urban-use" class that comprises industrial, commercial, and transportation uses was extracted using the method of reclassification (nominal data). From bands 3 and 4 of the Landsat TM data, the normalized difference vegetation index (NDVI) was computed for each cell (ratio data). From band 6, the thermal infrared band, surface temperature for each cell was also computed (interval data).

All the above data sets were rectified to the UTM coordinate system and aggregated to the census street block areal unit. Next, from the census data, the following variables were extracted at the census block group level: population density (ratio data), per capita income (ratio data), median home value (ratio data), and percent of college graduates (ratio data). Because these variables were recorded by census block groups in vector format, they had to be rasterized and registered with the Landsat TM image data layers mentioned earlier. Altogether seven map layers have been developed. From the urban map layer, the percent of urban use in each census block group was computed (ratio data).

All the attribute values of the seven map layers were then classified by equally dividing the range of values between the maximum and minimum into ten ranks (ordinal data) in ascending order. The higher the rank number (i.e., approaching ten), the more positive the attribute is for QOL. In this way, seven new map layers, showing attribute values in integers from 1 to 10 were created. These seven map layers were overlaid and added together. The higher the composite rank for a block group, the better was its QOL. The resultant map layer from the overlay displays the variations in QOL in the Athens-Clarke County by census block group (Figure 5.37). The

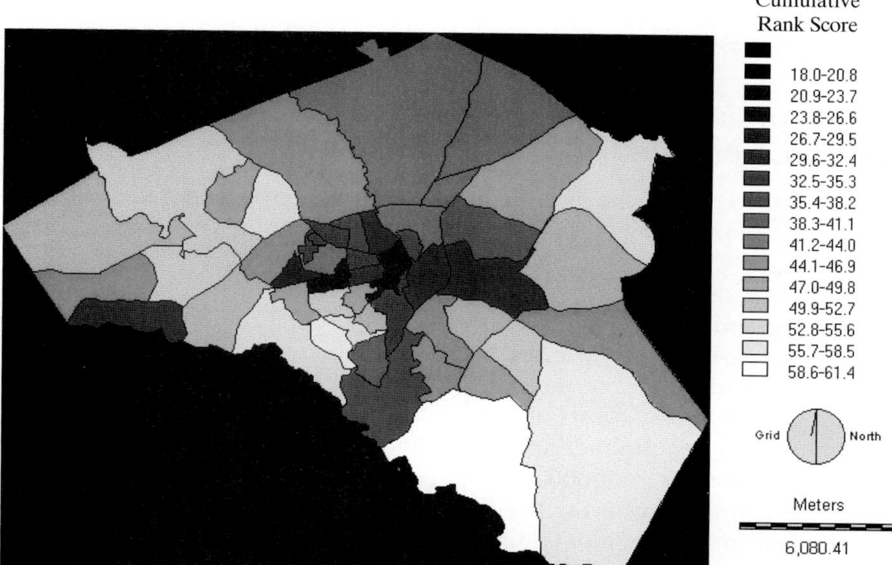

FIGURE 5.37
GIS overlay result of seven ranked map layers of biophysical and socioeconomic data for Athens-Clarke County, 1990, by census block groups. *(Source: Lo and Faber, 1997)*

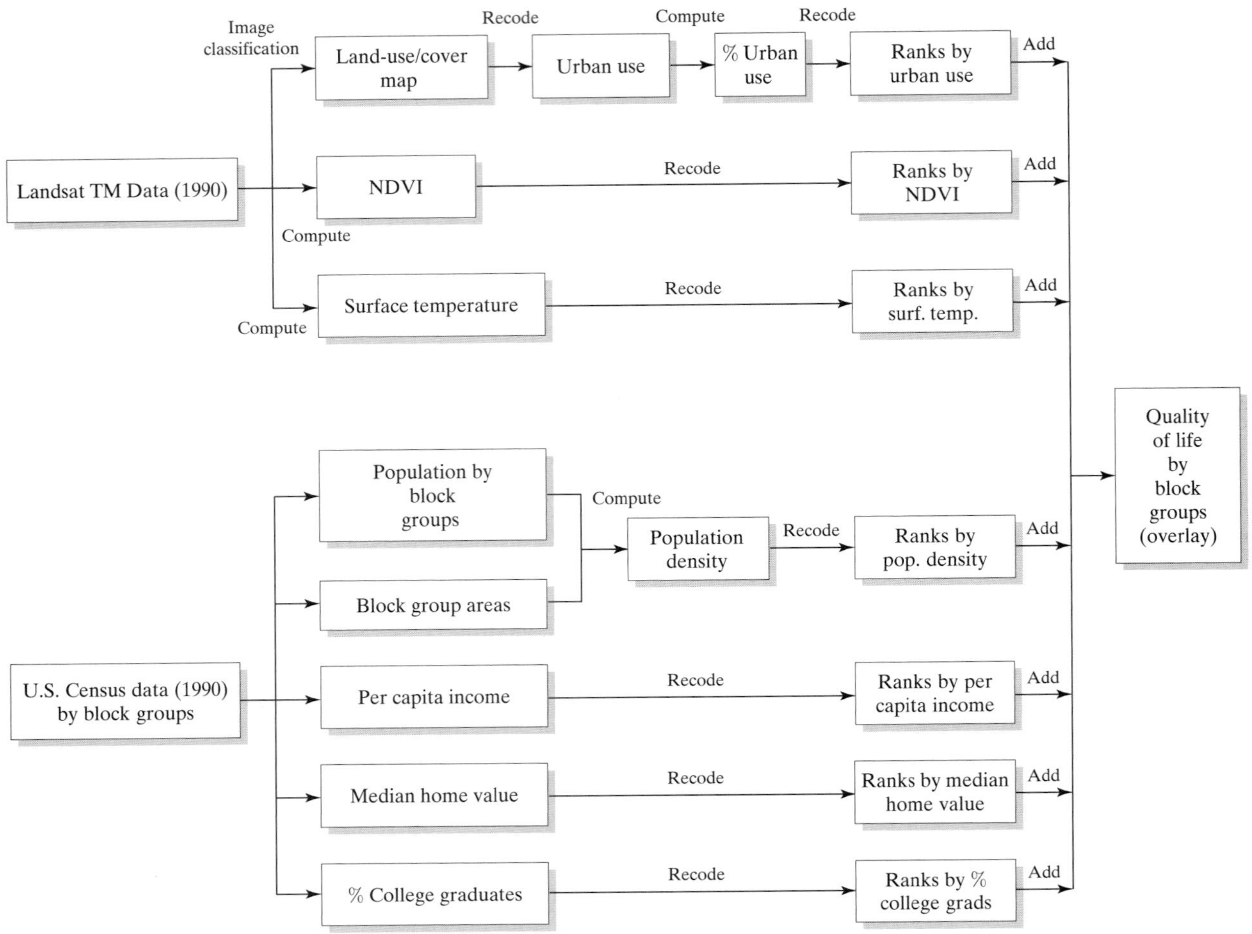

FIGURE 5.38
Flowchart showing modeling of quality of life by block groups in Athens-Clarke County, Georgia, 1990.

whole procedure of cartographic modeling is shown in the form of a flowchart in Figure 5.38.

5.5.3 AN EXAMPLE OF PRESCRIPTIVE CARTOGRAPHIC MODELING

The objective of this study was to predict the landslide probabilities of the mountain terrain in Nelson County, central Virginia (Gao and Lo, 1995). From 1:24,000-scale aerial photographs, 87 landslide scars within a 28.5 km^2 area were delineated and then field checked. These landslide scars were the results of heavy rainfall brought along by the remnants of hurricane Camille on the night of August 19–20, 1969. Any errors of the photo interpretation were corrected to produce a final landslide map of the study area. From a 1:24,000-scale USGS topographic map, the contours were digitized, and a DEM was constructed from a topographic map of the study area by interpolation using a dot grid with 24-m resolution. The DEM represented the terrain before the occurrence of the landslides.

There are three general groups of factors that have caused landslides: the topographic and environmental setting, such as hill-slope characteristics, soil moisture condition, and land cover; the internal properties of the earth materials, such as the geology and structure of the underlying bedrock; and independent external factors, such as rainfall intensity, earthquakes, and human interference. This research focused on the first group of variables at the microscale. At such a scale, the variation of bedrock lithology, structure, and rainfall intensity may be regarded as spatially uniform, thus leaving topographic conditions the main controlling factors for the landslides. To investigate the potentiality of the terrain to landsliding, the following terrain factors were believed to be important: elevation, slope gradient, slope configuration, and slope aspect. At a grid cell in a DEM, these factors can be incorporated in the following polynomial equation to model landslide potentials.

$$P(x, y) = W_h F_h + W_g F_g + W_c F_c + W_a F_a \quad \textbf{(5.6)}$$

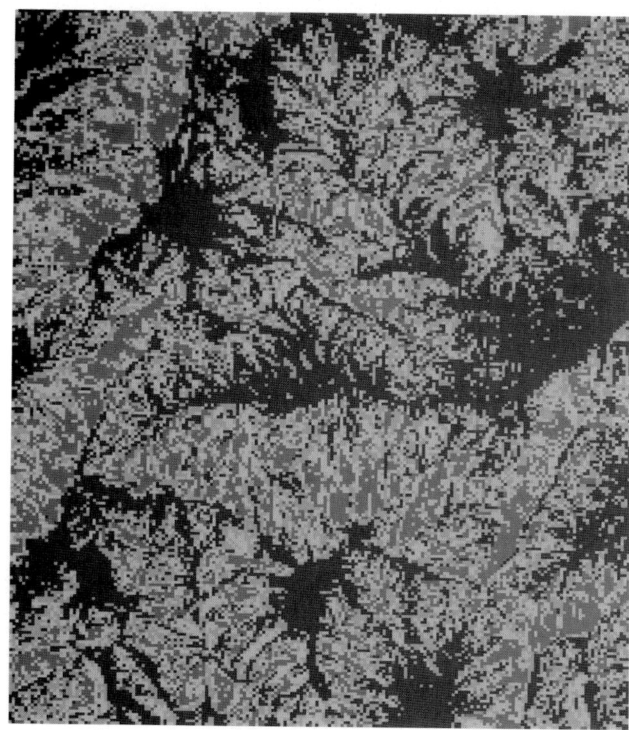

FIGURE 5.39
Modeled landslide susceptibility by integrating elevation, slope gradient, slope configuration, and slope aspect criteria. To view this photo in color, see the color insert. *(Source: Gao and Lo, 1995)*

where F_h, F_g, F_c, and F_a are the elevation, slope gradient, slope configuration, and slope aspect factors respectively, and W_h, W_g, W_c, and W_a are weights assigned to the factors of elevation, slope gradient, slope configuration, and slope aspect respectively.

From the DEM of the study area, the variables of elevation, slope gradient, and slope aspect were extracted to form three separate map layers (see Section 5.2.3). Slope configuration refers to the shape of the slope, which can be characterized as rectilinear, concave, or convex. A total of nine slope configurations may be recognized from the DEM by examining the combination of elevation changes along the *X* (cross section) and then along the *Y* (profile) directions. From this analysis, a reclassification function was used to produce a new map layer for slope configuration.

The landslide map produced from aerial photographic interpretation was converted to raster format and registered to the DEM. The landslide map layer was reclassified into two categories: affected terrain (1) and unaffected terrain (0). This newly created map layer was then overlaid with each of the four map layers on elevation, slope gradient, slope configuration, and slope aspect to determine the importance of the relationship between the landslide paths and each of the four factors. The percentage of cells affected by landslides according to each of the factors was computed and then employed to derive the weight for that factor. A new map layer for the weight for each factor was produced. These weights layers were then added on top of each other in a four-layer overlay to produce a final map layer of the overall landslide susceptibility of the terrain in five categories: least susceptible, less susceptible, average susceptible, more susceptible, and most susceptible. This final map layer therefore predicts the chance for a landslide to occur at a particular location of the grid cell (Figure 5.39). The flowchart for the cartographic modeling is shown in Figure 5.40.

5.5.4 ERROR PROPAGATION IN CARTOGRAPHIC MODELING

As noted in Chapter 4, errors occur in all forms of geographic data processing. Since cartographic modeling is concerned with both data and processes, error propagation, introduced in Chapter 4, is of particular interest and relevance. The objective of cartographic modeling is to generate a new map layer from existing map layers. The accuracy of the new map layer is obviously dependent on the type and level of errors that are present in the input map layers. Errors are also likely introduced during the transformation process. It follows that unless the behavior of the errors of the various input map layers during the transformation operations can be established, the resulting new raster layer will be used with an unknown degree of uncertainty. This will obviously diminish the usefulness of the new data and possibly defeat the whole purpose of using GIS. Therefore, error propagation should always be a key consideration in cartographic modeling.

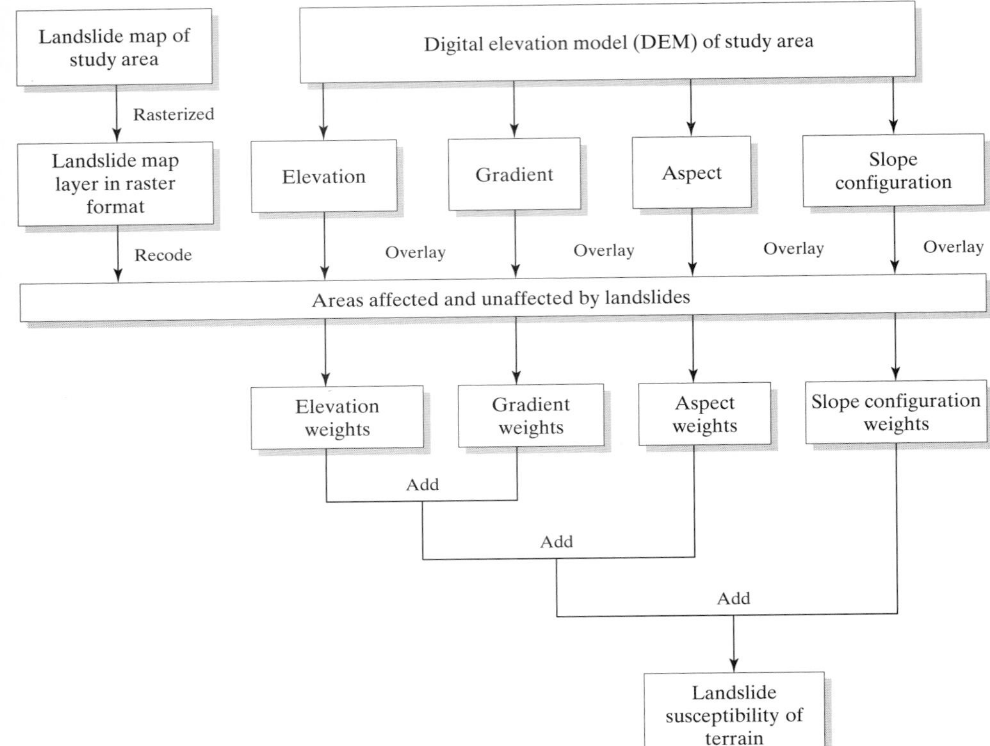

FIGURE 5.40
Flowchart showing modeling of landslide susceptibility in the mountains of Nelson County, Virginia.

As explained in Chapter 4, an error propagation function is a mathematical representation of the mechanisms whereby errors present in the input map layers are modified by a particular data transformation operation. This implies that the use of an error propagation function requires some a priori knowledge of the quality of the input map layers. The knowledge of data quality is usually based on the PCC (percent correctly classified) indices obtained from attribute accuracy tests of the source raster layers (see Chapter 4). It also assumes the ability to estimate the impacts of different data transformation operations on the errors of the input map layers. The impact of a particular operation can be estimated by considering a number of factors, which include the user's own experience, the nature of the operation as well as the theory of probability.

Lanter and Veregin (1992) have demonstrated error propagation in layer-based GIS data processing. The cartographic model (Figure 5.41) that they used involved three input map layers, namely, LANDUSE, PERMIT (grazing permits on private land), and VEGETATION. The objective of the application was to determine and identify lands in which oak regeneration would be at risk (a new map layer called AT-RISK) because of grazing activities. Three operations were used for transforming data in the input map layers of the application into the new map layer: UNION, which represents the logical OR operation; INTERSECT, which represents the logical AND operation; and RESELECT, equivalent to a reclassification process.

The percent correctly classified (PCC) for the LANDUSE and VEGETATION input map layers was based on the accuracy with which cover classes in these map layers were depicted. The PERMIT was a binary (Yes/No) input map layer. Its PCC was assumed to reflect uncertainty associated with the usual incompleteness in grazing permit records. The transformation of the PCC indices through the application was based on the assumption that the errors were uncorrelated across data layers and were distributed uniformly across classes of the thematic attributes. On the basis of this assumption, the impacts of the operations UNION, INTERSECT, and RESELECT were estimated. The UNION function created an output map layer containing the union of the features on two input map layers. A point needs to be correctly classified on only one of the input map layers in order to be considered accurate on the output map layer. In terms of the PCC index, this means that the PCC of the output map layer can never be lower than the PCC of the most accurate input map layer. On the other hand, the INTERSECT function creates an output map layer containing the intersection of the features on two input map layers. This means that classification for the output map layer is defined as the intersection of the correctly classified portions of the input map layers. A point has to be correctly classified on both input map layers in order to be considered

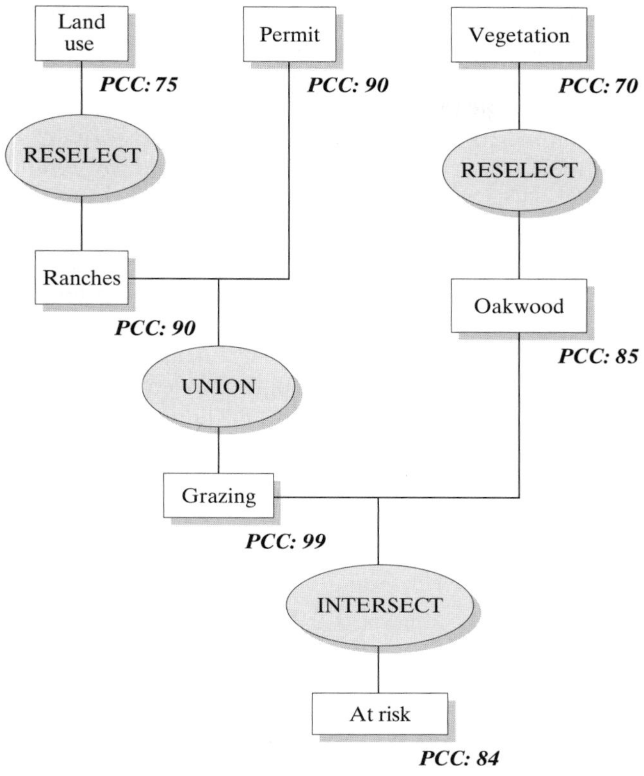

Land use

PCC: 75

RESELECT

Ranches

PCC: 90

Permit

PCC: 90

UNION

Grazing

PCC: 99

Vegetation

PCC: 70

RESELECT

Oakwood

PCC: 85

INTERSECT

At risk

PCC: 84

FIGURE 5.41
Error propagation in cartographic modeling. *(Based on Lanter and Veregin, 1992)*

accurate on the output map layer. In terms of the PCC index, this implies that the PCC of the output map layer can never be higher than the PCC of the least accurate input map layer. Finally, the RESELECT function involves the collapse of a larger number of classes to a smaller number of classes. This means that the PCC index tends to increase as a result.

As illustrated in Figure 5.41, the application went through these four operations to generate the required AT-RISK layer.

- RESELECT LANDUSE to create an intermediate binary layer called RANCHES (private ranches)
- RESELECT VEGETATION to create another intermediate binary layer called OAKWOOD (oak woodlands)
- UNION RANCHES and PERMIT to create an intermediate binary layer called GRAZING (private and public grazing lands)
- INTERSECT GRAZING and OAKLAND to create a binary layer called AT-RISK, which identifies lands in which oak generation would be at risk from grazing activities

As indicated in Figure 5.41, the output data from this application of cartographic modeling can be used with an uncertainty index of 0.84. This means that given the uncertainty of the input map layers and the estimate of the behavior of the errors during the transformation processes, 84% of the land identified as being at risk because of grazing activities can be correctly classified.

5.6 Summary

In this chapter, we have provided a comprehensive overview of the concepts and techniques of raster-based geographic data analysis, to ensure that users have a good understanding of the characteristics and sources of raster geographic data, what a typical raster-based GIS can do, and how it performs different types of spatial analysis and modeling tasks. It is important to note the compatibility of the raster GIS format with that of digital images in remote sensing, thus making remotely sensed image data an excellent source of data for GIS. However, raster GIS, while borrowing from the techniques of digital image processing (such as image-to-map rectification, image-to-image registration, and filtering) distinguishes itself by performing more varied operations on map layers at the local, neighborhood, and regional scales. At the local scale, raster GIS reclassifies, overlays, and computes aspects and slopes. At the neighborhood scale, raster GIS determines distances, buffers, and viewsheds. At the regional scale, raster GIS identifies regions and calculates areas, perimeters, and shapes of these regions.

As raster data are getting more accessible to users, often at increasingly high geometric quality and spatial resolution, raster-based GIS data processing is now widely used for spatial analysis and modeling. Raster-based methods are most suitable for applications for which the analysis and visualization of surface information, rather than high positional accuracy, are primary requirements. These applications include site identification and route selection, environmental mapping, natural resource management, forest resource inventory, landscape ecology and biodiversity studies, wildlife habitat analysis, wetland inventory and analysis, hydrological and geomorphological modeling, terrain analysis, predictive modeling, as well as land-use and land cover mapping.

In the past, methods of raster-based GIS data processing were carried out as distinct processes from vector-based processing. In recent years, advances in computer technology have made possible the integrated use of raster and vector data in many applications. The boundary between raster and vector geographic data processing is rapidly diminishing. The ability to integrate vector and raster data in a single working environment has become a standard rather than an optional feature of GIS software packages. This has totally changed the ways raster data are used in GIS. It must be noted, however, that there are still some areas of GIS applications where the use of raster-based techniques are not appropriate. These include network analysis, land parcel and title registration, resource inventories, and topographic mapping, all of which rely heavily on the use of linear boundaries and attributes of discrete spatial features.

This chapter also provided a detailed explanation of the use of color in computer graphics. Raster-based GIS produce hard-copy surface maps more efficiently than vector-based systems because these systems are generally equipped with excellent color manipulation capabilities. In recent years, there has been a phenomenal growth of interest in cartographic animation and multimedia GIS. These new branches of geographic information science depend heavily on the use of raster images. Therefore, the generation of raster images in support of applications using cartographic animation and multimedia GIS will be the most challenging objective of raster-based GIS in the future.

One of the most important features of raster-based GIS is cartographic modeling, by which multiple raster layers are mathematically combined or transformed in different ways to solve spatial problems. Cartographic modeling provides efficient data management, automates cartographic procedures, and permits dynamic simulation, thus making raster-based GIS a very powerful analytical tool. As Gao et al. (1996) pointed out, "raster GIS provide facilities to express almost any mathematical equations in the form of the map algebra and other tools, such as distance mapping, topographic feature extraction, and surface interpolation, to either generate input variables for modeling or perform the modeling itself."

REFERENCES

Acevedo, W., and Masuoka, P. (1997) "Time-series animation techniques for visualizing urban growth," *Computers and Geosciences*, Vol. 23, No. 4, pp. 423–435.

Afek, Y., and Brand, A. (1998) "Mosaicking of orthorectified aerial images," *Photogrammetric Engineering and Remote Sensing*, Vol. 64, No. 2, pp. 115–125.

ANA Tech (1997) *ANA Tech Scansmith Preditor User Guide* (Version 4). Littleton, CO: ANA Tech.

Anderson, J. R., Hardy, E. E., Roach, J. T., and Witmer, R. E. (1976) *A Land Use and Land Cover Classification System for Use with Remote Sensor Data*. Geological Survey Professional Paper 964. Washington, D.C.: United States Government Printing Office.

Berry, J. K. (1987) "Fundamental operations in computer-assisted map analysis," *International Journal of Geographical Information Systems*, Vol. 1, No. 2, pp. 119–136.

Berry, J. K. (1993a) "Cartographic modelling: The analytical capabilities of GIS," in *Environmental Modeling with GIS*, by Goodchild, M. F., Parks, B. O., and Steyaert, L. T. eds. New York: Oxford University Press.

Berry, J. K. (1993b) *Beyond Mapping: Concepts, Algorithms, and Issues in GIS*. Fort Collins, CO: GIS World Books.

Berry, J. K. (1995) *Spatial Reasoning for Effective GIS*, Fort Collins, CO: GIS World Books.

Bian, L., and Butler, R. (1999) "Comparing effects of aggregation methods on statistical and spatial properties of simulated spatial data," *Photogrammetric Engineering and Remote Sensing*, Vol. 65, No. 1, pp. 73–84.

Bian, L., and Walsh, S. J. (1993) "Scale dependencies of vegetation and topography in a mountainous environment in Montana," *Professional Geographer*, Vol. 45, No. 1, pp. 1–11.

Bonham-Carter, G. F. (1994) *Geographic Information Systems for Geoscientists*. New York: Elsevier Science Inc.

Chan, K. K. L., and White, D. (1987) "Map Algebra: An object oriented implementation," *Proceedings International Geographic Information System Symposium: The Research Agenda*, Vol. II, pp. 127–150. Washington, DC: The Association of American Geographers.

Cook, D., Majure, J., Symanzik, J., and Cressie, N. (1997) "Dynamic graphics in a GIS: More examples using linked software," *Computers and Geosciences*, Vol. 23, No. 4, pp. 371–385.

Eastman, J. R. (1997) *Idrisi for Windows*. Worcester, MA: Clark Laboratory for Cartographic Technology and Geographic Analysis, Clark University.

ERDAS (1992) *Writing Models in ERDAS IMAGINE Using the Spatial Modeler Language*. Atlanta, GA: ERDAS, Inc.

ERDAS (1999) *ArcView GIS Image Analysis Extension (Version 1.1)*, an ERDAS and ESRI White Paper. Atlanta, GA: ERDAS, Inc.

ER Mapper (1998) *Using and Distributing ECW V2.0 Wavelet Compressed Imagery*, Compression White Paper. San Diego, CA: Earth Resources Mapping Pty Ltd.

ESRI (1999) "Cell-based Modeling with GRID," in *ArcDoc (ArcInfo Version 7.1.2 On-line Help)*. Redlands, CA: Environmental Systems Research Institute.

Fellmann, J. D., Getis, A., and Getis, J. (1995) *Human Geography: Landscapes of Human Activities*. Dubuque, Iowa: W.C. Brown Publishers.

Fisher, P. F. (1993) "Algorithm and implementation uncertainty in the viewshed function," *International Journal of Geographical Information Systems*, Vol. 2, No. 7, pp. 331–347.

Fisher, P. F. (1994) "Viewshed operations: Probable and Fuzzy models," in *Innovations in GIS 1* by Worboys, M. F. ed. pp. 161–176. London: Taylor & Francis.

Flanagan, N., Jennings, C., and Flanagan, C. (1994) "Automatic GIS data capture and conversion," in *Innovations in GIS 1* by Worboys, M. F. ed. pp. 25–38. London: Taylor & Francis.

Gao, J., and Lo, C. P. (1995) "Micro-scale modelling of terrain susceptibility to landsliding from a DEM: a GIS approach," *Geocarto International*, Vol. 10, No. 4, pp. 15–30.

Gao, P., Zhan C., and Menon, S. (1996) "An overview of cell-based modeling with GIS," in *GIS and Environmental*

Modeling: Progress and Research Issues, by Goodchild, M. F., Steyaert, L. T., Parks, B. O., Johnston, C., Maidment, D., Crane, M., and Glendinning, S. eds. Fort Collins, CO: GIS World Books.

Giordano, A., Veregin, H., Borak, E., and Lanter, D. (1994) "A conceptual model of GIS-based spatial analysis," *Cartographica*, Vol. 31, No. 4, pp. 44–55.

Goodchild, M. F. (1992) "Analysis," in *Geography's Inner Worlds: Pervasive Themes in Contemporary American Geography* by Abler, A. F., Marcus, M. G., and Olson, J. M. eds. New Brunswick, NJ: Rutgers University Press.

Haggett, P., Cliff, A. D., and Frey, A. (1977) *Locational Models*. London: Edward Arnold.

Hodgson, M. E., and Gaile, G. L. (1998) "A cartographic modeling approach for surface orientation-related applications," *Photogrammetric Engineering and Remote Sensing*, Vol. 65, No. 1, pp. 85–95.

Jensen, J. R. (1996) *Introductory Digital Image Processing*, Upper Saddle River, NJ: Prentice Hall.

Kolbl, O., Best, M. P., Dam, A., Douglas, J. W., Mayr, W., Philbrik, R. H., Seitz, P., and Wehrli, H. (1996) "Photogrammetric scanners," in C. Greve ed. *Digital Photogrammetry: An Addendum to the Manual of Photogrammetry*, pp. 3–14. Bethesda, MD: American Society for Photogrammetry and Remote Sensing.

Lanter, D. P., and Veregin, H. (1992) "A research paradigm for propagating error in layer-based GIS," *Photogrammetric Engineering and Remote Sensing*, Vol. 58, No. 6, pp. 825–833.

Lillesand, T. M., and Kiefer, R. W. (2000) *Remote Sensing and Image Interpretation* 4th ed. New York: John Wiley and Sons.

Limp, W. F. (1999) "Image processing software—system selection depends on user needs," *GEOWorld*, Vol. 12, No. 5, pp. 36–46.

Lo, C. P., and Faber, B. J. (1997). "Integration of Landsat Thematic Mapper and census data for quality of life assessment," *Remote Sensing of Environment*, Vol. 62, pp. 143–157.

Lo, C. P., and Shipman, R. L. (1990) "A GIS approach to land-use change dynamics detection," *Photogrammetric Engineering and Remote Sensing*, Vol. 56, No. 11, pp. 1483–1491.

MacEachren, A. M., and Kraak, M. J. (1997) "Exploratory cartographic visualization: Advancing the agenda," *Computers and Geosciences*, Vol. 23, No. 4, pp. 335–343.

Mayo, T. (1994) "Computer-assisted tools for cartographic data capture," in *Innovations in GIS 1* by Worboys, M. F. ed. pp. 39–52. London: Taylor & Francis.

McHarg, I. L. (1969) *Design with Nature*. New York: Doubleday.

Menon, S., Gao, P., and Zhan, C. (1991) "GRID: A data model and functional map algebra for raster processing," *Proceedings*, pp. 551–561, GIS/LIS '91, Atlanta, GA.

MicroImages (1998) *Getting Good Color with TNTmips*. Lincoln, NE: MicroImages, Inc.

Mitchell, C. (1973). *Terrain Evaluation*. London: Longman.

Morain, S., and Baros, S. L. eds. (1996) *Raster Imagery in Geographic Information Systems*. Santa Fe, NM: OnWord Press.

Mullen, R. R. (1980) "Aerial mosaics and orthophotomaps," in *Manual of Photogrammetry* 4th ed. by Slama, C. C. ed. Falls Church, VA: American Society of Photogrammetry.

Peterson, M. P. (1995) *Interactive and Animated Cartography*. Englewood Cliffs, NJ: Prentice Hall.

Photometrics (1990) *Charge-coupled Devices for Quantitative Electronic Imaging*. Tucson, AZ: Photometrics Ltd.

Rhyne, T. M. (1997) "Going virtual with geographic information and scientific visualization," *Computers and Geosciences*, Vol. 23, No. 4, pp. 489–492.

Ritter, N., and Ruth, M. (1995) *GeoTIFF Format Specification Revision 1.8.1*. **http://home.earthlink.net/~nitter/geotiff/spec/geotiffhome.html**

Shapiro, M. (1993) "R.mapcal: Raster map layer calculator," *GRASS Reference Manual*.

Tomlin, C. D. (1983) *Digital Cartographic Modeling Techniques in Environmental Planning*, Unpublished doctoral dissertation, Yale University, New Haven, CT.

Tomlin, C. D. (1990) *Geographic Information Systems and Cartographic Modeling*. Englewood Cliffs, NJ: Prentice-Hall.

Tomlin, C. D. (1991) "Cartographic modelling," in *Geographic Information Systems, Vol. 1: Principles*, pp. 361–374, by Maguire, D. J., Goodchild, M. F., and Rhind, W. D. eds. Harlow, UK: Longman Scientific and Technical.

Triglav, J. (1999) "MrSID: A master of raster image compression," *Geoinformatics*, July–August, 1999 issue, pp. 36–41.

USGS (1997a) *DEM Data User's Guide Version 5*, Reston, VA: United States Geological Survey.

USGS (1997b) *Standards for Digital Terrain Models*, National Mapping Program Technical Instructions. Reston, VA: United States Geological Survey.

USGS (1999a) *Digital Orthophoto Quadrangles (Fact Sheet 129-95)*. Reston, VA: United States Geological Survey, Mapping Applications Center. **(http://mapping.usgs.gov/mac/isb/pubs/factsheets/fs12995.html)**

USGS (1999b) *Digital Raster Graphics (Fact Sheet 070-99)*. Reston, VA: United States Geological Survey, Mapping Applications Center. **(http://mapping.usgs.gov/mac/isb/pubs/factsheets/fs07099.html)**

USGS (2000) *Digital Orthoquadrangle and GeoTIFF*. Rolla, MO: United States Geological Survey, Mid-Continent Mapping Center. **http://mcmcweb.er.usgs.gov/stds/geotiff.html**

Vickers, E. W. (1993) "Production procedures for an oversize satellite image map," *Photogrammetric Engineering and Remote Sensing*, Vol. 59, No. 2, pp. 247–254.

Wolf, P. R. and Dewitt, B. A. (2000). *Elements of Photogrammetry with Applications in GIS*, 3rd ed. New York: McGraw-Hill.

CHAPTER

6

VECTOR-BASED GIS DATA PROCESSING

6.1 Introduction

Vector-based data processing is the manipulation and analysis of spatial features stored as points, lines, and polygons in a geographic database for spatial problem solving. It includes a wide spectrum of processes that range from digital data conversion, database creation, spatial data query and management, overlay and network analysis, to the development of sophisticated applications by computer programming. Since the vector data model is conceptually, logically, and physically dissimilar from the raster data model, the objectives and requirements of vector-based geographic data processing are distinctly different from those of raster-based processing that we learned in Chapter 5. This chapter examines the methods of geographic data processing that can be found in a typical vector GIS.

6.2 Characteristics of Vector-Based GIS Data Processing

Vector-based GIS data processing makes use of spatial data represented as discrete entities on well-defined layers in digital databases. Therefore, it is useful to revisit the properties of vector data and the vector layer and to explain the major characteristics of using these data and layers in geographic data processing.

6.2.1 Vector Data and the Vector Layer

Vector data in a digital geographic database depict individual spatial features as discrete entities using three basic graphical elements: points, lines, and polygons. In the database, these entities are identified as *feature classes*, each pertaining to a particular theme, such as drainage, vegetation, topography, transportation, land parcels, and administrative boundaries. Vector data, therefore, are used for GIS applications that focus primarily on the characteristics of individual or individual classes of spatial features in a particular geographic area of interest.

In a digital geographic database, feature classes are logically organized as *layers* (Figure 6.1). Since each layer is allowed to have only one type of graphical element, feature classes are also organized according to graphical element type by default. This means that features that pertain to a particular theme but are represented by different types of graphical elements must be identified separately according to their respective types of graphical elements. For example, although spatial features such as springs, rivers, and lakes are all associated with the same theme, that is, drainage, they must be identified as separate feature classes and stored on different layers, because they are point, line, and polygon features respectively. In this particular example, springs, rivers, and lakes are independent feature classes, and drainage is a *collection of feature classes*.

Conventionally, the topologically structured layer as explained in Chapter 3 has been the standard data format for vector GIS. In recent years, several other vector formats have emerged as a result of changing software design concepts and GIS application requirements. These include, for example, *shapefiles*, which is the nontopological data format of ArcView GIS and is commonly used not only in ESRI's own software products, but more generally as a de facto interchange format among GIS users (ESRI, 1997); spatially indexed and tiled files, which are created by map library functions of various GIS software packages, such as ArcInfo LIBRARIAN (ESRI, 1996c); database files, which are created by special spatial database management systems, such as ArcStorm (ESRI, 1996b); and *object-relational databases*, which are spatial database systems used by GIS software packages based on object-oriented technology, such as Oracle Spatial, ArcSDE, and ArcInfo 8 (ESRI, 2000; Zeiler, 1999).

Because the formats of these vector data types are different from the topologically structured data layers used by conventional vector GIS, spatial data processing using these data types tends to be done differently. It is not within the scope of this chapter to cover individual data-processing methods designed to work with these new vector data formats. However, although the actual methods for data processing using data in shapefile, map librarian, and object-relational database formats may be different, the concepts behind the construction of these methods are similar. In this regard, the vector-based functions to be discussed in this chapter will be generally useful to GIS users engaged in vector-based data processing, no matter what specific vector format their data are in.

6.2.2 Characteristics of Vector-Based GIS Data Processing

Vector GIS data processing is different from raster processing in several ways. One of the most important distinctions between these two approaches of GIS data processing is that whereas raster processing is based entirely on the layers, vector processing uses both data layers and individual features. As vector data are organized by layers, vector GIS data processing can be designed to work on layers, either individually or collectively in very much the same way as raster layers are processed. However, since spatial features are independently identified and represented on a vector layer, it is also possible to apply data-processing functions at the individual feature level. The use of layers and features in vector-based data processing is not

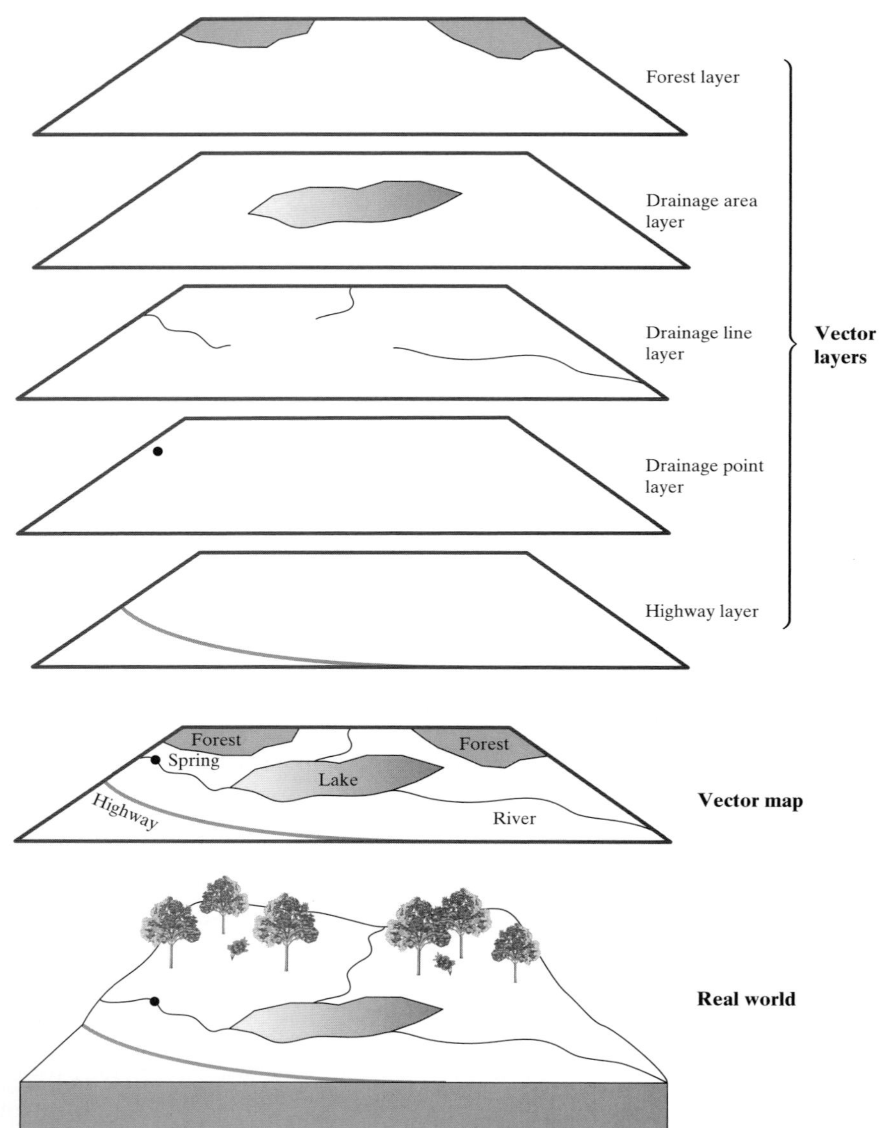

FIGURE 6.1
The vector method of representing real-world features.

mutually exclusive to one another. In fact, GIS users are free to mix layers and individual features in vector-based data processing, making it very flexible and intuitive to use in the construction of methods for spatial data analysis and modeling (Table 6.1).

Another important feature of vector-based GIS data processing is the reliance on topological relationships, as spatial problem solving always involves spatial relationships in addition to locations. Polygon processing, for example, relies on the topological relationships of adjacency (i.e., the left polygon and right polygon of a line) and containment (i.e., a point, line or polygon inside a polygon) to determine neighborhood information that will enable contiguous polygons to be identified, resampled, and merged. Similarly, using the topological relationship of connectivity (i.e., the from-node and to-node information) in a street database, network analysis algorithms are able to determine the shortest path between any two locations. The use of topology is made possible by the topological relationships that are either stored explicitly in the database or computed when the data are used.

Vector geographic data processing is performed using not only graphical data, but also attributes associated with these graphical data as well. The ability to integrate graphical data and their attributes, as noted in Chapter 1, is what distinguishes GIS from other forms of information systems. The need to integrate graphical and attribute data means that the use of a DBMS is often an integral part of vector-based geographic data processing. Conventionally, the role of DBMS in geographic data processing was confined mainly to attribute data management and database access. As the general role of DBMS is expanding from transaction processing to decision support, the role of DBMS in geographic data processing is also evolving toward the same direction,

Table 6.1

Examples of Mixing Layers and Individual Features in Vector-Based Processing

	Layer-Based Output	*Feature-Based Output*
Input by Layers	Land suitability analysis by topological overlay	Attribute database query to select individual features from a layer according to their characteristics
Input by Features	Creating a surface by interpolating point samples	Point-in-polygon search to determine whether a point falls within a particular polygon

leading to the introduction of data warehousing and data mining techniques that focus primarily on attribute data in the geographic database (ESRI, 1998; MapInfo, 1996).

Some vector-based geographic data-processing techniques are extremely demanding in terms of computational requirements, which means they require powerful computers and take time to complete. Notable examples include topology building and topological overlay analysis. In these two operations, the computer basically compares each line segment with every other line segment contained in a layer to determine whether or not they intersect; and if they do, the computer must create a node at the point of intersection. The computer then uses the node and line relationships to identify and construct the polygons on the vector layer. In order to maintain the integrity of the data, it is necessary to rebuild layer topology whenever a change has been made to the data. The maintenance of topological structure is the greatest overhead in vector geographic data processing and is probably the major reason behind the development and wide acceptance of the nontopological shapefiles format. For GIS applications that do not depend on the use of topological relationships, the shapefiles format provides a very useful and practical alternative to the conventional topologically structured layer.

Vector spatial data are normally recorded and processed using ground coordinates in double-precision. This sometimes creates the impression that results of vector data processing are more accurate than those obtained by using raster data (Berry, 1995). It must be reiterated, however, that any claim of accuracy in the use of vector data must be treated with care. This is because the collection of spatial data is subject to various factors that will introduce inherent and operational errors into the data (see Chapter 4). As a result, it is important for GIS users to realize that there is always a degree of uncertainty in the results of geographic data processing no matter whether vector or raster data are used. In this chapter, we will not deliberate on the issue of error propagation in vector GIS data processing in the same way we did in Chapter 5, on the understanding that the concepts and techniques of analyzing errors and their impacts on geographic data analysis discussed in Chapter 5 apply equally to vector GIS data processing.

6.2.3 CLASSIFICATION OF VECTOR-BASED GIS DATA-PROCESSING METHODS

In Chapter 5, we classified raster-based geographic data-processing functions according to their purposes into three categories: *input*, *analysis*, and *output*. Such a classification applies equally well to functions that are found in a typical vector-based GIS (Figure 6.2).

Generally speaking, geographic data are more readily available now than they were earlier. It is common for GIS to be implemented using spatial data obtainable from government agencies or commercial data suppliers. This has mostly eliminated the need for digital conversion of existing geographic data on maps and other hard-copy forms. However, data input remains a key function of vector data processing. This particular function includes three major processes: map digitizing, image scanning, and geometric transformation from data stored in another format. We will explain these data input processes in detail in Section 6.3.

The analysis function of a typical vector GIS consists of a variety of data-processing tasks used to detect the characteristics of spatial features in a particular geographic space of interest. Depending on the nature of the operations used, these functions can be classified into the same six categories as noted in Chapter 5. To recap, these categories are logical operations, arithmetic operations, overlay operations, geometric property operations, geometric transformation operations, and geometric derivation operations.

These six GIS functions accept vector data on one or more layers as input and produce a statistical report or a new vector layer as a result, much in the same way as raster-based functions work. However, whereas

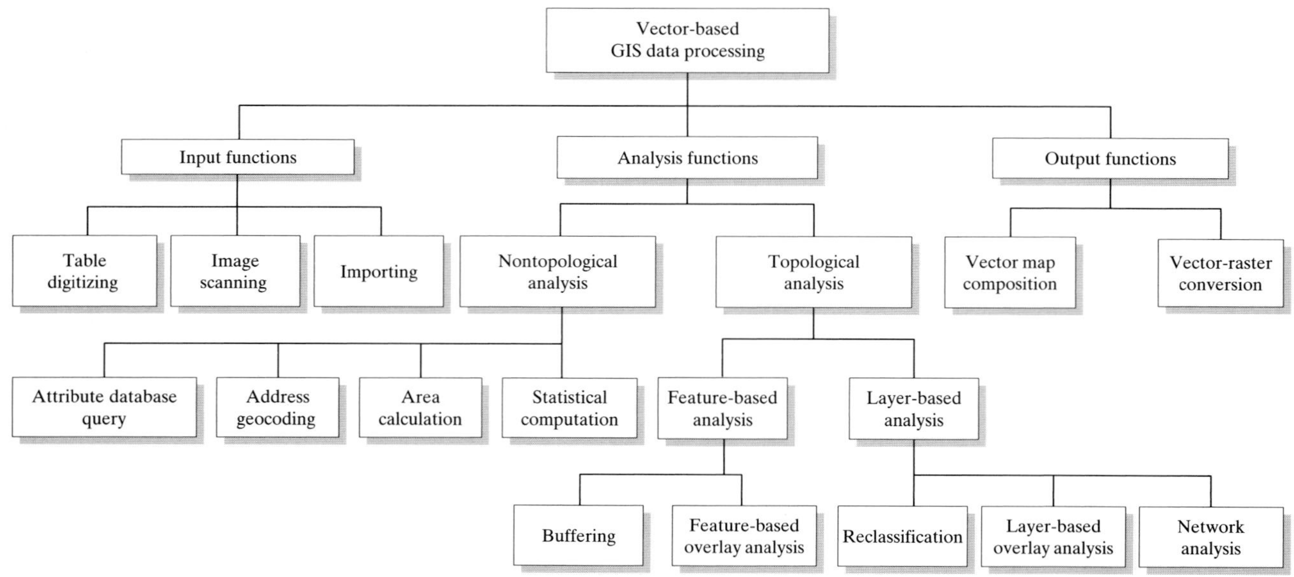

FIGURE 6.2
Classification of vector GIS data-processing methods.

raster-based methods work on raster cells on the layer, vector-based methods work on discrete spatial features. Also, some vector functions use attribute data only, others depend on the topological relationships among graphical elements on the input layers. These properties provide another way of classifying vector-based GIS functions into three categories: *nontopological functions*, which make use of attribute values and coordinates rather than stored or computed topological relationships to analyze spatial distribution, pattern, and trend implicit in one or more collections of geographic data; *feature-based topological functions*, which are performed by using the topological relationships between individual spatial features on one or more layers; and *layer-based topological functions*, which use all spatial features present on a layer in spatial analysis. In both the second and third categories of analysis functions, vector layers must be topologically structured in order that they can be used, and the output layers are topologically structured automatically during and on completion of these functions.

A matrix of vector-based GIS analysis functions can be created on these two approaches of classifying these functions (Figure 6.3). This matrix provides a useful way to illustrate the nature of commonly used vector-based GIS functions with respect to the characteristics of the operations that they use, as well as to the requirements of the data that are analyzed. We will use this classification scheme of vector-based GIS analysis functions to describe their properties, procedures, and applications in Sections 6.4 through 6.6.

	Nontopological	Topological	
		Featured-based	Layer-based
Logical operations	• Attribute database query		• Reclassification and aggregation
Arithmetic operations	• Change mapping • Summary statistics		
Overlay operations	• Address geocoding	• Overlay analysis	• Overlay analysis
Geometric property operations	• Calculation of areas, perimeters, and distances		• Network analysis
Geometric transformation operations	• Coordinate and geometric transformation • Surface interpolation		
Geometric derivation operations		• Buffering	

FIGURE 6.3
A matrix of vector-based geographic data-processing functions.

The output function of vector GIS data processing, which includes vector map composition and vector-raster conversion, aims to provide the results of a particular data analysis task to the end user in a form suitable for publication and distribution. We will explain the basic output methods of vector GIS data processing in Section 6.7. More detailed treatment of the concepts and techniques of presenting the results of geographic data processing will be given in Chapter 7.

6.3 VECTOR DATA INPUT FUNCTIONS

The input functions of vector-based GIS include different *digital data conversion* methods that aim to get spatial data ready for use in a computer system. In GIS, digital data conversion is made up of two components: *graphical data conversion*, which creates digital map layers; and *attribute data conversion*, which populates tabular data files associated with the graphical elements on a layer. The process of digital data conversion is made up of four sequential phases of work.

1. *acquisition* of digital data by digitizing existing maps, purchasing from government agencies or commercial data suppliers, or by collecting new data using field surveying, GPS-based attribute data logger, photogrammetric and remote sensing methods
2. *editing* to clean the acquired digital data, if necessary, in order to ensure that they are of an acceptable quality with respect to application objectives
3. *formatting* or *translating* to convert the digital data into the specific physical database format of the GIS where the data are used
4. *linking* graphical data to associated attribute data

The concepts and techniques of digital data conversion appear to be relatively simple and straightforward. In practice, however, data conversion is probably one of the most tedious, time-consuming, and error-prone processes in vector-based data processing. As we will demonstrate in the following discussion, digital data conversion embraces considerable technical details that must be well understood in order to build and maintain a geographic database cost-effectively.

6.3.1 DATA CONVERSION BY DIGITIZING EXISTING MAPS

Existing maps represent a very important source of geographic data in many organizations. Conventionally, map digitizing was done by means of a special data capture device called a *digitizer*. In recent years, the importance of the digitizer has diminished as a result of the growing use of scanning technology and the increasing availability of digital data from government agencies and commercial data suppliers. However, this method is still the handiest way of creating a geographic database in-house. Map digitizing is particularly useful for GIS applications that require project-specific data (i.e., data unlikely to be available from external sources).

6.3.2 THE DIGITIZER AND ITS SPECIFICATIONS

Digitizers are produced in different sizes. Large-format digitizers, with a minimum E-size (36″ × 44″ or 90 cm × 110 cm) surface area, are usually referred to as *digitizing tables*. Small-format digitizers, with a C-size surface area (17″ × 22″ or 43 cm × 56 cm), are commonly called *digitizing tablets*. A digitizer is made up of three components: a *table*, a *cursor*, and a *controller* (Figure 6.4). The table consists of a large printed circuit board of extremely fine grids that is embedded between flat fiberglass or plastic plates. The spacing of the measuring grids determines the *resolution* of the table, which, by definition, is the smallest distance separating two adjacent points. Most digitizers today have a resolution of 0.001 inch (0.025 mm). The cursor is a free-moving component connected to the controller by means of a thin cable. It has a tracking cursor around which a field coil is mounted. Whenever the cursor button is pressed, this field coil sends out a signal that is picked by the measuring grid to generate the coordinates of the cursor position on the table surface. The cursor button is one of the buttons (usually from four to sixteen in number) on the keypad of the puck. The controller is the interface between the digitizer and the host computer. It is a microprocessor that accepts the coordinates captured during the digitizing process and passes them to the host computer for display on the screen and for further processing.

In addition to the resolution, digitizers are also required to meet a number of technical requirements. These include (Masry and Lee, 1988):

1. *Stability* measures the amount the coordinate reading drifts when the cursor is fixed in a specific position, normally in the order of 0.001 inch (0.025 mm).
2. *Repeatability* is the ability of the digitizer to generate the same reading every time the cursor is placed at the same point on the table, normally within 0.001 inch (0.025 mm).
3. *Skew* is the measure of the orthogonality of the axes in the digitizing table, normally less than 0.002 inch (0.05 mm) in a 40-inch (one meter) distance.
4. *Accuracy* is the total sum of the errors resulting from the preceding sources, normally about ±0.004 inch (0.1 mm).

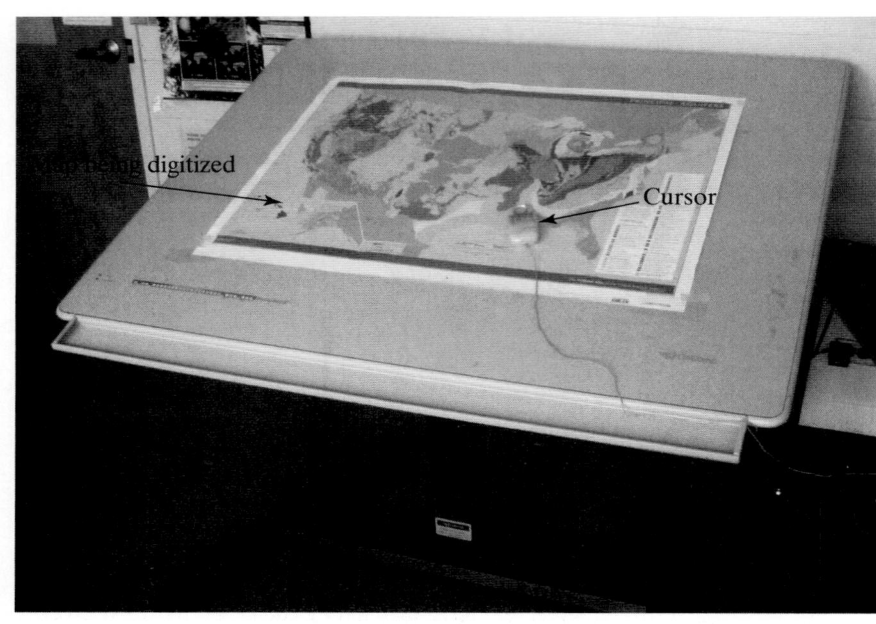

FIGURE 6.4
A large-format digitizing table.

6.3.3 MAP DIGITIZING PROCEDURES

Map digitizing is a multistep process (Figure 6.5). The typical map-digitizing procedures include: (1) preparation, (2) creating a digitizing template, (3) map digitizing, and (4) postdigitizing data processing.

Preparation for Digitizing Map digitizing starts by getting the map and the digitizer ready for the process.

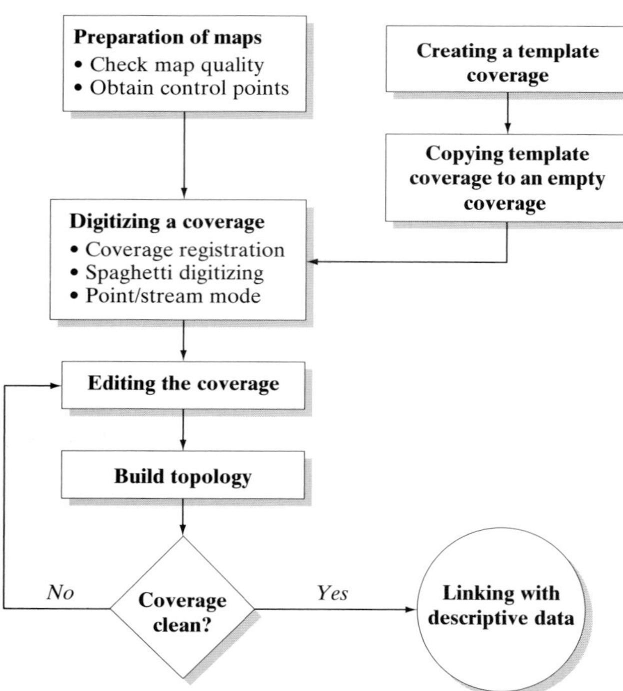

FIGURE 6.5
Graphical data conversion by map digitizing.

This includes checking the quality (i.e., accuracy, completeness, and currency) of the map data as well as identifying control points, sometimes known as *tic points*, for registering the output digital data to map coordinates. If the digitizer has not been in use for an extended period of time, it is necessary to perform an accuracy test to ensure that it is in good working condition (i.e., meets the specifications noted above). For map data in each map sheet, it is necessary to identify every one of the features (i.e., polygons, lines, and points) to be digitized and number them with a unique *identifier*.

Preparation also includes the establishment of the specifications of feature codes, line types, and approaches of data capture (Figure 6.6). The design of these specifications must follow an accepted data standard (e.g., corporate, industry, or national standards; see Chapter 4). The objectives of the digitizing specifications are to enforce data-processing standards, to allow for data sharing, and to ensure consistency of the data digitized by different operators.

Creating a Digitizing Template A digitizing template contains the tic points, map neat lines, and graphical elements that are common to all layers, such as boundaries of water bodies. For applications that require multiple layers, using a template enables these layers to be registered perfectly with one another. It also helps minimize the amount of work because graphical elements common to all layers need to be digitized only once (Figure 6.7).

The tic points and map neat lines of the digitizing template can be created either by actually digitizing them on the digitizer or by generating them from input coordinates. The latter method is always preferred because it uses exact coordinates. This practically eliminates

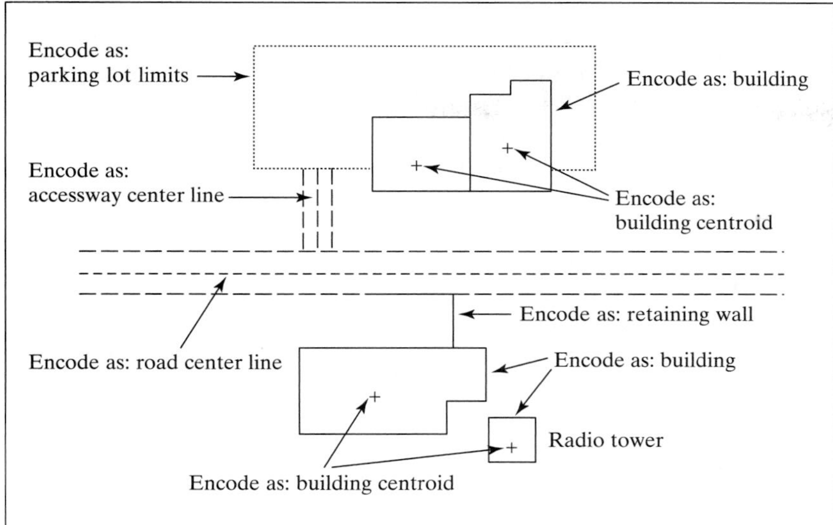

FIGURE 6.6
An example of data capture specifications. Data capture specifications provide the digitizing operators with very detailed instructions on what to digitize, how to digitize, and which feature code to use for each type of features. *(Source: Ontario Government Specifications for the Delivery of Digital Topographic Data and Cartographic Representation Products at Medium Scales [ONMR, 1992])*

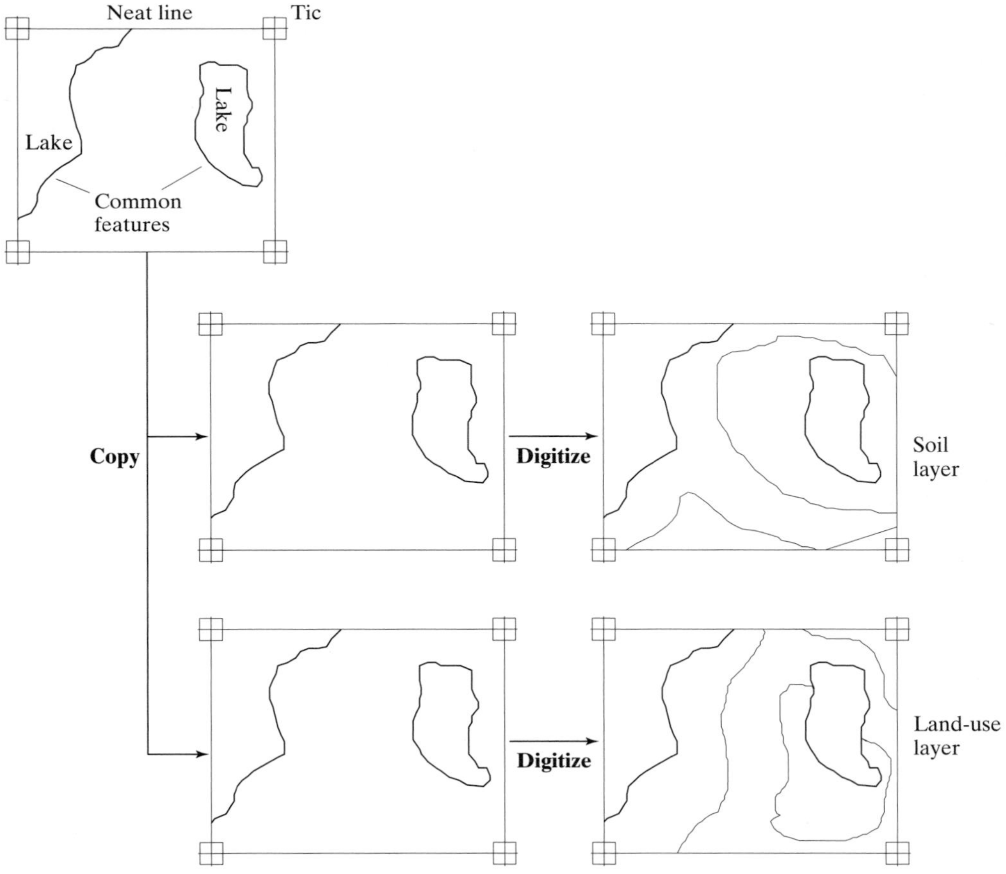

FIGURE 6.7
The method of using a digitizing template. Using a digitizing template eliminates the need to digitize common features on different layers and ensures perfect registration of these features when the layers are used for overlay analysis.

the possibility of errors that may occur when the points are digitized. The ways of generating tic points and map neat lines vary in different systems. In ArcInfo, for example, the GENERATE command and its suite of subcommands allow a blank layer containing only tic points and map neat lines to be created from an input ASCII file (ESRI, 1995a). The common graphical elements on the digitizing template have to be digitized from a topographic base map. The template is then edited and its topology built before it is used to provide the framework for digitizing the map layers.

Map Digitizing Map digitizing begins with the registration of the map mounted on the digitizing table to the digital map displayed on the screen. This is a transformation process that uses the grid coordinates obtained by digitizing the tic points of the map and the ground coordinates of corresponding points entered through the keyboard. On completion of this process, the coordinates for corresponding objects on the map and on the screen become identical (a coordinate transformation process, affine transformation, as explained in Chapter 2, Section 2.5.3, has been performed) (Figure 6.8).

During map digitizing, there is no need to follow a particular sequence (such as the order of the identifiers) by which graphical elements are digitized. There is also no need to digitize the locations at which lines intersect. This method of digitizing, usually referred to as *spaghetti digitizing*, enhances productivity by freeing the operator from the need to strictly follow a particular sequence of operation and to digitize every intersecting point. Intersecting lines will be automatically broken into line segments at their intersecting points during the topology building process later on. For maps that have complex contents, it is advisable to mark all objects (e.g., by crossing out their identifiers) when they are being digitized, in order to avoid missing or duplicating any features in the digitizing process.

Map digitizing may be carried out in *stream* or *point* modes. In the stream mode, the digitizer generates coordinates automatically at a predetermined time interval as the operator is tracing a line (which may be a line feature or the boundary of a polygon) on the map using the cross hair of the digitizer. In the point mode, the digitizer will generate coordinates only when the user presses a predefined button on the keypad of the digitizer. In practice, the point mode is more popular because it allows more operator control of the digitizing process by letting the operator determine which points are to be captured. The point mode of digitizing usually also results in a smaller data file because points are captured selectively rather than indiscriminately by sampling at very short time intervals.

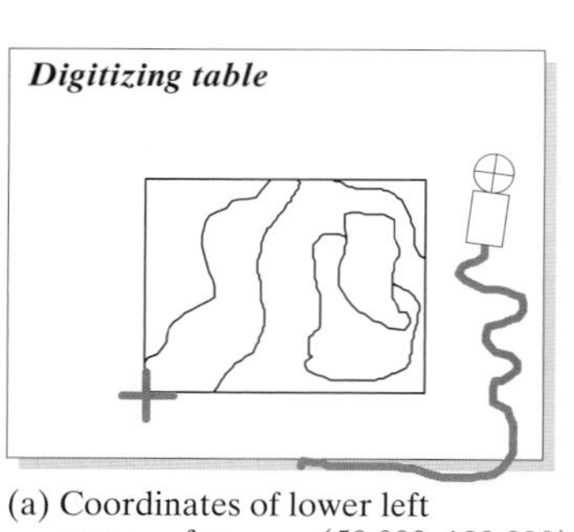

(a) Coordinates of lower left corner of map = (50,000, 100,000)

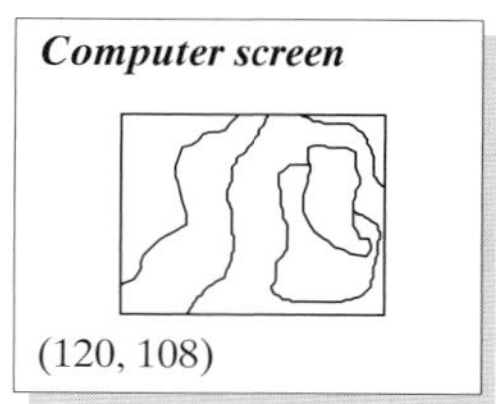

(b) Coordinates of lower left corner on screen without registration

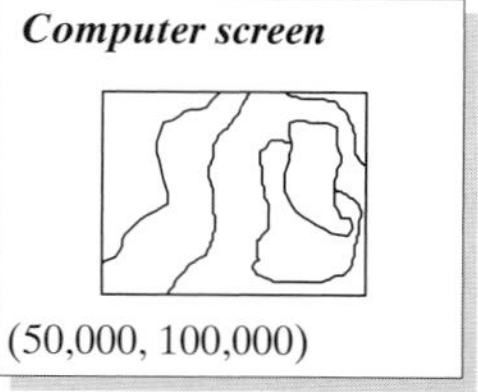

(c) Coordinates of same point on screen after registration

FIGURE 6.8
Coordinate registration in map digitizing. Without coordinate registration, data displayed on the screen are referenced to the arbitrary coordinate system of the digitizing table, but not the ground coordinate system of the map. After registration, all coordinates measured on the screen are referenced to the coordinate system of the map being digitized.

Postdigitizing Data Processing The steps of map digitizing described here represent only a portion of the process of spatial data conversion. Once the digitizing is complete and the data are saved to a file, the postdigitizing process as described in Section 6.3.8 will be carried out. Postdigitizing data editing should be perceived more as an extension to the digitizing process rather than as an independent process in itself. If omissions are detected during editing, it is sometimes necessary to return to the digitizing function. As a result, it is important to ensure that before the postdigitizing process is completed, the position of the map on the digitizing table must remain unchanged. Any movement of the position of the map requires the map registration procedure to be carried out again.

6.3.4 DATA CONVERSION BY SCANNING AND VECTORIZATION

Map data conversion by scanning and vectorization is often referred to as *screen digitizing* or *heads-up digitizing* to distinguish it from conventional map data conversion using a digitizing table. The use of this technology for map data conversion has grown considerably in the last few years as the result of improvement of hardware design, software capabilities, and data-compression techniques. This approach of digital data conversion is capable of converting a large number of maps in a relatively short time frame and at a cost comparable to or lower than the conventional method of map digitizing. As a result, the scanner has gradually replaced the digitizer as the means of digital data conversion in GIS projects (see Chapter 5).

It must be noted, however, that the process of scanning is only a very small part of the whole data conversion process (Figure 6.9). A considerable amount of work is needed to prepare the map for scanning. This includes separating the captured map data into different layers, touching up thin line work, as well as closing gaps in line objects. After scanning, it is necessary to convert the raster image to vector graphics, to build layer topology, and to link the resulting vector graphical elements to their associated attributes.

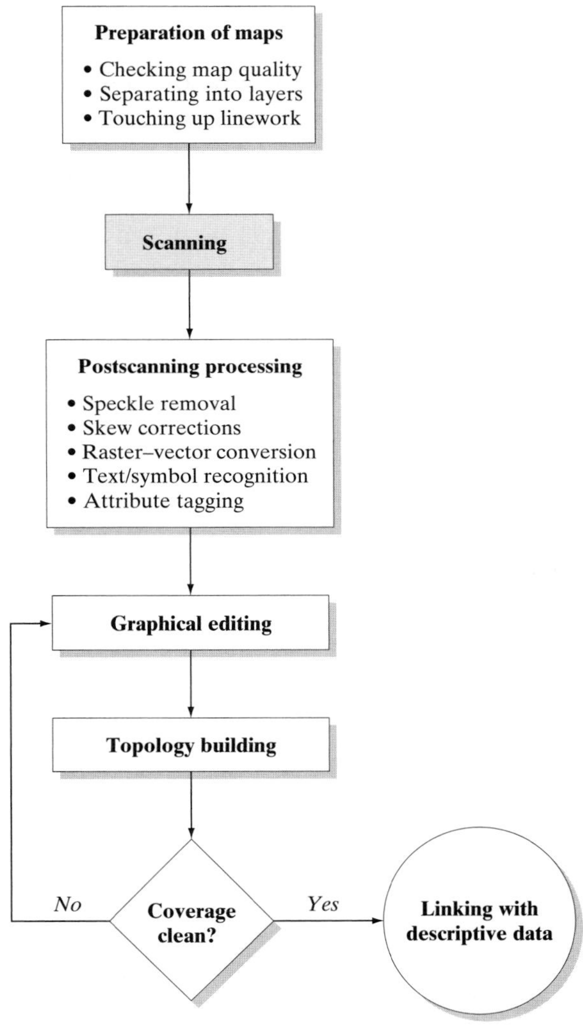

FIGURE 6.9
Graphical data conversion by scanning. Scanning of the map is only a small part of the whole data conversion process. Much work is needed to prepare the map before scanning and to postprocess the resulting vector data to make them useful for geographic database creation.

6.3.5 THE SCANNER AND ITS SPECIFICATIONS

For use in GIS data conversion, a scanner has to meet a number of technical specifications. These include (Masry and Lee, 1988):

- *Resolution*. This is a measure of the *sharpness* of the image, usually expressed in terms of *dots per inch* (dpi) or actual pixel size in microns. Experience has shown that a resolution of 300 to 600 dpi is sufficient for general GIS applications involving paper maps; scanning of aerial photographic images for digital mapping requires a resolution as high as 2000 dpi.
- *Accuracy*. The accuracy of scanners is expressed as a ratio between the dimensions of the raster image and the original document. A scanner in good working condition is expected to have an accuracy of ±0.1%.
- *Scan Width and Length*. A minimum scan width of 36 inches (91.4 cm) is required. Document length is not always a problem.
- *Output File Format*. It is essential for the output formats to be accepted by the vectorization software to be used.
- *Quality of the Accompanying Software*. This includes the ability of the software to integrate with the computer hardware platform, to perform batch- and icon-driven processing, to remove speckles, and to correct the skew of the resulting image.

6.3.6 POSTSCANNING DATA PROCESSING

Scanning is a nonselective data conversion process in the sense that every point on the map is captured. The resulting raster image is of limited use for GIS applications other than simple display on the computer screen. As the technology stands today, it is not yet possible to totally automate the vectorization of raster data on a scanned image. Postscanning data processing, using either computer-assisted or manual methods, is still an indispensable part of the data conversion process. The actual amount of work that is required for postscanning data processing depends on a variety of factors, including the quality of the original maps, the complexity of the map contents as well as the functionality of the vectorization software (ESRI, 1994c; Gisiger, 1998).

In general, postscanning processing is carried out as a sequence of processes that includes (1) raster–vector conversion, (2) raster text conversion, (3) raster symbol conversion, (4) graphical data editing, and (5) attribute data tagging.

Raster–Vector Conversion This process, which changes raster images into vector graphics, can be carried out manually or with computer assistance (automated or semi-automated vectorization). In recent years, there has been some success in automating the raster–vector conversion process using artificial intelligence and pattern recognition techniques (Flanagan et al., 1994; Mayo, 1994). The quality of the original maps seems to be the major limiting factor for the adoption of automated methods. For example, automated vectorization fails because of the presence of broken lines that cause gaps in the image. Lines that are too thin or do not have adequate contrast to be picked up during image scanning represent another common source of error. Special cartographic symbols and widely spaced characters in place names also cause failure in automated symbol and text recognition.

Raster Text Conversion This process recognizes characters in the raster image and changes them into alphanumeric data. Relatively well-developed character recognition programs can now be used to perform this process automatically. However, verification of the results of character recognition and the corrections of errors that may occur are largely manual tasks.

Raster Symbol Conversion This process recognizes cartographic symbols in the raster image and converts them into alphanumeric codes. The apparent lack of standards in the form, size, and codes of cartographic symbols has made automated symbol recognition a much more difficult task than character recognition. As a result, raster symbol conversion is largely a manual task in practice.

Graphical Data Editing This process cleans the graphics by removing data conversion errors in the same way as table digitizing errors are corrected. Graphics editing can be done using the editing functions of a GIS. Alternatively, it is possible to use the data-editing tools that come with raster–vector conversion software packages (Figure 6.10).

Attribute Data Tagging This process, which adds attribute data (e.g., feature identifiers, feature codes, and contour labels) to the graphical data after raster–vector conversion, is usually carried out interactively on the screen (see Section 6.3.8).

6.3.7 IMPORTING VECTOR DATA AND PROJECTION TRANSFORMATION

The increasing availability of digital spatial data from government agencies and commercial data suppliers means

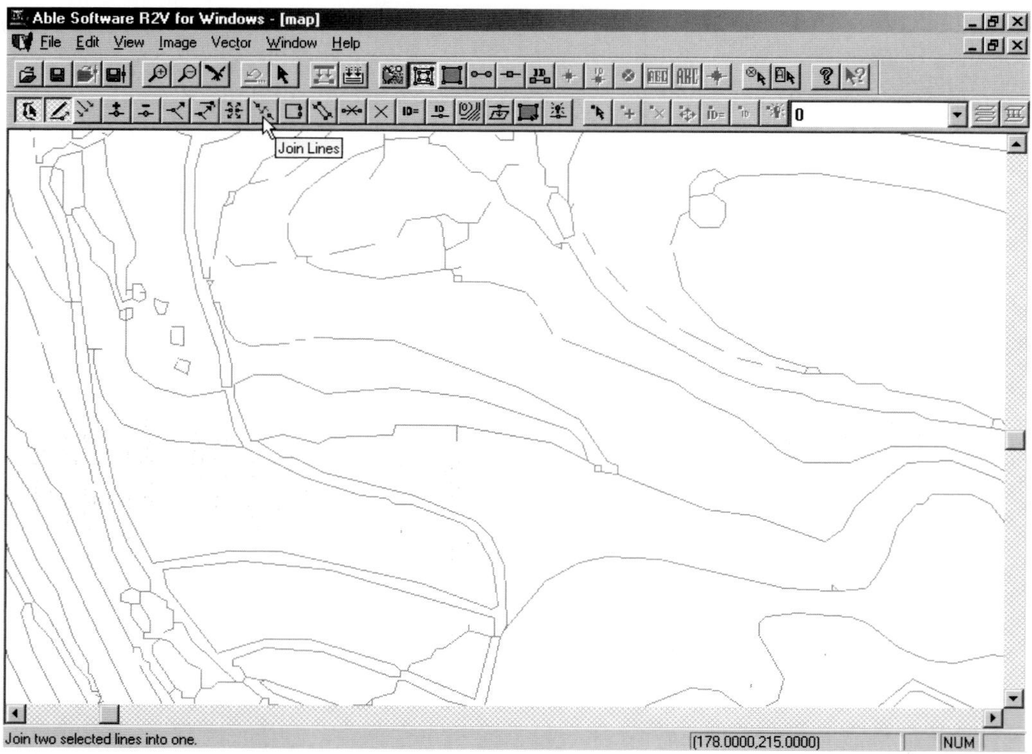

FIGURE 6.10
Graphical editing tools provided by a typical raster–vector conversion software package. Raster–vector conversion software packages in general have a rich set of data editing tools that allow the user to clean the resulting vector data before saving them to a specific GIS data file format.

that spatial data required by a particular GIS project can now be purchased rather than converted from existing maps. The ability to import digital data in different formats has now become a necessity in GIS. Data import in GIS is achieved by using a data translation program. Translation programs may be developed by GIS software vendors, third-party service bureaus and system consultants, or by public institutions such as government agencies and universities. As these data translators are basically all product-specific, the best sources of information are the user manuals of individual products.

In addition to importing digital data that are already in a particular proprietary or interchange format, GIS software packages are normally capable of importing data directly from field surveying using coordinate geometry (COGO) or GPS-based methods (ESRI, 1996a). Additionally, GIS software packages in general are able to accept and convert graphics files in CAD format, in particular the very popular AutoCAD DXF format. The function to import digital data in CAD formats are particularly important for users who are working in engineering and related areas, such as facility management.

An important data-processing function that is closely associated with data import is coordinate transformation. This function is required when the imported geographic data are not in the map projection or coordinate system that is adopted by a particular application, such as the Universal Transverse Mercator Coordinate System or the State Plane Coordinate System (see Chapter 2). Before the imported data can be integrated with existing data, they must be geometrically transformed to the common ground coordinate system. GIS software packages today often support a large set of standard projections and coordinate systems with predefined transformation parameters, such as spheroid, central meridian, and standard parallel. Alternatively, a stand-alone commercial coordinate transformation program can be used, such as the Geographic Transformer, produced by Blue Marble Geographics.

With extensive coordinate functions available in GIS software packages and obtainable from third-party software vendors, coordinate transformation has never been easier. However, GIS users must realize that map projections and coordinate transformation are relatively complex processes. In order to avoid the misuse of coordinate transformation functions in GIS, users must ensure that they clearly understand why they choose a particular projection and coordinate system, what options they have in a particular GIS software package or a commercial product, metadata such as datum, central meridian, and coordinates of the false origins, for both the input and output data sets.

6.3.8 GRAPHICAL DATA EDITING

Graphical data editing is a postdigitizing process that is designed to ensure the integrity of the data before they can be used in a geographic database. Integrity of the data in the context of a topologically structured layer means that they are free from errors as illustrated in Figure 6.11. In other words, this implies that

- lines intersect where they are expected to intersect (i.e., no undershoot or overshoot)
- nodes are created at all points where lines intersect
- all polygons are closed
- each polygon contains a label point
- the topology of the layer is built

Graphical data editing is a core function of GIS. The methods of graphical data editing applies to new data obtained by map and heads-up digitizing, as well as to data converted from other formats. Generally speaking, graphical data editing is made up of four basic phases of work, including (1) setting the editing environment, (2) topology building, (3) data editing and error corrections, and (4) joining adjacent layers.

Feature	Errors	Examples
Node	Dangling node	
	Pseudo-node	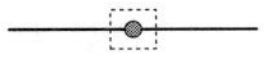
	Pseudo-node	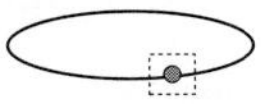
Arc	Dangling arc	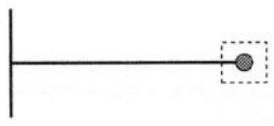
	Overshoot	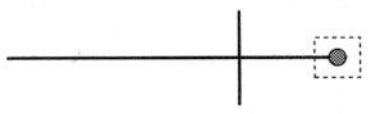
	Undershoot	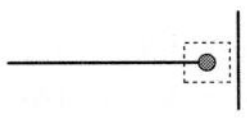
	Intersection (no node)	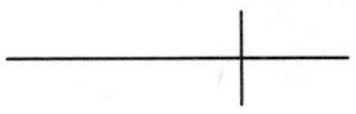
Label	Missing label	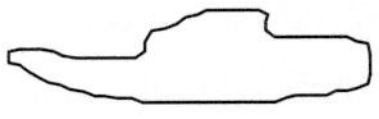
	Duplicate/multiple labels	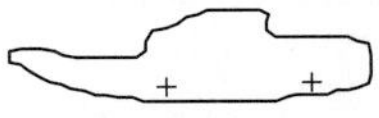

FIGURE 6.11
Examples of digitizing errors. Whether a digitizer feature is an error often depends on what it represents, rather than how it looks. For example, a cul-de-sac is always a dangling arc that must not be regarded as a digitizing error.

Setting the Editing Environment This process sets up the working environment for a data-editing session. Although the procedures used by an individual GIS may vary, the objectives of setting the editing environment are essentially the same. In ArcInfo, an important concept in setting the editing environment is *tolerance,* which is a user-defined measurement to facilitate the editing process. Common tolerances include *edit tolerance,* which enables the operator to select a graphical element more quickly by picking without having to actually click on it (i.e., the graphical element will be selected when the screen cursor is clicked at a small distance away from it) (Figure 6.12a); *weed tolerance,* which removes digitized arcs that are shorter than a specified length (Figure 6.12b); and *grain tolerance,* which filters out digitized points in curves, circles, and other smoothed arcs that are too close together (Figure 6.12c).

Topology Building This is probably the most important process in graphical data editing. As a postdigitizing process, however, topology building actually serves two interrelated purposes.

- *Building the Topological Structure and Relationships for the Graphical Elements on a Layer.* The actual process of topology building is dependent on the type of the graphical elements on a layer (Figure 6.13). In general, this includes the creation of point, line, and polygon topology by assigning an internal identifier to each graphical element identified and the creation of attribute tables.
- *Error Identification and Automated Corrections.* If digitizing errors exist in the digital data file, they will be highlighted by the topology building commands. These errors may also be corrected automatically by setting user-defined data-editing tolerances in the topology building process. ArcInfo, for example, uses the concept of *fuzzy* and *dangling tolerances* in conjunction with the CLEAN and BUILD commands (which invoke the topology building process) to resolve node problems resulting from digitizing (Figure 6.14). By specifying a fuzzy tolerance, nodes that are too close to one another will be resolved into a clean intersection. By specifying a dangling tolerance, small dangling nodes will be removed, thus eliminating the need to do this manually.

Topology building is normally a repetitive process. When the errors in a layer have been identified and corrected, either manually or automatically using fuzzy and dangling tolerances, topology must be rebuilt. If errors have been introduced or overlooked in the editing process, they will show up again in the new round of topology building. These errors must be removed, and topology rebuilt once again. The layer will be accepted only when it is shown to be totally free of digitizing and topological errors.

Data Editing and Error Corrections Graphical data editing includes procedures that enable the operator to select, delete, copy, and add graphical elements, as well as to change their properties, on a vector layer (Figure 6.15). Data editing is an integral part of the topology building process as noted earlier. It is also required when errors are detected as the result of quality assurance measures in data conversion (e.g., by verifying the data in a check plot).

Tolerance	Illustration
(a) Edit tolerance Cursor is able to select point *x* when it is clicked within the edit tolerance	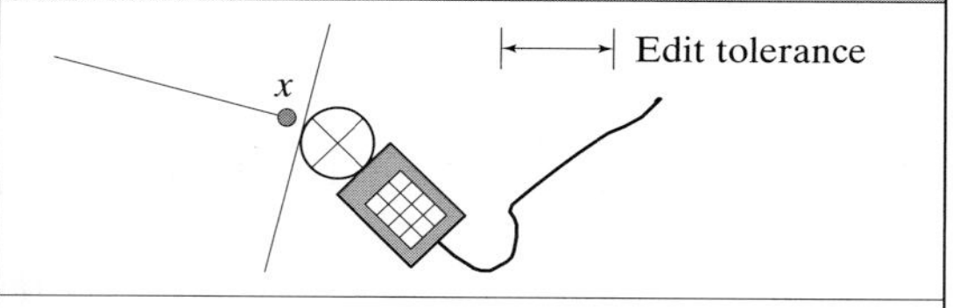 x Edit tolerance
(b) Weed tolerance Digitized arcs smaller than weed tolerance are filtered	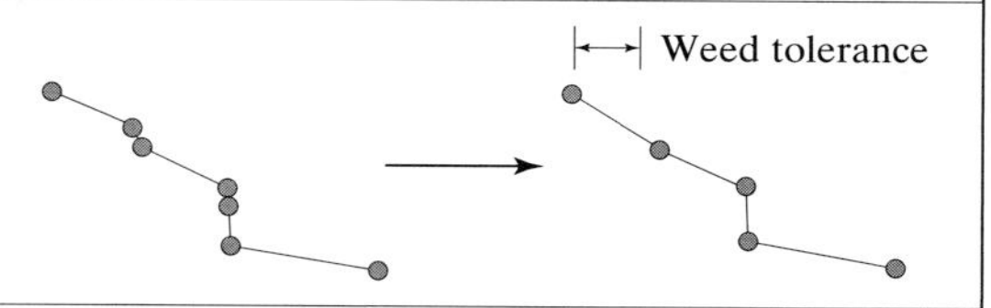 Weed tolerance
(c) Grain tolerance Digitized points too close to one another are removed	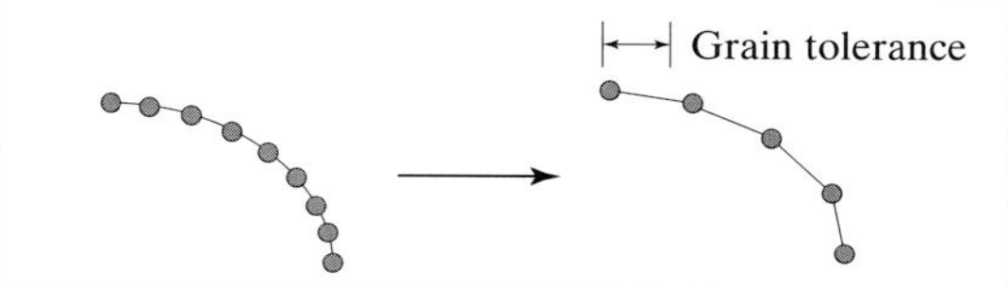 Grain tolerance

FIGURE 6.12
Edit, weed, and grain tolerances in graphical data editing.

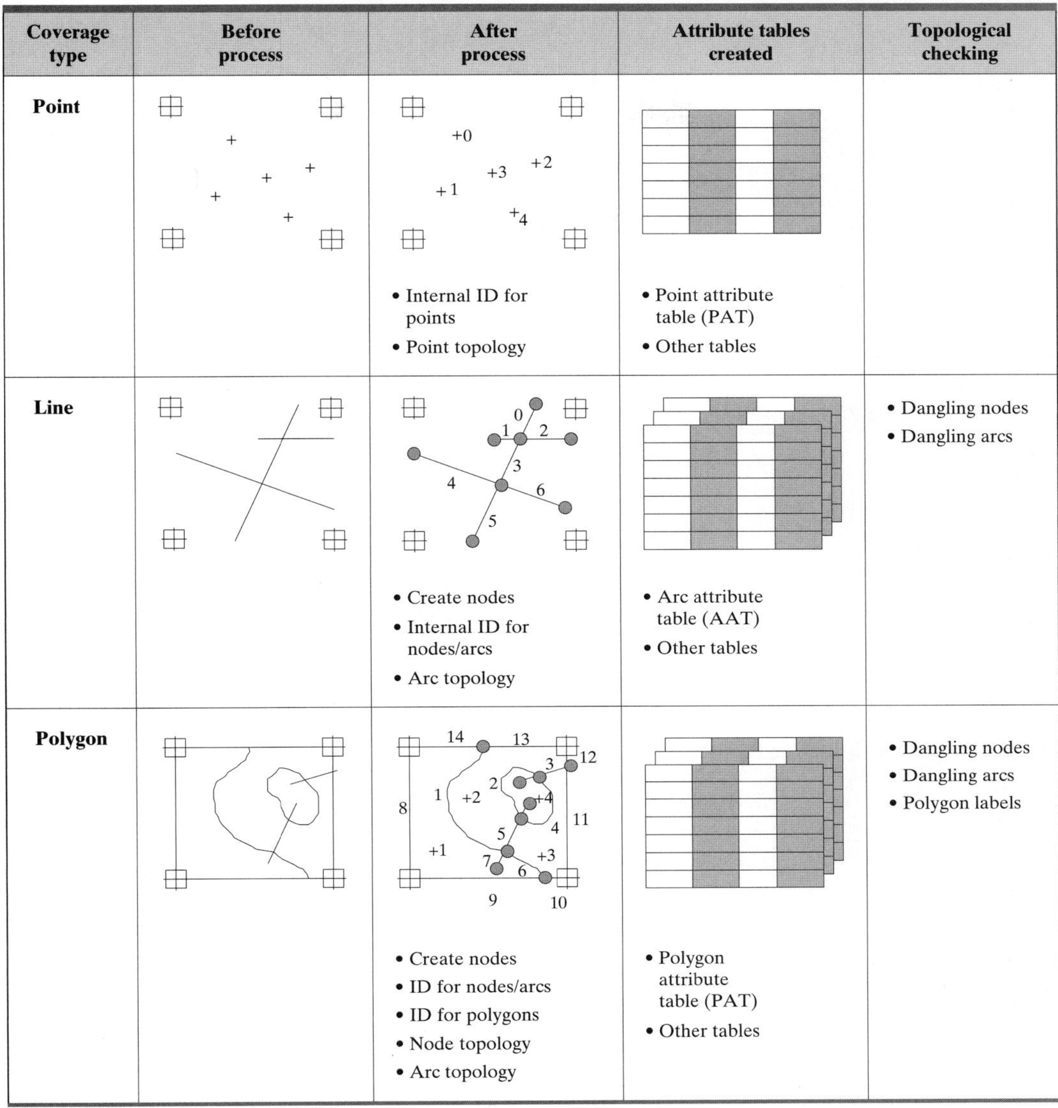

Coverage type	Before process	After process	Attribute tables created	Topological checking
Point		• Internal ID for points • Point topology	• Point attribute table (PAT) • Other tables	
Line		• Create nodes • Internal ID for nodes/arcs • Arc topology	• Arc attribute table (AAT) • Other tables	• Dangling nodes • Dangling arcs
Polygon		• Create nodes • ID for nodes/arcs • ID for polygons • Node topology • Arc topology	• Polygon attribute table (PAT) • Other tables	• Dangling nodes • Dangling arcs • Polygon labels

FIGURE 6.13
Tasks performed during topology building. The types of tasks performed in topology building depend on the types of features on a particular layer. Topology building on a polygon layer, for example, is far more complex than topology building on a point layer.

Data editing is an error-prone process, particularly when complex map layers are involved (e.g., forest resource inventory maps). It is good practice to back up the edit layer from time to time during a data-editing session. This allows data that have been deleted by mistake to be recovered by reloading the backup data, thus avoiding the more time-consuming process of redigitizing the data from the map. Data editing is also a time-consuming process because it must be performed layer by layer. It is therefore important to make use of every available method and software tool in the system to automate the process. These include the use of the various topology building and editing tolerances, editing macros as well as customized Graphic User Interface (GUI) described earlier (ESRI, 1995b).

Joining Adjacent Layers The procedure of data conversion described here applies to a single map with multiple layers. For applications that require more than one map, a further step of work is necessary to ensure that all the layers can be joined together to form a continuous geographic database. This usually involves three sequential processes that include:

- *Edge Matching*. The purpose of this step is to ensure that linear features across layer boundaries match with one another (Figure 6.16a, b and c).

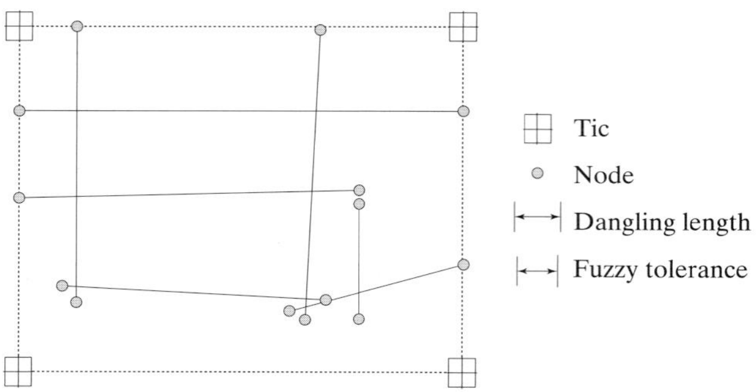

(a) Digitized graphics before applying CLEAN command

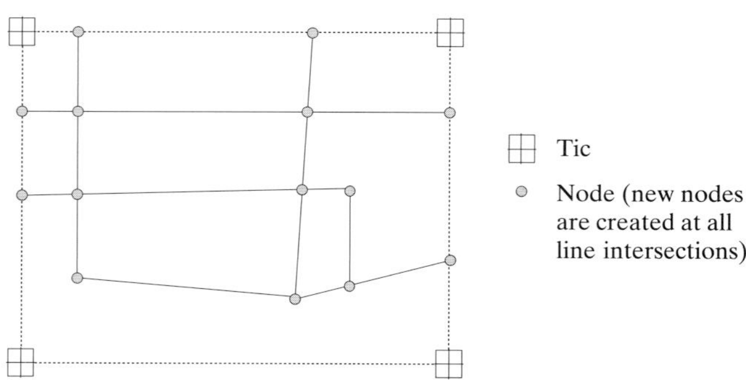

(b) Vector data after using the CLEAN command

FIGURE 6.14
Using fuzzy and dangling tolerances with the CLEAN command to automatically correct digitizing errors in ArcInfo. Dangling arcs smaller than the dangling tolerance are deleted and all nodes within the fuzzy tolerance are forced to become a single node.

Type of errors	Before correction	After correction
Missing label		+ Add polygon label
Missing arc		Add arc
Overshoot		Select and delete overshoot
Undershoot		Extend line to polygon boundary
Dangling node		Move node to close polygon
Wrong label ID	+ 1099	+ 1900 Correct polygon identifier

FIGURE 6.15
Examples of graphical editing to correct digitizing errors. Graphical editing includes a variety of processes such as selecting, adding, deleting, moving graphical elements, and editing of label text.

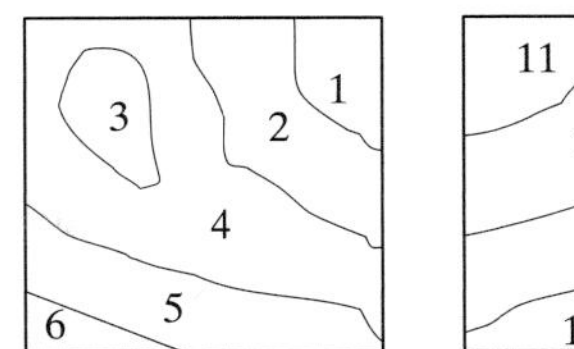

(a) GIS applications often require spatial data from more than one map.

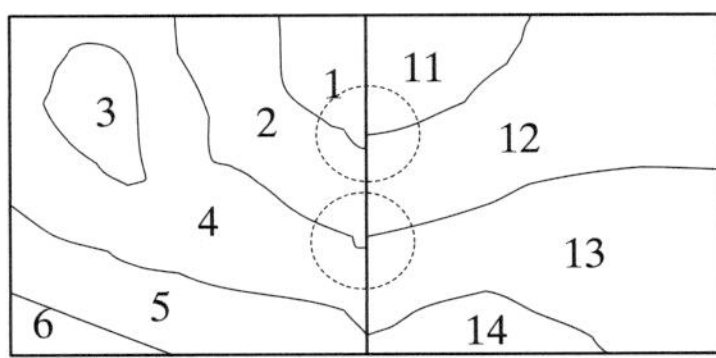

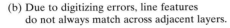

(b) Due to digitizing errors, line features do not always match across adjacent layers.

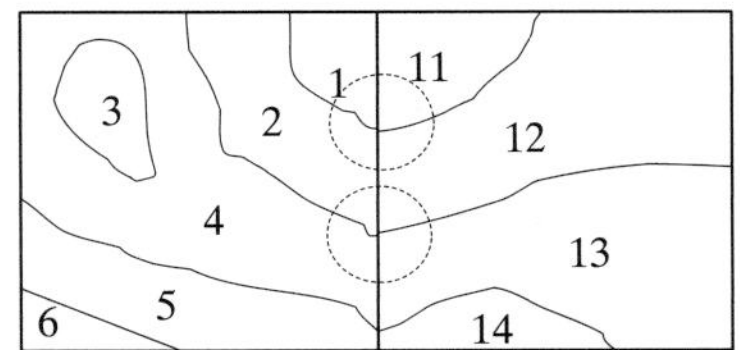

(c) Edge matching is the process by which line features on adjacent layers are forced to match with one another across layer boundary.

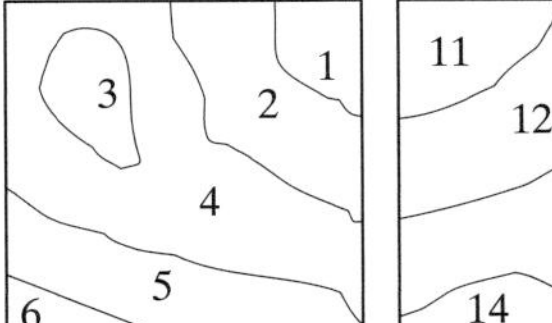

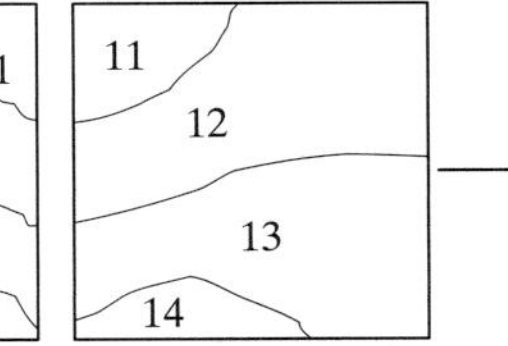

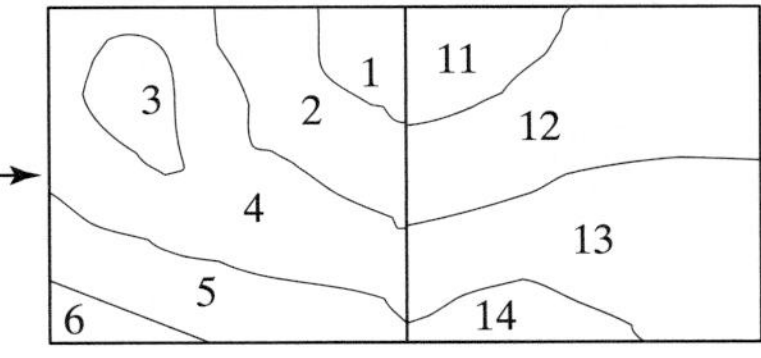

(d) After edge matching, the two original layers remain separate files. The process of map-join is to physically merge the two layers to create a single layer.

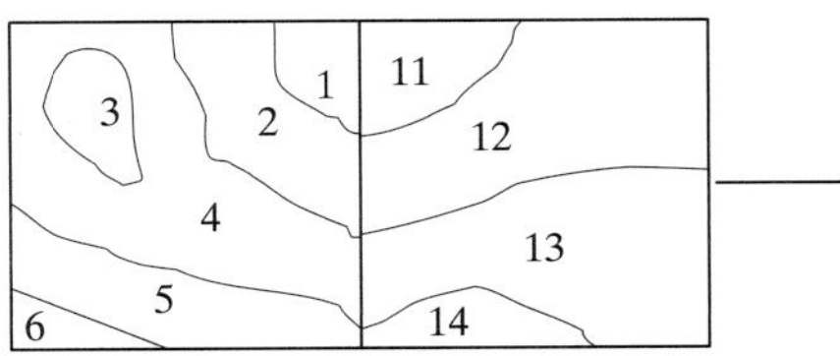

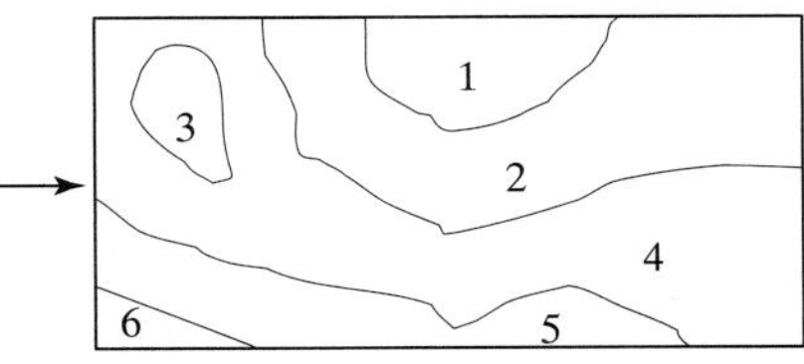

(e) The final step is to dissolve the neat line dividing the two original adjacent layers. The topology of the resulting layer is rebuilt to form a continuous layer.

FIGURE 6.16
Creating a continuous layer from individually digitized layers. This diagram illustrates how two adjacent layers are merged to form a single continuous layer. The same techniques can be applied to merge multiple adjacent layers in vector-based geographic data processing.

- *Map Joining.* The purpose of this step is to physically join adjacent layers to form one single layer (Figure 6.16d).
- *Dissolving Redundant Polygon Boundaries.* The aim of this process is to merge polygons on adjacent layers by removing their boundaries. The topology of the resulting layer is rebuilt to form a single continuous layer (Figure 6.16e).

6.3.9 ATTRIBUTE DATA CONVERSION

In geographic databases using the georelational or object-relational models, attribute data are stored in database tables. The objectives of attribute data conversion are to create the database tables containing descriptive data associated with the graphics and link them logically by means of the common feature identifiers in the feature attribute tables (Figure 6.17).

In GIS, descriptive attribute data are managed either internally as part of the database management component of the GIS itself or externally by a commercial database management system (DBMS). In ArcInfo, for example, internal database tables are INFO files. Unlike feature attribute tables, these files are not generated by the system automatically but must be created by the user in the INFO environment. Attribute data stored in external databases are processed and managed by a

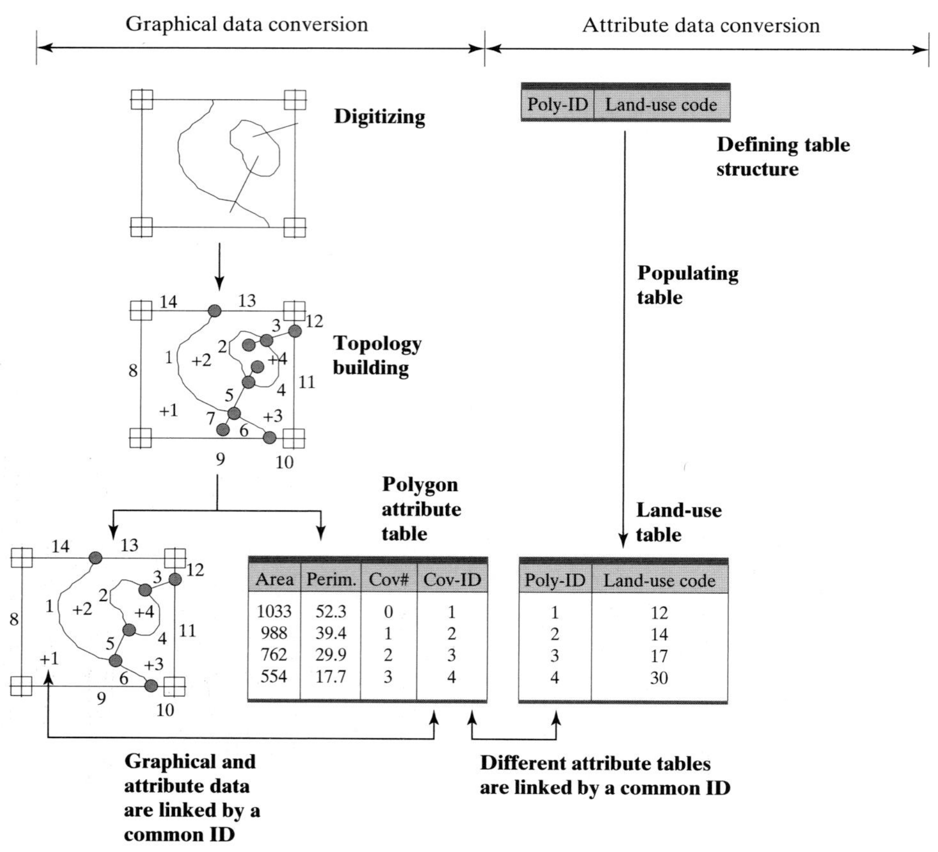

FIGURE 6.17
The relationship between graphical and attribute digital data. The graphical digital data conversion process generates an attribute file that contains the geometric properties of the graphical elements on the vector layer. The attribute digital data conversion process generates an attribute file that contains the descriptive properties of the graphical elements. These two types of attribute files are linked by the common identifiers of the graphical elements in the files.

DBMS supported by the GIS, which in the case of ArcInfo include Oracle, Sybase, DB2, and Informix. There are many reasons why an external commercial DBMS is used. These include, for example,

- to use existing data sets that are brought into the GIS as the result of enterprise data integration
- to share spatial data more effectively with other (non-GIS) applications
- to make use of the robust data management functions that are found in commercial DBMS packages, such as database security, data recovery, and data integrity checks using constraints
- to take advantage of the capabilities that are rarely found in the database management function of the GIS itself, for example, database query using SQL

Attribute data conversion is a three-step process: (1) defining the structure of the data file, (2) populating the data file (i.e., adding attribute data to the files), and (3) linking the file to its associated layer and other data files. Although the actual procedures may vary among different systems, this sequence of work applies to both internal and external database files.

Defining the Structure of a Data File The structure of a data file is defined by the characteristics of the items or columns in a relational database table. A table contains many items pertaining to the attribute properties that the table represents. Each item is defined by an *item name*, *data type* (i.e., character, numeric, logical, or date), an *item width* (the maximum number of characters or digits of an item), and for numerical data type, *the number of decimal places* (Figure 6.18). Both the items and their characteristics are determined in physical database design during the data modeling process (see Chapter 11).

Populating the Data File This is the process whereby attribute data are entered into a data file. There are many ways by which data files can be populated, such as by interactive data input using an input form on the

```
CREATE TABLE parcel
        (parcel_id        CHAR(30),
        reg_date          DATE,
        address_1         VARCHAR2(80),
        address_2         VARCHAR2(80),
        owner_name        VARCHAR2(50),
        zoning            CHAR(2),
        asst_value        NUMBER(10, 2),
        tax_rate          NUMBER(5, 4))
```

FIGURE 6.18
Specifying the structure of a data file using the Structured Query Language (SQL) in database creation. In this example, "CREATE TABLE" is the SQL command for creating a new data file. "parcel" is the name of the file to be created. The names of the attributes in the "parcel" file include: parcel_id, reg_date, address_1, address_2, owner_name, zoning, asst_value and tax_rate. The data types of these attributes are specified at the time when the data file is created: CHAR = fixed-length variables, the maximum size of which is given in brackets; VARCHAR2 = variable-length variables, the maximum size of which is given in brackets; NUMBER = numerical values, the first number in brackets represents the total number of decimal digits, and the second number is the number of digits to the right of the decimal point; DATE = date and time values of a format specific to a particular database system.

screen, and loading data from an existing file. Just like graphical data conversion, attribute data conversion is a time-consuming and error-prone process. Errors in attribute data conversion are hard to detect. As noted earlier, errors in graphical data conversion can always be detected automatically during topology building. Errors in attribute data conversion can be detected only by manual checking. Some systems allow the use of "permissible values" to filter out illegitimate input values. For example, if the permissible value of "age" is set at 1 to 150, zero and any number greater than 150 will not be accepted. Such a mechanism, however, cannot solve all error detection problems because it is not able to verify the correctness of data falling within the range of permissible values. For example, it is not possible to detect the error caused by entering "24" instead of "42." Also, the concept of permissible values does not apply to all items, such as names of people and places.

Linking to the Associated Layers Linkage between the data files created in attribute data conversion to their associated layers is done using the user-defined feature identifiers as the common key. This linkage can be physical (i.e., the attribute and graphical data files are joined together to form a single file) but more commonly it remains logical after the attribute data file has been populated. This means that the two types of files are physically separate files but are accessed as one single logical data file during data processing, using the common key.

6.4 Nontopological GIS Analysis Functions

Vector-based geographic data processing can be topological or nontopological in nature. Topological data processing, as the name implies, embraces those methods that rely on the use of topological relationships, rather than coordinates, to answer spatial questions. Nontopological data processing, on the other hand, refers to those methods that answer spatial questions without the need to use such relationships. These include a wide variety of processes that perform basic GIS functions such as spatial data display, spatial database query, attribute database query, and statistical computation. Nontopological data-processing functions can be used as stand-alone GIS analysis tools. In practice, however, they are more commonly used in conjunction with topological methods as building blocks to construct relatively sophisticated GIS applications.

6.4.1 ATTRIBUTE DATABASE QUERY

Attribute database query is an essential function of vector-based geographic data processing. The storage of descriptive data in the form of tables in the georelational and object-relational data models allows these data to be retrieved, updated, and analyzed by means of database query commands of the GIS software packages. For example, by using a combination of the selection commands and operators, as implemented in the INFO component of ArcInfo, GIS users can freely select items in INFO tables using the attribute values of one or more items as search indices. The selected items are then used to retrieve the graphical elements associated with them, either for screen display or for saving to a new data file for further analysis (Figure 6.19).

For geographic data stored in external DBMS, the Structured Query Language (SQL) is the standard query language. SQL is implemented using three key operations (Figure 6.20).

- *SELECT*. This operation extracts data items in specified rows of a table.
- *PROJECT*. This operation extracts data items in specified columns of a table.
- *JOIN* (also known as *RELATIONAL JOIN*). This is the operation that merges two tables on the basis of the values in the common columns of the tables.

SQL can be used interactively or as part of an application program (i.e., by embedding SQL statements within the codes of a computer program), making it extremely flexible to use. The syntax of a typical SQL query is

SELECT<attribute_name>FROM<table>
WHERE<condition_statement>

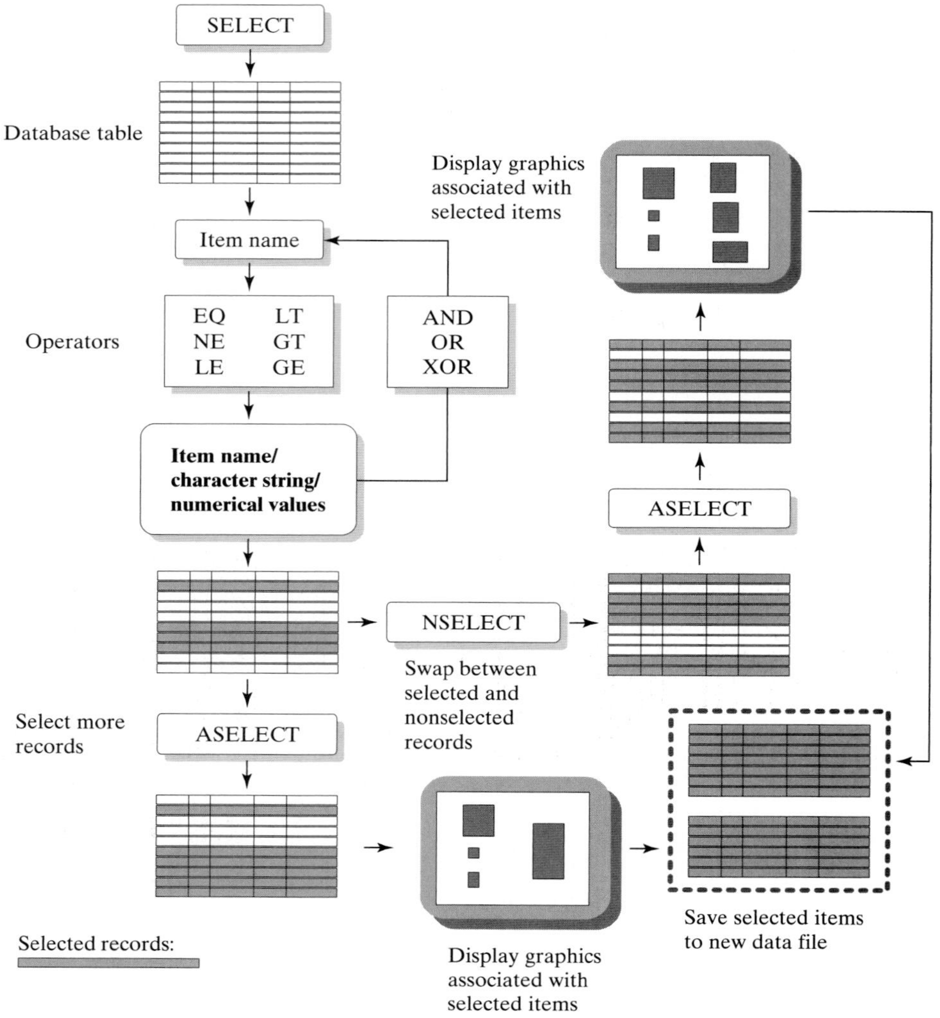

FIGURE 6.19
Attribute database query using SELECT, NONSELECT, and ADD-SELECT commands to manipulate the retrieval of attribute and graphical data in vector-based geographic data processing.

SQL can use the *relational* (GT, LT, EQ, GE, and LE), *arithmetic* (=, +, −, *, /), and *Boolean* (and, or, not, exclusive or) operators. Thus, to find the residential use represented by a code of 11 (according to the Anderson scheme) from a land-use map, the following SQL query can be used.

SELECT FROM lumap WHERE lucode = 11

To find land-use polygons that are greater than 100 hectares, for example, the above SQL query becomes

SELECT lucode FROM lumap WHERE area GT 100

Since SQL is very similar to the English language in syntax, it provides a very intuitive approach to attribute data query. The result of an SQL query is a table, which can be displayed on the computer screen, written to a data file, or printed as a hard-copy report. It should be noted that until very recently, the method of attribute data query by SQL has been used independently of spatial data query. Many attempts have been made to extend the capability of SQL for spatial data query. A typical spatially extended SQL query looks like this.

SELECT lucode FROM lumapWHERE area GT 100 and
WHERE location INSIDE orange county

The expected result of this search is a map showing all land-use polygons greater than 100 hectares that fall within Orange County in a land-use map covering many counties. Despite the apparent attraction of the idea and the practical need for spatially enabled SQL, there has been relatively limited success in the endeavors to spatially extend SQL. This is because of two reasons (Egenhofer, 1991 and 1992). One of these is the difficulty in incorporating the necessary spatial

Parcel table

PID	Zoning	Area
12345	C1	20000
23456	R2	15000
12332	R2	8000
88889	I2	35000

Owner table

Owner	PID	Tax
Smythe	12345	10,000
Burton	23456	3,500
Doe	12332	5,000
Jones	88889	8,300

(a) The original database tables

SELECT parcel where zoning = R2 (residential class 2)

Zoning	PID	Area
R2	23456	15000
R2	12332	8000

PROJECT parcel over zoning using PID

Zoning	Area
C1	20000
R2	23000
I2	35000

JOIN parcel and owner over PID (parcel identifier)

PID	Owner	Zoning	Area	Tax
12345	Smythe	C1	20000	10,000
23456	Burton	R2	15000	3,500
12332	Doe	R2	8000	5,000
88889	Jones	I2	35000	8,300

(b) The SELECT, PROJECT, and JOIN operations in database query

FIGURE 6.20
Using Relational DBMS Operations for Attribute Data Query. This example illustrates how descriptive data in two tables (Parcel and Owner) can be extracted by the conventional relational DBMS operations of SELECT, PROJECT, and JOIN. Note that in the PROJECT operation, the area values of R2 parcels have been summed to give the total area of that particular category of zoning.

concepts into SQL as graphical display and SQL's specification. The other is the inability within the relational framework to support qualitative answers, knowledge queries, and metadata queries, all of which are critical for spatial data processing.

With the recent emergence of new SQL standards (e.g., SQL3 and SQL Multimedia), the ability for GIS users to perform spatially extended attribute query will be greatly enhanced. For example, ESRI (1998) has introduced a spatially extended SQL API (application programming interface) to the SDE (Spatial Database Engine) software as part of its support of object- relational DBMS products. This API will allow users to perform SQL queries with spatial constraints (e.g., inside an administrative unit, within a certain distance from a point source of pollution, intersecting with a particular highway).

6.4.2 STATISTICAL COMPUTATION AND THEMATIC MAPPING

In essence, statistical computation is an extension of attribute database query. Standard statistical measures such as mean, maximum, minimum, and standard deviation can be computed for specified items in attribute tables. In ArcInfo, for example, there is a useful command called STATISTICS in the tables module, which generates the following statistics for a selected item: count (frequency of occurrence), minimum, maximum, sum, and mean. Another useful command in ArcInfo is FREQUENCY, which produces a list of the frequency of occurrences of a specified set of items by categories in a database data file together with useful information (to be specified by the user) on the total area of these occurrences by categories.

ArcView GIS has pushed the capability of statistical computation further by having a built-in function that is capable of aggregating attribute data and then displaying them on the fly for visualization and interpretation (ESRI, 1996b). With the aid of this function, the user is able to generate thematic maps that summarize the attributes of features (e.g., number of people in a particular income group) based on the polygons in another theme (e.g., census tracts) they fall within. This method not only works for aggregating point features as in the previous example, but also line features (e.g., roads within a traffic management district) and polygon features (e.g., land parcels within a county).

6.4.3 ADDRESS GEOCODING

Addresses are probably one of the most commonly used forms of geographic data. Many academic and research institutes, government agencies, and commercial organizations collect and maintain large address-based databases. These include census statistics, property tax records, electoral registers, social surveys, customer databases, and goods delivery locations. These types of data are known as "quasi geographic data," because using the addresses alone, it is not possible to pinpoint locations on a map and to integrate address-based data with other forms of geographic data for spatial analysis.

To locate addresses on a map, it is necessary to determine their grid references in an accepted georeferencing system. *Address geocoding*, also known as *address matching*, is the process by which grid references (e.g., latitude/longitude, UTM coordinates, and state plane coordinates) are added to point locations described by street addresses. In this sense, address geocoding is a nontopological overlay operation because it adds one layer of data onto another to create a new layer of spatial data. Address geocoding is a built-in function of many vector-based GIS, particularly those oriented toward applications in geodemographics, such as MapInfo, and those primarily designed for the presentation of spatial information, such as ArcView GIS (ESRI, 1996a).

The Concept of Address Geocoding

The objective of address geocoding is to match addresses in data files that have grid references (usually called the *reference theme*) to those addresses in data files that do not (usually called the *event table*) (Figure 6.21). The primary sources of reference theme data in the United States are the *TIGER* (Topologically Integrated Geographic Encoding and Referencing) files from the Bureau of the Census. In Canada, the Area Master Files (AMF) from Statistics Canada represent the most commonly used reference theme data in geocoding. In the United Kingdom, reference theme data can be obtained from either the *Address-Point* of the Ordnance Survey or the *Postcode Address File* (PAF) of the Royal Mail. Although the structures of different reference theme data files may be different from one another, they all contain the street network and the address ranges on both sides of the street necessary for geocoding. Addresses in the event table, on the other hand, are normally recorded as text strings that contain components such as house number, street name, city name, and postal code. The event table can be stored in the free format of a text file or the specific format of a database file such as dBASE dbf file.

The software module that performs geocoding is usually referred to as the *geocoding editor*. In order to compare the reference theme data with event table data in address geocoding, it is necessary for the two types of data to be in a compatible format. This always means the creation of a *geocoding index* from the graphical reference theme by the geocoding editor. The geocoding index is a database table. An associated *geocoding information file* is also created to record the characteristics of the geocoding index, such as address style and the address attribute fields. During the address geocoding process, the application software first reads the records in the event table one by one. It parses the address into separate components (i.e., street number, street name, city name, and postal code) and standardizes the components according to the specifications of a selected address style supported by the geocoding editor. It then determines the street segment where the particular address is found by matching the parsed components with data in the geocoding index. The actual grid reference of the address within the street segment is finally computed by interpolating between the two end points of the street segment according to the numerical

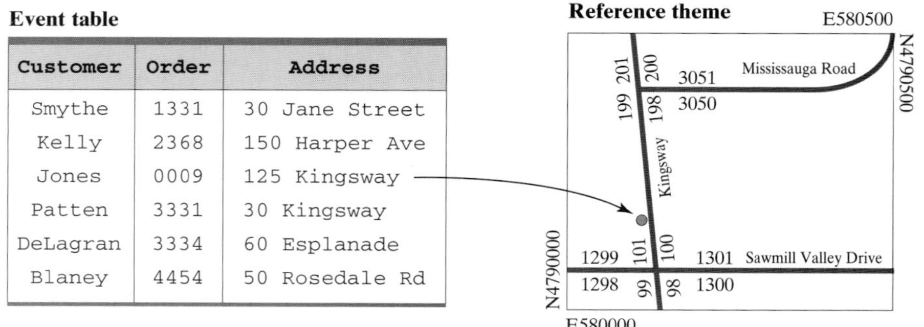

Event table

Customer	Order	Address
Smythe	1331	30 Jane Street
Kelly	2368	150 Harper Ave
Jones	0009	125 Kingsway
Patten	3331	30 Kingsway
DeLagran	3334	60 Esplanade
Blaney	4454	50 Rosedale Rd

FIGURE 6.21
Address geocoding concept. The process of address geocoding first matches the street name in both the event table and the reference theme, and then computes the coordinates of the addresses by determining whether a particular street number is on the odd- or even-number side of the street; estimating the distance of the address from a street intersection by proportion with reference to the range of the street numbers in the street segment concerned; and calculating the coordinate of the address using the coordinates of the street intersections.

relationship between the street number and the address range of the street segment.

The Process of Address Geocoding

Address geocoding is a multistep process that can be carried out in batch or interactive modes, or a combination of both. The procedure of address geocoding can be generalized into the following sequence of tasks (Figure 6.22).

1. *Data Preparation.* The objective of this step of work is to prepare the reference theme and the event table to be geocoded. For the reference theme, it is necessary to ensure that the address style used is one of the address styles supported by the geocoding editor, for example, U.S. street address with zone, U.S. street address without zone, ZIP+4 postal codes, and five-digit ZIP postal codes (ESRI, 1996a). For the event table, it is also necessary to ensure that all components of the street address are stored according to the format specified by the geocoding editor. If the geocoding editor supports the use of name aliases (e.g., City Hall, St. Michael's Hospital, Geoscience Building, Department of Geography) in addition to street addresses, then prepare a separate alias table for common place names.

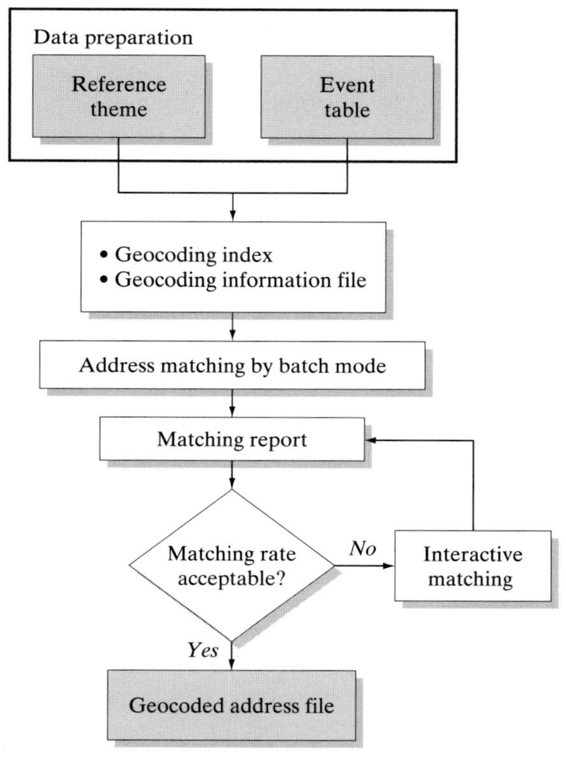

FIGURE 6.22
The procedure of address geocoding.

2. *Loading the Reference Theme and Event Table.* This step of work loads the reference theme and the event table into the geocoding editor, which then builds the geocoding index and the geocoding information file using information from these two input files.
3. *Geocoding.* The geocoding process can be performed in either batch or interactive mode. It is common practice to start the process in the batch mode, which lets the geocoding editor run until it reaches the last record of the event table. At the end of the batch run, a dialog box showing the number of matches that have been made is displayed. There are various reasons why a perfect match has not been achieved, such as spelling mistakes, incorrect house numbers, and outdated data in the reference theme or event table. At this point, it may be desirable to perform an interactive rematch by which each of the unmatched addresses will be examined and geocoded with the aid of the geocoding editor.
4. *Saving and Populating the Geocoding Data.* On completion of the interactive geocoding process, save the results to data files as required by the geocoding editor so that they can be used for further analysis.

Applications of Geocoding

Geocoding is normally an intermediate step in geographic data processing. Once the addresses in the event table have been geocoded, they may be integrated with other types of geographic data to create new map products and perform spatial analysis. A typical map product is an *electronic pin map*, which displays the locations of various events by address. An electronic pin map serves the same purpose as the conventional paper pin map, but it is easier to use and maintain. Geocoded addresses may also be used in advanced spatial query and analysis, such as location-allocation modeling to determine the optimal locations of new service centers or facilities, and point-in-polygon overlay analysis to find customers within a service or trading area (see Section 6.5.1).

6.4.4 CALCULATION OF AREAS, PERIMETERS AND DISTANCES

An important generic vector-based geographic data-processing function is to extract locations (x and y coordinates), distances, areas, and perimeter length on a layer. In the ArcPlot module of ArcInfo, for example, the command MEASURE is used in combination with three options: (a) WHERE (to show the coordinates of any point on the layer selected by the mouse cursor), (b) LENGTH (to join two or more points on the layer by a line and to determine its length), and (c) AREA (to determine the area of a polygon on the layer defined by three or more points chosen by the mouse cursor). The LENGTH option is a straight-line distance between two points computed using the Pythagorean theorem. It is also used to compute the perimeter of a polygon. The AREA computation is normally based on the coordinates of the points defining the figure of the area. The double area (2A) of any polygon of n points is given by the sum of abscissas (X) of a point multiplied by the difference of the ordinates (Y) of the following and the preceding point, counting clockwise (Methley, 1986). In general, this theorem can be written as

$$2A = \Sigma[X_n(Y_{n+1} - Y_{n-1})] \tag{6.1}$$

or

$$2A = \Sigma[Y_n(X_{n+1} - Y_{n-1})] \tag{6.2}$$

where Σ is summation over all values of n, X and Y are the coordinates of the points of the polygon taken in a clockwise order. In the equations given above, $n + 1$ means the polygon point ahead of and $n - 1$ means the polygon point preceding the polygon point. In the vector GIS database, the AREA and PERIMETER values computed for each polygon of a topologically structured layer are permanently stored as two attribute items.

Because distance and proximity are such important concepts in spatial analysis in GIS, most vector GIS provide different ways to extract these two attributes from the layer. In ArcInfo, there is a DISTANCE command in ArcEdit, which allows the coordinates of any two points on the layer to be read and the distance between these two points to be computed. Proximity is obtained from a command called NEAR, which computes the distance from each point in a layer to the nearest arc, point, or node in another layer. It is possible therefore to compute a distance matrix of points if the two input point layers are the same. In order to avoid computing an excessively large distance matrix, the program allows the user to specify the radius of search from each point in a layer to the features in another layer. Distance and proximity will be further discussed in Chapter 10 on spatial analysis.

6.5 FEATURE-BASED TOPOLOGICAL FUNCTIONS

Topological geographic data processing includes a variety of processes that rely on the use of topological relationships for spatial problem solving. As noted in Chapter 3 Section 3.5.2, there are three types of topological relationships: adjacency, containment, and connectivity. All methods of topological geographic data processing are constructed using one or more of these relationships. Just like nontopological methods, topological methods can be used individually as stand-alone spatial problem-solving tools. They can also be used in conjunction with other topological and nontopological methods in more complex GIS applications. Generally speaking, methods of topological data processing are more sophisticated than their nontopological counterparts, both conceptually and technically. These methods are among the most computationally demanding processes in vector-based geographic data processing.

Topological analysis functions can be applied to individual features or all features on a vector layer. In the following two sections, we will explain two important topological analysis functions that work on individual features. Topological analysis functions that work on entire vector layers will be discussed in Section 6.6.

6.5.1 FEATURED-BASED OVERLAY ANALYSIS

The aim of feature-based overlay analysis is to find the relationships between individual features of one layer and those of another layer. One of the most commonly used feature-based overlay functions is point-in-polygon matching. The objective of this particular function is to determine whether a given point feature falls inside a polygon. Different algorithms have been developed to address this spatial problem in computer-assisted cartography (Monmonier, 1982). The most straightforward test to determine the position of a point relative to a polygon is the *plumb-line algorithm*. Conceptually, this algorithm is relatively simple and involves dropping a line directly downward from the point and counting the number of times, if any, this line intersects with the polygon boundaries (Figure 6.23). If the number of intersections is odd, then the point is within the polygon; otherwise it is outside.

There are many GIS applications that require the use of the point-in-polygon test. Notable examples include the determination of customer locations within different sales territories in market analysis, location-allocation analysis of hospitals, fire stations, and other social or community facilities in planning, as well as the matching between surveyed sites of natural and scientific importance (e.g., nests of protected bird species) and forest types in resource management.

6.5.2 BUFFERING

By definition, a *buffer* is a zone with a specified width surrounding a spatial feature. The process of creating a buffer, or *buffering*, is an essential function in vector-based geographic data processing. Since the basic graphical elements of vector GIS are points, lines, and polygons,

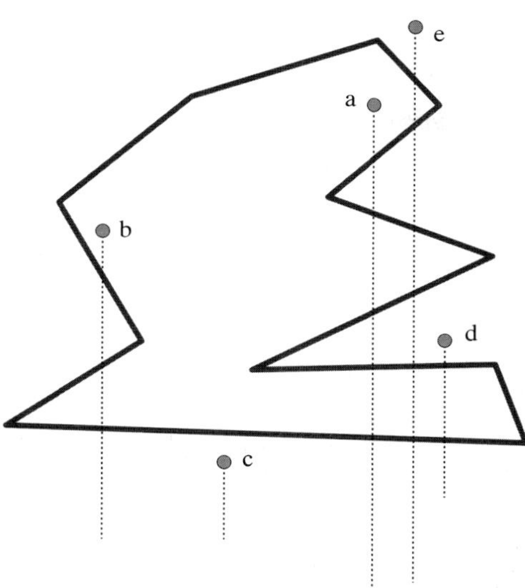

FIGURE 6.23
Point-in-polygon search by the plumb-line algorithm. In this method, a point is within a polygon if its plumb line has an odd number of intersections with the boundaries of the polygon, such as point a and b, which cross the polygon boundaries 5 and 3 times, respectively; otherwise the point is outside the polygon, such as c, d, and e, which cross the polygon boundaries 0, 2, and 6 times, respectively.

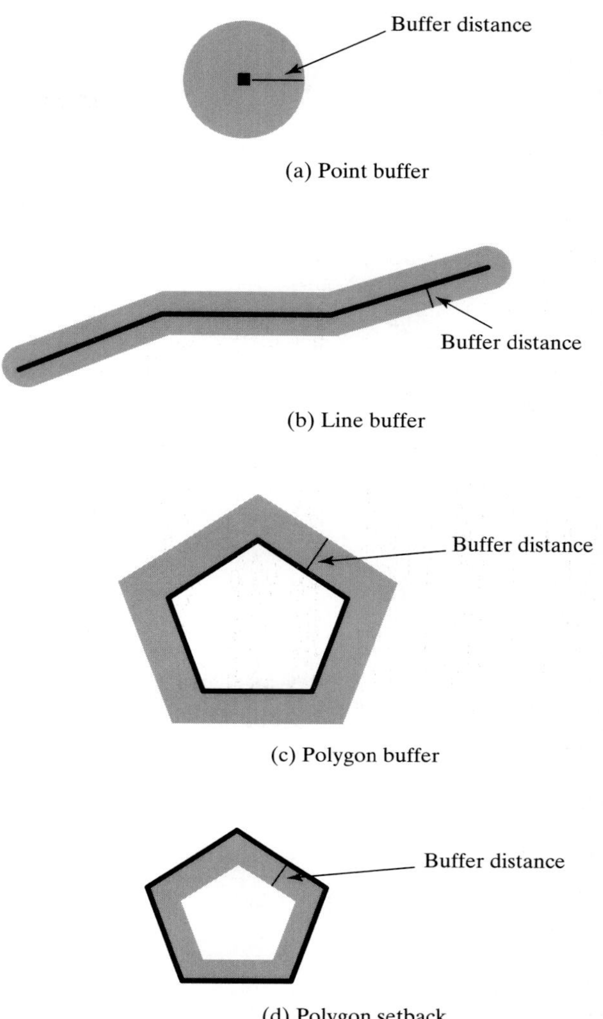

FIGURE 6.24
Point, line, and polygon buffers.

buffers can be generated around each of these elements. In each case, the buffer generated takes the shape from the feature. For a point buffer, it is a circle with a specified radius (as the buffer distance) drawn surrounding the point (Figure 6.24a). For a line buffer, it is a band with a specified distance created on both sides of the line conforming to the line's curve (Figure 6.24b). For a polygon buffer, it is a belt of a specified buffer distance from the edge of the polygon surrounding the polygon and conforming to its shape (Figure 6.24c). When the buffer zone is constructed inward from the polygon boundary, it is sometimes called a *setback* (Figure 6.24d).

Buffer zones are polygons, no matter whether they are generated from a point, line, or polygon feature. In geographic data processing, the buffer distance is a user-defined parameter that is used by the system to generate the buffer zones. For highly convoluted line features, such as a meandering river or complex polygon boundaries, buffer zones generated tend to overlap. However, boundaries of overlapping buffer zones are automatically dissolved to produce a single coherent buffer zone for the entire spatial feature (Figure 6.25).

Buffering is a very important function in vector-based geographic data processing. Like its raster-based counterpart, vector-based buffering is used primarily for neighborhood analysis that aims to evaluate the characteristics of an area surrounding a specific location or linear spatial feature. The objective of buffering is to delineate the area of interest for a neighborhood analysis, which is then used in conjunction with other geographic data-processing functions to search spatial features that fall inside or outside this area of interest and analyze their spatial patterns and interrelationships. Common applications of buffering include, for example, identification of properties within a certain distance of a zoning change application in urban development, delineation of areas around natural features where human activities are restricted in natural resource management, and determination of areas affected by location, movement, or spread of hazardous materials in environmental management.

In most GIS applications, it is sufficient to assume a uniform buffer distance. However, it is also possible to produce variable-distance buffer zones. Some applications require different buffer distances for different parts of the spatial features of interest. For example, Xiang (1993) conducted research on the use of variable-distance buffers for rivers for water quality management.

(a) Original spatial feature.

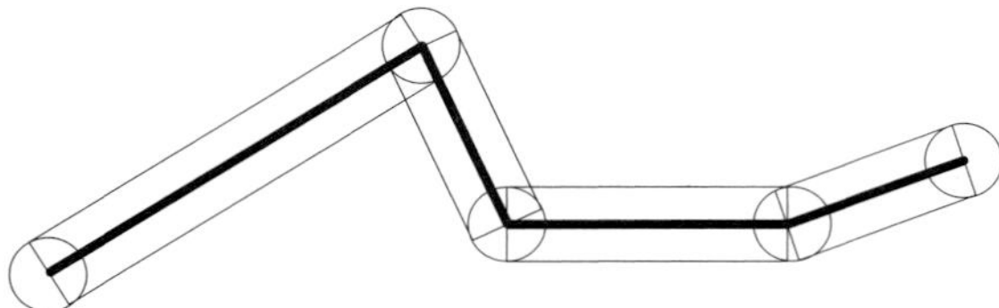

(b) Buffer zones are generated by each line segment and node of the spatial feature.

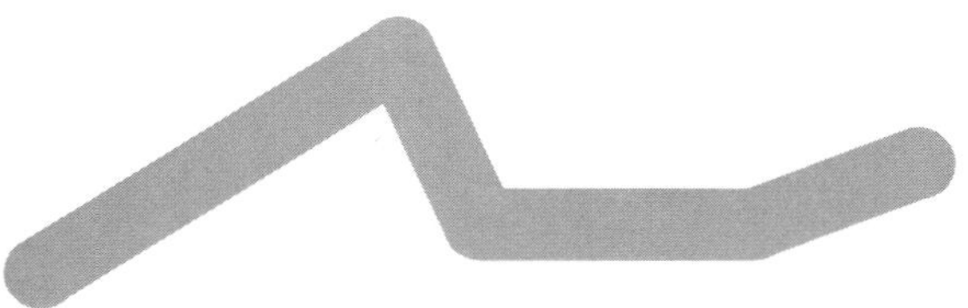

(c) Boundaries of individual buffer zones are dissolved to form a coherent buffer zone for the entire spatial feature.

FIGURE 6.25
The process of buffering a line feature.

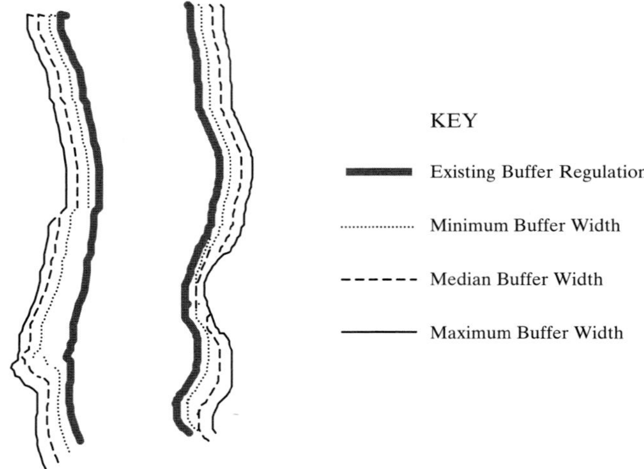

FIGURE 6.26
An example of variable-distance buffering: variable buffer zones along a river. *(Source: Xiang, 1993)*

Buffer distances are determined with reference to physical, ecological, and socioeconomic conditions of the region. He developed a formula to compute riparian buffer distance using a *detention time model* of buffer effectiveness developed by Phillips (1989), which incorporates soil, hydrological features, land cover, and topography. Xiang (1993) extracted soil, hydrologic, and topographic map data within an 805-meter distance from the river (using the CLIP command in ArcInfo). From the topographic data, slope angles were computed using a Triangulated Irregular Network (TIN) model. All these map layers were overlaid to form a new layer. Buffer distances were computed according to the riparian buffer distance formula he developed. These buffer distances were then applied to produce variable-distance buffer zones in different parts of the river (Figure 6.26).

6.6 Layer-Based Topological Functions

Layer-based topological analysis functions, as the name implies, are those that work on the entire layer rather than on individual spatial features. Map layers that are used as input to this particular class of geographic data-processing functions must be topologically structured. The output layers from these functions are topologically structured on completion of the function. Layer-based topological functions, as a rule, are computationally intensive in nature.

6.6.1 RECLASSIFICATION

Reclassification is a database simplification process that aims at reducing the number of categories of attribute data on a layer. This process is made up of two steps of work that involve a nontopological process using attribute data and a topological process using graphical data. The first step makes use of the nontopological methods of attribute data query to select attribute data of a particular value (or range of values) by an arithmetic operator (=) or a relational operator (EQ) and then assign a new value to them according to the new classification scheme (Figure 6.27).

After reclassification, some polygons will have the same attribute values as their neighbors. The common boundaries between polygons with identical attribute values become redundant and have to be removed in order to maintain database integrity. This is done in the second step in a topological process that makes use of the adjacency relationship to determine neighboring polygons with like attribute values and to remove the redundant boundaries in between. As the topology of the original layer has been changed after reclassification, it must be rebuilt to generate new attribute tables and to establish the topological relationships among the graphical elements in the new layer (Figure 6.27).

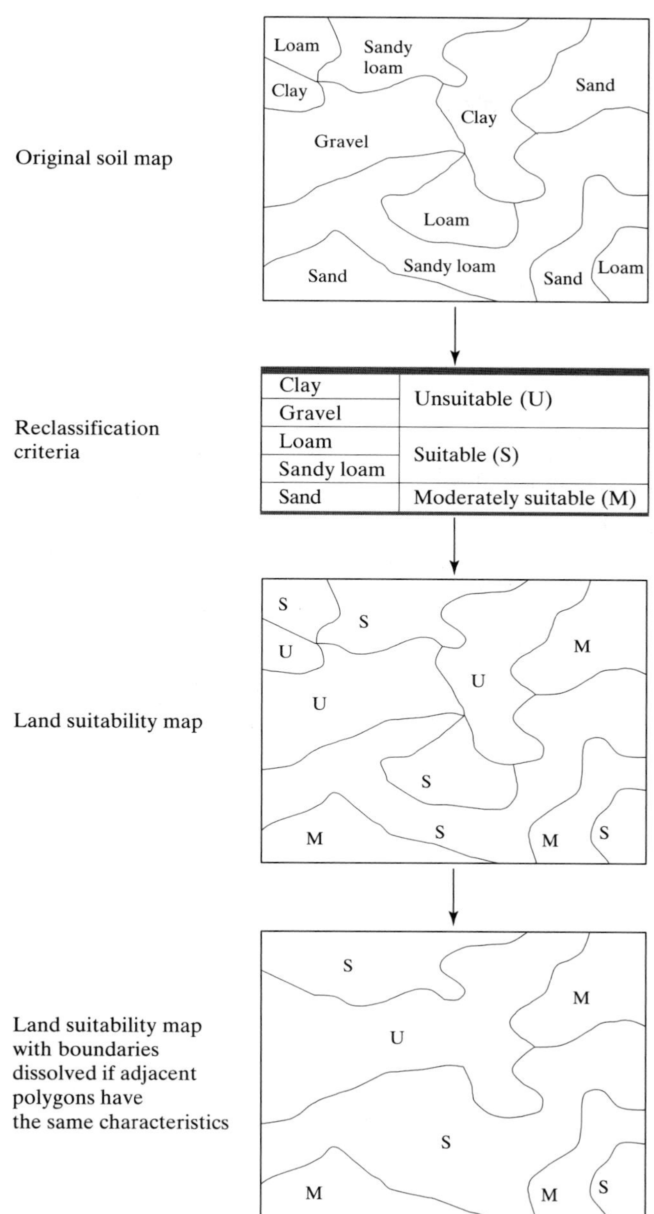

FIGURE 6.27
The process of reclassification.

6.6.2 LAYER-BASED OVERLAY ANALYSIS

Topological overlay is probably the most important function of vector-based geographic data processing. Functionally, it is the vector equivalent of cartographic modeling in raster GIS that we learned in Chapter 5 Section 5.5. Technically, however, topological overlay is much more complex and computationally demanding. The term "topological overlay" implies that the layers must have already been topologically structured, and after the overlay, the topology of the combined layer created from the overlay operation will have to be updated. The resulting attribute table for the new layer consists of the old and new attributes formed by logical or mathematical operation on the old ones (Figure 6.28).

The method of topological overlay can be most easily understood by explaining it from three perspectives, namely: (1) the types of overlay, (2) the topological overlay operators, and (3) the topological overlay process.

Types of Topological Overlay In topological overlay, features of two layers are intersected to create new output features, and attributes of the two layers are combined to describe the new output features. The new output features vary according to the nature of the two input layers. Because in vector GIS, layers can be built up of points, lines, or polygons, three types of topological overlay are possible: (a) overlaying a point layer on a polygon layer, (b) overlaying a line layer on a polygon layer, and (c) overlaying a polygon layer on a polygon layer (Figure 6.29).

Of these three types of topological overlay, the polygon-on-polygon overlay is the most complex. It is computationally intensive because of the need to compute the coordinates of the intersecting points and then to compare them with the list of coordinates of the two intersecting lines (Figure 6.30). This type of overlay is particularly useful for modeling applications. By overlaying soil types with land-use parcels, for example, it is possible to associate soil types with land use, and hence determine which types of land use have occupied fertile soil and where they are located.

Topological Overlay Operators For polygon-on-polygon overlays, different operators are used to combine two or more input layers into a single output layer. Topological overlay operators that are commonly implemented in a typical vector-based GIS such as ArcInfo include *UNION, INTERSECT, IDENTITY, CLIP, ERASECOV,* and *SPLIT* (Figure 6.29) (ESRI, 1994a). The definitions of these operators are as follows:

- *UNION* overlays polygons and keeps all areas in both layers.
- *INTERSECT* overlays polygons but keeps only those portions of the first input layer features falling within the second input layer polygon.
- *IDENTITY* overlays polygons and keeps all input layer features.
- *CLIP* cuts out a piece of the first input layer using the second input layer as a cookie cutter.
- *ERASECOV* erases a part of the first input layer using the second input layer.
- *SPLIT* divides the first input layer into a number of smaller layers based on the second input layer.

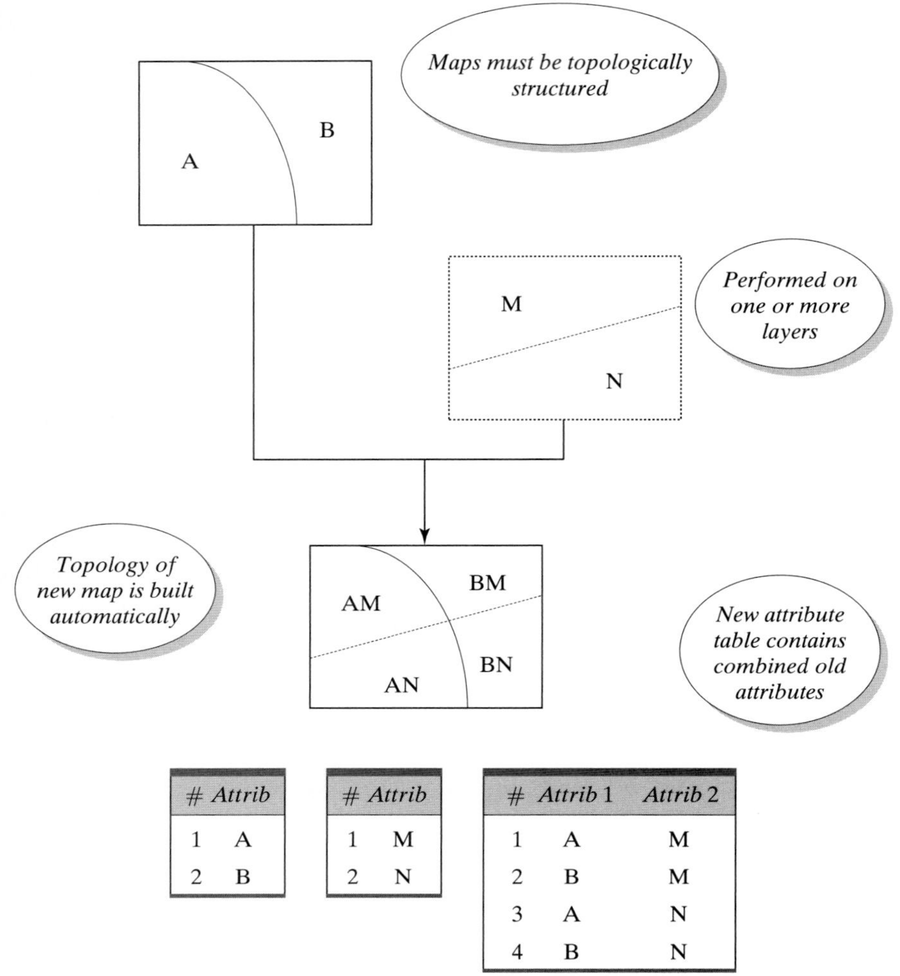

#	Attrib
1	A
2	B

#	Attrib
1	M
2	N

#	Attrib 1	Attrib 2
1	A	M
2	B	M
3	A	N
4	B	N

FIGURE 6.28
The topological overlay process.

Topological Overlay Process Topological overlay is a four-step process that includes (1) defining the objectives and data analysis criteria, (2) applying topological overlay operators, (3) postoverlay processing and interpreting the results, and (4) the removal of sliver polygons.

Defining the Objectives and Data Analysis Criteria For example, we may seek to create a land suitability map for agricultural development. Three layers are available: soil, topography (slope angles), and land cover (Figure 6.31a). The land assessment criteria are stipulated as follows:

- Areas with loam soil, two degrees of slope, and covered by grass are the most suitable for agricultural development (using a suitability code = 2).
- Areas with gravel, three degrees of slope, and covered by bush are unsuitable for agricultural development (using a suitability code = 0).
- Any other areas are suitable (using a suitability code = 1).

Applying Topological Overlay Operators Depending on the objective of topological overlay, the operators can be applied as explained. In this particular example, only the UNION operator is needed. It is applied two times as follows:

- to overlay soil (S) and topography (T) to produce a new layer S-T, creating a new attribute table combining soil and topography (slope angles) (Figure 6.31b)
- to overlay S-T layer with the land cover (L) layer to produce a new layer S-T-L and associated new attribute tables (Figure 6.31c)

Postoverlay Processing and Interpreting the Results This step of work requires the use of database management tools to change the structure of the attribute table by adding an item called "suitability code" to the S-T-L feature table (Figure 6.31d). It is possible to add the code to this item manually with SQL queries and joins, based on the criteria noted earlier. Another approach is to use attribute data manipulation commands to select from

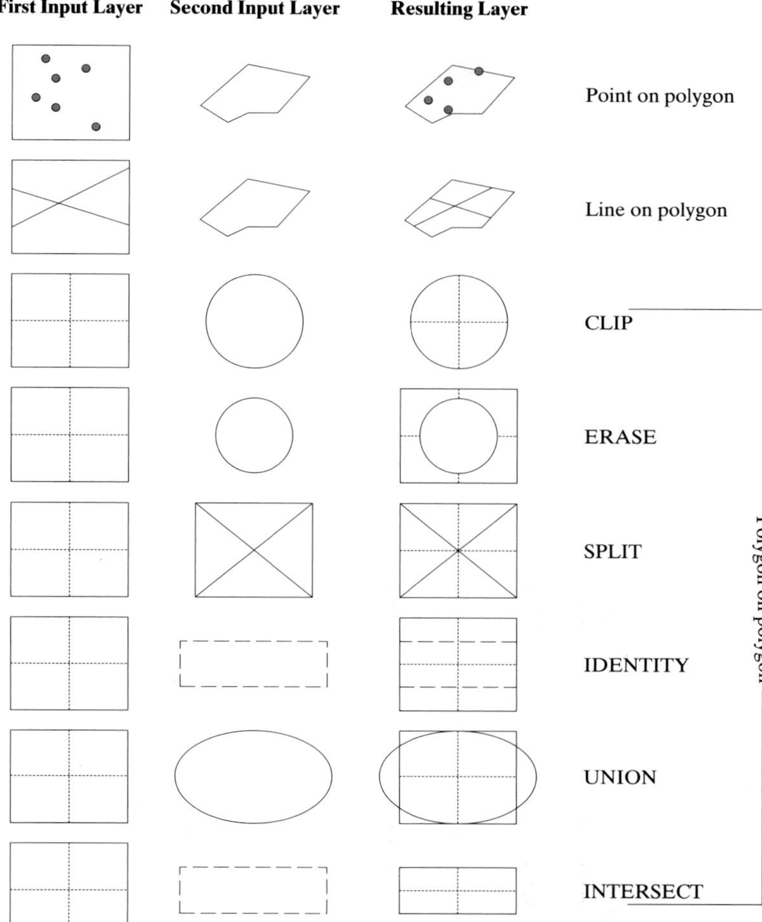

FIGURE 6.29
Types of topological overlay operations.

the attribute file of the S-T-L layer the type of soil, topography, and land cover that meet the stipulated criteria, and then add the correct suitability code. In ArcInfo, for example, this can be achieved using the *RESELECT* and *CALCULATE* commands in the TABLE module. The final attribute table will incorporate characteristics of all three input layers and an item of suitability code. A look-up table can be used to control the display of a land suitability map. It is also possible to extract only those polygons that are most suitable for agricultural development (suitability code = 2). Therefore, the end product of topological overlay is a land suitability map from which the most suitable locations may be selected. Using the function of statistical computation (see Section 6.4.2), a table showing the statistics of different categories of land-use suitability may also be generated.

Removal of Sliver Polygons *Sliver polygons,* also called *spurious polygons,* are very small polygons that are generated in topological overlay as a result of the mismatching of identical features on different layers (Figure 6.32). This can be due to a variety of factors, such as random errors in line digitizing and imperfection in registering the layers. There are different ways to eliminate these sliver polygons. These include:

- *Using Fuzzy Tolerance.* Fuzzy tolerance is the minimum distance between coordinates on the output layer. Setting a suitable fuzzy tolerance will cause the sliver polygon to disappear when coordinate points that are closer than the fuzzy tolerance are merged. To determine the correct fuzzy tolerance, make a hard copy of the overlay and measure the amount of sliver with a scale. This will give an initial estimate of the fuzzy tolerance to use in "cleaning" the overlay. If slivers persist, increase or decrease the fuzzy tolerance until the slivers are eliminated.
- *Using the Built-in Functions of GIS.* Many vector-based GIS software packages have a built-in function to eliminate the sliver polygons resulting from topological overlay. In ArcInfo, for example, the command ELIMINATE, which merges selected polygons by dropping the longest shared border between them, can be used to remove slivers based on one or more criteria. These criteria

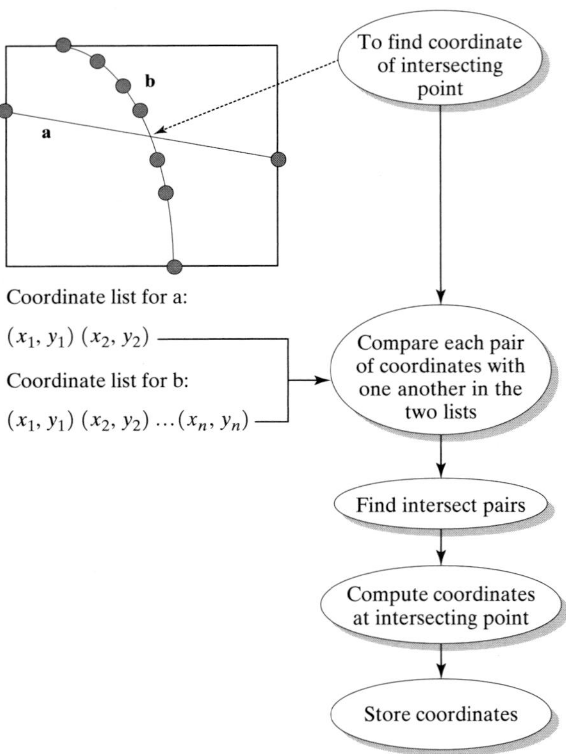

FIGURE 6.30
Computational intensity in polygon overlay.

can be the area (sliver polygons being very small), the shape (sliver polygons being very long and narrow), and the number of arcs (slivers having only two bounding arcs).

- *Using a Template in Digitizing.* In contrast to the two methods just described, which treat sliver polygons as a postoverlay process, this approach provides a preventive measure to ensure that no sliver polygon will occur in topological overlay. It takes advantage of the method of using a digitizing template as explained in Section 6.3.3. Since polygons (e.g., lakes) that are common to all layers in the overlay process are derived from the same digitizing template, they will match perfectly and the occurrence of sliver polygons is therefore avoided (Figure 6.33).

6.6.3 NETWORK ANALYSIS

In the context of GIS, a network is a set of linear features that are interconnected. Common examples of networks include highways, railways, city streets, rivers, transportation routes (e.g., transit, school buses, garbage collection, and mail delivery), and utility distribution systems (e.g., electricity, telephone, water supply, and sewage). Collectively, these networks form the infrastructure of modern society. They provide the means for the movement of people and goods, the delivery of services, the flow of resources

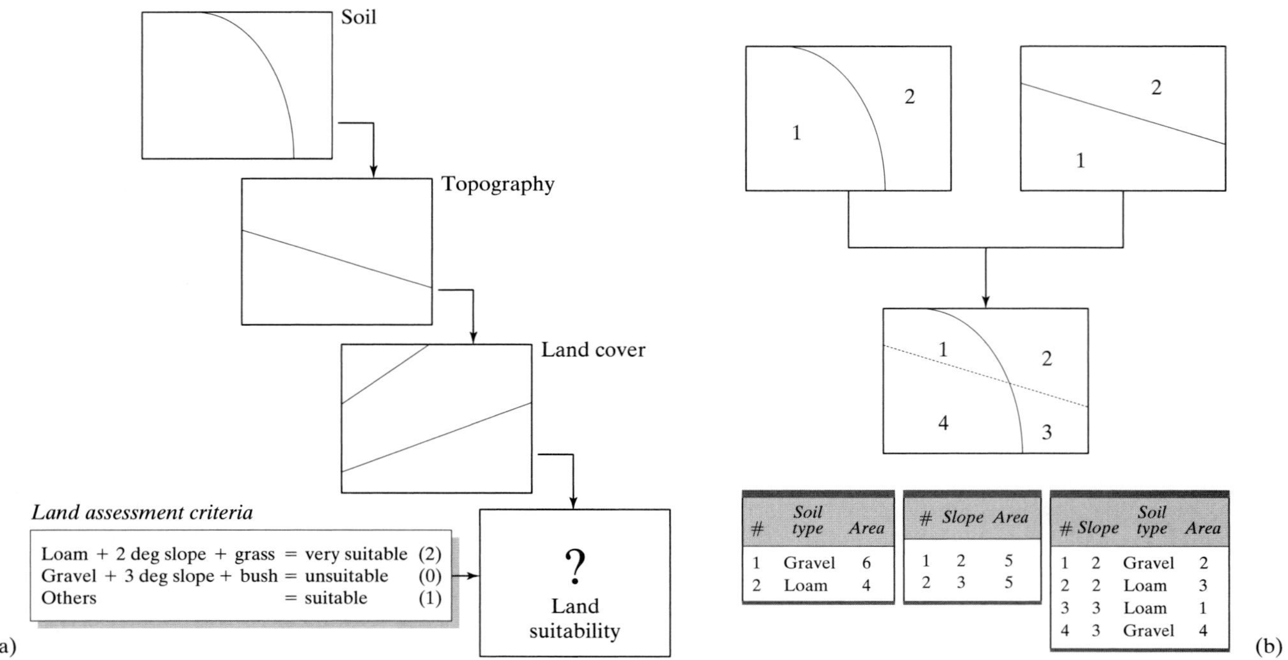

#	Soil type	Area
1	Gravel	6
2	Loam	4

#	Slope	Area
1	2	5
2	3	5

#	Slope	Soil type	Area
1	2	Gravel	2
2	2	Loam	3
3	3	Loam	1
4	3	Gravel	4

FIGURE 6.31
(a) An example of topological overlay. (b) Topological overlay: step 1. (c) Topological overlay: step 2. (d) Topological overlay: step 3.

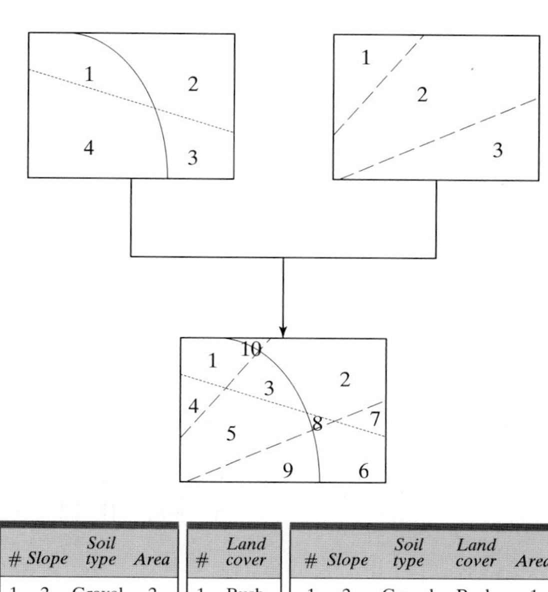

#	Slope	Soil type	Area
1	2	Gravel	2
2	2	Loam	3
3	3	Loam	1
4	3	Gravel	4

#	Land cover
1	Bush
2	Grass
3	Bush

#	Slope	Soil type	Land cover	Area
1	2	Gravel	Bush	1
2	2	Loam	Grass	2
3	2	Gravel	Grass	1
4	3	Gravel	Bush	1
5	3	Gravel	Grass	2
6	3	Loam	Bush	1
7	2	Loam	Bush	0.6
8	3	Loam	Grass	0.2
9	3	Gravel	Bush	1
10	2	Loam	Bush	0.2

(c)

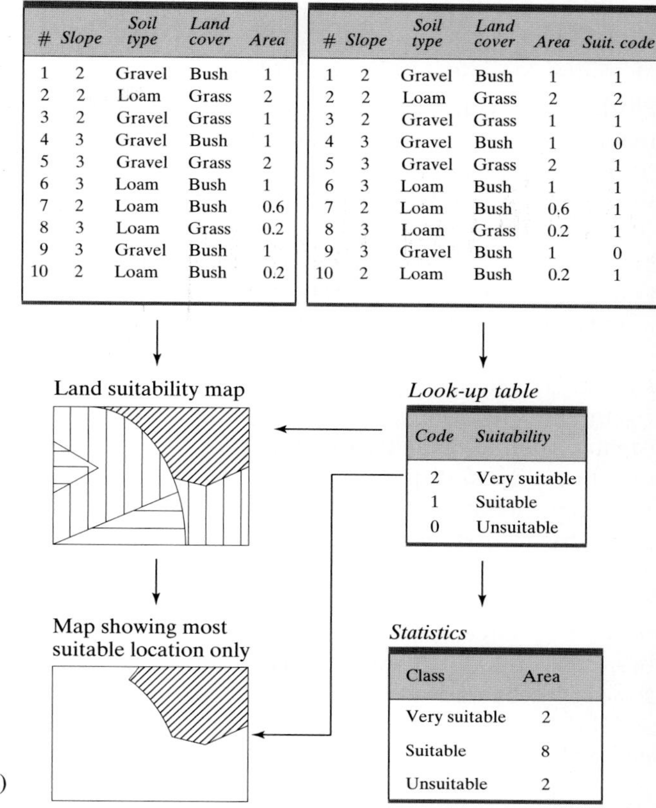

#	Slope	Soil type	Land cover	Area
1	2	Gravel	Bush	1
2	2	Loam	Grass	2
3	2	Gravel	Grass	1
4	3	Gravel	Bush	1
5	3	Gravel	Grass	2
6	3	Loam	Bush	1
7	2	Loam	Bush	0.6
8	3	Loam	Grass	0.2
9	3	Gravel	Bush	1
10	2	Loam	Bush	0.2

#	Slope	Soil type	Land cover	Area	Suit. code
1	2	Gravel	Bush	1	1
2	2	Loam	Grass	2	2
3	2	Gravel	Grass	1	1
4	3	Gravel	Bush	1	0
5	3	Gravel	Grass	2	1
6	3	Loam	Bush	1	1
7	2	Loam	Bush	0.6	1
8	3	Loam	Grass	0.2	1
9	3	Gravel	Bush	1	0
10	2	Loam	Bush	0.2	1

Code	Suitability
2	Very suitable
1	Suitable
0	Unsuitable

Class	Area
Very suitable	2
Suitable	8
Unsuitable	2

(d)

FIGURE 6.31 *(continued)*

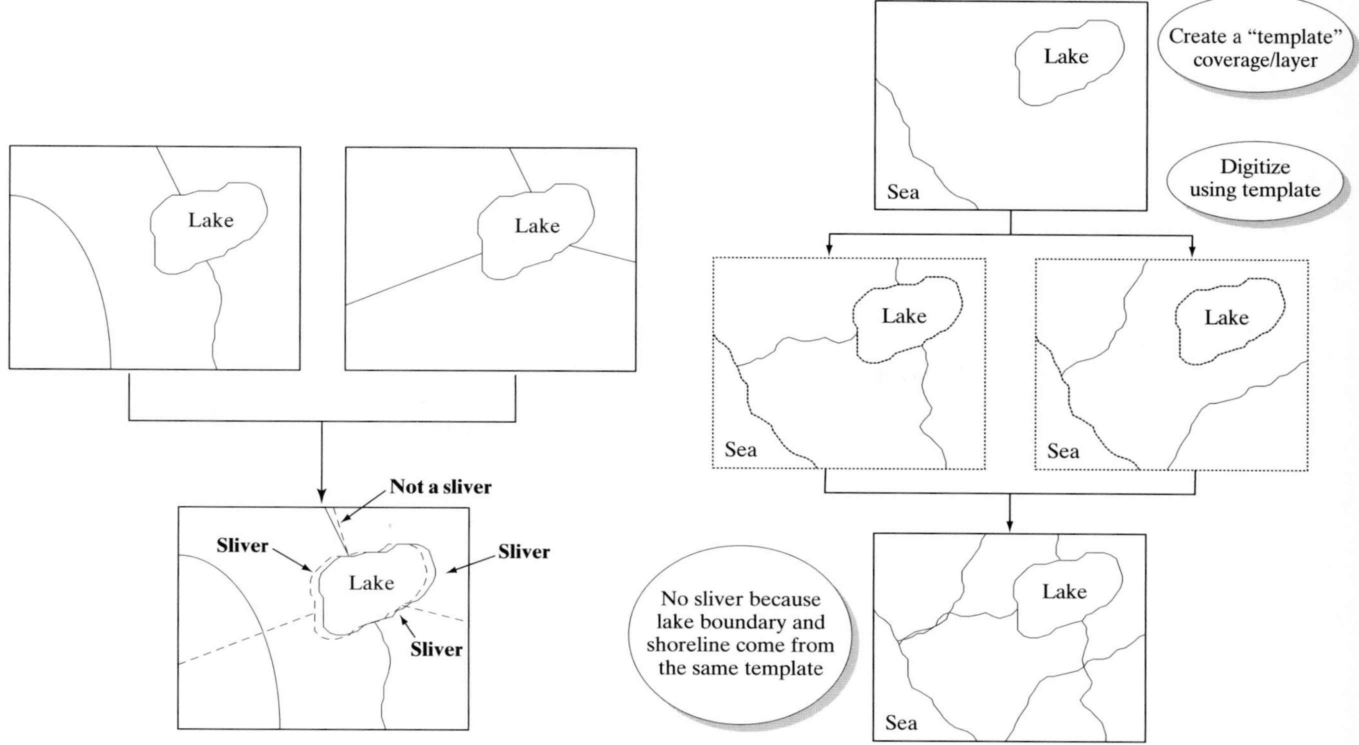

FIGURE 6.32
Creation of sliver polygons in overlay analysis.

FIGURE 6.33
The method of using a digitizing template to minimize sliver polygons in topological overlay.

and energy, as well as the communication of information (Haggett and Chorley, 1969; Kansky, 1963).

Despite the obvious geographic importance of networks, however, GIS in general is not optimized for network data processing. It is more oriented toward resource and environmental applications that emphasize the use of polygons. Dealing with networks requires sophisticated processes and data structure that may not be available in the core functionality of most GIS software packages. Network data-processing procedures are usually provided in the form of software extensions that are obtainable from the GIS vendors or third-party software developers. In ArcInfo, for example, the native commands and the georelational data model are not capable of handling network data processing other than simple graphical display and attribute data query. Advanced network processing can be done by using a special software extension called *ARC NETWORK* (ESRI, 1996a). ArcView GIS, similarly, provides an extension known as *Network Analyst* that allows the user to solve network-related spatial problems such as finding the best routes, defining travel costs, and modeling traffic flow (Ormsby and Alvi, 1999).

The concepts and techniques of network analysis in vector-based geographic data processing can be explained from three perspectives: (1) the network layer, (2) network analysis algorithms, and (3) applications of network analysis.

The Network Layer A network layer is made up of topologically structured *line segments, junctions,* and, optionally, *turns* (Figure 6.34). A line segment, also known as an *edge,* is the basic spatial unit of a network. It is identified by a segment identifier and is tagged by related attributes such as street name, length, and speed limit in a street network; and section number, length, and carrying capacity in a power transmission network. A junction, or an *intersection* as it is sometimes called, is a point or node in the network where line segments meet. It is identified by a junction identifier and the line segments leading to and from it in the network. A turn is a numerical value specifying the time or cost required to travel through a junction. In a street network, for example, a value of 5 seconds can be assigned to a turn if there is a stop sign at the particular intersection, and a value of 15 seconds is assigned if there is a traffic light. These values represent the average time that it will take for a car to turn from one street to another street in normal road and traffic conditions, and they can be set to different values for different conditions.

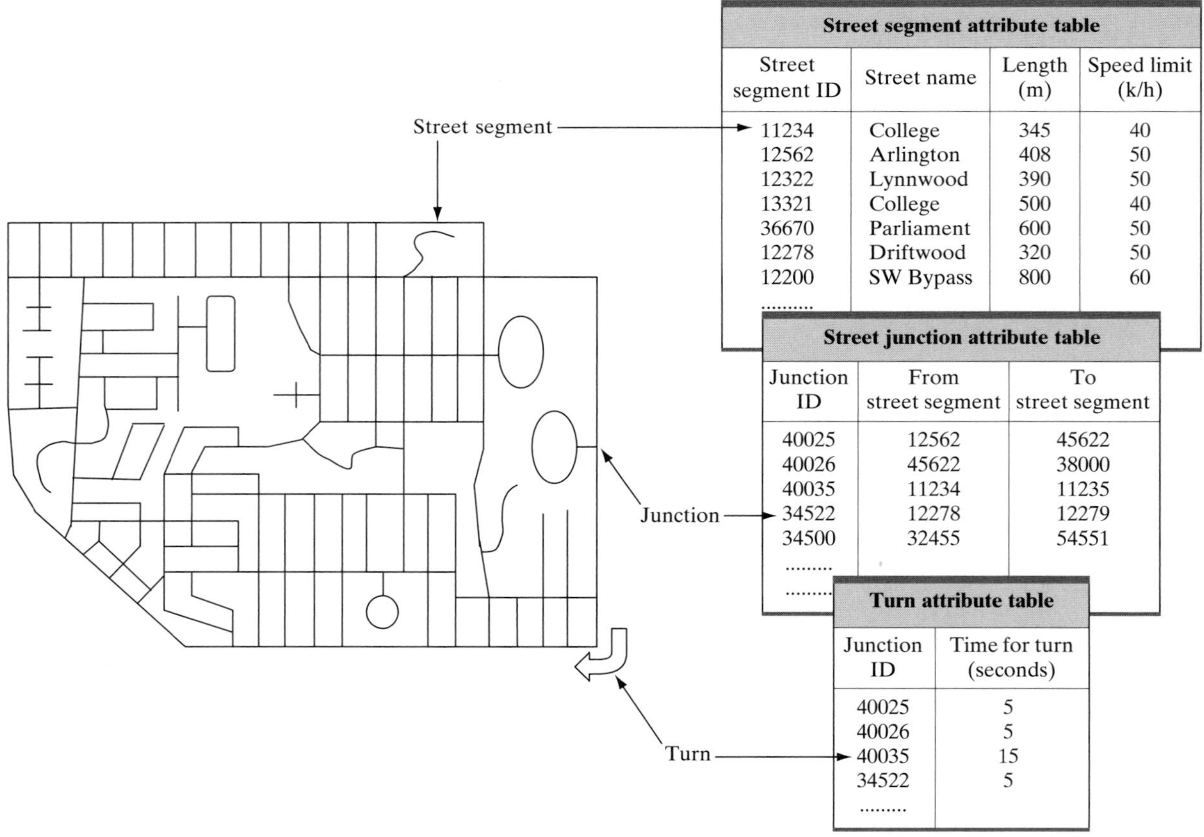

Street segment attribute table

Street segment ID	Street name	Length (m)	Speed limit (k/h)
11234	College	345	40
12562	Arlington	408	50
12322	Lynnwood	390	50
13321	College	500	40
36670	Parliament	600	50
12278	Driftwood	320	50
12200	SW Bypass	800	60
....			

Street junction attribute table

Junction ID	From street segment	To street segment
40025	12562	45622
40026	45622	38000
40035	11234	11235
34522	12278	12279
34500	32455	54551
....		
....		

Turn attribute table

Junction ID	Time for turn (seconds)
40025	5
40026	5
40035	15
34522	5
....	

FIGURE 6.34
Characteristics of a network layer.

Network Analysis Algorithms Network analysis algorithms are based on relatively sophisticated mathematics using graphs theories (see Chapter 10, Section 10.7). Even a simple spatial problem using networks, such as finding the shortest path between two points, requires a considerable amount of computation to solve. It is not within the scope of this book to discuss the different network analysis algorithms that have been used in GIS. However, it is useful to demonstrate the complexity of network analysis by explaining how the shortest path between two points is determined in GIS.

Generally speaking, it is relatively easy to visually find the shortest path between two points, A and B, on a map (Figure 6.35). In GIS-based network analysis, however, it is not so simple because the system has to compute and compare all possible paths between the two given points in order to arrive at an answer. Several algorithms have been developed to solve this problem mathematically. The best known of these algorithms, developed by E.W. Dijkstra, is used by the Network Analyst extension of ArcView GIS (Ormsby and Alvi, 1999). The algorithm begins with the starting point, A, by entering it as the first item in a "reached nodes" table (Figure 6.36a). It then considers all the junctions adjacent to this point and places them in a "scanned nodes"

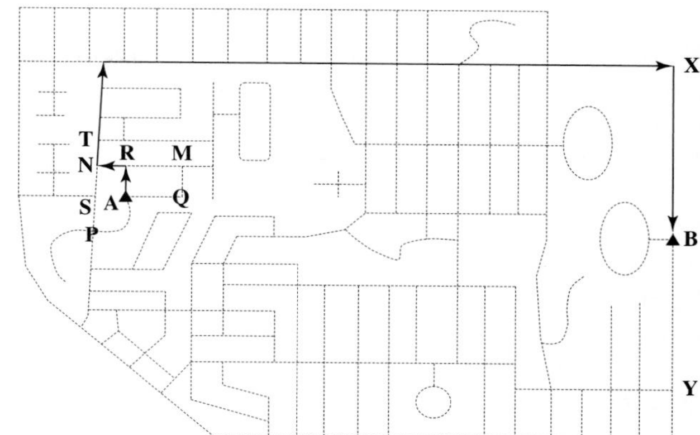

FIGURE 6.35
Determining the shortest path between A and B.

FIGURE 6.36
Changing values in the "reached nodes" and "scanned nodes" tables at different stages during the execution of Dijkstra's shortest-path algorithm.

Reached nodes table

Node	Cumulative distance	Previous node
A	0	--

Scanned nodes table

Node	Cumulative distance	Previous node

(a) Contents of tables at beginning

Reached nodes table

Node	Cumulative distance	Previous node
A	0	--
R	80	A

Scanned nodes table

Node	Cumulative distance	Previous node
P	120	A
Q	105	A
R	80	A

(b) Contents of tables after reaching P and R from A

Reached nodes table

Node	Cumulative distance	Previous node
A	0	--
R	80	A
N	130	R

Scanned nodes table

Node	Cumulative distance	Previous node
P	120	A
Q	105	A
M	184	R
N	100	R

(c) Contents of tables after reaching M and N from R

Reached nodes table

Node	Cumulative distance	Previous node
A	0	--
R	80	A
N	130	R
T	180	N
...		
...		
...		
B	1250	X

Scanned nodes table

Node	Cumulative distance	Previous node
B	1250	X
B	1565	Y

(d) Contents of tables after reaching B from X and Y

table. From this scanned nodes table, the system identifies the node with the shortest distance from A and moves it to the reached nodes table (Figure 6.36b). The process is repeated using the latest reached nodes (i.e., R, N, and so on) until the destination point B is reached by all possible paths (Figure 6.36c). On completion of this process, the reached nodes table will contain the cumulative distances from point A to all the nodes in the network, and the scanned nodes table will contain cumulative distances of all possible paths from point A to point B (Figure 6.36d). The shortest path between the starting and destination points is obtained by selecting the smallest cumulative distance in the scanned nodes table. In order to find the sequence of junctions that are included in the selected path, it is necessary to trace the node identifiers in the "previous node" column of the reached nodes table.

This description of the shortest-path algorithm probably represents the simplest scenario to which it is applied. The process will be much more complicated if the shortest travel time or least travel cost between the two points is required. Since travel time and travel cost are not necessarily proportional to distance directly due to, for example, different speed limits and different values of turns at different junctions, it is necessary to perform the analysis using the time and cost associated with each individual street segment and junction, rather than by simply multiplying the total distance by an average travel time or average cost.

The algorithm used in this example determines the shortest path between two specific junctions only. If the shortest path between any two points is located on line segments, rather than right at the junctions, it is necessary to create the junction by breaking the line segments concerned first and rebuilding the network topology before the algorithm is applied. It should also be noted that despite its complexity, the creation of the reached nodes and scanned nodes tables often represents only the first step in network analysis. Practically all the network-based spatial analysis tasks listed next require further processing using more sophisticated mathematical or statistical models.

Applications of Network Analysis There are many spatial problems that require the use of network analysis for their solution. These include, for example,

- *Pathfinding.* This analysis finds the shortest path (in terms of physical distance or least cost) that can be followed to visit a series of locations in a network.
- *Allocation.* This analysis assigns one or more portions of a network to be served by a facility or business location.
- *Tracing.* This analysis finds all portions of the network that are connected with the movement of a particular feature, for example, a truck.
- *Spatial Interaction.* This analysis aims to depict the accessibility of a location and the interactions that occur between different locations based on a technique known as *gravity modeling* that is widely used in economics, geography, engineering, and urban planning (see Chapter 10, Section 10.6).
- *Distance Matrix Calculation.* This analysis generates a distance matrix between different pairs of locations in the network.
- *Location-Allocation Modeling.* This is probably the most complex network analysis; it aims at determining simultaneously the locations of existing and planned facilities, as well as the allocation of demand to these facilities.

6.7 Vector-Based Output Functions

Output functions of vector-based GIS are data-processing procedures that generate maps, charts, and statistical reports, as well as transformation of digital data from one form or scale to another. The purpose of these functions is to provide the necessary means to present the results of the data analysis processes discussed earlier to end users in an effective manner. Vector-based output functions share many common concepts and techniques with raster-based methods explained in Chapter 5, in particular the use of color for displaying maps on the computer screen and producing hard-copy maps using printers and plotters (Section 5.4). However, vector-based output functions are also different from their raster-based counterparts in several ways. In the following sections, we will provide an overview of the characteristics of the output functions that are found in a typical vector-based GIS software package. More detailed explanation of GIS output functions from the broader perspective of information communication will be given in Chapter 7.

6.7.1 Methods of Presenting Vector-Based Spatial Information

Since GIS has its roots in cartography and mapping, it is invariably characterized by the ability to generate high-quality maps in both digital and hard-copy forms. Depending on the nature of the information produced, results of vector-based data-processing tasks can be presented in various forms. The most commonly used methods include the following (Figure 6.37):

- *Graduated Symbols.* Graduated symbols can be points drawn to show magnitude of data values at locations where they are measured or lines that show volumes or ranks for linear networks, such as roads, utility lines, and rivers.

Method	Example	Features	Advantages	Disadvantages
Graduated symbols	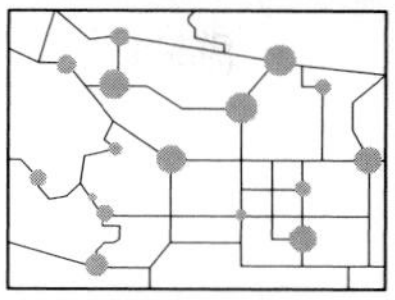 Accidents at intersections	• Locations • Lines • Areas	• Association between size of symbols and magnitude of spatial features • Easy to show spatial distribution	• Difficult to read if too many features are present
Graduated colors	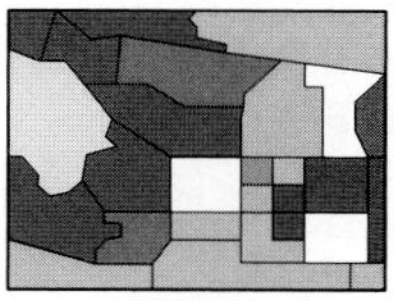 Population density	• Continuous phenomena • Areas	• Easy to recognize spatial patterns	• Difficult to associate colors with values they represent
Charts	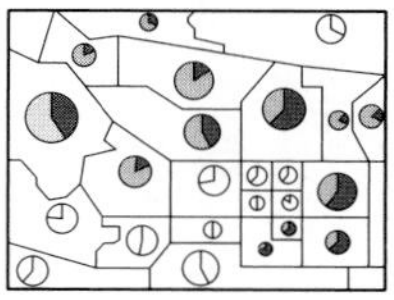 Percent of people with a college degree	• Locations • Areas	• Easy to show magnitude and categories of spatial features	• Difficult to identify spatial pattern and distribution of spatial features
Contours	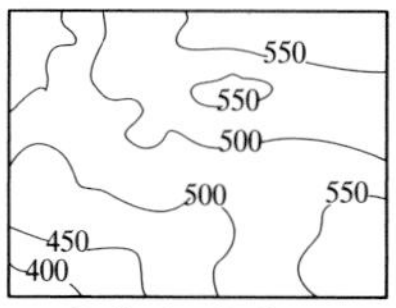 550 550 500 500 550 450 400 Pollution indices	• Continuous phenomena	• Easy to show rate of change and trend across an area	• Difficult to identify spatial pattern and distribution of spatial features
Three-dimensional perspective views	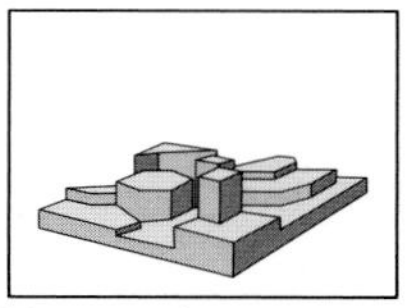 Crime rates by census tracts	• Continuous phenomena • Locations • Areas	• Easy to show rate of change • Association between 3-D values to quantities that they represent	• Difficult to read values of individual features • Difficult to construct

FIGURE 6.37
Different methods of presenting results of vector-based geographic data analysis.

- *Graduated Colors.* This method is most suitable for showing discrete areas by which data are summarized, using no more than seven to ten colors, with darker colors to mean more or greater, and lighter colors to mean fewer or smaller.
- *Charts.* This method is commonly used to display data summarized by area or discrete locations, in terms of both quantity and categories.
- *Contour Lines.* Contour lines are best for showing rate of change in data values across an area for spatially continuous phenomena, such as elevation, rainfall, temperature, and barometric pressure.
- *Three-dimensional Perspective Views.* This method is often used to show continuous surfaces or magnitudes of discrete areas by which spatial data are summarized.

6.7.2 MAP COMPOSITION

Although the methods of presenting vector-based spatial information are relatively simple in concept, the actual generation of the output products still requires a considerable amount of effort in planning and design. GIS software packages today generally provide a very rich set of tools that allow the user to compose a map to meet different data-processing needs (Figure 6.38). However, the quality of the end product is dependent not merely on the individual user's ability to master these map composition tools in GIS, but also the user's technical training in cartographic design. Therefore, users who do not have a cartographic background should take extra care when generating output products to ensure that these products are generated according to standard cartographic practices in terms of the use of

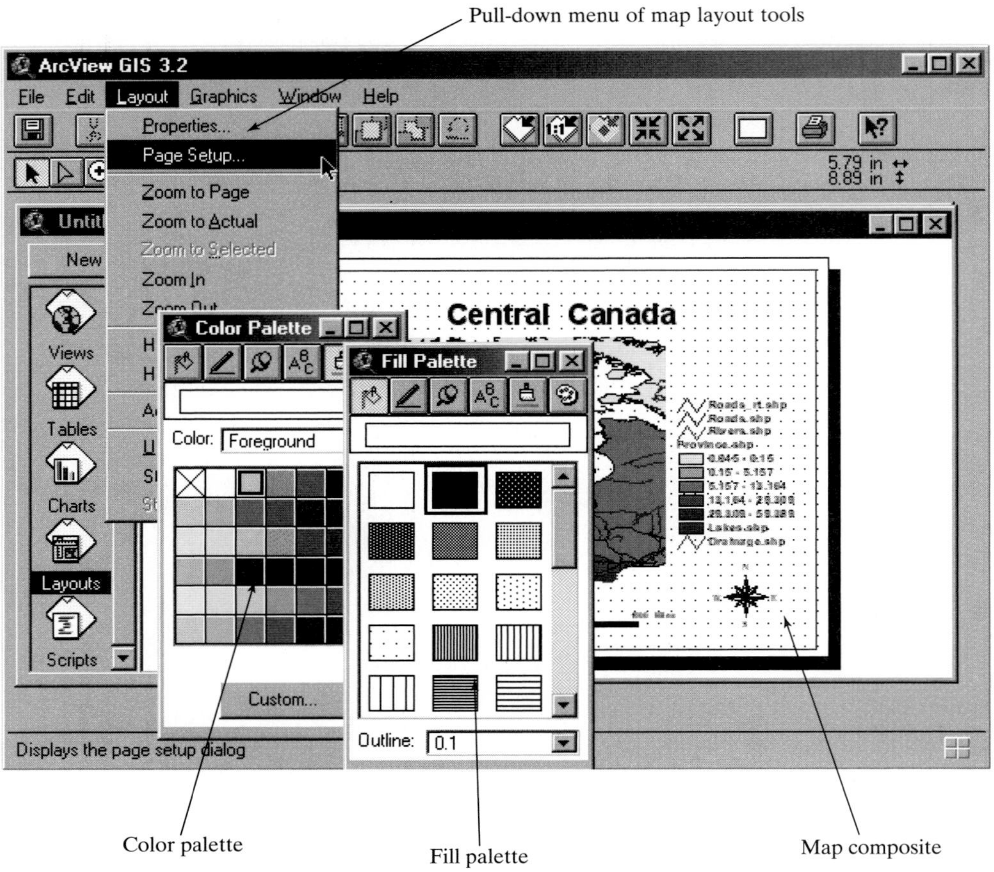

FIGURE 6.38
Map composition graphical interface and tools of ArcView GIS. To view this photo in color, see the color insert.

color, map layout, and symbology, as well as map scale and size.

From the data-processing perspective, map composition is a repetitive design process. Even experienced cartographers have to test different alternatives before arriving at the final result. In the digital cartographic environment of GIS, it is relatively easy for the user to try different ways of presenting the results of spatial information. As a general guideline, the user should pay particular attention to the following issues in the map composition process:

- *The Order of Layer Display or Plotting.* The normal sequence is to display or plot layers pertaining to the topography first, followed by layers showing man-made features, and finally annotation. This means that highways and buildings, for example, are often displayed or plotted on top of contour and drainage, and place names are displayed or plotted on top of all layers of graphical data.
- *Use of Fonts.* Make sure that font sizes are proportional to the cartographic prominence of the features that they describe and that the alignments of labels are appropriate to the orientation of their associated features.
- *Use of Symbols.* The symbols supplied by GIS software packages are sometimes not enough to represent all types of features to be displayed, particularly in thematic or special-purpose applications such as geological mapping. In such cases, the user should know where to obtain the necessary symbols and how to incorporate them into the GIS for deployment in map composition.

6.7.3 VECTOR–RASTER FORMAT CONVERSION

The increasing integration of GIS and other forms of computer-based applications, particularly desktop publishing and word processing, has led to the growing importance of vector–raster format conversion as a function in vector-based geographic data processing. Since vector GIS data most often use proprietary formats that are not accepted by desktop publishing and word-processing software packages, it is often necessary to convert vector-based output products into a commonly used raster format, such as GIF and JPEG, so that they can be incorporated into reports and presentation graphics.

Vector–raster format conversion is also required when output products of vector-based geographic data processing are disseminated to end users by means of the Internet. Up to now, spatial data on the Internet are largely transmitted and displayed in raster format, although there are standards and protocols for using vector graphics. Conversion of vector-based output products into raster format ensures that they can be viewed by end users directly on the Web browser, thus avoiding the need for the end user to install a product-specific program, called a plug-in or helper program, necessary for vector data.

There are several ways by which vector-based output products can be converted into raster format. These include, for example, using the built-in conversion function or extension of a GIS software package itself, using the data conversion facility of a commercial graphics processing package, or by means of a screen capture package. As the identities of individual spatial features are lost in the format conversion process, the resulting output products are good for display only. Their use as input for further geographic data analysis is rather limited.

6.8 Application Programming

Most GIS today use the "tool-box" approach to software design. The major advantage of this approach is the flexibility in application development. Since GIS software packages are developed as general-purpose tool kits for managing graphical and attribute data, they can be easily adapted to any application such as management of natural resources, municipal administration, and transportation planning. The trade-off, however, is that straight out-of-the box GIS are difficult to use. It is not easy for users to master the hundreds of application procedures that are found in a typical GIS. In order to make GIS easier to use, all software vendors deliver their products with programming capability that allows users to build their own applications. As a result, application programming is an essential part of vector-based GIS data processing today.

GIS application programming is a substantial topic that deserves an entire book in its own right. In the following sections, our discussion is limited to an introductory overview of the nature of GIS programming from the perspective of data processing. We will first explain the characteristics of macro programming using a software-specific scripting language. We will then examine the emerging concepts pertaining to software components and describe how they can be applied to geographic data processing with special reference to vector-based procedures.

6.8.1 Building Data-Processing Functions Using Scripting Languages

A scripting or macro language is an interpreted computer programming language. An interpreted language translates each command line, entered from either the keyboard or a program file, to machine language and executes it before moving to the next line. Scripting languages have simpler structure than conventional computer programming languages. This makes them relatively easy to learn and use. A basic function of application programming is to group frequently used command sequences into individual special-purpose tasks so that they can be used repeatedly. At a more advanced level, application programs may also be used to create prompts for data entry, validate input data, return error messages, build user-defined commands and menus, and develop a point-and-click interface for specific applications (Figure 6.39).

There are many advantages of using custom application programs in vector-based geographic data processing. These include, for example,

- enforcing the standardization of procedures by using a common interface, and thus making possible the development and deployment of custom applications on top of the host GIS software package
- improving user productivity by automating frequently used and repetitive procedures, as well as by reducing typing and mouse clicking by storing command sequences in a file
- minimizing the need to memorize commands and their functions, thus enabling less experienced users to master the skill of using the system more quickly
- reducing data input error by providing the user with possible options
- enhancing the native capabilities of the GIS software by developing application-specific commands, menus, and graphical interface windows
- enabling the integration of geographic data processing with computer-based applications

Scripting languages, therefore, serve a wide variety of purposes in vector-based geographic data processing. Conventionally, scripting was the only way by which GIS applications were developed. Advances in software development methodology have made available powerful programming tools by which GIS applications can be developed more cost-effectively. As a result, GIS application development is gradually moving

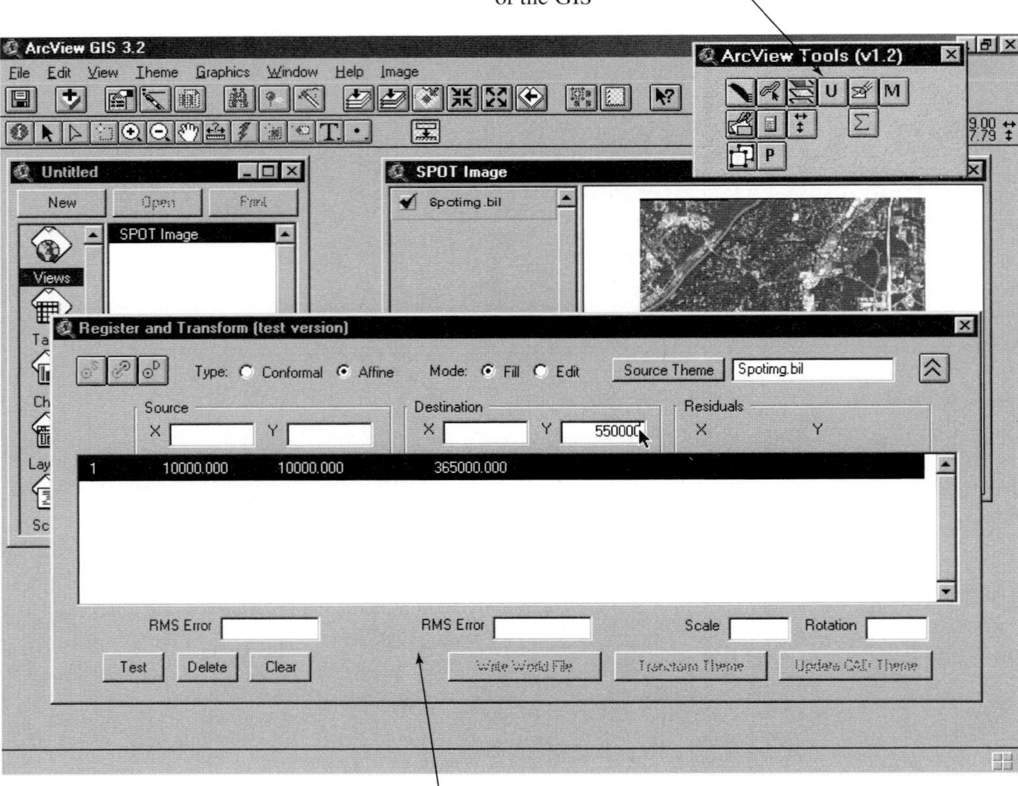

FIGURE 6.39
Example of using Avenue scripting language in ArcView GIS to enhance the standard data-processing tools and to automate repetitive tasks. Scripting languages allow users to develop new data-processing tools that are used in a particular project or by a specific group of users. The functionality of these tools ranges from relatively simple data editing procedures to very complex processes based on sophisticated mathematical solutions.

away from the use of scripting languages for particular GIS software packages, to the open system development approach using industry standards as described in the next section.

6.8.2 GIS APPLICATION DEVELOPMENT USING COMPONENT TECHNOLOGY

The advent of *component software* is one of the most important developments in software engineering in recent years. The concept of component software can be explained from two perspectives. First, component software is the product of a particular software development methodology. The component software development methodology evolves from the attempts to search for methods that can expedite application development and minimize development cost through the development of modular, reusable software products. Unlike conventional software development strategies that aim mainly at generating reusable source codes, the component strategy focuses instead on creating reusable preprogrammed software modules that are generally known as "components," "custom controls," or "automation objects." Components must be embedded within a *container program* in order to perform the data-processing functions for which they are intended (Figure 6.40). Components are created by low-level computer languages such as C and C++, but the container programs can be written in a variety of languages including Visual Basic, Visual C++, Delphi, PowerBuilder, and some others. Component software is therefore characterized by "binary (machine code) reusability," as opposed to "source code reusability" in conventional software application development (Anderson et al., 1996).

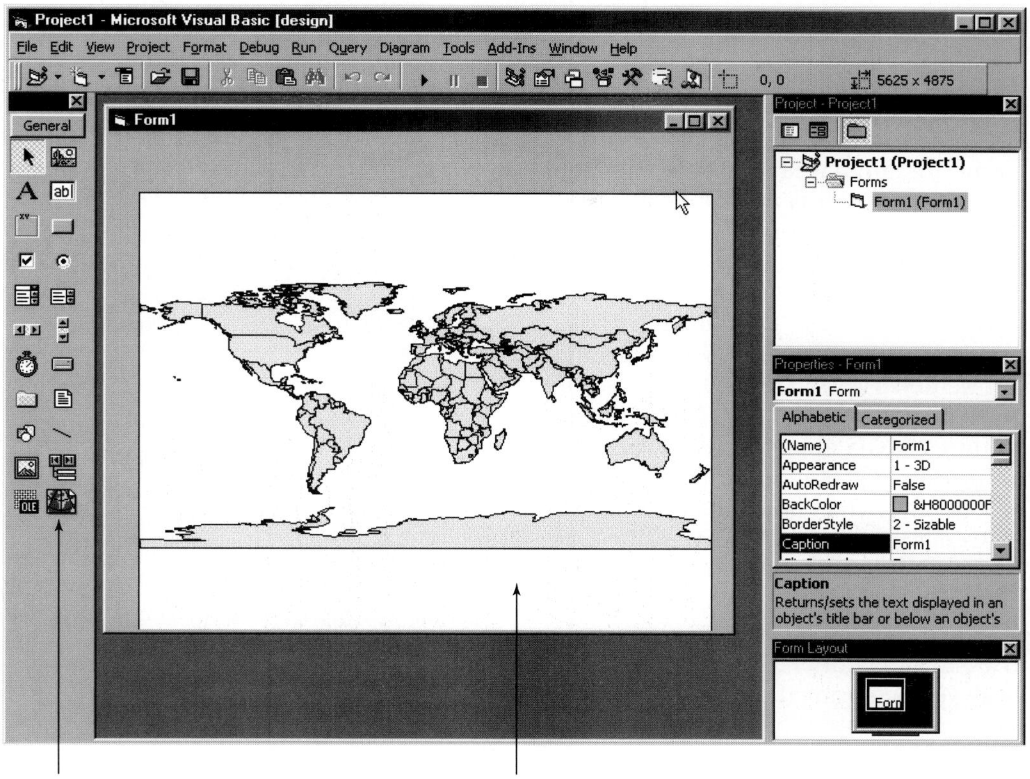

MapObjects map control loaded to the Visual Basic toolbox

Map added to a Visual Basic form

FIGURE 6.40
Example of using MapObjects components in Visual Basic to develop an application that uses map data. A form is the "drawing board" used in Visual Basic programming to design and create the graphical interface of Microsoft Windows applications. MapObjects provides the necessary software components that allow the user to load a map to a form of a Visual Basic program. It also provides the components to build the tools (e.g., zoom, pan, and edit) necessary to manipulate the data on the map when the Visual Basic program is run.

Second, component software is a software development architecture. Components are developed in compliance with industry-adopted programming standards that define both the interface among the components themselves and their behavior when embedded in application programs. The most widely accepted component architectures are Microsoft's *Object Linking and Embedding/Component Object Model* (OLE/COM, which has also been variously called *OLE controls* and *OCX;* it has now been renamed *ActiveX controls*) and Object Management Group's *Common Object Request Broker Architecture* (CORBA). The adherence to standard software architectures is the reason why components can run under different computer languages as noted above. This is also the underlying principle behind the concept of open GIS and the implementation of interoperable geographic data processing (see Chapter 1) (OGIS, 1998a, and 1998b).

Unlike macro programming, which is oriented toward both the end user and the application developer, component software is intended primarily for the application developer. There are two basic approaches to using component software in GIS application development (Hartman, 1997): *component suites* and *component parts*. Component suites are collections of components that are assembled into an integrated comprehensive system. Component technology has now been used by GIS software vendors to develop large-scale systems and to reengineer existing ones. ArcInfo Version 8, for example, has been designed and developed on the component software development methodology. This allows geographic data-processing functions of ArcInfo to be embedded within non-GIS in what is called the *Open Development Environment* (ODE) (ESRI, 1998). In Windows NT, the ODE is implemented by making ArcInfo functions (e.g., data editing and display) accessible via OCX using industry-standard software development languages (i.e., Visual Basic, Visual C++, Delphi, and PowerBuilder). In UNIX, ODE is implemented as shared libraries for development environments such as C, C++, Motif (an X-Windows–based graphical user interface tool kit, see Chapter 7) and Tcl/Tk (a cross-platform programming and graphical user interface tool kit (Zhuang and Engel, 1995). Previously, application programming could be developed only by using software-specific scripting languages such as Arc Macro Language (AML) of ArcInfo. It is now possible to use any computer language that supports the OLE software development architecture.

Component parts, on the other hand, are individual component software modules. Unlike component suites, component parts are not usually produced in the form of an assembled software package. Instead, they are designed to perform particular geographic data-processing procedures in GIS applications. A variety of geographic data-processing component parts are now available on the market, such as *GeoView* and *GeoObjects* of Blue Marble Geographics, *MapX* of MapInfo, *Sylvan/OCX* of Sylvan Ascent, and *MapObjects* and *ArcObjects* of ESRI. These component parts provide a very rich set of software tools for geographic data processing, particularly in the desktop application environment. Functions that can be developed using the above component parts include

- display of geographic information using maps, graphs, charts, and images
- browsing spatial information by panning and scrolling through multiple map layers
- spatial analysis and database query
- geometric manipulation of graphics elements
- coordinate transformation and format conversion
- address matching and geocoding
- real-time data collection such as event tracking using GPS

The availability of component software technology has opened a new horizon in geographic data processing. Component software makes possible the inclusion of geographic data-processing functions in computer applications by providing developers with plug-in GIS capabilities. Component software tends to be sold according to the low-cost, high-volume marketing philosophy. Application developers can purchase these tools at a relatively low one-time cost, use them to add geographic data-processing functionality to applications, and then deploy these applications to an unlimited number of end users at a fraction of the cost that would have otherwise been incurred.

The impact of component software technology on geographic data processing is probably most significantly felt in the area of Internet-based mapping and GIS applications. MapObjects, for example, has an optional extension called *MapObjects Internet Map Server* (IMS) that provides the developer the necessary tools to build Internet-based applications. The IMS extension lets the developer use MapObjects to create spatially enabled Web sites capable of delivering geographic data-processing functions in a distributed network environment (Figure 6.41). Component software technology is indeed in the process of changing the ways in which geographic data processing is carried out. It not only enables more users to use geographic data-processing functionality in their applications locally, but also to access large custodial geographic databases and powerful GIS applications via the Internet and other forms of communications networks.

6.9 Summary

This chapter explained the characteristics and nature of vector-based geographic data processing. Broadly speaking, vector-based geographic data processing serves three primary purposes: data conversion, data query and analysis, and address geocoding. Data conversion processes enable geographic data stored in different forms and available from different sources to be used by GIS. Despite the increasing availability of digital geographic data from government agencies and commercial data suppliers, in-house data conversion has remained an important task in many GIS projects. It is therefore essential for all users of GIS to have a sound understanding of the concepts and techniques behind all the processes involved.

Data query and analysis procedures represent the core functionality of GIS. These procedures can be classified as nontopological and topological. Nontopological processes provide the user with the basic functions to display geographic information and query the geographic database on the basis of coordinates. Topological processes, on the other hand, provide the user with the necessary software tools to solve spatial problems using the topological relationships of adjacency, containment, and connectivity among graphical elements in the geographic database. From the perspective of geographic data processing, topological processes are probably the most important functions in vector GIS. Methods such as buffering, network analysis, and topological overlay always form the major building blocks of developing GIS applications.

Descriptive data play a very important role in vector-based data query and analysis. Many processes are simply impossible without the use of attribute data. In the past, the capability of GIS to perform attribute database query was relatively limited. This was mainly because of the inherent weakness of relational database systems to manipulate graphical data. The establishment of SQL as the standard query language for commercial DBMS has greatly enhanced the functionality of GIS to access descriptive data during vector-

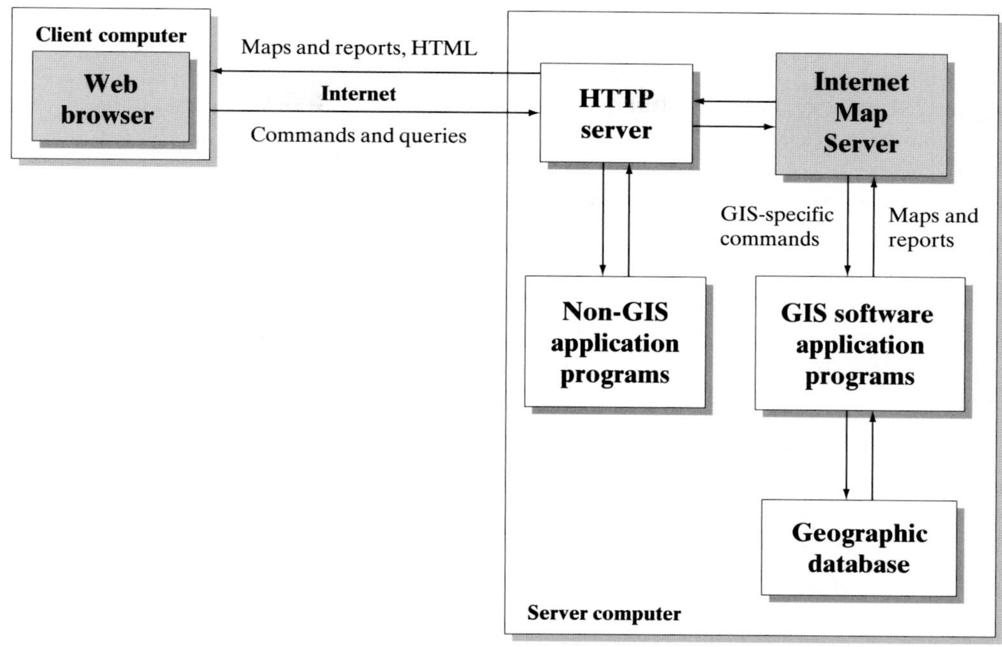

FIGURE 6.41
Geographic data processing via the Internet using component software technology. The function of the Internet Map Server is to process GIS-specific commands and graphical data, which the HTTP server is not capable of handling.

based geographic data processing. Until very recently, attempts to spatially enable SQL for GIS applications have met with rather limited success. Emerging database technology and SQL standards currently under development will remove the impediments experienced in the past. It is anticipated that access to descriptive data in vector-based processing in the future will not only be more powerful, but also more integrated as well.

In recent years, there has been significant development in the concepts and techniques of address geocoding. This is both the cause and effect of the phenomenal growth of the use of GIS in the social sciences and business that we are witnessing today. With the aid of address geocoding, descriptive data, which could not be used for spatial analysis in the past, can now be applied to a wide range of GIS applications in marketing, crime analysis, business planning and demographic studies. As a result, these new application areas are rapidly overtaking natural resource and environmental applications as the primary focus of vector GIS.

Conventionally, vector-based geographic data processing was carried out in the command line mode. This was not a very productive way of using GIS because of the amount of typing required. Many GIS software vendors tried to overcome this problem by offering the user the capability to use macro programming. This allows repetitive command sequences to be stored in program files so that they can be reused easily. Examples are Arc Macro Language (AML) and Simple Macro Language (SML) used by ArcInfo (ESRI, 1994b).

What is perhaps more challenging to the GIS community is the introduction of software component technology to GIS application development. Despite its relatively short history, software component technology has quickly become the predominant application development methodology for the GIS industry. Since this particular technology is based on computer industry standards, it allows GIS application development to be totally integrated with the enterprise computing environment. Very complicated GIS applications can now be developed using spatially enabled software components commercially marketed by software vendors. Software component technology makes possible geographic data processing using a totally custom-built GIS.

Raster-based and vector-based geographical data processing represent two very different approaches to spatial problem solving. Today's computer technology has made possible the integrated use of raster and vector data in spatial analysis and modeling. In this chapter, we explained the integration of raster and vector data from the perspective of data conversion in vector-based geographic data processing. In Chapter 8, we will examine in greater detail the issues of using these two types of data from the perspective of remote sensing and GIS integration.

REFERENCES

Anderson, D., Ledbetter, M., and Tepovich, D. (1996) "Piecing together component-based GIS application development," *GIS World*, Vol. 9, No. 6, pp. 62–64.

Berry, J. K. (1995) "Raster is faster, but vector is corrector," in *Spatial Reasoning for GIS*, pp. 91–94. Fort Collins, CO: GIS World Books.

Egenhofer, M. J. (1991) "Extending SQL for Graphical Display," *Cartography and Geographic Information Systems*, Vol. 18, No. 4, pp. 230–245.

Egenhofer, M. J. (1992) "Why not SQL!," *International Journal of Geographical Information Systems*, Vol. 6, No. 2, pp. 71–85.

ESRI (1994a) *PC ARC/INFO User Guides*, Redlands, CA: Environmental Systems Research Institute, Inc.

ESRI (1994b) *ARC Macro Language: Developing ARC/INFO Menus and Macros with AML*, Redlands, CA: Environmental Systems Research Institute, Inc.

ESRI (1994c) *Scanning Data Entry Solutions for ARC/INFO GIS*, an ESRI White Paper, Redlands, CA: Environmental Systems Research Institute, Inc.

ESRI (1995a) *Understanding GIS: The ARC/INFO Method*, Redlands, CA: Environmental Systems Research Institute, Inc.

ESRI (1995b) *ArcTools—Arc/Info Software's Graphical User*, an ESRI White Paper, Redlands, CA: Environmental Systems Research Institute, Inc.

ESRI (1996a) *Using ArcView GIS*, Redlands, CA: Environmental Systems Research Institute, Inc.

ESRI (1996b) *Arc/Info Extensions*, an ESRI White Paper, Redlands, CA: Environmental Systems Research Institute, Inc.

ESRI (1996c) *Arc/Info Subsystems*, an ESRI White Paper, Redlands, CA: Environmental Systems Research Institute, Inc.

ESRI (1997) *ESRI Shape File: A Technical Description*, an ESRI White Paper, Redlands, CA: Environmental Systems Research Institute, Inc.

ESRI (1998) *Spatial Data Warehousing*, an ESRI White Paper, Redlands, CA: Environmental Systems Research Institute, Inc.

ESRI (2000) *ARC/INFO 8: A New GIS for the New Millennium*, an ESRI White Paper, Redlands, CA: Environmental Systems Research Institute, Inc.

Flanagan, N., Jennings, C. and Flanagan, C. (1994) "Automatic GIS data capture and conversion," in *Innovations in GIS 1*, by Worboys, M. F. ed. pp. 25–38, London: Taylor & Francis.

Gisiger, A. (1998) "Hitachi Image Series" (Quick-take Review), *GIS World*, Vol. 11, No. 6, pp. 73–74.

Haggett, P., and Chorley, R. J. (1969) *Network Analysis in Geography*, London: Edward Arnold.

Hartman, R. (1997) *Focus on GIS Component Software: Featuring ESRI's MapObjects*, Santa Fe, NM: OnWard Press.

Kansky, K. J. (1963) *Structure of Transportation Networks*, Chicago, IL: Department of Geography, the University of Chicago.

MapInfo (1996) *MapInfo and the Data Warehouse*, a MapInfo White Paper, Troy, NY: MapInfo Corp.

Masry, S. E., and Lee, Y. C. (1988) *An Introduction to Digital Mapping*, Lecture Notes No. 56, Department of Surveying Engineering, University of New Brunswick, Fredericton, NB.

Mayo, T. (1994) "Computer-assisted tools for cartographic data capture," in *Innovations in GIS 1*, by Worboys, M. F. ed. pp. 39–52, London: Taylor & Francis.

Methley, B. D. F. (1986) *Computational Models in Surveying and Photogrammetry*. Glasgow: Blackie.

Monmonier, M. S. (1982) *Computer-assisted Cartography: Principles and Prospect*. Englewood Cliffs, NJ: Prentice Hall.

OGIS (1998a) *Open GIS Simple Features Specification for OLE/COM (Version 1.0)*, Wayland, MA: Open GIS Consortium, Inc.

OGIS (1998b) *Open GIS Simple Features Specification for CORBA (Version 1.0)*, Wayland, MA: Open GIS Consortium, Inc.

OMNR (1992) *Ontario Government Specifications for the Delivery of Digital Topographic Data and Cartographic Representation Products at Medium Scales (1 : 10000 and 1 : 20000)*, Version 5, Toronto: Ontario Ministry of Natural Resources.

Ormsby, T., and Alvi, J. (1999) *Extending ArcView GIS*. Redlands, CA: Environmental Systems Research Institute.

Phillips, J. D. (1989) "An evaluation of the factors determining the effectiveness of water quality buffer zones," *Journal of Hydrology*, Vol. 110, pp. 221–237.

Xiang, W.-N. (1993) "GIS method for riparian water quality buffer generation," *International Journal of Geographical Information Systems*, Vol. 7, No. 1, pp. 57–70.

Zeiler, M. (1999) *Modeling Our World: The ESRI Guide to Geodatabase Design*. Redlands, CA: Environmental Systems Research Institute.

Zhuang, X., and Engel, E. A. (1995) "Tcl/Tk GUI toolkit offers cross-platform application development," *GIS World*, Vol. 8, No. 7, pp. 58–60.

CHAPTER

7

Visualization of Geographic Information and Generation of Information Products

7.1 Introduction

Maps are the most effective and intuitive way of presenting geographic information. They are the natural outcome of geographic data processing. The production of maps, therefore, is one of the core functions of all GIS. This chapter aims to explain the concepts and techniques that GIS users must understand in order to make the best use of this important function in their work.

Producing maps using advanced digital technologies in GIS is very different from producing maps using conventional cartographic methods. In this chapter, we will explain the changing nature of cartography in the context of GIS. We will introduce the concepts of *human–computer interaction* and *scientific visualization* as they pertain to the access, communication, and analysis of geographic information in GIS. Finally, we will provide an overview of the principles of cartographic design in GIS and explain how they are used to generate *information products* that, by definition, include not only maps but also other forms of output such as statistical graphs, tabulated reports, and multimedia presentations.

7.2 Cartography in the Context of GIS

Cartography and GIS are linked by virtue of their common focus on maps. Traditionally, GIS users relied on concepts and techniques drawn from cartography to produce maps. However, since the technologies for producing maps in conventional cartography and GIS are quite different from one another, it is now generally recognized that conventional principles and techniques are no longer adequate for GIS applications. A new definition of cartography in the context of GIS is obviously in order.

7.2.1 The Renewed Interest in Cartography

GIS researchers and users have always been interested in cartography, but their interest has never been as great as it is today. This renewed interest in cartography was prompted primarily by the disappointment in the poor quality of many maps produced by GIS. Despite the use of advanced digital technologies, most GIS applications are still designed to present geographic information in a way that mimics paper maps in conventional cartography (Dykes, 1996). Since not all of today's GIS users are well trained in the use of graphics, typical design errors in conventional cartography continue to occur in maps produced by GIS (Kraak, 1995; Weibel and Buttenfield, 1992). There is no doubt that the use of digital technologies has removed much of the tedious and repetitive work in map production. However, it does not seem logical to assume that maps produced from these technologies should be identical to the traditional printed products and contain similar design deficiencies.

Many GIS users have come to the realization that it is the cartographic expertise of people, not computer functionality, that plays the most important role in ensuring good quality maps, even in a computer-based technology such as GIS. Weibel and Buttenfield (1992), for example, pointed out that although all GIS software packages include a standard function for map production, none of them has the mechanism to ascertain the correct design of maps. Within the discipline of cartography itself, there have been many attempts to automate the map design process (for example, by using expert systems), but up to now the achievement has been rather limited (Buttenfield and Mark, 1991; Fisher and Mackaness, 1987). The current interest of GIS users in cartography reflects the common understanding that it is not possible to rely entirely on the computer to produce maps, but they have to take personal responsibility themselves.

The interest of GIS users in cartography has also been stimulated by the changing application environment of GIS. The development of GIS has already reached a relatively mature stage that demands more sophisticated information presentation techniques not offered by conventional cartography. In addition to using GIS to answer the relatively straightforward question of "where," users now rely increasingly on these systems to answer a wide variety of highly complex spatial problems of "why," "when," "how," and "what-if." These problems require not only very sophisticated algorithms for their solution, but also very large volumes of data as well. The demand for maps as an aid to spatial problem solving has never been as great as it is today. In order to take advantage of the rapidly evolving computer technologies to present geographic information, GIS users started to explore the possibility of using maps within systems for scientific visualization, spatial decision support systems, hypermedia information systems, and virtual reality environments (Artimo, 1994). In this regard, the current thrust toward cartography not only represents the revival of an old interest, it is in fact an important area of GIS research and technology development as well.

7.2.2 The Changing Nature of Cartography

The nature of maps is now undergoing some of the most remarkable changes in the history of cartography. In terms of physical form, maps are no longer limited to conventional printed products. This change of the physical form of maps has fundamentally changed the ways by which geographic information is presented. Digital maps in the form of screen display (soft copies) have gradually replaced printed maps (hard copies) as the predominant method of viewing geographic information. Instead of the conventional static view of the paper map, geographic information can now be presented dynamically on computer screens. With the aid of appropriate application programs, geographic information can also be presented as a *perspective view* (a representation of an area observed from an oblique angle), an *animation* (a dynamic sequence of maps), a *fly-by* (a virtual reality representation of the real world, sometimes also called a *fly-through*), or a *simulation* (a *what-if* scenario that is not a representation of the real world). In addition, geographic information contained in digital maps can be abstracted or converted relatively easily into other forms of information products such as graphs, charts, statistical tables, and written reports.

From the perspective of functions, digital maps also differ considerably from printed maps in several ways. Conventionally, maps served a triple function simultaneously as a *data store*, a *data carrier* (i.e., medium of *data transmission*) and a mechanism for *information presentation*. In GIS, these functions are separated from one another. Computer databases are now the data store. Magnetic and optical media, and increasingly the computer network, have become the primary data carriers. This has left information presentation as the primary function of the map in the GIS environment. The removal of the data store and data carrier functions has effectively freed the map from many of its conventional design constraints and, consequently, made more room for new concepts, techniques, and media to be introduced in both its production and use (Clarke, 1995).

Central to the renewed function of information presentation of maps are two important and interrelated concepts: *human–computer interaction* (HCI) (see Section 7.3.1) and *visualization* as implied in the term *"scientific visualization"* (SciVis) (see Section 7.4.1). Within this new conceptual framework, the presentation of geographic information is dependent heavily on computer graphics, image processing, high-performance computing, and other areas of computer technology. This leads us to what can perhaps be described as the most important change of the nature of maps in GIS: the map as an *interface to geographic information*. People begin to realize that the efficient use of GIS relies not only on how geographic data are modeled and structured in the database, but also how users interact with the computer. While data modeling provides the conceptual construct that optimizes the *structuring* and *storage* of geographic data, human–computer interaction provides a working environment that optimizes the *communication* and *presentation* of geographic information.

As GIS applications become more sophisticated and interactive, the map is playing an increasingly important role in the human–computer interaction process. In this capacity, the map serves as the spatial index for accessing the geographic database. It provides the mechanism for the visualization of the results of spatial analysis. It also provides the means for users to view the intermediate results of spatial analysis and, on the basis of interpreting these results, to interactively change the parameters of the algorithms of spatial analysis. This capability, which is usually referred to as *computational steering*, is instrumental in making the map an integral part of spatial data analysis and modeling activities. This new function of the map as an interface to geographic information allows users to *communicate geographically* with one another. It is one of the cornerstones of the new model of GIS information communication to be developed in the next section.

7.2.3 THE MODEL OF GIS INFORMATION COMMUNICATION

The concept of communication is not new to cartography. Since the 1960s, the paradigm of communication has been a key element in the theories of cartography (Taylor, 1991). In the most basic form, a *model of cartographic communication* represents the way an average person views the assimilation of geographic information through the use of maps (Figure 7.1a). The source of information is the real world. The cartographer, acting as the encoder of information, employs a variety of techniques to represent selected spatial features in a two-dimensional space on a map. When the map is used, the map reader decodes the information by interpreting the meaning of and interrelationships among the spatial features depicted. In this model of communication, the map is basically a *data carrier* and the map user is a *passive receptor*.

The changing nature of maps in the GIS environment means that cartographic communication has to take a new form and a new importance (Figure 7.1b). As the context of information communication has changed from the use of the conventional map to the digital map in GIS, this new model of information communication should be more aptly called the *model of GIS information communication*. This model summarizes the nature of maps and their use within the GIS environment. It provides the conceptual framework for developing the methods of presenting geographic information to be discussed later in this chapter.

The characteristics of the model of GIS information communication can be best explained by comparison with the conventional model of cartographic communication. Although the word "communication" is used for both the conventional and the new models, it actually

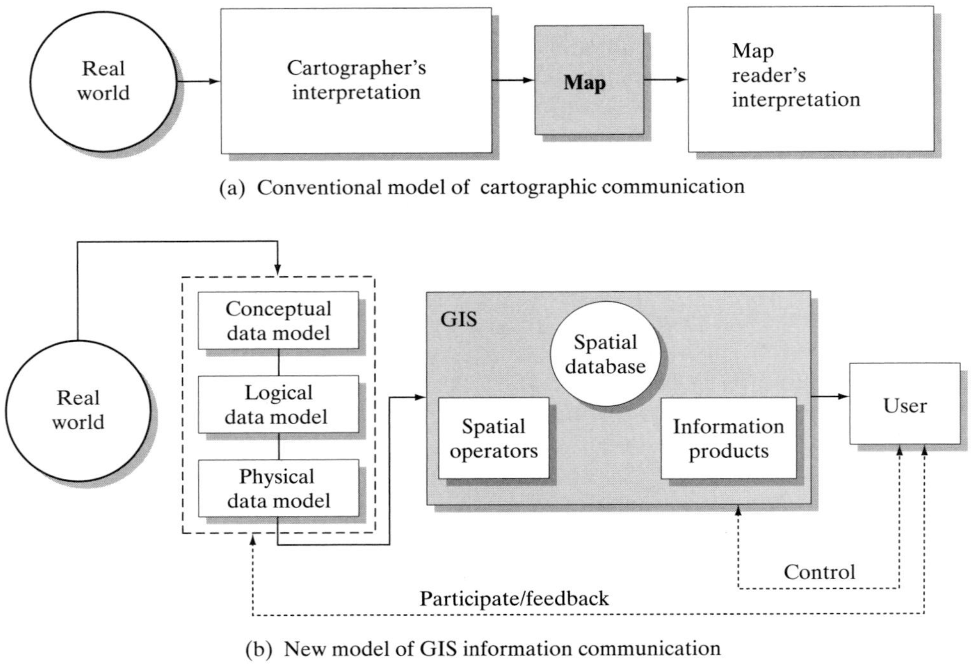

(a) Conventional model of cartographic communication

(b) New model of GIS information communication

FIGURE 7.1
Models of cartographic and GIS information communication. The new model of GIS information communication is different from the old model of cartographic information communication in terms of mode, objective, and scope of communication, as well as the level of interaction between the user and the map.

represents two very distinct modes of communication. As a rule in conventional cartography, an average map user had no control over the information contained in maps; he or she is simply a consumer of the information. The mode of communication is therefore largely a one-way flow of information. In contrast, information communication in GIS is multidimensional in the sense that information can be presented three-dimensionally, dynamically, and temporally. As the human perception of multidimensional electronic images is quite different from that of the static views of printed maps, GIS information communication is more sophisticated than conventional cartographic communication, both in concepts and methods.

GIS information communication can be described as a *user-oriented* approach, as opposed to the apparently *cartographer-centered* approach of the conventional cartographic communication model. Unlike map users, GIS users in general play a more active part in the information communication process. The role of specialist GIS users is particularly important. At different stages of the systems development life cycle of a GIS, these users help define the user's view in data modeling, create the geographic database, manage the system, and provide technical support to other classes of users (see Chapter 1). As maps in GIS can be generated on demand, they can be specifically customized to meet the application requirements of a particular user or group of users. Consequently, the map producer and the map user have a much more intimate relationship in GIS than in conventional cartography.

The objectives of GIS information communication also appear to be more diverse than those of conventional cartographic communication that aim almost exclusively to the provision of information about location, distribution, and pattern. With the aid of sophisticated data analysis tools, digital maps in GIS may be applied to many new application areas that are not possible with conventional paper maps. These include, for example, real-time decision support, animation of spatial processes, spatial simulation, and predictive modeling, as well as integration with nonspatial statistical analysis. Therefore, the new model of GIS information communication is different from the conventional model of cartographic communication not only in terms of complexity, but also in terms of scope as well.

7.3 Human–Computer Interaction and GIS

Human factors (also referred to as *ergonomics*, or *human engineering*) is a field of study that is concerned with human responses to the use of technology. *Human–computer interaction* (HCI) is a branch of human factors that deals particularly with the ways people use computers and how they can use computers more effectively. Although the primary focus of human–computer interaction is the interface between the user and the computer, the realm of human–computer interaction also includes the psychology of using machines, training requirements, workplace organization, and occupational health. In this regard, human–computer interaction is of interest not only to computer scientists and system developers, but also to psychologists, industrial engineers, human resource specialists, occupational health professionals, and practitioners in various fields of science and engineering. Human–computer interaction now plays a very important role in all aspects of the design of information systems, including GIS.

7.3.1 THE BEHAVIORAL APPROACH OF HUMAN–COMPUTER INTERACTION

There are different approaches to human–computer interaction (Gould, 1989). The prevailing approach is the *behavioral approach* (also called the *cognitive science approach*). This approach of human–computer interaction attempts to accommodate the interaction between the user and the computer as closely as possible to human behavior and thinking (Egenhofer and Frank, 1988). It encourages the design of systems that can be used efficiently with the minimum amount of training (which in turn implies the minimum need to modify the user's behavior). This is achieved by using three interrelated software design techniques.

1. *what-you-see-is-what-you-get* (WYSIWYG), which ensures that the interface always reflects the true state of the system
2. *direct manipulation*, which enables the user to use a pointing device, such as a mouse, to select an object on the screen, drag it across the screen, or click on it to activate a process
3. *graphical user interface* (GUI), which ensures graphical and operational consistency across different applications by using commonly accepted *widgets* (icons and menus) and *metaphors* (environments for doing things)

These three design concepts together make possible the development of an intuitive *windows-icon-menu-pointer* (WIMP) environment that allows the user to interact with the computer with the minimum need to learn and memorize any command syntax. There are many advantages to adopting the behavioral approach of human–computer interaction in GIS design. Application development by the behavioral approach of human–computer interaction involves the user in the iterative design process. It is therefore compatible with the *rapid application development* (RAD) approach of software construction that is now widely used in the computer industry (see Chapter 11). Adherence to the methods of the behavioral approach of human–computer interaction removes the barriers between the GIS user and the system. It helps popularize GIS technology because applications are easier to use despite the growing sophistication of the applications. As the dominant approach to human–computer interaction in the computer industry, the behavioral approach allows GIS to be integrated more easily with the mainstream enterprise computing environment. It also improves the overall productivity of using GIS by cutting down training requirements and by allowing users to apply their experience in other software packages directly to the use of GIS.

7.3.2 PRINCIPLES OF THE GRAPHICAL USER INTERFACE

The implementation of the behavioral approach to human–computer interaction depends on the use of *graphical user interfaces* (GUI). In its most generic form, a *user interface* is an application program that is designed to facilitate the communication between the user and the computer. In the early days of electronic data processing, the interface between the user and the computer took the form of *job control language* (JCL) commands on punch cards. When the keyboard and cathode ray tube (CRT) monitor were developed to replace punch cards as the mode of program and data input, the *command line interface* (CLI) became the standard method of communication between the user and the computer. Both the JCL and CLI were text-based interfaces.

The concept of GUI was pioneered by Xerox's Palo Alto Research Center (XeroxPARC) in California in the early 1980s (Sheldon et al., 1991). In essence, a GUI is an audiovisual metaphor on the computer's monitor representing the actions that the computer can perform. The means of interaction between the user and the computer is by the WIMP paradigm. The mouse is the pointing device for activating the processes represented by icons and options in pop-up or pull-down menus in one or more windows (Figure 7.2). The GUI for the Apple Macintosh, released in 1984, was the first one to become widely used (Apple Computer, 1987). The GUI method was quickly adopted by Microsoft, IBM, and other software vendors. It is now the standard method of human–computer interaction for PCs and UNIX workstations.

Several graphical user interface standards have been developed for the PC platform, including Microsoft *Windows*, IBM *OS/2 Presentation Manager*, Amiga *Workbench*, and a few others. Microsoft Windows is now the predominant

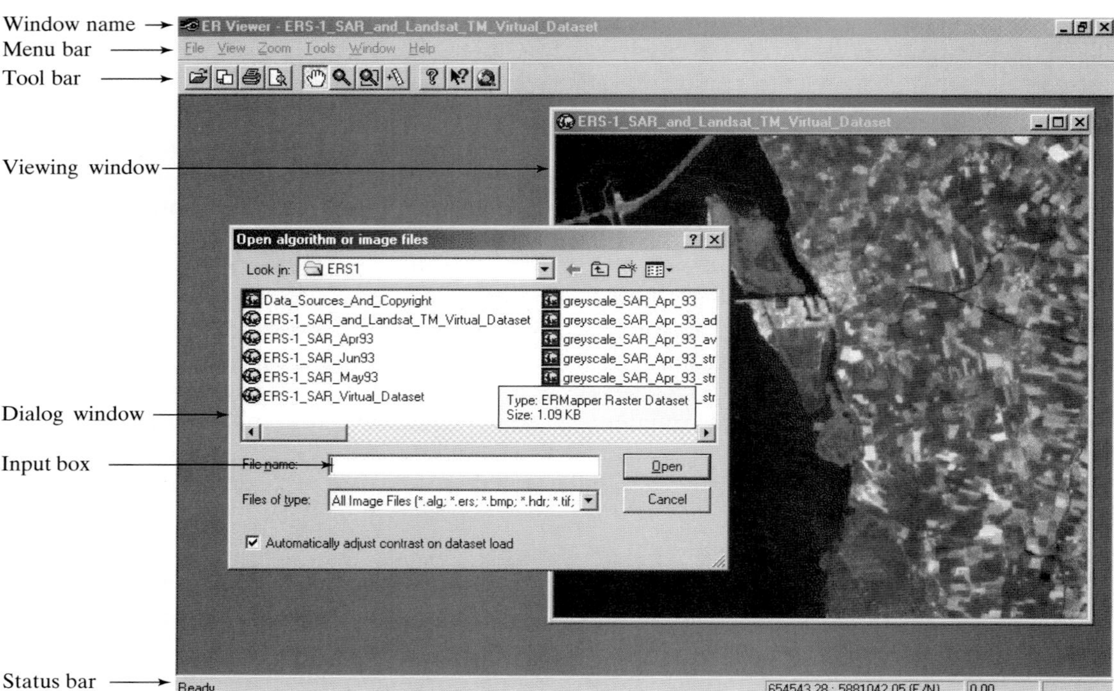

FIGURE 7.2
Features of a typical graphical user interface.

GUI for desktop applications. In the UNIX arena, the *X-Windows* system is the predominant windowing system. Unlike the Microsoft Windows system, the X-Windows system only provides the mechanism to enable the user to open windows and work with information between the keyboard, the mouse, and the computer. The actual look and feel of the graphical user interface is provided by one of two popular standards: *Open Look* and *Motif*. Open Look was jointly developed by Sun Microsystems, AT&T, Xerox, and others, whereas Motif was developed by the Open Software Foundation (OSF). The windows and icons of these two standards look the same on the screen but the controls and appearance of the two GUI environments (e.g., border, title line, menus, scroll bars, and message line) are different.

A typical graphical user interface is made up of three components: the *windowing system* that is a set of tools for creating windows and defining their characteristics, such as the X-Windows system; the *windows manager* that controls the drawing of the screen representation such as fonts and icons; and the *application program interface* (API) that acts as a link to a particular program and controls the feedback from its execution. As common user interactions, such as deletion and saving of files, have already been built into the GUI, it is not necessary for the same functions to be coded for individual applications. This gives the interfaces for different applications the same look and feel. Since these common user interactions are independent of a particular piece of hardware (e.g., printing device), they contribute to the concepts of resource sharing, system integration, and open systems design in application development.

The look and feel of graphical user interfaces is made consistent across different applications by adopting a standard set of elements for the graphical panel (i.e., title bar, option menus, scroll bar, and slider), a standard set of icons for the common procedures (e.g., opening, saving, printing, and copying files; cut-and-paste in editing; and the selection of font types and sizes), as well as standard window formats for pop-up and pull-down menus. At the application level, standard methods are used for manipulating the mouse, defining the sizes of viewing windows, and the selection of individual elements in textual and graphical documents. To cater to the differences in user profile and application requirements, the user is always allowed to exercise some degree of flexibility in organizing the GUI. This includes the display and hiding of particular icons on the screen, positioning of icons, and the creation of special interface elements by scripting or macro programming. For systems such as GIS where specific data-processing requirements cannot be always met by the generic application interfaces, this ability to customize the GUI is particularly important. It allows the graphical user interface to be tailored to the specific needs of a particular application while at the same time maintaining the overall consistency of the appearance of the interface.

7.3.3 GRAPHICAL USER INTERFACE FOR GIS

The graphical user interface is of great importance in using GIS. In the past, GIS with command line interface was invariably difficult to use. A considerable amount of training was required in order for a user to master the hundreds of commands and data-processing procedures in a typical GIS. Such a requirement was not practical for general users who were interested in using GIS only for specific applications but not in the general functionality of these systems. It was also a formidable task to train the growing number of desktop users whose only requirement was to use GIS for occasional spatial data viewing and query. These two classes of users had to rely on the specialist users (i.e., GIS managers, analysts, programmers, and application specialists) to provide the necessary support in their use of GIS applications. The difficulty of learning to use GIS on the command line mode was one of the major factors that deterred more people from using them before the advent of the graphical user interface.

In general, the majority of GIS users have very little interest in the structure of the database or the detail of the procedures required to complete a particular spatial analysis task. What they want is a user-friendly interface that allows them to perform a complex spatial analysis or query with minimum training. Today's GIS are practically all designed with this user expectation in mind, and as a result, the command line interface was quickly replaced by the graphical user interface. By using a combination of icons and menus, GIS is now much easier and more intuitive to use (Figure 7.3). The general user has largely been freed from the need for extensive training and can spend the time on solving the spatial problems instead. The ease of use with the graphical user interface definitely plays a significant role in popularizing GIS in the workplace as well as in homes.

The advent of the Internet brought a new dimension to the user interface of GIS. Internet-based GIS applications range from simple browsing of predrawn maps on a Web page, through interactive map generation,

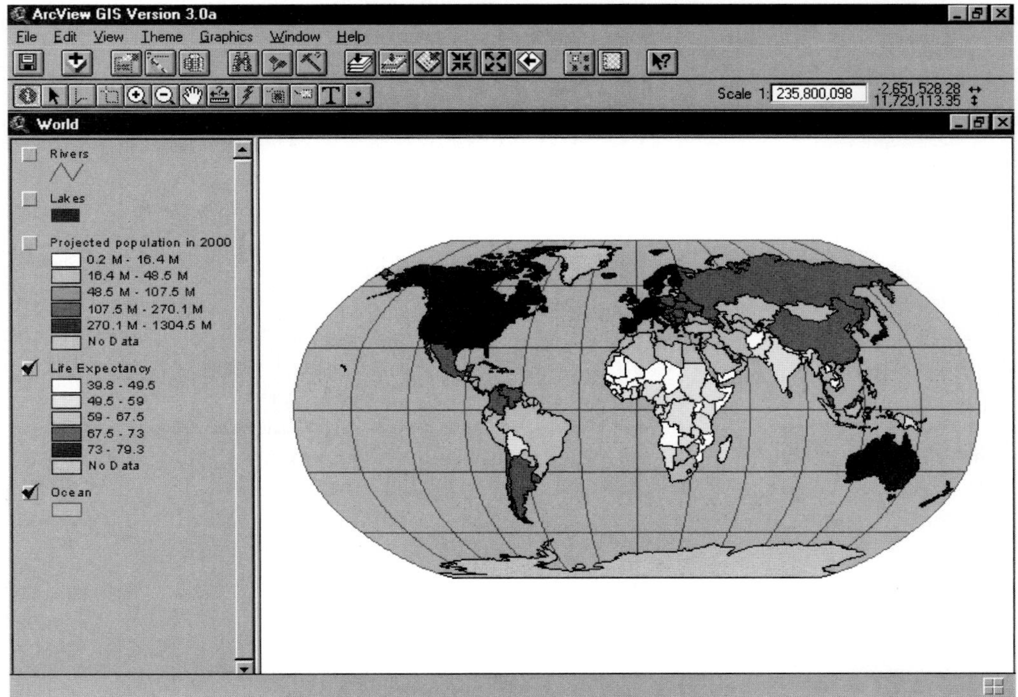

FIGURE 7.3
The graphical user interface (GUI) of ArcView GIS. ArcView GIS and similar packages are specially designed for use as a front-end application to display results of spatial analysis and modeling. They have very powerful data presentation capabilities that allow the user to interactively select and hide different layers of geographic information.

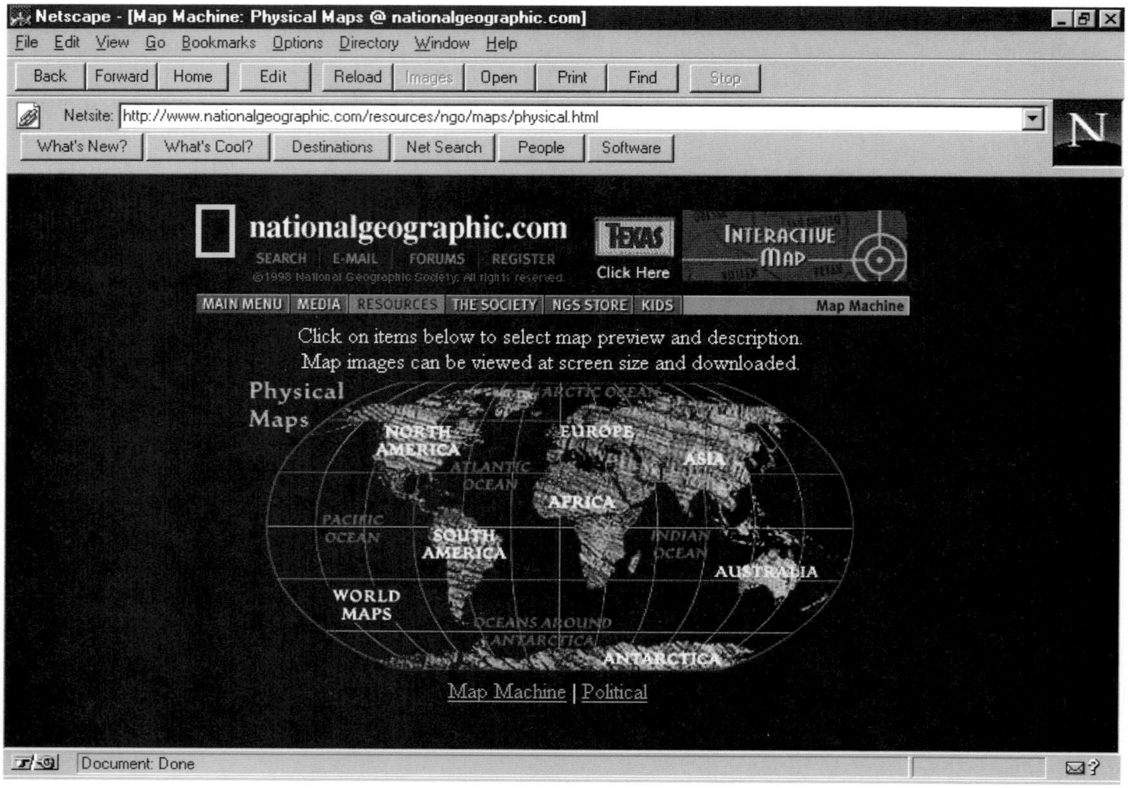

FIGURE 7.4
Graphical user interface of a distributed geographic information (DGI) application. A DGI interface allows the user to interact with a remote geographic database across the Internet (see Chapter 12 for further explanation of the nature and type of DGI applications). *(Source: National Geographic Web site)*

cartographic animation, spatial metadata search, to complex network-based collaborative GIS processing in which users at remote locations share common data and operations with one another in real time (Plewe, 1997). Such applications use the generic functions of the Web browser as the graphical user interface (Figure 7.4). Based on the client/server architecture and hypertext technology, distributed geographic information applications are highly interactive in operation. They are usually intended for a relatively large number of users, many of whom are nontraditional GIS users with different expectations of using the technology. The user interfaces for Internet-based GIS applications, therefore, have very different requirements from those for conventional GIS applications.

New software technology has now made possible the development of customized and integrated user interfaces that can concurrently access geographic data stored in the local memory and across the Internet. *ArcExplorer*, a product of ESRI (1998), is an example of such applications (Figure 7.5). When connected to the Internet, ArcExplorer can be used as a Web client application capable of browsing and downloading vector geographic data in the *Shapefiles* format (which is the file format for ArcView). Once the data have been downloaded, ArcExplorer becomes a local application interface for data visualization, spatial query, and map composition. By using the technologies employed to create ArcExplorer (i.e., Visual Basic and MapObjects components), similar custom-built GIS applications can be developed with Microsoft Windows-compliant graphical user interface (Liu et al., 1997; Luo, 1997; Ofenstein and Rauenzahn, 1997). User interfaces developed in this way can be designed to precisely meet the specific needs of a particular user group. They help popularize GIS by eliminating the usual heavy capital investment required for implementing a full-blown GIS. From the perspective of GIS development, these user interfaces represent the progression toward what can be described as *user-oriented GIS*.

7.3.4 GIS INTERFACE RESEARCH

The design of GUI is an important research topic in GIS (Mark, 1992). Earlier work in GIS user interface research focused mainly on graphical user interface design for efficient spatial data handling. As the WYSIWYG paradigm of user interface design was gaining popularity in the

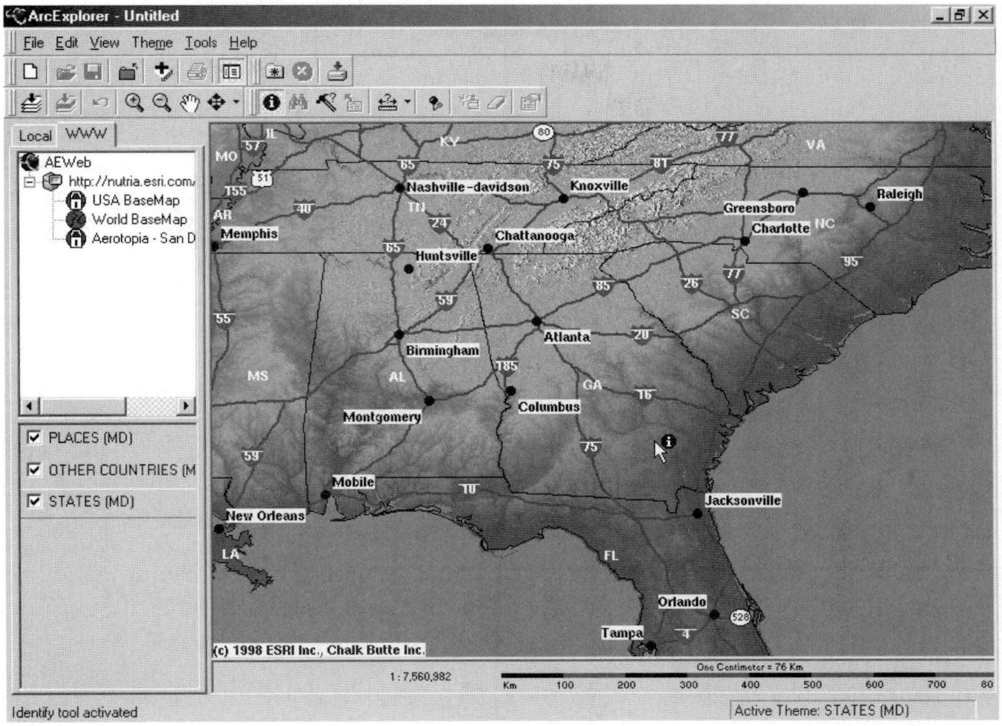

FIGURE 7.5
Graphical user interface of ArcExplorer. *(Source: ESRI)(ESRI Graphical User Interface is the intellectual property of ESRI and is used with permission. ©ESRI)*

computer industry, several GIS researchers started to look at the possibility of extending this concept to GIS. In a review of human–computer interaction as applied to GIS, Gould (1989) suggested that it was possible to improve GIS user interface design in three potential areas: (1) enhancement of the relatively simple process-response mode of graphical interaction to a more advanced mode of visual thinking, (2) the ability to handle fuzzy queries by parsing natural languages, and (3) the use of multimedia techniques of sound and video to construct spatial queries and responses.

All of these areas of potential improvement have received considerable attention in GIS user interface research in recent years. As a result, the emphasis of research gradually moved away from basic spatial data handling to include spatial query languages with particular reference to the suitability of Structured Query Language (SQL) as a GIS interaction language (Egenhofer, 1992; Egenhofer and Frank, 1988), cognitive structuring of space (Mark, 1989), the map overlay metaphor and direct manipulation using the method of visual and iconic programming (Egenhofer and Richards, 1993; Lee and Chin, 1995), as well as the requirements of user interfaces for special-purpose applications such as spatial decision support systems (Armstrong et al., 1991).

7.4 Visualization of Geographic Information

Cartographic cognition is the process by which the human brain recognizes spatial patterns and relationships. This is a mental process that was difficult to replicate by software in GIS (Taylor, 1991). Advances in the concepts and techniques of *scientific visualization* in the last few years have provided GIS users with the necessary tools to explore salient patterns and relationships in geographic data that are otherwise hard to comprehend. At present, scientific visualization is an important research topic in human–computer interaction as it is applied to GIS.

7.4.1 The Origin and Scope of Scientific Visualization

Scientists in many disciplines have a long tradition of using the method of *visual data analysis* in their research. The objective of visual data analysis is to use graphics to assist in data exploration and the development of ideas in scientific investigation (Unwin et al., 1994). Visual data analysis involves a suite of graphics-based techniques for interpreting the complex relationships in large volumes of data that are impossible by the use of mathematical formulas and equations. The term *scientific visualization* (SciVis), which is also commonly referred to as *visualization in scientific computing* (ViSC), has grown out of this established approach in scientific research (McCormick et al., 1987). New developments in data-acquisition technology and mathematical modeling have produced huge quantities of data pertaining to Earth's physical and biological environments that call for advanced techniques to transform them into useful information. This accounts for the spectacular growth of SciVis as an indispensable tool for data exploration in various branches of science and technology today.

The concept of *visualization* as implied in scientific visualization embraces three elements (Buttenfield and Mackaness, 1991): *computation,* which denotes visualization as a method of computing, a set of hardware and software tools, as well as the mechanisms that facilitate human–computer interaction; *cognition,* the human ability to develop mental representations, identify patterns, and create orders in data analysis; and *graphics design,* which includes the construction of visual displays using the principles of graphics communication. This concept of visualization is obviously much broader than the commonly held idea that visualization simply means "making visible" (MacEachren, 1995). It has been adopted by the cartographic community to explain the relationship between representation methods, spatial understanding and analysis, and the use of digital technologies in modern cartography (Figure 7.6) (Taylor, 1994).

The emphasis of visualization is on the *exploratory analysis* of information by means of symbolic-to-geometric transformation (i.e., turning *nonvisible* digital data into *visible* graphical representations), interactive data manipulation and display, graphical rendering, as well as time-series animation. From the perspective of GIS, the objective of visualization is to produce visual images that can effectively exploit the ability of the human visual system to identify spatial patterns and processes in spatial problem solving and decision making (Robertson, 1988). Defined in this way, visualization basically excludes standard GIS functions such as map creation and spatial data viewing because no element of data exploration is involved.

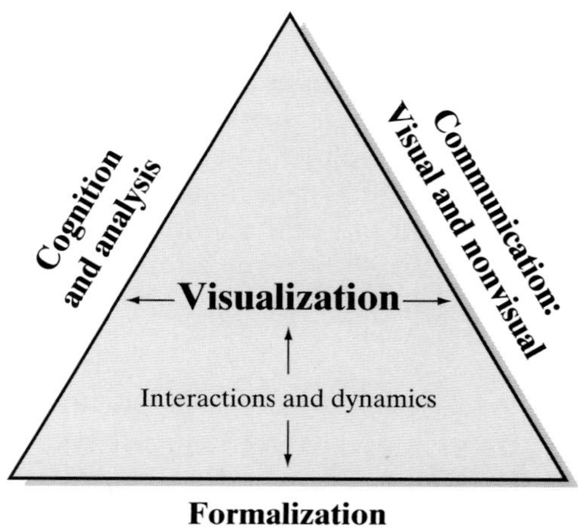

FIGURE 7.6
Concept of visualization in the context of cartography. *(Source: Taylor, 1994)*

The use of visualization can be divided into two categories (Gordin et al., 1996): *interpretive* and *expressive.* In the interpretive use, the user is basically a "reader" who is attempting to extract the meaning of the data by visualization (i.e., the cognition aspect of visualization). In expressive use, the user is an "author" who is attempting to convey the meaning of the data through visualization (i.e., the communication or graphics design aspect of visualization). In practice, however, these two uses of visualization are complementary with, rather than exclusive to, one another. The process of visualization usually starts with interpretative use by displaying the data through color coding to reveal gradations or fluctuations. The image (or the *scene,* as it is sometimes called) is then superimposed on objects such as maps (to obtain spatial references) and/or displayed in three dimensions (to make it resemble the natural three-dimensional world for easier interpretation). If necessary, images showing intermediate results at different stages of data analysis can be produced. This allows the user to understand not only the final outcome of data analysis and modeling, but also the characteristics and behavior of intermediate processes. It also enables the user to identify features that are critical to the final outcomes and the factors that can serve as predictors of later events. What is more, by varying the values or relative weightings of the factors in the analysis, the user can obtain and visually compare the different scenarios that may occur as a result. Once the interpretive visualization processes are completed, the user then applies expressive visualization techniques (e.g., using graphs, three-dimensional diagrams and animation) to communicate the findings to others. In this way, visualization is perceived as the core technology that has made possible the use of the digital map as an interface to geographic information. It is the key mechanism by which GIS users can communicate geographically with one another as noted in the model of GIS information communication (see Section 7.2.2).

In essence, visualization is a technology-driven computer application. As the technology evolves, the scope of visualization changes correspondingly. Consequently, visualization as we know it today is very different from what it was ten years ago. From the perspective of technology, it is now possible to compress a large amount of data into one single visual image, view the data interactively in real time, change the simulation parameters and obtain the results immediately, and perform highly complex time-dependent simulations in a distributed computing environment. From the perspective of applications, visualization techniques are now increasingly used as an operational tool in environmental and resource management, in addition to their traditional role as a data exploration tool in scientific research (Buckley et al., 1998). This has led to more stringent demands on the quality of the geographic data used for visualization. At the same time, it also calls for more sophisticated visualization capabilities to be built into current GIS software products.

7.4.2 TECHNIQUES FOR VISUALIZATION OF GEOGRAPHIC INFORMATION

Visualization of geographic information can be done using a variety of techniques. Masuoka et al. (1996) identify five categories of visualization techniques, namely, (1) *two-dimensional plots*; (2) *three-dimensional plots*; (3) *two-dimensional planimetric views*; (4) *three-dimensional perspective views*; and (5) *animation*. These techniques can be used in different combinations, depending on the objectives of the applications and the types of data being visualized. The characteristics and areas of applications of these techniques are given here and in Figure 7.7.

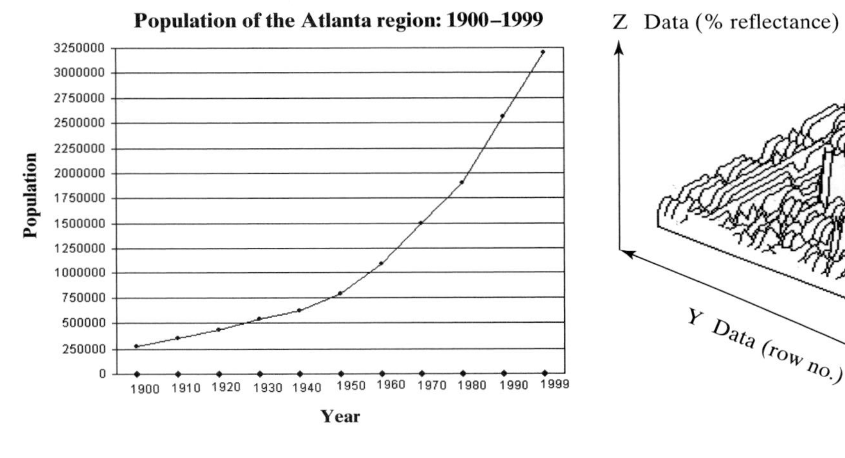

(a) 2-D Plot

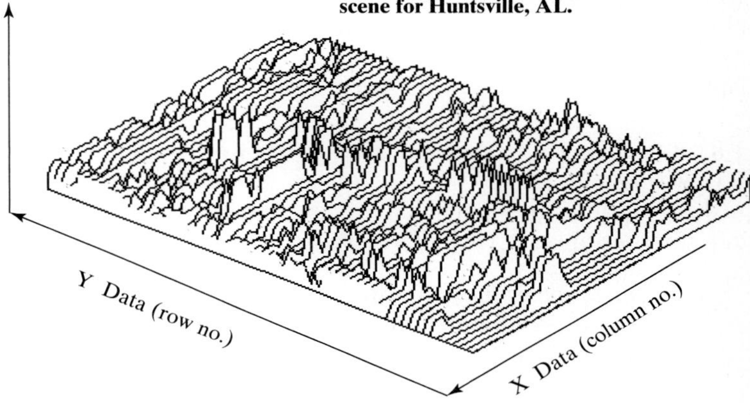

(b) 3-D Plot

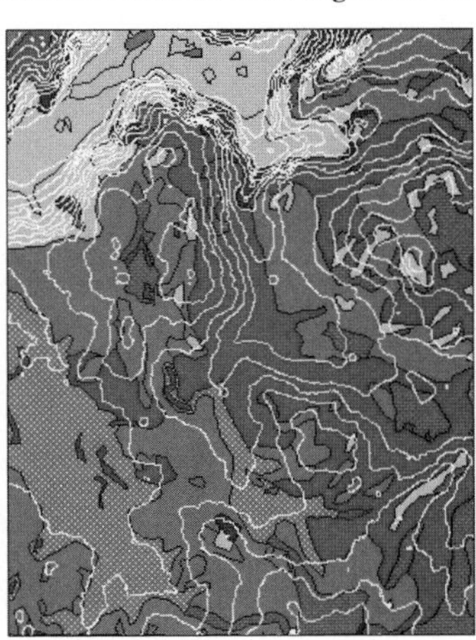

(c) 2-D Planimetric view

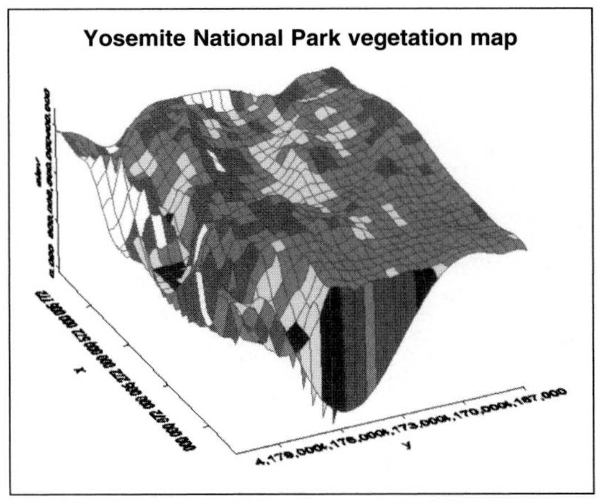

(d) 3-D Perspective view

FIGURE 7.7
Examples of four techniques for visualization of geographic information. ([c] and [d] Courtesy of Insightful Corporation, formerly MathSoft)

Two-Dimensional Plots

These plots are relatively simple to construct and are useful for visualizing the relationship between two numerical variables (e.g., population against time as shown in Figure 7.7a for the city of Atlanta).

Three-Dimensional Plots

These plots (sometimes referred to as *surface plots*) are used to visualize the relationships among three numerical variables (e.g., Figure 7.7b, which shows the percentage of reflectance of the land cover in a Landsat MSS scene at each pixel position for a part of Huntsville, Alabama).

Two-Dimensional Planimetric Views

A two-dimensional planimetric view is the technique used in conventional cartographic visualization. Spatial variation and patterns resulting from data analysis can be effectively depicted using different colors and symbology (e.g., Figure 7.7c, which shows a map of vegetation laid on top of the contours in Yosemite National Park).

Three-Dimensional Perspective Views

A three-dimensional perspective view can be created in a number of ways (see also Table 7.1).

- *Geometric modeling* generates views ranging from simple three-dimensional perspective drawings of the landscape to relatively complex wire-frame models capable of showing local details at large scales.
- *Video imaging* produces photorealistic quality views that can faithfully depict the conditions of the landscape.
- *Geometric video imaging* is a hybrid approach combining geometric modeling with video-imaging techniques (so that it can take advantage of the relative merits of the parent techniques, i.e., ability to depict small changes and high image quality).
- *Image draping* is a well-established function in GIS and image-processing software, involving the draping of an image, such as a digital orthophoto or a classified satellite image, onto a three-dimensional perspective view created by geometric modeling. Figure 7.7d is a graphic representation in three dimensions of Figure 7.7c, that is, with the vegetation draped on top of a digital elevation model (DEM).

Animation

Animation is the computer graphics technique for visualizing time-dependent spatial data in sequence. Two-dimensional planimetric views have been most commonly used for animation. However, two-dimensional and three-dimensional plots can also be usefully employed to visualize temporal changes of spatial data. When an animation sequence is produced in three dimensions, it is called a fly-by or fly-through. In its most sophisticated form, an animation sequence can be visualized in *virtual reality* (VR, also called *virtual world* or *virtual environment*).

7.4.3 APPROACHES TO VISUALIZATION OF GEOGRAPHIC INFORMATION

Various approaches to visualizing geographic information have been developed, with different degrees of sophistication and complexity in the underlying technologies. These approaches include (1) using the generic display function of GIS software, (2) using GIS in con-

Table 7.1
Comparison of Characteristics of Visualization Techniques

Visualization Techniques	*Data Requirements*[1]	*Level of Realism*[2]	*Operational Complexity*[3]	*Data Integrity*[4]
Geometric modeling	High	Low to moderate	Moderate to high	High
Video imaging	Low	High	Moderate	Low to moderate
Geometric video imaging	Moderate to high	High	Moderate to high	Moderate to high
Image draping	Low to moderate	Moderate	Moderate to high	Moderate

Notes: [1]High data requirements means either large volume or detailed data are needed. [2]High level of realism refers to photographic or near-photographic quality. [3]High operational complexity means that the technique is difficult to use. [4]High data integrity refers to the ability of the technique to represent small changes.
Source: McGaughey, 1997.

junction with CAD systems, (3) linking GIS to statistical software, (4) linking GIS to visualization software, (5) using stand-alone spatial visualization software, and (6) dynamic visualization via the Internet. The characteristics of these approaches, together with typical examples of applications, are summarized as follows.

Using the Generic Display Function of GIS Software

Presentation of spatial information is a generic function of GIS software. In general, however, this function is not adequate for the purpose of visualization as defined in Section 7.4.1. The display function of GIS software is designed primarily for static two-dimensional views. Three-dimensional display capability is either absent or very limited. The capability to represent flows or movements, another essential requirement for visualization, is also largely unavailable. However, these constraints of the generic display function do not mean that it cannot be used for visualization. As Batty (1994) has demonstrated, it is possible to use macro programs, such as the Arc Macro Language (AML) of ArcInfo, to develop very powerful visualization applications (Figure 7.8). In this approach, the display function of the GIS (which in Batty's example is ArcPlot of ArcInfo) is used to display maps quickly and efficiently. The macro programs serve as the link between the graphics and the modeling routines. The advantages of using GIS include the use of its own data sources (and therefore removing the need for data conversion), the generality of the user interface, as well as the flexibility in coupling with the data analysis and modeling modules (Batty and Xie, 1994). The major drawbacks to such an approach include the limitation of the generic display function to two-dimensional static views, as well as the requirement for the user to have access to a high level of proficiency in macro programming.

Using GIS in Conjunction with CAD Systems

Visualization for landscape analysis and terrain modeling requires three-dimensional capability that is seldom offered as a built-in function of GIS software. Some GIS packages offer software extensions for processing and displaying three dimensions (e.g., the triangulated irregular networks [TIN] extension of ArcInfo). However, the application of the extended function is usually subject to a variety of constraints, including the processing required for generating the triangulated irregular networks, the vertical distortion that must usually be applied to the z-dimension, and the inability to place three-dimensional objects on the model

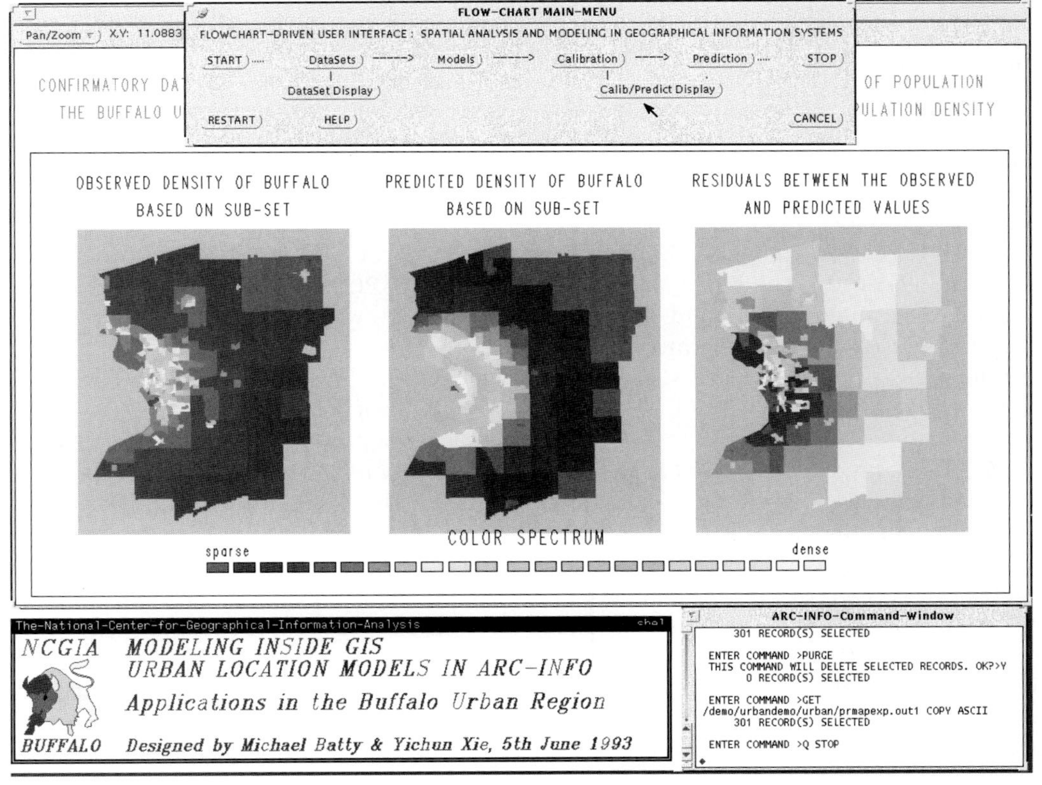

FIGURE 7.8
Visualization using generic display function of GIS. *(Courtesy: Michael Batty, University College London)*

FIGURE 7.9
Landscape visualization using GIS and CAD. *(Courtesy of Prof. G. B. Hall, University of Waterloo)*

surface (Mayall and Hall, 1994). One practical way to make up for the deficiency of GIS for three-dimensional modeling is to take advantage of the three-dimensional modeling capability of CAD systems. For example, Mayall and Hall (1994) propose a method that combines GIS and CAD technologies for visualizing two-dimensional landscape data in three-dimensional scenes (Figure 7.9). This method may be enhanced for visualizing actual changes in the landscape, as well as the impacts of changes in building codes and engineering standards on the landscape. Predictive modeling of changes in economic behavior, ecological succession, and wave erosion may also be implemented.

Linking GIS to Statistical Software

This approach combines the use of a GIS and a statistical software package. In such a system, the GIS serves as the graphical front end for the display of visualization data and the statistical package is the data exploration engine. Logically in between these two components is an *application programming interface* (API) that is established to facilitate the transfer of data and commands between them. A commercial product built on the basis of this approach is *S+ ArcView for GIS* developed by MathSoft (1998) (Figure 7.10). This package is based on S-PLUS, a popular statistical package for exploratory data analysis and modeling of three broad classes of spatial data: geostatistical data, point patterns, and lattice data. Using an interface known as S+API, the statistical and graphical functions of S-PLUS are integrated with the spatial visual capabilities of ArcView. The user can seamlessly export ArcView data to S-PLUS for spatial statistical analysis and then return the graphics and analytical result to ArcView for display. Within the ArcView environment, the user can also directly access S-PLUS objects to obtain information to plot a map with residual value, add a table, or add an event theme.

Linking GIS to Visualization Software

A variety of scientific visualization tools have been developed by the research community and made available to the public (Buckley et al., 1998). While many of these tools provide basic three-dimensional rendering capabilities, most are not tightly integrated with GIS software that will enable them to use georeferenced data in the operational environment in, for example, forest management. *Virtual Forest*, developed by Innovative GIS Solutions, Inc., is one of the attempts to integrate the advanced rendering capabilities of SciVis software with the geographic data management functions of a GIS for forestry applications (Buckley et al., 1998). This product is based on the use of ArcInfo as the repository of geo-

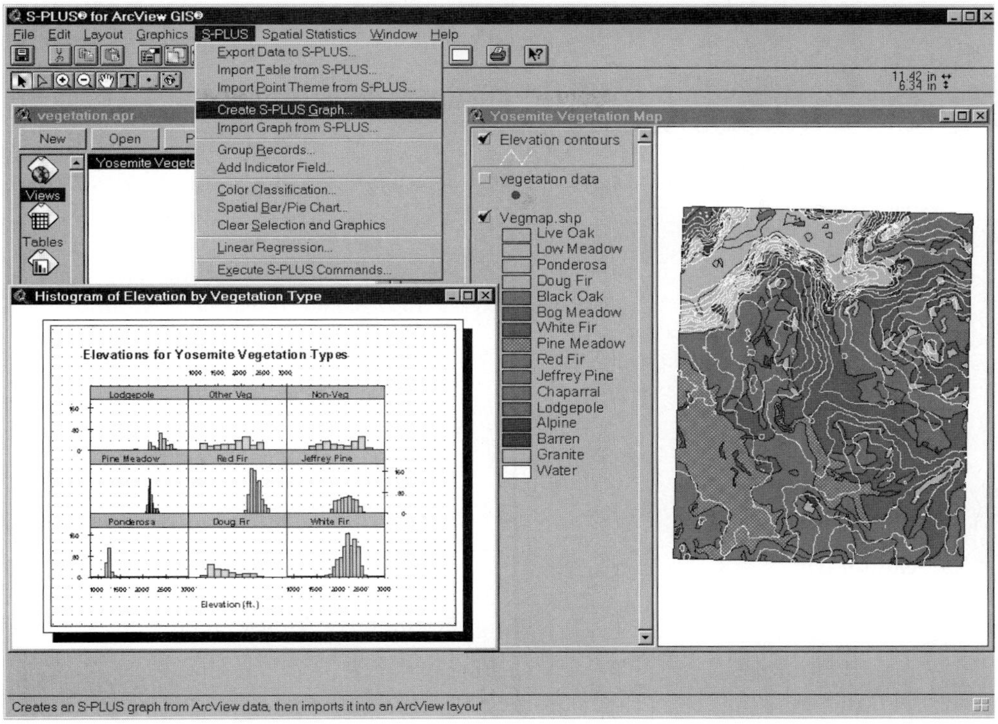

FIGURE 7.10
Visualization by linking GIS (ArcView) to a statistical analysis package (S-PLUS). *(Source: Insightful Corporation, formerly MathSoft)*

graphic data that is tightly coupled with two independent visualization interfaces: the *Tree Designer* and the *Landscape Viewer*. The Tree Designer allows the user to interactively create and edit tree objects. The Landscape Viewer, on the other hand, provides the tools for rendering three-dimensional visualization using the three-dimensional objects created by the Tree Designer. The rendering techniques provided by the Landscape Viewer include the definition of a DEM view, DEM surface texturing, tree-stand boundaries, tree rendering, the conditions of the light source and the sky, as well as atmospheric effects. Through a separate interface, Virtual Forest provides the capability for the user to define landscape views with separate themes denoting specific visualization events. This allows the user to represent multiple landscape visualization scenes such as different alternatives for harvest cutblock design, and temporal events for time-series analysis.

Using Stand-alone Visualization Software

There is now a host of spatial data visualization software packages on the market. Some of these are commercial products and others are public-domain products developed by academic institutions or government agencies. These products run on different hardware platforms and are designed for different application areas, particularly for landscape analysis in forestry management (Table 7.2). In most cases, these visualization software products are capable of generating high-quality scenes that serve the interpretive use of visualization well. However, these products in general also suffer from a number of deficiencies, such as poor or weak linkage between different applications of visualization, the absence of mechanisms for testing the validity and reliability of the visualization scenes, as well as the lack of integration with GIS data models and databases.

Dynamic Visualization via the Internet

This is the newest and most exciting development of the visualization of geographic information. Advances in computer technologies in general, and display technologies in particular, have led to the development of several prototype systems that attempt to approach functional integration between visualization and GIS. On the basis of the technologies used, these systems can be described as *virtual GIS* (Rhyne, 1997). These systems provide the means for visualizing terrain models consisting of elevation and imagery data in conjunction with GIS raster layers, protruding features, buildings, vehicles, and other landscape objects. The user is totally immersed in a three-dimensional virtual world that allows for navigation through GIS databases (Neves et al., 1997). The initial problem with these systems was the requirement for high-performance computer systems, which was prohibitively expensive and prevented the popular use of the technology.

Table 7.2
Examples of Stand-Alone Landscape Visualization Software

Package Name	*Visualization Technique*	*Hardware Platform*	*Description and Additional Information*
UTOOLS and UVIEW	Geometric modeling	PC—Windows	USDA Forest Service, Pacific Northwest Research Station; free distribution *http://forsys.cfr.washington.edu/utools.html*
SmartForest	Geometric modeling	UNIX	Imaging Systems Laboratory, University of Illinois at Urbana-Champaign, IL; free distribution *http://www.imlab.uiuc.edu/SF/SF_II.html*
Landscape Management System (LMS)	Geometric modeling	PC—Windows	College of Forest Resources, University of Washington, Seattle, WA; free distribution *http://silvae.cfr.washington.edu/lms/lms.html*
USFS Southern Research Station Visualization System	Geometric modeling	UNIX	A mix of public domain and custom software, using GRASS GIS; free distribution *http://so4702.usfs.auburn.edu/research/prob4/standviews.html*
Visual Explorer	Geometric modeling and image draping	PC—Windows	Commercial product for landscape visualization; free demonstration version *http://www.woolleysoft.co.uk*
CLRview	Geometric modeling	Silicon Graphics IRIX	Centre for Landscape Research, University of Toronto, Toronto, ON; free distribution *http://www.clr.utoronto.ca/CLRVIEW/cvmain.html*
World Construction Set	Image draping	PC—Windows	Commercial product for landscape visualization by fly-through *http://www.questarproductions.com*
Truflite	Image draping	PC—Windows	Commercial product for landscape visualization by fly-through; free demonstration version *http://www.truflite.com*

Source: Buckley et al., 1998

Recent development in the Internet provides a less expensive alternative tool, which is called the *Virtual Reality Modeling Language* (VRML) for building virtual GIS (see Chapter 9). The Virtual Reality Modeling Language is an open, extensible, industry-standard scene description language for three-dimensional scenes (which are called "worlds" in virtual reality jargon) on the Internet. With the VRML, it is possible to create and view distributed, interactive worlds that are rich with text, images, animation, sound, and video. Since the Internet is based on the client/server architecture, all CPU-intensive data analysis and rendering processes in virtual GIS can be carried out on the server side, and the resulting three-dimensional worlds can be viewed on any client computer with a Web browser that supports the VRML standard (Figure 7.11).

7.4.4 VISUALIZATION OF UNCERTAINTY IN GIS

In Chapter 4, we defined uncertainty as unknown quality of data. The presence of uncertainty is the rule rather than the exception for geographic data. GIS users have long recognized this problem, but the high cost of accuracy evaluation has forced most, if not all, users to ignore the potential impacts of uncertainty on GIS applications. Even when information of uncertainty is available, the lack of adequate reporting mechanisms has stopped the users from taking advantage of the information. However, new concepts and techniques of visualization have provided the necessary tools by which information about uncertainty can be usefully employed in practice. Since the primary form of information presentation of GIS is the visual display, the best solution to

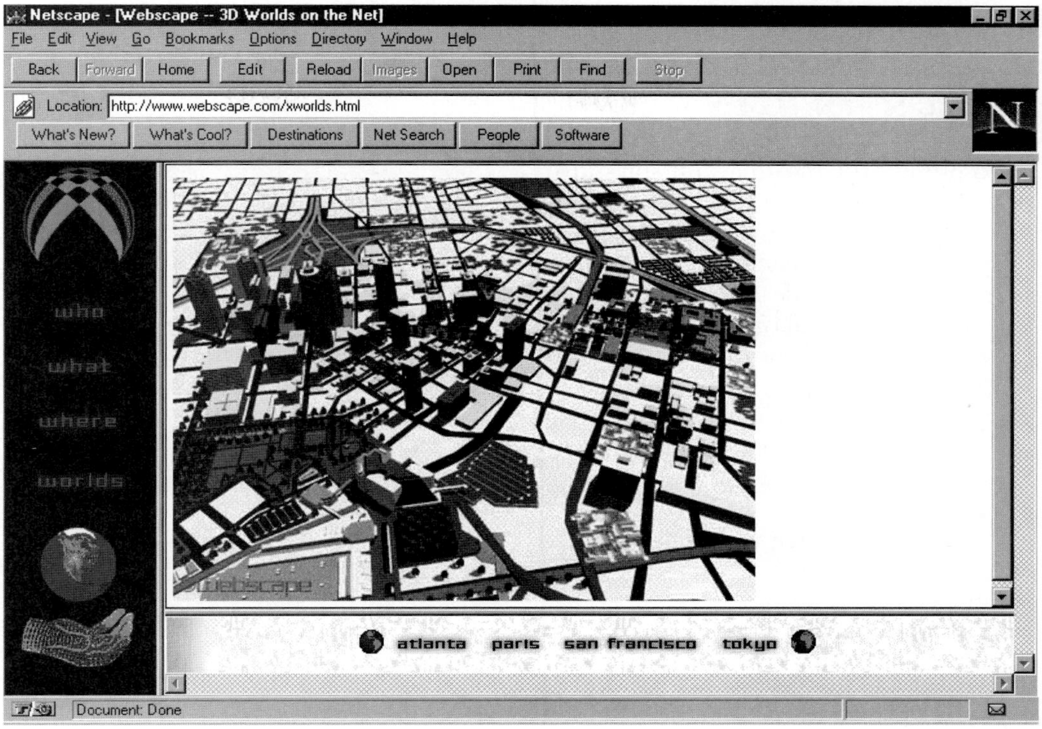

FIGURE 7.11
Presentation of geographic information by virtual reality (City of Atlanta, GA). *(Courtesy: Webscape)*

the problem of uncertainty is to use a visual approach. In general, techniques for the visualization of uncertainty may be divided into two broad classes (Davis and Keller, 1997): *static maps* and methods derived from using static maps, and *animation*. The characteristics of these two classes of techniques are explained as follows.

Using Static Maps of Uncertainty

A static map of uncertainty records the spatial variation of uncertainty across a particular area (Kraak and Ormeling, 1996). The simplest way of using a static map is to place it side by side with the map it represents. The user visualizes the uncertainty by directly comparing the two maps. Another way is to generate a composite map (manually or with the aid of the computer) that combines map data with uncertainty data in a single map (Berry, 1995; Kraak and Ormeling, 1996). This represents a more advanced form of visualization of uncertainty because the resulting composite maps can be further combined to compute the *joint probability* for the uncertainties of two variables, thus giving further insight into the original data. In general, static maps are easy to produce, and the methods of map comparison are easy to understand. However, the methods of static maps are relatively limited in their ability to demonstrate variability in a theme (Davis and Keller, 1997). Therefore, advanced users may prefer the methods of dynamic visualization by animation explained next.

Dynamic Visualization of Uncertainty by Animation

This class of visualization methods includes a variety of systems of different degrees of sophistication, ranging from the relatively simple method of toggling between the actual data and uncertainty information, to a highly complex interactive manipulation that allows the user to alter parameters and view results in real time (Davis and Keller, 1997). Ehlschlaeger et al. (1997), for example, describe a method of dynamic visualization that tests the impact of spatial uncertainty in elevation data on a corridor location algorithm. The uncertainty of this particular investigation is introduced as the result of using DEM data sampled at an interval of 3 arc seconds for an application that requires elevation data obtained at 30-m intervals, which is about three times finer than the 3 arc-second data. By analyzing the hundreds of potential realizations of the elevation surface and the cost surface using animation, the researcher is able to answer numerous questions that would otherwise be difficult to understand because of the complexity and the massive amount of data involved. It is also found that animation is an effective way of communicating the results of the research to a wider audience. This type of application will probably open a new horizon for visualization of geographic information in environmental impact assessment in which public hearing is always a mandatory requirement.

7.5 Principles of Cartographic Design in GIS

As visual images represent the most natural and intuitive way to understand geographic space, maps play a critical role in spatial problem solving using GIS. The generation of maps is a relatively complex process that usually starts with a cartographic design phase. In general, the principles of conventional cartographic design that are found in textbooks (e.g., Slocum, 1999; Robinson et al., 1995; Dent, 1993; Keates, 1989) are applicable to the design of maps in GIS. There are, however, fundamental differences between the two working environments of conventional cartography and GIS (see Section 7.2.1). In the following sections, we will review the principles of cartographic design with special reference to their implementation in GIS.

7.5.1 TECHNICAL CONSIDERATIONS IN CARTOGRAPHIC DESIGN

Unlike conventional cartography, maps in GIS can be generated both as soft-copy products on the screen of the computer monitor (see Section 7.6.1) and hard-copy products using a plotter or a printer (see Section 7.6.2). Since no printing press is used, maps in GIS can be produced on demand when they are needed. This allows a high degree of customization of the contents of individual maps. However, customization also demands considerable knowledge of the GIS user in the design and generation of graphics. In order to master the skills of using the map production functions of a GIS, the user has to understand the major concepts in cartographic design, including (1) use of color, (2) use of text, (3) symbols and symbol sets, and (4) map-to-page transformation.

Use of Color

The primary function of color is to make information on a map visually distinguishable. Any color on a map can be described by three dimensions (Figure 7.12) (Slocum, 1999): *hue*, the dominant wavelength, which is what we usually think of as "color," such as red, green, or blue; *value* (or *lightness*), which is the description of how light or dark a color is when holding hue constant; and *saturation*, which is the purity of a hue or the range of wavelengths reflected (the narrower the range the purer the saturation). Cartographers utilize the systematic changes along these dimensions to show different types of relationships between data. As a general rule, changes in hue are used to indicate *qualitative* or *nominal* differences (e.g., land cover types and administrative units), whereas changes in value and saturation are used to represent *quantitative* or *hierarchical* differences (e.g., population density and amount of rainfall). In practice, color can be used either as a graphics variable in itself or in combination with the use of text and symbols in cartographic design, as we will explain.

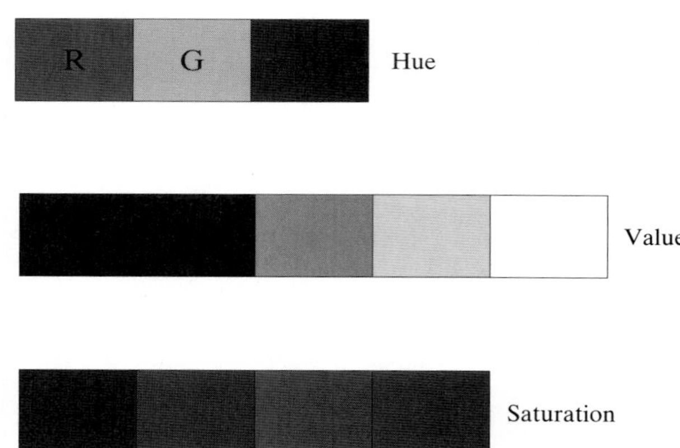

FIGURE 7.12
Dimensions of color. To view this photo in color, see the color insert.

There are several factors that influence the perception of color for presenting geographic information on the screen and using a plotter or printer. The number of colors that can be used is governed by the *bit depth* of the computer's video system for screen displays (see Section 7.6.1) and by the number of colors that a plotter or a printer is designed to support for hard-copy products (see Section 7.6.2). It is important to note that the specifications of color in different monitors may be slightly different. The actual colors displayed on the screen may also be slightly different as the result of the age of the monitor, the length of time it has been turned on, as well as the setting of the brightness and contrast of the screen. Colors on the screen are displayed by the *illuminant* or *additive mode*, which is essentially different from colors printed on paper by the *object* or *subtractive mode*. The translation between these two modes of color display is not straightforward. This is why it is impossible to exactly replicate a color that is displayed on a map and on the computer screen (see Section 5.4.1 in Chapter 5).

Use of Text

Descriptive text is used to give a map its title and to explain the legends. It is also used to label features by distinguishing their relative importance nominally (for qualitative data) and hierarchically (for quantitative data). Text can be used in the forms of different character sets that are described by three typographical characteristics (Figure 7.13), including *family*, which refers to a set of typefaces with variations based on design; *face* or *style*, which describes the specific variation based on weight, width, and angle; and *font*, which refers to a character set with a particular face and at a specific size. In addition to these three basic characteristics,

- **Family**: variation based on design

  Times New Roman **Impact** CourierNew Script Albertville

- **Face**: variation based on weight, width, and angle

  Wide Latin **Brittanic Bold** *italic*

- **Font**: specific size and style

  Times New Roman (16) Times New Roman (24)

FIGURE 7.13
Characteristics of text.

most GIS allows users to specify the color to enhance the cartographic appearance of the map. Some systems also allow for the use of different text qualities so that check plots can be produced more cost-effectively by drawing the text in *draft quality*.

Text is stored in different ways. In ArcInfo, for example, text can be stored as symbols of a coverage, or alternatively, as a coverage of its own, in which case the coverage is called an *annotation coverage*. When text is stored as symbols of a coverage, it is a temporary feature that appears only when the text display commands are issued in ArcPlot. Annotation, on the other hand, is a permanent feature class that can be displayed in ArcPlot. It can also be edited in ArcEdit. Properties and commands for text and annotation are similar in terms of syntax, but the purposes of their applications are distinct.

Symbols and Symbol Sets

A symbol is a graphic pattern that is used to represent a feature on a map. According to the types of features they represent, symbols in ArcInfo are classified into four categories (Figure 7.14): *marker symbols*, representing point and node features; *line symbols*, representing arcs, routes, and sections using lines of different colors, types, and widths; *shade symbols* for filling polygons and regions using solid color or shade patterns; and *text symbols*, descriptive text used to label features that can be points, lines, and polygons. Each type of symbol spans a set of properties. These properties vary between different systems, but in general they include color, size, angle, pattern, and, in the case of text symbols, font size and style.

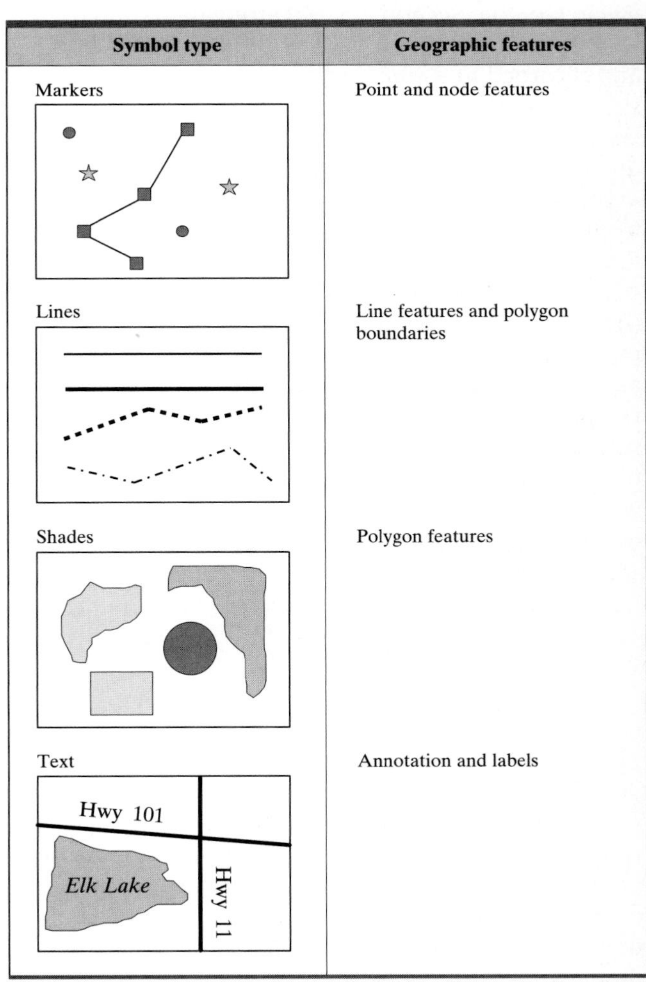

FIGURE 7.14
Symbols and symbol sets.

The creation of symbols is a relatively complex task since each symbol must be programmed individually. Because of this, symbols are usually supplied as part of the GIS software package in the form of *symbol sets*. ArcInfo, for example, is delivered with four symbol sets: *markersets, linesets, shadesets,* and *textsets*. They are computer files with the extensions of .MRK, .LIN, .SHD, and .TXT, respectively. High-end systems such as the workstation version of ArcInfo also include symbol editors that allow the user to change the properties of individual symbols to suit the purpose of specific applications. These symbol editors are particularly beneficial to users whose requirements cannot be satisfied by the generic symbol sets supplied by the GIS software vendor. They may also be used to create symbol sets to optimize the output of a particular application or the output of a particular type of plotter or printer.

Map-to-Page Transformation

Map-to-page transformation is the placement of coverage features onto an output medium of a specific size. This is an important concept pertaining to *map composition* in GIS (see Section 7.5.2). During map composition, the graphics are simultaneously drawn on the screen and recorded in a *graphics file* as a series of drawing instructions. This graphics file can be saved for use at a later time, both for display on the monitor and for producing a hard-copy product. When the monitor is used as the display device, the map-to-page relation is hidden from the user. However, when a *graphics file* is to be used for plotting or printing on a hard-copy output device, it is necessary to perform a map-to-page transformation so that the map can be properly positioned on the output medium. The method of map-to-page transformation is based on the relationships between the following terms (Figure 7.15) (ESRI, 1995a): *physical page*—the actual surface of the medium on which the map is displayed or plotted; *graphics page*—the area of the physical page where the map is drawn; *map limits*—the area on the graphics page where features in a coverage are drawn; and *map extent*—the rectangular limits (i.e., the bounding maximum and minimum *x* and *y* coordinates), in actual ground measurements, that define the area of Earth's surface to be displayed.

In order to produce a hard-copy information product from a graphics file, three basic operations are required to fit the coverage (i.e., the map) onto a page of the output medium: rotation, scaling, and translation. Different systems tend to perform these operations in different ways. In ArcInfo, for example, map-to-page transformation is a relatively straightforward process using a sequence of commands that enable the user to define the units of measurements (PAGEUNITS and MAPUNITS), physical size of output (PAGESIZE), range of page units (MAPLIMITS), and the shift, scale, extent, and position of the coverage to be displayed (MAPSHIFT, MAPSCALE, MAPEXTENT, and MAPPOSITION) (ESRI, 1995a).

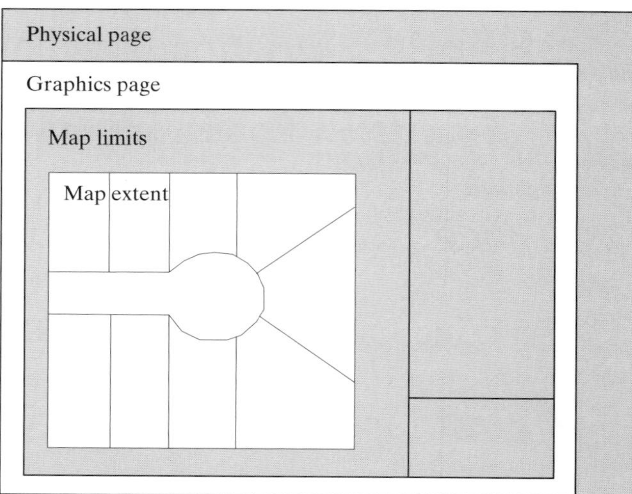

FIGURE 7.15
Relationships between physical page, graphics page, map limits, and map extent.

7.5.2 MAP LAYOUT AND GEOGRAPHIC CONTENT IN CARTOGRAPHIC DESIGN

Map composition is the process by which maps in GIS are produced. Conventionally, this process was done in the command mode whereby graphics elements were drawn one by one to form the end product. The process was highly interactive in the sense that the user was allowed to determine the position of the map on the output medium (see map-to-page transformation), change graphics variables of map contents, and define the positions surrounding map elements. In many GIS packages today, map composition commands are grouped together to form a special functional module. In ArcInfo, for example, such a functional module is called ArcPlot. Map composition in ArcPlot can be carried out in one of three modes: (1) command mode; (2) macro programming using Arc Macro Language (AML, for workstation and Windows NT versions) or Simple Macro Language (SML, for PC version); and (3) using the Map Tools function of the ArcTools graphical user interface (for workstation and Windows NT versions) (ESRI, 1995b).

No matter which mode is used, map composition can be perceived as being made up of three components: (1) map layout design; (2) geographic contents; and (3) label placement. The basic considerations pertaining to these components are explained here.

Map Layout Design

The process of map composition usually starts with an initial layout. There is no single standard to a map layout design, but all output products should include the

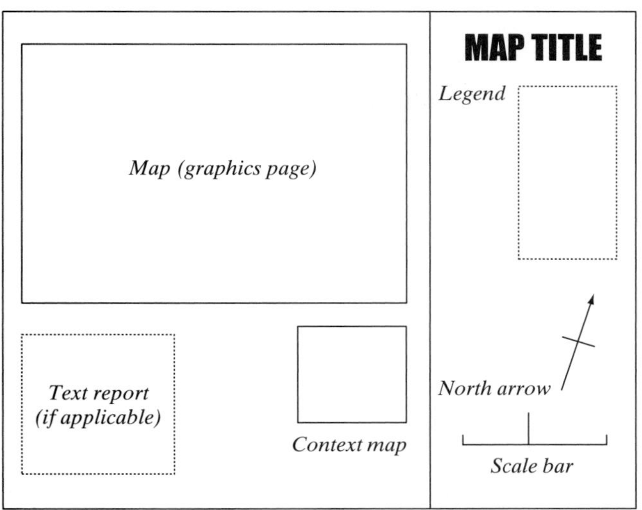

FIGURE 7.16
Elements of map layout.

basic elements as shown in Figure 7.16. Once the layout design has been determined, map surround elements will be added. These include *neat lines, title blocks* and *title text, north arrow, scale bar, legend,* as well as the *logo* and *name* of the producer of the map. These elements are ephemeral objects that are not stored in coverage. Some of the elements, such as neat lines and north arrows, can be drawn using routines and symbols supplied by the GIS software vendor. Other elements, such as the scale bar and legend, are unique to a specific map or map series. These elements have to be created by macro programming. Since map layout design is a relatively complex and time-consuming process, it is always advisable to store a completed design as a template so that it can be reused for other applications. It should be understood that a layout design is an iterative process. Before finalizing and adopting a certain design, it is advisable to generate several alternatives and to test them against cartographic constraints as well as the application requirements of the end user.

Geographic Contents of the Map

This particular component of cartographic design is concerned with the selection of features to be included in a particular map. In essence, the geographic contents of any map are basically governed by three factors: the theme or related themes to be presented, the area to be covered, and the scale of the presentation. For most maps, this part of the design process is relatively easy and straightforward. Problems usually occur when it comes to the use of color, text, and symbology to represent the features on the map, and to make judgment for label placement and map generalization. Although there are attempts to develop guidelines to assist in the decision making for these tasks, the procedures still involve a considerable amount of subjectivity and intuition. In general, a good design for geographic contents is characterized by

- *visual balance* with other map components (e.g., title, legend, scale bar, and other marginal information) to facilitate the reading of the map
- *visual clarity* by resolving spatial conflicts between map elements (this usually involves one or more of the operations of generalization, see Section 7.5.3)
- *visual hierarchy* by the appropriate use of font and symbol sizes
- *visual contrast* by the appropriate use of color and shade patterns
- provision of suitable *context* or *reference information* (this involves the selection of appropriate layers from the topographic base map)
- appropriate positioning of text and annotations by following the guidelines for label placement as explained next

Label Placement

Labels on maps provide the attribute data associated with graphical map elements. Label placement is an important component of cartographic design because it directly affects the readability of the map. A good label placement design enables the reader to associate labels with the map elements that they describe. In contrast, a poor design causes difficulty and uncertainty in using the map. Just like other components of cartographic design, label placement is also a very complex process because of the variety of factors involved. Basically, label placement is governed by the characteristics of the map element that a label describes, including type of the element (i.e., whether it is a point, a line, or a polygon), shape of the element, orientation of the element, position of the element on the map, and position of the element relative to other map elements.

Of the many attempts to automate cartographic design, label placement has had the greatest success. Automation of label placement is based on the development of guidelines which, when enforced, will enable map labels to be positioned properly during computer processing. Different types of map elements have different placement requirements and consequently different guidelines (Table 7.3). Although these guidelines are intended for computer-assisted label placement, they can be equally applicable to label placement by manual methods.

7.5.3 CARTOGRAPHIC GENERALIZATION

Cartographic generalization is the process by which map data are abstracted and transformed into a representation at a reduced scale. Broadly speaking, generalization

Table 7.3
Guidelines for Label Placement

General Guidelines
- Easily perceived and unambiguous feature-label association
- Avoidance of overlap (among names, or among names and point features), but overlap for name and line features allowed (e.g., elevation value on contour lines)
- Conformance to shape and extent of area features
- Avoidance of excessive clustering of names
- Conformance to the curvature of the parallels for horizontal placements
- Conformance to applicable standards and conventions

Guidelines for Labeling Area Features
- Following dominant shape of boundary of the feature
- Preference of horizontal placement over vertical placement
- Positioning name baseline closely parallel to the dominant centerline of the area
- Having nonhorizontally placed names curved, preferably starting along the horizontal and then curving away, rather than curving toward the horizontal
- Having neighboring area-feature names of the same category curved similarly unless the shapes of the features are significantly different
- Permission of placement outside the confines of the feature if it has a severely curving shape or consists of two or more disjoint parts separated by water

Guidelines for Labeling Point Features
- Close proximity between name and feature
- Preference of placement above feature to placement below feature
- Placement horizontally, reading from left to right
- Avoidance of spreading out of characters
- Hyphenation and abbreviation permissible

Guidelines for Labeling Line Features
- Conformance to curvature of feature
- Using the feature as placement baseline
- Repetition permissible for long features
- Avoidance of spreading out of characters
- Reading upward in left half of a map; downward in the right half

Source: Freeman, 1991

in cartographic design is based on the consideration of three factors (McMaster and Shea, 1988): (1) the intrinsic objective of *why* it is necessary to generalize; (2) an assessment of the situation which indicates *when* to generalize; and (3) an understanding of *how* to generalize using spatial and attribute transformation operations to effect the requisite changes.

Why Generalize?

The answer to the why question is relatively simple. When geographic data are collected at a scale that is larger than the scale at which they are presented, it is necessary to reduce the complexity of the data to make the resulting map look aesthetically more pleasing. Generalization ensures that geographic data are presented at a scale appropriate to the purpose of the map and the application requirements of its users. In the digital working environment of GIS, generalization is particularly important because it allows faster response in data processing and, consequently, reduces the occurrence of screen flicker (see Section 7.6.1). It also minimizes data collection and storage requirements because data at one particular scale can be used for different applications where different map scales are required.

When to Generalize?

Shea and McMaster (1989) identified six conditions that may necessitate cartographic generalization.

- *congestion*, which occurs when too many features are packed into a limited geographic space
- *coalescence*, which occurs when features touch one another as the result of inadequate resolution or symbolization
- *conflict*, a situation in which the spatial representation of a feature is incompatible with its background

- *complication,* which results from using data from different sources, at different scales, or automated with different tolerance levels
- *inconsistency,* which refers to a set of generalization decisions applied nonuniformly across a given map
- *imperceptibility,* a situation resulting from a feature falling below a minimum representation size of a map

How to Generalize?

Generalization can be carried out in a number of ways, depending on a variety of factors such as the nature of the data, the reason for generalization, and the information requirements of the end user. Broadly speaking, the following operations have been identified to effect scale and data reduction in cartographic generalization (Figure 7.17) (Shea and McMaster, 1989).

- *simplification,* which involves the selection of a subset of the original coordinate pairs
- *smoothing,* which relocates or shifts coordinate pairs in order to plane away any small perturbations
- *aggregation,* which groups point features into a higher order class
- *amalgamation,* which joins small features into a larger map element
- *merging,* which is usually applied to parallel line features
- *collapse,* which decomposes area or line features to point features
- *refinement,* which discards smaller features from among a cluster of features
- *typification,* which is similar to refinement, but uses a representative pattern of features or symbols
- *exaggeration,* which amplifies the shape or size of features to meet the specific requirements of a map
- *displacement,* which shifts the position of features to gain clarity
- *classification,* which groups together objects into categories of features sharing identical or similar characteristics

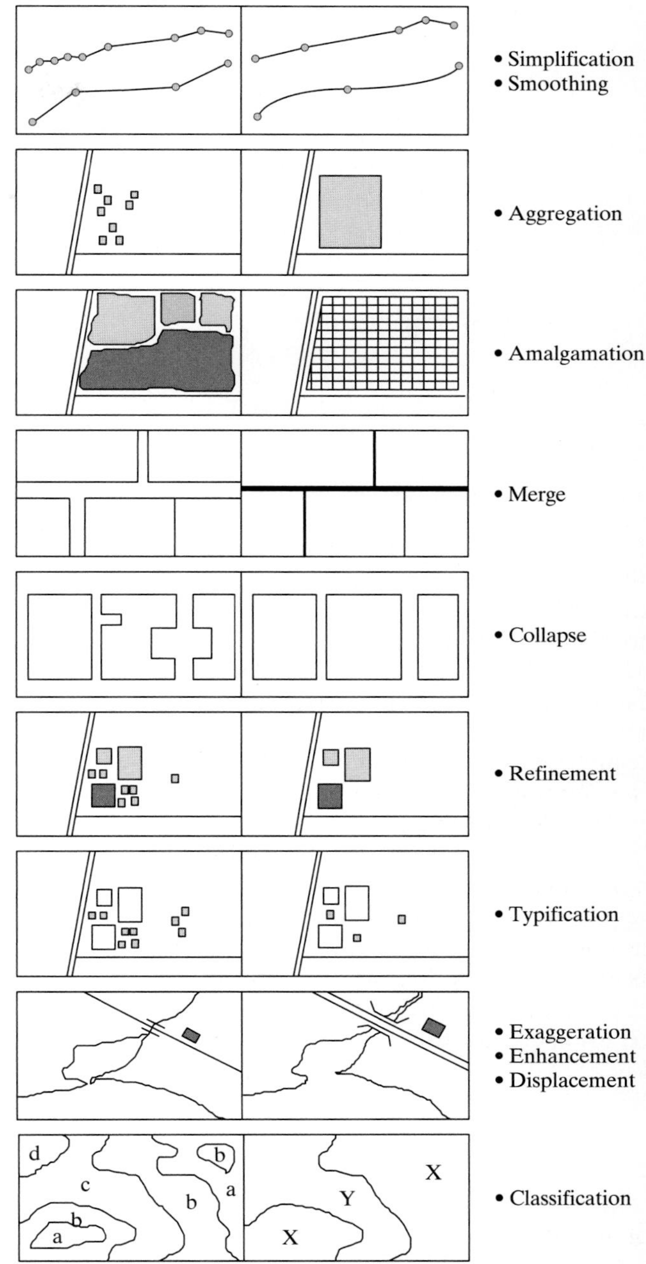

FIGURE 7.17
Methods of cartographic generalization. *(Source: Shea and McMaster, 1989)*

Generalization is one of the most complex tasks in cartography because considerable decision making is involved. As a result, there has not been much success in attempts to automate the process. In individual GIS software packages, however, there are usually functions that assist in line generalization. In ArcInfo, for example, line generalization can be effected at different points during data processing, including:

- *Edit Tolerances.* There are two edit tolerances: *grain tolerance* and *weed tolerance*. The grain tolerance (the GRAIN command) sets the closest distance between adjacent vertices and is used for modifying existing arcs with ArcEdit commands such as SPLINE and for adding new curved arcs with ADD. The weed tolerance (the WEEDTOLERANCE command) also sets the closest distance between adjacent arc vertices but it is used when digitizing with the ArcEdit ADD command and when modifying selected arcs with the GENERALIZE command.
- *Draw Tolerance.* The WEEDDRAW command reduces the number of vertices that are used to represent an arc from a line or polygon coverage while preserving its basic shape and orientation.

This process increases the speed with which arcs are drawn, but at the same time leads to a consequential loss of resolution and detail. The WEEDDRAW command does not weed nodes, and the reduction in the number of vertices is purely for cartographic purposes; that is, it does not alter the stored coverage data.

- *Permanent Line Generalization.* The GENERALIZE command permanently reduces the number of vertices that represent line features (i.e., it changes the stored coverage data).

7.6 Generation of Information Products

Conventionally, the map is the standard method for presenting information in GIS. However, as GIS applications become more diverse and sophisticated, other forms of information products have also been used. These include charts, graphs, statistical tables, textual reports, perspective views, as well as animated presentations. The methods for the generation of new GIS information products are largely the results of advances in computer graphics and related technologies. It is important for GIS users to understand the current state of these technologies and to take full advantage of them accordingly. In the context of our discussion, "presentation of information" refers not only to the *display* of geographic information on the computer screen or the printed map, but also the *dissemination* and *publication* of the information using different forms of multimedia technologies for the purpose of mass communication.

7.6.1 Generation of Soft-Copy Information Products

Soft-copy information products are visual presentations generated on the computer monitor (also referred to as the *visual display unit,* or VDU). Computer monitors are constructed using different technologies and can be classified into one of the following categories: *cathode ray tube* (CRT), *liquid crystal display* (LCD), and *light-emitting diodes* (LED). Since the majority of desktop monitors used today are CRTs, our discussion will focus on the characteristics of these monitors with particular reference to the PC hardware platform.

There are two types of cathode ray tube monitors: *storage tube* and *refresh tube*. Storage tube monitors were old technology used in the early years of computer graphics. All monitors used today are of the refresh tube type. A refresh tube monitor is made up of four components (Figure 7.18).

1. the *electron gun*: a system of electrodes (cathode and anode) for producing a beam of high-velocity electrons
2. the *electronic lens*: a mechanism by which the electron beam is focused to a sharp point on the screen
3. the *plates* or *coils*: a deflection system for sweeping the electron beam
4. the *screen*: the viewing surface of the monitor with phosphor coating on one side

In the refresh tube monitors, images are formed by directing a beam of electrons against the phosphor coating on the screen horizontally line by line, from top to bottom. In order to keep the images visible, the computer has to "refresh" the screen by constantly repeating the image-forming process. It is necessary to refresh the screen approximately 30 times per second to avoid the occurrence of *flicker*. This rate is called the *refresh rate* of the monitor. The simplest implementation of refresh tube technology is the monochrome display, which is obtained by varying the intensity of the electron beam. Colored displays are produced by equipping the monitor with three electron guns and three sets of phosphors, one for each of the three primary colors: red, green, and blue. In typical raster graphics processing, three bytes are used to represent the intensity of the colors in

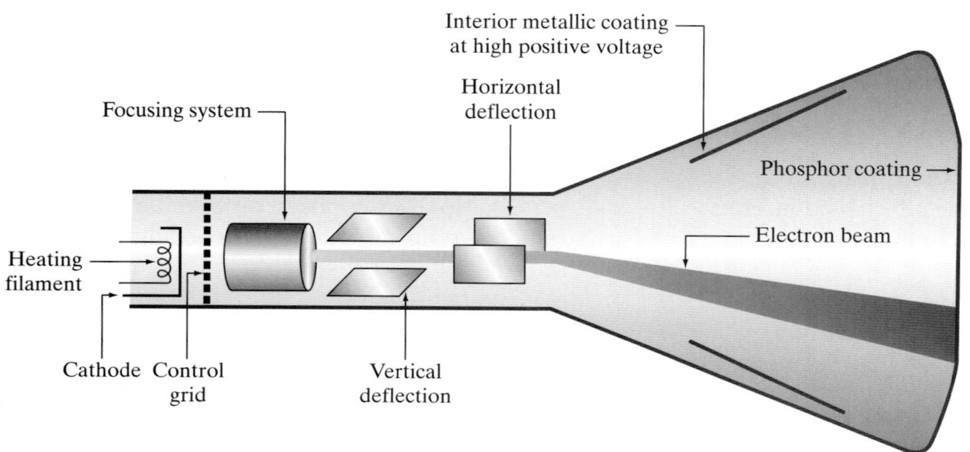

FIGURE 7.18 Components of a CRT monitor.

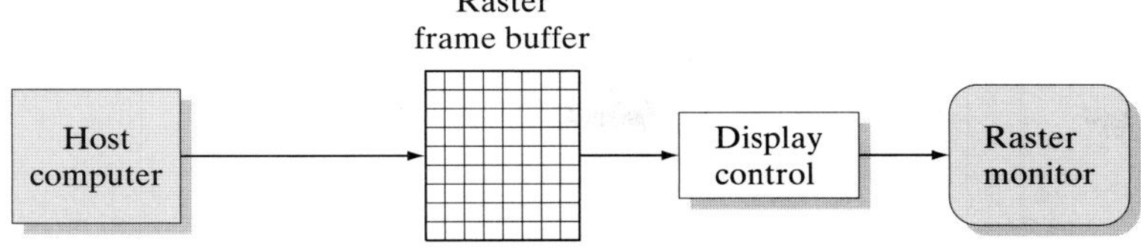

(a) Memory organization for raster display

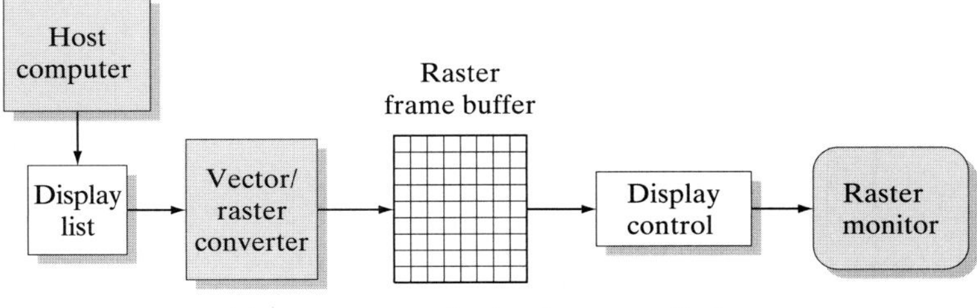

(b) Memory organization for vector display

FIGURE 7.19
Memory organization for displaying raster and vector data on a raster CRT monitor.

each pixel. Since one byte represents 2^8 (8-bit) digital values, 256 levels of intensity of red, green, and blue can be displayed for each pixel on the screen. By varying both the intensity and the combination of the three beams, it is possible to obtain up to 2^{24} (24-bit) (i.e., over 16.5 million) colors (see Section 5.1.1).

Refresh tube monitors can be used for both raster and vector graphics. In raster data processing, the data to be displayed are stored in the computer's *frame buffer* as an array of numbers representing the brightness of individual screen pixels. The frame buffer is located between the local memory (where graphics processing takes place) and the display controller (Figure 7.19a). Since the data in the local memory and the frame buffer are both in raster format, the display of the images simply involves the mapping of the pixels to the dots on the screen. Images formed in this way are therefore usually called *bit-map graphics*. When a bit map is enlarged, it becomes a pixelated image (i.e., the detail of the image is lost). Bit-map graphics also tend to create large files. However, the relatively simple mechanism by which images are formed on the screen means that graphics processing can be done quickly. The ability to use three separate image planes for the three basic colors in the frame buffer also allows the use of millions of colors for the display. The methods of bit-mapped graphics are used by raster-based GIS applications and remote sensing image processing.

In vector processing, on the other hand, the data to be displayed are stored as drawing instructions for individual graphics elements in a *display list*. A *display list processor* continuously loops through the list and converts it into pixels in the frame buffer (Figure 7.19b). Technically, therefore, vector display is a much more complex process than raster display. It requires constant vector–raster conversion between the data in the local memory (where vector processing takes place), the display list (where drawing instructions are stored), and the frame buffer (which stores the screen image in raster form). To minimize the requirement for data conversion, refreshing of the screen is done by reading the frame buffer. Vector display usually makes use of only one image plane. With an 8-bit image plane, 256 colors can be used to produce the screen display. This number of colors should be more than enough for most GIS and mapping applications.

The visual quality of images on the screen is governed by the characteristics of the computer monitor, which include (1) screen resolution, (2) color, (3) size, (4) dot pitch, (5) refresh rate, and (6) refresh mode. The implications of these characteristics with special reference to GIS applications are explained below.

Screen Resolution

Screen resolution is the degree of sharpness, or clarity, of the images displayed on the monitor. The resolution of computer monitors is measured by the number of dots, or pixels, of color that can be displayed across and down the screen. For example, the 800 × 600 resolution means that there are 800 dots of color across all 600 rows down the screen. Standard screen resolution in common use in today's desktop monitors include 640 × 480, 800 × 600, 1024 × 768, 1280 × 1024, 1600 × 1200, 1800 × 1440 (Poor, 1997). The resolution of a monitor is related to, and governed by, the standards of graphics cards, which are defined in terms of monitor resolution and number of colors. For example, the *Super VGA* monitor, an industry display standard maintained by the *Video Electronics Standards Association* (VESA), can display at resolutions ranging from 640 × 480 to 1280 × 1024 with up to 16 million colors. If a Super VGA monitor is used in conjunction with a graphics card that can display only at 800 × 600, the effective resolution of the screen will be that of the graphics card (i.e., 800 × 600). It is not possible to take advantage of the highest possible resolution of the monitor. In practice, therefore, it is important for the resolution of the monitor to be compatible with the resolution supported by the computer's video system (as determined by the graphics card).

Color

The number of colors that can be displayed depends on the graphics card of the computer. Depending on the amount of on-board random-access memory (RAM), graphics cards vary with respect to their ability to support different combinations of screen resolutions and colors. With 256K of on-board RAM, for example, a Super VGA card can support a screen resolution of 800 × 600 with 16 colors, or a resolution of 640 × 480 with 256 colors. The number of colors and the resolution can be increased by adding more on-board memory. When one MB of on-board RAM is available, a Super VGA card can display at the 1024 × 768 resolution with 256 colors, or 640 × 480 with 32768 colors. To take full advantage of the maximum resolution (1280 × 1024) and maximum number of colors (16.5 million) that can be displayed by a Super VGA monitor, 4 MB of on-board RAM is needed. It is generally accepted that for multimedia applications, which include presentation of information in GIS, a screen resolution of 640 × 480 with 256 colors is the minimum requirement.

Size

The size of a monitor refers to the *diagonal length* of its cathode ray tube, but not to the measurement of the actual screen. This is because the plastic rim that holds the monitor masks a small portion of the cathode ray tube. The portion of the screen that can be seen is called the *viewable image size,* which may be up to an inch (2.54 cm) smaller than the monitor size measurement. The smallest PC monitors measure 14 inches. This size is adequate for information viewing purposes at 640 × 480 or 800 × 600. When the resolution increases, it is necessary to increase the monitor size for more comfortable viewing. When working at a 1280 × 1024 resolution, for example, a 17-inch monitor is recommended. For applications that require simultaneous multiple windows, a 20- or 21-inch monitor is strongly recommended.

Dot Pitch

The dot pitch of a monitor refers to the physical size of a pixel. Monitors with large dot pitches are cheaper to produce, but the screen displays are fuzzy and can cause eyestrain easily. For best viewing results, the dot pitch of the monitor must measure no more than 0.28 mm.

Refresh Rate

These are measures, expressed in terms of *hertz* (Hz), that denote the number of times the screen is redrawn each second. Low refresh rates cause the screen to flicker, which in turn leads to eyestrain and fatigue when using the monitor for an extended period of time. Most monitors today are constructed with a 75 Hz refresh rate, which is slightly lower than the 85 Hz recommended by the Video Electronics Standards Association (VESA). Large high-end monitors usually have refresh rates of 85 or 100 Hz.

Refresh Mode

The refresh mode of a monitor can be *interlaced* or *noninterlaced.* A noninterlaced monitor redraws every line of information on the screen with each refresh, while an interlaced monitor refreshes only every other line. Because of the reduced workload, interlaced monitors are less expensive, but the quality of the output is not as clear and crisp as that of a noninterlaced monitor.

7.6.2 GRAPHICS PERFORMANCE AND ACCELERATION

Screen display of geographic information can be tediously slow unless the computer's graphics performance is accelerated. This can be done either by using a faster central processing unit (CPU) or adding a special graphics card to the video system of the computer, or preferably both. There are different types of graphics cards that have been developed for different purposes. From the perspective of displaying geographic information, the most important graphics card function is *graphics acceleration*. A *graphics accelerator* increases graphics performance by taking some of the load off the CPU and by using different types of random-access memory (RAM), such as *video memory* (VRAM) and *synchronous graphics RAM* (SGRAM). VRAM is different from conventional system RAM in that it is dual-ported. This means that it can both send and receive data simultaneously, while system RAM can do only one task at a time (and is therefore slower). SGRAM, similarly, is a very fast type of RAM that operates at the same speed as the graphics card itself. Specially designed for use on graphics accelerators, it improves performance on such tasks as painting a large block of color. Another important graphics card function is *three-dimensional acceleration* that aims to enhance the display of information in three dimensions. An example is the *OpenGL card.* OpenGL is the three-dimensional graphics standard of the Windows NT operating system. The OpenGL card is a three-dimensional accelerator with OpenGL procedures hardcoded onto a specialized processor. It works fast because there is a specialized processor using its own dedicated memory and executing the OpenGL commands directly.

Graphics processing performance of the CPU can be improved by additional RAM. As GIS processing and display are extremely memory-intensive, 64 MB of RAM is the minimum requirement and 128 MB is always recommended for the CPU. If RAM is important, *cache memory* is even more so. Cache memory is a small but very high-speed memory installed as a buffer between the CPU and the main memory (the hard drive). It stores instructions and data that are most frequently and repetitively used by the CPU. This expedites processing by effectively minimizing the traffic between the CPU and the main memory. As a general rule, 256 KB of cache is the minimum requirement, but 512 KB is preferred.

Computer display technologies have advanced spectacularly in the last few years. From the perspective of presenting geographic information, the ability to perform three-dimensional graphics and animation on the PC platform is particularly important. Conventionally, these methods were limited exclusively to very expensive workstations. At present, sophisticated three-dimensional and animated presentations can be performed on PCs at an affordable price. This technological breakthrough came as the result of the convergence of a number of developments, including the implementation of OpenGL standard by Microsoft as part of the Windows NT operating system, the introduction of very fast microprocessors by Intel and its competitors, and the production of very powerful three-dimensional graphics processing engines, such as the GLiNT 500TX processor by 3DLabs and RealiZM by Intergraph. These developments, taken together, created the momentum that has greatly expanded and enhanced the display function of GIS from the generation of digital maps to the visualization of spatial information through three-dimensional and animated presentations.

7.6.3 GENERATION OF HARD-COPY INFORMATION PRODUCTS

Despite the increasing use of the screen display as the predominant means for information presentation, hard-copy outputs remain a requirement for many applications in GIS. Unlike the conventional printed products, GIS information products are usually produced on demand using a graphics plotter or graphics printer. There are now many types and classes of plotters/printers in the market. These devices use a variety of technologies and exhibit a wide range of capabilities (Table 7.4).

Table 7.4
Characteristics of Graphics Plotters and Graphics Printers

	Graphics Plotters	*Graphics Printers*
Technology	Pen (vector) Drum Plotter Flatbed Plotter	Electrostatic (Raster) Ink-jet Laser Thermal
Output Size	D-size (22″ × 34″; 55.88 cm × 86.36 cm) E-size (34″ × 44″; 86.36 cm × 111.76 cm)	D-size (22″ × 34″; 55.88 cm × 86.36 cm) E-size (34″ × 44″; 86.36 cm × 111.76 cm) Letter size (8.5″ × 11″)
Resolution	Drum plotter: 0.001 inch (0.0254 mm) Flatbed plotter: 0.0003 inch (0.0076 mm)	Electrostatic: 200–400 dpi (0.127 mm–0.0635 mm) Ink-jet: 600 dpi (0.0423 mm) to 1200 dpi (0.0212 mm) Laser: most commonly 600 dpi (0.0423 mm), but can be up to 1200 dpi (0.0212 mm) Thermal: 300 dpi (0.0846 mm)
Use of Colors	Generally 4–6 colors	Electrostatic and ink-jet: monochrome or color Laser: monochrome or color Thermal: color
Speed	Ball point pen 40 inches/sec (101.6 cm/sec) Ink: 2 inches/sec (5 cm/sec)	Large-format electrostatic and ink-jet: less than 1 ppm (page per minute) Desktop ink-jet: 4–8 ppm monochrome; 2–4 ppm color Laser: typically 2 ppm monochrome; 6 ppm color for personal laser printer; up to 100 ppm for high-end workgroup and departmental laser printers Thermal: less than 1 ppm

Technically speaking, the generation of hard-copy information products is the weakest link in GIS technology. Extremely advanced and sophisticated spatial analysis and modeling can now be performed by the computer and displayed dynamically on the monitor. However, hard-copy output devices always impose constraints on the production and quality of the resulting information products. For example, although the monitor is capable of using literally millions of colors, plotters/printers can normally use no more than a dozen. Graphics plotters/printers are also relatively limited in the use of fonts and symbology (fonts and symbols must be individually coded). They are comparatively much slower than the monitor in generating graphics, and the cost of using them is usually higher (hardware and consumables such as ink and paper are expensive).

Installing a plotter/printer is not a simple plug-and-play task. In order for a plotter/printer to work properly, it must be able to communicate with the host computer. Conventionally, plotters/printers use a computer language to jointly control and process a plotting/printing job with the operating system of the computer. The most commonly used plotter/printer languages include Hewlett-Packard *HPGL* and *PCL*, IBM *Proprinter*, Adobe *Postscript*, Epson *ESC/P*, and others that we noted in Chapter 3. Within each of these languages, there can be different versions that may not be totally compatible. The lack of standards and compatibility can sometimes be a serious problem in installing plotters/printers. The plotters/printers will simply not work if they are not using the same language for the drawing instructions generated by the host computer. The incompatibility of different versions of the same language, on the other hand, may affect the quality of the resulting products (e.g., lowering the resolution of output). There are now host-based plotters/printers that aim to remove these problems. These plotters/printers rely on a *graphics device interface* (GDI) in the computer to translate and send all the commands to the plotter/printer for processing. By using the processor of the computer to generate the image, this approach removes the need for installing sophisticated and expensive controllers, processors, and memory on the plotters/printers themselves. As a result, the prices of the plotters/printers can be reduced considerably.

Graphics plotters are required for the production of large-format high-quality information products. These plotters draw by moving a drafting tool across a medium (such as paper or mylar) in much the same way as in manual drafting. The most commonly used drafting tool is a ball point or ink pen. This is the reason why these plotters are usually referred to as *pen plotters*. The use of a ball point pen gives good drafting speed, but the quality of the output is not as high as the use of an ink pen. However, the use of an ink pen limits the drawing speed because the ink takes time to flow.

Large-format plotters can be constructed in two distinct ways: drum plotting and flat-bed plotting. With the *drum plotter*, the drafting tool is moved across the paper (i.e., along the axis of the drum) to achieve movements in the x-direction, and the paper is moved over a rotating drum on which it is mounted to achieve movements in the y-direction. The drafting head can then be moved to any position by the simultaneous movements of the drafting tool and the drum. The output of drum plotters can be D-size or E-size (which are approximately equivalent to the metric sizes of A1 and A0, respectively). In practice, however, the length of the output can be as long as 150 feet (45.72 m). The typical resolution for drum plotters is dependent on the drafting tool used, and can be as fine as 0.001 inch (0.0254 mm). The drafting head is usually equipped with 4 to 6 pens for changing color and line widths. Drum plotters are usually used for generating check plots or large maps for use as exhibits in public presentations.

With the *flat-bed plotter*, the medium is held in a fixed position on a table top while the drafting pen is moved over its surface in both the x and y directions (Figure 7.20). The flat-bed plotter is usually used for very high-quality plots, probably on scribe-coat and photographic film rather than on paper. The best line work, at a resolution as fine as 0.000005 inch (0.000127 mm), is obtained by using a photohead drafting onto a film. The photohead is a miniature projector that exposes cartographic images onto the film photographically. Since there is no direct physical contact between the drafting tool and the medium, photographic drafting can be done at a relatively high speed. The process is made even more efficient because it does not require operator intervention (e.g., changing pens) for a long operating time. However, the high cost of the equipment and the need to work in a darkroom environment imply that the use of this method of drafting can be only justified when extremely high-quality plots are required, for example, in the preparation of printing separates of topographic maps.

A conventional alternative to the vector-based pen plotter is the wide-format *electrostatic printer*. Electrostatic printers are raster-based devices and, therefore, are based on entirely different technologies. Instead of using a drafting head to draw on paper or mylar, the image is obtained by electrostatically charging specially treated paper that is moved over a bar containing very fine wires or nibs. A toner is then applied to the paper. Images are formed where the toner adheres to the charged areas and when excess toner is removed by a suction device. Earlier generations of electrostatic printers had a relatively poor resolution of 100 dpi. Colored electrostatic printers today can achieve a resolution up to 400 dpi. Colored electrostatic printing is obtained by using what is called the

Typical RF	1:1000 1:5,000 1:10,000	1:20,000 1:50,000 1:100,000	1:1,000,000 1:2,500,000
Description	LARGE-SCALE	MEDIUM-SCALE	SMALL-SCALE
Characteristics	• Depict small features • Show geometric shapes	• Small features disappear • Generalize geometric shapes • Good compromise between map detail and extent of map coverage	• Symbolize features, e.g., areas represented by point or line symbols • Show macro features, e.g., climatic zones

Large: 1:2,400 scale map of Athens, GA

Medium: 1:24,000 scale map of Athens, GA

Small: 1:2,500,000 scale map of Athens, GA

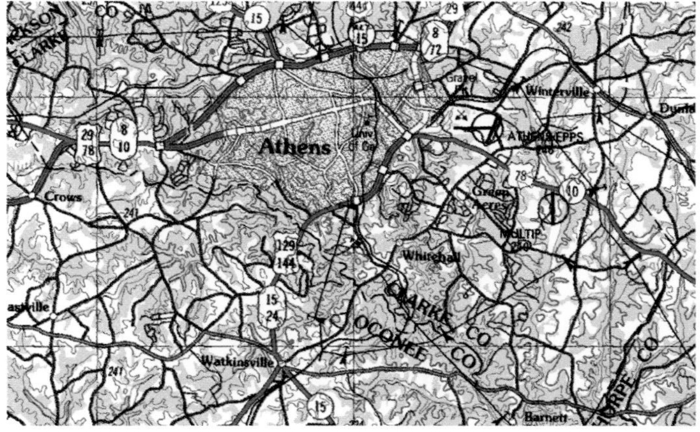

FIGURE 2.2
Continuum of large-, medium-, and small-scale maps, and their characteristics.

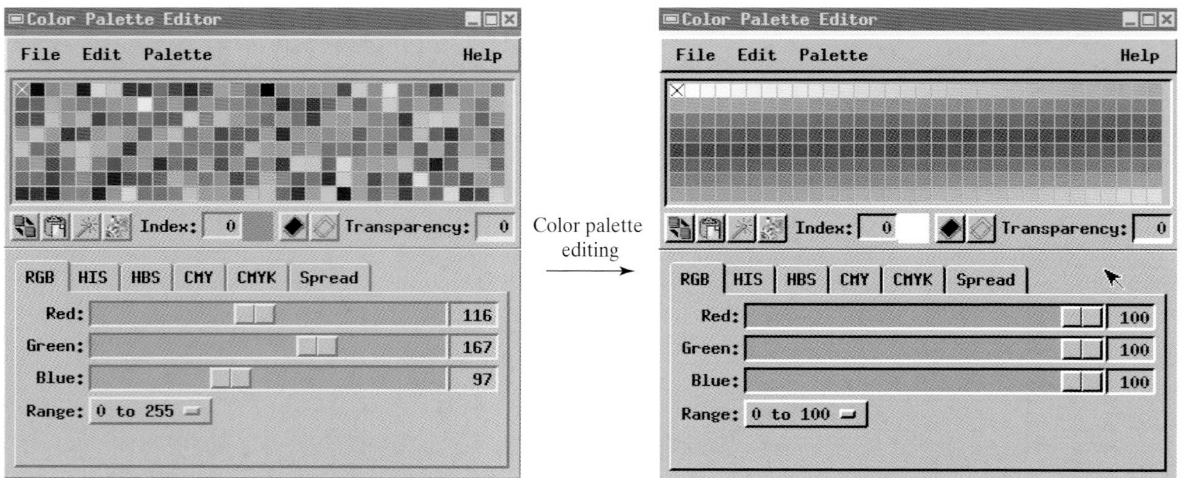

FIGURE 5.32
The color palette editor of TNTmips. A raster image displayed by the application-specific palette will show smooth transition from one class of cell values to another because the change of the colors in the color cells is gradual, rather than random as shown in the default color palette. *(Source: Microimages, 1998)*

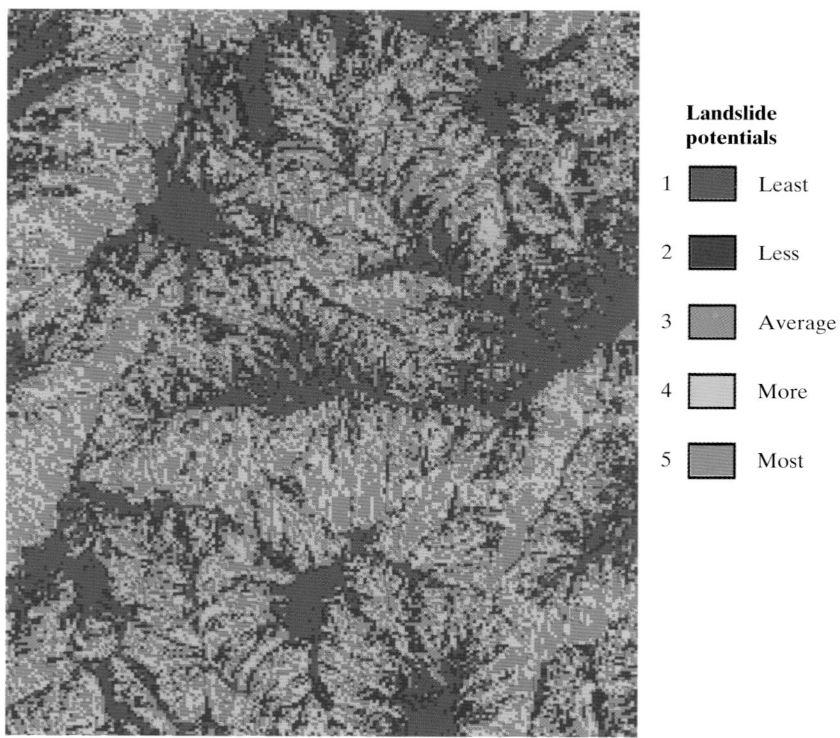

FIGURE 5.39
Modeled landslide susceptibility by integrating elevation, slope gradient, slope configuration, and slope aspect criteria. *(Source: Gao and Lo, 1995)*

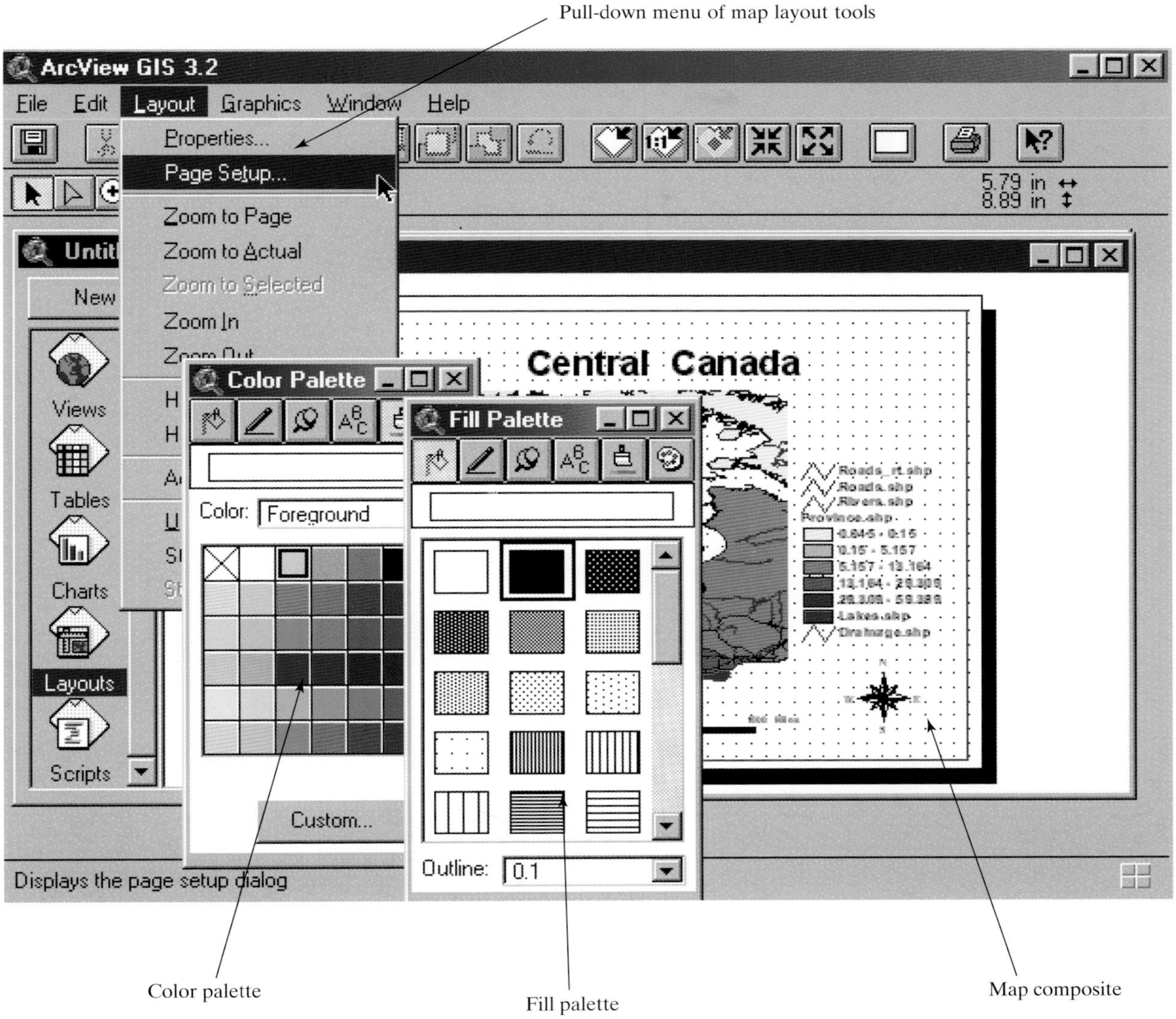

Pull-down menu of map layout tools

Color palette

Fill palette

Map composite

FIGURE 6.38
Map composition graphical interface and tools of ArcView GIS.

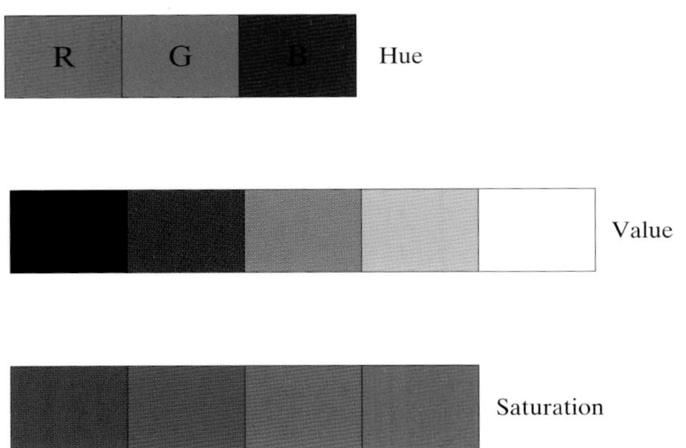

FIGURE 7.12
Dimensions of color.

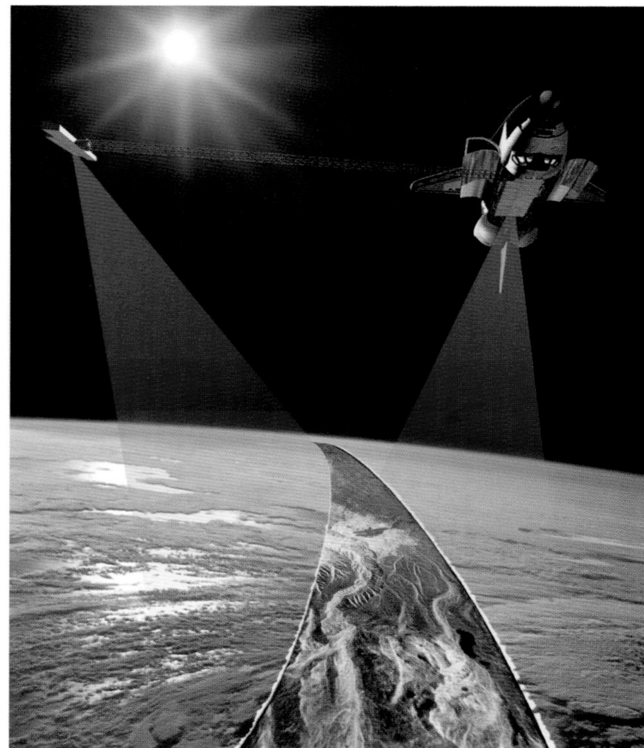

PLATE 8.3
The Shuttle Radar Topography Mission (SRTM) acquires radar images of Earth from two antennas (one on the payload and one at the end of the 60-m mast), thus allowing 3-D radar images to be produced to map Earth's terrain heights. *(Courtesy: NASA/JPL/Caltech)*

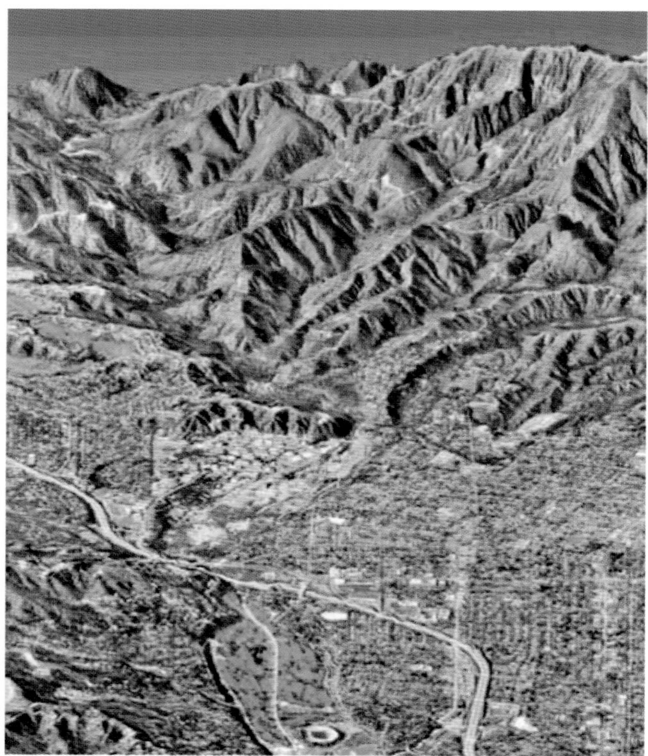

PLATE 8.4
This perspective view of Pasadena, California, is formed from three datasets: (1) Shuttle Radar Topography Mission (SRTM) for elevation, (2) Landsat data from the land surface color, and (3) USGS digital aerial photography for image detail. *(Courtesy: NASA/JPL/Caltech)*

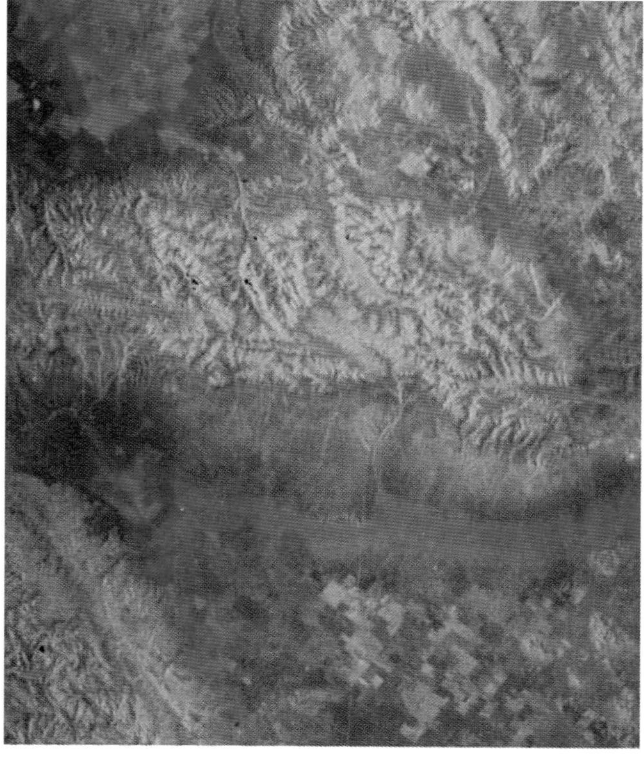

PLATE 8.5
San Andreas fault, Southern California. The image combines two types of data from the Shuttle Radar Topography Mission: (1) The image brightness corresponds to the strength of the radar signal reflected from the ground, and (2) the contourlike color bands show the elevation as measured by interferometry. *(Courtesy: NASA/JPL/Caltec)*

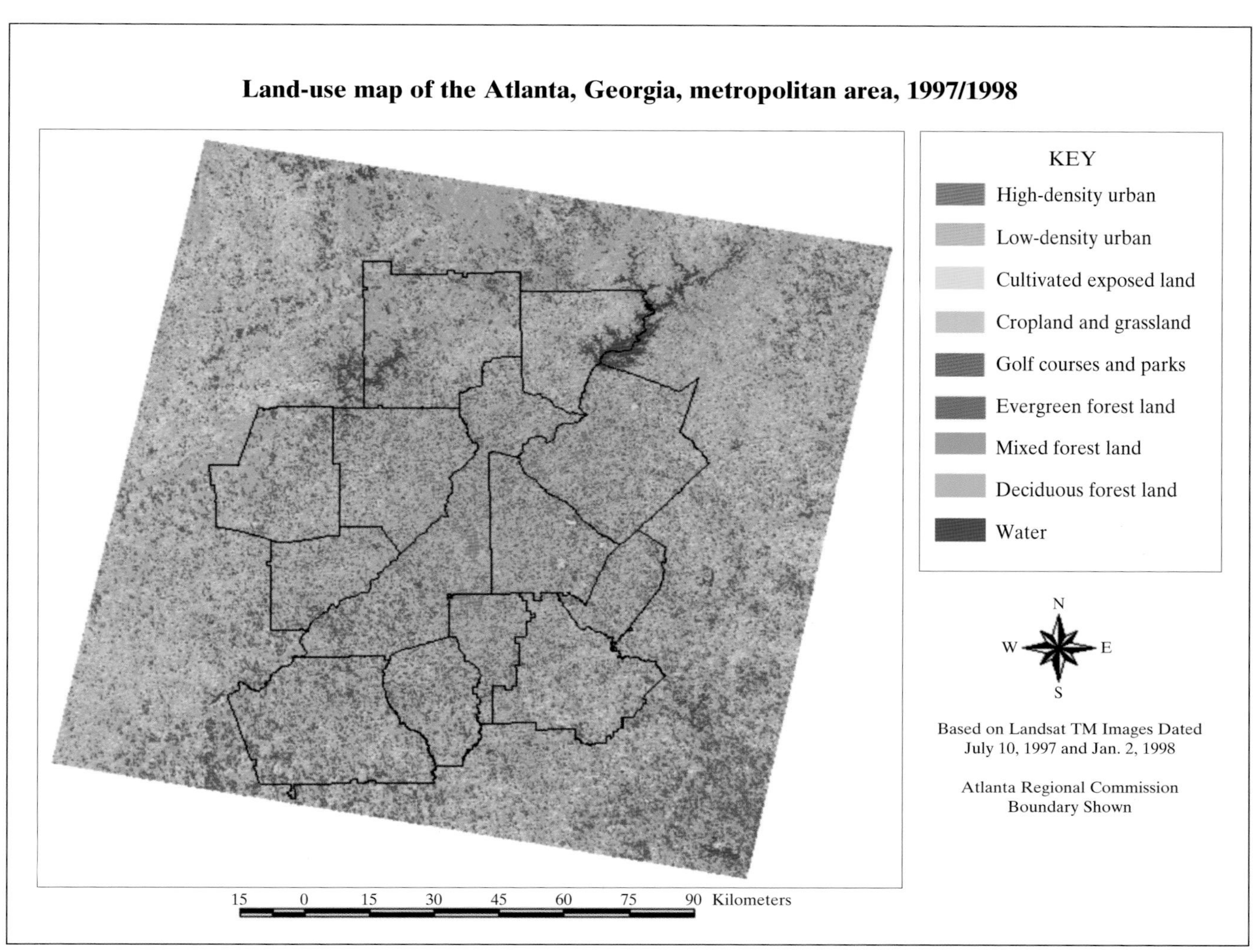

PLATE 8.7

A 1997 land-use/cover map of the Atlanta region produced from an unsupervised image classification approach (ISODATA) based on the Landsat TM image (Plate 2.4). This map was produced using ArcView GIS software. The boundaries of the 13 urban counties of Atlanta (vector) were overlaid on top of the land-use/cover classification (raster) so that land-use/cover statistics may be extracted for each county.

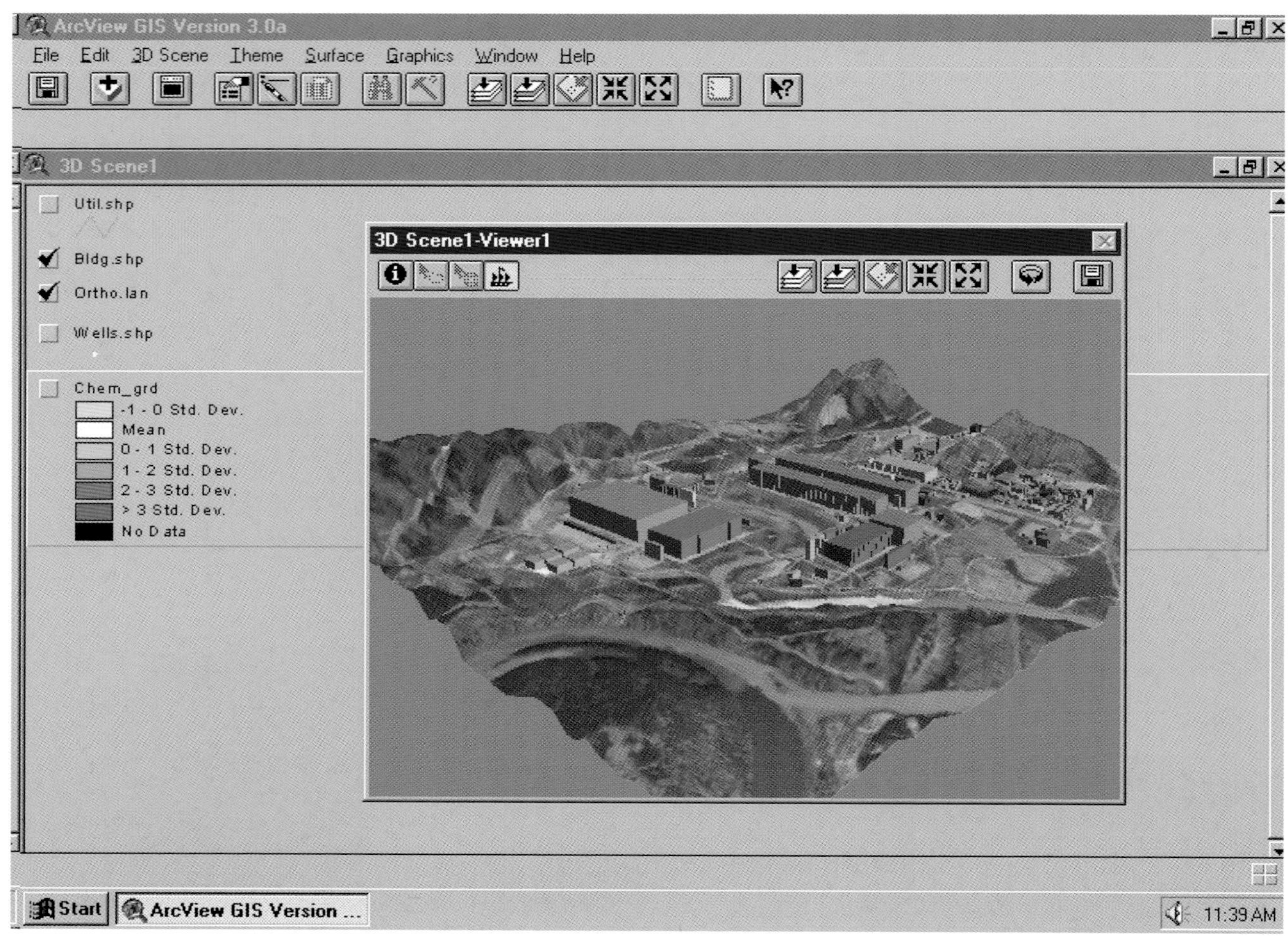

FIGURE 9.26
True three-dimensional terrain visualization using three-dimensional analysis of ArcView GIS. *(Courtesy: ESRI, Redlands, CA)*

(a) Landscape scene before the construction of the reservoir

(b) Landscape scene when the reservoir starts storing water

(c) Landscape scene when the reservoir is at full capacity

FIGURE 9.28
Selected scenes from a three-dimensional visualization application showing how a valley is flooded when a reservoir is built. *(Courtesy: Questar Productions)*

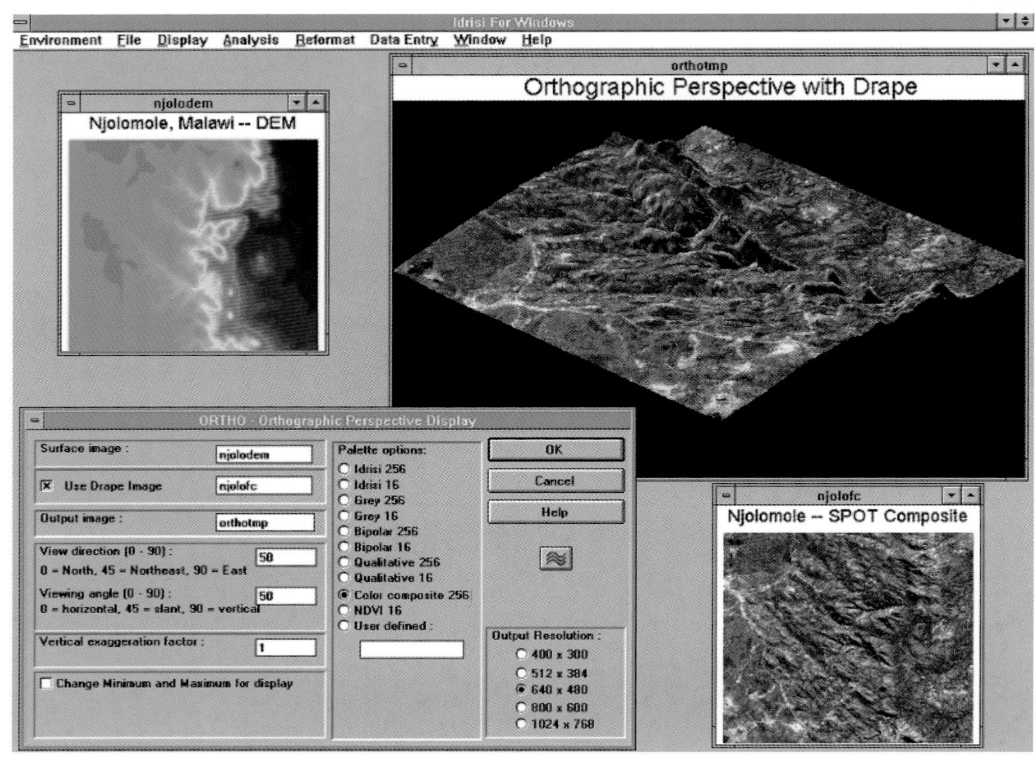

FIGURE 9.30
GUI of Idrisi showing its three-dimensional visualization capability. *(Courtesy: Clarke University)*

FIGURE 9.33
Application of three-dimensional terrain visualization in combat pilot training Virtual world construction programs are now commonly used to construct multimedia applications designed to train military pilots in the identification of military vehicles, including animated aircrafts, ground vehicles, and ships. *(Courtesy: Stage 22 Imaging, Winter Park, FL)*

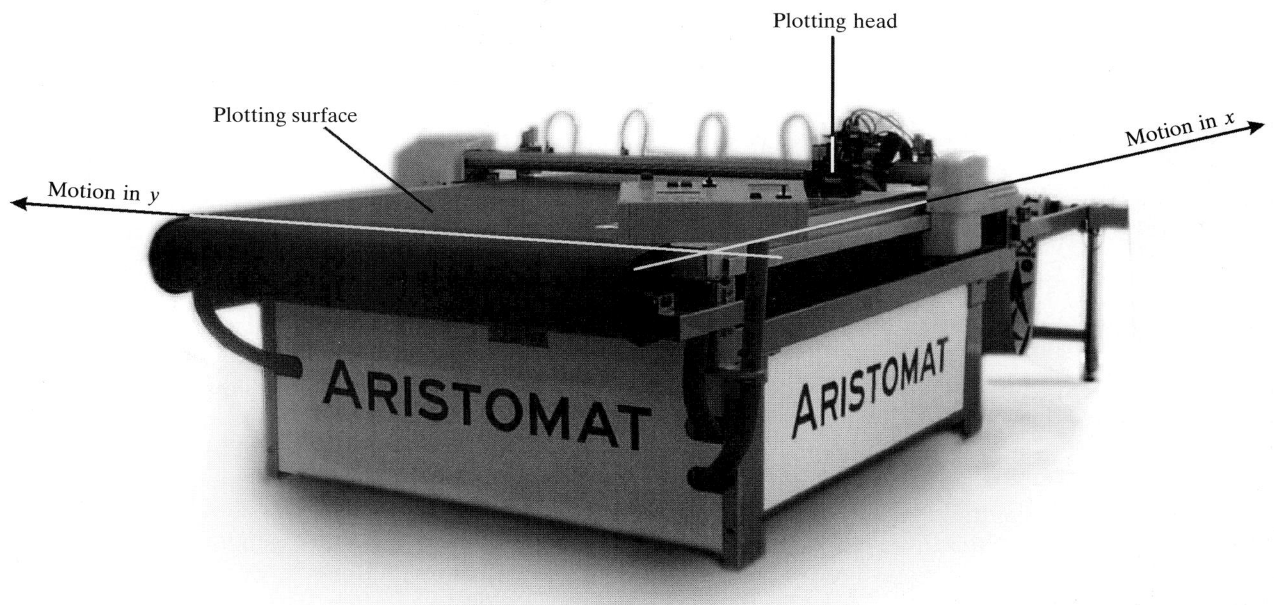

FIGURE 7.20
A flat-bed plotter. *(Courtesy: Aristomat)*

four-color printing process. During the process, the paper is subject to four passes whereby the three subtractive primary colors (i.e., yellow, magenta, and cyan) and black are applied separately. The primary advantage of the electrostatic printers over the pen plotter is the drafting speed. Since the output is built up by evenly spaced rows of images (i.e., output is formed line by line), the time required to complete a plot is not affected by the complexity of its content.

In recent years, wide-format ink-jet printers (Figure 7.21) have gradually overtaken the conventional electrostatic printers for raster plotting. This type of printer creates images on paper by spraying streams of ink droplets through tiny holes. The droplets are electrically charged when they pass through charged plates in the process. This allows them to be either attached to or deflected from the paper. Earlier generations of ink-jet printers had a resolution similar to that of the conventional electrostatic printer. New ink-jet printers such as the Hewlett-Packard DesignJet 3000CP capable of printing at 600 dpi or higher resolution with millions of colors are now available. This new generation of ink-jet printers is capable of creating inexpensive color proofs and provides a relatively simple way of generating big colored GIS information products for public display purposes.

For desktop GIS applications, GIS users can choose from a variety of small-format printers based on ink-jet, laser, and thermal printing technologies (Figure 7.22a). Desktop *ink-jet printers* appear to be the most popular for personal use because of their low cost and wide acceptance as a standard peripheral of PCs. However, they are generally slow, and colored output is not always good enough for publishing. A new type of desktop ink-jet printer, known as a *photo-centric printer*, offers photo-quality printing at 1200 dpi resolution. These

(a)

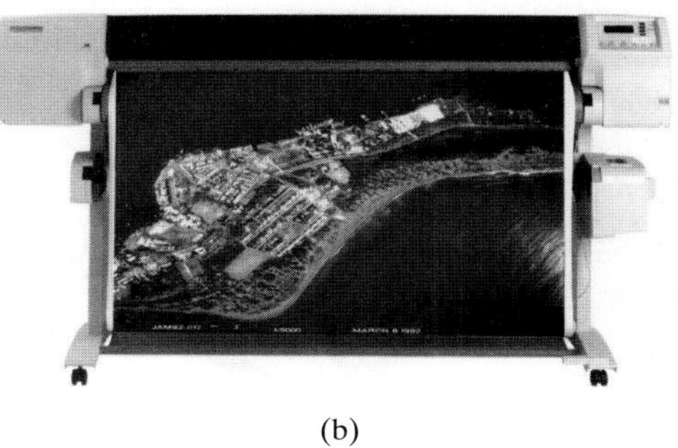

(b)

FIGURE 7.21
Two examples of wide-format ink-jet printers used in GIS (a) HP DesignJet 1050C and (b) HP DesignJet 3000CP. *(Source: Hewlett Packard and ESRI)*

(a) Epson ink-jet printer (b) HP laser printer

FIGURE 7.22
Desktop ink-jet and laser printers.

printers are more suitable than the regular ink-jet printers for GIS applications. The major disadvantage is that they are more expensive. The requirement to use high-quality inks and special glossy papers adds additional costs to their use in the production environment.

Laser printers work on a principle similar to electrostatic printing, except that a laser beam directed by a rotating drum mirror is used to charge the paper (Figure 7.22b). They are in general more expensive than ink-jet printers, but the production cost per page (toner, developer, and drum) is about 25–50% lower. Laser printing is fast and produces crisp, clean output. The best result is obtained when an *image setter* is used. This laser-based device can generate images with a resolution of 1200 dpi on paper and 2400 dpi on photographic film. The ability to generate high-resolution prints at high speed on regular paper makes laser printers the best choice for workgroup or departmental GIS applications in a network environment. From the perspective of generating GIS information products, laser printers are most suitable for printing a large quantity of maps, graphs, and charts for distribution to the public and for inclusion in technical reports and manuals. If color is a critical requirement (e.g., displaying results of visualization), the use of *thermal printers* should be considered. This particular type of printer makes use of temperature-sensitive pigments or dyes to generate colored prints of photo-quality. The obvious drawbacks are the high costs of both the hardware and the media, which have prevented them from popular use for GIS applications.

7.6.4 GENERATION OF MULTIMEDIA INFORMATION PRODUCTS

The idea of using nonprint media for geographic information has been around for quite a while. Its history can probably be traced to the *Domesday Project* that took place in the United Kingdom in the 1980s (Rhind et al., 1988). Key components of the resulting system of this project were two video disks containing a collection of 21,000 spatial digital data sets, 24,000 Ordnance Survey topographic maps at various scales, statistical tabulations, time series, picture libraries, and TV film clips. Since that time, the interest in *spatial multimedia* has gradually grown as multimedia technologies evolve and incorporation of multimedia data types into GIS increases (Groom and Kemp, 1995). The use of multimedia technologies has now irrevocably become an integrated part of GIS. Using multimedia technologies is the most convenient and effective way of integrating geographic information from different types and sources. It provides the essential concepts and techniques for many new GIS applications such as visualization and spatial decision support systems. It also helps popularize geographic information by delivering it to more people and in more understandable ways.

The mechanisms by which geographic information can be presented by multimedia vary considerably from one application to another. In terms of the complexity of the technologies used, the applications of spatial multimedia can be grouped into the following categories: (1) publication of geographic information by CD-ROM, (2) production of multimedia information products, (3) electronic atlases, (4) real-time geographic information display by Web cams, and (5) hypermaps and digital spatial libraries. The characteristics of and technologies used by these applications are explained as follows.

Publication of Geographic Information by CD-ROM

Other than the conventional printed products, CD-ROM has now become the most popular medium by which geographic information is disseminated and published. According to a customer survey conducted by the USGS in 1995, CD-ROM was the preferred distribution media of 72% of the respondents (Gillespie and Snyder, 1995). As a media for data delivery, CD-ROM has many advantages, including

1. high-capacity (650 MB)—suitable for distribution of large volumes of data
2. efficiency of use—the ability of holding a large volume of data in one single disc minimizes or eliminates the need to swap discs frequently, as with floppy discs
3. easy handling and shipment by mail—CD-ROM is less bulky than the stack of floppy discs or magnetic tapes for similar amount of data
4. accessibility by large audience—CD-ROM is now a standard peripheral of PCs, making data more accessible to more users than the old method of using magnetic tapes that require special equipment
5. stability of the medium secured by standards such as ISO 9660

6. direct use from a CD drive (i.e., no need to transfer data to hard drive)
7. low per-unit production and distribution cost

The generation of CD-ROM information products is a relatively straightforward process. It involves the creation of a master that is used for mass duplication under computer control. The price of CD-ROM mastering and duplication technologies has been reduced considerably in recent years. There are now desktop CD-ROM duplicators, complete with hardware and software (authoring programs) that allow the direct transfer of selected files from the hard drive to a recordable compact disc. This makes information dissemination and publication by CD-ROM not only more affordable, but also more practical and flexible. CD-ROM information products can now be generated on demand, and their contents can also be specifically customized according to the particular needs of individual users.

Production of Multimedia Information Products

Multimedia allows for the use of audio, graphical, textual, and video files in an integrated information communication environment. For the presentation of geographic information, these media can be used individually or in different combinations. A full-blown multimedia system for geographic information is called a *hypermap*, which we will explain later. In the following discussion, we will briefly look at the characteristics and applicability of individual electronic media.

- *Audio*. Sound is used as a common addition to multimedia presentation of geographic information. In some electronic atlases, sound is used to enhance graphics (e.g., Microsoft's *Encarta World Atlas* contains 4000 sound clips). Sound is particularly useful for environmental impact assessment involving noise nuisance when noise levels are calibrated to match visual observation conditions (Hughes, 1996; Krygier, 1994). It is relatively easy to create a digital audio file that can be used to support visual presentation of geographic information. Analog sound recorded by an ordinary recorder can be encoded into the digital form using the *Sound Recorder* that comes with Windows (Figure 7.23) on any PC equipped with a sound card. Sound Recorder is capable of capturing sounds in 8-bit mono at 11 KHz. For more advanced sound recording and editing (e.g., 16-bit stereo), it is necessary to use special audio application software (such as Sound Blaster's *WaveEditor*). Digital sound files can be stored in a variety of formats, including *AIFF* (Audio Interchange File Format), Macintosh *SND*, Windows *Wave*, and *MPEG* (Moving Picture Experts Group) *Audio*.
- *Video*. Digital video (usually referred to as a *movie* in multimedia terminology) is extremely useful for showing the landscape of an area of

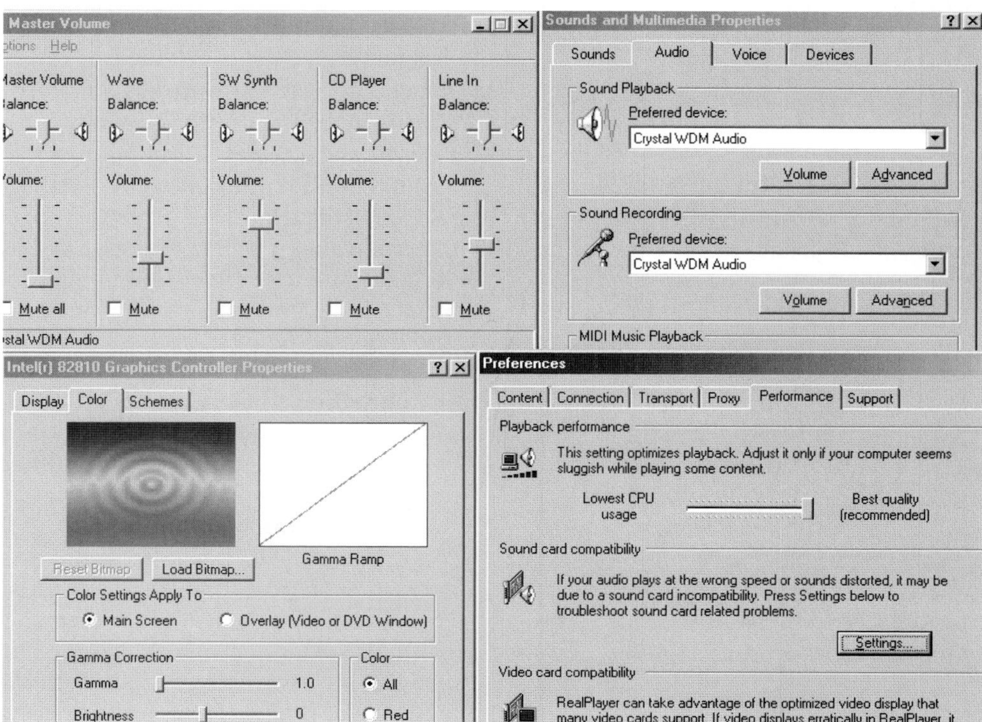

FIGURE 7.23
Interfaces for recording and playing multimedia files in Windows.

interest. Video images have been widely used for pavement management in highway engineering and also for real estate and tourism promotion. The process of capturing video into the computer, similar to audio capture, is easy with the aid of a video capture card, image capture software, and necessary hook-up to the camcorder or camera. As video files are usually of very large size, digital video data are always stored in compressed form. Decompression occurs on the fly when the files are played back. The three video file formats accepted by the computer industry at large are *QuickTime* (QT), *Video for Windows* (VfW), and *MPEG* (Moving Picture Experts Group) *Video*.

- *Virtual Reality (VR), or Virtual Environment (VE)*. This is a collective name for a variety of techniques designed to allow the user to interact with spatial information in three dimensions. Depending on the hardware and interface capabilities, virtual reality can take several forms (Fairbairn and Parsley, 1997). *Desktop VR*, which makes use of the standard computer monitor and conventional PC software to create and view virtual worlds, is probably the most important for geographic information because of its low cost, compatibility, and availability on the Internet.
- *Animation*. The creation of animated maps has long been a cartographic practice. Advances in multimedia have made the creation of animated maps much easier than ever before. Animation can be done by means of special software programs called animators.

Electronic Atlases

An electronic atlas, in simple terms, can be defined as an atlas that is developed for use on electronic media. It permits real-time interactions between the atlas and its user. Earlier attempts to produce electronic atlases aimed primarily at making data viewable on a computer screen and transferable in digital form. Developments in GIS and multimedia technologies have injected many new concepts into the design and production of electronic atlases. These include advanced methods of spatial data indexing and query, hyperlinks (i.e., nonsequential access to data; see explanation of *hypermap* later in the text), as well as analytical functions. As a result, today's electronic atlases are very much different from what they were previously in terms of both functions and forms.

The production of electronic atlases requires careful design and planning. A good understanding of user expectations and requirements is essential. This helps define the contents of the electronic atlas (i.e., the relevance and depth of the treatment of the subject matter). The design of the GUI is another critical consideration because it determines how easy or difficult the user can navigate back and forth when interacting with the data. In order to properly set up the hyperlinks between the data sets, it is necessary to ascertain at the planning stage the relationships between the maps and associated documents (e.g., photographs, imagery, articles, and statistical tables), as well as the relationships among the maps themselves (e.g., topographic maps, thematic maps, and three-dimensional maps).

Many electronic atlases are now available in the market. Some of these are produced by commercial organizations. Others are produced by universities and government agencies. Examples of commercial electronic atlases include Rand McNally's *New Millennium World Atlas*, Compton's *Interactive World Atlas*, and Microsoft's *Virtual Globe*. These products are invariably characterized by the use of CD-ROM as the media of distribution, intuitive user interface, multimedia presentation, and a myriad of data analysis functions such as map comparison and data aggregation. For users with access to the Internet, there are also direct links to the wealth of information available in the World Wide Web, as exemplified by the interface of the *New Millennium World Atlas* in Figure 7.24.

An example of electronic atlases produced by universities is the *Interactive Atlas of Georgia* (Hodler et al., 1994), which replaced the traditional hard copy *Atlas of Georgia* produced in 1986 (Hodler and Schretter, 1986). This interactive atlas allows users to make maps from the data in the atlas using any class intervals desired. Users can also query the maps for locations of places and data.

One of the best examples of electronic atlases produced by a government agency is the *National Atlas Information System* (NAIS) of Canada (Figure 7.25). This product was first conceptualized in the late 1970s (Groot, 1979). Its development since then has undergone considerable changes in response to changes in both technologies and user requirements (Siekierska et al., 1990). In its present implementation, which is delivered through the Internet *(http://atlas.gc.ca/english)*, the National Atlas Information System contains several components. These include educational resources for geography and cartography, community and natural resource mapping, links to information resources provided by different government agencies and departments, election mapping, as well as Canadian heritage sites.

In 1998, the USGS started work on the *National Atlas of the United States* (Figure 7.26). This is a cooperative project between USGS and its partners in private industry. The national atlas is designed to promote greater geographic awareness through the development and delivery of products that provide easy-to-use, maplike views of the natural and socioeconomic landscapes of the United States. It is intended to serve the interests

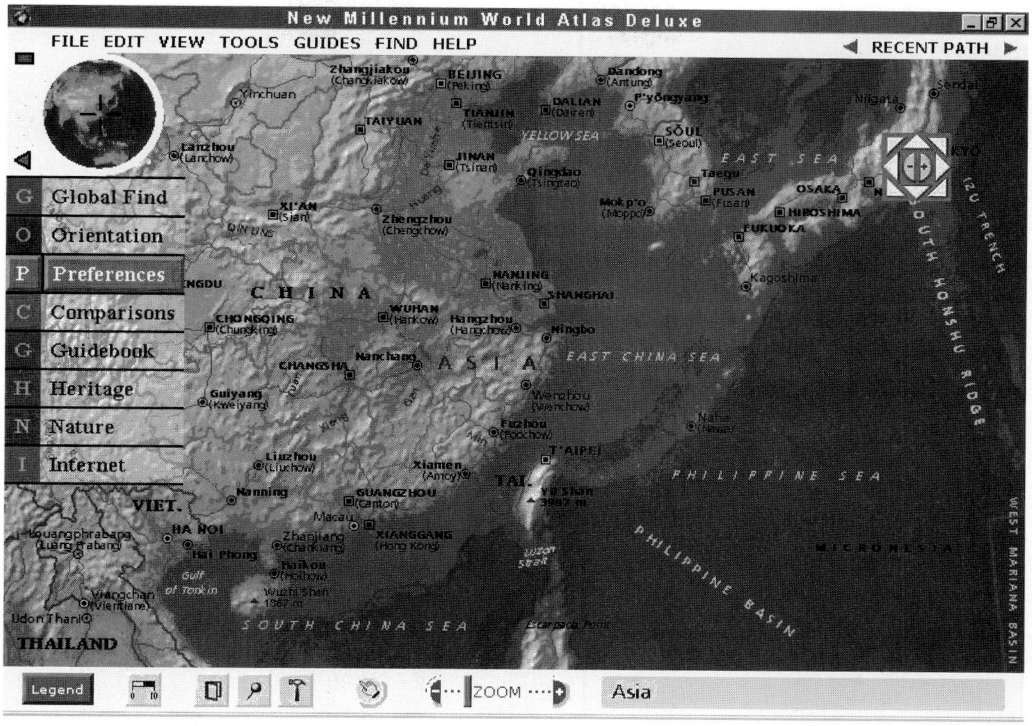

FIGURE 7.24
An example of an electronic atlas. *(Source: New Millennium World Atlas)*

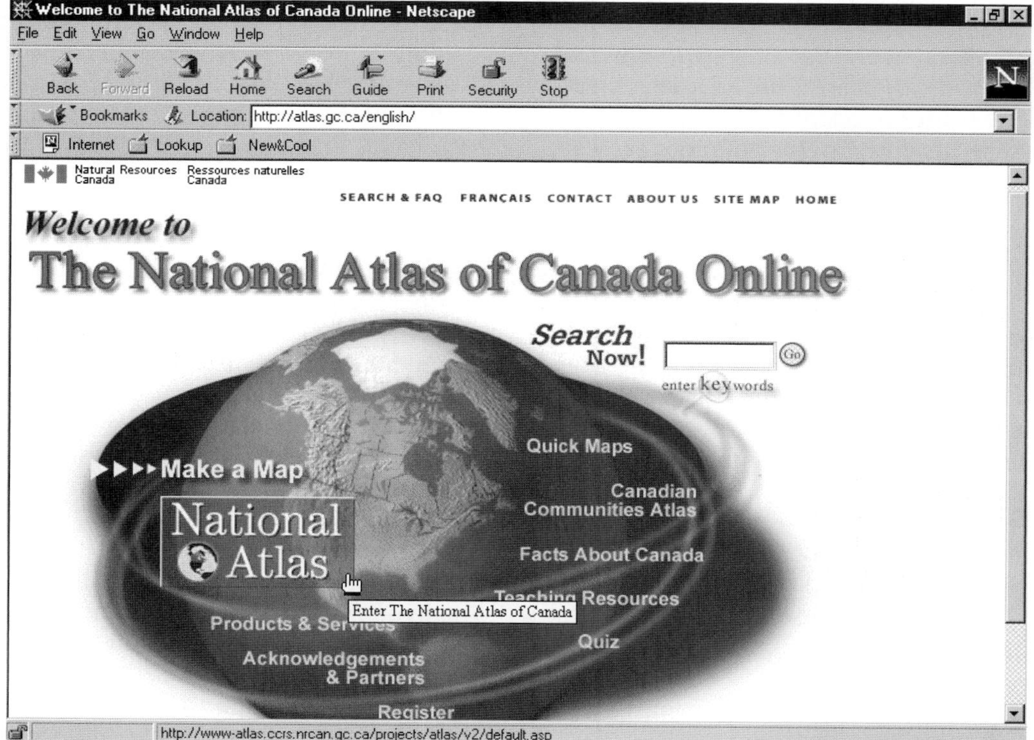

FIGURE 7.25
Interface of the Canadian National Atlas Information System. *(This is based on information taken from the National Atlas of Canada web site http://atlas.gc.ca ©2001. Her Majesty the Queen in Right of Canada with permission of Natural Resources Canada)*

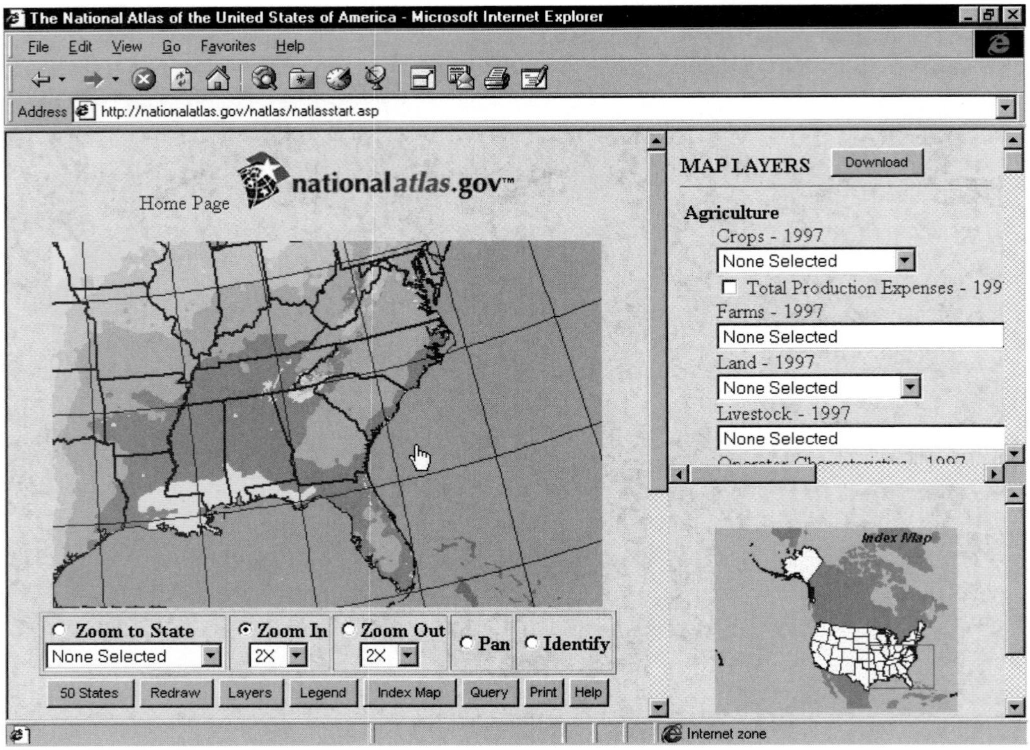

FIGURE 7.26
National Atlas of the United States. *(Source: US Geological Survey)*

and needs of a diverse populace in different ways: as a basic reference, a framework for information discovery, an education tool, a research aid, as well as a reliable and accurate source of scientific information. The U.S. National Atlas, like its Canadian counterpart, is accessed by the Internet (http://nationalatlas.gov/).

Real-Time Geographic Information Display by Web Cams

A *Web cam* is basically a video camera that is connected to a Web page on an Internet server. In a sense, a Web cam is both a data capture and a data display device, depending on the purpose of the user. In the last couple of years Web cams have become a popular tool for geographic information dissemination of the tourism industry. The Resorts Sports Network, for example, has live Web cams at many resorts in the United States. People planning a ski trip can obtain real-time information about the depth of the powder and the length of the lift lines at any of these locations. Another fast growing application of Web cams is traffic management. A series of Web cams can be installed in key locations (e.g., intersections) along major corridors to form a real-time highway information system (Figure 7.27). Such a system can be used by traffic managers to monitor road conditions and by road users to plan their routes. Web cams have also been used to monitor scientific instruments and events. Examples include (Thoen, 1996) the NBC Seismo Cam, which provides a live shot of a seismograph in southern California; the Web cam that was installed to observe the eruption of Mt. Ruapehu in New Zealand in 1995; and the Web cam set up by the Center for Coastal Geology of the USGS to record cliff erosion along the southern shore of Lake Erie in Ohio.

Conventionally, the capture and presentation of time as the fourth dimension of geographic information has been a difficult and time-consuming task. It appears that Web cams have proved to be a viable solution. Many Web cam sites offer downloadable MPEG video files that allow users to see the passage of time at those locations. This technology has in fact been increasingly used by many weather-related sites to capture and present daily and seasonal weather conditions.

Hypermaps and Digital Spatial Libraries

The hypermap probably represents the most advanced and complex use of multimedia for geographic information. Practically speaking, a hypermap is more a concept, rather than a physical information product. Its objectives are used to structure individual multimedia components in respect to one another and to the map, and to allow the user to navigate the data. The concept of a hypermap is based on the methods of *hypertext* and *hypermedia* (Laurini and Thompson, 1992). Hypertext is

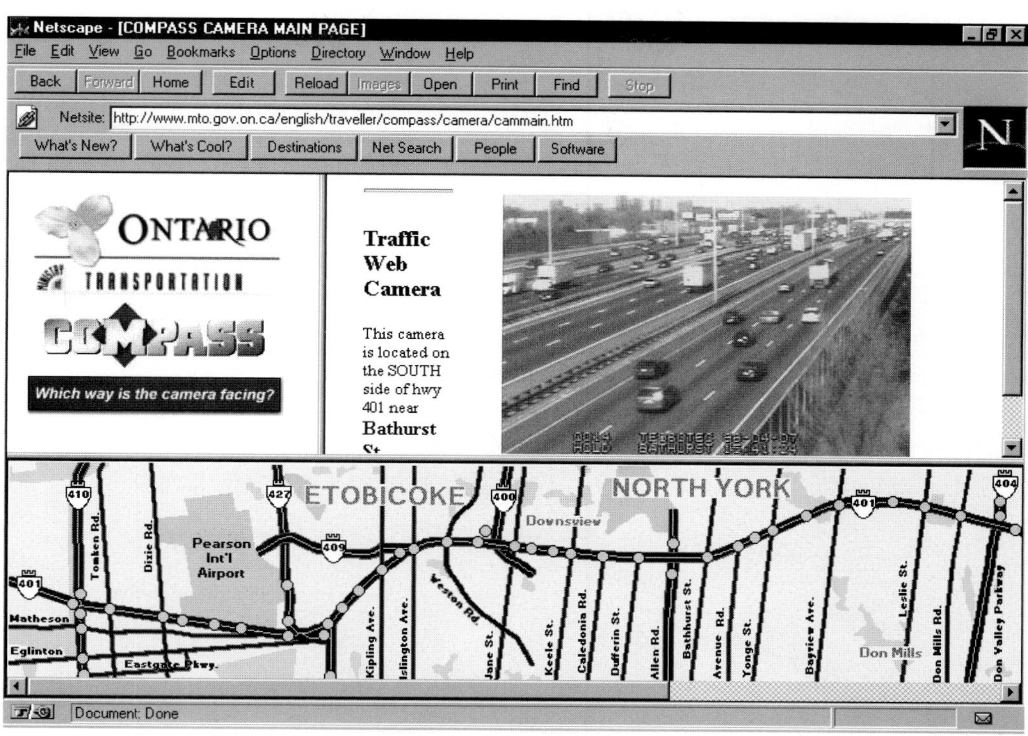

FIGURE 7.27
Example of a Web cam. *(©Queen's Printer for Ontario, 2001. Reproduced with permission)*

the method by which links, called *hyperlinks* or *hotlinks*, are incorporated in a document to enable the seamless referencing to other documents and retrieving information from them. The hyperlinks form a web of logically connected files that allows nonsequential access to information at different locations, in much the same way as the human brain links associated pieces of information (Figure 7.28). Hypermedia is the extension of hypertext when other forms of media, such as graphics, sound, and digital movies, are supported. By using hyperlinks, textual, graphics, audio, and video files may be seamlessly linked together for the presentation of information in what has come to be known as the multimedia environment.

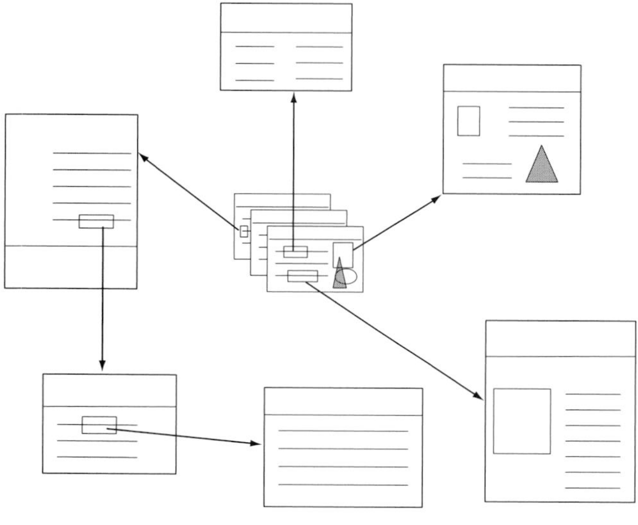

FIGURE 7.28
Principle of hypertext.

The inclusion of a map as a component of a hypermedia system gives it a new spatial dimension and a new identity called the *hypermap*. A hypermap is distinguished from other forms of hypermedia systems by its capability to navigate data sets not only by theme but also by geography. In a way, this is an implementation of the idea of the map as an interface to geographic information that we noted in Section 7.2.1. Spatial data navigation in a hypermap always starts with a search window on a digital map that covers the area of interest (Kraak and van Driel, 1997) (Figure 7.29). This search window can be defined in a number of ways, such as keyboard input of minimum and maximum coordinates, using the mouse to create a window on the screen, and entering a key word or geographic name such as name of a country, state, city, or neighborhood. Time may also be entered to define a specific time frame for the search. By using the vector-based GIS processing method of overlay (see Chapter 6), spatial objects falling within the search window can be extracted from the geographic database. Information relating to a particular spatial object can then be accessed through its hyperlinks. For example, hyperlinks

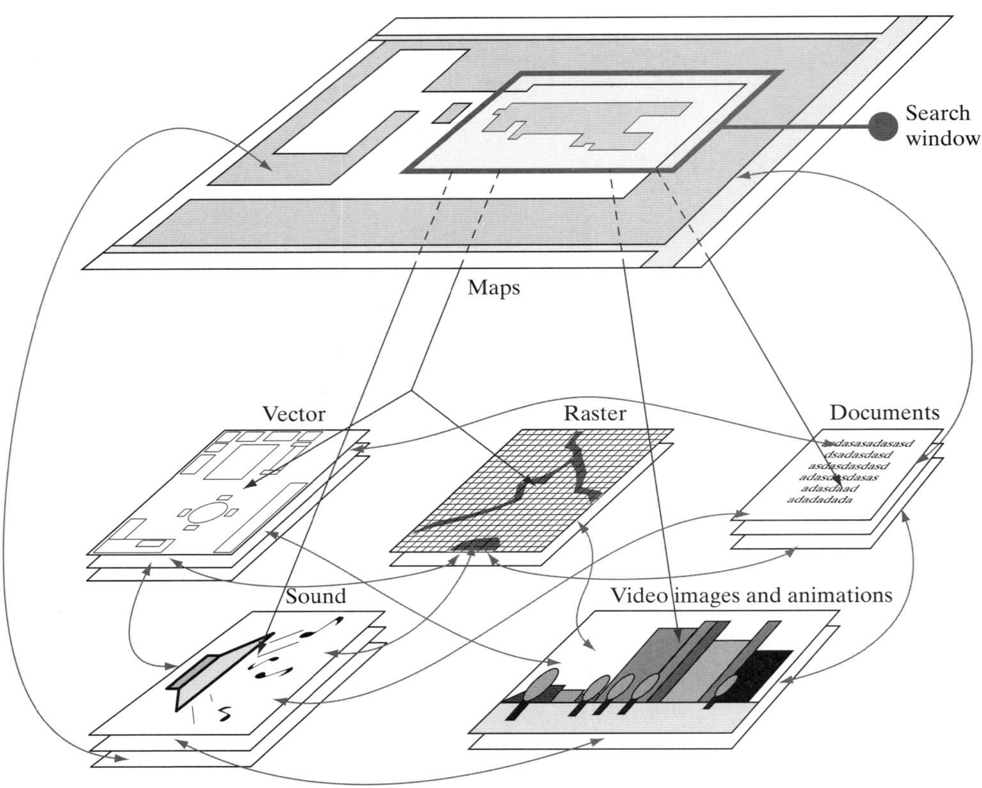

FIGURE 7.29
The concept of hypermap. A hypermap makes use of multimedia technology to access geographic information in a variety of visual and audio forms. *(Source: Kraak and Ormeling, 1996)*

for a particular building can provide access to its floor plans, photographs or video files showing its appearance and structural conditions, zoning and engineering plans of the surrounding areas, transit routes passing through the street, as well as social and community information pertaining to the neighborhood. These documents themselves may have hyperlinks to other documents that lead to other sources of information. These may include, for example, property assessment files of the building, schedules for individual transit routes, gazetteers governing land-use planning, telephone number and address of the community center, and census statistics of the area.

Over the last several years, a variety of systems have been developed using more or less the concept of hypermaps as described here (Kraak and van Driel, 1997). In recent years, the *digital spatial library* has gradually evolved to become a reality. This particular concept can be regarded as an extension of the hypermap. At the same time, it is also an extension of the *digital library* (Wiederhold, 1995). As its name implies, a spatial digital library is a digital library that is specially designed for spatial information. A spatial digital library can be constructed in different ways, but typically it is made up of four components.

1. *Geographic data* include map data from government agencies, statistical data from government and research institutes, and data sets associated with particular GIS projects.
2. *Metadata* are descriptions of the digital library's geographic data holding for use by potential users to determine if the data are relevant to their applications or not.
3. *Metadata index* is the mechanism for the users to search through the metadata for a data set that meets desired application requirements.
4. *Search interface* is the front end for the metadata index that allows the user to enter queries in an intuitive manner.

As digital spatial library is a relatively new concept, there are not many operational implementations around. A good example is the *Alexandria Digital Library* (ADL) at the University of California, Santa Barbara (Goodchild, 1995) (Figure 7.30). ADL was established in 1994 as one of the six *Digital Library Initiative* (DLI) projects funded by the National Science Foundation (NSF) and other government agencies. The goal of ADL is to design, implement, and deploy a digital library for spatially indexed information, which includes three types

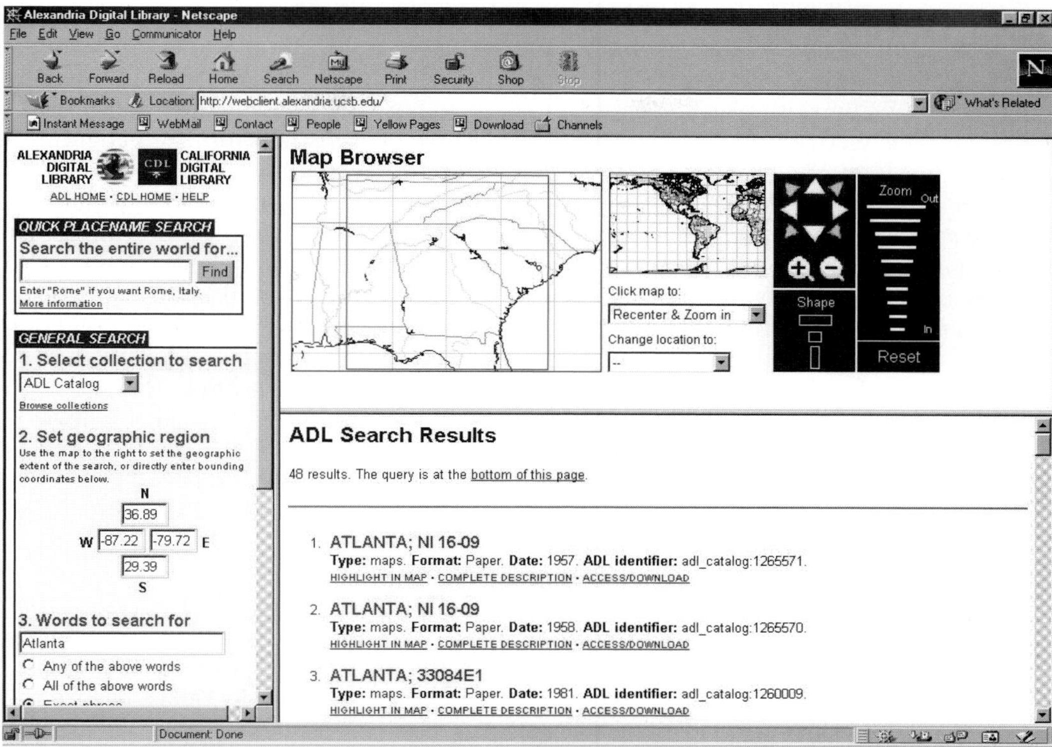

FIGURE 7.30
Spatial query GUI of Alexandria Digital Library (ADL). The view is divided into three parts: the Map Browser, the Search Box, and the Search Results Box. The map can be zoomed in and out, and the map extent box can be changed. The latitudinal and longitudinal extent of the box is also displayed *(Courtesy: Department of Geography, University of California, Santa Barbara).*

of information objects: *spatial objects*, such as maps, images or photographs; *spatially referenced objects*, such as books and audio and video files that can be positioned in space and for which spatial reference is a useful basis for retrieval; and *general objects*, which have no spatial reference. The first prototype of ADL was completed in early 1995 and was available on CD-ROM. A prototype World Wide Web implementation was completed later in the same year. ADL has demonstrated that digital spatial libraries can greatly expand the traditional role of map libraries as a repository of printed maps and related materials. Like all other types of digital libraries, digital spatial libraries are perceived not only as a public and educational service for the advancement and dissemination of spatial knowledge, but also as an important component of the information technology infrastructure of modern society. The next stage of the project, known as Alexandria Digital Earth Prototype (ADEPT), was also funded by NSF for the period 1999–2004. The Digital Library has now expanded into a Digital Earth, which is a multiresolution, three-dimensional representation of Earth into which we can put in vast amount of georeferenced data. Using a variety of multimedia tools, we can now explore Earth through space and time by means of three-dimensional visualization.

7.7 Summary

In this chapter, we have examined two important issues pertaining to the information presentation function of GIS: visualization and generation of information products. The application environment of GIS today is very different from what it was earlier. This has necessitated the development of a new model of GIS information communication that describes how geographic information is assimilated and presented. One of the key features of this new model of information communication is the emerging role of the map as an interface to geographic information. As a result, the use of GIS now requires considerably more human–computer interaction (HCI) than ever before. The adoption of sound HCI methods for graphical user interface (GUI) design has made GIS easier to use despite the increasing sophistication of the applications. This has also stimulated GIS users to explore the opportunities of using new technologies in computer graphics to meet their diverse application requirements. In essence, this is the major driving force behind the current thrust toward the use of scientific visualization and multimedia technologies in GIS.

Advances in the methods of scientific visualization (SciVis) have given the presentation of geographic information a new meaning and a new horizon of applications. It has freed the presentation of geographic information from the static two-dimensional world. Using SciVis techniques in conjunction with related technologies such as the World Wide Web protocol, digital spatial library technologies, and high-speed communications networks, GIS users are now able to access, retrieve, merge, and analyze complex spatial data sets dynamically in three dimensions and in real time. It enables users to detect and understand spatial relationships that would otherwise be impossible using conventional methods.

The diversification of GIS applications and the use of multimedia technologies for the presentation of geographic information imply that the map is no longer the sole output product of GIS. Geographic information can now be presented in both printed and nonprinted forms, using a variety of media and for a multitude of purposes that include conventional mapping, spatial problem solving, and the distribution of spatial knowledge. Visualization of geographic information does not mean merely making the information visible, and generation of information products does not only involve pressing the button of a plotter or a printer. Both of these tasks require leading-edge computer technologies, innovative thinking, and advanced knowledge of graphics design. In the context of modern GIS, cartography is an integral part of GIS processing and analysis and, as we have demonstrated, one of the major driving forces behind the development of GIS technology today.

REFERENCES

Apple Computer (1987) *Human Interface Guidelines: The Apple Desktop Interface*, Cupertino, CA: Apple Computer Inc.

Armstrong, M. P., Densham, P. J., and Lolonis, P. (1991) "Cartographic visualization and user interface in spatial support systems," *Proceedings*, GIS/LIS '91, Vol. 1, pp. 321–330, Atlanta, GA.

Artimo, K. (1994) "The bridge between cartographic and geographic information systems," in *Visualization in Modern Cartography* by MacEachren, A. M., and Taylor, D. R. F. eds. pp. 45–61, Oxford: Pergamon Press.

Batty, M. (1994) "Using GIS for visual simulation modeling," *GIS World*, Vol. 7, No. 10, pp. 46–48.

Batty, M., and Xie, Y. (1994) "Urban analysis in a GIS environment: Population density modelling using ArcInfo," in *Spatial Analysis and GIS* by Fotheringham, S., and Rogerson, P. eds. pp. 189–219, London: Taylor and Francis.

Berry, J. K. (1995) "Topic 4—Toward an honest GIS: Practical approaches to mapping uncertainty," *Spatial Reasoning for Effective GIS*, Fort Collins, CO: GIS World.

Buckley, D. J., Ulbricht, C., and Berry, J. K. (1998) "The Virtual Forest: Advanced 3-D visualization techniques for forest management and research," *Conference Proceedings*, GIS '98/RT '98, pp. 382–389, Toronto, ON.

Buttenfield, B. P., and Mackaness, W. A. (1991) "Visualization," in *Geographical Information Systems, Vol. 1: Principles*, by Maguire, D. J., Goodchild, M. F., and Rhind, D. W. eds. pp. 427–443, Harlow, UK: Longman Scientific and Technical.

Buttenfield, B. P., and Mark, D. M. (1991) "Expert systems for cartographic design," in *Geographic Information Systems: The Microcomputer and Modern Cartography* by Taylor, D. R. F. ed. pp. 129–150, Oxford: Pergamon Press.

Clarke, K. C. (1995) *Analytical and Computer Cartography*, Englewood Cliffs, N.J.: Prentice Hall.

Davis, T. J., and Keller, C. P. (1997) "Modelling and visualizing multiple spatial uncertainties," *Computers and Geosciences*, Vol. 23, No. 4, pp. 397–408.

Dent, B. D. (1993) *Cartography: Thematic Map Design*, Dubuque, IA: Wm. C. Brown Publishers.

Dykes, J. (1996) "Dynamic maps for spatial science: A unified approach to cartographic visualization," in *Innovations in GIS 3* by Parker, D. ed. pp. 177–187, London: Taylor & Francis.

Egenhofer, M. J. (1992) "Why not SQL!," *International Journal of Geographical Information Systems*, Vol. 6, No. 2, pp. 71–85.

Egenhofer, M. J., and Frank, A. U. (1988) "Designing objected oriented query languages for GIS: Human interface aspect," *Proceedings*, Symposium on Spatial Data Handling, pp. 79–96, Sydney, Australia.

Egenhofer, M. J., and Richards, J. R. (1993) "The Geographer's Desktop: A direct-manipulation user interface for map overlay," *Proceedings*, AutoCarto 11, pp. 63–71, Minneapolis, MN.

Ehlschlaeger, C. R., Shortridge, A. M., and Goodchild, M. F. (1997) "Visualizing spatial data uncertainty using animation," *Computers and Geosciences*, Vol. 23, No. 4, pp. 387–396.

ESRI (1995a) *Understanding GIS: The ArcInfo Method* (Version 7 for UNIX and Open VMS), 3rd ed. Redlands, CA: Environmental Systems Research Institute, Inc.

ESRI (1995b) *ArcTools—Arc/Info Software's Graphic User Interface*, an ESRI White Paper, Redlands, CA: Environmental Systems Research Institute, Inc.

ESRI (1998) *Using ArcExplorer*, Redlands, CA: Environmental Systems Research Institute, Inc.

Fairbairn, D., and Parsley, S. (1997) "The use of VRML for cartographic presentation," *Computers and Geosciences*, Vol. 23, No. 4, pp. 475–482.

Fisher, P. F., and Mackaness, W. A. (1987) "Are cartographic expert systems possible?," *Proceedings*, AutoCarto 8, pp. 530–534, Baltimore, MD.

Freeman, H. (1991) "Computer name placement," in *Geographical Information Systems, Vol. 1: Principles*, by Maguire, D. J., Goodchild, M. F., and Rhind, D. W. eds. pp. 445–456, Harlow, UK: Longman Scientific and Technical.

Gillespie, S., and Snyder, G. (1995) "U.S. Geological Survey customers speak out," *GIS World*, Vol. 8, No. 9, pp. 48–50.

Goodchild, M. F. (1995) *Alexandria Digital Library* (Report on a Workshop on Metadata held in Santa Barbara, California, November 8, 1995), National Center for Geographic Information and Analysis, University of California, Santa Barbara, CA.

Gordin, D. N., Edelson, D. C., and Gomez, L. M. (1996) "Scientific visualization as an interpretive and expressive medium," *Proceedings*, Second International Conference on the Learning Sciences, pp. 409–414, Charlottesville, VA.

Gould, M. (1989) "Human factor research and its value to GIS user interface design," *Proceedings*, GIS/LIS '89, Vol. 2, pp. 542–550, Orlando, FL.

Groom, J., and Kemp, Z. (1995) "Generic multimedia facilities in geographic information systems," in *Innovations in GIS 2* by Fisher, P. ed. pp. 189–200, London: Taylor & Francis.

Groot, R. (1979) "Canada's National Atlas program in the computer era," *Cartographica*, No. 23, pp. 41–52.

Hodler, T., and Schretter, H. A. (1986) *The Atlas of Georgia*, Athens, GA: Institute of Community and Area Development, University of Georgia.

Hodler, T., Lawson, N., Schretter, H., and Torgeson, J. (1994) *The Interactive Atlas of Georgia*, Athens, GA: Institute of Community and Area Development, University of Georgia.

Hughes, J. R. (1996) "Technology trend mark multimedia advances," *GIS World*, Vol. 9, No. 11, pp. 40–43.

Keates, J. S. (1989) *Cartographic Design and Production*, 2nd ed. Harlow, U.K.: Longman.

Kraak, M.-J. (1995) "The map beyond geographical information systems," in *Innovations in GIS 2* by Fisher, P. ed. pp. 163–168, London: Taylor & Francis.

Kraak, M.-J., and Ormeling, F. (1996) *Cartography: Visualization of Spatial Data*, Harlow, Essex: Longman.

Kraak, M.-J., and van Driel, R. (1997) "Principles of hypermaps," *Computers and Geosciences*, Vol. 23, No. 4, pp. 457–464.

Krygier, J. (1994) "Sound and cartographic visualization," in *Visualization in Modern Cartography*, by MacEachren, A. M., and Taylor, D. R. F. eds. Oxford: Pergamon Press.

Laurini, R., and Thompson, D. (1992) *Fundamentals of Spatial Information Systems*, London: Academic Press.

Lee, Y. C., and Chin, F. L. (1995) "An iconic query language for topological relationships in GIS," *International Journal of Geographical Information Systems*, Vol. 9, No. 1, pp. 25–46.

Liu, G., Ayers, R., Murillo, B., and Tam, W. (1997) "Using Visual Basic and MapObject to develop mapping applications," *Proceedings* (in CD-ROM), Paper #464, ArcInfo Users Conference, San Diego, CA.

Luo, Y. (1997) "Spatial analysis education and GIS componentware technology," *Proceedings* (in CD-ROM), Paper #448, ArcInfo Users Conference, San Diego, CA.

MacEachren, A. M. (1995) *How Maps Work: Representation, Visualization and Design*, New York: Guilford Press.

Mark, D. M. (1989) "Cognitive image schemata for geographic information: Relations to user views and GIS interfaces," *Proceedings*, GIS/LIS '89, Vol. 2, pp. 551–560, Orlando, FL.

Mark, D. M. (1992) *Research Initiative 13 Report on the Specialist Meeting: User Interface for Geographic Information Systems*, Technical Report 92-3, National Center for Geographic Information and Analysis, University of California at Santa Barbara, Santa Barbara, CA.

Masuoka, P., Acevedo, W., Fifer, S., Foresman, T., and Tuttle, M. J. (1996) "Techniques for visualizing urban growth using a temporal GIS database," *Proceedings*, Vol. 3, pp. 89–100, ASPRS/ACSM Annual Convention and Exhibition, Baltimore, MD.

MathSoft (1998) *S + ArcView for GIS*, Cambridge, MA: MathSoft, Inc.

Mayall, K., and Hall, G. B. (1994) "Information systems and 3-D modeling in landscape visualization," *Proceedings*, pp. 796–804, URISA Annual Conference.

McCormick, B. H., Defanti, T. A., and Brown, M. D. (1987) "Visualization in scientific computing," *SIGGRAPH Computer Graphics Newsletter* (Special Edition), Vol. 21, No. 6.

McGaughey, R. J. (1997) "Techniques for visualizing the appearance of timber harvest operations," paper presented at 20th Annual Meeting of the Council on Forest Engineering, Rapid City, SD.

McMaster, R. B., and Shea, K. S. (1988) "Cartographic generalization in a digital environment: A framework for implementation in a geographic information system," *Proceedings*, GIS/LIS '88, Vol. 1, pp. 240–249, San Antonio, TX.

Neves, J. N., Goncalves, P., Muchaxo, J., and Silva, J. P. (1997) "Interfacing spatial information in virtual environments," *Computers and Geosciences*, Vol. 23, No. 4, pp. 483–488.

Ofenstein, W. T., and Rauenzahn, K. A. (1977) "Development of a MapObject-based spatial data viewer and analysis tool for remote sensing applications," *Proceedings* (in CD-ROM), Paper #211, ArcInfo Users Conference, San Diego, CA.

Plewe, B. (1997) *GIS Online: Information Retrieval, Mapping and the Internet*, Santa Fe: NM.

Poor, A. (1997) "Big screen, low prices," *PC Magazine*, Vol. 16, No. 17, pp. 106–157.

Rhind, D., Armstrong, P., and Openshaw, S. (1988) "The Domesday machine: A nationwide geographical information system," *Geographical Journal*, Vol. 154, No. 1, pp. 56–68.

Rhyne, T. M. (1997) "Going virtual with geographic information and scientific visualization," *Computers and Geosciences*, Vol. 23, No. 4, pp. 489–492.

Robertson, P. K. (1988) "Choosing data representation for the effective visualization of spatial data," *Proceedings*, 3rd International Symposium on Spatial Data Handling, pp. 243–252, Sydney, Australia.

Robinson, A. H., Morrison, J. L., Muhrcke, P. C., Kimerling, A. J., and Guptill, S. C. (1995) *Elements of Cartography*, 5th ed. New York: Wiley.

Shea, K. S., and McMaster, R. B. (1989) "Cartographic generalization in a digital environment: when and how to generalize," *Proceedings*, AutoCarto 9, pp. 56–67, Baltimore, MD.

Sheldon, K. M., Barron, J. J., and Smith, B. (1991) "Windows Wars," *BYTE*, June 1991, pp. 124–134.

Siekierska, E., Katznelson, P., and Adomaitis, V. (1990) "Recent developments in the implementation of the Electronic Atlas of Canada," *Proceedings*, pp. 1624–1635, National Conference (GIS for the 1990s), Ottawa, ON.

Slocum, T. A. (1999) *Thematic Cartography and Visualization*, Upper Saddle River, NJ: Prentice Hall.

Taylor, D. R. F. (1991) "Geographic information systems: The microcomputer and modern cartography," in *Geographic Information Systems: The Microcomputer and Modern Cartography*, by Taylor, D. R. F. ed. pp. 1–20, Oxford: Pergamon Press.

Taylor, D. R. F. (1994) "Perspectives on visualization and modern cartography," in *Visualization in Modern Cartography*, by MacEachren, A. M., and Taylor, D. R. F. eds. pp. 333–342, Oxford: Pergamon Press.

Thoen, B. (1996) "Web cams take you here, there and everywhere," *GIS World*, Vol. 9, No. 4, pp. 64–66.

Unwin, D., Dykes, J. A., Fisher, P. F., Stynes, K., and Wood, J. (1994) "WYSIWYG? Visualization in the spatial sciences," paper presented at AGI Conference, Birmingham, U.K.

Weibel, R., and Buttenfield, B. P. (1992) "Improvement of GIS graphics for analysis and decision-making," *International Journal of Geographical Information Systems*, Vol. 6, No. 3, pp. 223–245.

Wiederhold, G. (1995) "Digital libraries, value and productivity," *Communications of the ACM*, Vol. 38, No. 4, pp. 85–86.

CHAPTER

8

Remote Sensing and GIS Integration

8.1 Introduction

The success of any GIS application depends on the quality of the geographic data used. Collecting high-quality geographic data for input to GIS is therefore an important activity. Traditionally, environmental data can be collected directly in the field, using in situ (ground survey) methods. This type of data collection normally makes use of an instrument that measures a phenomenon directly in contact with the ground, such as the pH value of soil, the temperature of the water in a lake, the angle of a slope, or the height of a building. Data collected in the field are normally regarded as high quality, although human and instrument errors do occur (see Chapter 4, Section 4.2.3). In situ data collection can be expensive because it is labor-intensive and time-consuming.

The term *remote sensing* was coined by geographers in the Office of Naval Research of the United States in the 1960s to refer to the acquisition of information about an object without physical contact. The term usually refers to the gathering and processing of information about Earth's environment, particularly its natural and cultural resources, through the use of photographs and related data acquired from an aircraft or a satellite (Simonett, 1983). Today, remote sensing is the preferred method to use if environmental data covering a large area are required for a GIS application.

Remote sensing data can be analog or digital in form as well as small or large in scale, according to the type of sensor and platform used for acquiring the data. In some usage, remote sensing refers only to imagery acquired by sensors using electronic scanning, which detects radiation outside the normal visible range (0.4–0.7 μm) of the electromagnetic spectrum, such as microwave, radar, and thermal infrared. In this chapter and throughout this book, the term "photograph" is used to refer to the picture acquired by a conventional camera in the visible region of the electromagnetic spectrum and is analog in form, while the word "image" or "imagery" refers to nonphotographic pictures acquired by electronic detectors operating in the invisible portion of the electromagnetic spectrum and is digital in form. One should, however, note that near-infrared radiation between 0.8 and 1.2 μm is photographically actinic, which means that it can be recorded on near-infrared film using an ordinary camera.

One notable characteristic of remote sensing is that it is not just a data-collection process. Remote sensing also includes data analysis: the methods and processes of extracting meaningful spatial information from the remote sensing data for direct input to the GIS. In the era of digital computers, remote sensing has been predominantly digital in nature. Even aerial photographs in analog form have been converted to digital form by highly accurate photogrammetric scanners as input to the computer for the production of orthophoto images (Kolbl et al., 1996). In digital form, remote sensing data are compatible with the raster-based GIS data model and can be readily integrated with other types of raster GIS data. In fact, the raster GIS data-processing functionality explained in Chapter 5 originates from the digital analysis of remote sensing data (see Section 5.1.3).

The advantage of remote sensing is the bird's-eye view or synoptic view it provides, so that environmental data covering a large area of Earth can be captured instantaneously and can then be processed to generate maplike products. Another advantage of remote sensing is that it can provide multispectral and multiscale data for the GIS database. However, the adoption of remote sensing does not eliminate in situ data collection, which is still needed to provide ground truth to verify the accuracy of the remote sensing data collected. It helps minimize the amount of in situ data collection.

As our Earth constantly changes, remote sensing from an orbital platform in space provides the easiest means to keep the GIS database up to date. Because remote sensing images and aerial photographs are nonselective in nature, they have to be interpreted or analyzed before useful data can be added to the GIS database. A common application of remote sensing images and aerial photographs is land-use and land cover mapping because human activities are best revealed by the use of the land and the changes to the land cover that have taken place. Land-use and land cover data are descriptive in nature (categorical with the use of a nominal measurement scale). In addition to this application, remote sensing can be used to collect information on several important biophysical variables, which include planimetric location (x, y), topographic/bathymetric elevation (x, y, z), color and spectral signature, vegetation chlorophyll absorption characteristics, vegetation biomass, vegetation moisture content, soil moisture content, temperature, and texture/surface roughness (Jensen, 1983a). All these variables are metric in nature.

8.2 Principles of Electromagnetic Remote Sensing

Both photographic and nonphotographic remote sensing systems record data of reflection and/or emission of electromagnetic energy from Earth's surface (Figure 8.1). The major source of electromagnetic energy is the sun, although Earth itself can emit geothermal and man-made energy. Electromagnetic radiation is a form of energy derived from oscillating magnetic and electrostatic fields (Figure 8.2) and is capable of transmission through

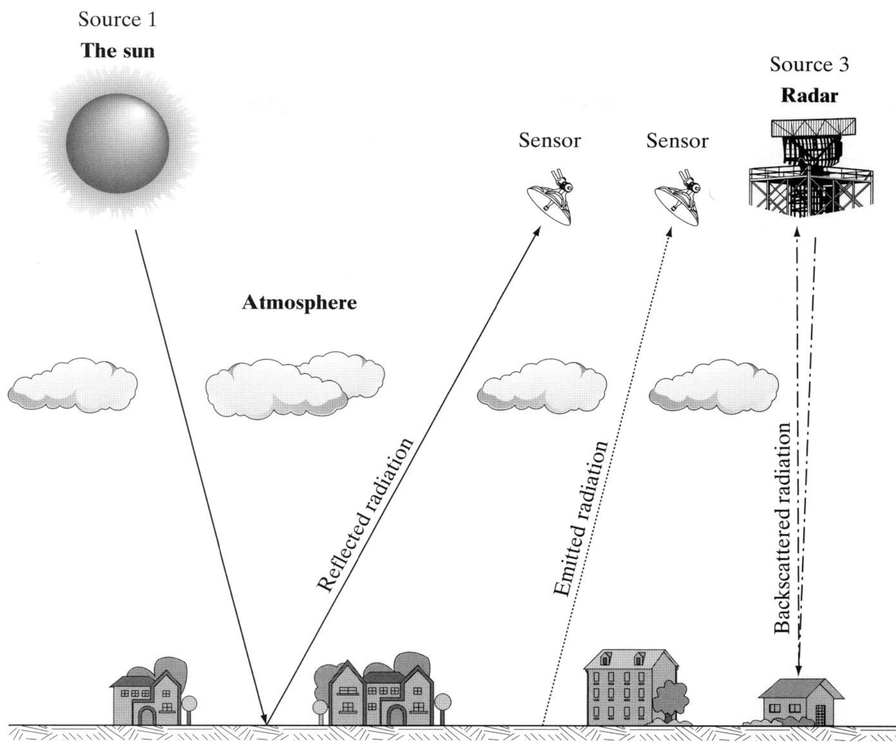

FIGURE 8.1
Sources of electromagnetic energy in remote sensing. *(Modified from Curran, 1985)*

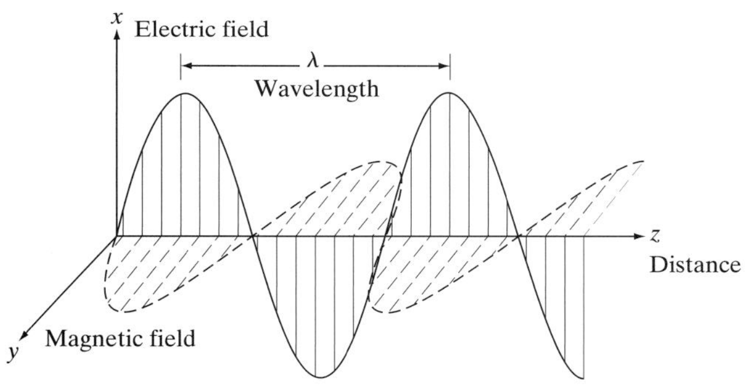

FIGURE 8.2
Electromagnetic wave pattern. *(Source: Lo, 1986)*

empty space in a plane harmonic wave pattern at the velocity (c) of light (3×10^8 m s^{-1}) (Lo, 1986). The frequency of oscillation (f) is related to the wavelength (λ) by the standard wave equation (Monteith, 1973).

$$c = \lambda f \tag{8.1}$$

Electromagnetic radiation occurs as a continuum of wavelengths and frequencies from short wavelength, high-frequency cosmic waves to long-wavelength, low-frequency radio waves (Figure 8.3) (Curran, 1985). This is known as the *electromagnetic spectrum*. Although the visible portion of the electromagnetic spectrum (0.4−0.7 μm) is very narrow, it is important in remote sensing because aerial photography directly records the electromagnetic radiation in this region, using a camera-filter-film combination. Aerial photography is still flown at regular intervals (such as every 5 years) to update land-use/cover information in each county in the United States.

Electromagnetic energy generated from the sun is seriously attenuated by its passage through the atmosphere to Earth. The atmosphere contains aerosol particles and gas molecules that scatter or absorb the electromagnetic energy according to its wavelength. Aerosol particles such as water droplets, dust, and smoke in the atmosphere tend to scatter the electromagnetic energy. Scattering causes change in the direction and intensity of radiation. Generally, scattering decreases with an increase in the wavelength of electromagnetic radiation. Therefore, ultraviolet radiation, which adjoins the blue end (0.4–0.5 μm) of the visible spectrum, is scattered much more than the radiation in the longer visible wavelengths, giving rise to the blue sky on a clear day. The gas molecules, notably water vapor (H_2O), carbon dioxide (CO_2), oxygen (O_2), and ozone (O_3), absorb electromagnetic radiation. However, absorption is selective by wavelengths. Electromagnetic

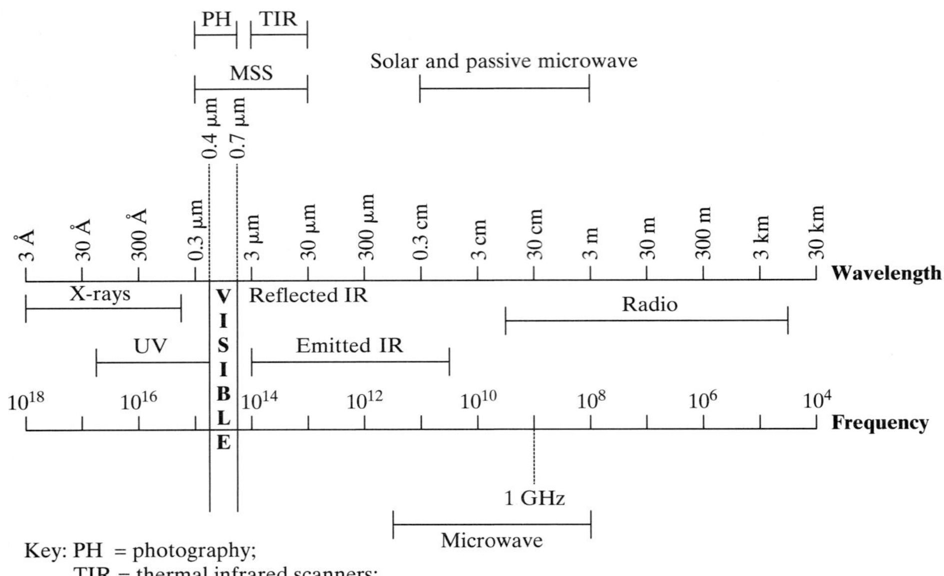

FIGURE 8.3
Electromagnetic spectrum.
(Source: Curran, 1985)

radiation with wavelength shorter than 0.3 μm is completely absorbed by the ozone (O_3) in the upper atmosphere, whereas water particles in clouds absorb and scatter electromagnetic radiation at wavelengths less than about 0.3 cm. There are certain "transmission windows" in the atmosphere through which the electromagnetic energy of certain wavelengths can be fully transmitted, such as the 3–5-μm and 8–14-μm transmission windows for thermal infrared energy. The windows in the 1-mm to 1-m wavelength region allow radar and microwave energy to be transmitted (Figure 8.4). These transmission windows have been exploited for use in nonphotographic remote sensing.

Once the electromagnetic energy reaches Earth, it is further modified through interacting with features on the surface of Earth. The energy may be *reflected, refracted, transmitted,* or *absorbed*. Energy absorbed by an object will be given out again in the form of *emitted* energy by that object. A remote sensing system can detect reflected and emitted energy from Earth's surface (Figure 8.1). The reflection of the radiation energy depends on the surface roughness and the nature of the material. A rougher surface (by which the height variation of the surface is greater than the wavelength of the incident energy) gives rise to more *diffused* and brighter reflection than a smoother surface. A very smooth surface

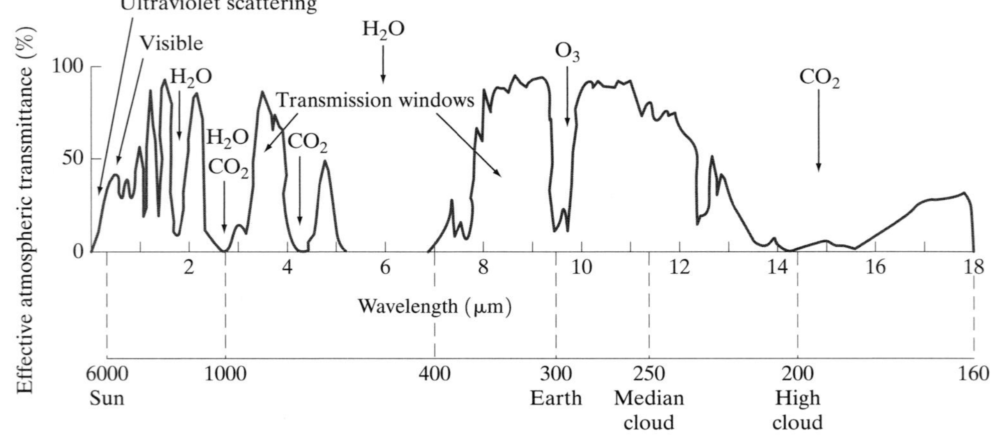

FIGURE 8.4
Atmospheric attenuation of electromagnetic energy and transmission windows. *(Source: Lo, 1986)*

such as a lake will give rise to total reflection away from the remote sensor (known as *specular*, or *mirrorlike*, reflection), thus explaining why a lake surface is normally dark toned in images acquired in the visible as well as the radar and microwave regions. Emitted energy from Earth's surface is invisible and can be detected only by using detectors sensitized to the wavelength region of emission. A good example is thermal infrared (TIR) detectors (such as an indium antimony TIR detector or a mercury-cadmium-telluride TIR detector) sensitized to the TIR wavelength regions of 3–5 μm and 8–14 μm, where transmission through the atmosphere is maximum (Figure 8.4). Hotter surfaces emit more heat energy and are brighter in tone than cooler ones as recorded on the TIR image.

In summary, electromagnetic remote sensing involves the use of a camera or a detector to record the reflection and/or emission of the electromagnetic radiation from the land cover on the surface of Earth. For example, a photographic camera records reflection of the electromagnetic energy in the visible wavelength region (0.4–0.7 μm). The variation in the tone of an object on the photograph reveals its surface roughness and characteristics. It is believed that each object has its own "tonal signature" revealing its atomic or molecular structure and allowing it to be separated from other objects. This provides the basis for the extraction of information from photographs and images.

8.3 Remote Sensing System Classifications

Basically, remote sensing systems can be classified into two types: *passive* and *active*. Passive remote sensing systems sample emitted and reflected radiation from ground surfaces when the energy source is independent of the recording instrument. Good examples are the camera and TIR detectors. In other words, without illumination from the sun, no photographs can be taken with the camera. On the other hand, active remote sensing systems can send out their own electromagnetic radiation at a specified wavelength to the ground and then sample the portion reflected back to the detecting devices. A good example is imaging radar.

Passive remote sensing systems can be further subdivided into *analog* and *digital* types. An important passive analog remote sensing system is the aerial camera, which can produce high-quality aerial photographs for topographic and thematic mapping at varying scales. Aerial photographs remain a very important source of data for GIS applications. Another type of passive analog remote sensing system is videography obtained with a video recorder. Both the aerial photographs and video data can be converted to digital form for input to the computer. As mentioned before, highly accurate photogrammetric scanners can be used to digitize the aerial photographs.

Passive digital remote sensing systems include multispectral scanners, linear and area array scanners, and spectroradiometers. Although digital cameras are available, they are not commonly used in aerial photographic missions for topographic mapping purposes because the camera format is smaller and spatial resolution is much poorer than those of the analog aerial cameras designed for photogrammetric applications.

Active remote sensing systems may also be analog and digital. In the past, *side-looking airborne radar* (SLAR) images that were processed optically were in analog form. Today, synthetic aperture radar (SAR) data are digitally processed, and the resulting images are therefore digital in nature.

Another way to classify remote sensing systems is based on the type of imaging platform used. Both aerial and space platforms have been used to acquire geographic data of Earth. Aerial platforms normally make use of small aircraft such as the Cessna type, while space platforms include space shuttles and satellites. Aerial photography has been commonly used to cover a small area such as a county area for map revision purposes. The photographic scale is usually large, normally not smaller than 1 : 25,000. Space shuttles have been used to acquire both analog and digital data of the earth. On October 5, 1984, NASA Shuttle Mission 41G carried the analog Large Format Camera (LFC) into space and took a total of 2160 photographs of Earth from an altitude of 237 km. Each photograph, covering an area of 178 km × 356 km, has a scale of 1 : 778,000 (Doyle, 1985). Similarly, the space shuttle has also been used as a platform to image Earth using C-band synthetic aperture radar digitally processed (known as Spaceborne Imaging Radar-C or SIR-C mission, flown on April 9–20, 1994, and September 30–October 11, 1994, at an altitude of 215 km in a circular orbit). However, in order to continuously monitor Earth, an unmanned polar orbiting satellite carrying a digital multispectral scanner system is preferred so that it can image the whole Earth from pole to pole at a fixed interval of time. The satellite altitude varies from 700 to 900 km, which determines the frequency of the repeat cycle of the satellite. A good example is the Landsat series of satellites: Landsat 1 through 3 (known as first-generation Landsat), Landsats 4 and 5 (the second-generation Landsat), and most recently, Landsat 7 (the third-generation Landsat), which use multispectral scanner technology to acquire images of Earth. Each Landsat scene covers an area of 185 km × 185 km (see Section 2.9.2 and Plates 2 and 3 in Chapter 2). Unlike aerial photography, satellite images cover a very large area of Earth in very small scale, thus revealing large spatial patterns in the landscape. Aerial photography and satellite remote sensing therefore complement each other in data collection.

8.4 Imaging Characteristics of Remote Sensing Systems

The success of data collection using remote sensing technology requires an understanding of the basic characteristics of remote sensing systems. Remote sensing imaging systems, whether analog or digital, possess four major resolution characteristics that determine the type of geographic data that can be detected: (1) spatial resolution, (2) spectral resolution, (3) radiometric resolution, and (4) temporal resolution (Simonett, 1983).

8.4.1 Spatial Resolution

Spatial resolution is by far the most important (but also conceptually complex) characteristic of a remote sensing system. It determines the ability of a remote sensing system in recording spatial detail. In analog photography, the spatial resolution of the photograph refers to the sharpness of the image. This is the combined result of the camera lens and the film used. The resolving power of the camera lens can be determined by means of a resolution test pattern made up of numerous sets of parallel black lines of varying thickness separated by white spaces of the same thickness, known as *line pairs (lp)*. An example of the resolution target used by the USGS for camera calibration is shown in Figure 8.5 (Light, 1992). The resolution target is photographed with the camera lens using very high-resolution film emulsion (i.e., very fine grain film) under the low-contrast lighting condition of aerial photography (normally a ratio of line-to-space luminance of 1.6 to 1). The number of line pairs per millimeter that can be detected visually from the photographed target under magnification is the resolution of the camera lens (Light, 1992). The standard aerial photogrammetric camera with 23 cm × 23 cm (9″ × 9″) format is capable of producing aerial photographs with a resolution ranging between 20 and 40 lp/mm for the low-contrast target (1.6 : 1 contrast ratio). Modern photo-grammetric cameras equipped with forward motion compensation (FMC), gyro-controlled stabilized mounts, and high-resolution film, such as the Wild RC 30 and Zeiss RMK-TOP, can produce aerial photographs with a resolution as high as 60 lp/mm.

The number of line pairs per millimeter does not take into account the scale of the aerial photograph. The ground resolution (GR), which combines the effects of the scale and resolution together, can be computed by using the following formula (Lillesand and Kiefer, 2000):

$$GR = w^{*}SF \qquad \textbf{(8.2)}$$

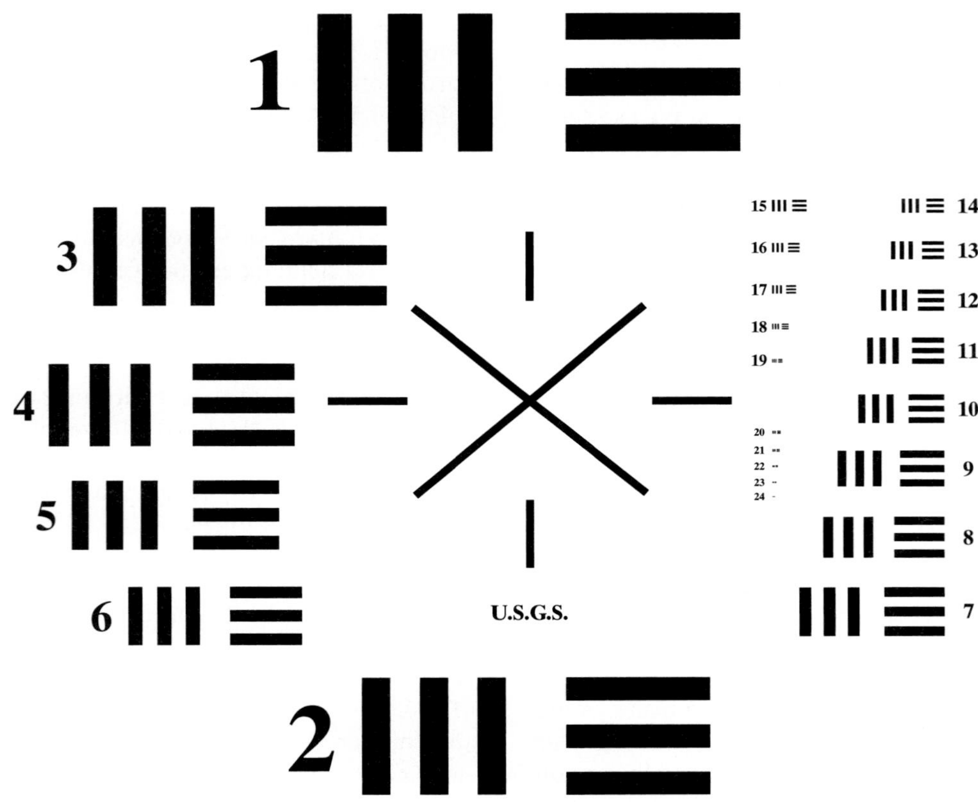

FIGURE 8.5
Reticle and resolution target used by the United States Geological Survey to test photogrammetric camera. *(Source: Light, 1992)*

where w is the width in millimeters for one line pair of the photographic system and SF is the scale factor of the aerial photograph. Thus, for an aerial photograph taken at the scale of 1 : 10,000 with the most modern photogrammetric camera with 60 lp/mm, $w = 1/60$ mm, and $SF = 10{,}000$. Therefore,

$$GR = 10{,}000/60 = 166.7 \text{ mm} = 0.17 \text{ m}$$

For electro-optical scanning systems, which produce digital images, spatial resolution is usually described as the instantaneous field of view (IFOV) or the solid angle through which the detector is sensitive to the electromagnetic energy, and it is measured in terms of milliradians, a linear measure of angles. Obviously, IFOV is affected by the height of the imaging platform, the size of the detector element, and the focal length of the optical system. To calculate the spatial resolution of the image, the following formula is used (Curran, 1985).

$$D = H^*\beta \tag{8.3}$$

where D is the ground dimension of the detector element (in meters), H is the flying height (in meters), and β is the IFOV (in milliradian). For Landsat MSS detectors, the ground dimension of the detector element (i.e., spatial resolution) is normally cited as 79 m, but because of changes in the orbital altitude (H) of the satellite (880–940 km), the ground dimension of the Landsat MSS detector element varied from 76 to 81 m (Freden and Gordon, 1983). This means that a Landsat MSS image is made up of picture elements (or pixels) with a nominal dimension of 79 × 79 m. Each pixel records the electromagnetic energy reflectance in the form of digital numbers. One should note that spatial resolution of digital images is not the same as that of an analog aerial photograph. It was determined that for electro-optical remote sensing systems, two pixels are required to present the same content as one line pair on an aerial photograph (Petrie, 1997). Therefore, for the standard aerial photographs with a spatial resolution of 20 to 40 lp/mm to be converted to digital form, the equivalent pixel size is between 25 μm (1016 dpi) and 12.5 μm (2032 dpi). For the 60 lp/mm spatial resolution of the analog aerial photograph produced from the most advanced photogrammetric camera, the pixel size equivalent is 8.5 μm (2988 dpi). This is why a high-quality scanner with linear measuring resolutions of 1 to 2 μm and accuracies (RMSE) of ±3 to 5 μm on each axis is needed for converting aerial photographs into digital form (Petrie, 1997). It is also interesting to note that a standard 9″ × 9″ aerial photograph with 60 lp/mm spatial resolution, when converted into digital form, will have a data volume of 26892 pixels by 26892 pixels or over 723 megapixels (equivalent to 723 Mb). If a pair of aerial photographs are converted for stereoviewing, as commonly is the case in photogrammetric applications, the total volume of data in terms of pixels will be well in excess of 1.5 Gb!

In remote sensing, it seems that everybody is chasing after the highest spatial resolution possible. This is particularly the case with satellite imaging systems. Space Imaging, a commercial company, has successfully launched its satellite IKONOS-2, which images Earth with one panchromatic band (0.45–0.90 μm) at 1-m spatial resolution and four multispectral bands at 4-m spatial resolution, thus surpassing SPOT's 10-m spatial resolution of its panchromatic band (Jensen, 2000) (Plate 8.1). Even Landsat 7 has included a 15-m panchromatic band (0.52–0.90 μm) in its Enhanced Thematic Mapper Plus (ETM+) sensor. The Advanced Spaceborne Thermal Emission and Reflection Radiometer (ASTER) flown on NASA's EOS (Earth Observing System) Terra satellite, successfully launched on February 24, 2000, images Earth in its three visible to near-infrared bands with 15-m spatial resolution (King and Greenstone, 1999; *http://terra.nasa.gov*) (Plate 8.2). In reality, the major consideration should be whether the spatial resolution of a remote sensing system is appropriate for the task at hand. The nature of the data to be extracted from remotely sensed images dictates the spatial resolution requirement. It is well known that a higher spatial resolution is needed to extract data from the urban environment (Welch, 1982). Jensen and Cowen (1999) indicated that a spatial resolution of ≤5 meters by 5 meters is needed to extract urban and suburban information from remotely sensed images. Research conducted by Lo (1997a) also confirmed that a spatial resolution of about 5 meters or better is needed to map the urban land cover in Hong Kong. On the other hand, for global scale phenomena such as vegetation pattern, the coarse 1-km spatial resolution of the AVHRR (Advanced Very High Resolution Radiometer) data flown on NOAA satellites is quite adequate. Therefore, spatial resolution is closely related to scale (Lam and Quattrochi, 1992). Smaller features on earth, such as potholes and caves, require high spatial resolution images (i.e., large-scale) to extract. Larger geographic phenomena, such as landforms and geological structures, require smaller scale images with lower spatial resolution to detect. The Anderson land-use/cover classification scheme for use with remote sensor data (Anderson et al., 1976) makes use of a multilevel land-use/cover classification system to meet the different spatial resolutions of the remote sensing systems (Table 8.1). One important issue in GIS is the integration of data with different spatial resolutions, which can give rise to the problem of "ecological fallacy", i.e., inference made at one geographic scale is not applicable to a different geographic scale. For example, before overlaying images of different spatial resolutions, it is always necessary to resample one image to conform to the spatial resolution of the other (to be discussed in Section 8.6.2; also see Section 5.2.2).

PLATE 8.1
Space Imaging 1-m spatial resolution IKONOS-2 image of the Washington, D.C., area. The Washington Monument is clearly visible at the top middle edge of the image. *(Courtesy: Space Imaging)*

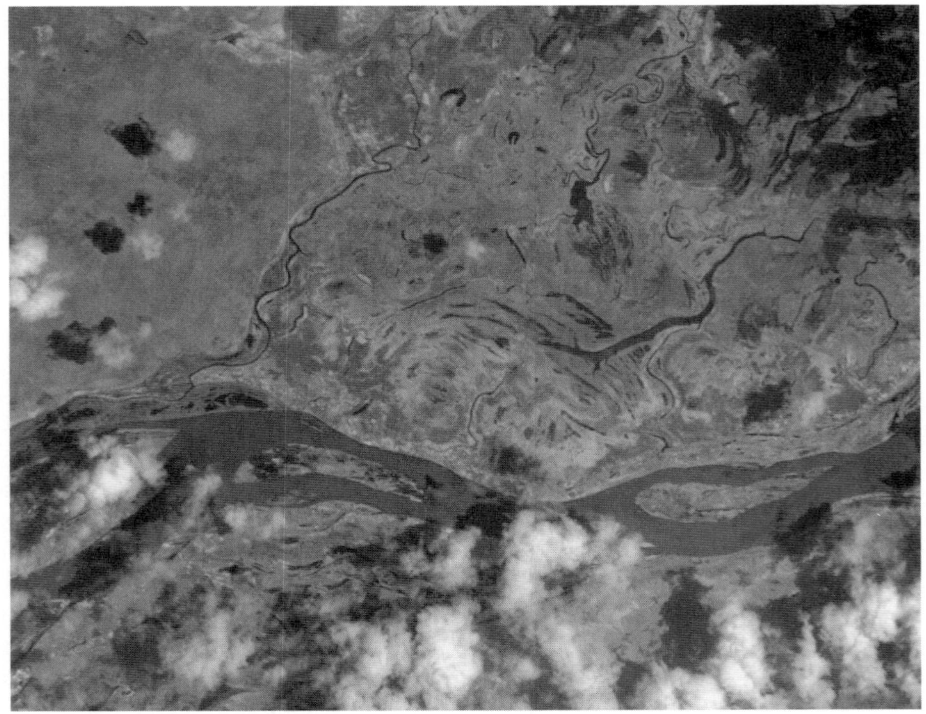

PLATE 8.2
The ASTER image of the San Francisco River region in Brazil recorded in visible/near-infrared band. The image covers an area of about 20 × 20 km with a ground resolution of 15 meters. *(Courtesy: NASA)*

Table 8.1
Spatial Resolution and Classification Level

Classification Level	Typical Data Characteristics
I	Landsat MSS data
II	High-altitude data at 40,000 ft (12,400 m) or above (less than 1:80,000 scale)
III	Medium-altitude data taken between 10,000 and 40,000 ft (3,100 and 12,400 m)
IV	Low-altitude data taken below 10,000 ft (3,100 m) (more than 1:20,000 scale)

Source: Anderson et al., 1976

8.4.2 SPECTRAL RESOLUTION

Spectral resolution refers to the electromagnetic radiation wavelengths to which a remote sensing system is sensitive. There are two components: the number of wavelength bands (or channels) used and the width of each wave band. A larger number of bands and a narrower bandwidth for each band will give rise to a higher spectral resolution. An aerial photograph obtained by an aerial camera with a film sensitized to the 0.4–0.7 μm (the visible spectrum) has only one wide band, which gives the name panchromatic to the film. The spectral resolution is therefore very low. In the case of Landsat MSS images, there are four spectral bands: 0.5–0.6 μm (green), 0.6–0.7 μm (red), 0.7–0.8 μm (reflected infrared), and 0.8–1.1 μm (reflected infrared). On the other hand, Landsat Thematic Mapper (TM) and Landsat Enhanced Thematic Mapper Plus (ETM+) all detect in seven bands: 0.45–0.515 μm, 0.525–0.605 μm, 0.63–0.69 μm, 0.75–0.90 μm, 1.55–1.75 μm, 10.40–12.50 μm, and 2.09–2.35 μm. Both Landsat TM and ETM+ data therefore have a much higher spectral resolution than Landsat MSS. The bandwidth within each band is also narrower than that of Landsat MSS. The use of narrower bandwidths allows more unique spectral signatures of objects (such as crops and vegetation) to be recorded, thus helping to discriminate more subtle differences among these objects.

As in spatial resolution, it is important to select the correct spectral resolution for the type of information to be extracted from the image. This involves deciding which wave bands are to be used. For example, the spectral reflectance of a vegetation canopy varies with wavelengths according to pigmentation, physiological structure, and water content of the leaves of the vegetation (Curran, 1985). The 0.4–0.7-μm wavelength region (blue, green, and red) shows pigment absorption, the 0.7–1.3-μm region (the near-infrared) exhibits high reflectance, and 1.3–1.6-μm region (the middle infrared) absorbs water (Figure 8.6).

Recent developments in satellite remote sensing have shown interest in greatly enhancing the spectral resolution of the remote sensing system. For example, the Moderate Resolution Imaging Spectrometer (MODIS) flown on NASA's EOS Terra satellite senses in 21 wave bands within 0.4–3.0 μm and 15 within 3–14.5 μm with very narrow bandwidths (King and Greenstone, 1999; Jensen, 2000; *http://terra.nasa.gov*) (Plate 8.2). This is known as *hyperspectral* remote sensing. These wave bands have been designed to provide simultaneous observations of cloud cover, sea surface temperature and chlorophyll, land cover, land-surface temperature, and vegetation properties. Since 1992, the Jet Propulsion Laboratory in California has been experimenting with a hyperspectral sensor developed by NASA known as Airborne Visible/Infrared Imaging Spectrometer (AVIRIS) capable of collecting 20-m spatial resolution image data in 224 bands (0.38–2.5 μm) from a NASA ER-2 aircraft at a height of 20 km (Jensen, 2000). These hyperspectral data are collected for the purpose of application in ecology, geology, coastal and inland waters, snow and ice hydrology, atmospheric aerosols and gasses, clouds, biomass burning, hazards assessments, as well as satellite calibration and validation. Analysis of hyperspectral data requires the use of sophisticated digital image processing techniques, including the need for radiometric correction and the conversion of the data to percent reflectance.

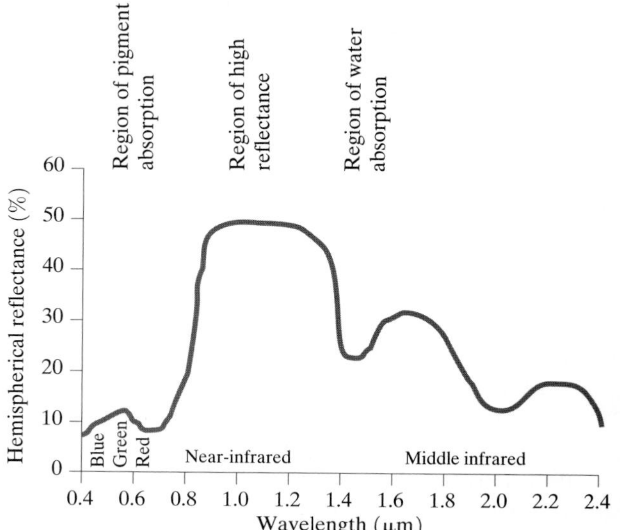

FIGURE 8.6
Hemispherical reflectance of a rhododendron leaf. *(Source: Curran, 1986)*

8.4.3 RADIOMETRIC RESOLUTION

Radiometric resolution is the smallest difference in radiant energy that can be detected by a sensor, and it is applicable to both photographs and digital images. In photography, radiometric resolution is inversely proportional to the contrast of the film, so that a higher contrast film will be able to resolve smaller differences in exposure (Lillesand and Kiefer, 2000). In other words, the film will be able to resolve subtle changes in gray tones. For digital images, radiometric resolution refers to the number of discrete levels into which a signal may be divided during the analog-to-digital conversion. This is also known as the *quantization* level. For Landsat MSS detectors, the reflected radiant energy recorded is in 6 bits (2^6, or 64 quantizing levels, giving digital numbers from 0 to 63 to the pixel). Later, the quantization level of three of the Landsat MSS bands was expanded to 7 bits (2^7, or 128 quantizing levels, giving digital numbers from 0 to 127 to the pixel). Landsat TM data are recorded with 8 bits (2^8, or 256 quantizing levels, giving digital numbers from 0 to 255). With the increased radiometric resolution, Landsat TM detectors are more sensitive to small radiant energy changes than Landsat MSS detectors. AVIRIS data are also superior to other types of data in radiometric resolution because of their use of 12-bit (from 0 to 4095) quantizing.

8.4.4 TEMPORAL RESOLUTION

Temporal resolution is the frequency of data collection. Obviously, more frequent remote sensing data acquisition will capture changes in environmental phenomena that occur daily, monthly, seasonally, and yearly. These environmental phenomena can be physical and cultural in nature. Remote sensing of vegetation and crops can benefit from a high temporal resolution. Vegetation shows phenological changes, while crop cultivation follows the seasons. In these cases, temporal resolution, which uses time as the discriminating factor, can compensate poor spatial and spectral resolutions to help identify vegetation and crop types. An important application of remote sensing is change detection, which is possible only with good temporal resolution. For satellite remote sensing, temporal resolution refers to repeat coverage at the Equator, or the number of days required for an orbiting satellite to return to the same point on Earth. Landsats 1, 2, and 3 had a repeat coverage at the Equator of 18 days, while Landsats 4, 5, and 7 required only 16 days. Thus, Landsats 4, 5, and 7 have better temporal resolution than Landsats 1, 2, and 3. Currently, having Landsats 5 and 7 orbiting Earth has improved the temporal resolution to shorter than 16 days because an image of a point on Earth can be acquired by either satellite. More frequent coverage of Earth by satellites will provide a better chance for cloud-free images to be acquired. Most important, temporal resolution of the remotely sensed data should be capable of matching the frequency of change of the landscape or phenomena to be mapped (Jensen and Cowen, 1999).

8.5 EXTRACTION OF METRIC INFORMATION FROM REMOTELY SENSED IMAGES

Metric information is basically topographic in nature and includes planimetric location (x, y), height (z), and slope angles. These variables are in ratio or interval measurement scales. Extracting these topographic variables requires an understanding of the geometry of the remotely sensed images in photographic and digital form. The basis for this understanding comes from photogrammetry.

8.5.1 FUNDAMENTALS OF PHOTOGRAMMETRY

Photogrammetry is traditionally defined as the "art or science of obtaining reliable measurements by means of photography." This definition has been extended to cover nonphotographic and digital images (Colwell, 1997). Today, digital (or soft-copy) photogrammetry uses computer and digital images to produce accurate topographic maps or orthoimages. It totally replaces analog photogrammetry of the past, which depended on complicated mechanical or optical stereoplotting machines to make maps.

Photogrammetry for topographic applications is normally applied to a stereopair of vertical aerial photographs. Although a vertical aerial photograph has a map-like view, it does not have a consistent scale as on a map. An aerial photograph is a central perspective projection, implying that a terrain object will be displaced away from the center of the photograph according to its height and location on the photograph (Figure 8.7). In Figure 8.7, terrain heights AA' (h_A) and BB' (h_B) are imaged as aa' and bb' respectively through the camera lens O (a fixed point). aa' and bb' are known as *relief displacements*. On the other hand, a map is an orthographic projection, by means of which terrain heights are all projected to one common horizontal datum so that the scale is correct throughout the map.

Photographic Scale

For a vertical aerial photograph, the *nominal* scale (S) can be determined from the following formula:

$$S = f/H \tag{8.4}$$

where f is the focal length of the camera, and H is the flying height. For an aerial photograph taken by a cam-

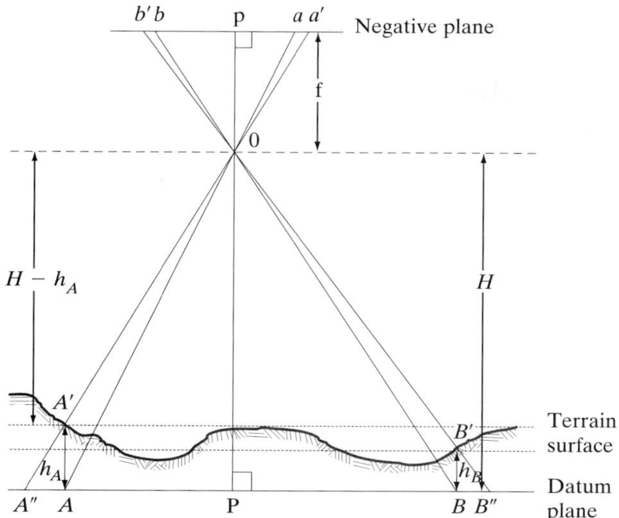

FIGURE 8.7
Relief displacements caused by central perspective projection in an aerial photograph. *(Source: Lo, 1986)*

era with a focal length of 6 inches at a height of 10,000 feet, the nominal scale is 1 : 20,000. However, this nominal scale does not take into account changes in terrain elevation in the area covered by the photograph. Normally, the average terrain elevation (h_{av}) as determined from a topographic map is used, and the average photographic scale (S_{av}) becomes

$$S_{av} = f/(H - h_{av}) \qquad \textbf{(8.5)}$$

Aircraft Tilts

An aerial photograph is subject to scale error caused by tilts of the aircraft at the time of photography. These tilts can be visualized geometrically as rotations around the x, y, and z axes of the aircraft. A rotation around the x-axis of the aircraft (which parallels the flight direction) gives *omega* (ω), or *roll*; a rotation around the y-axis of the aircraft (wings) gives *phi* (ϕ), or *pitch*; and a rotation around the z-axis (the height line) of the aircraft gives *kappa* (κ), or *yaw* (Wolf and Dewitt, 2000). The tilts cause scale errors that increase radially from the center of the aerial photograph. These *tilt displacements* affect the planimetric positions (x, y) of objects and can be removed by means of a rectification procedure using a minimum of four ground planimetric control points. The rectification can be done instrumentally or analytically. The instrumental approach uses a film projector to project the photographic image (enlarged) on to a plane (the easel) on which the ground control points are marked. By rotating the plane in the x and y directions and the film projector in the z direction, the projected image positions of the ground control points are matched up with their correct positions on the plane. Once this is achieved, an image of the rectified negative is exposed on a coated film placed on the plane under darkroom conditions. Today, with the use of computers, the analytical approach to rectification is preferred; it involves solving the following two-dimensional projective transformation equations for each of the ground control points (Wolf and Dewitt, 2000),

$$X = \frac{a_1x + b_1y + c_1}{a_3x + b_3y + 1} \qquad \textbf{(8.6a)}$$

$$Y = \frac{a_2x + b_2y + c_2}{a_3x + b_3y + 1} \qquad \textbf{(8.6b)}$$

where X and Y are ground coordinates, x and y are photo coordinates, and the a's, b's, and c's are eight parameters of the transformation.

Relief Displacement and Stereoscopic Parallax

To produce a geometrically correct topographic map from aerial photographs, relief displacements caused by the terrain (Figure 8.7) have to be corrected also. This can be achieved only by using a stereoscopic pair of aerial photographs, which allows the heights of terrain objects to be accurately measured using the theory of stereoscopic parallax. From Figure 8.8, it can be seen that the x-parallax, which is defined as the apparent image displacement created along the x-direction (flight direction) of the aerial photograph, is related to the parallax angle (α), which varies with the height of an

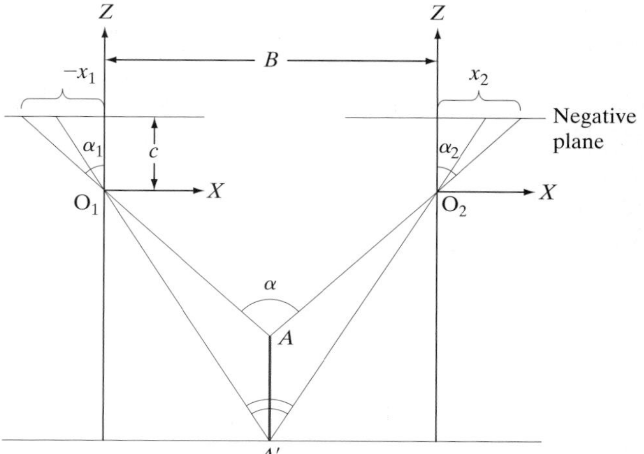

FIGURE 8.8
The definition of x-parallax in a stereoscopic pair of aerial photographs. *(Source: Lo, 1986)*

object imaged in a stereopair of aerial photographs. The x-parallax is really an algebraic difference in the x-coordinates of a point as imaged on the left and right photographs. Based on this concept, a parallax formula was developed to compute the height of an object, as follows,

$$\Delta h_{ba} = \frac{\Delta p_{ba}(H - h_a)}{p_a + \Delta p_{ba}} \qquad \textbf{(8.7)}$$

where Δh_{ba} is the difference in height between two terrain points a and b; Δp_{ba} is the difference in x-parallax between the same two terrain points; p_a is the x-parallax of point a; h_a is the height of point a; and H is the flying height measured from the datum plane (normally, mean sea level).

To measure the stereoscopic parallax accurately, the *floating mark* has to be used (Figure 8.9). When the stereopair of aerial photographs is viewed under a stereoscope, a stereomodel revealing the third dimensions of terrain objects is perceived by the human operator. A small mark is placed on an object point on the left photograph, and an identical small mark is placed on the conjugate position of the object point on the right photograph. The two small marks will be fused as one single mark and appear to float in the human operator's mind when viewed stereoscopically. By changing the separation of the two small marks in the x-direction (i.e., along the flight direction of the photographs) while maintaining the stereoscopic view, the human operator will be able to "float" the mark and adjust it to sit on the terrain object. This will measure the x-parallax of the point, and by using the parallax equation (8.7), differences in height between points can be obtained. The floating mark can also be used to locate points on the stereomodel. This floating mark is invaluable and is still used today in digital photogrammetric workstations.

Orientation of Stereomodel

Before accurate metric data can be extracted from a stereopair of aerial photographs, a photogrammetrist will carry out the following steps to set up the stereomodel correctly: (1) *inner orientation*, which sets the correct focal length and ensures that the center of each photograph is correctly determined; (2) *relative orientation*, which creates a tilt-free stereomodel from the pair of aerial photographs; and (3) *absolute orientation*, which relates the stereomodel to the correct terrestrial coordinate system so that any measurement on the stereomodel will give

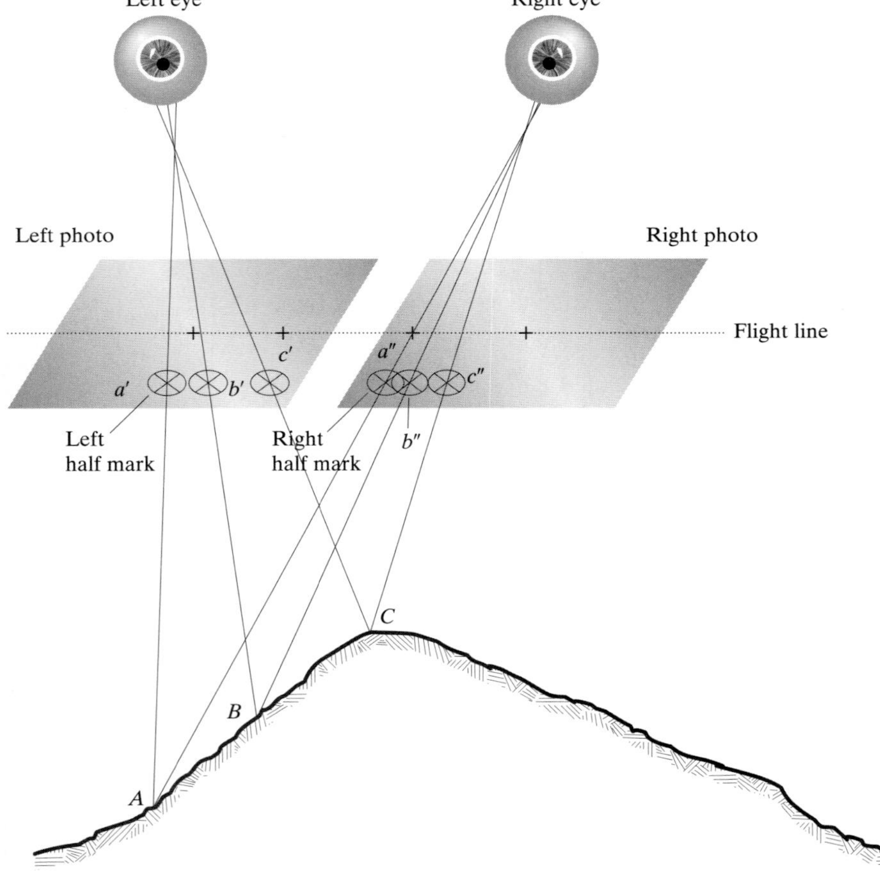

FIGURE 8.9
The principle of the floating mark.
(Source: Wolf and Dewitt, 2000)

the correct x, y, and z coordinates. Absolute orientation requires knowledge of high-quality ground control points (X, Y, and Z coordinates), which can be obtained from existing maps, aerial triangulation, or GPS (see Chapter 2). All these steps can be achieved with an analog, an analytical, or a digital approach, depending on the available instrumentation (Petrie, 1990). After the stereomodel has been properly set up, all measurements made on positions, heights, and slopes (based on the height difference between two points and its horizontal distance) will be accurate.

Digital (or soft-copy) photogrammetry has made it possible to carry out the orientation procedures much more quickly and accurately using a suite of computer programs developed based on the photogrammetric principles. All these steps merely involve mathematical transformation of coordinates: Inner orientation makes use of affine transformation to center the photographs; relative orientation makes use of a linearized form of collinearity condition equations; and absolute orientation uses three-dimensional conformal transformation between model and ground coordinates (Wolf and Dewitt, 2000). All these can be achieved using expensive photogrammetric workstations or low-cost personal computers (Petrie, 1997).

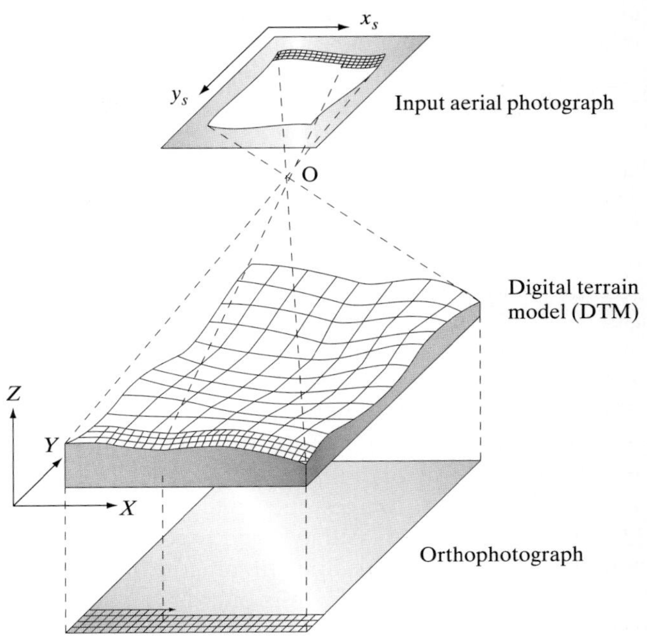

FIGURE 8.10
Principle of digital orthophoto production. *(Source: Wiesel and Behr, 1987)*

Orthophotograph and Orthoimage Production

Digital photogrammetry has been used to produce *orthophotographs* and *orthoimages*, which can be easily integrated in the GIS database. An orthophotograph is a reprojection of an aerial photograph in which all tilts and relief displacements have been removed, and the photograph has a consistent scale just like a map. An orthoimage is the digital counterpart of an orthophotograph and may be produced from a stereoscopic pair of scanned aerial photographs or from a stereopair of satellite images (such as those from SPOT or ASTER). All rely on the use of a digital elevation model (DEM) properly registered to the stereomodel to provide the correct height data in a regular matrix of points to carry out *differential rectification* of the image (Jensen, 1996a). After the digital images that form the stereomodel have been properly oriented (as explained earlier), the DEM is then registered and resampled to the same resolution as that of the digital images. For each pixel at the stereo overlap of the digital images, the correct height as given by the DEM at the same position is set and the pixel of the digital image is transformed using the collinearity condition equations. In this way, each pixel of the digital image is orthorectified, hence the term "differential rectification" (or bit-by-bit rectification) (Figure 8.10). After the differential rectification, the central perspective projection of the digital image has been transformed to the orthogonal projection.

The DEM for orthoimage production can be extracted from the contours of the topographic maps by digitizing or acquired from the USGS. It is also possible to generate the DEM from the stereopair of digital images using the method of *digital image correlation* (or *stereoimage matching, digital image matching,* or *stereocorrelation*) (Moffitt and Mikhail, 1980; Wolf and Dewitt, 2000). After the stereoimages have been properly rectified, each line of pixels on the left image should be properly aligned with the corresponding line of pixels on the right image. In photogrammetry terminology, this is called an *epipolar line*. In stereoimage matching, the digital values of pixels along the epipolar line of the left image are correlated with the corresponding epipolar line of the right image using the following formula for a one-dimensional correlation,

$$C_{uv} = \frac{\sum (u_i - u_m)(v_i - v_m)}{\sqrt{\left[\sum (u_i - u_m)^2 \sum (v_i - v_m)^2\right]}} \quad \textbf{(8.8)}$$

where C_{uv} is the normalized correlation coefficient, u and v are digital image values along the left and right epipolar lines respectively, $u_m = \Sigma u_i/N$ (or the mean of the image values along the left epipolar line), $v_m = \Sigma v_i/N$ (or the mean of the image values along the right epipolar line), and N is the number of sequential image elements used in the matching. Any lack of correlation between the image elements of the left and right epipolar lines represents a displacement in the x-direction, or the x-parallax, which is related to terrain elevation. In this way, the height at each image element (pixel) is obtained, and after the whole image is stereo matched, a DEM is generated.

A two-dimensional digital image correlation is also possible by using a correlation window of a reference image space, which is moved over all possible locations within the search window of the search image space, as demonstrated by Ehlers and Welch (1987). Using a better base-to-height (B/H) ratio, such as with a stereopair of panchromatic SPOT images with 10-m spatial resolution, an RMSE for height (Z) of ±6 m to ±18 m can be achieved, thus allowing the generation of DEMs and orthoimages at scales of 1:50,000 (Ehlers et al., 1989). One should, however, note that stereocorrelation is not perfect when tall structures and trees occur in large-scale aerial photographs and images, and manual editing of the orthoimage is required (Jensen, 1996a). For satellite images, the problem is less severe. Orthoimages are increasingly used by cartographers, remote sensing scientists, and GIS practitioners as base maps upon which thematic information is overlaid.

8.5.2 PHOTOGRAMMETRIC CONSIDERATIONS OF NONPHOTOGRAPHIC REMOTE SENSING SYSTEMS

Nonphotographic remote sensing systems are those operating in the invisible portions of the electromagnetic spectrum. These systems normally require indirect acquisition of images and make use of scanning in the cross-track direction to acquire the electromagnetic energy reflected or emitted from the terrain. Unlike aerial photography, the resulting image is a continuous strip, not frame by frame. The image does not display the same central perspective projection of the aerial photograph, and, based on the scanner geometry of the image, special formulas have to be developed to extract topographic information. Two particularly useful systems in this category are thermal infrared and radar, a passive and an active system, respectively.

Thermal Infrared Sensor A thermal infrared (TIR) detector array is incorporated as band 6 (10.40–12.50 mm) of the Landsat Thematic Mapper (TM) with a spatial resolution of 120 m. The Landsat 7 Enhanced Thematic Mapper Plus (ETM+) data have the same TIR band but a much-improved spatial resolution of 60 m. Nevertheless, the main developments in TIR remote sensing occurred with the use of aerial platforms. The most noteworthy are the TIR sensors developed by NASA: *Thermal Infrared Multispectral Scanner* (TIMS) and, most recently, *Advanced Thermal and Land Applications Sensor* (ATLAS) (Table 8.2).

TIMS is a calibrated, aircraft-mounted thermal scanner with six channels in the wavelength regions of 8.2 to 12.2 μm. A precision of 0.2° C can be obtained over the temperature range of 10 to 65° C. The usable swath width is ±30° of nadir. The sensor's IFOV is 2.5 milliradian, and

Table 8.2
ATLAS System Specifications

Channel	*Bandwidth Limits* (μm)	*NER* mW/cm² (μm)	*NEΔT* °C	*MTF at* 2 mrad	*Cooling*
1	0.45–0.52	<0.008	N/A	0.5	Ambient
2	0.52–0.60	<0.004	N/A	0.5	Ambient
3	0.60–0.63	<0.006	N/A	0.5	Ambient
4	0.63–0.69	<0.004	N/A	0.5	Ambient
5	0.69–0.76	<0.004	N/A	0.5	Ambient
6	0.76–0.90	<0.005	N/A	0.5	Ambient
7	1.55–1.75	<0.05	N/A	0.5	77 K
8	2.08–2.35	<0.05	N/A	0.5	77 K
9	3.35–4.20	N/A	<0.3	0.5	77 K
10	8.20–8.60	N/A	<0.2	0.5	77 K
11	8.60–9.00	N/A	<0.2	0.5	77 K
12	9.00–9.40	N/A	<0.2	0.5	77 K
13	9.60–10.2	N/A	<0.2	0.5	77 K
14	10.2–11.2	N/A	<0.2	0.5	77 K
15	11.2–12.2	N/A	<0.3	0.5	77 K

Key: N/A = not applicable NER = Net Energy Radiance NEΔT = Noise Equivalent Temperature Change MTF = Modulation Transfer Function for an IFOV of 2 mrad (milliradians)
Source: Lo et al., 1997

the spatial resolution can be varied from a 5–30 m pixel size, according to the altitude of the aircraft (Luvall et al., 1990). On the other hand, ATLAS is a more advanced 15-channel imaging system that incorporates the bandwidths of the Landsat TM with additional bands in the middle reflective infrared and TIR bands. The thermal infrared bands are identical to the six bands in TIMS (Table 8.2). ATLAS may be regarded as an improvement over TIMS. Before the TIR data acquired by TIMS and ATLAS can be used, they have to be calibrated using atmospheric data collected at the time of data acquisition, and the 8-bit digital numbers of the pixels have to be converted to 32-bit radiance in units of $W\ \text{cm}^{-2}\ sr^{-1}\ \mu\text{m}^{-1}$ (Lo et al., 1997). TIR images acquired by these airborne systems have been applied to map surface temperatures of water (Anderson, 1992), tree canopy (Luvall et al., 1990), and the urban heat island (Lo et al., 1997). Surface temperature is an important quantitative biophysical variable. TIR image data can be acquired during daytime and nighttime (Figure 8.11). A comparison of the day–night TIR images reveals the day–night temperature contrast of land cover types, such as lakes and forests.

Geometrically, the TIR scanner imagery suffers from severe scale distortion caused by the scanning movement of the imaging sensor, which collects environmental data by scanning across the track (y-axis) with the aid of rotating

(a) Day

(b) Night

FIGURE 8.11
ATLAS day and night thermal infrared images [Channel 13 (9.60–10.20 μm)] of downtown Atlanta, GA, acquired on May 11, 1997.

mirrors. Because of the scanning motion, a nadir line of pixels is formed below the sensor, and scale distortion along the y-axis increases away from this nadir line. In theory, there should be no distortion of scale along the flight direction (x-axis). All these combine to give rise to a "quasi-panoramic distortion" (Figure 8.12) (Masry and Gibbons, 1973). Although scale errors inherent in the geometry of TIR imagery can be corrected using the rectification method mentioned earlier, it is not a favored form of image for the production of topographic maps or for the extraction of topographic information. The use of TIR images is to extract nontopographic biophysical variables: surface temperatures of objects on earth, including the surface soil temperature using the 10.5–12.5-μm region (Jensen, 1983a), forest canopy temperature changes (Luvall et al., 1990), and the urban heat island effect (Lo et al., 1997). In an application using the TIR band of Landsat TM data, Nichol (1994, 1995) mapped the surface temperatures of the forested central catchment area and the high-rise housing estates of Singapore. Roth et al. (1989) employed the TIR data (10.5–11.5 μm) of the *Advanced Very High Resolution Radiometer* (AVHRR) data with a spatial resolution of 1 km to assess the urban heat island intensities of several cities on the west coast of North America. Thus, both high and low spatial resolution TIR image data from air

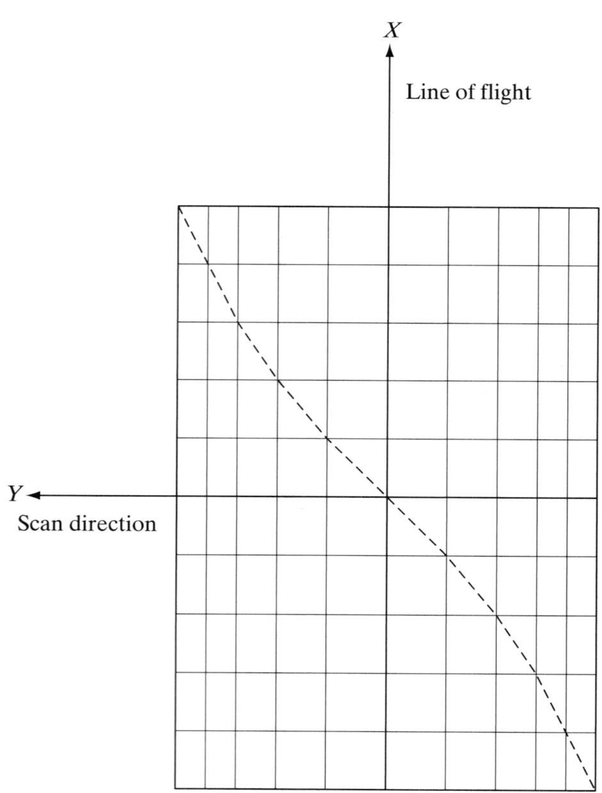

FIGURE 8.12
Panoramic distortion of thermal infrared imagery. *(After Masry and Gibbons, 1973)*

and space platforms have been used to extract surface temperatures of land cover. All these applications require proper calibration and conversion of the image data to radiance or radiant temperatures.

Interpreting the TIR images to extract surface temperature requires an understanding of the following basic physical laws.

The *Stefan–Boltzmann law* states that the energy emitted by a full radiator is proportional to the fourth power of its absolute temperature.

$$B = \sigma T^4 \qquad \textbf{(8.9)}$$

where B is $(W\text{m}^{-2})$ is the flux emitted by unit area of a plane surface into an imaginary hemisphere surrounding it, σ is the Stefan–Boltzmann constant $(5.57 \times 10^{-8}\, W\text{m}^{-2}\,\text{K}^{-4})$, and T is the absolute temperature in kelvins (Monteith, 1973).

Wien's law states that the maximum energy per unit wavelength should be emitted at a wavelength λ_m given by

$$\lambda_m = 2897/T\ \mu\text{m} \qquad \textbf{(8.10)}$$

The *quantum equation* indicates that the energy of a quantum is given by (Lillesand and Kiefer, 2000)

$$Q = hf \qquad \textbf{(8.11)}$$

where Q is the energy of a quantum in joules, h is Planck's constant $(6626 \times 10^{-34}$ joule sec), and f is the frequency of the energy. By substitution from equation 8.1, equation 8.11 can be written as

$$Q = hc/\lambda \qquad \textbf{(8.12)}$$

which states that the longer the wavelength, the lower the energy content and vice versa.

Finally, one should note that the amount of radiation emitted from an object is affected by *emissivity* (e), which is a ratio comparing the amount of energy radiated from an object at a given temperature with that radiated from an ideal object known as a blackbody at the same temperature. A blackbody is a hypothetical body that absorbs the entire radiation incident upon it and emits the maximum amount of radiation at all wavelengths. The value of emissivity varies from 0 to 1. For example, the average emissivity of clear water at the 8–14 μm wavelength region is 0.98–0.99, while that of brick is 0.93–0.94. Aluminium foil has very low emissivity: 0.03–0.07. Usually, the higher the reflectivity of the object, the lower is its emissivity.

Side-Looking Airborne Radar (SLAR) Radio detection and ranging (radar) is an active imaging system that sends out a signal to the target and then receives the reflectance of the signal from the target. Radar operates at the very long wavelength region (0.8–100 cm), and as a result, radar energy is less affected by aerosols in the atmosphere. Radar imaging can be carried out both during the day and night, independent of the weather. Radar belongs to the microwave region of the electromagnetic spectrum, and frequencies are usually used together with wavelengths to designate the bands. Because radar was a military sensing device, microwave frequencies have been arbitrarily assigned to bands identified by an alphabet. The most commonly used bands for imaging radar are: *X-band* (2.4–3.8 cm in wavelengths, or 12.5–8 GHz in frequencies), *C-band* (3.8–7.5 cm, or 8–4 GHz), *S-band* (7.5–15 cm, or 4–2 GHz), and *L-band* (15–30 cm, or 2–1 GHz).

Radar consists of an antenna that can transmit and receive pulses of energy to and from a target. In side-looking airborne radar (SLAR), an antenna is mounted at the belly of an aircraft. As the aircraft is flying, energy pulses are sent out sideways (in the *y*-direction) from the antenna at the desired angle (look angle) toward the ground. The reflected signals are received by the antenna and then processed to produce an amplitude/time video signal, which modulates the intensity of a cathode ray tube to expose an image line on the photographic film (Figure 8.13) (Lillesand and Kiefer, 2000). As the aircraft moves forward (*x*-direction), a radar image is built up line by line.

Radar imagery is very unlike conventional aerial photography because the geometry is totally different. As a side-looking system, radar imaging gives an oblique view of the earth. In fact, the oblique view is necessitated by the fact that radar is a distance-measuring (or ranging) device. Return signals from objects located at the same distance from the aircraft are received at the same time, and as a result, these objects cannot be distinguished. To avoid ambiguity in imaging, the terrain vertically below the aircraft is never imaged. The SLAR antenna will not accept return pulses from an incidence angle of less than 45° (Figure 8.13). For a *brute force radar* (or *real aperture radar*), scale errors of the SLAR image increase toward the nadir line of the aircraft. A major distortion to note in radar images is *feature foreshortening*, which is the appearance of compression of features tilted toward the radar, thus giving rise to a brighter appearance of the affected slope in the image. When the reflected energy from the upper portion of a feature is received before the return from the lower portion of the feature, a phenomenon called layover occurs because the top of the feature will be displaced, or "laid over" relative to its base (Figure 8.14). In other words, relief displacement is radial toward the nadir line (aircraft) in a radar image.

Another important parameter affecting the appearance of terrain objects in a radar image is the *incidence angle*, which is the angle between the radar illumination and the ground surface (Figure 8.13). The incidence angle will change from the near range (near the aircraft) to the far range (away from the aircraft) along the *y*-direction. Thus, even identical terrain forms located

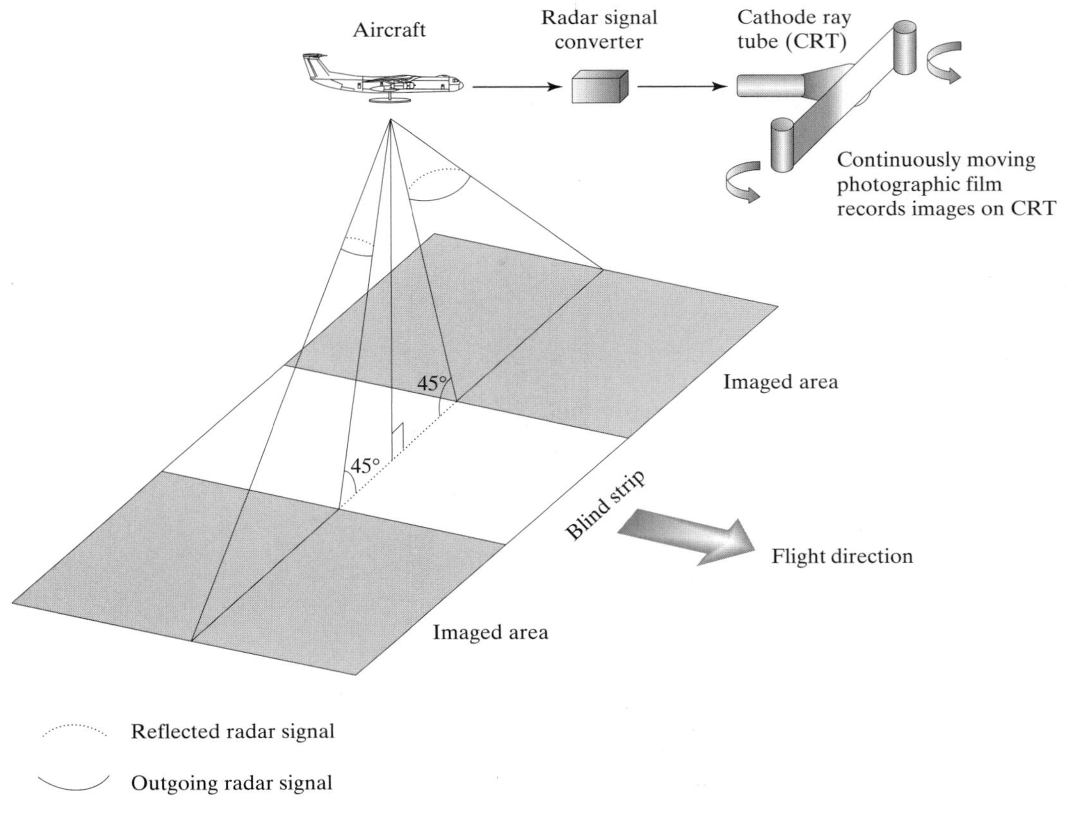

FIGURE 8.13
Acquiring SLAR (side-looking airborne radar) images. *(Modified from Cooke and Harris, 1970)*

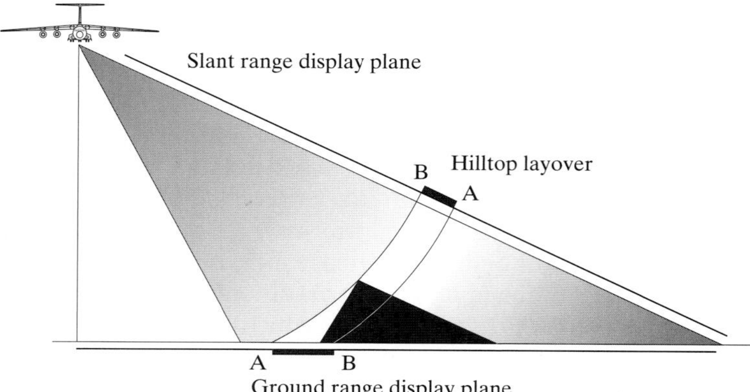

FIGURE 8.14
Radar layover: reflected energy from the top of the hill (A) is received before the return from the base of the hill (B). The resulting radar image (BA) shows the top of the hill displaced, or "laid over," relative to its base. *(Source: RADARSAT International)*

in different directions from the aircraft will appear differently. A more uniform illumination across the radar swath can be achieved by raising the flying height of the radar imaging system. Therefore, a satellite platform is better suited for radar imaging than the aircraft platform.

The spatial resolution of the real aperture radar is complicated by the fact that it is different between the range (y) and azimuth (x) directions. The *range resolution* is determined by the pulse length used in the radar while azimuth resolution is affected by the physical length of the antenna that defines the beam width. In order to minimize the geometric distortion and spatial resolution discrepancies, *synthetic aperture radar* (SAR) is more commonly used for environmental data extraction. SAR uses a short physical antenna (e.g., 2 m) to receive radar signals reflected from the terrain and then stores the phase and amplitude of the signals on film or tape. They are later processed to synthesize the effect of a very long antenna (600 m). SAR therefore produces a synthetic beam with constant width in the azimuth direction, and the azimuth resolution is usually made equal to range resolution. SAR is the preferred system used in space platforms, such as Seasat, SIR-A, SIR-B, SIR-C, ERS-1, and RADARSAT. (See Figure 10.4 for an SIR-A image of the North China Plain.)

Despite its oblique view, radar imaging is the non-photographic system of greatest potential for topographic mapping, and hence the extraction of terrain-related biophysical variables, such as elevation, slope lengths, and angles. It is possible to use the shadow of a single SLAR image to determine the heights of terrain objects as follows:

$$h = H \times L/D \tag{8.13}$$

or

$$h = L \times \sin \beta \tag{8.14}$$

where h is the height of an object above the datum plane, H is the flying height above the datum plane, L is the length of the radar shadow in slant range, D is the slant range distance from the antenna to the object, and β is the depression angle of the antenna. If radar shadows are invisible, relief displacement (layover) can be used to determine the height of an object. Stereopairs of SLAR images may also be acquired through special flight configuration. An area is imaged twice from two different look-directions, either from the same side or opposite side at the same height or different heights (Figure 8.15). This creates *radar parallax*. To avoid the sidelighting effect,

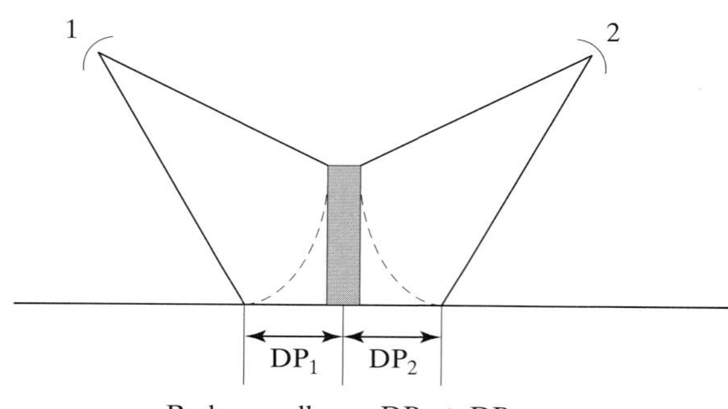

(a) Opposite side configuration

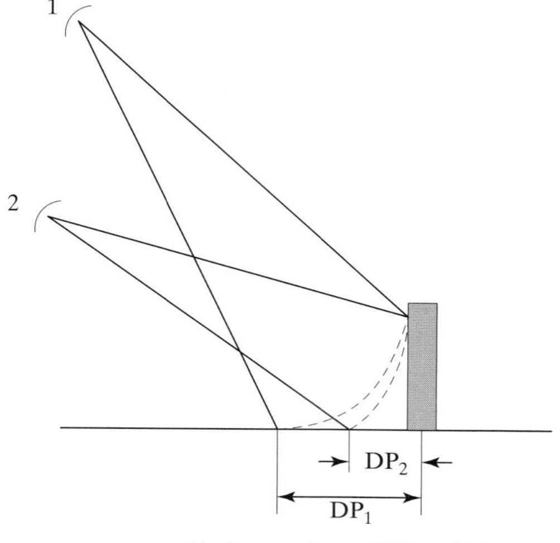

(b) Same side configuration

FIGURE 8.15
Radar parallax obtained from two different flight configurations. *(Source: Simonett and Davis, 1983)*

stereo radar imagery is best flown using the same side and the same height coverage (Simonett and Davis, 1983). Stereoviewing is possible as in conventional aerial photography using a mirror stereoscope, and the radar parallax can be used to determine heights of terrain objects and for topographic mapping. It was found that for areas of low to moderate relief, airborne radar permits mapping with a planimetric accuracy of 100 meters and a height accuracy of 20 meters (Simonett and Davis, 1983).

Because SLAR is an active remote sensing system, the type of signal to be transmitted and received can be controlled through *polarization*, which is the direction of vibration of the electrical field vector of electromagnetic radiation. There are two types of polarization in SLAR: either horizontal or vertical. It is possible for SLAR to transmit the signals with horizontal polarization and receive the return signals with horizontal polarization (HH). Similarly, radar signals can be transmitted with vertical polarization and received with vertical polarization (VV). Both HH and VV are called *like polarization*. Alternatively, the radar signals can be transmitted with horizontal polarization and received with vertical polarization (HV), or just the reverse (VH). Both HV and VH are known as *cross polarization*. Radar return signals are usually strong in like polarization but weak in cross polarization. VV polarization is more sensitive to vertical objects such as tree trunks and plants because of their vertical orientation, while HH polarization is more sensitive to physical and cultural surfaces that exhibit a dominant horizontal configuration.

SLAR is also a useful source of other biophysical variables, notably soil moisture and surface roughness/texture. Research has shown that active radar imaging systems in the C-band (6.3 cm) with HH polarization operating at 10° incidence angle and L-band (9–21 cm) at various incidence angles exhibit sensitivity to soil moisture (Jensen, 1983a). An SLAR image exhibits a peppery appearance, and texture is clearly emphasized. This makes it easy to extract surface roughness, which is defined as the average vertical relief of small-scale irregularities of the terrain surface measured in centimeters (Sabins, 1997). According to Peake and Oliver (1971), the following formulas for smooth and rough surfaces are respectively

$$h_{smooth} < \lambda/(25 \sin \beta) \quad \textbf{(8.15)}$$

$$h_{rough} > \lambda/(4.4 \sin \beta) \quad \textbf{(8.16)}$$

where λ is the radar wavelength in centimeters, and β is the depression angle of the antenna (Sabins, 1996).

Interpretation of SLAR images requires a knowledge of the radar system parameters of wavelength, incidence angle, and polarization of the incident wave; and the terrain parameters of slope angle, surface roughness, and the dielectric constant. The dielectric constant is an electrical property of the matter that influences the radar returns. For example, water added to soil can affect the dielectric constant of the soil and differentiate it from drier soil in the SLAR images. SLAR images tend to show linear or geometric features such as man-made structures very well because of the shadowing effect. On the whole, SLAR images are a good source of geological and geomorphological information, both metric and descriptive. However, applications to land-use/cover mapping, population, and human settlement studies have also been attempted (Lo, 1986; Lo, 1998; Henderson and Xia, 1998).

8.5.3 PHOTOGRAMMETRIC CONSIDERATIONS OF SATELLITE IMAGING SYSTEMS

Both photographic and nonphotographic systems discussed in the previous two sections have been used for remote sensing of Earth from space. Aerial cameras have been placed on spacecraft to photograph Earth, including the Large Format Camera (LFC) from a space shuttle by NASA, and the KFA-1000, KFA-3000, KVR-1000, and TK-350 cameras on satellites by the Russian Space Agency (Jensen, 2000). Radar has now become an important all-weather imaging system in space as Canada's RADARSAT (see Section 2.9.2). TIR imaging has been featured in Landsat TM and ETM+ sensors and has played an increasingly important role in NASA's Terra sensors (notably ASTER, MODIS, and MISR)*(http://terra.nasa.gov)*.

Space platforms have the advantage that they provide a wider global coverage than that of aerial platforms. Satellite platforms allow images of Earth to be acquired at a regular interval. As a result, there are now many land remote sensing satellites orbiting the earth (Table 8.3). One of the major purposes of satellite remote sensing is to speed up world topographic mapping. Unfortunately, this has not been achieved. In 1994, it was reported that the annual progress of world topographic mapping at 1 : 25,000 and 1 : 50,000 was only 2.8% and 1.1%, respectively (Konecny and Schiewe, 1996). Also, updating of topographic maps is essential. It was found that maps of 1 : 25,000 and 1 : 50,000 scale were 20 and 40 (or more) years out of date, respectively.

There are stringent requirements to meet for topographic mapping, which include planimetric accuracy, height accuracy, and completeness of details (Petrie, 1970). As soon as Landsat images became available in 1972, numerous attempts were made by photogrammetrists to produce topographic maps that can meet these three requirements. The satellite platform has many advantages over the aerial platform from the photogrammetric point of view. The satellite platform, which is subject to much less turbulence than in the atmosphere because space is a vacuum, is more or less tiltfree. The very high orbiting altitude of the satellite means that relief displacement is not as severe as that on

Table 8.3
Land Remote Sensing Satellite Systems

Country	Sponsoring Program	Sensor	No. of Bands $\Delta\lambda(\mu m)$	IFOV	Swath Width (km)	Launch Date	Stereo
USA	NOAA-12 to -14	AVHRR	5 0.58–12.5	1.1 km	2400	1991	N/A
	NASA Landsat-5	Thematic Mapper (TM)	7 0.45–12.5	28.5 m VNIR,SWIR 120 m TIR	185	1984	N/A
	NASA Landsat-7	Enhanced TM Plus (ETM+)	7 0.45–12.5 1 pan 0.52–0.90	30 m 28.5 m VNIR,SWIR 60 m TIR 15 m	185	1999	N/A
	EOS-Terra	ASTER	14 0.5–12.0	15 m VNIR 30 m SWIR 90 m TIR	60	2000	AT
		MODIS	36 0.4–14.5	250 m, 500 m, 1 km	2330 × 10	2000	N/A
		MISR	4 0.425–0.886	275 m, 550 m, 1.1 km	360	2000	AT
	Space Imaging	IKONOS-2	1 pan 4 MSS 0.45–0.90	1 m pan 4 m MSS	11	1999	AT and CT
	EarthWatch	Quickbird-2	1 pan 4 MSS	0.61 m pan 2.5 m MSS	20–40	2001	AT and CT
France	SPOT 3	HRV	1 pan, 3 XS 0.5–0.89	10 m pan 20 m XS	60	1993	CT
	SPOT 4	HRV	1 pan, 4 XS 0.5–1.75	10 m pan 20 m XS	60	1997	CT
	SPOT 5	HRV	1 pan, 4 XS 10 m VNIR	5 m pan 20 m SWIR	60	2002	CT
ESA and Canada	ERS-1,2	SAR	C-band	25 m	100	1991, 1995	Interf.
Japan	JERS-1	OPS SAR	7 0.52–2.40 L-band	18 m	75	1992	AT
	ADEOS	AVNIR	4 0.42–0.89	8 m pan 16 m VNIR	80	1996	CT

(continues)

Table 8.3 (Continued)
Land Remote Sensing Satellite Systems

Country	Sponsoring Program	Sensor	No. of Bands $\Delta\lambda(\mu m)$	IFOV	Swath Width (km)	Launch Date	Stereo
India	IRS-1A,1B	LISS I, II	4	72 m, 36 m	148	1988, 1991	N/A
	IRS-1C/D	LISS II	4	5.8 m pan	148	1995, 1996	CT
			0.52–1.7	23.5 m VNIR 70.5 m SWIR			
Canada	RADARSAT	SAR	C-band HH	10–50 m	50–500	1995	Interf.
Germany	MOMS-02/D2	pan, M/S	7	4.5 m HR pan	37	1993	AT
			0.44–0.81	13.5 m VNIR	78		
	MOMS-02/P	pan, M/S	7 0.44–0.81	6.0 m pan 18.0 km VNIR	50 HR 105 VNIR	1996	AT

Key: N/A = not applicable
AT = along track
CT = cross track
Interf. = interferometry
Source: Konecny et al., 1996; Carlson and Patel, 1997; Jensen, 2000

an aerial photograph. Wong (1975) demonstrated that a general polynomial model could be used to model geometric distortions of the Landsat MSS images, and with four or more control points for distortion correction, a Landsat MSS image could be processed to meet the National Map Accuracy Standards for topographic mapping at 1:500,000 scale or smaller. By applying digital image rectification using a nearest-neighbor resampling method and an affine transformation to Landsat MSS images covering the United Kingdom, Dowman and Mohamed (1981) found that an RMSE of ±52 meters in eastings and ±38 meters in northings could be achieved. They also found that for a stereomodel of Landsat MSS, the RMSE for height was ±69 meters.

While some progress has been made, many of these efforts in the 1970s and 1980s have failed to use satellite images to produce topographic maps at 1:50,000 scale or larger. It became clear that the spatial resolution of the satellite images available at that time was inadequate to provide all the planimetric details required for the topographic maps, particularly at large scales (Wong, 1975). The high orbital altitude has also limited the base-to-height (B/H) ratio, a measure that relates the separation of two imaging (exposure) stations (known as the air base) with the flying height of the sensor. This affects the accuracy of triangular fix of positions as well as the stereoscopic parallax for height measurement, using a floating mark. Thus, for large-scale topographic mapping, these two aspects have to be improved.

On October 5, 1984, the LFC was flown using the space shuttle as the platform, and a total of 2160 frames of space photographs in the format of 23 cm × 46 cm (9″ × 18″) were acquired from altitudes of 352 km, 272 km, and 222 km. The camera has a focal length of 30.5 cm and a spatial resolution of about 80 lp/mm on high-resolution aerial film at a contrast ratio of 2:1 (Malhotra, 1989). It was found that the LFC could be used to produce topographic maps at a scale of 1:50,000 and smaller, with a planimetric accuracy of 14 meters and spot height accuracy of 15 meters (Derenyi and Newton, 1987). The use of a large format was intended to give a more favorable base-to-height ratio. The high spatial resolution photographs should allow more details to be detected. Unfortunately, an analog camera with film on the space shuttle platform is not an economical remote sensing system. The use of LFC did not go beyond the experimental stage.

With the arrival of the so-called high spatial resolution digital satellite imaging systems, namely, Landsat *Thematic Mapper* (TM) with 30 m resolution and SPOT (Systeme

Probatoire d'Observation de la Terre) panchromatic (with 10 m resolution) and multispectral (with 20 m resolution) in the 1980s, a more automated approach to topographic mapping using satellite images to meet National Map Accuracy Standards at 1 : 50,000 scales and larger became a possibility. The panchromatic images from France's commercial satellite SPOT are particularly suited for this purpose because of their very high spatial resolution and the stereoscopic coverage made possible by off-nadir imaging (across-track imaging), which gives a high base-to-height ratio. Geometrically, a SPOT image is less distorted than that of a Landsat TM image because of its use of a linear detector array and push-broom scanning in image acquisition. Push-broom scanning does not use rotating mirrors to scan; instead it pushes a linear array of tiny detectors along to cover an area on Earth. Evaluation by Gugan and Dowman (1988a) has demonstrated that SPOT is a potential source of imagery for 1 : 50,000 and smaller scale mapping tasks using an analytical photogrammetric approach, with planimetric accuracy of ≥20 meters and height accuracy of ≥8 meters. It was further confirmed that based on height and planimetric accuracy and completeness evaluation, SPOT images can be used for topographic mapping at 1 : 50,000 scale with 20-m contours, and that if the image quality is very good and the ground control sufficient, 1:25,000 scale mapping is also possible (Theodossiou and Dowman, 1990; Gugan and Dowman, 1988b).

With the use of a digital photogrammetric approach, DEMs on a 30-m grid to an accuracy of ±10 meters can be generated from the panchromatic SPOT stereoimages. These DEMs can be used to produce orthoimages and for topographic mapping at 1:25,000 scale (Dowman, 1994). The major weakness of satellite images still lies in the restricted spatial resolution (Dowman and Peacegood, 1989). However, the future looks bright for the application of satellite remote sensing to topographic mapping and a data source for GIS. As has been shown in Section 8.4.1, the Landsat 7 ETM+ has a 15 m panchromatic band (0.52–0.90 μm). NASA's ASTER on the Terra satellite has three visible and near-infrared (VNIR) bands with 15 m spatial resolution. ASTER is particularly suited to topographic mapping because it is equipped with two telescopes: one looking vertically downward with a three-band detector and another looking backward with a single-band detector. Together they produce along-the-track stereoscopic coverage of Earth in the near-infrared band (0.76–0.86 μm) using the more advanced push-broom scanning. ASTER has the greatest potential to elevate the accuracy of topographic mapping from space. The superior 1 m spatial resolution of Space Imaging's IKONOS-2 will no doubt contribute to significant improvement of the accuracy of topographic mapping, particularly in horizontal positions of mapped objects and completeness of map details (Plate 8.1). More high spatial resolution imaging satellites similar in design to IKONOS-2 will be launched by commercial corporations, such as EarthWatch, which plans to launch QuickBird-2 in October 2001 to image Earth at 0.61 meter for the panchromatic band and 2.5 meters for the multispectral bands, with stereoscopic capability, making it the highest spatial resolution civilian imaging satellite system.

In other parts of the world, Germany has successfully developed a space imaging system called the Modular Optoelectronic Multispectral Scanner-02 (MOMS-02) flown on the Second German Spacelab Mission (D-2) on Space Shuttle flight STS-55 in April/May 1993, which used one 4.3 m spatial resolution panchromatic channel to combine with four multispectral visible and near-infrared channels with 13 m spatial resolution to obtain stereoimages of Earth along the flight direction (i.e., along the track) (Konecny and Schiewe, 1996). MOMS-02 was later integrated onboard the Russian Mir station's PRIRODA module in a "preoperational" mission from 1996 to 1999 as MOMS-2P. The MOMS system will be extended to be operational on an Earth Observing (EO) satellite within the so-called EarthMap program to become MOMS-3D. The Indian Remote Sensing Satellites IRS-1C and IRS-1D feature a panchromatic sensor (0.50–0.75 μm) capable of producing a cross-track stereoimages of 5.8 m resolution. SPOT-5, to be launched in 2002, will provide multispectral data of 10 m resolution and panchromatic data of 2.5 m resolution in an along-track stereo mode. The along-track stereo mode, usually achieved with three sensors—one pointing forward, one pointing vertically downward, and one pointing backward—will allow stereoimages to be acquired at the same satellite orbit with identical atmospheric conditions, and the base-to-height ratio can also be improved. Some countries, notably Japan, China, and India, will launch 2–3 m spatial resolution imaging satellites in 2002 and 2003. All these developments will certainly improve the positional accuracy and completeness of detail in topographic mapping using satellite images. Digital photogrammetry has helped lower the cost of topographic mapping and the extraction of height and planimetric information. Even low-cost PCs can be used for this purpose, thus promoting the integration of remote sensing with GIS.

As mentioned in Section 2.9.2, because of its all-weather capability, radar is used on satellite platforms. Imaging radar has the greatest potential for use in topographic mapping. Indeed, in areas of perennial cloud cover, radar has been used to produce topographic maps (the most notable of such an effort being Project RADAM—Radar Amazon—of the 1970s) (Roessel and Godoy, 1974). Canada launched *RADARSAT-1* on November 4, 1995, as a commercial venture. It uses SAR with C-band wavelength (5.6 cm), HH polarization, and a right-looking, steerable antenna to acquire radar images of the earth from an altitude of 798 km. Stereo coverage may also be obtained. The spatial resolution may be varied from 10 to 100 meters, depending on the operational beam mode. RADARSAT stereoimages may be used to extract terrain heights, which will allow DEMs to be generated for subsequent application in topographic mapping (Mercer et al., 1998). Radar images may also be merged with SPOT or Landsat TM or ETM+ images. There is a plan to launch RADARSAT-2 in 2003. Apart from RADARSAT, there are other satellite radar imaging systems, which include the European Space Agency's (ESA) *ERS-1* and Japan's *JERS-1* (Table 8.3). Radar satellite imaging systems will complement the high spatial resolution and high spectral resolution visible and near-infrared imaging systems nicely in many areas of applications.

The Shuttle Radar Topography Mission (SRTM) on the Space Shuttle Endeavour on February 11, 2000, collected over 224 hours of three-dimensional radar images of Earth between latitudes 60° N and 54° S from an altitude of 233 km (145 miles) using C-band and X-band interferometric synthetic aperture radars (IFSARs) during the 11-day mission, thus acquiring topographic data for over 80% of Earth's land mass. The unique characteristic of this mission is that single-pass interferometry was used, which means that two radar images were obtained at the same time, one from the radar antenna in the shuttle's payload bay, and the other from the radar antenna at the end of a 60-m (200-ft) mast extending from the shuttle. Together they formed a baseline from which a swath 225 km (140 miles) wide swept across Earth like a push broom (Plate 8.3). The antenna at each end of the baseline sent signals to and received the signal returns from Earth. This is similar to the method of obtaining stereoscopic radar images explained in Figure 8.15. By combining the two images acquired by the antenna at each end of the baseline, a single 3-D image was produced. These 3-D image data will allow digital topographic map products of the world at 30 m × 30 m spatial sampling with ≤16 m absolute vertical height accuracy to be produced, thus meeting the National Map Accuracy Standards for topographic mapping *(http://www.jpl.nasa.gov/srtm/)*. The Shuttle Radar Topography Mission was sponsored by NIMA, NASA, the German Aerospace Center, and the Italian Space Agency. Plate 8.4 shows an example of using the Shuttle Radar Topography Mission data to produce the elevation data on which the Landsat image and digital aerial photographs were combined. Plate 8.5 shows a Shuttle Radar Topography Mission C/X SAR interferogram, displaying the contourlike colored bands. The Shuttle Radar Topography Mission will be used to produce accurate DEM data of the world, replacing the use of GTOPO30 data (see Chapter 9, Section 9.4.4).

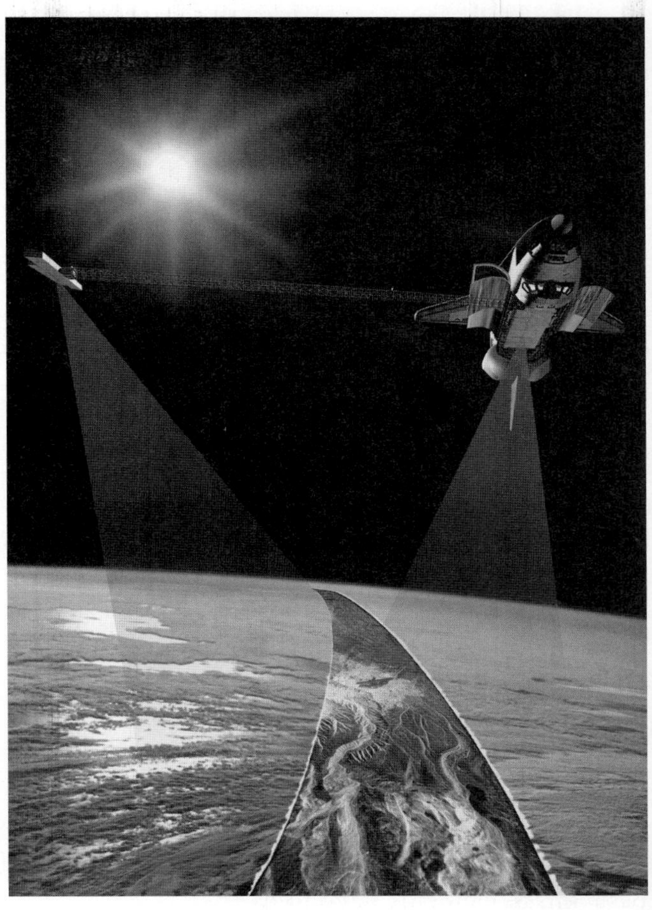

PLATE 8.3
The Shuttle Radar Topography Mission (SRTM) acquires radar images of Earth from two antennas (one on the payload and one at the end of the 60-m mast), thus allowing 3-D radar images to be produced to map Earth's terrain heights. To view this photo in color, see the color insert. *(Courtesy: NASA/JPL/Caltech)*

8.6 Extraction of Thematic (Descriptive or Attribute) Information from Remotely Sensed Images

As has been mentioned, all forms of remotely sensed images are nonselective in nature and cannot be directly integrated into a GIS database. An interpretative process is necessary before useful thematic information relating to the environment can be extracted from these images. This information is descriptive in nature and uses the nominal scale of measurement (e.g., land use or land cover). The extraction can be achieved either through visual interpretation or by computer-assisted image analysis.

8.6.1 PRINCIPLES OF PHOTOGRAPHIC INTERPRETATION

Photographic interpretation is defined by the American Society for Photogrammetry and Remote Sensing as "the act of examining photographic images for the purpose of identifying objects and judging their significance" (Colwell, 1997). The process of photographic interpretation can be divided into five stages as illustrated in Figure 8.16: detection, identification, analysis and deduction, classification, and theorization—verification or falsification of hypotheses. If the hypothesis is rejected, the knowledge of the human interpreter is improved. He/she will then reformulate the hypothesis and repeat the photographic interpretation until the hypothesis is accepted. Thus, photographic interpretation is an iterative process. The photographic interpretation process as explained here adopts deductive logic: the specification of a null hypothesis, observation, and verification or falsification (Curran, 1987). The deductive logic also implies an approach of interpreting the aerial photograph from the general to the specific. The human interpreter examines on the aerial photograph the general geographical setting of the study area first before coming down to examine specific details. The human interpreter will then proceed from the known to the unknown.

In assisting the identification of objects, at least nine basic image elements have to be used (Figure 8.17). These are, in order of complexity: tone/color, size, shape, texture, pattern, height, shadow, site, and association (Estes et al., 1983, Teng et al., 1997).

PLATE 8.4
This perspective view of Pasadena, California, is formed from three datasets: (1) Shuttle Radar Topography Mission (SRTM) for elevation, (2) Landsat data from the land surface color, and (3) USGS digital aerial photography for image detail. To view this photo in color, see the color insert. *(Courtesy: NASA/JPL/Caltech)*

Tone/color is the most important element used in photographic interpretation. Tone/color is the record of the amount of light reflection from the land surface onto the film. The amount of reflection is dependent on the nature of land cover, notably the surface roughness, color, the atomic structure of the materials, and the wavelength used (see Section 8.2). On black-and-white panchromatic film, tone varies from black through gray to white in continuous steps, thus making use of the spectral and radiometric resolution of the film. In general, the more light reflected by the object, the lighter its tone on the photograph. For true-color and color infrared photography, the term "color" is used. Vegetation is green and water is blue in true-color photographs, while vegetation is usually pink, red, or purple in color infrared photographs. It is the difference in tone or color between objects or between an object and its background that is important (Plate 8.6).

Size refers to the dimension of an object, which should be known from the scale of an aerial photograph. By comparing with an object of known dimensions, such as a car, one should be able to know the size of an object. *Shape* is the general form of an object as it appears from a vertical view and provides a useful clue to identifying an object. Both size and shape represent geometric arrangements of the tone or color of the elements making up the object or phenomenon (Plate 8.6).

Texture is the frequency of tone change within the image that arises when a number of small features are viewed together. It gives the visual impression of the roughness or smoothness of an object. Texture is scale-dependent and is affected by the spatial resolution of the photograph. Obviously, the texture of forest (small scale) is different from that of distinct tree canopies (large scale) (Plate 8.6).

Pattern is the spatial arrangement of objects and is also highly dependent on scale. Geographers often refer to drainage patterns, field patterns, and settlement patterns in their study. The pattern is the result of a process, or a driving force, which will shed light on physical and human activities in the study area. One can describe a pattern as regular or irregular; concentrated or dispersed (Plate 8.6).

Height provides the third dimension of the object. The ability to see the third dimension of objects is a major advantage when aerial photographs are viewed stereoscopically under a mirror stereoscope in the photographic

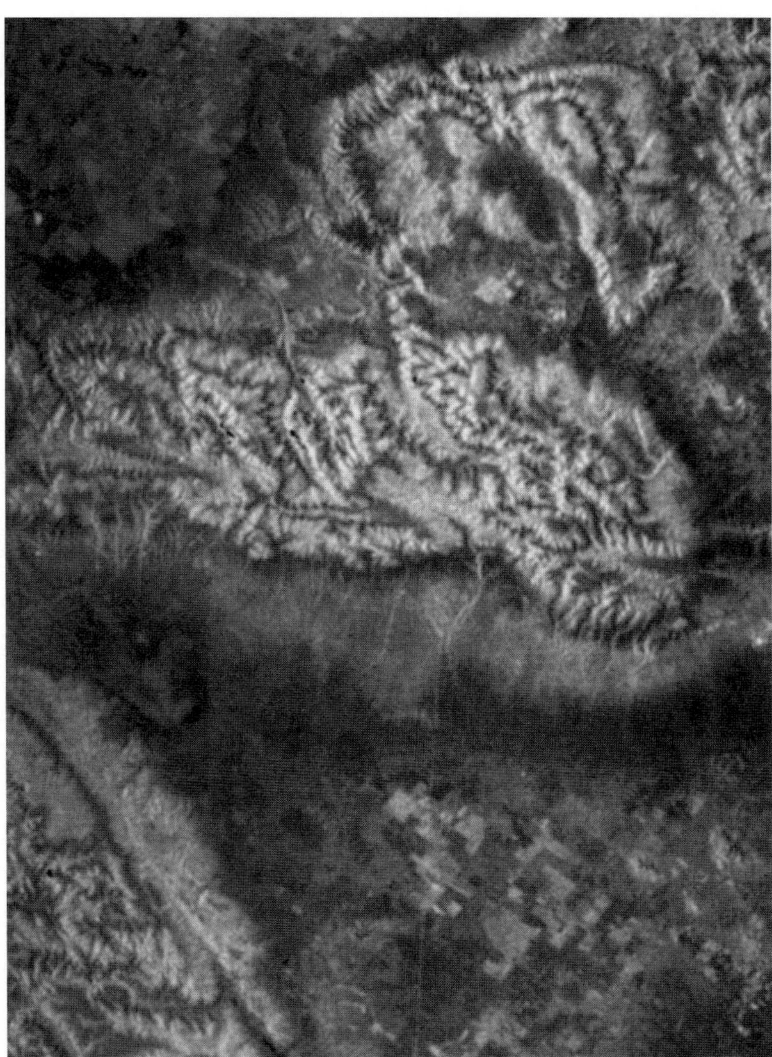

PLATE 8.5
San Andreas fault, Southern California. The image combines two types of data from the Shuttle Radar Topography Mission: (1) The image brightness corresponds to the strength of the radar signal reflected from the ground, and (2) the contourlike color bands show the elevation as measured by interferometry. To view this photo in color, see the color insert. *(Courtesy: NASA/JPL/Caltech)*

interpretation process. Heights of trees and buildings are a useful clue to vegetation and building types (Plate 8.6).

Shadow can provide some idea of the height of an object if no stereoscopic viewing is possible. Shadows tend to emphasize linear features and shapes. However, shadows also obscure features. Aerial photography is normally taken within two hours of noon in order to avoid excessive shadows.

Site refers to the locational characteristics of objects, which include such terrain conditions as topography, soil, vegetation, and cultural features. For example, certain trees prefer a marshy environment to grow. *Association* is the spatial relationship of objects and phenomena. The identification of one object or phenomenon will point toward the occurrence of associated objects and phenomena. For example, a playground and an athletic field are associated with a school building.

8.6.2 FUNDAMENTAL CONCEPTS IN COMPUTER-ASSISTED IMAGE CLASSIFICATION

The computer-assisted approach is used to extract thematic information (attributes) from digital images, such as those acquired by satellites (Landsat MSS, Landsat TM, Landsat ETM+, SPOT, and IKONOS). The same process of interpretation is applicable to image

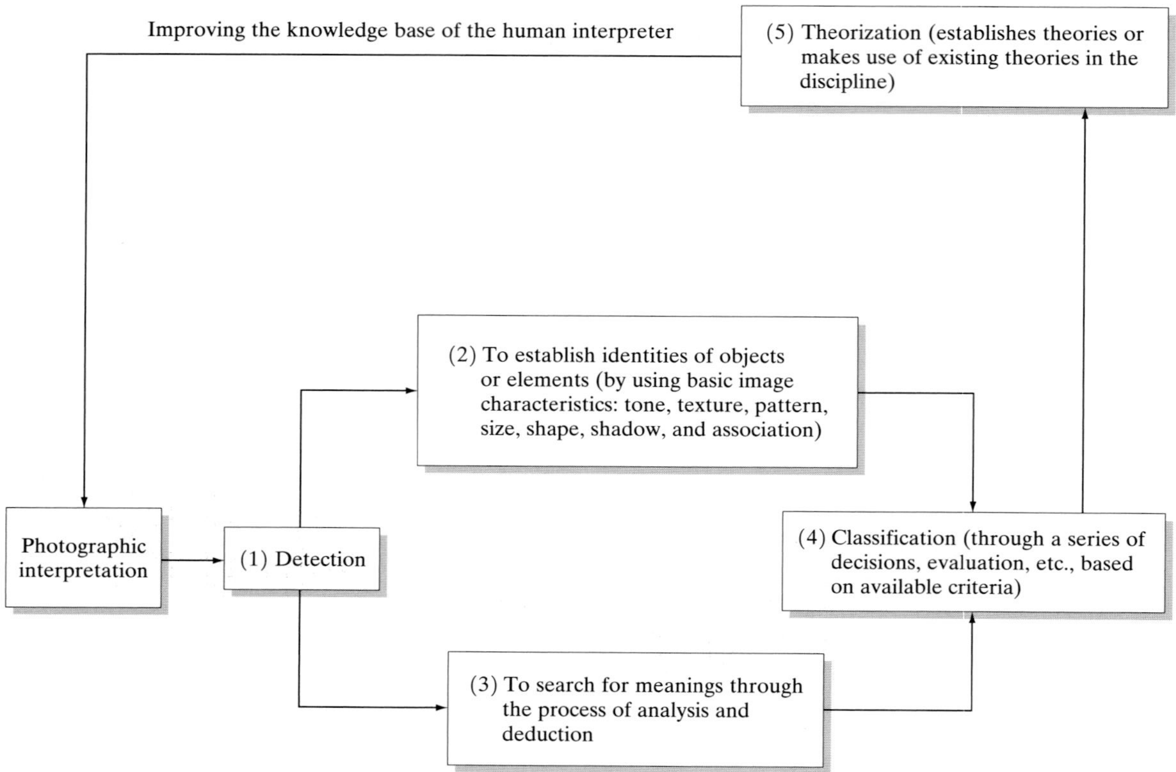

FIGURE 8.16
Processes of photographic interpretation. *(Source: Lo, 1976)*

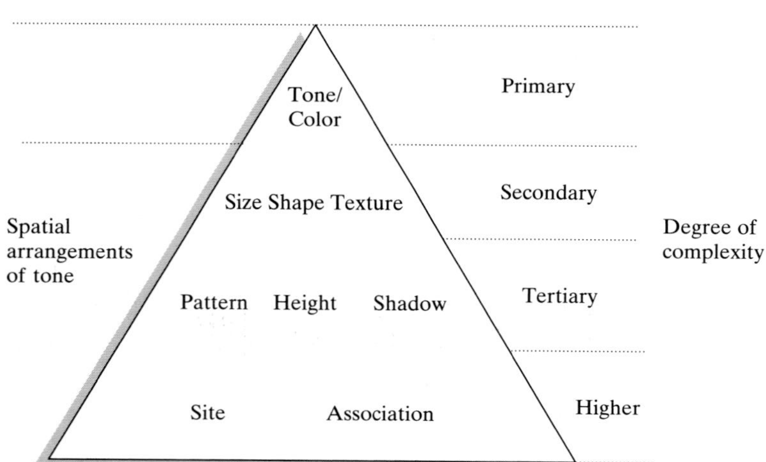

FIGURE 8.17
Primary ordering of image elements fundamental to photographic interpretation and image analysis. *(Source: Estes et al., 1983)*

classification. Although the nine basic image elements were developed with reference to aerial photography in analog form, they are readily applicable to computer-assisted analysis of digital images. Tone, which is represented as a digital number in each pixel of the digital image, is obviously the most important element used in digital image classification. Size, shape, texture, and pattern represent spatial arrangements of tone and color. Texture has been used in digital image classification too (Haralick et al., 1973; Franklin and Peddle, 1989), and it is particularly useful in the analysis of cultural features (Webster, 1996). For radar imagery, texture is as important as tone in visual and digital image classification. Site and association provide the basis for the development of contextual image classifiers to improve image classification accuracy (Wharton, 1982; Gong and Howarth, 1992).

PLATE 8.6
All the basic image elements for photographic interpretation. Tone variations, textural differences (cf. A, B, and D), the pattern of the forest plots, the size of trees (cf. A, C, and D), the heights of trees (compare A, B, and D), the site and association suggest that this is a tree nursery.

Data Preprocessing

Before digital image data can be classified, they need to be preprocessed because of errors caused by the remote sensing systems and the atmospheric conditions at the time of image acquisition. When satellite images are used, these errors are even more severe. *Preprocessing* attempts to compensate for these errors, including radiometric and geometric corrections. This is raster data preprocessing in GIS (see Chapter 5, Section 5.2.3 for discussion on raster data processing).

Radiometric Correction This is to correct (1) errors in the detectors of the remote sensing system, such as missing scan lines, striping or banding (caused by a detector out of adjustment), or a line-start problem (scanning system not able to collect data at the beginning of a scan line); (2) atmospheric attenuation caused by scattering and absorption in the atmosphere; and (3) topographic attenuation caused by slopes and aspects of terrain (Jensen, 1996b). If land-use/cover change characterization is to be carried out using digital images acquired at different times, radiometric correction is necessary; otherwise the same land cover types on images of different years will not give the same tonal reflectance. Absolute radiometric correction requires knowledge of the atmospheric conditions at the time of imaging, which is often not easily available. Relative normalization is a more economical approach by which a reference image in a time sequence of images is chosen, and digital values of the other images within the full range from very dark to very light targets are then related to this reference image (Yuan and Elvidge, 1996). While some of the methods of relative radiometric normalization are mathematically complicated, standard image-processing software packages such as ERDAS Imagine provide a function called *histogram matching,* which matches the histogram of brightness values of one image to that of another image, so that the apparent distribution of brightness values in the two images are made as close as possible (Richards, 1986) (see Section 5.2.4 and Figure 5.10). Another commonly used method of relative radiometric normalization is *image regression,* by which the brightness value of each pixel of the subject image is related to that of the reference image band by band to produce a linear regression equation (Jensen, 1996b).

Geometric Correction This is to correct the systematic and nonsystematic errors in the remote sensing system during the process of image acquisition. Systematic errors include scan skew caused by the forward motion of the platform during the time for each mirror sweep, irregular mirror-scan and platform velocities, while nonsystematic errors include variations in flying heights (or orbital altitudes) and attitude errors (i.e., roll, pitch, and yaw) (Jensen, 1996b). Geometric correction involves a mathematical transformation of the coordinates as explained in Chapter 2 (Section 2.5.3). In the case of digital images, which are in raster format, the coordinates are pixels fixed by columns and rows rather than *x* and *y*. For geometric correction, the method of image-to-map rectification, followed by attribute interpolation by resampling, as explained in Chapter 5, Section 5.2.2 on georeferencing of raster data, is used.

Image Classification: Approaches

Image classification is usually used to extract thematic information (attributes) from multispectral images. There are three approaches to digital image classification: supervised, unsupervised, and hybrid.

Supervised Classification This is characterized by the need to use training areas to specify to the computer algorithm the brightness values that will represent one category of land use or land cover in each band of the digital image. In this way, the algorithm is trained to identify different categories of land use/cover based on the brightness values of pixels in different spectral bands. This approach requires knowledge of ground truth on land use/cover that can be obtained from fieldwork or from large-scale aerial photographs. The selection of training areas is a very important step because it will affect the accuracy of the final classification. As far as possible, pixels of homogeneous brightness values are selected for each category of land use/cover to be classified. However, in reality, it is difficult to have total homogeneity, and confusion between some land-use/cover categories will arise.

After the training stage, the image is then classified based on the input of training areas. There are three commonly used "hard" classifiers for multispectral image classification: parallelpiped, minimum distance to means, and maximum likelihood (Jensen, 1996b). These are all per-pixel classifiers, that is, individual pixels are used as the basic unit for classification.

Parallelpiped classifier is also known as a box classifier because the upper and lower limits of brightness values in each band for each category of land use/cover (as determined from training) are fixed as a box, and the brightness value of each pixel is then compared to each of these boxes to determine in which box the pixel's brightness value should belong. There are always some pixels that cannot be classified because their brightness values do not fall in any of these boxes (Figure 8.18a).

The *minimum distance to means classifier* computes the Euclidean (or straight-line) distance between a pixel and the means of the clusters of land-use/cover categories derived from the training areas in a feature space defined by the number of spectral bands (Figure 8.18b). The means of the clusters are averaged brightness values of these clusters in each band. The Euclidean distances between a pixel and the cluster means are computed using the Pythagorean theorem as explained in Section 2.5.2 (Equation 2.6). A pixel is assigned to the class of the closest cluster. However, because clusters are different in size, a pixel can be easily misclassified. A spectrally less homogeneous cluster that is large tends to be "closest" to a pixel than more homogeneous clusters that are smaller.

The *maximum likelihood classifier* computes the probability that a pixel belongs to a class. Data from the training sets are assumed to be normally distributed, which allows the *mean vector* and the *covariance matrix* of the spectral cluster of each category of brightness values to be computed (Lillesand and Kiefer, 2000). Based on these results, the statistical probability of a pixel's brightness value belonging to each category of land use/cover can be computed. The pixel is assigned to the category with the highest probability value (Figure 8.18c). This classifier usually assumes equal probability of occurrence of each category of land use/cover. If known, one can specify the probability of occurrence of each land-use/cover category before carrying out the image classification (known as a priori maximum likelihood classification). This makes use of weight factors for particular categories. This variation of the maximum likelihood decision rule is known as the Bayesian deci-

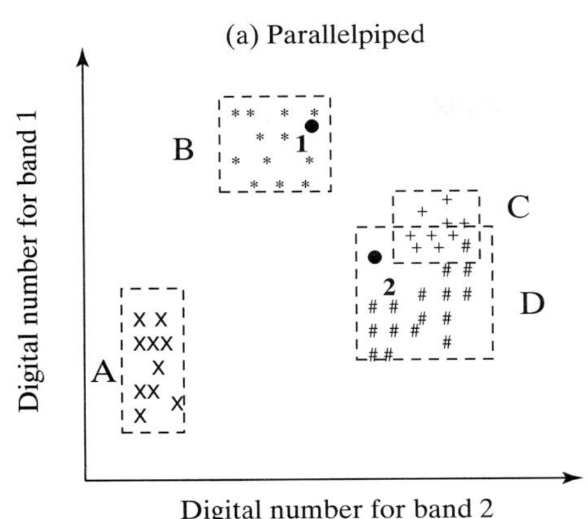

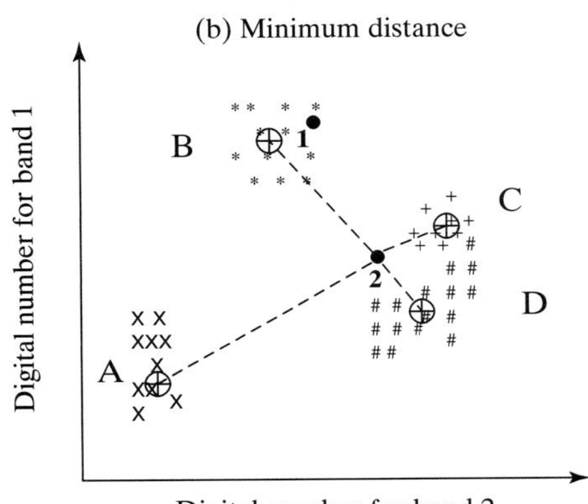

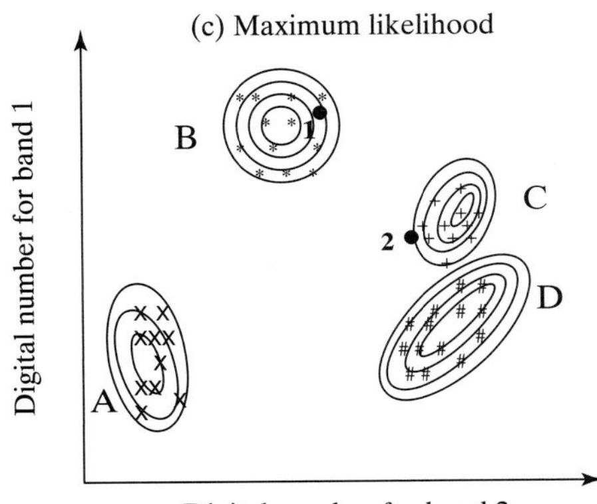

FIGURE 8.18
Three different types of classifiers. (a) A parallelpiped creates a box with an upper and lower bounds. A pixel falling within the box belongs to that class. (b) Minimum distance: Pixel #2 being closest to the centroid of pixel cluster C is assigned to class C. (c) Maximum likelihood: Equiprobability contours indicate that pixel #1 most likely belongs to class B, and pixel #2 most likely belongs to class C. *(Modified from Lillesand and Kiefer, 2000)*

sion rule. Although the maximum likelihood method is computationally intensive, it is often regarded as the best classifier to use.

Unsupervised Classification This approach uses the method of cluster analysis to produce natural clusters of pixels of similar brightness values from the multispectral image data without the need to train the algorithm. These natural clusters of pixels are then related to actual land-use/cover categories after consulting the ground-truth data. Cluster analysis normally makes use of distances to form clusters. The number of clusters to be produced has to be specified. Too large a number of clusters will produce too many homogeneous clusters, and association with actual land-use/cover categories may be more difficult. Too few clusters will produce heterogeneous clusters and land-use/cover categories will intermingle. The optimal number of clusters can be determined only by trial and error. A clustering algorithm known as Iterative Self-Organizing Data Analysis Technique (ISODATA) is very widely used (Jensen, 1996b). The unsupervised classification is very popular among GIS agencies for database maintenance because of the very fast clustering programs now available, and the unsupervised classification is easier to learn.

Hybrid Approach The unsupervised and supervised approaches to classification can be mixed together to produce a *hybrid* approach. After unsupervised classification has been completed, training areas are extracted based on these natural clusters. A maximum likelihood classifier is then applied to classify the whole scene. Such a hybrid approach is quite commonly used today to extract land-use/cover information from satellite image data.

Image classification results in the production of a map in hard- and soft-copy forms, usually with the aid of GIS functionality. Plate 8.7 is a 1997 land-use/cover map of Atlanta, Georgia, produced from the Landsat TM images shown in Plate 2.4 using the ISODATA algorithm (unsupervised classification approach). By overlaying and properly registering a vector file of county boundaries over the resulting raster map, land-use/cover statistics of each county can be extracted using GIS functionality.

Image Classification: Types of Classifiers

Soft Versus Hard Classifiers The methods of classification explained previously make use of Boolean logic, that a pixel can belong to only one class (1) or nothing at all (0). It is clear that a pixel is not 100% pure, and at the boundary between two or more types of land-use/cover, the mixed-pixel problem arises. In the case of *hard* classification, a line is drawn to separate

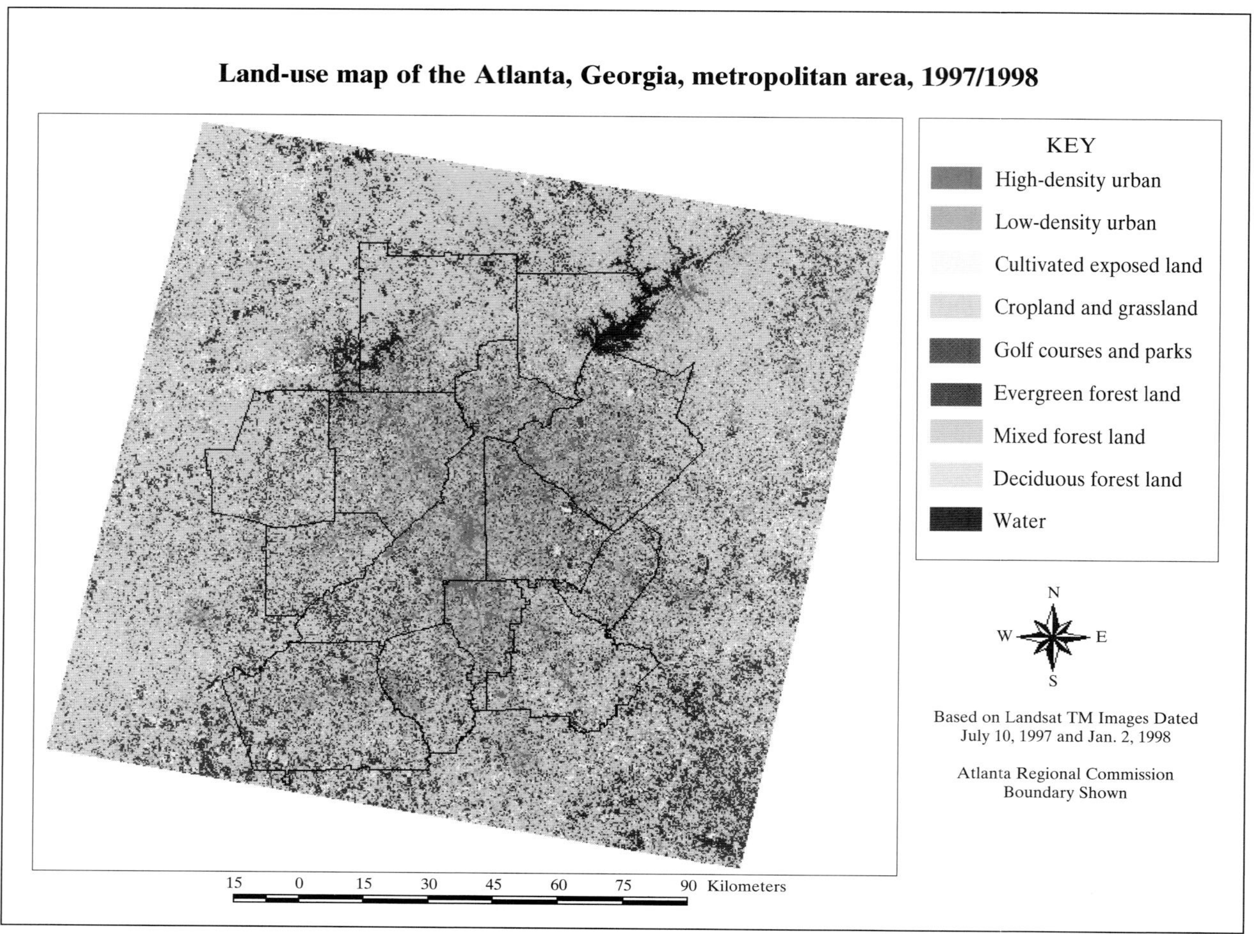

PLATE 8.7
A 1997 land-use/cover map of the Atlanta region produced from an unsupervised image classification approach (ISODATA) based on the Landsat TM image (Plate 2.4). This map was produced using ArcView GIS software. The boundaries of the 13 urban counties of Atlanta (vector) were overlaid on top of the land-use/cover classification (raster) so that land-use/cover statistics may be extracted for each county. To view this photo in color, see the color insert.

the different categories of land use/cover. However, a more appropriate approach is to determine the different proportions of land use/cover in the pixel, for example, 0.75 forest, 0.03 water, and 0.22 cropland, which are known as the membership grade values, based on fuzzy logic (Wang, 1990). The sum of the membership grade values will be 1. This method of classification does not assign a definitive class to the pixel but gives a combination of classes instead, and so is known as *soft* classification (Figure 8.19). If soft classification is used, the traditional method of representing land-use/cover polygons with sharp boundaries is no longer appropriate. The result can affect the vector representation of geographic boundaries in GIS. Wang and Hall (1996) recommended the use of fuzzy representation of geographic boundaries, in which boundaries describe not only the location but also the rate of change of environmental phenomena.

Contextual and Neural Network Classifiers The classifiers explained earlier segment images on a pixel-by-pixel basis. Because of the mixed-pixel problem, another class of classifiers has been developed that do not just classify pixel by pixel. A contextual classifier makes use of contextual information among neighboring pixels in order to locate spectrally homogeneous regions of pixels (Wharton, 1982). This is a zone-based method of classification also known as the *cover-frequency method of classification* (Gong and Howarth, 1992). This approach involves two stages: The multispectral image is first classified into its land-use/cover classes using either a supervised or unsupervised approach, followed by moving a pixel window (normally 3×3) over the resultant land-use/cover image to determine the cover frequency within the window. These frequencies are assigned to the central pixel of the window. Figure 8.20 is a 3×3 window showing the class of land cover in

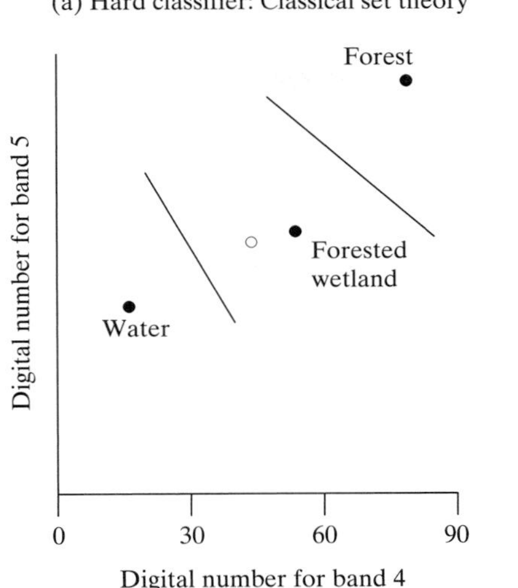

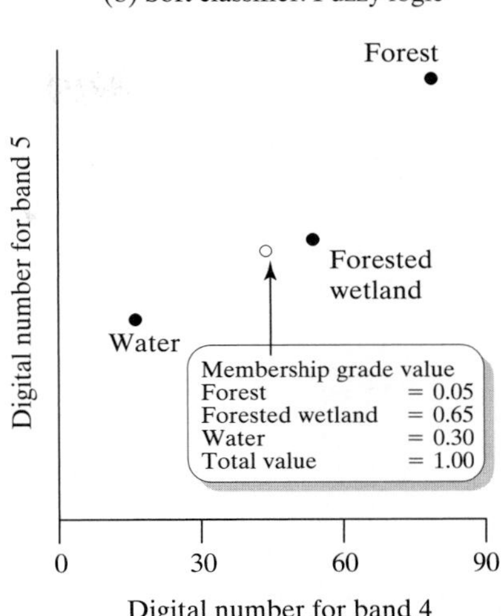

FIGURE 8.19
Feature space showing mean class vectors for three types of land cover (black circles). In (a) the unknown measurement vector (white circle) is assigned to "forested wetland," based on classical set theory, while in (b) fuzzy logic is used to show the membership grade values of the unknown measurement vector, indicating how close it is from each mean-class vector. *(Source: Jensen, 1996b)*

each pixel. Assume that there are four classes of land cover represented by numbers 1, 2, 3, and 4; the cover frequency displayed is 5 (1), 3 (2), 0 (3), and 1 (4). This can be represented in the form of a frequency vector for the given window as $(5, 3, 0, 1)^T$. This vector is assigned to the center pixel of the window. After the moving window has completely covered the whole image, these frequency vectors extracted are classified using cluster analysis to produce a final land cover map. The size of the pixel window can affect the accuracy of the classification by land cover class. In other words, land cover classes that are more homogeneous are better classified using a small pixel window, while a large pixel window is best for more heterogeneous land cover classes.

1	2	2
2	1	1
4	1	1

FIGURE 8.20
Contextual classifier. For four possible classes, the vector of component frequencies corresponding to the given window is $(5, 3, 0, 1)^T$. *(Source: Wharton, 1982)*

Another type of classifier that does not classify on a pixel-by-pixel basis makes use of the *artificial neural network*, which allows ancillary data to be included in image classification (Civco, 1993). Artificial neural networks are models that attempt to simulate the functionality and decision-making processes of the human brain. Neural network features corresponding to the *synapses*, *neurons*, and *axons* of the brain are *input weights*, *processing elements*, and *output paths*. The processing elements are the same as the human brain's biological neuron. There are many input paths in the processing elements (analogous to the brain's dendrites) and the information transferred along these paths is combined by a mathematical function, commonly simple summation. The combined input is modified by a transfer function before being passed on to other connected processing systems, whose input paths are usually weighted. Commonly, the function is a sigmoid or hyperbolic tangent function that has the effect of smoothing the processing element's internal values. In applying to multispectral image classification, neural networks consist of three or more layers: an *input layer*, an *output layer*, and one or more *hidden layers* (Figure 8.21). The input layer consists of one or more processing elements (the training data), and the output layer consists of one or more processing elements that store the results of the network. In land cover classification of a Landsat TM image, the input layer consists of the vector of the brightness values for each spectral band (seven altogether). The input layer may also have ancillary data (such as elevation, slope, soil type). The hidden layers

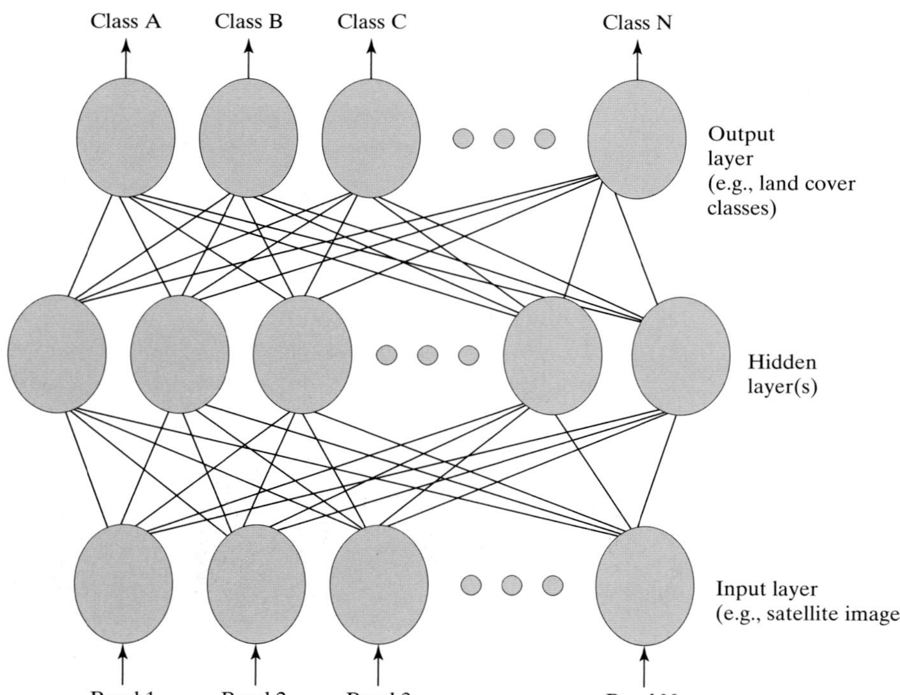

FIGURE 8.21
Topology of an elementary artificial neural network. *(Source: Civco, 1993)*

consist of a number of processing elements that translate the input data into output information (i.e., land cover). The number of hidden layers depends on the nature and structure of the input and output data. The neural networks need to learn the characteristics of each output response using the training data. These characteristics are then recalled to classify unknown pixels. Hepner et al. (1990) observed that artificial neural networks could produce results superior to those using the conventional methods of supervised classification. There are many examples of using artificial networks to classify satellite images for land use/cover (Yoshida and Omatu, 1994; Bischof et al., 1992; Kanellopoulos et al., 1992; Benediktsson et al., 1990).

Collateral Material

To be successful in both photographic interpretation and image analysis, collateral material that includes literature, maps, and laboratory and field data must be used. Field checking by sampling is essential after an interpretation or analysis has been completed in order to develop an error matrix, from which thematic accuracy is determined (see Chapter 4).

8.6.3 DIGITAL CHANGE DETECTION USING REMOTELY SENSED IMAGES AND GIS

One of the major functions of GIS is its analysis of changes between two or more layers of geographic data (see Chapters 5 and 6). In remote sensing, change detection is the process of identifying differences in the state of an object or phenomenon by observing it at different times (Singh, 1989). Digital change detection in remote sensing has been greatly facilitated by the use of GIS, particularly in the following two approaches: (1) map-to-map comparison and (2) image-to-image comparison (Green et al., 1994).

Map-to-Map Comparison

This is also known as "postclassification comparison change detection" (Jensen, 1996b). If satellite images for two different years are used, the two images need to undergo geometric rectification and registration so that they can be matched up exactly. Each image will then be classified using one of the approaches explained in Image Classification (Section 8.6.2). The overlay function of GIS is used to compare the two maps pixel by pixel. A change detection matrix, which is a cross-tabulation between the two maps, is produced. A change map may also be generated. Figure 8.22a and b are two land-use maps in raster format of East Tuen Mun in the New Territories of Hong Kong in 1976 and 1987 respectively. The two maps were overlaid and subtracted pixel by pixel. The resulting map would show zeros for pixels with no change. Other pixels where land-use change had occurred would have positive or negative values. Using reclassification, all these positive and negative values were recoded to become "ones." In this way, a binary mask of "ones" and "zeros" was produced that vividly depicted the areas where land-use change had occurred (Figure 8.22c). A change matrix revealed the dynamics of land-use change

(a) 1976

(b) 1987

(c) Binary change mask

FIGURE 8.22
Land-use change analysis using GIS. (a) 1976 land-use map of East Tuen Mun; (b) 1987 land-use map of East Tuen Mun; (c) land-use change map produced by subtracting the 1976 from the 1987 map, and then recoding to form a binary change mask, with black indicating areas where land-use change has occurred. *(Source: Lo and Shipman, 1990)*

in the study area (Table 8.4). The map-to-map comparison may also be performed on the basis of polygon types rather than pixels. In this case, the two land-use maps extracted from digital images need to be converted to vector format for this purpose. Vector GIS software may be used to overlay the two vectorized maps ("UNION"). The polygons that have undergone land-use change can then be extracted from the overlay using the appropriate GIS functions.

Image-to-Image Comparison

The radiance value of each pixel of the two satellite images of two different dates is compared by this approach (Jensen, 1996b). Because any change in radiance value is assumed to reflect a genuine change of the land cover, the two satellite images will have to be calibrated in order to minimize sensor calibration effects and standardize data-acquisition effects. Because data on the conditions of the atmosphere at the time of satellite overpasses required for radiance value correction are not always available, the relative normalization method of radiometric correction approach mentioned in data preprocessing is used (Caselles and Garcia, 1989). Also, the two images have to be geometrically rectified and accurately registered. The two images are compared either by means of subtraction (known as image differencing—Equation 5.5c in Section 5.3.2) or by band ratioing (i.e., by dividing two layers, as shown in Equation 5.5e in Section 5.3.2). These are raster GIS overlay functions. In the subtraction case, the results can be negative or positive. Commonly, a constant value is added to convert all negative values into positive values. Green et al. (1994) used a constant of 100 in their application of multidate Landsat TM images for vegetation change detection. Thus, values of 100 in the resulting image difference file indicate "no change" in reflectance value between the two dates; areas of increased reflectance are indicated by values greater than 100; and areas of decreased reflectance are indicated by values less than 100. This method will allow the areas of gain or loss to be mapped and quantified. In band ratioing, the same spectral bands of the images for the two years are divided, with the result that changes will have values of either greater than or less than 1, thus revealing the direction of land cover reflectance change for a pixel between the two years (Lo, 1997b).

Because overlay functionality is involved in both the map-to-map and image-to-image comparison approaches, the accuracy of the change map produced is adversely affected by the accuracy of each map that forms the overlay. As has been observed in Chapter 4, the accuracy of the overlay of two maps each with an accuracy of 95% will be reduced to 90%. Therefore, the more maps to overlay, the less will be the accuracy of the final overlay.

8.7 Integration of Remote Sensing and GIS

A fundamental technical requirement for integrating remote sensing with GIS data is the need to have both types of data in the same georeferencing system. This involves coordinate transformation and resampling. Coordinate transformation can be either *rectification* if the spatial dataset is to be transformed to a specific georeferencing system, such as the UTM, or *registration* if the coordinates of one spatial dataset are transformed to those of another spatial dataset without specific

Table 8.4

Land-Use Change Matrix Based on an Interpretation of the 1976 and 1987 Land-Use Maps of Tuen Mun as Shown in Figure 8.22

From 1976 / *to 1987*	*LDU*	*HDU*	*TRSP*	*CULT*	*AGLN*	*WOOD*	*RESER*	*BAYS*	*BADL*	*BARR*	*TRANS*	*Total (1987)*
LDU	**0**	304	70	271	0	1195	75	114	1203	620	136	3988
HDU	1842	**0**	432	857	2	426	81	2725	267	973	849	8454
TRSP	892	182	**0**	346	169	109	10	301	105	796	114	3024
CULT	815	29	21	**0**	24	765	0	0	341	147	275	2417
AGLN	701	0	119	1130	**0**	408	79	0	589	114	0	3140
WOOD	1064	0	58	408	99	**0**	3	20	10,657	1617	100	14,026
RESER	0	0	0	0	0	3	**0**	0	0	0	0	3
BAYS	217	95	127	2	0	0	0	**0**	6	5	113	565
BADL	360	0	0	758	0	6917	0	0	**0**	787	12	8834
BARR	569	0	0	25	0	2142	0	167	3628	**0**	0	6531
TRANS	1254	9	19	681	0	181	0	798	90	61	**0**	3093
Total (1976)	7714	619	846	4478	294	12,146	248	4125	16,886	5120	1599	54,075

Note: All values are given in numbers of pixels. Each pixel is a 20 m × 20 m cell on the land-use maps.
Key: LDU = low-density urban; HDU = high-density urban; TRSP = transportation; CULT = cultivated land; AGLN = Agricultural; WOOD = woodland; RESER = reservoirs and ponds; BAYS = bays and estuaries; BARR = barren land; TRANS = transitional land.
The zero diagonal in the matrix indicates no change.
Source: Lo and Shipman, 1990

reference to a georeferencing system. Resampling is required for the interpolation of pixel values after the rectification or registration for raster data. The principles of these processes have already been explained in Chapters 2 (Section 2.5.3) and 5 (Section 5.2.2) as well as Section 8.6.2 of this chapter. The major concern of rectification, registration, and resampling is the accuracy achievable and the extent to which these processes can be fully automated to deal with the large volume of high-resolution satellite image data (Ehlers, 1997).

As satellite image data of different spectral bands and spatial resolutions become available, the problem of efficient storage and processing of these data arise. GIS technology has been regarded as the best way to solve the problems (Ehlers et al., 1989). However, GIS technology has to be modified, adapted, and extended to meet this objective. In other words, there is an urgent need to integrate remote sensing with GIS. Ehlers et al. (1989) presented three evolutionary stages in the integration of remote sensing and GIS as follows (Figure 8.23).

Stage I: Separate but Equal There are two separate systems: one is the GIS and the other is image processing (Figure 8.23). The two systems are linked using some sort of data exchange format that allows data exchange to take place. For this first level of integration, simultaneous display of GIS (usually in vector format) and remotely sensed images (in raster format) is possible in one of the following ways: the results of low-level image processing (e.g., a thematic map) may be moved to the GIS allowing for the assignment of attribute values to the theme; the results of GIS overlays may be moved between images and vector data to image analysis software, thus facilitating geometric registration of images to a common coordinate system; and the results of GIS spatial analyses may be moved to image analysis software for support and validation of the image.

Stage II: Seamless Integration Both the GIS and image-processing system that share the same user interface are still separate but complementary to each other and interwoven (Figure 8.23). Seamless integration possesses

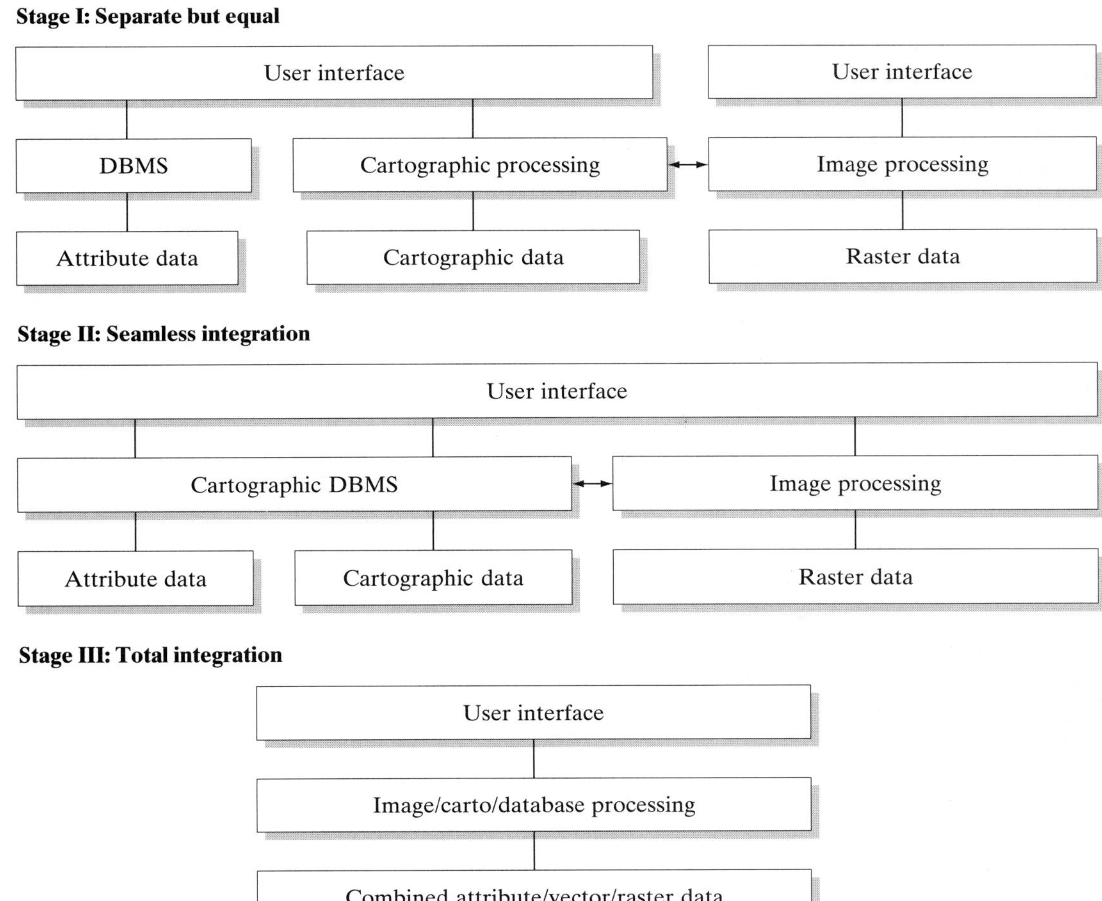

FIGURE 8.23
Three stages in the integration of image analysis with GIS technology. *(Source: Ehlers et al., 1989)*

the following capabilities: control over remote sensing image components (themes); incorporation of GIS (vector) data directly into image processing (through rasterization); accepting spatially, radiometrically, spectrally, and temporally heterogeneous data input in a coherent manner; handling time, accommodating hierarchical entities (e.g., house, block, city—three different levels); analysis of random and systematic errors; and generating simulations cartographic and image data together with temporal evolution.

Stage III: Total Integration Both GIS and image processing become one system and are no longer separable (Figure 8.23). The vector–raster dichotomy no longer exists and is valid only at low processing levels. At higher processing levels (i.e., higher levels of abstraction), a vector representation is used, while at lower levels a raster representation is adequate. In total integration, the object-based and phenomena-based representation of geographic space can be handled flexibly. Both raster and vector data may be accommodated at different hierarchical levels. In a total system, remote sensing becomes a part of the data input functionality.

There have been some major technical and institutional impediments to total integration. The raster–vector dichotomy was one of the technical impediments. Remote sensing is a raster-based data collection and analysis system, while GIS tends to be vector-oriented (although raster-based GIS also exists). Each of the methods of data representation has its advantages and disadvantages as was explained in Chapter 3. Computer programs that convert between raster and vector formats can introduce significant errors after the conversion (Lunetta et al., 1991). Figure 8.24 clearly demonstrates the difference between a raster and vector representation of the same shape. Another technical impediment is the problem of data uniformity. GIS data are clearly defined before collection, while remote sensing data, which contain a wealth of information about the environment, need to be interpreted before they can be used. The user does not necessarily know what to extract from the remote sensing data. GIS data may come from many different sources, and as a result, it is not always known how they are collected. Tracking errors of GIS data will be difficult. On the other hand, remote sensing data combine data collection and data processing together, and it is often easier to determine the errors. This lack of data uniformity made integration of GIS and remote sensing data difficult in the past.

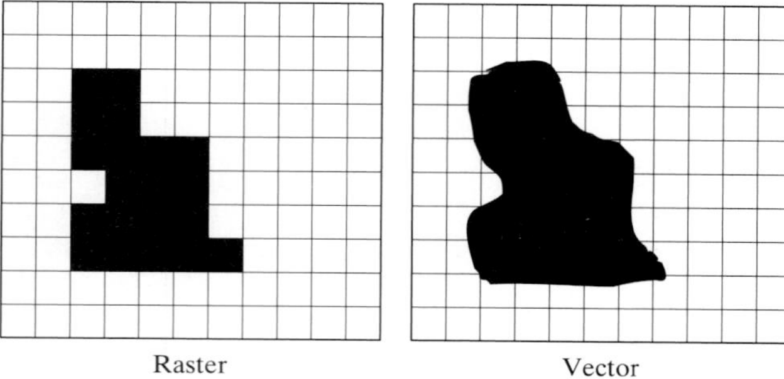

FIGURE 8.24
Differences in a raster and vector representation of the same shape. *(Source: Lunetta et al., 1991)*

Institutional impediments refer to the groups of users with different needs who are involved with GIS and image processing. The decision makers who are concerned with strategic development and management of natural resources generally work with a restricted area (e.g., city, county, or country) or with a restricted task (e.g., providing building permits). On the other hand, research scientists may require a much wider view of Earth. The third group of users is the general public who need a system sophisticated enough to suggest to them which are the important or relevant relationships. It is difficult to design an integrated GIS and remote sensing system that meets all these diverse needs.

Today, much progress has been made toward better integration of remote sensing and GIS. Advances in computer hardware and software have permitted an expansion of the current GIS/remote sensing capabilities in dealing with data structure conversion. High-speed computers make possible more complicated data processing and analysis in a short time. Networking of computers allows high-speed data transfer, real-time data compression and decompression, and real-time interactive spatial analysis (Faust et al., 1991). Improvements in the GUI have made GIS/remote sensing systems easier to use (see Chapter 5). New display technology facilitates GIS and remote sensing data to be displayed in higher resolution and better color fidelity. Techniques of multimedia communication using voice, video, and graphics may also be employed. The display technology that allows stereoviewing of digital images, fly-bys, and the draping of a vector data set on a raster image (e.g., land-use polygons draped on top of an orthoimage) assists the integration of GIS and remote sensing.

An important area of integration of remote sensing and GIS lies in combining vector information in image classification for the selection of training areas. The system is capable of performing a raster–vector intersection query (Ehlers et al., 1989). In essence, this is to find out which pixels fall within which polygon, given an image and a polygon file, without the need for data format conversion. A fully integrated system requires two-way flow of data between the raster image and the vector dataset. Image statistics within polygons are generated and then returned directly to the GIS database as attributes of the polygons (Hinton, 1996).

Because artificial neural networks incorporate multisource data in the input layers for image classification (Section 8.6.2.) without the need to depend on any statistical model of the dataset concerned, they have helped to integrate remote sensing with GIS (Wilkinson, 1996).

Finally, a new model that will not treat GIS/remote sensing integration as a vector–raster dichotomy is needed if total integration is to be achieved. Ehlers et al. (1989) presented a model to show the different representations of spatial information in geographic information and image processing systems. In a representation space defined by three orthogonal axes of aspects of representation—concept of space, scale, and level of abstraction (Figure 8.25)—a representation is shown as a point (S1 or S2). Representations can vary in several aspects. Some transformations from one realization of representation to another (e.g., from S1 to S2) may be reversible. Other axes that can be added to this space are level of uncertainty and accuracy, or temporal abstraction to create higher order representation spaces. In this model, vector and raster data represent different levels of abstraction. New computer technology and high-resolution (submeter) satellite image data will blur the distinction between vector and raster data.

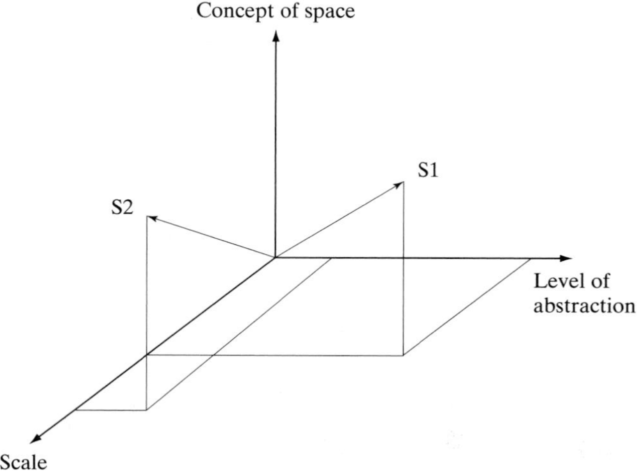

FIGURE 8.25
A conceptual model for different representations of spatial information in Geographic Information and Image Processing systems. The representations (such as S1 and S2) can vary in the concept of space, level of abstraction, or scale. Other presentation axes (e.g., level of uncertainty and accuracy) can be added. *(Source: Ehlers et al., 1989)*

8.8 Summary

Remote sensing data, particularly satellite image data, are an important source of data for GIS. Technologically advanced satellite remote sensing systems capable of imaging Earth with high spatial and spectral resolutions have been successfully launched. Landsat 7 not only improves upon the multispectral sensing of the Landsat MSS (Multispectral Scanner) and Thematic Mapper (TM) sensors with its Enhanced Thematic Mapper Plus (ETM+) sensors, it also adds a 15-m panchromatic band to acquire high spatial resolution images of our earth. The 1-m spatial resolution IKONOS-2 from the commercial company Space Imaging makes it possible to produce very accurate topographic maps at medium and large scales. NASA's Earth Observing System (EOS) Terra satellite comprising ASTER, MODIS, MISR, CERES, and MOPPETT sensors provides a constant stream of data about Earth's atmosphere and its terrestrial environment. Satellite imaging systems similar to IKONOS-2 will be launched by additinal commercial companies. NASA also launched the EOS Aqua satellite systems that cover the marine environment. Inevitably, all these satellite data will form a part of an immense GIS database. Radar and thermal infrared sensors have also grown in importance. Therefore, a good understanding of the fundamentals of remote sensing is essential for GIS users.

In this chapter, a brief survey of remote sensing, photogrammetry, and satellite remote sensing systems in relation to GIS was provided. The physical principles of remote sensing were explained, followed by an examination of some important imaging characteristics of remote sensing systems, namely, spatial, spectral, radiometric, and temporal resolutions. Aerial photography is still an important form of remote sensing, particularly if detailed metric and descriptive information is to be extracted. Metric information, most notably heights and positions, can be obtained from stereopairs of aerial photographs using the principles of photogrammetry. Photogrammetry emphasizes reconstructing the stereomodel at the time of photography through inner, relative, and absolute orientation so that the tilts and relief displacements inherent in the aerial photographs may be removed and so that the topographic information produced may be related to a proper terrestrial coordinate system. Photogrammetry may be applied to digital image data, such as those from satellites, to produce digital elevation models (DEM) and orthoimages, thus integrating seamlessly with GIS.

Extraction of descriptive (thematic) information from aerial photographs makes use of the principles of photographic interpretation and requires the application of the basic image elements of tone, texture, pattern, shadow, shape, height, site, and association. Again, the principles of photographic interpretation form the basis of computer-assisted image classification when digital images are used. Both the supervised and unsupervised approaches to image classification are computer algorithms that classify the pixels based on the intensity of their values. Raster GIS functionality, such as reclassification and overlay, has played an important part in digital image classification and, in particular, change detection using the methods of map-to-map comparison and image-to-image comparison. Digital image data are easy to correct geometrically and radiometrically. Rectification and registration followed by resampling are common operations to both GIS and remote sensing, particularly if two or more images are to be matched up for overlay in the case of change detection.

Great progress has been made toward the *total integration* of GIS and remote sensing largely because of the rapid advancement in computer hardware and software technology. The raster–vector dichotomy has become blurred as powerful high-speed computer and ingenious computer programming have made the data structure conversion almost transparent. New display technology, which features fly-bys or drapes a vector data set on a raster image, promotes the integration of remote sensing and GIS. Finally, a model of integration is presented in which the vector and raster geographic data representations are viewed as two different levels of abstraction, one complementing the other. The emergence of very high-resolution (submeter) satellite images and high-speed computer power will spearhead the complete integration of remote sensing and GIS.

REFERENCES

Anderson, J. E. (1992) "Determination of water surface temperatures based on the use of thermal infrared multispectral scanner data," *Geocarto International*, Vol. 7, No. 3, pp. 3–8.

Anderson, J. R., Hardy, E. E., Roach, J. T., and Witmer, R. E. (1976) *A Land Use and Land Cover Classification System for Use with Remote Sensor Data*, Geological Survey Professional Paper 964, Washington, D.C.: United States Government Printing Office.

Benediktsson, J. A., Swain, P. H., and Ersoy, O. K. (1990) "Neural network approaches versus statistical methods in classification of multisource remote sensing data," *IEEE Transactions on Geoscience and Remote Sensing*, Vol. 28, No. 4, pp. 542–552.

Bischof, H., Schneider, W., and Pinz, A. J. (1992) "Multispectral classification of Landsat-images using neural networks," *IEEE Transactions on Geoscience and Remote Sensing*, Vol. 30, No. 3, pp. 482–490.

Carlson, G. R., and Patel, B. (1997) "A new era dawns for geospatial imagery," *GIS World*, March, pp. 36–40.

Caselles, V., and Garcia, M. J. L. (1989) "An alternative simple approach to estimate atmospheric correction in multitemporal studies," *International Journal of Remote Sensing*, Vol. 10, pp. 1127–1134.

Civco, D. (1993) "Artificial neural networks for land-cover classification and mapping," *International Journal of Geographical Information Systems*, Vol. 7, No. 2, pp. 173–186.

Colwell, R. N. (1997) "History and place of photographic interpretation," in Philipson, W. R. (1997) *Manual of Photographic Interpretation*, 2nd ed. Bethesda, MD: American Society for Photogrammetry and Remote Sensing.

Curran, P. (1985) *Principles of Remote Sensing*, London: Longman.

Curran, P. J. (1987) "Remote sensing methodologies and geography," *International Journal of Remote Sensing*, Vol. 8, No. 9, pp. 1255–1275.

Derenyi, E. E., and Newton, L. (1987) "Control extension utilizing Large Format Camera photography," *Photogrammetric Engineering and Remote Sensing*, Vol. 53, No. 5, pp. 495–499.

Dowman, I. J. (1994) "Satellite imagery—a guide to mapping applications," *SW*, Vol. 2, No. 3, pp. 20–22.

Dowman, I. J., and Mohamed, M. A. (1981) "Photogrammetric applications of Landsat MSS imagery," *International Journal of Remote Sensing*, Vol. 2, No. 2, pp. 105–113.

Dowman, I. J., and Peacegood, G. (1989) "Information content of high resolution satellite imagery," *Photogrammetria*, Vol. 43, pp. 295–310.

Doyle, F. J. (1985) "The Large Format Camera on Shuttle Mission 41-G," *Photogrammetric Engineering and Remote Sensing*, Vol. 51, No. 2, p. 200.

Ehlers M. (1997) "Rectification and registration," in Star, J. L., Estes, J. E., and McGwire, K. C., eds. *Integration of Geographic Information Systems and Remote Sensing*. Cambridge: Cambridge University Press, pp. 13–36.

Ehlers, M., Edwards, G., and Bedard, Y. (1989) "Integration of remote sensing with geographic information systems: A necessary evolution," *Photogrammetric Engineering and Remote Sensing*, Vol. 55, No. 11, pp. 1619–1627.

Ehlers, M., and Welch, R. (1987) "Stereocorrelation of Landsat TM images," *Photogrammetric Engineering and Remote Sensing*, Vol. 53, pp. 1231–1237.

Estes, J. E., Hajic, E. J., and Tinney, L. R. (1983) "Fundamentals of image analysis: Analysis of visible and thermal infrared data," in R. N. Colwell ed., *Manual of Remote Sensing*, 2nd ed. Falls Church, VA: American Society of Photogrammetry, pp. 987–1124.

Faust, N., Anderson, W. H., and Star, J. L. (1991) "Geographic information systems and remote sensing future computing environment," *Photogrammetric Engineering and Remote Sensing*, Vol. 57, No. 6, pp. 655–668.

Franklin, S. W., and Peddle, D. R., (1989) "Spectral texture for improved class discrimination in complex terrain," *International Journal of Remote Sensing*, Vol. 10, pp. 1437–1443.

Freden, S. C. and Gordon, Jr., F. (1983). "Landsat satellites," in Colwell, R. N. ed. *Manual of Remote Sensing*, 2nd ed. Falls Church, VA: American Society of Photogrammetry, pp. 517–570.

Gong, P., and Howarth, P. J. (1992) "Land-use classification of SPOT HRV data using a cover-frequency method," *International Journal of Remote Sensing*, Vol. 13, No. 8, pp. 1459–1471.

Green, K., Kempka, D., and Lackey, L. (1994) "Using remote sensing to detect and monitor land-cover and land-use change," *Photogrammetric Engineering and Remote Sensing*, Vol. 60, No. 3, pp. 331–337.

Gugan, D. J., and Dowman, I. J. (1988a) "Topographic mapping from SPOT imagery," *Photogrammetric Engineering and Remote Sensing*, Vol. 54, No. 10, pp. 1409–1414.

Gugan, D. J., and Dowman, I. J. (1988b) "Accuracy and completeness of topographic mapping from SPOT imagery," *Photogrammetric Record*, Vol. 12, No. 72, pp. 787–796.

Haralick, R. M., Shanmugam, K., and Dinstein, I. (1973) "Textural features for image classification," *IEEE Transactions on Systems, Man and Cybernetics*, SMC, Vol. 3, No. 6, pp. 610–621.

Henderson, F. M., and Xia, Z. G. (1998) "Radar applications in urban analysis, settlement detection and population estimation," in Henderson, F. M., and Lewis, A. J. eds. *Principles and Applications of Imaging Radar, Manual of Remote Sensing*, 3rd ed. Vol. 2, American Society for Photogrammetry and Remote Sensing, New York: John Wiley and Sons, pp. 733–768.

Hinton, J. C. (1996) "GIS and remote sensing integration for environmental applications," *International Journal of Geographical Information Systems*, Vol. 10, No. 7, pp. 877–890.

Jensen, J. R. (1983a) "Biophysical remote sensing," *Annals of the Association of American Geographers*, Vol. 73, pp. 111–132.

Jensen, J. R. (1983b) "Urban/suburban land use analysis" in Colwell, R. N. ed. *Manual of Remote Sensing*, 2nd ed. Falls Church, VA: American Society of Photogrammetry, pp. 1571–1666.

Jensen, J. R. (1996a) "Issues involving the creation of digital elevation models and terrain corrected orthoimagery using soft-copy photogrammetry," in Greve, C. ed. *Digital Photogrammetry: An Addendum to the Manual of Photogrammetry*, Bethesda, MD: American Society for Photogrammetry and Remote Sensing, pp. 167–179.

Jensen, J. R. (1996b) *Introductory Digital Image Processing: A Remote Sensing Perspective*. Upper Saddle River, NJ: Prentice Hall.

Jensen, J. R. (2000) *Remote Sensing of the Environment: An Earth Resource Perspective*. Upper Saddle River, NJ: Prentice Hall.

Jensen, J. R. and Cowen, D. C. (1999) "Remote sensing of urban/suburban infrastructure and socioeconomic attributes," *Photogrammetric Engineering and Remote Sensing*, Vol. 65, No. 5, pp. 611–622.

Kanellopoulos, I., Varfis, A., Wilkinson, G. G., and Meger, J. (1992) "Land-cover discrimination in SPOT HRV imagery using an artificial neural network—a 20-class experiment," *International Journal of Remote Sensing*, Vol. 13, No. 5, pp. 917–924.

King, M. D., and Greenstone, R. (1999) *1999 EOS Reference Handbook: A Guide to NASA's Earth Science Enterprise and the Earth Observing System*, Greenbelt, MD: Goddard Space Flight Center, National Aeronautics and Space Administration. **(http://eos.nasa.gov/)**

Kolbl, O., Best, M. P., Dam, A., Douglass, J. W., Mayr, W., Philbrik, Seitz, P., and Wehrli, H. (1996) "Photogrammetric scanners," in Greve, C. ed. *Digital Photogrammetry: An Addendum to the Manual of Photogrammetry*, Bethesda, MD: American Society for Photogrammetry and Remote Sensing, pp. 3–14.

Konecny, G., and Schiewe, J. (1996) "Mapping from digital satellite image data with special reference to MOMS-02," *ISPRS Journal of Photogrammetry and Remote Sensing*, Vol. 51, pp. 173–181.

Lam, N. S. N., and Quattrochi, D. A. (1992) "On the issues of scale, resolution, and fractal analysis in the mapping sciences," *Professional Geographer*, Vol. 44, No. 1, pp. 88–98.

Light, D. L. (1992) "The new camera calibration system at the U.S. Geological Survey," *Photogrammetric Engineering and Remote Sensing*, Vol. 58, No. 2, pp. 185–188.

Lillesand, T. M., and Kiefer, R. W. (2000) *Remote Sensing and Image Interpretation*, 4th ed. New York: John Wiley and Sons.

Lo, C. P. (1986) *Applied Remote Sensing*, London: Longman.

Lo, C. P. (1976) *Geographical Applications of Aerial Photography*, New York: Crane, Russak and Company.

Lo, C. P. (1997a) "Resolution and urban mapping from satellite images," in Au, K. N. and Lulla, K. eds. *Hong Kong and the Pearl River Delta as Seen from Space Images*, Hong Kong: Geocarto International Centre, pp. 111–117.

Lo, C. P. (1997b) "Detection of Hong Kong's inner harbour change using SPOT HRV data," in Au, K. N., and Lulla, K. eds., *Hong Kong and the Pearl River Delta as Seen from Space Images*, Hong Kong: Geocarto International Centre, pp. 75–82.

Lo, C. P. (1998) "Applications of imaging radar to land use and land cover mapping," in Henderson, F. M., and Lewis, A. J. eds. *Principles and Applications of Imaging Radar, Manual of Remote Sensing*, 3rd ed. Vol. 2, American Society for Photogrammetry and Remote Sensing, New York: John Wiley and Sons, pp. 705–732.

Lo, C. P., and Shipman, R. L. (1990) "A GIS approach to land-use change dynamics detection," *Photogrammetric Engineering and Remote Sensing*, Vol. 56, No. 11, pp. 1483–1491.

Lo, C. P., Quattrochi, D. A., and Luvall, J. C. (1997) "Application of high-resolution thermal infrared remote sensing and GIS to assess the urban heat island effect," *International Journal of Remote Sensing*, Vol. 18, No. 2, pp. 287–304.

Lunetta, R. S., Congalton, R. G., Fenstermaker, L. K., Jensen, J. R., and Tinney, L. R. (1991) "Remote sensing and geographic information system data integration:

Error sources and research issues," *Photogrammetric Engineering and Remote Sensing*, Vol. 57, No. 6, pp. 677–687.

Luvall, J. C., Lieberman, D., Lieberman, M., Hartshorn, G. S., and Peralta, R. (1990) "Estimation of tropical forest canopy temperatures, thermal response numbers, and evapotranspiration using an aircraft-based thermal sensor," *Photogrammetric Engineering and Remote Sensing*, Vol. 56, pp. 1393–1401.

Malhotra, R. C. (1989) "Potential of Large Format Camera photography," *Photogrammetric Engineering and Remote Sensing*, Vol. 55, No. 2, pp. 183–189.

Masry, S. E., and Gibbons, J. G. (1973) "Distortion and rectification of IR," *Photogrammetric Engineering*, Vol. 39, pp. 845–849.

Mercer, J. B., Tennant, K., and Thornton, S. (1998) "Operational DEM production from airborne interferomentry and from RADARSAT stereo technologies." *1998 ASPRS Annual Conference Proceedings*, Tampa, Florida, April 1998, CD-ROM.

Moffitt, F. H., and Mikhail, E. M. (1980) *Photogrammetry*, New York: Harper and Row.

Monteith, J. L. (1973) *Principles of Environmental Physics*, New York: Elsevier.

Nichol, J. E. (1994) "A GIS-based approach to microclimate monitoring in Singapore's high-rise housing estates," *Photogrammetric Engineering and Remote Sensing*, Vol. 60, No. 10, pp. 1225–1232.

Nichol, J. E. (1995) "Monitoring tropical rain forest microclimate," *Photogrammetric Engineering and Remote Sensing*, Vol. 61, No. 9, pp. 1159–1165.

Peake, W. H., and Oliver, T. L. (1971) *The Response of Terrestrial Surfaces at Microwave Frequencies*. Ohio State University Electroscience Lab Technical Report AFAL-TR-70–301. Columbus, OH: Ohio State University Electroscience Lab.

Petrie, G. (1990) "Developments in analytical instrumentation," *ISPRS Journal of Photogrammetry and Remote Sensing*, Vol. 45, pp. 61–89.

Petrie, G. (1997) "Developments in digital photogrammetric systems for topographic mapping applications," *ITC Journal*, 1997-2, pp. 121–135.

Petrie, G. (1970) "Some considerations regarding mapping from earth satellites," *Photogrammetric Record*, Vol. 6, No. 36, pp. 590–624.

Richards, J. R. (1986) *Remote Sensing Digital Image Analysis: An Introduction*, Berlin: Springer-Verlag.

Roessel, J. van, and Godoy, R. de (1974) "SLAR mosaics for Project RADAM," *Photogrammetric Engineering*, Vol. 40, No. 5, pp. 583–595.

Roth, M., Oke, T. R., and Emery, W. J. (1989) "Satellite derived urban heat islands from three coastal cities and the utilization of such data in urban climatology," *International Journal of Remote Sensing*, Vol. 10, No. 11, pp. 1699–1720.

Sabins, F. (1997) *Remote Sensing Principles and Interpretation*, New York: W.H. Freeman.

Simonett, D. S. (1983) "The development and principles of remote sensing," in R. N. Colwell ed. *Manual of Remote Sensing*, 2nd ed. Falls Church, VA: American Society of Photogrammetry, pp. 1–35.

Simonett, D. S., and Davis, R. E. (1983) "Image analysis—active microwave," in R. N. Colwell, ed. *Manual of Remote Sensing*, 2nd ed. Falls Church, VA: American Society of Photogrammetry, pp. 1125–1181.

Singh, A. (1989) "Digital change detection techniques using remotely-sensed data," *International Journal of Remote Sensing*, Vol. 10, No. 6, pp. 989–1003.

Teng, W. L. (1997) "Fundamentals of photographic interpretation," in W. R. Philipson, ed. *Manual of Photographic Interpretation*, 2nd ed. Bethesda, MD: American Society for Photogrammetry and Remote Sensing, pp. 49–113.

Theodossiou, E. I., and Dowman, I. J. (1990) "Heighting accuracy of SPOT," *Photogrammetric Engineering and Remote Sensing*, Vol. 56, No. 12, pp. 1643–1649.

Wang, F. (1990) "Improving remote sensing image analysis through fuzzy information representation," *Photogrammetric Engineering and Remote Sensing*, Vol. 56, No. 8, pp. 1163–1169.

Wang, F., and Hall, B. (1996) "Fuzzy Representation of Geographical Boundaries in GIS," *International Journal of Geographical Information Systems*, Vol. 10, No. 5, pp. 573–590.

Wang, F., Hall, G. B., and Subaryono (1990) "Fuzzy information representation and processing in conventional GIS software: Database design and application," *International Journal of Geographical Information Systems*, Vol. 4, No. 3, pp. 261–283.

Webster, C. J. (1996) "The potential of urban texture measures in monitoring urbanisation from space," in Fung, T., Lai, P. C., Lin, H., and Yeh, A. G. O. eds. *GIS in Asia*, Singapore: GIS Asia Pacific, pp. 309–321.

Welch, R. A. (1982) "Spatial resolution requirements for urban studies," *International Journal of Remote Sensing*, Vol. 3, pp. 139–146.

Wharton, S. W. (1982) "A context-based land-use classification algorithm for high-resolution remotely sensed data," *Journal of Applied Photographic Engineering*, Vol. 8, No. 8, pp. 46–50.

Wiesel, J. W., and Behr, F. J. (1987) "Digital orthophoto generation using the Kern DSR-11 analytical stereo restitution instrument." *Technical Papers, ASPRS-ACSM Annual Convention*, American Society for Photogrammetry and Remote Sensing, pp. 216–225.

Wilkinson, G. G. (1996) "A review of current issues in the integration of GIS and remote sensing data," *International Journal of Geographical Information Systems*, Vol. 10, No. 1, pp. 85–101.

Wolf, P. R., and Dewitt, B. A. (2000) *Elements of Photogrammetry with Applications in GIS*, New York: McGraw-Hill Book Company.

Wong, K. W. (1975) "Geometric and cartographic accuracy of ERTS-1 imagery," *Photogrammetric Engineering*, Vol. 41, No. 5, pp. 621–635.

Yoshida, T., and Omatu, S. (1994) "Neural network approach to land cover mapping," *IEEE Transactions on Geoscience and Remote Sensing*, Vol. 32, No. 5, pp. 1103–1109.

Yuan, D., and Elvidge, C. D. (1996) "Comparison of relative radiometric normalization techniques," *ISPRS Journal of Remote Sensing*, Vol. 51, pp. 117–126.

CHAPTER

9

Digital Terrain Modeling

9.1 Introduction

Terrain information is essential to applications in many disciplines: geography, land surveying, civil engineering, landscape architecture, earth and environmental sciences, as well as resource planning and management. The history of using digital methods to represent and analyze the characteristics of topography may be traced to the 1950s (Weibel and Heller, 1991). Since then, digital terrain modeling has been a major focus of research and application in various fields of science and technology. The increasing availability of digital terrain data and advancements in computer technology in the last several years have resulted in a spectacular growth in the use of this particular component of geographic data processing.

Digital terrain modeling may be approached in different ways. The mathematical approach aims to develop algorithms for terrain representation, contour interpolation and three-dimensional visualization. The nonmathematical approach, on the other hand, tends to focus more on the applications of digital terrain modeling techniques. In this chapter, we will examine digital terrain modeling mainly from the perspectives of data in the context of GIS. The emphasis of our discussion will be on the sources and characteristics of digital terrain data, the principles of terrain data processing including terrain modeling software and the use of digital terrain modeling in various fields of science and technology such as surveying and mapping, earth and environmental science, engineering design, as well as military operation and training.

9.2 Definitions and Terminology

It has been generally accepted that the term "digital terrain model" (DTM) was first introduced by Miller and La Flamme (1958), who defined it as "a statistical representation of the continuous surface of the ground by a large number of selected points with known *X*, *Y* and *Z* coordinates in an arbitrary coordinate field." Several terms have been coined to describe methods and processes pertaining to digital terrain data. These include: *digital elevation model, digital terrain elevation data, digital terrain model,* and several others. Although many people assume these terms to be synonymous, Petrie and Kennie (1991) suggested that different terms often refer to different concepts and products as explained below.

- *Digital Elevation Model (DEM).* The word "elevation" in DEM is the measurement of height above a datum. It implies the absolute altitudes or elevations of the points contained in the data. The term "digital elevation model" and its acronym DEM are particularly widely used in North America to refer to data sets containing regularly spaced elevations in the form of a lattice or square grid (see Section 9.3.1).
- *Digital Terrain Elevation Data (DTED).* This term is used primarily as a product name for elevation data in the grid format that are generated by the former United States Defense Mapping Agency [now a part of the National Imagery and Mapping Agency (NIMA)].

The term "digital terrain modeling" (DTM) as it is used in this chapter refers generally to the concepts and techniques of acquiring and using digital elevation data. In the context of geographic data processing, DTM is a multistep process that is made up of the following sequence of tasks (Figure 9.1) (Weibel and Heller, 1991).

- *Digital terrain data sampling* is the structuring and acquisition of digital terrain data by photogrammetric, cartographic, and field survey methods.
- *Digital terrain data processing* is the manipulation of digital terrain to ensure their usability by GIS.
- *Digital terrain data analysis* involves the use of algorithms and procedures that restructure digital terrain data into useful geographic information.
- *Digital terrain visualization* entails the development of algorithms and methods that will allow the effective display of the terrain to assist in spatial problem solving and decision making.
- *DTM applications* comprise the practical use of DTM in different fields of science and technology.

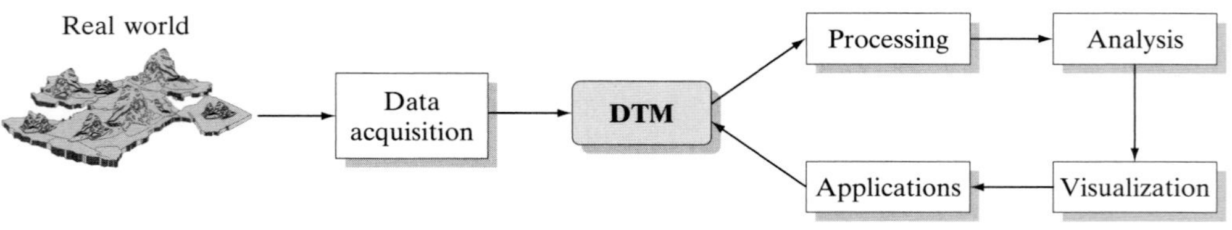

FIGURE 9.1
The main tasks of digital terrain modeling in GIS.

The relationships between DTM and geographic data processing may be understood from two perspectives: functions and data. From the perspective of functions, DTM is a critical component of geographic data processing. The concepts and techniques of DTM are now used not only for the representation and analysis of topographic surfaces, but also for other forms of geographic phenomena such as climate, meteorology, pollution, land cover, natural resources, engineering design, and distribution of the socioeconomic variables. From the perspective of data requirements, however, DTM is distinct from conventional geographic data processing. This is because terrain modeling is three-dimensional in nature, which requires special data structure and software functionality to handle. Conventionally, GIS software packages have very limited or no generic capability for digital terrain modeling. A functional extension, such as *ARC TIN* in ArcInfo (ESRI, 1994) and *3D Analyst* in ArcView GIS, has to be developed for DTM purposes. There are also numerous stand-alone DTM packages specially designed for terrain modeling and visualization in the market (see Section 9.5.4).

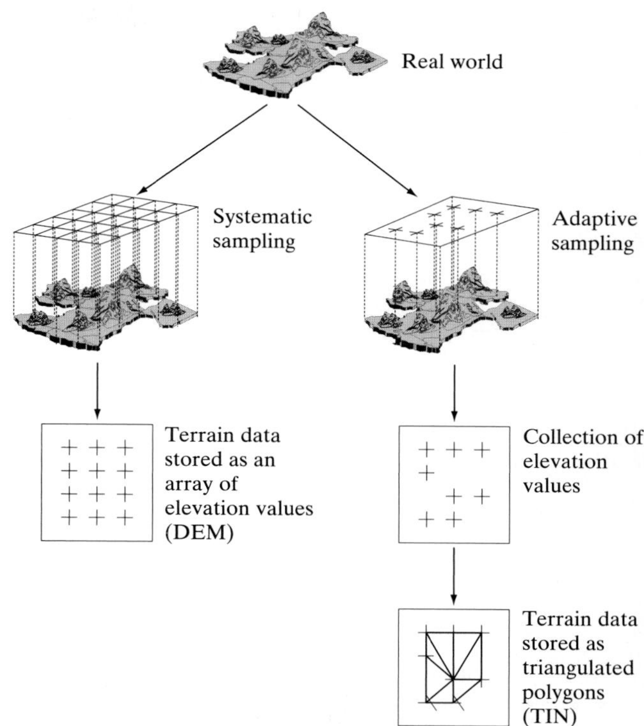

FIGURE 9.2
Approaches to digital terrain data sampling. Terrain data may be sampled either at regularly spaced intervals to form a digital elevation model (DEM) or selectively at salient ground points to form a triangulated irregular network (TIN).

9.3 Approaches to Digital Terrain Data Sampling

Acquisition of terrain data is a sampling process because it is impossible to record each and every point on Earth's surface. There are two approaches to digital terrain data sampling (Figure 9.2): *systematic* and *adaptive*. In systematic terrain data sampling, elevation points are measured at regularly spaced intervals. The result is a matrix of elevation values that is usually referred to as a *digital elevation model* (DEM). When the adaptive sampling method is used, elevation measurements are made at selected points that are assumed to be representative of the terrain. The result is a collection of irregularly distributed elevation values that must be properly structured before they can be used for further processing. Since the method of triangulation is used to build the spatial framework for storing the elevation values, the data collected by this approach are referred to as a *triangulated irregular network* (TIN).

The DEM and TIN approaches to terrain data sampling are not mutually exclusive. There are now well-established methods for the conversion of DEM and TIN data to and from one another (see Section 9.5.1). Most digital terrain modeling systems today accept both formats of data as input (see Section 9.5.4). In practice, the choice between DEM and TIN is governed more by the considerations explained next, rather than by the particular GIS or DTM application software to be used. These considerations include:

- *The Nature of the Terrain*. The TIN approach is more suitable for representation of complex terrain where local changes are significant, because systematic DEM sampling is unable to ensure that characteristic points in the terrain can be included.
- *The Purpose of Modeling*. It is easier to perform spatial analysis (e.g., cartographic overlay) with raster-based DEM data than with vector-based TIN data, but the latter method tends to produce more accurate results (see Chapters 5 and 6).
- *The Needs of Specific Applications*. Some applications, such as the production of orthophotographs, work more effectively with DEM data, whereas others, such as the generation of shaded relief maps, work better with TIN data.
- *The Method of Data Acquisition*. The DEM approach is most suited to terrain data acquisition by automated photogrammetric digitizing, but the TIN approach allows terrain data to be collected more efficiently by map digitizing and field survey methods.

9.3.1 CHARACTERISTICS OF A DEM

The idea of sampling terrain elevation data in the form of a lattice or grid square by the DEM approach is identical to the raster data model (see Chapter 3). A DEM may be described by three elements (Figure 9.3): *block, profile,* and *elevation point*. According to the United States Geological Survey (USGS), a block (the same as a tile) is used to describe the physical extent of a DEM (USGS, 1997b). It is usually tied to a particular topographic map series but does not necessarily always cover the same geographical extent. For example, among the different DEM products of the USGS, the 7.5-minute DEM quadrangles have a one-to-one relationship with their corresponding map quadrangles, whereas there are four 15-minute DEM blocks in one 30-minute quadrangle and two 1-degree DEM blocks in each 1- by 2-degree quadrangle (see USGS Web site for further information: *http://edcwww.cr.usgs.gov/glis/hyper/guide/usgs_dem)* (USGS, 1997a).

A profile is a linear array of sampled elevation points. The spacing between the profiles represents one dimension of the spatial resolution of the DEM. The other dimension is the spacing between the elevation points. There are three types of elevation points: *regular points, first points* along a profile, and *corner points*. Of these three types of points, coordinates are stored only for the first points along a profile and the corner points. These coordinates are used to tie the DEM block to an accepted georeferencing system. They are also used to calculate the spacing of the profiles (i.e., Δx and Δy) as well as the coordinates of the regular elevation points.

A DEM is usually georeferenced either to the geographic (latitude/longitude) or the UTM coordinate system. When it is georeferenced to the geographic coordinate system, the spacing of the profiles is expressed in terms of *arc seconds* or *arc minutes* (Figure 9.4). One arc second is approximately equivalent to 30 meters of linear measurement (depending on the latitude). The USGS 30-minute DEMs, for example, have a 2- by 2-arc second spacing, whereas the 1-degree DEMs have a 3- by 3-arc second spacing. The use of arc seconds makes the computation of the positions of the regular elevation points relatively complicated because it involves the conversion of an angular measurement to a linear measurement (USGS, 1997b). It should also be noted that in geographic coordinates, each DEM block is a *geographic rectangle* described by the positions of its corner elevation points. This means that the DEM block

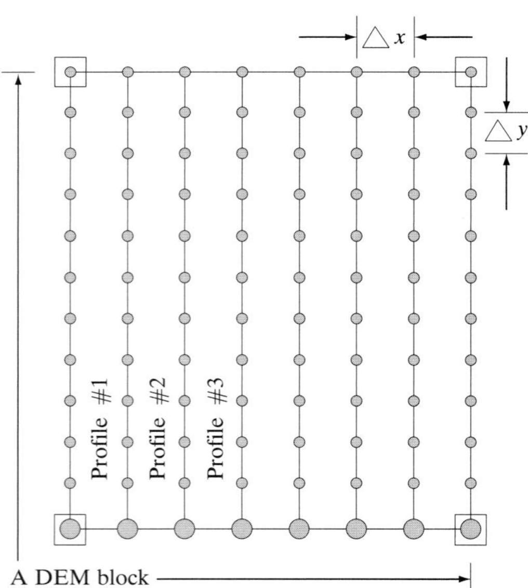

FIGURE 9.3
Elements of a USGS DEM block.

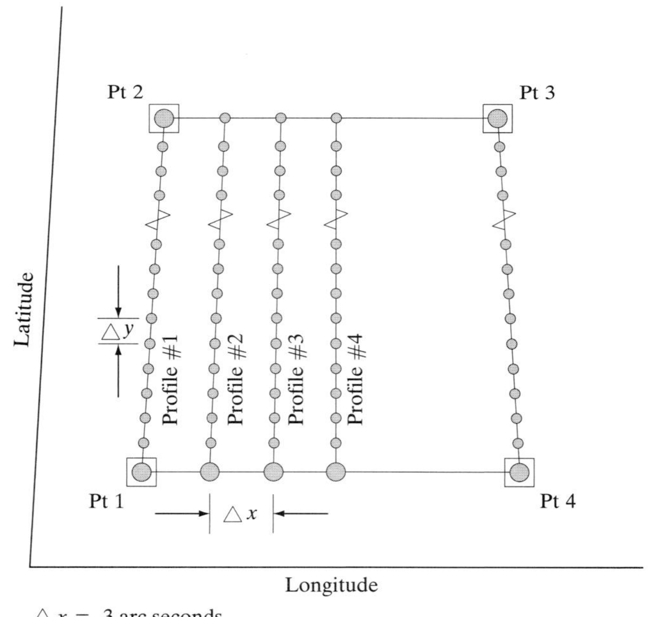

FIGURE 9.4
Characteristics of a USGS 1-degree DEM block georeferenced to the geographic coordinate system. *(Source: USGS, 1997b)*

is not a regular rectangle but one that exhibits a slight convergence of the meridians (i.e., the northern bounding latitude is shorter than the southern bounding latitude in the Northern Hemisphere).

When a DEM block is georeferenced to the UTM coordinate system, the spacing of the profiles is expressed in meters (Figure 9.5). Of the five types of DEMs produced by the USGS, the 7.5-minute DEMs are georeferenced to UTM. In these DEMs, the profiles are clipped to the straight-line intercept between the four geographic corners of the blocks. This represents an approximation of the neat lines of the corresponding 7.5-minute map quadrangles. The resulting area of coverage for the DEM is a quadrilateral, the opposite sides of which are not quite parallel. Because of the systematic change in orientation of the quadrilateral in relation to the UTM grid, profiles intersect the east and west neat lines as well as the north and south neat lines as shown in Figure 9.5. This means that profiles in UTM-based DEMs do not always have a uniform number of elevation points like those that are georeferenced to the geographic coordinate system.

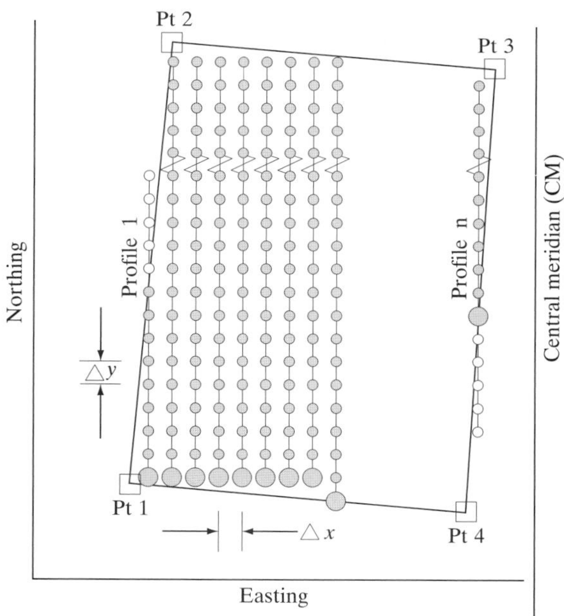

FIGURE 9.5
Characteristics of a USGS 7.5 minute DEM block georeferenced to the UTM coordinate system. *(Source: USGS, 1997b)*

Like many other forms of raster-based geographic data, DEMs are usually stored as arrays of ASCII characters or as binary numbers. The data structure of these data files may be described simply as a grid-based elevation matrix. Since this data structure reflects the array data storage structure of programming languages, DEM data are relatively simple to handle. The method of elevation point sampling using regular grids makes DEM particularly suitable for digital terrain data acquisition using automated techniques such as photogrammetric digitizing (see Section 9.4.2). However, the regular grid pattern obviously cannot precisely adapt to the complexity of the terrain. An excessive number of elevation points are needed in order to represent the terrain surface to a specific level of accuracy. This always poses the problem of data redundancy in areas where there is little variability in the terrain. A common solution is to collect terrain data automatically using the DEM model. The data are then used to build TINs, which are supposed to be more efficient in terms of data storage as explained in the next section.

9.3.2 CHARACTERISTICS OF THE TIN

In the TIN data model, the terrain is recorded as a continuous surface made up of a mosaic of nonoverlapping triangular facets formed by connecting selectively sampled elevation points using a consistent method of triangle construction (Figure 9.6). For the purpose of simpler mathematical interpolation, most TIN models assume planar triangular facets defined by three *edges* (sometimes referred to as *legs*). Each edge is bounded by two *vertices* (sometimes referred to as *nodes*). In a TIN, these edges depict linear terrain features (e.g., breaks, ridges, and river channels) while the vertices describe nodal topographic features (e.g., peaks, pits, and passes). In addition to the difference in the method of sampling elevation points, the TIN data model is distinct from the DEM data model in two important ways.

- Each and every sample point in a TIN has an (x, y) coordinate and an elevation, or z value (the locations of elevation points in DEM are implicit in the data model).
- The TIN data model may include explicit topological relationships between points and their proximal triangles.

Topological relationships play a significant role in the TIN data model. By building these topological relationships

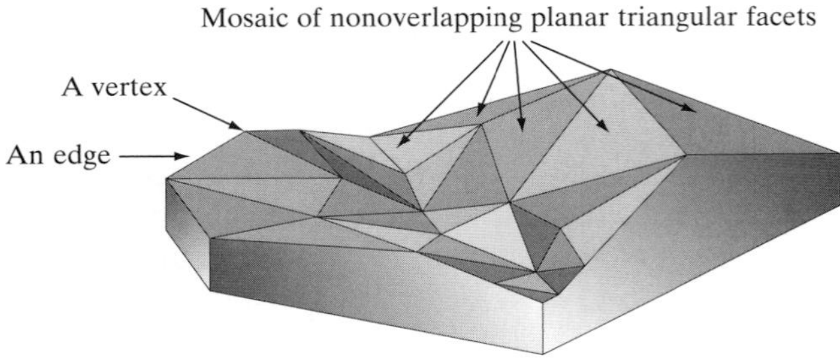

FIGURE 9.6
Elements of a Triangulated Irregular Network (TIN).

using the method of triangulation, the totally unstructured elevation points as they are collected are turned into a properly organized geographic database suitable for terrain modeling applications. The process of triangulating thousands of discrete points is no trivial task, both conceptually and computationally. As illustrated in Figure 9.7, a given set of elevation points may be triangulated in many ways. If the resulting TINs are used for contouring, very drastically different maps will be generated. The primary requirement in the TIN data model, therefore, is to develop the necessary procedures that will ensure the production of a unique TIN for a given set of elevation points (Petrie, 1991a; Tsai, 1993).

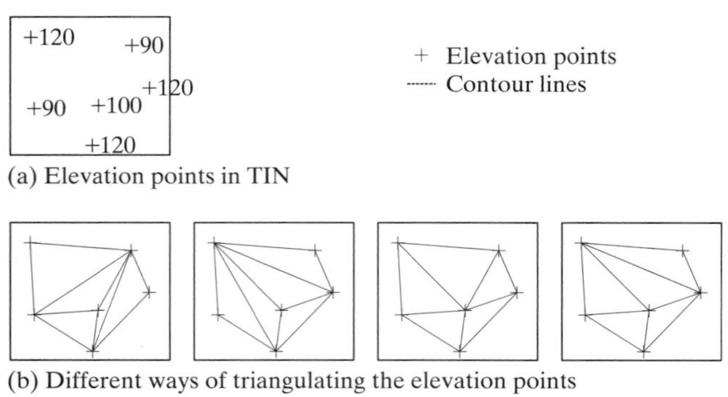

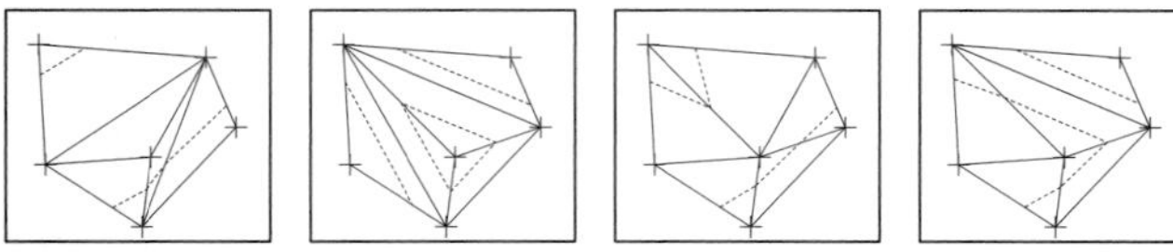

FIGURE 9.7
The impact of the ways of triangulation on the application of a TIN for contour interpolation. The same set of elevation points in (a) can be triangulated in different ways as shown in (b). When these triangulated networks are used to interpolate contours, drastically different results are obtained as shown in (c).

Delaunay triangulation is the most commonly used method of creating a network of triangles from sampled points over a particular geographic space for a TIN because it allows well-shaped "fat" triangles to be generated. Delaunay triangulation is closely associated with *proximal regions* that are also called *nearest-neighborhood regions, Thiessen polygons*, or *Voronoi polygons* (Laurini and Thompson, 1992). Proximal regions are created as the result of subdividing a particular geographic domain or space encompassing a set of points into a set of convex polygons. Given three points P_1, P_2, and P_3 in a plane S (Figure 9.8a), for example, it is possible to partition the plane using the perpendicular bisectors to the segments P_1P_2, P_2P_3, and P_1P_3 into three proximal regions V_1, V_2, and V_3. Each of these regions contains one of the original points called *anchor points* (Figure 9.8b). By definition, a proximal region is a polygon within which every point is closer to its own anchor point than the anchor points of all other regions. This means that any point located in V_1, for example, is closer to P_1 than to P_2 and P_3. The process of systematically subdividing a geographic domain in this way is called *Dirichlet tessellation* in computational geometry. The resulting surface is known as a *Voronoi diagram* or *Thiessen polygons* (Figure 9.8c).

If all the pairs of anchor points sharing a common edge of a polygon in the Voronoi diagram are connected, a network of triangles may be obtained (Figure 9.8d). The process of creating a network of triangles in this way is known as Delaunay triangulation. The triangles obtained by this method have a unique property, i.e., the circumcircle that passes through the vertices of a particular triangle contains no other point in its interior (Figure 9.8e). This property, which is called the *empty-circle criterion*, is used as a mechanism to automatically construct a TIN from a set of points in Delaunay triangulation (Petrie, 1991a; Tsai, 1993). Delaunay triangulation may be applied to two-dimensional data as well as three-dimensional data. It may also be extended to include predefined limiting factors (e.g., noncrossing break lines) in what is called "constrained" Delaunay triangulation (Tsai, 1993). This will make a TIN fit even better than the terrain surface that it represents.

In essence, the TIN data model uses the vector data model. Therefore, TIN data may be stored using the method of relational attribute tables in the georelational data model (Figure 9.9). A TIN data set contains three basic attribute tables:

- *arc attribute table*, which contains the length, from-node, and to-node of the edges of all the triangles
- *node attribute table*, which contains the (x, y) coordinates and elevation (z) of individual vertices

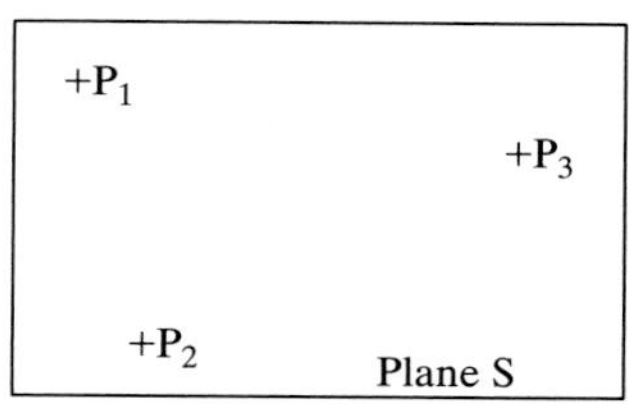

(a) Points in a plane

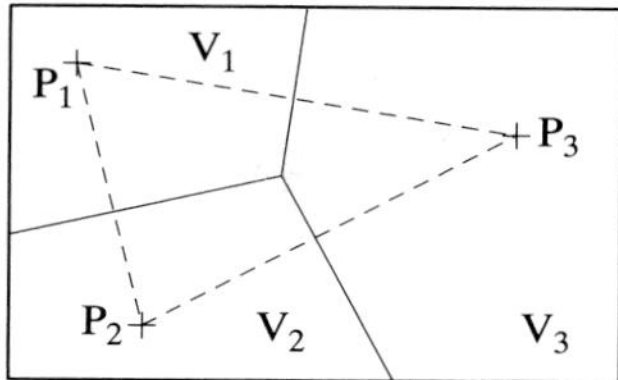

(b) Subdivision into proximal regions

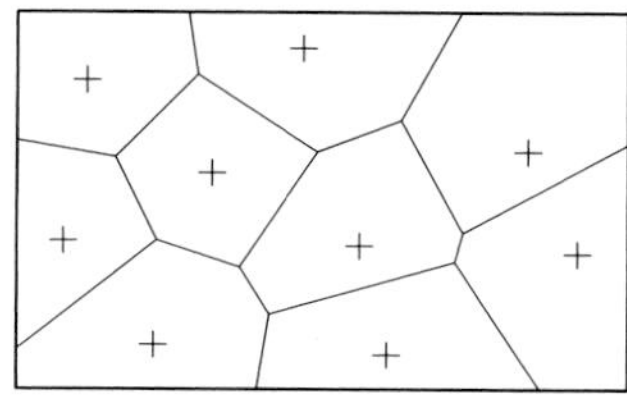

(c) A Voronoi diagram

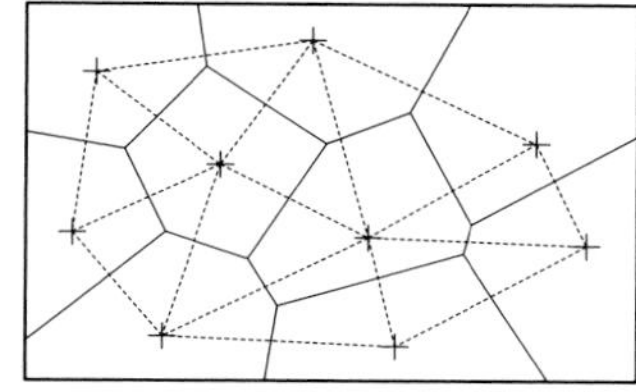

(d) Delaunay triangulation

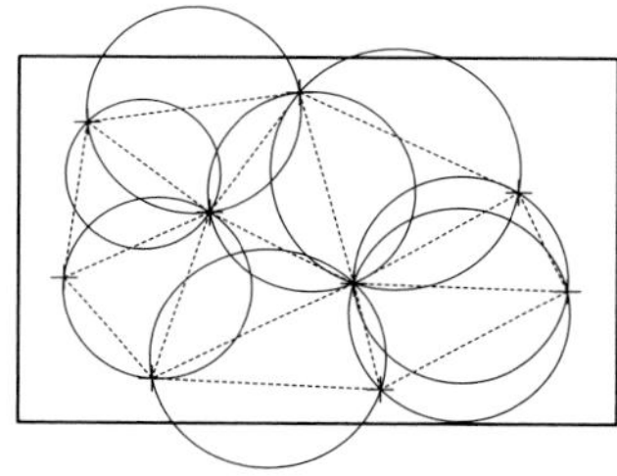

(e) Delaunay triangles and the empty-circle criterion

FIGURE 9.8
The concepts of proximal regions, Voronoi diagram, and Delaunay triangulation. Delaunay triangulation is the process of subdividing a particular geographic domain into proximal regions and systematically connecting the anchor points that share a common edge of a polygon in the resulting Voronoi diagram.

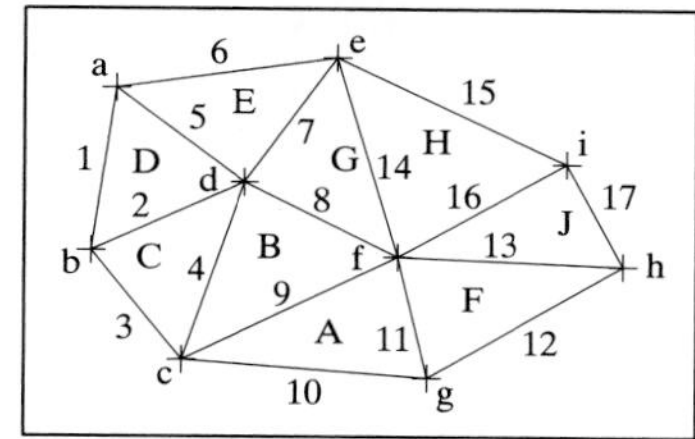

Arc Attribute Table

Edge ID	Length	From-node	To-node
1	124	b	a
2	108	b	d
3	119	c	b
4	178	c	d
5	120	d	a
...			
...			

Node Attribute Table

Vertex ID	*X*	*Y*	Elevation
a	780,005	33,036	105
b	780,003	33,020	92
c	780,016	33,005	85
d	780,021	33,028	120
e	780,048	33,040	95
...			
...			

Polygon Attribute Table

Triangle ID	Area	Edge 1	Edge 2	Edge 3	Neighbors
A	7800	9	10	11	B, F, –
B	8210	9	4	8	A, G, C
C	6504	2	4	3	D, –, B
D	6890	1	2	5	–, E, C
E	7650	5	6	7	G, D, –
...					
...					

FIGURE 9.9
Data structure of a triangulated irregular network (TIN).

- *polygon attribute table,* which contains the areas of the triangles, the identification numbers of the edges that form individual triangles, and the identifiers of the adjacent polygons

Storing data in this way allows terrain to be manipulated and queried with the aid of standard data base management system (DBMS) tools. It also eliminates redundancy in data storage because it is necessary to store all vertices and edges only once even though they are used for more than one triangle. Other terrain parameters such as slope and aspect can be computed on the fly when the data are used. With the explicit storage of topological relationships (i.e., the from-nodes and to-nodes in the arc attribute table and the neighboring polygons in the polygon attribute table), it is relatively simple to use the TIN models to tackle problems involving spatial topology. In addition, since TIN data are based on the topological arc-node structure of the georelational data model, they may be applied relatively easily to vector-based geographic data processing such as automated contouring, three-dimensional landscape visualization, volumetric and cut-and-fill computation in engineering design, surface characterization, as well as site visibility studies. The selective method of elevation point sampling always ensures that all the important topographic characteristics (e.g., peaks, pits, break lines, and fault lines) are included in the TIN model. As a result, the

TIN approach of digital terrain modeling can accurately describe more complex terrain surfaces with a far smaller number of elevation points than the DEM approach.

9.4 Acquisition of Digital Terrain Data

Digital terrain data may be acquired by a variety of methods, depending on factors such as the location and size of the area of interest, the purpose of terrain modeling, and the technical resources available (Table 9.1). Generally speaking, ground survey methods are most suitable for large-scale terrain modeling for engineering and mining applications (Kennie, 1991). At smaller scales covering larger geographic areas, photogrammetric methods are always used. However, as a vast amount of terrain data are already in existence in topographic maps, many national mapping agencies tend to acquire digital terrain data by digitizing existing maps. Digital terrain data sets obtained in this way are usually of a small scale and have a national or regional coverage. In the following sections, we will explain the current technologies used for acquiring digital terrain data by field surveying, photogrammetric, and cartographic methods. We will also examine the sources and characteristics of DEM data that are available in the United States, Canada, the United Kingdom, and Australia, with a view to demonstrating how different government mapping agencies produce and treat digital terrain data as a standard cartographic product.

9.4.1 Terrain Data Collection by Field Surveying Methods

Conventionally, terrain data were obtained in field surveying by grid leveling and stadia tacheometry. These methods have been replaced by the new generation of survey instruments such as the *electronic tacheometer* and the *Global Positioning System* (GPS).

Acquisition of Digital Terrain Data by an Electronic Tacheometer

An *electronic tacheometer* is an instrument that is capable of electronically measuring both angles and distances and performing basic computation (e.g., reducing slope distances to the horizontal and determining coordinates from bearings and distances). Some electronic tacheometers are also equipped with an internal memory or an external data recorder for temporary storage of data. Because of the ability of this category of instruments to perform multiple functions in field surveying, they are sometimes referred to as *total stations* (Plate 2.1).

When conducting field measurements using the electronic tacheometer, two sampling strategies may be adopted: at predetermined regular grid intervals or at locations where salient terrain characteristics occur. The time required to set out a grid and the possibility of failing to depict terrain variability between the sampling points have precluded the general application of the first approach. In the second approach, the surveyor is free to select elevation points that are deemed to be representative of the characteristics of the terrain. These

Table 9.1
Methods of Acquiring Digital Terrain Data

Methods	*Technology*	*Accuracy*	*Areal Coverage*	*Typical Applications*
Ground survey	Total or semitotal stations	Very high	Limited to specific sites	Small-area site planning and design
Photogrammetry	Stereoplotting machines (with or without correlators)	High if from spot heights; low if from contour	Large-area projects, especially in rough terrain	Large engineering projects, e.g., dams, reservoirs, highways, open cast mines
Cartography (existing maps)	Manual digitizing; Semi-automated line following; Scanning	Low	Nationwide at small scales	Aircraft flight simulation, landscape, visualization, landform study, military battlefield simulation

Source: Petrie, G. (1991b)

include changes of slope, peaks, ridges, and the base of river channels. The result of the survey will be a collection of irregularly distributed elevation points. The method of Delaunay triangulation as described in Section 9.3.2 is used to generate a TIN before the terrain data can be used for further processing.

Acquisition of Digital Terrain Data by GPS

As explained in Chapter 2 (Section 2.9.3), the GPS is a satellite-based surveying system that allows the user to obtain highly accurate digital terrain data electronically. There are two basic methods of field measurements using GPS: *static* and *differential* (Figure 2.32). Static GPS surveying is used to determine positions of survey control points in areas where geodetic control is lacking or unreliable. It is also used to measure the distance between two points accurately. Distance measurements require the use of two or more dual-frequency GPS receivers that record measurements to the GPS satellites simultaneously for about six hours. Static GPS surveying is used mainly for geodetic control and the measuring of national and international networks. It is not intended for ordinary digital terrain data acquisition. Differential GPS surveying makes use of existing or newly established control points as a reference to determine the positions and heights of ground points. It may take one of the following forms, all of which may be used to acquire elevation data for digital terrain modeling:

- *A kinematic GPS surveying* method is used to make measurements along a line in open areas. The technique requires the use of a reference station (a GPS receiver set up at a point of known position and elevation) and a GPS rover unit (Plate 2.7). Before the start of the measurement process, both the GPS receiver and the rover unit must be initialized. This is done by switching on both machines and holding the rover unit in a fixed position for about 5 to 10 minutes. The rover unit is then moved, taking measurements at a preset interval (typically 1 or 2 seconds). Discrete points may also be recorded if necessary. In a sense, kinematic GPS surveying is a direct and 1 : 1-scale digitizing of ground points in the stream mode. It is used mainly for surveying road edges and centerlines as well as digital data in the form of profiles and cross sections.
- *Stop-and-go GPS surveying* is used to survey detail points that are relatively close together in an open area. The instruments and initialization procedure for this method are identical to those for kinematic GPS surveying. However, instead of sampling ground points in the stream mode, the operator has the discretion to select the points to be measured. Stop-and-go GPS surveying is usually used for detail surveys, engineering surveys, and collection of salient points for digital terrain modeling.
- *A real-time GPS surveying* configuration is made up of a reference station and one or more rover units. The reference station includes a GPS receiver, a controller, and a radio modem. Its function is to track the satellites and retransmit the satellite data to the rover units through the radio modem. The rover unit of a real-time system is similarly made up of a GPS receiver, a controller, and a radio modem. During the measurement process, the rover unit receives data from the GPS satellites directly and the reference station through the radio modem. These two sets of data are then processed in the rover controller. The real-time method of GPS surveying gives highly accurate results. It is most suitable for detail surveys, engineering surveys, open pit mine surveys, and data collection for digital terrain modeling.

In theory, GPS is an ideal technology for acquiring digital terrain data because it is an all-weather 24-hours-a-day system that may be used anywhere on Earth's surface with sky visibility. In practice, however, the use of GPS is subject to several limiting factors:

- *Errors of Measurements.* Different types of errors may occur when making GPS measurements. These include noise error of the satellite signals; *geometric dilution of precision* (GDOP), which is a ranging error magnified by the range vector difference between the receiver and the satellites; and blunders in measurement and data processing. An error known as Selective Availability (SA), the intentional degradation of the satellite signals carried out by the United States Department of Defense (DOD), has been dropped from the list because SA was set to zero at midnight of May 2, 2000, by the order of President Bill Clinton, thus boosting GPS receiver horizontal accuracy from 100 meters to 15–25 meters (Divis, 2000). The idea behind the methods of differential positioning described here is to correct bias errors at the sampled locations with measured bias errors at a known position. Other forms of errors that remain are treated by different methods.
- *The Availability of Control.* In order to acquire data that meet survey specifications, it is necessary to perform GPS measurements in the differential mode. This requires the presence of horizontal and vertical control in or near the area of interest where simultaneous measurements are made. If such a condition cannot be satisfied, the accuracy

of the terrain data obtained will be less. Although geodetic control points may be established by static GPS surveying, such measurements are very time-consuming as explained previously.

- *The Observation Window.* GPS measurements require a minimum of three satellites for two-dimensional coordinates and four satellites for three-dimensional coordinates. GPS is not usable unless this number of satellites with good geometry can be tracked with sufficient accuracy. Different locations on Earth's surface have different optimal observation windows during which the required number of satellites may be detected by the GPS receiver.
- *The Effects of Local Canopy Cover, Urban Canyons, and Mountain Valleys.* A dense canopy cover reduces the observation window and, consequently, the operational effectiveness of using GPS in forested areas (Rempel and Goadsby, 1992). For the same reason, tall buildings in the urban environment and deep mountain valleys may cause difficulty in obtaining accurate GPS fixes.

GPS is a very productive tool for acquiring digital terrain data where the necessary control is available and the sky is visible. Additionally, it provides a useful means by which the accuracy of existing digital terrain data may be evaluated (Adkins and Merry, 1994). It also provides a less expensive alternative to the conventional method of aerotriangulation to densify control for the photogrammetric methods of acquiring digital terrain data as explained in the next section.

9.4.2 TERRAIN DATA COLLECTION BY PHOTOGRAMMETRY

Photogrammetric methods are used when the terrain of interest is too extensive or too rugged for field survey methods to be applied practically. With the appropriate choice of photographic scale, flying height, base to height ratio (i.e., the geometry of the overlapping photographs), and equipment, photogrammetry can be used for the acquisition of digital terrain data over a wide range of map scales and degrees of accuracy (Petrie, 1991b). Digital terrain data may be obtained by any of these conventional photogrammetric methods: (1) using an analog stereoplotter equipped with encoders, (2) using an analytical plotter, and (3) using digital photogrammetry. The rapid development of *digital* (or *soft-copy*) *photogrammetry* in recent years has resulted in new tools and techniques by which digital terrain data can be extracted from aerial photographs and spaceborne remotely sensed images.

Digital Terrain Data Acquisition Using an Analog Stereoplotter

This is the conventional photogrammetric method to obtain three-dimensional data from aerial photographs. Three-dimensional coordinates of elevation points in the stereomodel may be digitized by equipping the stereoplotter with linear and rotary encoders. A variety of digitizing units may be attached to the stereoplotter to convert the electronic signals from the encoders into numerical coordinate values (Petrie, 1981). The digitizing unit may be connected either to a dedicated microcomputer or to a mainframe computer on a time-sharing basis. This method of photogrammetric digitizing is largely manual. The operator has complete control over the sampling of elevation points, and the strategy of photogrammetric sampling is an important operating consideration in the data acquisition process (Figure 9.10) (Petrie, 1991b).

Digital Terrain Data Acquisition Using an Analytical Plotter

An analytical plotter (Figure 9.11) uses a mathematical or numerical solution in photogrammetric measurements. Using the *X*, *Y* and *Z* terrain coordinates of a point measured by the operator, the main processor of the analytical plotter computes the corresponding photographic coordinates of the same points on each photograph of the stereopair in real time and moves the positions of the photographs accordingly, so that the

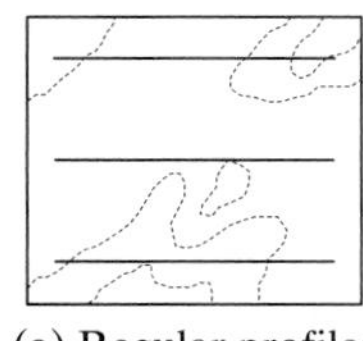
(a) Regular profiles

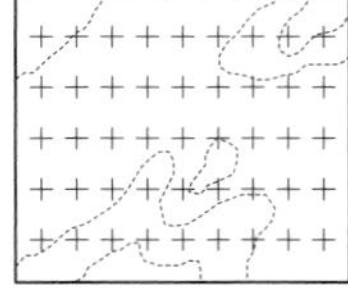
(b) Regular grid

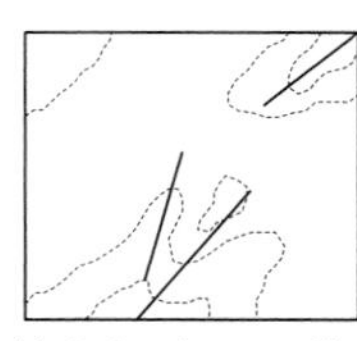
(c) Selective profiles

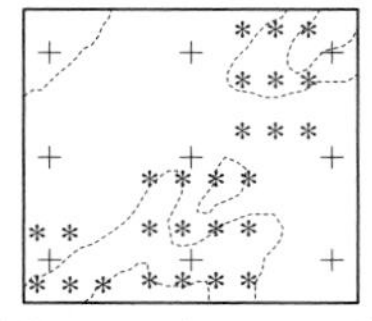
(d) Progressive sampling

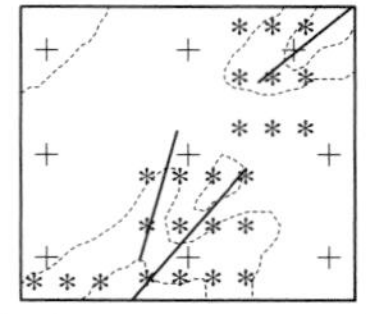
(e) Composite sampling

— Profiles
\+ Regular sampling points
* Additional sampling points

FIGURE 9.10
Photogrammetric sampling strategies. Elevation points in the stereomodel can be sampled in different ways. In order to overcome the shortcomings of systematic sampling methods (a, b, and c), it is possible to use the methods of progressive and composite sampling (d and e) in areas where important terrain characteristics occur. *(After Weibel and Heller, 1991, with minor modifications)*

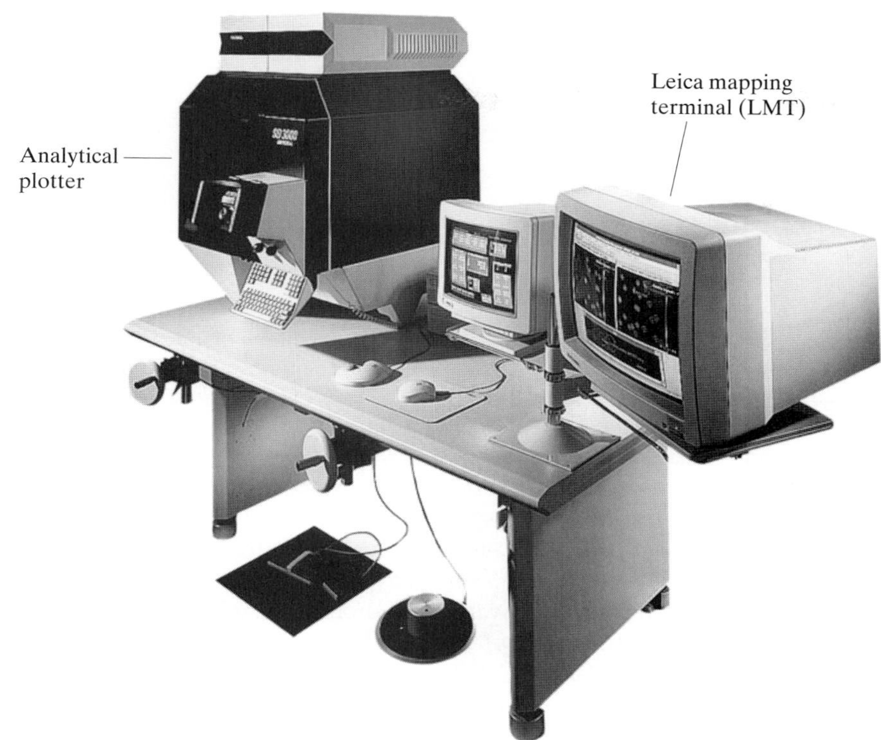

FIGURE 9.11
Leica SD2000/3000 analytical photogrammetric workstation.
(Courtesy: Leica Geosystems, Inc.)

measured point can be viewed stereoscopically (i.e., in three dimensions). The movement of the measuring mark of the analytical plotter may be programmed so that it can be driven to any desired position in the stereomodel. This makes it extremely easy to implement a predetermined sampling strategy for digital terrain data. As soon as the elevation of a point has been measured, the computer automatically moves the measuring mark to the next sampling point. This particular capability of the analytical plotter has made it possible to develop a sampling strategy, using the methods of *progressive* and *composite sampling,* that overcomes the inherent shortcomings of systematic sampling (Makorovic, 1973, 1977). Instead of sampling at regular grid intervals or along regularly spaced profiles, the density of the sampling points may be programmed to vary in different parts of the stereomodel according to the nature of the local relief. This allows the system to pick up more elevation points in areas of rugged topography and fewer points in areas of low relief variability (Figure 9.10d and e).

Digital Terrain Data Acquisition Using Digital Photogrammetry

Digital photogrammetry is a technology that has changed the methods of acquiring topographic data from aerial photographs. Instead of using conventional photographic diapositives, digital photogrammetry uses digital images obtained either by scanning aerial photographs at extremely high resolutions (1200 dpi or higher) or by the use of remote sensing imaging systems (see Chapter 5). A digital photogrammetric system includes the hardware and software that make up the *digital photogrammetric workstation* (DPWS) and peripheral devices for data conversion between the analog and digital formats (Figure 9.12) (Helpke, 1995). These devices include digital cameras and film scanners for analog–digital conversion as well as film makers and plotters for digital–analog conversion.

A digital photogrammetric workstation (Figure 9.13) is an integrated high-end workstation that is equipped with the following hardware components (Helpke, 1995):

- a fast CPU with a vast amount of both main and cache memories
- a graphics subsystem, including a true-color (24-bit per pixel) imagery memory, a graphics accelerator, a nondestructive overlay, and a fast processing unit for on-line computation and display
- a high-resolution stereo color monitor, typically 1920 × 1080 resolution
- fast buses to ensure high-speed data transfer
- a three-dimensional measuring device (a three-dimensional mouse) for interactive stereo measurements
- high-capacity disks, network access, keyboard, printer, plotter, data backup, and other peripheral devices

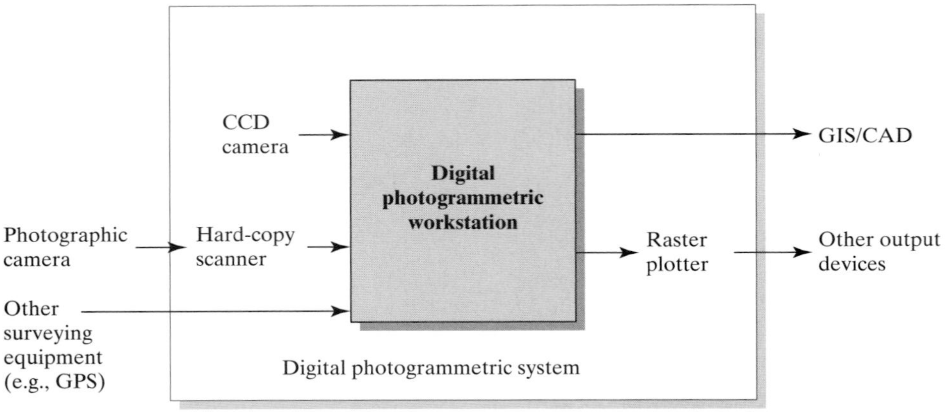

CCD camera
Hard-copy scanner
Photographic camera
Other surveying equipment (e.g., GPS)
Digital photogrammetric workstation
GIS/CAD
Raster plotter
Other output devices
Digital photogrammetric system

FIGURE 9.12
Components of a digital photogrammetric system. *(Source: modified from Helpke, 1995)*

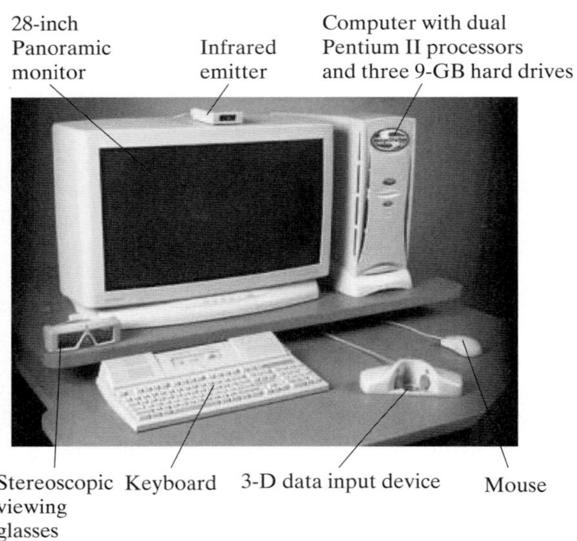

FIGURE 9.13
Intergraph ImageStation ZII digital photogrammetric workstation. *(Courtesy: Intergraph Corporation)*

Stereoscopic viewing in digital photogrammetry may be achieved in a number of ways. In most systems, temporal separation (i.e., alternative display of two images) in connection with polarized light is used in one of the following two ways (Helpke, 1995):

- By mounting a polarization screen in front of the monitor, which polarizes the emitted light in synchronization with the display of the images, images may be viewed stereoscopically with the aid of a set of passive polarizing glasses.
- By integrating the polarization screen into passive polarizing glasses to make them active. Synchronization between the screen and the glasses is ensured by using an infrared beam.

The same image on the screen may be viewed by more than one user. For interactive stereo measurement, however, a three-dimensional data input device is required. Subpixel accuracy is achievable. From the hardware perspective, it is this ability of achieving accurate interactive stereo measurement that distinguishes a DPWS from a general-purpose workstation. In terms of software implementation, a DPWS is equipped with a suite of computer programs that enables a substantial part of the photogrammetric operations to be automated (e.g., interior, relative and absolute orientations, point transfer, DTM extraction, and geometric image transformation). However, the user is always allowed to visualize, verify, and edit the data collected interactively. The DPWS is an extremely powerful tool for acquiring digital terrain data by photogrammetry. There are now digital photogrammetric systems that are especially designed for DTM data generation (Helpke, 1995).

The advent of digital photogrammetric technology has made it possible to use high-resolution stereoscopic satellite imagery for generating digital terrain data. Al-Rousan et al. (1997), for example, reported the successful use of SPOT level 1B imagery to extract a DEM using the EASI/PACE image-processing package. Level 1B imagery is one of the varieties of SPOT imagery formats that has been subject to substantial geometric processing, such as correction for Earth's rotation and tilts. Test results indicated that the average RMSE for the residual errors in both the ΔE and ΔN directions was around ± 6 meters; and the RMSE for errors in elevation ΔH was in the ± 4.4 to ± 7.7 m range. In another investigation, Maas and Kersten (1997) demonstrated the feasibility of generating DEM data from high-resolution still-video imagery obtained using a CCD (solid-state) camera, with a resolution of 1524×1012, mounted on a helicopter. By employing self-calibration techniques, accuracies of 2 centimeters for planimetric coordinates and 5–6 centimeters for height coordinates were obtained in digital aerotriangulation using imagery at 1:20,000 scale, and an accuracy of 0.03% of the flying height above ground could be achieved for digital terrain data.

9.4.3 TERRAIN DATA COLLECTION BY DIGITIZING EXISTING MAPS

Existing topographic maps contain a wealth of terrain data that may be used for digital terrain modeling. As the acquisition of digital terrain data by field survey and photogrammetric methods is a very costly and time-consuming undertaking, digitizing existing maps has often been adopted as a practical alternative. Digital terrain data for a large area may be obtained within a relatively short time at a modest cost. At present, most digital terrain data collected at national or state/provincial levels have been converted from existing topographic maps rather than acquired by carrying out new field or photogrammetric surveys. However, Petrie (1991c) cautioned that digital terrain data obtained from existing maps should be used with care. He noted that the typical accuracy of contours is only about one-third of that of the spot heights even when both are obtained from the same aerial photography. This is because contour measurement in a stereoplotter is conducted in a dynamic mode, as opposed to the static mode of spot height measurement that allows the operator more time to ensure accuracy. Contours on maps at smaller scales may be of even lower accuracy because errors might have been introduced during the process of generalization when the maps were produced. According to the National Map Accuracy Standard of the United States (Table 4.3), no more than 10% of the elevation points tested on contour lines can be in error by more than one-half the contour interval (see Chapter 4, Section 4.3.2). Clearly, even meeting this standard, the accuracy of the contours on the USGS topographic maps is not very high.

Without proper data structure, digitized contours are no more than a collection of coordinated elevation points (Figure 9.14a and b). They cannot be used

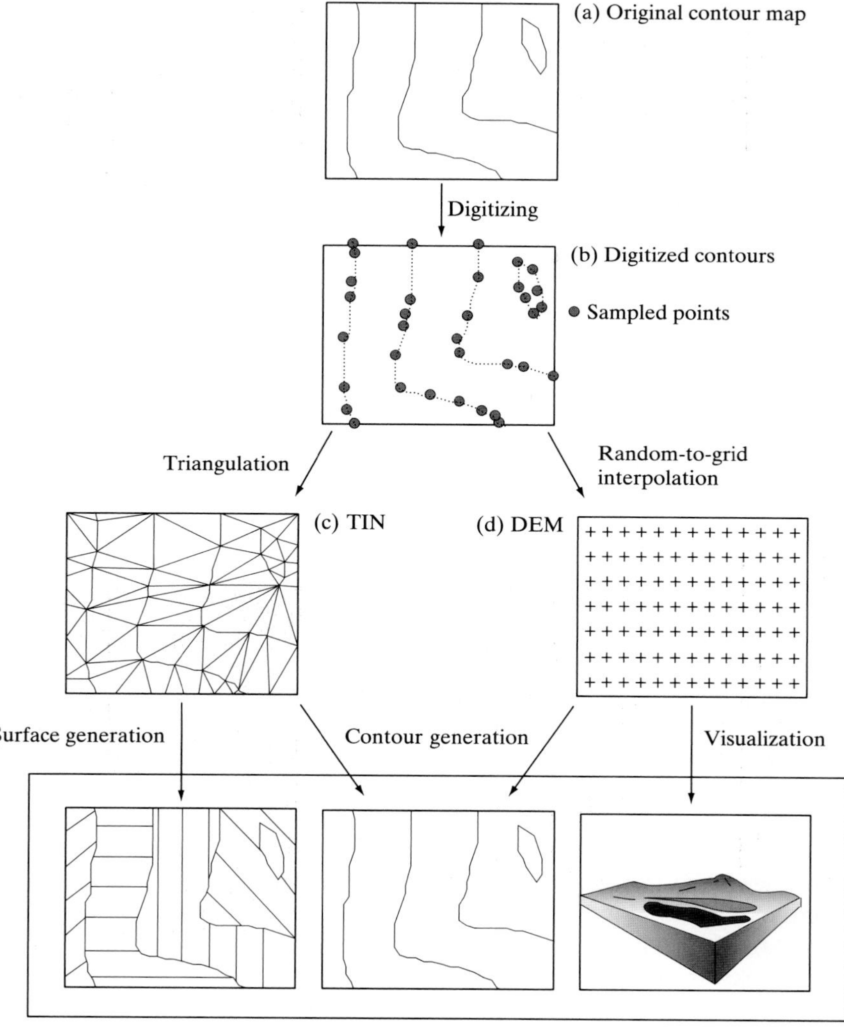

FIGURE 9.14
Relationships between contour maps, DEM/TIN, and digital terrain modeling.

for any useful GIS application other than regenerating the original contours themselves. This is the primary reason why surveying and mapping agencies seldom store terrain data as digitized contours. For the digitized contour data to be used for digital terrain modeling, it is necessary to carry out a series of postdigitizing processes whereby sampled points are turned into a TIN or a DEM (Figures 9.14c and d). To produce a DEM, the digitized contours are "gridded" at the desired interval (which fixes the resolution), and the elevation at each of the grid points is interpolated (using the method of inverse distance or kriging, as explained in Section 9.5.2). A TIN may be generated by using specially written algorithms to select significant break points from the DEM. Contour lines and other forms of cartographic products may be generated relatively easily from DEMs and TINs when required for digital mapping and DTM purposes (Figure 9.14e).

Digitizing of maps may be carried out using the same technology and procedures for map automation as described in Chapter 6. The method of scanning with automatic or semiautomatic vectorization has also been popular. This is the method used by the USGS to supplement manual table digitizing and photogrammetric methods in the production of DEMs. Generally speaking, map scanning and vectorization account for a only small portion of the work involved. Just like other forms of map digitizing, the most time-consuming part of the task is concerned with feature coding and attribute tagging, that is, attaching height values to the graphics (see Chapter 6). Quality control, which includes checking the data and the correction of any errors that may be found, also requires considerable effort in postdigitizing processing of the data (see Section 9.4.5).

9.4.4 SOURCES OF EXISTING DEM DATA

GIS users who require digital terrain data may obtain them from government mapping agencies or commercial data suppliers. Gittings (1996) maintains a very comprehensive catalog providing details of the sources and characteristics of digital terrain data in different countries. Such information may be obtained by consulting this catalog that is posted on the Internet. Information regarding digital terrain data marketed by commercial data suppliers may be obtained from advertisements in trade journals, industry service directories, and yearbooks. In this section, we introduce major digital terrain data sets produced in the United States, Canada, the United Kingdom, and Australia (Table 9.2).

Digital Terrain Data in the United States

Of the many digital terrain data sets produced by agencies of the United States federal government, three are of special interest to scientists in general and GIS users in particular because of their worldwide coverage. These are *ETOPO5*, *GTOPO30*, and the *Digital Chart of the World DEM* (DCW-DEM). ETOPO5 is a mosaic of a variety of different terrain data sources produced in the mid-1980s by the National Geophysical Data Center (NGDC) *(http://www.ngdc.noaa.gov/mgg/global/etopo5.html)*. It consists of a 2160 × 4320 geographic (latitude/longitude) centroid-registered grid. Although it is presented at 5 arc minutes, a large portion of the land areas were actually resampled from the 10 arc minute DEMs produced by the Fleet Numerical Oceanography Center (FNOC) of the United States Navy. The marine component, however, is based on the *Digital Bathymetric Data Base 5-minutes* (DBDB5) of the United States Naval Oceanographic Office. Local data were also used for those parts that cover Australia and Europe. ETOPO5 is available free via the USGS *Global Land Information System* (GLIS) interactive query system in the Internet. It is also available as part of the *Global Ecosystem Database* produced by the NGDC.

GTOPO30 (Global 30-Arc-Second Elevation Data Set), completed in late 1996, was developed over a 3-year period through a collaborative effort led by the USGS EROS Data Center. There were contributions and funding from organizations such as the National Aeronautics and Space Administration (NASA), the United Nations Environment Programme/Global Resource Information Database (UNEP/GRID), the United States Agency for International Development (USAID), the Instituto Nacional de Estadistica Geografica e Informatica (INEGI) of Mexico, the Geographical Survey Institute (GSI) of Japan, Manaaki Whenua Landcare Research of New Zealand, and the Scientific Committee on Antarctic Research (SCAR) *(http://www1.gsi-mc.go.jp/ gtopo30/gtopo30.html)*. TOPO30 is based on data derived from eight sources of elevation data, including vector and raster data sets, namely, Digital Terrain Elevation Data (DTED), Digital Chart of the World (DCW-DEM), USGS DEM, U.S. Army Map Service maps, International Map of the World, Peru map, New Zealand DEM, and Antarctic Digital Database. Elevations in GTOPO30 are regularly spaced at 30 arc seconds (approximately 1 kilometer) of latitude and longitude, resulting in a DEM having dimensions of 21,600 rows and 43,200 columns. The horizontal coordinate system is decimal degrees of latitude and longitude referenced to WGS84. The vertical units represent elevation in meters above mean sea level. GTOPO30 data may be obtained

Table 9.2
Examples of DEM Data Sets

Country/ Producing Agency	Product Name	Scale/Resolution	Coverage	Sources of Data
United States National Geophysical Data Center	ETOPO5	5- by 5-arc minutes	Whole world	Various
United States U.S. Geological Survey (EROS Data Center)	GTOPO30	30- by 30-arc seconds	Whole world	Various
United States U.S. Geological Survey (EROS Data Center)	DCW-DEM	30- by 30-arc seconds	Africa, N. America, Europe, to cover world eventually	Various, but mainly from contour, points, and hydrology layer of Defense Mapping Agency 1 : 1M maps
United States U.S. Geological Survey (National Mapping Program)	7.5-minute DEM	30- by 30-m	Cont. U.S., Hawaii, Puerto Rico	Various, including contours from 1 : 24,000 and 1 : 100,000 maps; National Aerial Photography Program (NAPP) quad-centered photographs
	30-minute DEM	2- by 2-arc sec		
	1-degree DEM	3- by 3-arc sec	U.S.	
	7.5-minute Alaska DEM	1- by 2-arc sec	Alaska	
	15-minute Alaska DEM	2- by 3-arc sec	Alaska	
Canada Department of Natural Resources	Canadian Digital Elevation Data (CDED)	1:250,000 3-arc sec to 12-arc sec depending on latitude	Whole country	Contours of 1 : 250,000 NTS maps, supplemented by geodetic and air survey data
United Kingdom Ordnance Survey	Land-Form PANORAMA	50 m	Whole country	Contours of 1 : 50,000 maps
	Land-Form PROFILE	User-defined, typically 10 m		Contours of 1 : 10,000 maps
Australia Australian Surveying and Land Information Group (AUSLIG) and others	GEODATA 9 SECOND DEM	9- by 9-arc seconds	Whole country	Contours, spot heights, coastlines, and hydrography from the 1 : 250,000 topographic maps; elevation from the national gravimetric database and geophysical surveys
Australia Australian National University	Australian DEM	3- and 1.5-arc minute 1-arc minute for individual states	Whole country	Various

electronically, at no cost, by anonymous *File Transfer Protocol* (FTP) from the EROS Data Center.

Digital Chart of the World (DCW) is a vector cartographic data set digitized under contract from the former United States Defense Mapping Agency (DMA) (now NIMA). The source was the *Operational Navigation Chart* (ONC) map series at a scale of 1 : 1,000,000. Sponsored by the USGS National Mapping Division, DCW-DEM was created by resampling elevation data from the contour, point heights, and hydrology layers of DCW at regularly spaced intervals of 30 by 30 arc seconds (approximately 1 kilometer). Supplementary terrain data of higher accuracy were used for the DEMs covering North America and Europe. The horizontal datum is WGS84. Elevation values are expressed in feet above mean sea level. Just like ETOPO5 and GTOPO30, DCW-DEM data may be downloaded at no cost over the Internet.

The three-dimensional radar image data collected by the Shuttle Radar Topography Mission (SRTM) on the Space Shuttle Endeavour mentioned in Chapter 8 (Section 8.5.3) are used to generate topographic map products. These three-dimensional image data will allow DEM of the world at 30 m × 30 m spatial sampling with ≤16 m absolute vertical height accuracy to be produced, thus meeting the National Map Accuracy Standards for topographic mapping *(http://www.jpl.nasa.gov/srtm/)*. These DEM data represent a great improvement in accuracy and will replace the GTOPO30 data as the major source of digital terrain data of the world.

The United States Geological Survey (USGS) is the lead federal agency for the collection and dissemination of digital cartographic data. It produces five primary types of DEM products in its *National Mapping Program* (USGS, 1997b).

- *7.5-Minute DEM Covering the Contiguous United States, Hawaii, and Puerto Rico.* These are provided as 7.5- by 7.5-minute blocks having the same coverage as a standard USGS 7.5-minute quadrangle. The resolution is 30 by 30 meters and the DEMs are on the UTM coordinate system.
- *30-Minute DEM Covering the Contiguous United States and Hawaii.* Two 30-minute DEMs provide the same coverage as a standard USGS 30- by 60-minute quadrangle. The resolution is 2 by 2 arc seconds and the DEMs are based on geographic coordinates.
- *1-Degree DEM Covering the Entire United States.* Two DEM blocks (three in Alaska) provide the same coverage as a standard USGS 1- by 2-degree quadrangle. The resolution is 3 by 3 arc seconds, and the DEMs are based on geographic coordinates.
- *7.5-Minute Alaska DEM.* Each of these DEMs has the same coverage of a 7.5-minute quadrangle. The resolution is 1- by 2-arc second latitude by longitude. This type of DEM is georeferenced to geographic coordinates.
- *15-Minute Alaska DEM.* The coverage of each of these DEMs corresponds to a 1:63,360-scale quadrangle. The resolution is 2- by 3-arc second latitude by longitude. These DEMs are based on geographic coordinates.

Digital Terrain Data in Canada

In Canada, the Department of Natural Resources (NRCan) is the federal agency responsible for national surveying and mapping. The Center for Topographical Information (CTI) of NRCan, in conjunction with the Ontario Region of the Canadian Forest Service (CFS), produces the nationwide coverage of digital terrain data called the *Canadian Digital Elevation Data* (CDED) (NRCan, 1997). These data are based on the *National Topographic System* (NTS) maps at the scale of 1 : 250,000. Other complementary sources include elevation data, aerotriangulation control points, and geodetic control points. The coverage of every CDED file corresponds to half an NTS map sheet, which means that there are western and eastern parts to the CDED for every NTS map at the 1 : 250,000 scale. The grid spacing is based on geographic coordinates, with a minimum resolution of 3 arc seconds and a maximum resolution of 12 arc seconds, depending on latitude. Elevation data are referenced to the NAD 83 horizontal datum and are expressed in meters above the mean sea level (the Canadian Vertical Geodetic Datum).

At the provincial level, various government mapping agencies also produce DEMs. Alberta and British Columbia, for example, have created DEMs by converting contours of the federal government's National Topographic System (NTS) maps. The British Columbia government is also in the process of producing DEMs at a large scale (1 : 20,000) in its *Terrain Resource Information Management* (TRIM) program. In Ontario's Basic Mapping program (OBM), a DEM layer is created for new digital map compilation for the *Digital Topographic Database* (DTDB). Automation of existing hardcopy maps maintains a contour layer.

Digital Terrain Data in the United Kingdom

The Ordnance Survey (OS) is the national mapping agency of the United Kingdom. It maintains two major types of digital terrain data products: *Land-Form PANORAMA* and *Land-Form PROFILE*. Land-Form PANORAMA is produced from contours at 10-m vertical intervals on OS Landranger maps, which have a scale of 1 : 50,000. It contains both contour data in vector format and DEM data with a 50-m resolution. Elevation values of the DEM are mathematically interpolated from the contours on the Landranger maps, rounded to the nearest meter. Land-form PROFILE, on the other hand, is created by digitizing contours at 5-m vertical intervals (10 m in mountainous areas) on 1 : 10,000 maps. Both contour and DEM data are maintained. The former is in vector format, whereas the latter can be supplied at a user-defined resolution, typically at 10 m intervals. The DEM data are calculated directly from the contour data including all spot heights and tide lines.

Digital Terrain Data in Australia

Several DEM data sets have been produced in Australia. *GEODATA 9 SECOND DEM* is a cooperative data set produced by the Australian Surveying & Land Information Group (AUSLIG), the Australian Geological Survey Organization (AGSO), the Australian Heritage Commission, and the Centre for Resource and Environmental Studies at the Australian National University. This DEM data set has a grid spacing of 9 arc seconds (approximately 250 m) in latitude and longitude. It was produced by combining the massive data sets of elevation and topographic information of AUSLIG and AGSO, including contours, spot heights, coastlines, and hydrography from the 1:250,000 topographic maps, as well as elevation from the national gravimetric database and geophysical surveys.

The *Australian Digital Elevation Model* is another continental-scale digital terrain data set. It is developed by the Centre for Resource and Environmental Studies at the Australian National University. The data are available at two spatial resolutions: 3 arc minutes and 1.5 arc minutes. For individual states, data at a finer resolution (1 arc minute) are also available. The 1.5-arc minute DEM data were calculated by a specific gridding technique that ensures the presence of a connected drainage structure by automatically removing spurious pits or sinks (Hutchinson, 1988, 1989). It also includes trigonometric points that serve to ascertain that most of the principal peaks are incorporated into the data.

9.4.5 QUALITY AND STANDARDS OF DIGITAL TERRAIN DATA

Quality and standards are key considerations in digital terrain data acquisition. The objective of quality control and enforcement of standards is to ensure that the errors inherent in the data sets are within an acceptable limit and quantifiable so that they may be used with confidence. The quality and standards of digital terrain data may be explained from three perspectives: (1) factors governing data quality, (2) quality control criteria and procedures during data acquisition, and (3) DEM data standards.

Factors Governing the Quality of Digital Terrain Data

The quality of digital terrain data is affected by a number of interrelated factors.

- *Methods of Data Acquisition and Model Production.* Of the different methods of digital terrain data acquisition described in the previous sections, field surveying methods produce digital terrain data of the highest quality, followed by photogrammetric methods and then cartographic map digitizing. Large DEM data sets are usually produced using different methods and with different types of source data. Such data sets probably contain all kinds of inherent errors that may have occurred and accumulated during the acquisition of the source data sets (see Chapter 4).
- *Types of Source Data.* In a research study aiming to evaluate the accuracy of DEMs, Bolstad and Stowe (1994) compared elevation values in a 7.5-minute DEM obtained from the USGS (produced using a Gestalt Photomapper II and National High Altitude Photography (NHAP) 1:40,000-scale, leaf-off, panchromatic aerial photographs) with those in a DEM of the same area generated from SPOT images (using an image correlation algorithm and a panchromatic stereopair obtained on different dates, one partially leaf-off and the other total leaf-off). The two DEMs reported the same elevation for only 3.5% of the cells in the study area, and differences of up to 82 meters were observed. This indicates that even with similar production methods, different types of source data may generate digital terrain data of very different qualities.
- *Nature of the Terrain.* In the research study of Bolstad and Stowe, it was found that the greatest discrepancies occurred in high-relief forested areas. Image correlation algorithms often exhibit the poorest fit, and hence lowest accuracies, in forested areas because of the lack of clearly defined topographic features such as roads, buildings, and vegetation boundaries.
- *Methods of Interpolation.* Different methods of interpolation used to generate DEMs and TINs employ different algorithms (see Sections 9.5.1 and 9.5.2). They may or may not include the mechanism that performs drainage enforcement to overcome the problem of spurious sinks and maintain a connected drainage pattern (Hutchinson, 1988, 1989). As a result, the quality of DEMs and TINs produced by different methods of interpolation may vary considerably.

Quality Control Criteria and Procedures

Different organizations have developed different criteria and procedures for quality control in digital terrain data acquisition. The quality control system at the USGS, for example, includes the following principal items (USGS, 1997b).

- *Statistical Accuracy Tests.* This includes testing of the general accuracy of individual data sets to ensure compliance with mapping specifications. All USGS DEMs are tested and assigned a vertical *root mean square error* (RMSE). This is computed using a minimum of 28 test points per DEM. Of these test points, 20 are to be selected from interior points and 8 from edge points. The edge points are used for edge testing, which evaluates how well elevation values along adjacent DEM edges match.

- *Data Editing.* The objective of editing is to ensure the conformance of the data to criteria governing constraints to data for water bodies (e.g., type, size, elevation), hydrography (e.g., drainage, lakes, swamps, and shorelines) and slopes (e.g., natural break points).
- *Verification of Logical and Physical Formats of Data Files.* The logical and physical formats of all DEM data are validated by a special program as part of the database entry procedure.
- *Visual Inspection.* Using a computer program called the *DEM Editing System* (DES), this process aims at identifying blunders such as irregularly gridded data and wrong tagging of elevation values, which may require verification with original source data.

DEM Data Standards

USGS DEMs are classified into three levels according to data quality. Data sets that fail to satisfy the standards for the classification are not normally released for distribution. The specifications for the three levels of classifications are as follows (USGS, 1997b).

- *Level 1*, which represents the lowest level, includes 7.5-minute DEMs or an equivalent derived from photogrammetric stereo profiling or image correlation using high-altitude aerial photography. The desired accuracy standard is a vertical RMSE of 7 meters or less, with 15 meters as the permissible maximum. A 7.5-minute DEM at this level has an absolute elevation error tolerance of 50 meters for blunders for any given node when compared to the measured elevation. Within each array of points, there must be no more than 49 contiguous elevations in error by more than 21 meters. The 30-minute DEMs produced from level 1 or level 2 7.5-minute DEM data are classified as level 1. They carry a computed RMSE, but no maximum value is set for this value because minimum accuracy requirements are assumed to have been satisfied in the production of the source data.
- *Level 2* DEMs have been processed or smoothed for consistency and edited to remove identifiable systematic errors. These include elevation data sets derived from hypsographic and hydrographic data digitizing, either by photogrammetry or by map digitizing, that have passed the review process on a DEM editing system. An RMSE of one-half contour interval is the maximum permissible error tolerance, with no errors greater than one contour interval.
- *Level 3* DEMs, which represent the highest accuracy of the three levels, are derived from *digital line graph* (DLG) data by using selected elements from both hypsography (i.e., contours and spot heights) and hydrography (i.e., lakes, shorelines, and drainage). Ridge lines and hypsographic effects of major transportation features are also included in the data. An RMSE of one-third of the contour interval is the maximum permissible error tolerance, with no errors greater than two-thirds of the contour interval.

9.5 Data Processing, Analysis, and Visualization

The core activities of digital terrain modeling involve three typical phases: processing terrain data to ensure that they are optimized for storage and application; performing analysis to convert topographic attributes (elevation, slope, aspect, profile curvature, and catchment area) derived from DEMs or TINs into useful terrain information; and presenting the terrain information to the user in an easily understandable manner. In the discussion that follows, we will explain these processes in detail and introduce the different classes of software tools that have been developed for such purposes.

9.5.1 MESH SIMPLIFICATION

Mesh simplification is the process by which a TIN model is constructed from DEM data. The objective is to extract from a DEM the topographically important elevation points to form a TIN with the minimum number of points possible, while at the same time preserving the maximum amount of information about terrain structure. This is an essential function in digital terrain data processing because it allows the user to take advantage of both the DEM and TIN models. As noted earlier, the DEM approach is more suitable than the TIN approach for automated digital terrain data sampling. On the other hand, the TIN approach is noted for its efficiency in data storage and ability to generate visually realistic topographic surfaces. With the aid of mesh simplification techniques and powerful computers, it is now possible to acquire digital terrain data with the DEM approach. The data are then processed to form TINs to optimize storage and modeling efficiency. It has been shown that the number of elevation points in DEMs obtained from the USGS may be reduced significantly (to less than 5% of the original) in TINs with only a very minor decrease in model quality (MCVL, no date).

Although the algorithms used by different methods of mesh simplification may be quite different from one another, the underlying principles are apparently very similar. They are all commonly characterized by an iterative process comprising two phases (Lee, 1991):

1. Determine the *surface significance* of elevation points in the DEM data by using the spatial relationships between an individual point and its neighbors. An elevation point is considered "important" or "surface-specific" when its elevation cannot be closely interpolated using values of its neighbors.
2. Establish the rules or tolerances to terminate the point selection process. These rules may be based on a quantitative criterion (i.e., when the number of selected points reaches a predetermined number) or a qualitative criterion (i.e., when there is no point whose computed importance exceeds a predefined level, such as median or maximum errors).

Lee (1991) identified four categories of methods of mesh simplification.

- *The skeleton method* is a two-phased process that first selects a set of surface-specific structural points (i.e., peaks, pits, passes, ridge lines, and river channels) and then a set of support points progressively in locations where differences in elevation exceed a predefined vertical tolerance (Fowler and Little, 1979).
- *The filter method,* which is also referred to as the *Very Important Points* (VIP) method, evaluates the importance of a given point by comparing its measured elevation value with an estimated value computed from the elevation values of its eight neighbors (Chen and Guevara, 1987). Elevation points are retained or discarded on the basis of either a predetermined significance level or a desired number of points to construct the TIN.
- *The hierarchy method,* proposed by De Floriani et al. (1984), approximates the terrain surface progressively to finer levels of detail by hierarchical or successive triangulation. It is in fact both a method of terrain point selection and a data structure for triangular tessellation of topographic surfaces.
- *The drop heuristic method* is in essence based on the technique of optimization by which points of least importance are selectively dropped until a predetermined threshold is reached (Lee, 1989).

There are advantages and disadvantages of these methods (Table 9.3). It is hard to say categorically which method is superior to the others. The choice of the

Table 9.3

Advantages and Disadvantages of Different Methods of Mesh Simplification

Method	*Advantages*	*Disadvantages*
Skeleton method	Works well on certain types of terrain, especially those with sharp breaks along ridges or channel lines	Complex and difficult to program
Filter method	Conceptually simple All points considered Flexibility in controlling the number of points used to construct TIN	No attempt to test global fitness between surfaces described by DEM and TIN Not performing well with curved surfaces and on topographic features with more gentle slopes
Hierarchy method	Hierarchical data structure that allows successive refinement of surface approximation and rapid data retrieval Particularly suitable for applications that require the use of spatial relationships	Tendency to generate long and thin triangles
Heuristic method	Defining important points in a global context Flexibility in using user-defined number of points or tolerance level	Computationally very demanding

Source: Compiled from Lee, (1991)

method to use depends on a variety of factors, such as the nature of the terrain, the purpose of a particular application, the geographic extent of the area of interest, as well as the computing resources available. Mesh simplification is not only a topic of interest in GIS, but more generally in computer graphics as well. It is always advisable for GIS users to keep an eye on new algorithms and methods that may appear in the technical literature in the fields of computer graphics and elsewhere. There are good possibilities for these new developments to be usefully employed for digital terrain modeling in the context of GIS (Polis and McKeown, 1992).

9.5.2 SURFACE REPRESENTATION AND ANALYSIS

Digital terrain modeling may be used to answer a variety of spatial questions by employing one or more of the following classes of analytical techniques: (1) *interpolation and surface fitting*, (2) *creating new DEM data sets*, (3) *extraction of topographic attributes and landscape features*, and (4) *visibility and viewshed analysis*.

Interpolation and Surface Fitting

Interpolation is the process by which elevation values of one or more points in geographic space are used to produce estimated values for positions where elevation information is required. It is used for contouring and for the generation of DEMs from selectively or randomly sampled elevation points. Interpolation may be computed by directly using elevation values in the DEM or TIN without the assistance of a surface fitting algorithm. In Figure 9.15, for example, the estimated elevation values for m (z_m) and n (z_n) in the DEM and TIN may be calculated as

$$z_m = (z_{3b} + z_{4b} + z_{3c} + z_{4c})/4$$

where z_{3b}, z_{4b}, z_{3c}, and z_{4c} are elevation values for grid intersections 3b, 4b, 3c, and 4c respectively; and

$$z_n = (z_c + z_b + z_e)/3$$

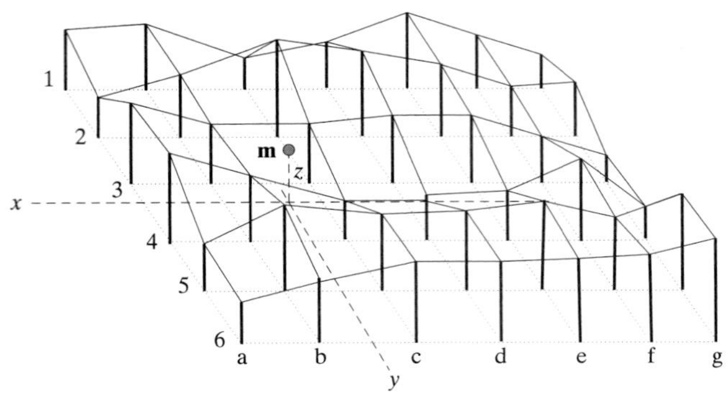

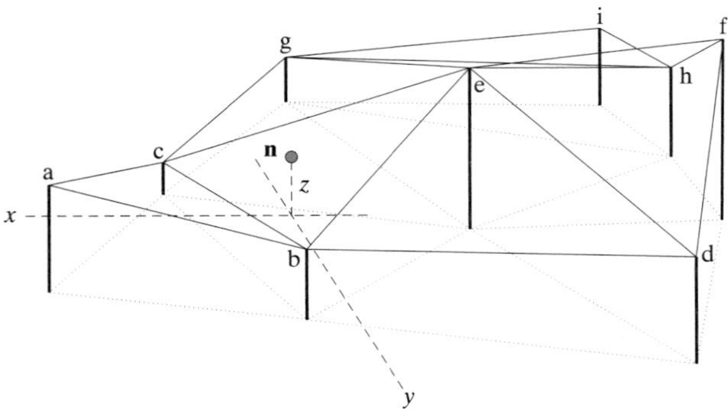

FIGURE 9.15
Interpolation in DEM and TIN. The objective of interpolation is to answer questions such as: "What are the elevation values at points m and n?" and "Where is the 100-m contour line running?"

where z_c, z_b, and z_e are elevation values for vertices c, b, and e respectively.

In practice, however, a surface-fitting algorithm is commonly used to improve the result of terrain modeling. A surface-fitting algorithm may be linear or nonlinear, depending on the order of the polynomial equations used for estimating the elevation of the required point (Petrie, 1991d). The general polynomial equation for surface fitting is given in Table 9.4, where z_i is the

Table 9.4
General Polynomial Equations for Interpolation

Individual Terms	*Order of Terms*	*Descriptive Terms*	*No. of Terms*
$z = a_0$	Zero	Planar	1
$+ a_1x + a_2y$	First	Linear	2
$+ a_3x^2 + a_4y^2 + a_5xy$	Second	Quadratic	3
$+ a_6x^3 + a_7y^3 + a_8x^2y + a_9xy^2$	Third	Cubic	4
$+ a_{10}x^4 + a_{11}y^4 + a_{12}x^3y + a_{13}x^2y^2 + a_{14}xy^2$	Fourth	Quartic	5
$+ a_{15}x^5 + \cdots$	Fifth	Quintic	6

Source: Petrie, 1991d

elevation value of an individual point i; x_i and y_i are its coordinates, and a_0, a_1, a_2, ... are coefficients of the polynomial. To find the elevation value for point m in Figure 9.15 using linear interpolation, for example, the first step is to set up a set of equations to solve for the coefficients a_0, a_1, and a_2.

$$z_{3b} = a_0 + a_1x_{3b} + a_2y_{3b}$$
$$z_{4b} = a_0 + a_1x_{4b} + a_2y_{4b}$$
$$z_{3c} = a_0 + a_1x_{3c} + a_2y_{3c}$$
$$z_{4c} = a_0 + a_1x_{4c} + a_2y_{4c}$$

Since there are more equations than the number of unknowns, this is a least-squares solution. Once the values of a_0, a_1, and a_2 have been determined, the elevation value of m can be found by

$$z_m = a_0 + a_1x_m + a_2y_m$$

where x_m and y_m are known coordinates of point m.

This method of interpolation is an example of what is known as the *local* approach. Local interpolators apply an algorithm repeatedly to a small portion of the total area in order to ensure a good fit. This approach of interpolation is conceptually simple and computationally easy to implement. However, since the interpolated elevation value (and consequently the fitted surface) at a point depends only on neighboring data, there is no consideration of the characteristics of terrain continuity and smoothness in nature. The *global* approach of interpolation removes such a shortcoming of the local approach by utilizing all or most of the elevation data to characterize the surface at a point, thus allowing prediction to be made on the trend of the surface. Figure 9.16 illustrates conceptually the difference between these two approaches of interpolation. For the terrain point p, the estimated elevation is h_1 when interpolated locally using only the values of the two neighboring points 3 and 4 (Figure 9.16a). If interpolated globally using the values of points 1 through 6 (Figure 9.16b), the estimated elevation value of p becomes h_2, which is greater than h_1. The estimated elevation values of p will be the same in Figures 9.16b and 9.16c if interpolated locally, assuming that the neighboring data points used in both cases have the same elevation values. If the effects of non-neighboring points are taken into account in global interpolation, the resulting elevation values of p in the two cases will be different. The examples of Figures 9.16b and 9.16c clearly demonstrate that the estimated elevation value of a terrain point is affected not only by its immediate neighbors, but also by those points that are located away from it. Therefore, global interpolation approaches obviously have the advantage of preserving terrain continuity and smoothness. However, as terrain modeling always involves a very large number of points, global methods are computationally very demanding.

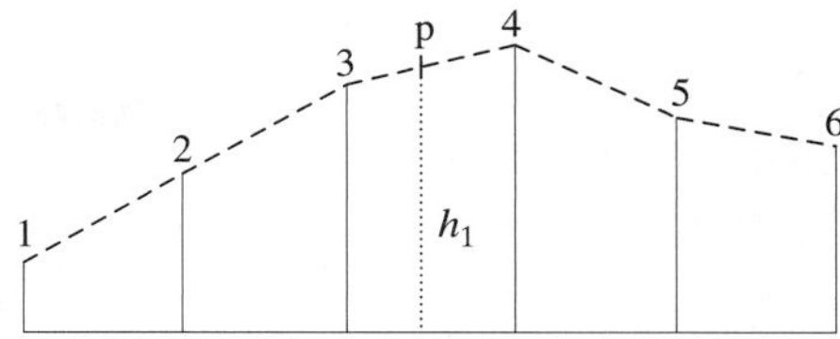

(a) Local interpolation, elevation of point p = h_1

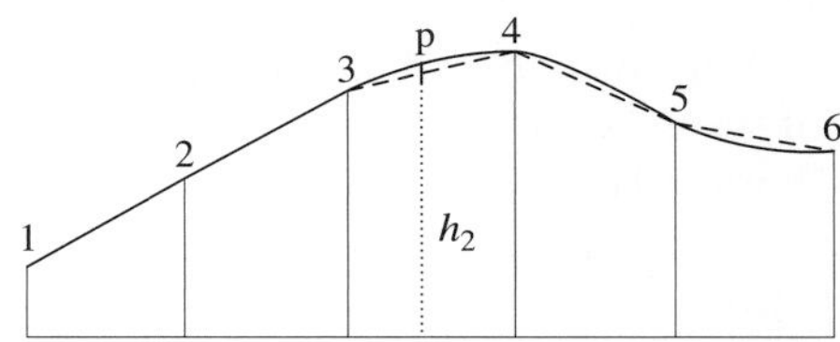

(b) Global interpolation with surface fitting, elevation of point p = h_2, where $h_2 > h_1$ in (a)

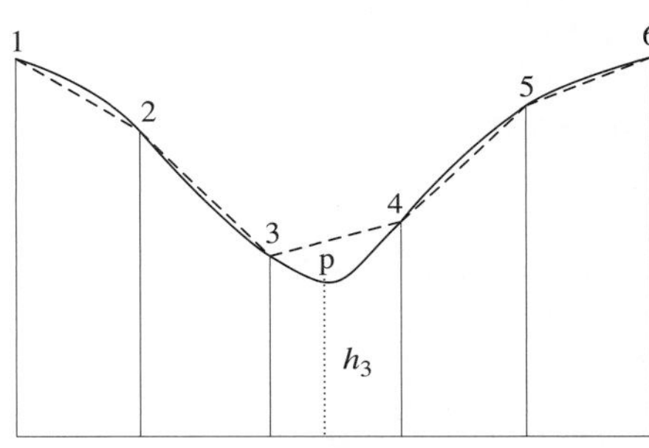

(c) Global interpolation with surface fitting, elevation of point p = h_3, where $h_3 < h_1$ in (a)

- - - - - - - - Locally interpolated surface
———— Globally interpolated surface

FIGURE 9.16
The impact of the methods of interpolation on the estimated elevation values. The estimated elevation values (h_1, h_2, and h_3) of point p in the three cases vary according to the methods of interpolation used, assuming that the heights of points 3 and 4 are the same in (a), (b), and (c).

A widely used global surface-fitting method is *trend surface analysis* (Mather, 1976; Unwin, 1981). In this method, the terrain surface is approximated by a polynomial expansion of the coordinates of the sampled points. The coefficients of the polynomial function are determined by solving a set of simultaneous equations, which include the sums of powers and cross products of the *x*, *y*, and *z* values, by the method of least squares. This ensures that the sum of the squared deviations from the trend surface is a minimum. Once the coefficients have been computed, the polynomial function can be evaluated at any point within the geographic space under investigation. Commonly used polynomials are as follows:

$$\textit{linear}: Z = a + bX + cY$$

$$\textit{quadratic}: Z = a + bX + cY + dX^2 + eXY + fY^2$$

$$\textit{cubic}: Z = a + bX + cY + dX^2 + eXY + fY^2 + gX^3 + hX^2Y + iXY^2 + jY^3$$

where Z is the estimated height, X and Y are geographic coordinates, and a, b, c, . . . , j are polynomial coefficients (Figure 9.17). Trend surface analysis is also a useful data exploratory tool. More detailed discussions of the implications of these equations for spatial analysis can be found in Chapter 10, Section 10.5.

It is relatively simple to create a grid matrix of values by substituting the coordinates of the grid nodes into the polynomial and calculating an estimate of the surface for each node.

Inverse distance weighted (IDW) interpolation is the simple local interpolation method most commonly used. It is usually implemented in the form of a moving window to define the *zone of influence* that we learned in raster-based neighborhood analysis in Chapter 5. In vector-based processing, a circular moving window is often used instead. This method is based on the assumption that the influence of each input point on the interpolated value at the center of the window is inversely proportional to a power (p) of its distance from the center as expressed in the following formula (Burrough and McDonnell, 1998) (Figure 9.18).

$$z(u_0) = \frac{\sum_{i=1}^{n} z(u_i) d_{ij}^{-p}}{\sum_{i=1}^{n} d_{ij}^{-p}} \quad \textbf{(9.1)}$$

where $z(u_0)$ is the estimated value of the point at an unsampled location, $z(u_i)$ are the data points, d_{ij} is the distance between each data point to the point at an unsampled location, and p is a parameter. Typically, $p = 2$; in other words, the weights are usually inversely proportional to the squared distance between the data point and the point at the unsampled location. The quality of the resulting surface, as indicated by the fitting between the surface and the input points, is dependent on the density, distribution, and accuracy of the input points. A shortcoming of this interpolation method is that it does not reproduce the local shape implied by the data and tends to produce local extrema at the data points (Mitas and Mitasova, 1999). Just like trend surfaces, IDW surfaces seldom go through the input points. In other words, IDW is not an exact interpolation method.

Finally, an optimum local interpolation method, known as *kriging*, originally developed for ore reserve estimation in geostatistics, is now widely used in DTM software packages because it is flexible and can handle any type of data (Cressie, 1988; Cressie, 1990; Oliver and Webster, 1990). Kriging, named after the pioneering work of Danie Krige (1951), is a generic name for a family of least-squares linear regression algorithms that are used to estimate the value of a continuous at-

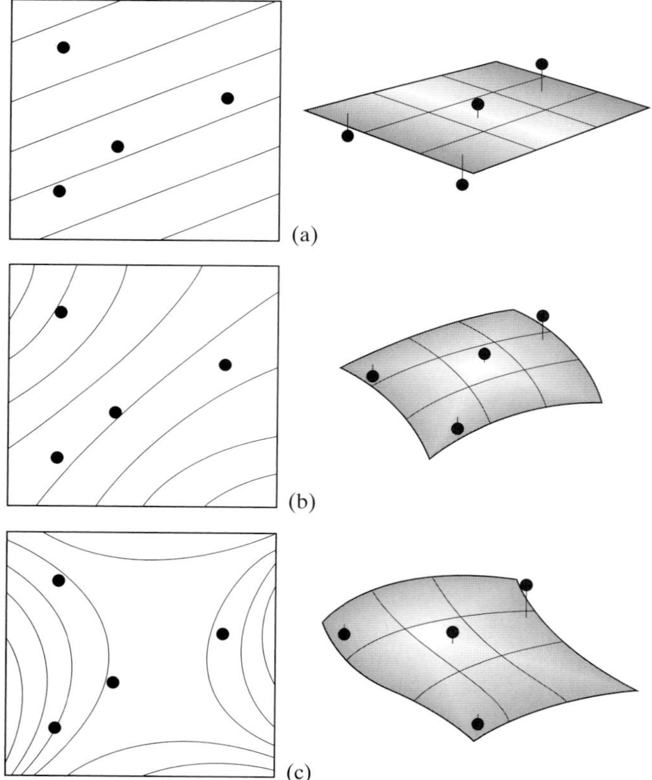

FIGURE 9.17
Graphical representation of the three forms of trend surfaces produced by the (a) linear, (b) quadratic, and (c) cubic polynomial equations. *(Source: Burrough, 1986)*

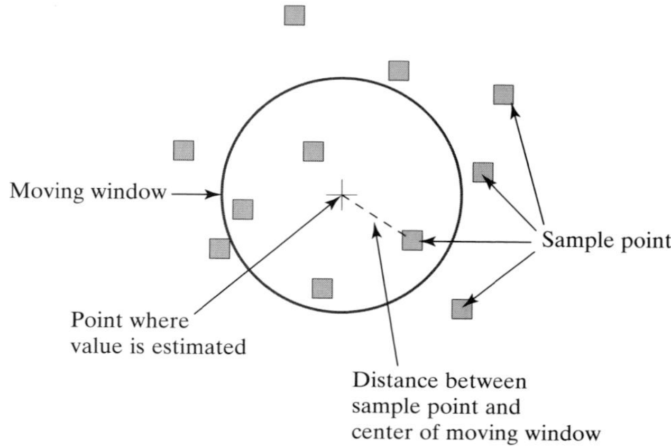

FIGURE 9.18
Moving average interpolation using inverse distance weighting.

tribute (such as terrain height) at any unsampled location using only attribute data available over the study area (Goodvaerts, 1997). Kriging treats the continuous attribute to be interpolated as a *regionalized variable*. The estimated value of the unsampled point is obviously affected by its distances from the surrounding points of known attribute values. There is spatial autocorrelation, that is, a dependence between sample data values, which decreases as the distance increases. There may also be a certain degree of directional effect on the variations of the sample data values. These characteristics of regionalized variables are quantified in the form of an autocovariance matrix (expressing the relationship between the covariance of the sample points and their distance), from which estimates of values at unsampled locations are determined (Lam, 1983). However, different assumptions on the regionalized variables may be made. In one approach, known as *simple kriging*, it was assumed that the expected value of a random variable at a specified location (i.e., the estimate) equals the mean value in the study area. In this case, a variogram, which represents the relationship between the mean-square difference between sample values and their distances, can be used to determine the value at the unsampled location. Mathematically, the variogram ($2r$) or semi-variogram (r) is defined by (Lam, 1983).

$$r = \frac{N}{2} \sum_{i-1}^{N} [z(u_i + d) - z(u_i)]^2 \qquad \textbf{(9.2)}$$

where d is the distance between two sample points, $z(u_i)$, $i = 1, 2, \ldots, N$ represent the z data available over the study area, say, the N data, and N is the total number of data points. Graphically, this can be represented in Figure 9.19. Another approach, known as *ordinary kriging*, assumes that the mean value is constant but unknown and the local variation of the mean needs to be accounted for. The variogram may be used to determine weights needed for local interpolation (Burrough and McDonnell, 1998). Because the assumptions for simple kriging are not always realistic, another approach called *universal kriging* (or *kriging with a trend model*) is used (Bonham-Carter, 1994). This considers that the unknown local mean value varies smoothly within each local neighborhood and the whole study area, so that the local trend over the whole study area can be modeled by a polynomial function such as a linear or quadratic trend (shown in the section on trend surface analysis).

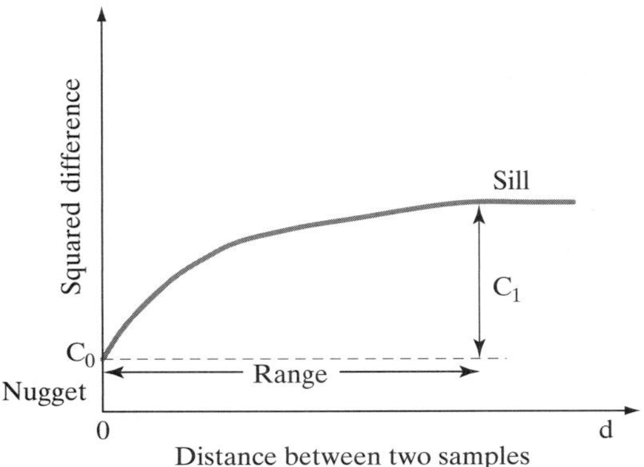

FIGURE 9.19
An example of a simple variogram. *(Source: Lam, 1983; Burrough and McDonnell, 1998)*

The various approaches of kriging mentioned earlier estimate values at points. The resulting map based on results from point kriging displays a lot of sharp spikes and pits at the data points (Burrough and McDonnell, 1998). This may be overcome by estimating an average value over a block of land, hence the name *block kriging*. Values within the block of land are estimated using the point kriging method, and the average value over the block is then calculated. Finally, another useful approach to kriging is called *co-kriging*. At one sampled location, data for more than one attribute may be available. If the data for each attribute are spatially correlated, it is possible to use the spatial variation of one attribute to help map the other attribute. This applies when the data for one particular attribute are more difficult or too expensive to obtain. Therefore, the other less expensive set of attribute data becomes the surrogate for the more expensive set. There are many more forms of kriging (e.g., factorial kriging, dual kriging, probability kriging). More discussion can be found in Goovaerts (1997) and Burrough and McDonnell (1998).

Creating New DEM Data Sets

An important application of the interpolation methods explained above is to create new DEM data sets from point data. New digital terrain data are sometimes required because existing data may not be of a resolution or quality compatible to a certain application. The production of new DEMs may also be warranted because of the availability of new source data. Instead of creating a completely new data set, it may be more economical to refine an existing DEM by incorporating these new data. With the aid of interpolation using a surface-fitting algorithm, a DEM of a higher resolution may be constructed from an existing one of a lower resolution.

The creation of new DEMs from existing data is by no means a trivial and straightforward task. It requires a good understanding of the nature and requirements of both the old and new DEM data sets. If, for example, one of the sources of data to be incorporated is of a higher resolution than the target DEM data set, the method of aggregation or generalization is applied instead of the method of interpolation. There are many ways to aggregate a multitude of high-resolution grid cells into one generalized grid cell at a reduced resolution, such as:

- *the statistical approach*, which uses a representative statistic such as mean or median of the original cells
- *the resampling approach*, which accomplishes resolution reduction by geometric transformation (see Chapter 8, Section 8.6.2)

For situations that require the use of surface fitting, the choice of the method depends on the geographic extent of the new DEM. DEMs that cover a small area may be generated using either a local or global approach. For continental or national scale DEMs, which may require millions of elevation points, global approaches such as kriging may be computationally impractical. On the other hand, local approaches are not feasible alternatives because of their sensitivity to the positions of the data points, which may result in the generation of spurious edge effects. In the development of a continentwide DEM for Australia, Hutchinson (1993) developed an iterative finite difference interpolation technique that is said to have the computational efficiency of a local interpolation method without sacrificing the major advantages of global methods (e.g., assurance of terrain continuity and freedom from spurious features). This iterative technique uses a simple multigrid strategy that calculates grid points at successively finer resolutions, starting with a coarse initial grid and continuously halving the grid spacing until the final user-defined resolution is achieved. For each resolution, grid points are computed by an iterative method known as the Gauss-Seidel method. Iteration for each resolution stops when the maximum number of iterations (typically 30) is reached.

The example of the Australian DEM also serves to illustrate how a drainage enforcement algorithm may be used to make the resulting DEM suitable for hydrological applications and related analysis. This algorithm, described in detail by Hutchinson (1989), attempts to remove all sink points in the fitted DEM that have not been identified as such in input sink data. It may also incorporate stream line data if such data are available. The drainage enforcement algorithm is run concurrently with the iterative interpolation procedure described here (Hutchinson, 1993). For each grid resolution, the DEM is periodically inspected for sinks and their accompanying lowest saddle points. Constraints affecting drainage clearance are then applied to the DEM by inserting ordered chain conditions that lead from each spurious sink, via the lowest associated saddle point, to a data point or existing ordered chain on the other side of the saddle. The integration of the drainage enforcement algorithm into the interpolation process has led to significantly improved results.

Extraction of Topographic Attributes and Landscape Features

Topographic attributes, which are defined as numerical descriptions of the terrain, may be classified as *primary* and *secondary attributes* (Moore et al., 1991). Primary topographic attributes are those geomorphometric parameters that may be directly calculated from digital terrain such as elevation and slope. Secondary attributes, also referred to as *compound attributes*, are formed by combining the primary attributes with other environmental indices that characterize the spatial variability of specific processes occurring in the landscape (e.g., soil water content distribution). Of the primary topographic attributes used to describe landforms, the following may be easily estimated from DEM data using automated methods (Moore et al., 1991).

- *elevation*: the altitude or height of terrain points
- *slope*: the gradient of the land
- *aspect*: azimuth of the maximum local slope
- *specific catchment area*: upslope area draining across a unit width of contour
- *flow path length*: maximum distance of water flow to a point in a catchment
- *profile curvature*: curvature of slope profile
- *plan curvature*: curvature of contour

Further parameters pertaining to geomorphological analysis may also be extracted, including local relief, drainage density, and statistics of slope and convexity (Weibel and Heller, 1991). The extraction of topographic attributes and geomorphometric parameters is usually an intermediate process in digital terrain modeling. The data obtained are used more as input for further terrain modeling (e.g., multivariate classification of landforms, trafficability studies, and soil erosion modeling) rather than as an end product in themselves.

A wide variety of landscape features may be extracted from DEMs and TINs. Conventionally, most of the landscape features extracted are associated with drainage and hydrological applications as a result of the thrust toward the development of automated computer-assisted channel network analysis in the last two decades (see Section 9.6.2). However, landscape features are now extracted increasingly for studies in different branches of science and technology in a broader context of using GIS. Numerous algorithms have been developed for extracting landscape features from digital terrain data. For the purpose of our discussion, we classify these algorithms by the three primary types of landscape features they are designed to extract: *surface-specific points*, *linear* or *network features*, and *areal features*. We will explain the working principles of selected examples of each type of algorithm accordingly.

Surface-specific points include topographic features such as peaks, pits, passes, and saddles. The extraction of these features from DEM is a relatively straightforward process that involves the comparison of elevation differences in a local neighborhood (i.e., local relief). In a DEM, peaks are represented by local elevation maxima, whereas pits, passes, and saddles are represented by local elevation minima. A detailed description of the methods for deriving surface-specific points from digital terrain data may be found in a classic paper by Peucker and Douglas (1975).

The delineation of linear topographic features (e.g., river channels and ridges) is a much researched topic area in digital terrain modeling. For obvious reasons, the algorithms for extracting linear features from DEMs are different from those for TINs. Mark (1988) identified several algorithms for extraction from DEMs, but not all of these are computationally practical to implement (Moore et al., 1991). At present, the most commonly used method appears to be the so-called "hydrological approach" proposed by Mark (1984). In this method, the "drainage area" of each DEM elevation grid (i.e., the number of elevation grids that drain into that grid) is first determined by climbing recursively through the DEM (Figures 9.20a and 9.20b). This process results in a matrix, called the "drainage area transform" (Figure 9.20c), that contains the drainage areas for all the grids in the DEM. The information in the drainage area transform is then used to trace the "channel pixels," as identified by those grids with large drainage areas. Channels are recursively followed upstream until there is no more point that exceeds a minimum threshold (Figure 9.20d). Once the drainage network has been delineated, ridges may be delineated

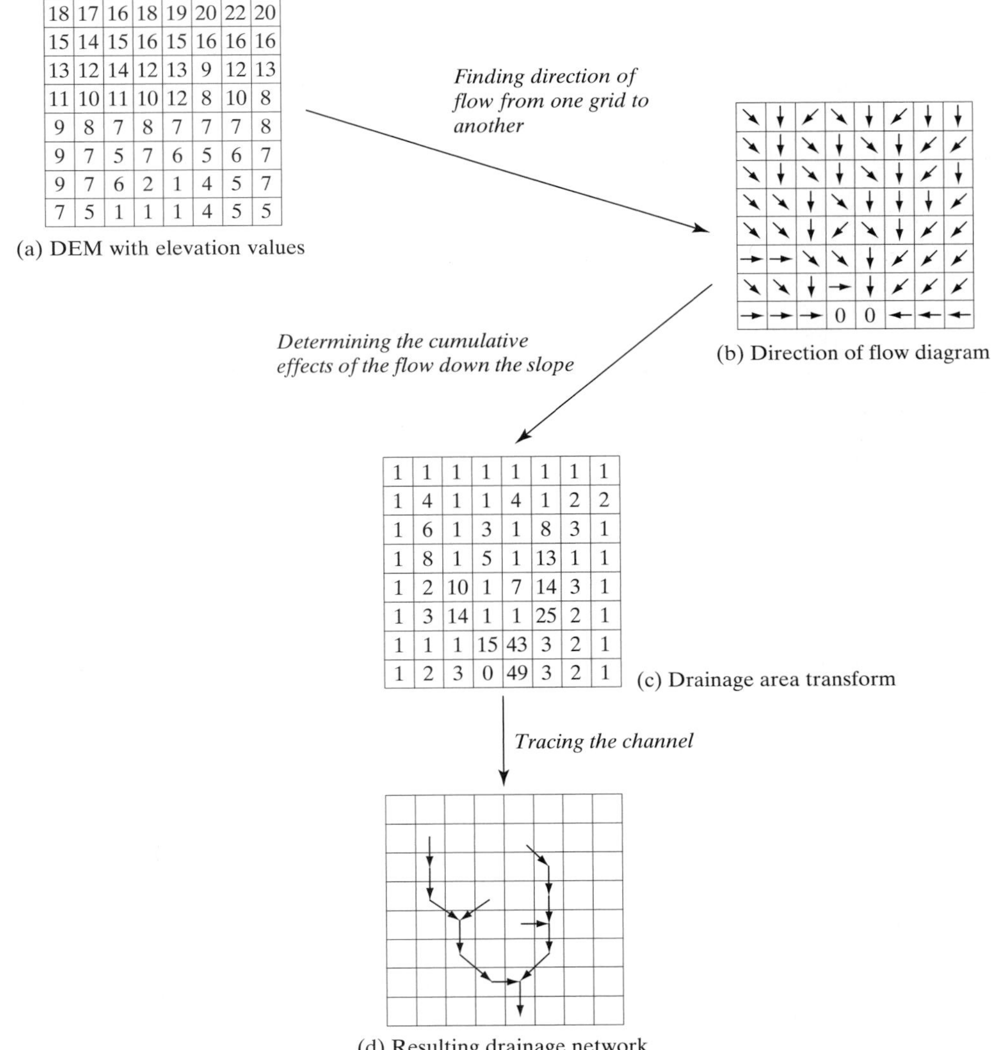

FIGURE 9.20
Concept of the "hydrological approach" of delineating drainage network from a DEM.

either by gray-scale thinning of all nonchannel pixels or by tracing the boundaries of the catchment area. The channels and ridges delineated may be noncontinuous. It is necessary to carry out an interpolation process to connect the broken line segments into a properly connected drainage network. On the other hand, if the channels and ridges consist of multiple adjacent lines, a thinning process is required to turn them into continuous lines of one grid width.

These methods for delineating drainage networks and ridges form the basis for the delineation of areal hydrological features (e.g., catchments and subcatchments) as the final stage of landscape feature extraction from DEMs. This is done by using variants of the recursive DEM climbing algorithms to identify those grids belonging to a specific channel or ridge. The process, described in detail by Band (1989), starts by first identifying each channel junction as a divide, which is used to anchor the divide graph to the channel network. Individual subcatchments are then defined for each channel link. Finally, the subcatchments are joined to form the catchment or drainage basin of the entire drainage network.

Visibility and Viewshed Analysis

The objective of visibility analysis in digital terrain modeling is to determine all the positions that can be seen from a specific point (Figure 9.21). In a DEM, two pixels, p and p′, are said to be mutually visible if a straight line can connect them without intersecting any part of the surface in between (Lee, 1994). The pixels p and p′ are usually referred to as "viewpoints." For each viewpoint in a DEM, a visibility matrix, V, can be constructed. In this matrix, $v_{ij} = 1$ when p_{ij} is visible from the viewpoint, and $v_{ij} = 0$ if otherwise. All the pixels that are visible from a viewpoint constitute its "visibility region," ΣV_p (Lee, 1994). The determination of the mutual visibility between p and p′ is referred to as *line-of-sight analysis*. The tasks associated with the computation and interpretation of the visibility matrices and visibility regions are collectively known as *viewshed analysis*.

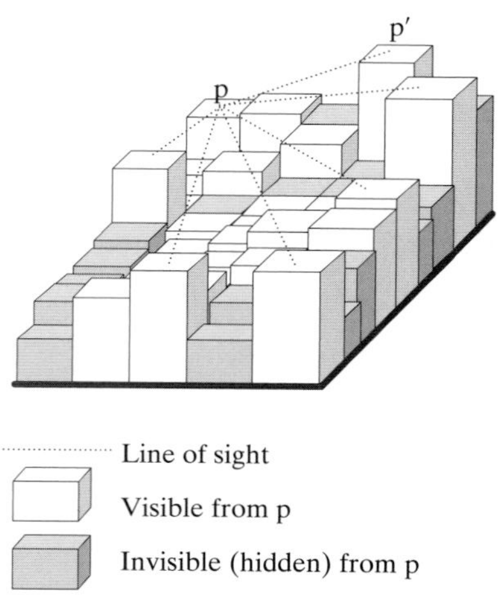

FIGURE 9.21
Determining the visibility in a digital elevation model.

The construction of the visibility matrix is a computationally demanding task, as a line of sight has to be constructed between the viewpoint grid cell and every one of the grid cells in the DEM. Earlier solutions were based mainly on the algorithms proposed by Yoeli (1985) and Mark (1987). More recently, Lee (1994) recommended the use of an algorithm known as the simple *digital differential analyzer* (DDA) as described by Newman and Sproull (1979). Lee (1994) also suggested this algorithm should be used in conjunction with data filtering that helps select "visually dominant" points for the analysis. This speeds up the computation since a smaller number of elevation points are involved.

If visibility analysis using DEMs is described as computationally intensive, visibility analysis using TINs is even more so. Many algorithms have been developed for three-dimensional visualization in computer graphics. However, there is an essential distinction between three-dimensional visualization and visibility analysis. Three-dimensional visualization algorithms are designed to display an image of a model from a given viewpoint *outside* the model, but not at a viewpoint *within* the model as required in visibility analysis. Therefore, these algorithms are not appropriate for the purpose of visibility analysis in digital terrain modeling. As part of the effort to automate landscape feature extraction in TINs, De Floriani et al. (1986) developed a method for visibility analysis using TIN data. In brief, this method makes use of the vertices of the TIN as the optimal set of viewpoints. For each of these viewpoints, a "visibility region" is constructed by considering the intersections of edges with planes defined by other edges and the current viewpoint. To reduce the number of edge-pairs to be considered, edges that cannot block the visibility of any other edges are eliminated at the outset, and the rest are sorted according to their elevations with respect to the viewpoint. When all the visibility regions have been delineated, the entire area covered by the TIN is decomposed into mutually exclusive and completely exhaustive set of polygons by repeated application of an intersection operator. The visibility relations are then stored in a matrix, V, where v_i represents the individual viewpoints, and v_j the set of elementary visibility regions. If the value of $v_{ij} = 1$, it signifies that region j is visible from viewpoint i. If the value of $v_{ij} = 0$, then region j is not visible from viewpoint i. Once the matrix is created, a viewshed map using any one of the vertices in the TIN may be constructed relatively easily.

9.5.3 DIGITAL TERRAIN VISUALIZATION

The ultimate aim of DTM is to present relevant terrain information about a given geographic space that results from the analysis of the characteristics of its topography and related spatial phenomena. Visualization is therefore an integral component of DTM, from the perspectives of both process and technology. As noted in Chapter 7, visualization in the context of GIS modeling serves two primary purposes: to communicate geographic information and to provide a means for data exploration and hypothesis refinement. The same objectives apply to digital terrain visualization.

There are numerous well-developed techniques for digital terrain visualization. New techniques are being developed as technology evolves. For the purposes of describing what these techniques are and explaining what methods and technologies are involved, a convenient way is to classify them according to the dimension of the graphical display: two-dimensional, two-and-a-half-dimensional, three-dimensional, and multidimensional.

Two-Dimensional Terrain Visualization

Contour lines are the most conventional, and probably still the most commonly used, method of digital terrain visualization. To visualize digital terrain by contours, it is necessary to generate and plot contour lines, either on the computer screen or on a hard-copy medium, using any of the methods described in McCullagh (1991). Using contour lines is a quantitative way of representing three-dimensional terrain in two dimensions because numerical measurements of elevation may be readily made on the display. From the perspective of visualization, there is one major drawback of contour lines: they are difficult to comprehend by inexperienced map users. As a result, different methods have been proposed to create a three-dimensional impression of relief by contours. An example is the "shaded contours" method, which draws contours by varying their widths according to illumination brightness (Peucker, 1972). This method has not been in popular use in GIS.

Analytical or automated hill shading is another widely used method of two-dimensional terrain visualization. Hill shading is based on the principles of applying illumination to the surface normal vector (i.e., the vector perpendicular) for the DEM or TIN. Figure 9.22 explains how it works on the triangular facets of a TIN. Hill shading using DEMs is similar in principle but different in algorithms. Figure 9.23 is an example of a hill shaded relief image generated from a USGS DEM. Hill shading has been described as a qualitative method of two-dimensional terrain data visualization. Although it readily creates a good impression of topographic form, there is no way by which numerical elevation values may be measured from the display.

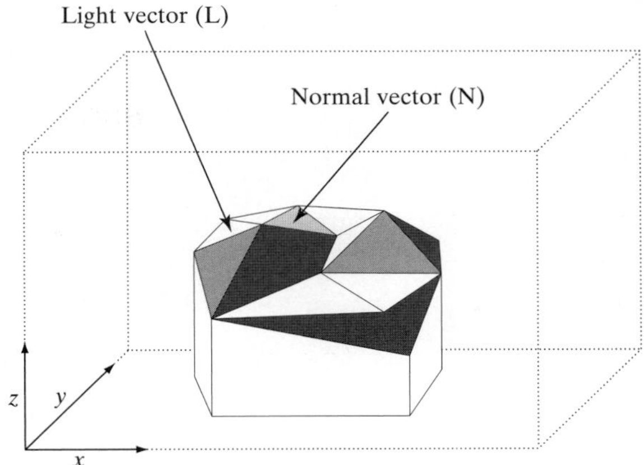

L × N = 0.0 < illumination values < 1.0

FIGURE 9.22
The concept of hill shading. The illumination value of an individual triangular facet, which will be translated into a gray scale for display and plotting, is computed from its normal vector (N) and a vector representing the light source in respect to the observer (L). The light source is conventionally placed in the northwest at an angle of 45° to the viewer's position.

Two-and-a-Half-Dimensional Terrain Visualization

A two-and-a-half-dimensional display is basically an *isometric* model. In such a model, the z attribute associated with an x, y location is projected onto an x, y, z coordinate reference system. This transforms the map of z attributes for an x, y position so that each z attribute defines a position on the z axis, creating a surface that is perceived as three-dimensional (Figure 9.24). Since all the principal axes are equal in scale in all three directions and they make equal angles with one another, two-and-a-half-dimensional visualization does not create the impression of visual depth (i.e., distant objects are smaller than closer ones). Two-and-a-half-dimensional models are relatively easy to implement from the programming point of view. They may be used together with other kinds of images that are draped on to them to create a more realistic perspective view of the terrain (Figure 9.25). From the user's perspective, two-and-a-half-dimensional visualization represents a great improvement over two-dimensional techniques. It has been the standard method of terrain visualization until relatively recently. However, two-and-a-half-dimensional visualization has now been rapidly superceded by three-dimensional techniques made possible by the advent of extremely powerful computer graphics hardware and software in the last few years.

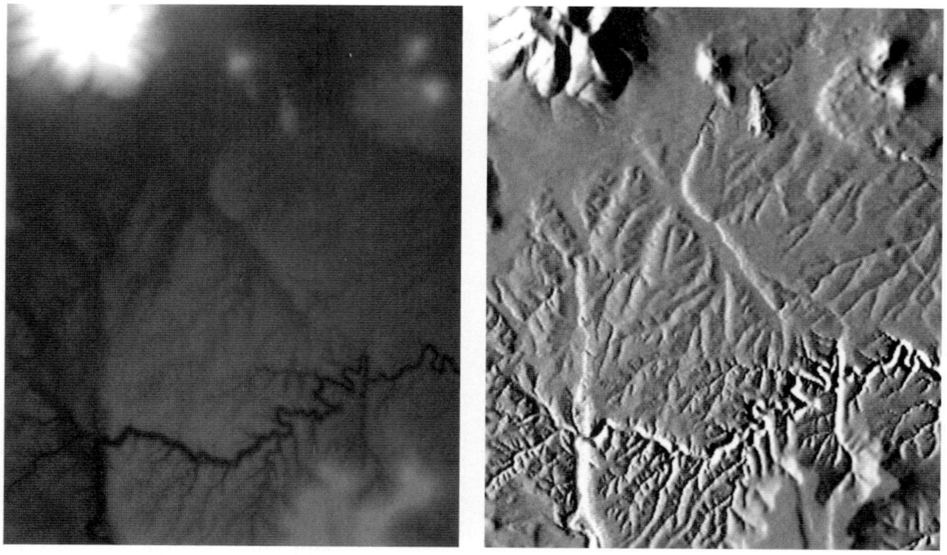

(a) Original DEM

(b) Relief image

FIGURE 9.23
Example of a hill shaded relief image (b) generated from a DEM (a). Note how the drainage, faults, and joint fractures have been enhanced in the relief image. *(Source: USGS)*

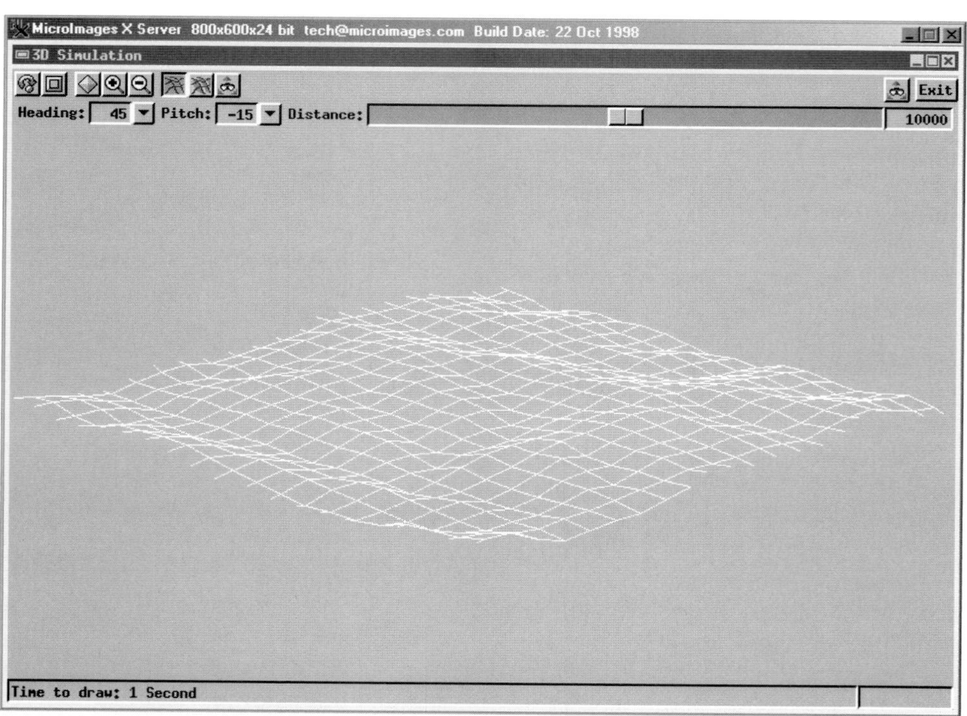

FIGURE 9.24
A two-and-a-half-dimensional isometric terrain model.

Three-Dimensional Terrain Visualization

A three-dimensional terrain model is a solid model in which many *x, y, z* data points are used to form a solid structure that may be visualized in a perspective view. Unlike a two-and-a-half-dimensional view, which presents only a pseudo perspective of the terrain, a three-dimensional terrain model is an analog for the physical space in nature as perceived by an observer (Figure 9.26). Three-dimensional terrain modeling allows

FIGURE 9.25
Draping a remote sensing image and vector data onto a two-and-a-half-dimensional isometric terrain model.

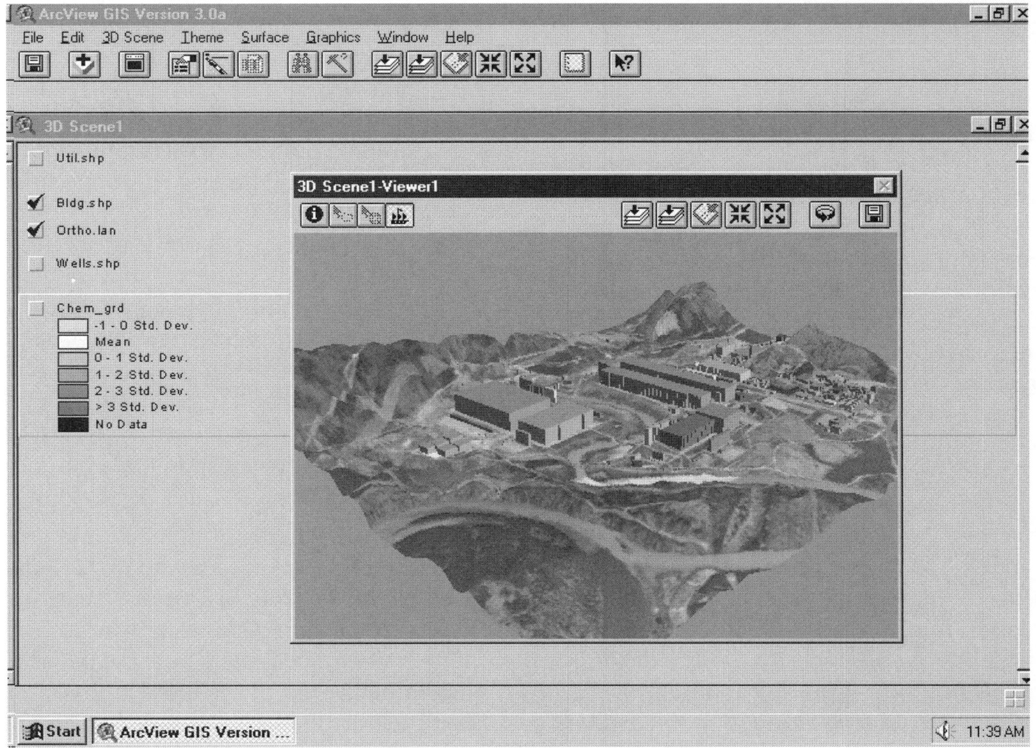

FIGURE 9.26
True three-dimensional terrain visualization using 3D Analyst extension of ArcView GIS. To view this photo in color, see the color insert. *(Courtesy: ESRI, Redlands, CA)*

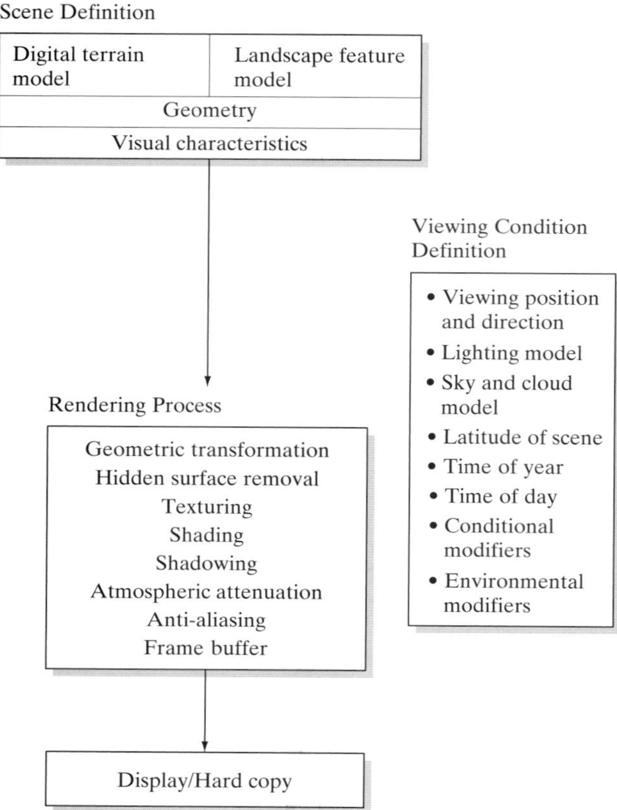

FIGURE 9.27
Rendering techniques for three-dimensional terrain visualization. *(Source: Kennie and McLaren, 1991)*

the full specification of three-dimensional operations on the objects and phenomena within the constraints of the geometrical model used. It represents one of the most exciting developments in GIS technology.

The concepts and techniques of three-dimensional terrain visualization have been drawn heavily from computer graphics (Foley et al., 1990; Newman and Sproull, 1979). By nature, three-dimensional terrain modeling is conceptually complex and computationally intensive. It is a rendering process that is made up of a combination of sophisticated computer graphics techniques that include (Figure 9.27) (Kennie and McLaren, 1991):

- *Geometric transformation* is a perspective transformation that maps three-dimensional terrain data into a two-dimensional space. The transformation results in the resizing of objects in the image space by scaling them inversely to their distances from the assumed position of the observer.
- *Depth cueing* is the technique that is used to match the perceived image on the computer screen to the "natural" visual cue model of a human observer. Depth cuing is usually done by increasing the three-dimensional interpretability (i.e., decreasing or compensating for the loss of three-dimensional information) of the surface, such as using the maximum DEM and image resolution possible.
- *Hidden edge and surface removal* involves various algorithms that are capable of improving the interpretability of terrain surfaces by computing and removing hidden images from being displayed. This means that objects and surfaces that are obscured by other objects or surfaces in the terrain model will not be seen by the observer.
- *Antialiasing* is the technique used to remove distortions (called *aliasing artifacts*) in the visual image caused by sampling in the original data.
- *Shading* provides a sense of realism of the terrain surface. Many factors are considered in shading the terrain surface, including light source and intensity of illumination, the condition of the intervening atmosphere, the surface characteristics (e.g., color, reflectance, and texture), position and orientation of the surface relative to the light source, other surfaces, and the observer.
- *Ray tracing* is an enhancement to the shading algorithm by giving additional consideration to the effects of, for example, refraction illumination and multiple light sources.
- *Shadows* provide further enhancement to the shading algorithm by considering whether an object or surface can be "seen" from a certain light source.
- *Surface texture detail* is the technique that adds details to the terrain surface by, for example, draping a digital photograph or other form of remote sensing imagery onto the surface of a terrain model.
- *Atmospheric attenuation* is the technique that modifies the appearance of the displayed terrain model by taking into account the effect of atmospheric or hazing effects (which may vary according to the time of a day, season, weather conditions, and level of air pollution).

Multidimensional Terrain Visualization

Many of the three-dimensional terrain visualization techniques in the market today are functionally so sophisticated that it is probably an oversimplification to label them as three-dimensional. It is more appropriate to call them multidimensional. These include the various methods that make use of the technique of animation to present the terrain in sequences of "scene rendering," forming the frames of a movie (see Chapter 7). Animation techniques may be used to visualize changes in the terrain over time from a static viewpoint (i.e., a tempo-

(a) Landscape scene before the construction of the reservoir

(b) Landscape scene when the reservoir starts storing water

(c) Landscape scene when the reservoir is at full capacity

FIGURE 9.28
Selected scenes from a three-dimensional visualization application showing how a valley is flooded when a reservoir is built. To view this photo in color, see the color insert. *(Courtesy: 3D Nature LLC, Arvada, CO)*

ral dimension is added to the visualization) (Figure 9.28). More advanced animation techniques allow the user to change the viewpoint continuously, creating what has become known as the fly-by, or fly-through, effect of visualization (i.e., a spatiotemporal dimension is added to the visualization) (Figure 9.29). Fly-by techniques are the foundation of DTM applications such as flight simulation in pilot training (see Section 9.6.6), route planning, and computer games. NASA has produced many interesting fly-by movies. (See an impressive Mt. Fuji fly-by developed using ASTER image data in NASA's EOS Terra satellite Web site:*http://terra.nasa.gov/Gallery/browse.php3?query=5).*

Digital terrain data of a multidimensional nature are best managed in *virtual environments* (or *virtual worlds*). Virtual environments may be perceived in different ways: plane images, two-and-a-half-dimensional models in a conventional computer screen, and a true three-dimensional space in an immersed environment (i.e., virtual reality, or VR). Virtual environments may also include information with more than three dimensions, such as temporal information (Jacobson, 1994). As noted in Chapter 7, the virtual reality modeling language (VRML) is a standard language for describing interactive three-dimensional objects. VRML 2.0 is the World Wide Web standard for the delivery of virtual reality on the Internet. VRML supports two types of object models suitable for digital terrain data: *ElevationGrid* and *IndexFaceSet*. These are in fact the VRML equivalents of the DEM and TIN models, respectively. They have been designed specifically to build terrain models and perform geometric modeling in VRML, although many do not support coordinates. It is anticipated that

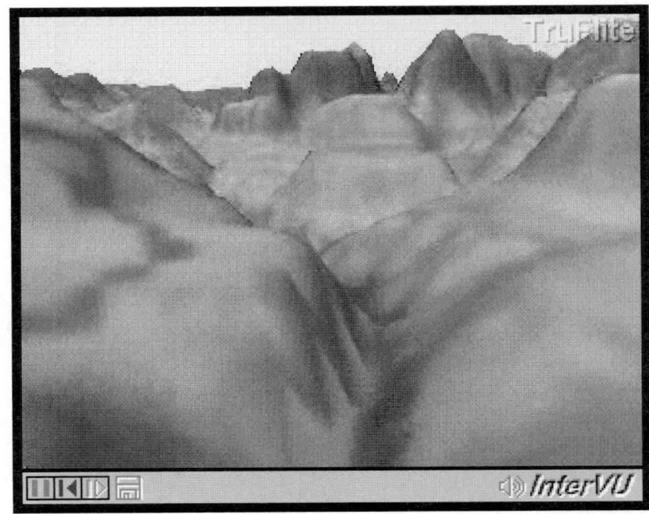

FIGURE 9.29
A scene captured from a fly-by animation application.
(Courtesy: The Martin D. Adamiker LLC, Vienna, Austria)

VRML will have a significant impact on the development of terrain visualization techniques in the future. The language itself has a strong user base, and it does not require any expensive proprietary modeling software. Working in the interactive mode, VRML applications allow users to perform dynamic query on the database. As we will see in the next section, many three-dimensional modeling software systems and some GIS now support some forms of VRML export. An extension of VRML called GeoVRML has been developed to improve on the coordinate system for three-dimensional representation of georeferenced data on the Web (Rhyne, 1999). This means that terrain visualization will be accessible to more GIS users at a lower cost and serving more purposes than the simple communication of terrain information.

9.5.4 SOFTWARE FOR DIGITAL TERRAIN DATA PROCESSING

Generally speaking, conventional GIS is weak in DTM functions. Although geographic data are obviously three-dimensional in nature, practically all GIS software packages have until now been designed on two-dimensional data models. Digital terrain data are geographic data in every sense of the term, but they have specific requirements in terms of data structure, processing procedures, and methods of information presentation. Software used for digital terrain data processing may be divided into four categories according to their designed functions: (1) terrain model construction software, (2) functional extensions to GIS software, (3) terrain data visualization software, and (4) contouring software. The characteristics of typical examples of these four categories of terrain data processing software are explained here.

Terrain Model Construction Software

Terrain model construction software is specifically for generating DEMs or TINs from unstructured input data. There are few stand-alone terrain model construction software packages around. The terrain model building process is usually handled in GIS software packages as a shared function between its generic data entry function (for terrain data input and editing) and the DTM functional extension (for creating the DEM or TIN). *ANUDEM*, developed by the Centre for Resource and Environmental Studies at the Australian National University, Canberra, Australia, is perhaps the best known and most widely used terrain model building package (Hutchinson, 1998). It is used not only in Australia, but also in the United States, Canada, and many other countries. ANUDEM accepts as input irregularly sampled elevation point data, contour line data, and stream line data, and generates a regular grid of discretized smooth surface that fits the original data. The program imposes a global drainage enforcement algorithm that automatically removes spurious sinks where possible (see Section 9.5.2). This particular feature has been found to be extremely useful in practice. It can significantly increase the accuracy, especially in terms of drainage properties, of DEMs interpolated from sparse sets of surface specific data. Originally developed for use on UNIX workstations, it was released in a Windows 95/NT version in 1998.

DTM Functional Extensions to GIS Software

For most GIS packages on the market today, DTM is offered as an optional software extension rather than as a generic function. There are two notable exceptions. One of these is Idrisi (Figure 9.30), developed by Clark Labs for Cartographic Technology and Geographic Analysis of Clark University, Worcester, MA. DTM is an integrated function of this particular software package. The other is the *Mini Image Processing System* (MIPS), a public-domain software suite developed by the USGS (1998) for Digital UNIX and Digital VMS operating systems on the Alpha computer platform. In both of these software packages, DTM is one of the many digital image and vector data-processing functions in a modular suite of programs.

A GIS software extension is an add-on to a host package that is designed to serve a particular data-processing function or a special data type. Software extensions are usually purchased separately as an option when the host GIS software is installed. As a result of the growing importance of DTM in geographic data processing, most GIS vendors now offer a software extension for terrain modeling applications (Table 9.5). The actual capabilities of individual software extensions vary considerably from one package to another. Some are complete with data input, editing, model creation, and visualization functions (e.g., ARC TIN, GeoTerrain). Others are designed primarily for model building and visualization. They rely on the host system for data editing-functions (e.g., 3-D Analyst for ArcView GIS). Usually, individual packages are capable of accepting a wide variety of input data formats, including DEM, TIN, ASCII text files, and raster images of different formats. The examples in Table 9.5 allow standard two-dimensional and three-dimensional visualization, but their capabilities to produce animation vary. No package has fly-by capability, and only a few of them (IMAGINE VIRTUAL GIS and FLY!) are able to export visualization scenes in standard movie formats.

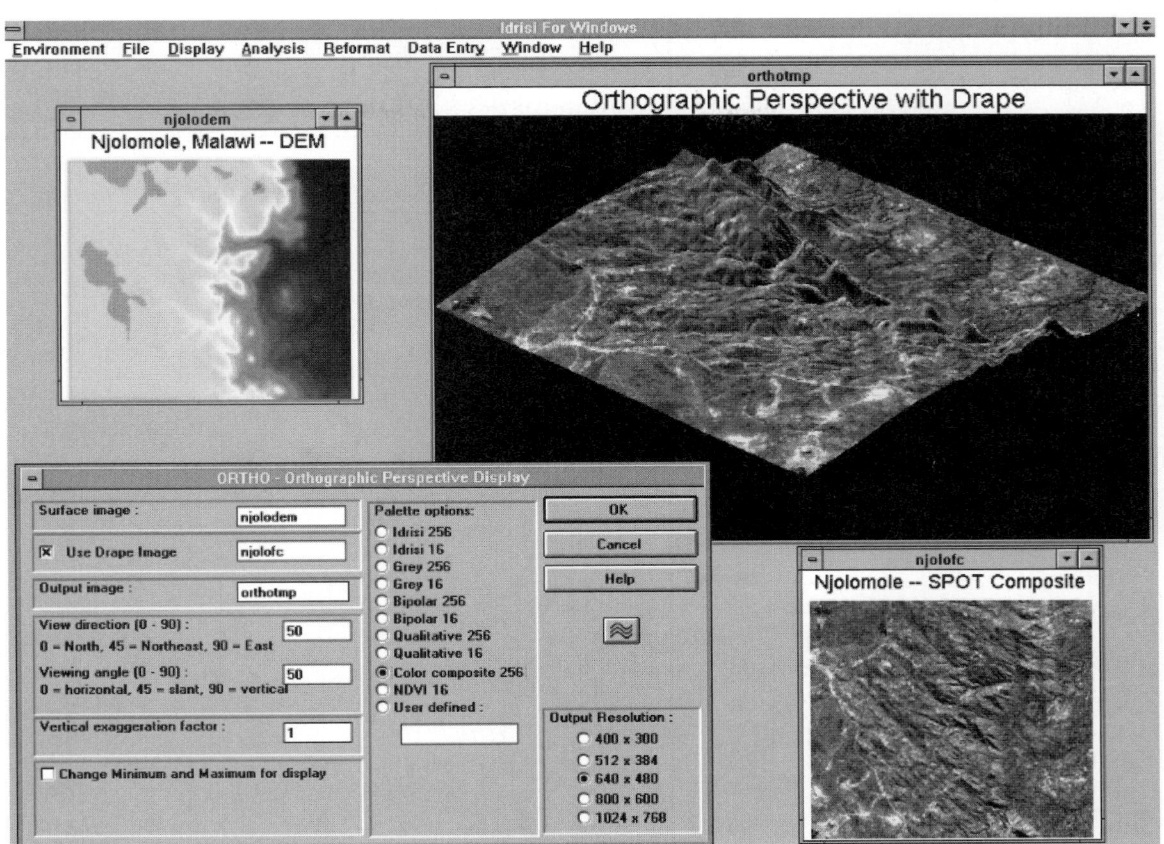

FIGURE 9.30
GUI of Idrisi showing its three-dimensional visualization capability. To view this photo in color, see the color insert.
(Courtesy: Clarke University)

Terrain Data Visualization Software

There are now many stand-alone software packages in the market specially designed for terrain analysis and modeling (Table 9.6). Until relatively recently, DTM could be implemented only on the mainframe, mini, and workstation classes of computers. Advancements in computer graphics, both in terms of hardware and software, have made DTM on the PC platform a reality. These stand-alone packages, as exemplified by those listed in Table 9.6, are generally characterized by relatively limited data automation and editing functionality. They have to rely mostly on packaged data sets obtainable from government agencies and commercial data suppliers. Some packages have very sophisticated user interfaces designed to allow the user considerable control on the parameters governing visualization (Figure 9.31). Unlike software extensions that are subject to the constraints of the technologies used by their respective host package, these stand-alone software packages have more flexibility in the use of new technologies (e.g., OpenGL and VRML support). There are two obvious shortcomings common to most packages. One is the lack of ability or difficulty in using georeferenced data. The other is the lack of database management capability. These have limited the usefulness of the packages to visualization functions with relatively little possibility for integrated geographic data processing in a true GIS environment.

Contouring Software

There are now many software packages in the market specially designed for automated contouring using digital terrain data of various types and formats. Since contouring is probably the most popular method of visualizing terrain data and it is not as computationally demanding as other forms of visualization, this category of packages may be implemented easily on the PC platform. With a few exceptions, software vendors tend to develop their applications within the working environment of an existing GIS or CAD package. This allows them to minimize development cost and time by making use of the display functionality of the host software. In a survey of Windows-based contouring software packages, Thoen (1996) noted that these packages

Table 9.5

DTM Functional Extensions to GIS/RS Software Packages

Product Name/Vendor	*Host GIS Software*	*Input/Output Formats*	*Interface Type*	*Special Notes*
ARC TIN ESRI, Redlands, CA	ArcInfo	DEM, TIN, point coverages, images, others	GUI, command line	Special ARC and ARCPLOT commands; AML macros
3-D Analyst ESRI, Redlands, CA	ArcView GIS	DEM, TIN, others; output to 3-D shapefiles	GUI	Totally integrated into ArcView GIS environment; tools for creation, analysis; display of 3-D data
GeoTerrain GEOPAK, North Miami Beach, FL	MicroStation GeoEngineering family of software by Bentley Systems, Exton, PA	DEM, TIN, others	GUI	Tools for data editing and display; merging and clipping of DTMs; thematic mapping; volumetric computation
MGE Terrain Analyst, Intergraph, Huntsville, AL	Modular GIS Environment (MGE) family of software	DEM, TIN, images, others	GUI, command line	Produces both raster and vector output
IMAGINE VIRTUAL GIS ERDAS, Atlanta, GA	ERDAS IMAGINE	DEM, TIN, images; output to movie formats	GUI	Movie creation capability for screen, output to several movie formats
MOSSpro Infrasoft, Boston, MA	MX family of engineering; site-design software	TIN, contour	GUI	Fly-by, automated cross section, volumetric computation
FLY! PCI, Richmond Hill, ON	EASI/PACE. OrthoEngine	DEM, TIN, images, other GIS/CAD formats	GUI	Fly-by and playback capability, export data to movie formats

Table 9.6
Stand-alone Terrain Analysis and Visualization Packages

Product Name/ Vendor	Platform	Input/Output Formats	Interface	Special Notes
LandForm Rapid Imaging Software, Albuquerque, NM	PC	DEM, ASCII, various image formats	GUI	OpenGL development environments; VRML support for Internet applications
TruFlite Martin D. Adamiker, Anif, Austria	PC	DEM, ASCII, ERDAS DEMs, IDRISI DEMs, others	GUI	
World Construction Set (WCS) 3D Nature LLC, Arvada, CO	Pentium or DEC Alpha AXP	WCS DEM format converted from USGS DEMs, images, ASCII files	GUI	Internal database to manage DEM files and vector objects
Visual Explorer 98 Wolleysoft, Balquhidder, Scotland	PC	DEM, ASCII, DXF, others	GUI to Vistapro, MapInfo, AutoCAD, and other programs	OpenGL development environments; support for VRML and Apple's Quicktime VR
TAPES Australian National University Canberra, Australia	UNIX workstations	DEM, contour	X-windows	TAPES-G uses grid data; TAPES-C uses contour data
3D Geographer Synoptech, Montreal, QC	PC	DCW, DEM, DXF, DGN, prefiltered geographic database of ETOPO5 and CIAWDBII	GUI	Ability to use georeferenced data and internal database
GWN-DTM Scientific Software Group (SSG), Washington, DC	PC and UNIX workstations	DEM, TIN, CAD formats	GUI	One of the components of an integrated set of programs for mapping, engineering, geoscientific applications

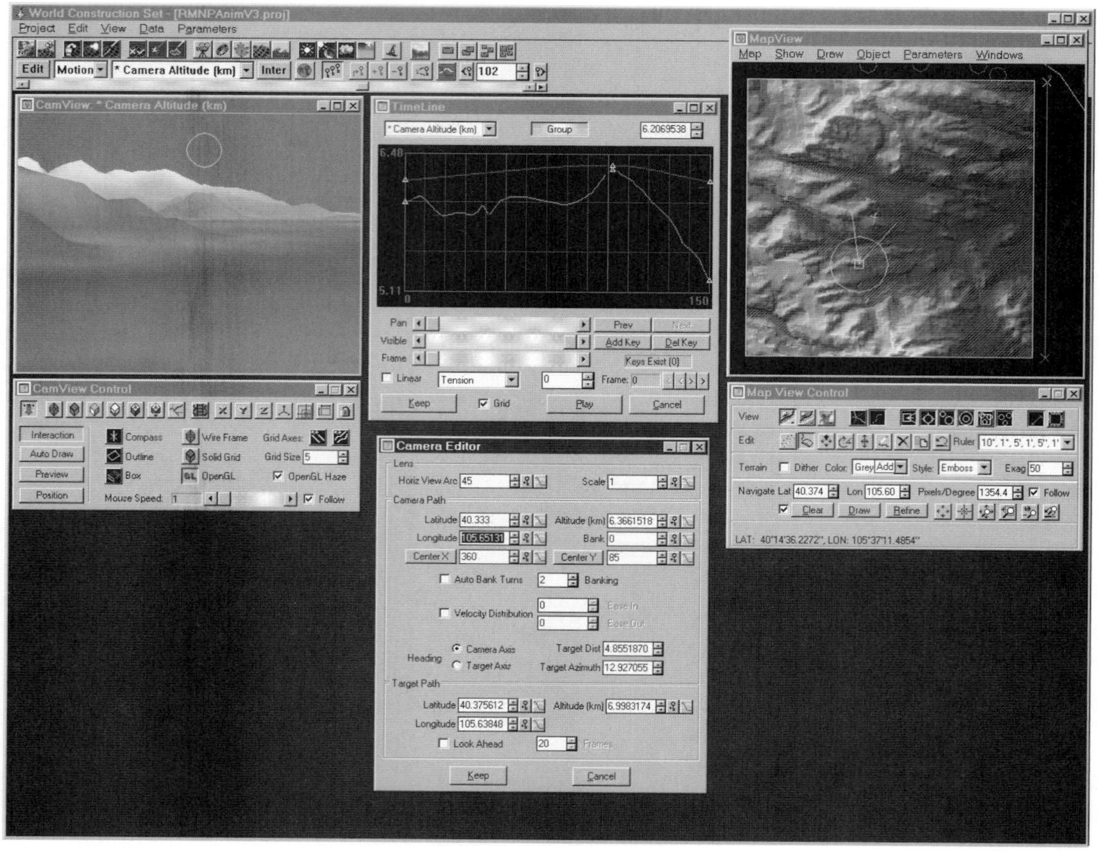

FIGURE 9.31
The GUI of World Construction Set. The GUI of terrain visualization software packages sometimes have relatively sophisticated functions that allow the user to control the ways by which the terrain scenes are rendered. *(Courtesy: 3D Nature LLC, Arvada, CO)*

usually accept a variety of input formats, including ASCII text files, DEM, TIN, and DXF (Table 9.7). He also found that although the primary function of these packages is automated contouring, most of them have some kind of three-dimensions rendering capability as well. In general, however, these packages have only limited GIS analysis capabilities. Not many of them, for example, are able to delineate watersheds from digital terrain data.

9.6 Applications of Digital Terrain Models

In the last several years, there has been a tremendous growth in the application of DTMs, not only in the traditional fields of geography, surveying and mapping, and earth and environmental sciences, but also in landscape design, biodiversity analysis, environmental impact analysis and site selection for telecommunication facilities (e.g., radio, television, and mobile telephone transmission towers). This is partially due to the increasing availability of digital terrain data from government agencies, academic and research institutions, as well as commercial geographic data suppliers. At the same time, advancing computer technology has helped popularize the use of DTMs. As noted, using a DTM is both data- and computation-intensive. Until relatively recently, it was simply impossible to implement DTM except by using powerful mainframes, mini-computers, and workstations. Rapid developments in hardware and software technologies in the last few years have made DTM widely affordable in the PC environment.

Several application domains may be identified for DTM: surveying and mapping (including cartography and photogrammetry), hydrology and fluvial geomorphology, geoscience, civil engineering, resource planning and management, as well as military operations. This list is certainly not exhaustive, and the boundaries between some of the application domains are not always clear. However, it provides a convenient framework to summarize the status of using DTM in different fields of science and technology.

Table 9.7
Examples of Contouring Software Packages

Product Name/ Vendor	*Host Software*	*Platform*	*Typical Input Format*	*3-D Rendering*	*Shaded Relief*	*GIS Analysis**
3-D Mapper	MapInfo	PC	ASCII points, DEM, DLG, DFX	Yes	Yes	Limited
IsoMap	Geographix Exploration System	PC	ASCII points, DLG	Yes	Yes	Limited
Surfer for Windows		PC	ASCII points, DEM, DLG, DFX	Yes	Yes	Limited
TroCon		PC	ASCII points, DEM, DFX	No	No	Limited
QuickSurf	AutoCAD	UNIX workstation, PC	ASCII points, DEM, DFX	Yes	Yes	Limited
SiteWorks	AutoCAD Microstation	UNIX workstation, PC	ASCII points, DEM, DFX	Yes	Yes	Yes
Vertical Mapper	MapInfo	PC	ASCII points, DFX	Yes	No	Limited

*GIS analysis functions include watershed delineation, flow lines, slope, aspect, ridge-line detection, watershed generation, as well as volume and surface area calculation
Source: Thoen, 1996

9.6.1 SURVEYING AND MAPPING APPLICATIONS

Cartographers have a long tradition of and interest in using terrain data to represent relief in the form of hill shading, slope profiles, and block diagrams (Imhof, 1982; Clarke, 1995). The development of the principles and techniques in DTM, together with the advent of advanced graphics-oriented hardware and software technologies, has provided cartographers an extremely powerful set of tools to depict relief more quickly and effectively and in more innovative ways. Cartographic representation of the terrain may now be done interactively and dynamically. It may also incorporate a vast amount of data that are not traditionally used, such as remote sensing and digital photographic imagery. DTM use is therefore not only a production tool but also a data integration tool in modern cartography.

The DEM and TIN models are standard data structures in digital cartography. The adoption of the DTM approach has drastically changed the ways land surveyors and photogrammetrists collect elevation data for the production of contour maps. In the past, the acquisition of elevation data and the interpolation of contour lines were labor-intensive, time-consuming, and error-prone processes. With the aid of DTM, high-quality contour maps may now be produced more quickly and economically. Errors that occurred during the data acquisition and map production processes may also be detected more easily when the data are examined visually in three dimensions.

A growing use of DEM is in the production of orthophotographs and orthoimages. The different methods of production have already been explained in Section 8.5.1. DEMs provide the terrain elevation information needed for differential rectification of digital aerial photographic images using an analytical or digital approach, thus greatly facilitating the production of orthophotographs and orthoimages. Orthophotographs and orthoimages may be displayed directly on a computer screen for use in heads-up digitizing or as a background for the draping of other types of spatial information, such as land use and land cover. They may also be used to drive a *photowriter* that prints them onto an unexposed file to create a negative, from which a print of the orthophotograph may be produced.

9.6.2 HYDROLOGICAL AND GEOMORPHOLOGICAL APPLICATIONS

The growing awareness of the deficiencies of the old methods of hydrologic analysis and the increasing availability of digital computers since the 1960s has led to the rapid development of numerical methods in hydrology and fluvial geomorphology. According to Moore et al. (1991), the period from about 1960 to 1975 was the era of "hydrologic modeling," in which mathematical descriptions of fluvial processes were developed and incorporated into hydrological models. Most of these models were concerned with predicting water quantities (e.g., runoff volumes and discharge) at a catchment or subcatchment outlet. These models were described as "lumped parameter models," which means that there was very little or no consideration for spatially variable processes and catchment characteristics.

The decade between 1975 and 1985 witnessed a change in the emphasis of hydrologic modeling. The growing concern with the environment, including management of pollution, resulted in the development of what have been commonly known as "transport models." These models, using the hydrologic models developed in the 1960s as the flow component, were perceived as the best way to predict water pollution. Just like their earlier counterparts, transport models also poorly accounted for the effects of space and topography on catchment hydrology.

Since the mid-1980s, however, there has been an increasing recognition of the need to predict spatially variable hydrologic processes at a fine resolution. This has led to the era of "spatial modeling" in the science of hydrology. Digital terrain data and remote sensing catchment characteristics (e.g., vegetation cover) are now considered as crucial data input to the new generation of hydrologic and water quality models. Different forms of digital terrain data (i.e., DEM-based, TIN-based, and contour-based) are used in different models to provide the spatial component of the analysis. An excellent review of typical examples of topologically-based hydrological and geomorphological models may be found in Moore et al. (1991). The issues of using GIS in modern hydrologic modeling (including status, data structure, and the effective coupling between GIS and hydrologic models) have been explained in detail by Maidment (1993).

9.6.3 GEOSCIENTIFIC APPLICATIONS

Practitioners in geoscience (i.e., geologists, geophysicists, mining engineers, and hydrogeologists, etc.) have a long tradition of using graphical techniques for terrain representation and analysis. These techniques, which include surface maps, multiple layers, fence diagrams, and stereoimages, have been largely limited to three-dimensional displays in the two-dimensional format of a computer screen or hard-copy output from a plotter (Van Driel, 1989). GIS and DTM together have provided geoscientists with the tools necessary to improve their productivity. The impact of DTM on geoscientific data processing may be explained from these two somewhat interrelated perspectives.

- *Data Management and Integration.* Geoscientific modeling usually requires the use of data from a variety of sources. Many applications require nonstandard solutions to database access and three-dimensional representation because of the range of data types, collection specifications, spatial data structures, and access keys, as well as the complex geometry of the subsurface domains. The adoption of GIS and DTM techniques appears to have provided the data structure and data management tools for the formulation of a more unified approach to geoscientific data processing. It has also provided the mechanism for integrating data obtained in different ways, at different scales, and with different cartographic specifications.
- *Data Analysis and Visualization.* Manual processing and analysis of data represented a major constraint in geoscientific investigations in the past. Individual geoscientific data sets are characteristically incomplete in nature. The creation of graphics from numerical data was both time-consuming and error-prone. The ability to rapidly create, retrieve, and integrate geological, geophysical, and related topographic data allows geoscientists to

understand the surface and subsurface environments by collaborating information from different sources. By using true three-dimensional visualization techniques, sometimes in real-time or near real-time situations, the ability of geoscientists to analyze problems and develop solutions may be greatly increased.

DTM techniques, by themselves, cannot solve geoscientific problems entirely. They must be interfaced with other forms of numerical models and analytical tools. However, it is obvious that the concepts and techniques of DTM now play a central role in geoscientific data processing. They support both the management of the spatial component of geoscientific data and the three-dimensional visualization tasks that until very recently have been notably lacking (Turner, 1989).

FIGURE 9.32
Three-dimensional visualization for evaluating highway design. *(Courtesy: Poitra Visual Communications, Denver, CO)*

9.6.4 ENGINEERING APPLICATIONS

Civil engineers have always had considerable interest in the use and development of DTM (Petrie and Kennie, 1991). In fact, many of the DTM software packages we noted earlier are oriented toward engineering applications. In general, engineering applications of DTM make use of digital terrain data of a large scale, are site-specific, and have an emphasis on precision of measurement. Digital terrain data for engineering applications are, therefore, more likely acquired by field surveying and photogrammetric methods rather than by map digitizing.

Three major areas can be identified for engineering applications of DTM (Craine, 1990).

- *Design.* In engineering projects, contouring is the perfect method for checking irregularities and smoothness in designed surfaces. The objective of using DTM is to identify the geometric skeleton of the engineering design and to create combined horizontal and vertical alignment strings that define the major features or structures.
- *Analysis Including Computation.* DTM software packages, particularly those developed for engineering applications, contain a range of functions for the analysis of digital terrain data. Typical functions include calculation of areas, volumes, slopes, and aspects; generation of longitudinal and cross sections; computation of setting-out information; triangulation of surfaces and contouring; and the delineation of watersheds and viewsheds.
- *Visualization.* This function allows the engineer to evaluate the design before it is finalized. Both wire-line and hidden-line perspective views can be used. More sophisticated packages allow the draping of image data to produce photorealistic views. The use of visualization is particularly popular in road design (Figure 9.32). Many highway design systems now incorporate terrain visualization capabilities. These allow the design to be subjectively evaluated and refined for safety and visual intrusion into natural environment along a highway corridor.

9.6.5 NATURAL RESOURCE MANAGEMENT APPLICATIONS

DTM plays a significant role in natural resource planning and management. As the relative magnitudes of many biophysical processes that occur in the natural landscape are sensitive to topographic characteristics (e.g., slope and aspect), digital terrain data are central to climatic and biological modeling pertaining to natural resources. DTM in this particular domain may be applied at a global or local scale. It may also be applied in the operational environment as well as in predictive modeling. In the past, these applications were seriously hampered by the lack of digital terrain data. With the increasing availability of global and national digital terrain data, biophysical models for natural resource inventory, analysis, and planning may now be developed using landscape processes as a basis.

Digital terrain data are used in these landscape process-based models to estimate solar radiation, moisture, and nutrient regimes that are highly dependent on the topographic attributes of elevation, slope, and aspect. As vegetation diversity and biomass production are closely related to these biophysical parameters, digital terrain data indirectly contribute to the estimation of the distribution and variation of land cover over space. Additionally, as animal habitats are governed by vegetation types and topography, digital terrain data are essential for habitat supply analysis and modeling in natural resource management. The use of DTM in spatial biophysical modeling has significantly enhanced the ability of resource managers and planners to develop more realistic strategies for the sustainable use of natural resources. Besides, as DTM is carried out mainly in the interactive mode, it helps to create a dynamic modeling environment that is more akin to the dynamic nature of the forest and wildlife that we endeavor to characterize, understand, and manage.

Environmental assessment (EA) is an integral component of natural resource management (Wray et al., 1994). An EA is an in-depth evaluation of the impact of human activities on the environment. In many jurisdictions, EA is governed by legislation and is mandatory prior to the approval of proposed activities or undertakings concerning natural resources. EA is usually performed by identifying alternatives to the proposal and alternative methods of carrying out the proposal. It provides the opportunity for public input by presenting these alternatives in open hearings. DTM plays a crucial role in developing EA alternatives. Digital terrain data are required for locating and estimating soil erosion (such as in the Universal Soil Loss Equation, or USLE), water quality (such as in Agriculture Non-Point Source Pollution System, or AGNPS, or BASINS), and wildlife habitats. They are also used for two-dimensional and three-dimensional visibility analysis that allows the public and the governing authorities to visually examine the impact of the proposed undertakings on the "values" of existing resources and facilities. Clear-cuts in timber harvest, for example, will negatively affect the values of tourist resorts if visible from the resorts. A buffer zone where no timber harvest is allowed must then be created to provide visual protection. Determining a visually acceptable buffer zone was a very laborious task in the past. With the use of DTM, it is not only relatively easy to generate one buffer zone as necessary, but to produce a number of alternatives for decision makers to choose from.

There is now a trend in natural resource management toward more detailed design using small treatment areas scattered across a large landscape (Berry et al., 1998). There is also growing public concern about the stewardship of the environment and natural resources. These factors have led to tighter guidelines for natural resource management strategies. This has in turn necessitated the adoption of more accurate planning and evaluation procedures in the decision-making process. In this regard, DTM is an indispensable tool for natural resource management because it allows the resource planners to work at a relatively detailed level of the landscape, typically at the watershed scale. It also provides an extremely powerful set of tools to visualize the impacts of resource management strategies before and after their implementation by the method of simulation.

9.6.6 MILITARY APPLICATIONS

The military establishments in many countries are important producers and users of digital terrain data and DTM applications. This accounts to a great extent for the intimate involvement of military mapping agencies in the specification of digital terrain data and the production of DEMs, such as the Shuttle Radar Topography Mission (SRTM) mentioned in Section 9.4.4. Initially, most military applications of DTM were oriented toward the automation of contour map production. New computer and military engineering technologies have greatly expanded the scope of DTM applications to include a wide variety of application areas ranging from relatively simple battlefield management to sophisticated missile guidance as summarized here (Griffin, 1991):

- *Visibility Studies.* This is probably the most conventional and commonly used military application of DTM. The purpose is to provide military personnel the means to assess and determine the observer-to-target intervisibility in battlefield planning and management.
- *Unmanned Vehicle Guidance.* Unmanned vehicles may be used in the battlefield for a myriad of purposes, such as surveillance, mine laying or clearing, rescue operations, tactical and terrain surveys, and offensive action. These vehicles must interact with and navigate over the terrain. DTM plays a crucial part in the research and development and the actual operation of military unmanned vehicles.
- *Artificial Intelligence in Route Planning.* DTM may be used in conjunction with onboard artificial intelligence (AI) software in vehicles or missiles for route planning purposes. The route planning software is usually based on various heuristic algorithms that can define routes with reference to the stored digital terrain data. This allows military vehicles and missiles to find the optimal routes over "unprepared" terrain.

FIGURE 9.33
Application of three-dimensional terrain visualization in combat pilot training. Virtual world construction programs are now commonly used to construct multimedia applications designed to train military pilots in the identification of military vehicles, including animated aircrafts, ground vehicles, and ships. To view this photo in color, see the color insert. *(Courtesy: Stage 22 Imaging, Winter Park, FL)*

- *Pilot Training.* Terrain visualization using the principles and techniques of DTM is now widely used in the training of military pilots in the rehearsal of battlefield operations. This is done by overlaying animated aircraft, ground vehicles, ships, and other military structures on the terrain model in a variety of realistic settings (Figure 9.33).

9.7 Summary

Until relatively recently, DTM has been commonly regarded as a special application of GIS that is of interest to a particular group of users only. Increasing availability of digital terrain data and rapid advances of computer technology in the last several years have greatly expanded the utilization of DTM in geographic data processing. The concepts and techniques of DTM are now used not only for the representation and analysis of topographical data, but also for other forms of surface-based geographic phenomena, such as climate, vegetation, environmental pollution, and distribution of socioeconomic activities. As we have demonstrated in this chapter, DTM is certainly one of the most important applications of geographic data processing today.

DTM may be carried out using two basic data models: DEM and TIN. These two data models are not mutually exclusive. Instead, they are always used to complement one another in practice. It is essential for GIS users to understand how DEM and TIN are constructed, to know their relative merits and limitations, and to be able to decide when one model should be used in favor of the other in different DTM applications. These two data models are closely associated with the techniques of digital terrain data acquisition. Although we covered all methods of terrain data collection in relatively great detail, we placed special emphasis on the use of digital photogrammetry and the Global Positioning System (GPS) because of their growing importance. At present, digitizing of existing contour maps is still the primary way of obtaining digital terrain data at the continental and national levels. However, as the demands for high-quality digital terrain data at larger scales grow, the use of digital photogrammetry and GPS increases correspondingly. Orthophotographs have become an important source of data for GIS, but their production requires the use of DTMs to provide accurate terrain elevation information for differential rectification. The production of orthophotographs and orthoimages is closely related to digital photogrammetry.

GIS users should be familiar with the concepts and techniques of digital terrain data processing, particularly those pertaining to the creation of new data sets, interpolation (notably, the method of kriging), and visualization (notably, animation and fly-by). Users should also be familiar with software resources, including the types and functionality of software packages from commercial vendors, government agencies, and educational institutions. Conventionally, DTM software products have been developed independently of GIS. Many of these special-purpose terrain modeling software packages are available. However, most GIS software vendors have now made DTM functions available in their products in the form of software extensions. As a result, DTM may now be carried out conveniently within the GIS environment.

The concepts and techniques of DTM have been drawn largely from computer graphics and those disciplines associated with the application domains noted earlier. This does not mean, however, that DTM is not a worthwhile topic area in the context of GIS research and development. There are many problem areas of DTM that deserve further investigations from the perspective of geographic data processing. One of the most imminent problems is the integration between DTM and geographic data processing. Although terrain modeling and visualization may now be done in most GIS, many special-purpose DTM software packages are still unable to accept terrain data directly from GIS. Another imminent problem is concerned with the quality and standards of digital terrain data. Without proper mechanisms to assure and quantify the quality of digital terrain data in use, the validity of the results obtained by DTM will inevitably be compromised. Besides, there are also considerable research opportunities in the integration of terrain visualization with spatial decision support, as well as the impacts of new computer graphics techniques, such as the VRML and virtual reality modeling, on the future development of DTM. Terrain modeling has become an indispensable tool of GIS.

REFERENCES

Adkins, K. F., and Merry, C. J. (1994) "Accuracy assessment of elevation data sets using the Global Positioning System," *Photogrammetric Engineering and Remote Sensing,* Vol. 60, No. 2, pp. 195–202.

Al-Rousan, N., Cheng, P., Petrie, G., Toutin, Th., and Valadan Zoej, M. J. (1997) "Automated DEM extraction and orthoimage generation from SPOT Level 1B imagery," *Photogrammetric Engineering and Remote Sensing,* Vol. 63, No. 8, pp. 965–874.

Band, L. E. (1989) "A terrain-based watershed information system," *Hydrological Processes,* Vol. 3, pp. 151–162.

Berry, J. K., Buckley, D. J., and Ulbricht, C. (1998) "Visualize realistic landscapes: 3-D modeling helps GIS users envision natural resources," *GIS World,* Vol. 1, No. 8, pp. 42–47.

Bolstad, P. V., and Stowe, T. (1994) "An evaluation of DEM accuracy: elevation, slope, and aspect," *Photogrammetric Engineering and Remote Sensing,* Vol. 60, No. 11, pp. 1327–1332.

Bonham-Carter, G. F. (1994) *Geographic Information Systems for Geoscientists: Modelling with GIS,* Kidlington, U.K.: Elsevier Science Limited.

Burrough, P. A. (1986) *Principles of Geographical Information Systems for Land Resources Assessment,* Oxford: Clarendon Press.

Burrough, P. A., and McDonnell, R. A. (1998) *Principles of Geographical Information Systems,* Oxford: Oxford University Press.

Chen, Z.-T., and Guevara, J. A. (1987) "Systematic selection of very important points (VIP) from digital terrain models for constructing triangular irregular networks," *Proceedings,* AutoCarto 8, pp. 50–56, Baltimore, MD.

Clarke, K. C. (1995) *Analytical and Computer Cartography,* Englewood Cliffs, N.J.: Prentice Hall.

Craine, G. S. (1990) "Application of the MOSS system to civil engineering projects," in *Terrain Modelling in Surveying and Civil Engineering,* by Petrie, G., and Kennie, T. J. M. eds. pp. 217–226, New York: McGraw-Hill.

Cressie, N. (1988) "Spatial prediction and ordinary kriging," *Journal for the International Association for Mathematical Geology,* No. 20, pp. 405–421.

Cressie, N. (1990) "The origin of kriging," *Journal for the International Association for Mathematical Geology,* No. 22, pp. 239–252.

De Floriani, L., Falcidieno, B., Nagy, G., and Pienovi, C. (1984) "A hierarchical structure for surface approximation," *Computer and Graphics,* No. 8, pp. 183–193.

De Floriani, L., Falcidieno, B., Pienovi, C., Allen, D., and Nagy, G. (1986) "A visibility-based model for terrain features," *Proceedings,* Second International Symposium on Spatial Data Handling, pp. 235–250, Seattle, WA.

Divis, D. A. (2000) "SA no more. GPS accuracy increases 10 fold," *Geospatial Solutions,* Vol. 10, No. 6, pp. 18–20.

ESRI (1994) *ARC TIN—Surface Modeling Extensions for ArcInfo GIS,* an ESRI White Paper, Redlands, CA: Environmental Systems Research Institute, Inc.

Foley, J. D., van Dam, A., Feiner, S. K. and Hughes, J. F. (1990) *Computer Graphics: Principles and Practice*, 2nd ed. Reading, MA: Addison-Wesley.

Fowler, R. J., and Little J. J. (1979) "Automated extraction of irregular network digital terrain models," *Computer Graphics*, No. 13, pp. 199–207.

Gittings, B. (1996) *Digital Elevation Data Catalogue*, Department of Geography, University of Edinburgh, Edinburgh, Scotland. **http://www.geo.ed.ac.uk/home/ded.html**

Goovaerts, P. (1997) *Geostatistics for Natural Resources Evaluation*, New York: Oxford University Press.

Griffin, M. W. (1991) "Military applications of digital terrain models," in *Terrain Modelling in Surveying and Civil Engineering*, by Petrie, G. and Kennie, T. J. M. eds, pp. 277–289, New York: McGraw-Hill.

Helpke, C. (1995) "State-of-the-art of digital photogrammetric workstations for topographic applications," *Photogrammetric Engineering and Remote Sensing*, Vol. 61, No. 1, pp. 49–56.

Hutchinson, M. F. (1988) "Calculation of hydrologically sound digital elevation models," *Proceedings*, Third International Symposium on Spatial Data Handling, pp. 117–133, Sydney, Australia.

Hutchinson, M. F. (1989) "A new procedure for gridding elevation and stream line data with automatic removal of spurious pits," *Journal of Hydrology*, No. 106, pp. 211–232.

Hutchinson, M. F. (1993) "Development of a continent-wide DEM with applications to terrain and climate analysis," in *Environmental Modeling with GIS*, by Goodchild, M. F., Parks, B. O., and Steyaert, L. T. eds. pp. 392–399, New York: Oxford University Press.

Hutchinson, M. F. (1998) *ANUDEM Version 4.6.2* (Last update, April 1998), Canberra, Australia: The National University of Australia. **http://cres.anu.edu.au/software/anudemtxt.html**

Imhof, E. (1982) *Cartographic Relief Presentation*, New York: W. de Gruyter.

Jacobson, R. (1994) "Virtual world capture spatial reality," *GIS World*, Vol. 7, No. 12, pp. 36–39.

Kennie, T. J. M. (1991) "Field data collection for terrain modelling," in *Terrain Modelling in Surveying and Civil Engineering*, by Petrie, G., and Kennie, T. J. M. eds. pp. 4–16, New York: McGraw-Hill.

Kennie, T. J. M. and McLaren, R. A. (1991) "Visualization for planning and design," in *Terrain Modelling in Surveying and Civil Engineering*, by Petrie, G. and Kennie, T. J. M. eds. pp. 252–261, New York: McGraw-Hill.

Krige, D. G. (1951) A Statistical Approach to Some Mine Valuations and Allied Problems at the Witwatersrand. Master's Thesis, University of Witwatersrand, South Africa.

Lam, N. S. N. (1983) "Spatial interpolation methods: A review," *The American Cartographer*, Vol. 10, No. 2, pp. 129–149.

Laurini, R., and Thompson, D. (1992) *Fundamentals of Spatial Information Systems*, London: Academic Press.

Lee, J. (1989) "A drop heuristic conversion method for extracting irregular networks from digital elevation models," *Proceedings*, GIS/LIS '89, Vol. 1, pp. 30–39, Orlando, FL.

Lee, J. (1991) "Comparison of existing methods for building triangular irregular network models of terrain from grid digital elevation models," *International Journal of Geographical Information Systems*, Vol. 5, No. 3, pp. 267–285.

Lee, J. (1994) "Digital analysis of viewshed inclusion and topographic features on digital elevation models," *Photogrammetric Engineering and Remote Sensing*, Vol. 60, No. 4, pp. 451–456.

Maas, H.-G., and Kersten, T. (1997) "Aerotriangulation and DEM/Orthophoto generation from high-resolution still-video imagery," *Photogrammetric Engineering and Remote Sensing*, Vol. 63, No. 9, pp. 1079–1084.

Maidment, D. R. (1993) "GIS and hydrologic modeling," in *Environmental Modeling with GIS*, by Goodchild, M. F., Parks, B. O., and Steyaert, L. T. eds. pp. 147–167, New York: Oxford University Press.

Makarovic, B. (1973) "Progressive sampling for digital terrain models," *ITC Journal*, 1973-3, pp. 397–416.

Makarovic, B. (1977) "Composite sampling for DTMs," *ITC Journal*, 1977-3, pp. 406–433.

Mark, D. M. (1984) "Automated detection of drainage networks from digital terrain models," *Cartographica*, No. 21, pp. 168–178.

Mark, D. M. (1987) "Recursive algorithms for the analysis and display of digital elevation data," *Proceedings*, First Latin American Conference on Computers in Cartography, pp. 375–397, San Jose, Costa Rica.

Mark, D. M. (1988) "Network models in geomorphology," in *Modelling Geomorphological Systems*, by Anderson, M. G. ed. pp. 73–79, Chichester, UK: John Wiley.

Mather, P. M. (1976) *Computational Methods of Multivariate Analysis in Physical Geography*, New York: Wiley.

MCVL (Multimedia Communications and Visualization Laboratory) (No date) *Fly-by Mesh Simplification*, Department of Computer Engineering and Computer Science, University of Missouri, MO. **http://meru.cecs.missouri.edu/mvl/flyby/simplification/index.html**

Miller, C., and La Flamme, R. A. (1958) "The digital terrain model—theory and applications," *Photogrammetric Engineering*, Vol. 24, No. 3, pp. 433–442.

Mitas, L., and Mitasova, H. (1999) "Spatial interpolation," in Longeley, P. A., Goodchild, M. F., Maguire, D. J., and Rhind, D. W. eds. *Geographical Information Systems, Vol. 1: Principles and Technical Issues*, 2nd ed. New York: John Wiley, pp. 481–492.

Moore, I. D., Grayson, R. B., and Ladson, A. R. (1991) "Digital terrain modelling: A review of hydrological, geomorphological and biological applications," *Hydrological Processes*, Vol. 5, pp. 3–30.

Newman, W. M., and Sproull, R. F. (1979) *Principles of Interactive Computer Graphics*, New York: McGraw-Hill.

NRCan (1997) *Standards and Specifications—Canadian Digital Elevation Data*, Sherbrooke, QC: Centre for Topographic Information, Natural Resources Canada. **http://www.ccg.nrcan.gc.ca/ext/html/english/products/cded/stdcded.html**

Oliver, M. A., and Webster, R. (1990) "Kriging: a method of interpolation for geographical information system,"

International Journal of Geographical Information Systems, Vol. 4, No. 3, pp. 313–332.

Petrie, G. (1981) "Hardware aspects of digital mapping," *Photogrammetric Engineering and Remote Sensing*, Vol. 47, No. 3, pp. 307–320.

Petrie, G. (1991a) "Modelling, interpolation and contouring procedures," in *Terrain Modelling in Surveying and Civil Engineering*, by Petrie, G., and Kennie, T. J. M. eds. pp. 112–127, New York: McGraw-Hill.

Petrie, G. (1991b) "Photogrammetric methods of data acquisition for terrain modelling," in *Terrain Modelling in Surveying and Civil Engineering*, by Petrie, G., and Kennie, T. J. M. eds. pp. 26–48, New York: McGraw-Hill.

Petrie, G. (1991c) "Terrain data acquisition and modelling from existing maps," in *Terrain Modelling in Surveying and Civil Engineering*, by Petrie, G., and Kennie, T. J. M. eds. pp. 85–111, New York: McGraw-Hill.

Petrie, G. (1991d) "Modelling, interpolation and contouring procedures," in *Terrain Modelling in Surveying and Civil Engineering*, by Petrie, G., and Kennie, T. J. M. eds. pp. 112–127, New York: McGraw-Hill.

Petrie, G., and Kennie, T. J. M. eds. (1991) *Terrain Modelling in Surveying and Civil Engineering*, New York: McGraw-Hill.

Peucker, T. K. (1972) *Computer Cartography*, Commission on College Geography Resource Paper 17, Washington, DC: Association of American Geographers.

Peucker, T. K., and Douglas, D. H. (1975) "Detection of surface specific points by local parallel processing of discrete terrain elevation data," *Computer Graphics and Image Processing*, No. 4, pp. 375–387.

Polis, M., and McKeown D. (1992) "Iterative TIN generation from digital elevation models," in *Proceedings*, IEEE Conference on Computer Vision and Pattern Recognition, pp. 787–790, Urbana-Champaign, IL.

Rempel, R., and Goadsby, J. M. (1992) "An evaluation of GPS under Boreal forest canopy cover," *Proceedings*, Geographic Information Seminar, Ontario Ministry of Natural Resources, pp. 164–172, Toronto, ON.

Rhyne, T.-M. (1999) "A commentary on GeoVRML: A tool for 3D representation of georeferenced data on the web," *International Journal of Geographical Information Sciences*, Vol. 13, No. 4, pp. 439–443.

Thoen, B. (1996) "Exploring the third dimension—a comparative review of contouring products for Windows 95," GIS World, Vol. 9, No. 12, pp. 51–59.

Tsai, V. J. D. (1993) "Delaunay triangulation in TIN creation: an overview and a linear-time algorithm," *International Journal of Geographical Information Systems*, Vol. 7, No. 6, pp. 501–524.

Turner, A. K. (1989) "3D GIS for hydrological applications," in *Three Dimensional Applications of Geographic Information Systems* by Raper, J. ed. pp. 115–127, London: Taylor and Francis.

Unwin, D. (1981) *Introductory Spatial Statistics*, London: Methuen.

USGS (1997a) *DEM Data User's Guide Version 5*, Reston, VA: United States Geological Survey.

USGS (1997b) *Standards for Digital Terrain Models*, National Mapping Program Technical Instructions, Reston, VA: United States Geological Survey. **http://rockyweb.cr.usgs.gov/nmpstds/demstds.html**

USGS (1998) *The USGS Mini Image Processing System*, Reston, VA: United States Geological Survey. **http://TerraWeb.wr.usgs.gov/TRS/software/mips/**

Van Driel, I. N. (1989) "Three dimensional display of geologic data," in *Three Dimensional Applications of Geographic Information Systems* by Raper, J. ed. pp. 1–10, London: Taylor and Francis.

Weibel R., and Heller, M. (1991) "Digital terrain modelling," in *Geographical Information Systems, Volume 1: Principles*, by Maguire, D. J., Goodchild, M. F., and Rhind, D. W. eds. pp. 269–297, Harlow, U.K.: Longman Scientific and Technical.

Wray, D., Curlew, C., Glassford, P., and Kapron, J. (1994) "Utilizing spatial technologies to support the Megisan Lake area environmental assessment," in *Proceedings of the 17th Annual Geographic Information Seminar and Resource Technology '94 Symposium (Decision Support—2001)*, Vol. 2, pp. 1078–1089, Toronto, ON.

Yoeli, P. (1985) "The making of intervisibility maps with computer and plotter," *Cartographica*, Vol. 22, No. 3, pp. 88–103.

CHAPTER

10

SPATIAL ANALYSIS AND MODELING

10.1 Introduction

One of the criticisms of GIS is that because it was initially developed as a tool for the storage, retrieval, and display of geographic information, the capabilities for spatial analysis are limited (Rogerson and Fotheringham, 1994). What is termed spatial analysis and modeling in GIS is basically map data manipulation such as polygon overlay and buffering (Fischer et al., 1996). It may also refer simply to the spatial *summarization* of data. Spatial data are retrieved selectively for a region of interest. Basic summary statistics are then computed and mapped.

In this chapter, spatial analysis will be used in its original sense, which is the manipulation of spatial data into different forms in order to extract additional meaning (Fotheringham et al., 2000). The major concerns are to investigate the patterns in spatial data and to discover possible relationships between such patterns and other attributes within the study region. The ultimate goal is to model such relationships for the purpose of understanding and prediction (Bailey, 1994).

Fischer et al. (1996) suggested four major areas where statistical spatial data analysis techniques may strengthen current GIS practices: (1) sampling objects from the database and the choice of an adequate spatial scale of analysis, (2) data rectification to compare variables for a different and incompatible set of zones, (3) exploratory spatial data analysis aiming to explore and exploit the GIS database to arrive at new insights and an understanding of the data characteristics, and (4) confirmatory or explanatory spatial data analysis concerned with systematic analysis of data and hypothesis testing based on some specified assumptions.

Because the GIS environment has been regarded as data-rich but theory-poor (Openshaw, 1990), it is advantageous to develop general data exploratory procedures in spatial analysis to help explore and exploit the GIS database to arrive at new insights. These procedures include the search for data characteristics such as trends, spatial outliers, spatial patterns, and associations (Scholten and LoCascio, 1997) and will be discussed in more detail in this chapter.

It should be noted that linking the spatial analytical functions to standard GIS may be achieved by three major approaches (Goodchild, 1992): (1) full integration of the spatial analysis procedures within the GIS, (2) close coupling between the spatial analysis software and GIS by sharing the same data structures, and (3) loose coupling in which an independent spatial data analysis module relies on a GIS for its input data and for functions such as graphic display via the import and export of the data in a common format. The first approach is not yet realized although Idrisi for Windows developed by Professor Ron Eastman of Clark University has integrated a number of useful spatial statistical procedures in the program, which are further enhanced in Idrisi32 (Eastman, 1997, 1999). The second approach is more realistic, but it requires a good user interface in the linkage. The S+ SPATIALSTATS is an add-on module to the S-PLUS system for data analysis and graphics. It develops specially written commands to transfer spatial data back and forth between ArcInfo and S-PLUS (Kaluzny et al., 1998). The third approach is used most often and has the advantage of using fully the advantages of the GIS and the statistical packages. It is quite common to use statistical packages such as SAS or SPSS or Minitab with the ArcInfo program for this purpose (Griffith, 1989). In the following, the discussion of the spatial analysis procedures will also indicate which type of linking approach has been used.

10.2 Descriptive Statistics

As an exploratory tool, spatial analysis facilitates visualization of the geographic data one variable at a time as maps, histograms, bar charts, box plots, or pie charts. The method of visualization chosen is a function of both the data and the objective of a particular analysis. The purpose is to discover outliers in the spatial data. Obviously, a map is the most effective way to reveal any spatial clustering of similar or dissimilar values, or a pattern of randomness. These are closely related to the concept of spatial autocorrelation, already briefly mentioned in Chapter 4, Section 4.4.3 in connection with the evaluation of attribute accuracy and in Chapter 9, Section 9.5.2 in connection with a discussion on kriging. For geographic data in interval and ratio scales, descriptive statistics may enhance the visualization by summarizing and revealing the degree of complexity of the geographic data under study. These descriptive statistics include measures of central tendency and dispersion. All these descriptive statistical techniques are fully integrated into the GIS.

10.2.1 Central Tendency

The three commonly used measures of central tendency are the *mean*, the *median*, and the *mode*, applicable only to data in interval or ratio scales. The mean can be found by summing all the values under one variable in the geographic data and divided by the total number of values. The median may be found by first arranging all the values in order. The middle value is the median if the set contains an odd number of values. In the case of an even number, the mean of the two central values is the median. The median is superior to the mean in that it is not affected by extreme values. The mode is the most frequently occurring value. The mean, median, and mode may also be found for grouped data in which the individual values are grouped into a number of classes and the frequency of occurrence in each class recorded (Griffith and Amrhein, 1991).

10.2.2 DISPERSION

Dispersion allows the spread in values of the geographic data and the pattern of the spread to be obtained. Commonly, three measures are used: standard deviation, skewness, and kurtosis.

Standard deviation (σ) is the most important measure of dispersion, and may be obtained from the following formulas:

$$m = \frac{\sum_{i=1}^{N} X_i}{N} \tag{10.1}$$

$$x_i = X_i - m \tag{10.2}$$

$$\sigma = \sqrt{\frac{\sum_{i=1}^{N} x_i^2}{N - 1}} \tag{10.3}$$

where x_i is the difference between the mean (m) and the individual values (X_i), and N is the total number of values. The square of standard deviation, or σ^2, is the variance.

Skewness measures the degree of asymmetry in a frequency distribution around a measure of central tendency and is expressed in terms of the relationship of the mean to the median divided by the standard deviation, or

$$\text{skewness} = (\text{mean-median})/\sigma \tag{10.4}$$

When the median is greater than the mean, the skewness is negative, indicating that the tail of the distribution extends closer to the small values. When the median is smaller than the mean, the skewness is positive, suggesting that the tail of the distribution extends at the larger value end. When skewness has a value of 0, the distribution is symmetrical.

Kurtosis indicates the peakedness or the spread of the frequency distribution of the geographic data variable and can be computed from the following formula:

$$\text{Kurtosis} = \frac{\sum_{i=1}^{N} (x_i)^4/N}{(\sigma^2)^2} \tag{10.5}$$

If kurtosis is less than 3, it suggests a pointed curve; if it is bigger than 3, a flat or less peaked curve; and if the value equals 3, it is a normally distributed curve.

10.3 SPATIAL AUTOCORRELATION

In addition to the descriptive statistics mentioned in Section 10.2, which shed light on the nature of the statistical distribution of the geographic data, spatial autocorrelation examines the spatial ordering of the geographic data. In addition, it explores the spatial covariance structure in attribute data. Therefore, spatial autocorrelation deals with both the attributes and locations of spatial features (Goodchild, 1986). In other words, spatial autocorrelation indicates whether adjacent or neighboring values in the geographic data vary together, and if so, how. Spatial autocorrelation statistics make it possible to measure interdependence in a spatial distribution and to use formal statistical methods to test hypotheses about spatial interdependence (Odland, 1988). It is important to note that spatial autocorrelation is affected by the scale of the spatial pattern. The same spatial pattern at different scales may produce different spatial autocorrelation results.

In general, spatial autocorrelation is present when similar values cluster together on a map. Spatial autocorrelation allows use of formal statistical procedures to measure the dependence among nearby values in a spatial distribution, test hypotheses about geographically distributed variables, and develop statistical methods of spatial patterns (Odland, 1988). There are two commonly used measures of spatial autocorrelation: Geary's index (c) and Moran's index (I).

10.3.1 GEARY'S INDEX (C)

This index was developed by Geary (1968) as a measure of spatial autocorrelation for area objects with interval attributes. As such, it is a suitable measure for use in the analysis of data aggregated by statistical reporting zones (such as census tracts). The index measures the similarity of *i*'s and *j*'s attributes, c_{ij}, which can be calculated as the squared difference in value. Thus,

$$c_{ij} = (z_i - z_j)^2 \tag{10.6}$$

where z_i is the value of the attribute of interest for object *i*; z_j is the value of the attribute of interest for object *j*; *i*, *j* are any two of the objects measured on an interval scale (Goodchild, 1986).

A locational similarity measure (the similarity of *i*'s and *j*'s locations), w_{ij}, was also used by Geary who defined it in a binary fashion, with $w_{ij} = 1$ if *i* and *j* shared a common boundary, and $w_{ij} = 0$ if not. Geary's index (c) is expressed as follows

$$c = \frac{\sum_i \sum_j w_{ij} c_{ij}}{2 \sum_i \sum_j w_{ij} \sigma^2} \tag{10.7}$$

where σ^2 is the variance of the attribute z values, or

$$\sigma^2 = \frac{\sum_i (z_i - \bar{z})^2}{(n - 1)} \tag{10.8}$$

where

$$\bar{z} = \frac{\sum_i^n z_i}{n}$$

The value of c computed will be largest if large values of w_{ij} coincide with large values of c_{ij}. If the value of $c = 1$, the attributes are distributed independently of location. If the value of $c < 1$, similar attributes coincide with similar locations. If the value of $c > 1$, attributes and locations are dissimilar.

The measure of the similarity of attributes (c_{ij}) given in Equation (10.6) is for interval data only. However, it is possible to design ways to measure the similarity of attributes in a manner appropriate to the data measurement scale used. For nominal data, the usual approach is to set $c_{ij} = 1$ if i and j have the same attribute value, and $c_{ij} = 0$ if not the same. For attributes in ordinal scale, similarity is based on a comparison of the ranks of i and j.

10.3.2 MORAN'S INDEX

Another measure of spatial autocorrelation was provided by Moran (1948), which has the advantage of giving a more logical result, with positive value implying that nearby areas tend to be similar in attributes, negative value implying dissimilar; a zero value indicates uncorrelated, independent, and random arrangement of attribute values. Moran's index (I) also involves the use of c_{ij} and w_{ij}, as explained in the section on Geary's index, and is defined as follows

$$c_{ij} = (z_i - \bar{z})(z_j - \bar{z}) \tag{10.9}$$

and

$$I = \frac{\sum_i \sum_j w_{ij} c_{ij}}{s^2 \sum_i \sum_j w_{ij}} \tag{10.10}$$

where s^2 denotes the sample variance,

$$s = \frac{\sum_i (z_i - \bar{z})^2}{n}$$

The w_{ij} terms represent the spatial proximity of i and j and can be computed in any suitable way, as explained in Section 10.3.1.

Both the Geary and Moran indices as developed were for area objects, but they may be equally applied to points, lines, and raster objects, provided an appropriate method can be developed to measure the spatial proximity of a pair of objects (w_{ij}). For point objects, one can compute distances between pairs of points and use the inverse distance weighting to compute similarity. Another approach is to transform the points into areas by partitioning the study area into Thiessen polygons (see Chapter 9, Section 9.3.2). For line objects, if the lines represent links between nodes with attributes, c_{ij} will measure the similarity of the attributes of each pair of nodes while w_{ij} will measure the links between them. We can set $w_{ij} = 1$ if a direct link exists between i and j, and 0 if otherwise. On the other hand, if the links carry attributes, w_{ij} will be a measure of proximity between two links and c_{ij} will measure the attribute similarity between the links. Finally, for raster data, w_{ij} may be defined in terms of boundary sharing. If a pair of raster cells share a common boundary, $w_{ij} = 1$ and 0 otherwise.

Spatial autocorrelation is fully integrated in Idrisi for Windows with a module called AUTOCORR, which computes the first lag spatial autocorrelation of an image using Moran's I. It is not available in ArcInfo, but may be loosely coupled to vector GIS. Ding and Fotheringham (1991) developed a spatial analysis module, SAM, for use in ArcInfo to compute spatial autocorrelation. The S+ SPATIALSTATS package may be coupled to ArcInfo (and ArcView) to compute the Moran and Geary statistics of spatial autocorrelation (Kaluzny et al., 1998).

Spatial autocorrelation is typically applied to examine spatial distribution patterns of phenomena. A good example is provided by Can (1993) in research on residential quality assessment in the city of Syracuse, New York. To construct the residential quality score, Can employed the following seven variables to capture the demographic dimensions of socioeconomic variations: (1) percentage of non-white persons, (2) percentage of female-headed single-parent households with children under 18 present, (3) percentage of occupied housing units that lack complete plumbing facilities, (4) percentage of occupied housing units that have 1.01 or more persons per room, (5) percentage of vacant housing units, (6) the median value of specified owner-occupied housing units, and (7) the median contract rent of specified renter-occupied housing units. These data were obtained at two spatial scales: census tracts and census block groups. The seven variables at each spatial scale were subjected to factor analysis, and standardized component scores were obtained. The first component accounted for 75% of total variation at the census tract level and 68% at the census block group level. The components were mapped (Figure 10.1). Visual inspection indicates a clustering of the densest shading (disadvantaged) stretching out from the city

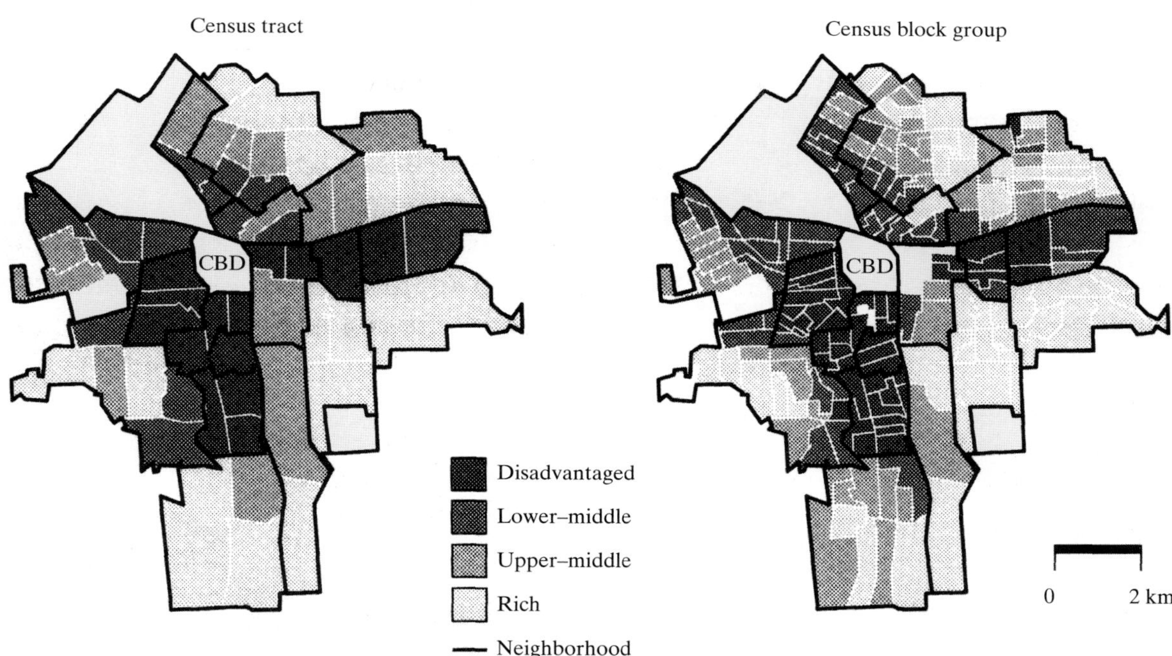

FIGURE 10.1
Spatial patterns of principal components scores of residential quality in the city of Syracuse, New York, at the census tract (left) and census block group (right) levels. *(Source: Can, 1993)*

center to the south and east of the city. However, an important question arises: Which spatial level should be used for this type of analysis? The level of aggregation in the data collected can obscure the underlying spatial pattern. To find the answer, Can calculated Moran's I to measure the extent of spatial clustering among census tracts and census block groups with respect to residential quality scores determined by factor analysis. The w_{ij} term (spatial proximity) was calculated with the aid of ArcInfo data files for coverage: Arc Attribute Table (AAT) and Polygon Attribute Table (PAT). The AAT file provides contiguity information (based on its topology), while the PAT file contains the attribute for spatial autocorrelation statistics. The results of Moran's I for the two spatial scales were 0.5101 (with a sample variance of 0.0082205) for the census tract level and 0.7552 (with a sample variance of 0.0023878) for the census block group level. The conclusion was that the extent of clustering was much stronger at the census block group level than at the census tract level. Aggregation at the census tract level would cause bias. This led to the conclusion that the census block group should be used in the formation of neighborhoods. Spatial autocorrelation has found a lot of application in epidemiological research (Mayer, 1983; Gatrell and Loytonen, 1998) and in the analysis of the spatial pattern of errors in the classification of remotely sensed images to determine whether these errors are spatially clustered (Congalton, 1988).

10.4 Quadrat Counts and Nearest-Neighbor Analysis

In a spatial distribution made up of points, it is necessary to describe its pattern. In real-life situations, these points could be birds' nests, persons, trees, cities, or galaxies. The spatial pattern is the result of a spatial process, which reveals both physical and socioeconomic forces at work. A spatial pattern of points may be described as regular or irregular. Under each type, one can further subdivide into *clustered, random,* or *anticlustered* (Figure 10.2) (Cole and King, 1968). There are several approaches to the analysis of the spatial pattern of points. The best-known ones are the quadrat counts and nearest-neighbor analysis. In this type of analysis, the main concern is with the location characteristics of the points rather than their attributes.

10.4.1 Quadrat Counts

The quadrat counts require covering the spatial pattern with a quadrat, or a sampling area of any consistent size and shape, which can be circular, rectangular, square, or so forth (Getis and Boots, 1978; Thomas, 1977). Usually, square quadrats are used. The number of points contained in each quadrat is counted, and the

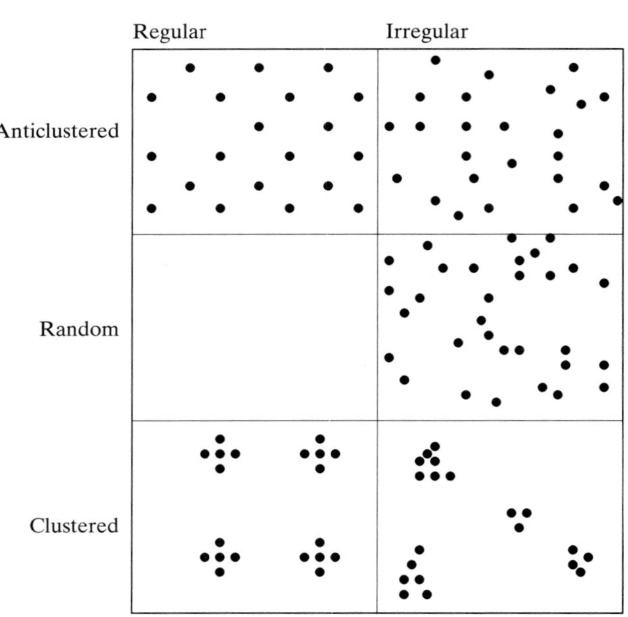

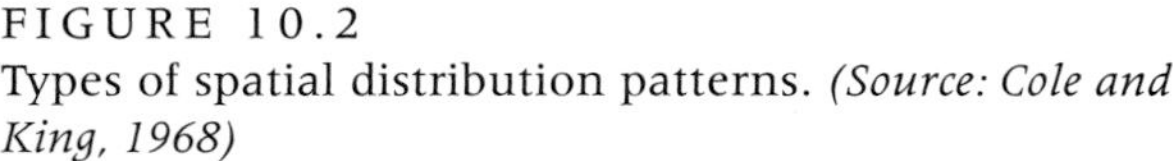

FIGURE 10.2
Types of spatial distribution patterns. *(Source: Cole and King, 1968)*

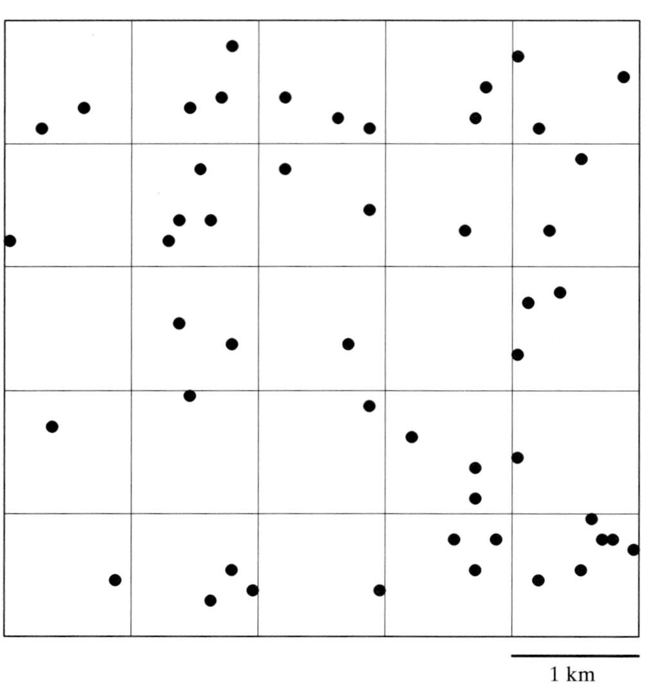

FIGURE 10.3
A lattice of 25 square quadrats containing 50 points placed by a Poisson process. *(Source: Getis and Boots, 1978. Reprinted with the permission of Cambridge University Press)*

observed frequencies of occurrence of points is tabulated (Table 10.1 and Figure 10.3). To determine whether the spatial pattern observed is random, the following Poisson distribution model is used:

$$P(x; \lambda) = \frac{e^{-\lambda}\lambda^{x}}{x!} \quad \textbf{(10.11)}$$

for, $x = 0, 1, 2, \ldots$

where x is a random variable whose values are the number of points per quadrat and λ is the mean number of points per quadrat, which can be estimated by counting the total number of points in the study area divided by the total number of quadrats used to cover the area, that is, an estimate of the mean number of points per quadrat. The Poisson distribution model is used to compute the expected frequency of

Table 10.1
Computation of the Poisson Process Model Based on the Observed Frequency of Points in Figure 10.3

No. of Points per Quadrat		*Observed Frequency*		*Expected Frequency*	*Expected Proportion*
0		{2		{3.375	0.135
1	11*	{9	10.15*	{6.775	0.271
2		5		6.775	0.271
3		{7		{4.500	0.180
4	9*	{1	8.075*	{2.250	0.090
5		{0		{0.900	0.036
≥6		{1		{0.425	0.017
N = Total		**25**		**25**	**1.000**
$\chi^2 = 0.642$		$df = 1$		$\chi^2_{0.20} = 1.642$	

*Aggregate value by combining frequency less than 5 into one single category
Source: Getis and Boots, 1978. Reprinted with the permission of Cambridge University Press.

occurrence of points (Table 10.1). A chi-square (χ^2) one-sample test is then applied to compare the observed and expected frequencies of points per quadrat as follows:

$$\chi^2 = \sum_{i=1}^{k} \frac{(O_i - E_i)^2}{E_i} \quad \textbf{(10.12)}$$

where O_i is the observed number of cases in the ith category and E_i is the expected number of cases in the ith category of the model, and k is the total number of categories. It is important to note that each category should not have a value less than 5. If this occurs, two categories have to be collapsed together to form one single category. The number of degrees of freedom of the chi-square (χ^2) test = (k − 1). The number of degree of freedom of the chi-square (χ^2) test = (k − 1), where k is the number of comparisons made. However, an additional degree of freedom is lost for the use of the parameter λ in calculating the expected frequencies. Therefore, in reality, the number of degree of freedom should be $k - 2$ for the chi-square (χ^2) test. The hypothesis that there is no difference between the observed and expected frequencies is tested. If the calculated χ^2 value falls below the χ^2 tabulated figures for a predetermined level of significance, the two sets of frequencies are statistically in accordance (i.e., a good fit), and the hypothesis of no difference is accepted. The implication is therefore that the spatial pattern was produced by a Poisson process (i.e., random). Table 10.1 illustrates the calculation based on quadrat counts in Figure 10.3 conducted by Getis and Boots (1978).

The main problem of the quadrat counts method in mapped-point pattern analysis is that the size of the quadrat may affect the results of the χ^2 test. According to ecologists, a rule of thumb to determine an appropriate quadrat size is twice the study area divided by the total number of points. Some geographers feel that for geographical studies, this is too large, and they advocate that an ideal quadrat size should be equal to the study area divided by the total number of points.

The quadrat counts method is implemented in Idrisi for Windows for raster data employed in GIS. However, it computes the mean number of points per cell (quadrat), the variance in cell totals, the variance : mean ratio, the degrees of freedom, the t statistic and significance level in comparing the variance : mean ratio with a value of 1.0. Variance : mean ratio values close to 1.0 suggest a random point pattern, although this can be substantiated only by checking for the absence of autocorrelation using the spatial autocorrelation module (AUTOCORR). Values smaller than 1.0 suggest an anticlustered pattern while values greater than 1.0 suggest a clustered pattern.

Quadrat analysis has found applications in both social and physical areas. There are two types of point patterns that quadrat analysis will deal with: the contagious processes in human populations and the structural features of the cultural and physical landscapes in homogeneous regions (Thomas, 1977). The study of diseases, innovation diffusion, and social contacts on voting behavior all involve the contagious processes. Structural applications include analysis of human settlement pattern, karst depressions in limestone regions, factory and shop distributions in urban areas, distribution of herb remains in savannas after burns, dispersal of seeds, the number of diseased trees per hectare, the number of sightings of birds, and many others. Following is an example carried out by one of the authors to determine the settlement pattern in the Baoding area of the North China Plain as revealed by the 1 : 500,000 scale Shuttle Imaging Radar-A (SIR-A) image (Lo, 1984, 1986) (Figure 10.4). The image was enlarged to 1 : 130,000 scale to facilitate the identification and digitization of settlements (SIR-A being a film product). The digitized coordinates of the settlements were

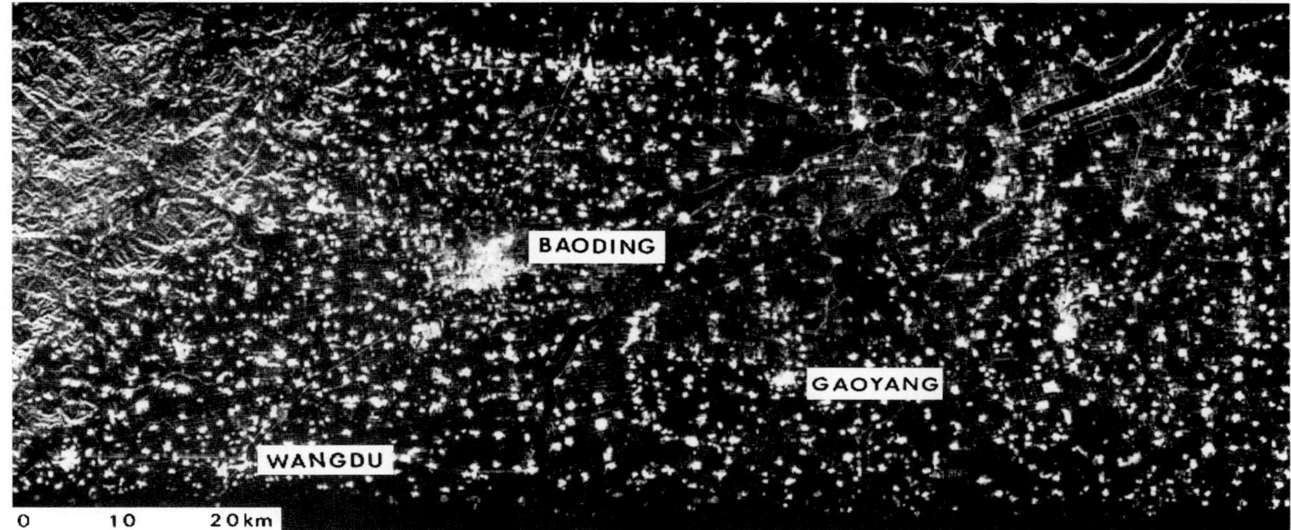

FIGURE 10.4
Shuttle Imaging Radar-A image of the Baoding area in Hebei Province located in the heart of the North China Plain. The strong backscatter clearly reveals the settlements. *(Source: Lo, 1986a)*

Table 10.2

Results of Quadrat Analysis and Nearest-Neighbor Analysis of the Baoding Area in the North China Plain

	Poisson Distribution		*Negative Binomial Distribution*	
Points per Quadrat	*Expected Frequency*	*Observed Frequency*	*Expected Frequency*	*Observed Frequency*
0	167.6	305	270.3	305
1	319.4	319	296.5	221
2	304.4	233	225.6	233
3	193.4	172	146.3	172
4	92.1	104	86.7	104
5	35.1	54	48.5	54
6	11.2	20	26.0	20
7	3.0	15	13.5	15
8	0.7	3	6.9	3
Total	1126.9	1127	1120.3	1127
Total χ^2 235.1			Total χ^2 36.3 Value of D 0.030 Critical D value at 0.2 level 0.032	

Note: Number of points = 2148, Number of quadrats = 1127, Area of map (km^2) = 7129.3, Mean density of points (per quadrat) = 1.906, Expected nearest-neighbor = 0.697, Observed nearest-neighbor = 0.687, Nearest-neighbor statistics = 0.986

Source: Lo, 1984

transformed to the map coordinates of the 1:250,000 scale topographic map of this part of China. The direct linear distances between these settlements were computed using the Pythagorean theorem (see Equation 5.8 in Section 5.3.3). A total of 2148 settlement points were obtained, and quadrats of 2.5 × 2.5 km dimension were interpolated by the computer. The results of the quadrat analysis are shown in Table 10.2. It was found that the negative binomial distribution seemed to fit the pattern much better than the Poisson distribution. The implication is that the settlement pattern for the Baoding area of the North China Plain was produced by a mixed random-clustering process. Because of the homogeneous nature of the terrain and the rural nature of these settlements, a generalized process was postulated, which implied that the clustering of the settlements was the result of some basic affinity among them. In research conducted by Davies et al. (1999), this affinity could be the ordering of settlements in a hierarchy following Christaller's Central Place Theory (1966).

10.4.2 NEAREST-NEIGHBOR ANALYSIS

To avoid the problem of quadrat size determination in quadrat counts, nearest-neighbor analysis is the alternative approach for mapped-point pattern analysis. This method may be easily integrated in both vector and raster GIS because it makes use of distance measurements between points to generate the statistic.

Nearest-neighbor analysis measures distances between sample points and their nearest neighboring points. The mean of the nearest-neighbor distance measurements (observed) for the study area is then compared with the expected mean distance. To derive the Poisson probability distribution for mean nearest-neighbor distance, one assumes that the sample area is now a circle with radius r, so that its area $a = \pi r^2$. In Equation 10.11, $e^{-\lambda}$ is the probability that a randomly chosen area of size a will contain no points. This is the same as the proportion of distances to nearest neighbor greater than or equal to r (Getis and Boots, 1978). The mean of r, called $E(r)$, is the expected mean distance to nearest neighbor in a Poisson process generated pattern (Clark and Evans, 1954:

$$E(r) = \frac{1}{2\sqrt{\lambda}} \tag{10.13}$$

where λ is the number of points per unit area (density). The standard error of $E(r)$ is given as:

$$\sigma_r = \frac{0.26136}{\sqrt{N\lambda}} \tag{10.14}$$

where N is the number of distance measurements made. To test the significance of departure from the Poisson model, the following equation is used:

$$Z = \frac{\bar{r} - E(r)}{\sigma_r} \qquad \textbf{(10.15)}$$

where Z is the standard variate of the normal curve and $\bar{r}$ is the observed mean distance to nearest neighbor. The rejection region of a hypothesis of no difference at 0.975 level of confidence is greater than or equal to 1.96.

The measurement of distances to nearest neighbor may take two approaches: (1) sampled location to nearest point and (2) point to nearest point (Figure 10.5). The "sampled-location-to-nearest-point" approach randomly selects a location in the study area and measures the distance to the nearest point (Figure 10.5a). The "point-to-nearest-point" approach randomly selects one point from all the points and measures the distance to the nearest neighbor (Figure 10.5b). Each point has the equal chance of being selected. For the nearest-neighbor method, there are a number of problems to note. Because of the need to know the density of points in the study area, the size of the area relative to the number of points is important and will affect the nearest-neighbor value. A cluster of points in a large area will produce low point density, while the same cluster of points in a small area will produce a high point density. To avoid this problem, the sample area under study must be positioned exactly within the total pattern of points. Another problem is the edge effect. Around the edge of the study area, there may be other points outside the boundary of the study area. The nearest-neighbor distances measured around the edge will be biased. Therefore, nearest-neighbor distances should be made to points even outside the boundary of the study area. Another approach is to eliminate all distance measurements to nearest-neighbor points outside the boundary of the study area before carrying on with the computation.

In applying nearest-neighbor analysis in the GIS environment, there is a shorthand measure called the K function developed by Ripley (1976) that captures the spatial dependence between different regions of the point process. It is defined as

$$K(d) = \lambda^{-1} E\left[\begin{matrix} \textit{number} \\ \textit{of points} \end{matrix} \leq \begin{matrix} \textit{distance d of} \\ \textit{an arbitrary point} \end{matrix}\right], d \geq 0,$$

where λ is the density of points in an area, and $E[\,]$ denotes the expectation. The advantage of the K function is that the theoretical value for $K(d)$ is known for several useful models of spatial point processes. The K function for a homogeneous process with no spatial dependence (complete spatial random) is πd^2. If there is clustering, there will be more points within short distances, leading to $K(d) > \pi d^2$. If the pattern is regular, $K(d) < \pi d^2$. The K function may be estimated with this formula:

$$\widehat{K}(d) = n^{-2}|A| \sum \sum_{i \neq j} w_{i,j}^{-1} I_d(d_{i,j}) \qquad \textbf{(10.16)}$$

where n is the number of points in region A with area $|A|$; $d_{i,j}$ is the distance between the ith and jth points; $w_{i,j}$ is the proportion of the circle with center at i and passing through j that lies within A; and $I_d(d_{i,j})$ is an indicator function, which is 1 if $d_{i,j} \leq d$. This estimate includes an adjustment for edge effects (Kaluzny et al., 1998).

The implementation of nearest-neighbor analysis in GIS should be straightforward because distances are used for proximity analysis by GIS. Distances from one

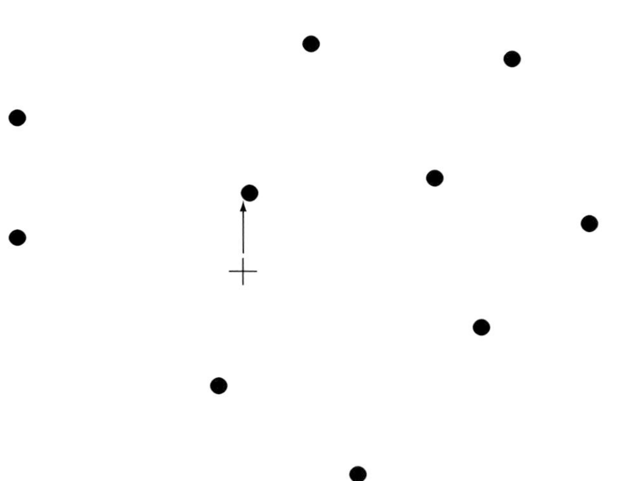

(a) Sample location to nearest point

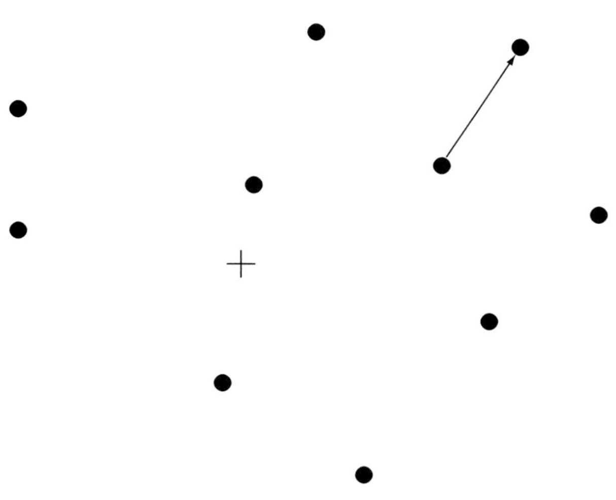

(b) Point to nearest point

FIGURE 10.5
Two types of distance measurement: + denotes a sample location. *(Source: Ripley, 1981)*

point to other sample points may easily be calculated. In the ArcInfo program, there is a command called ADDXY, which adds the X and Y coordinates to each point in a point coverage (after the topology has been correctly built). These coordinates will allow distances between points to be computed. Also, in ArcInfo, the command NEAR is suitable for the purpose of obtaining nearest-neighbor distance measurements. In ArcView, the Spatial Analyst extension has a distance mapping function that finds how far each cell is from the nearest source. A source is anything that the user selects. In Idrisi for Windows designed for raster data analysis, the command modules DISTANCE and ALLOCATE may be used to compute the nearest-neighbor distances. From all these measurements, the user still has to compute the required statistics to decide whether the spatial pattern is random, clustered, or anticlustered. The S+ SPATIALSTATS program has spatial analysis routines to compute a number of nearest-neighbor statistics for testing the hypothesis of complete spatial random (CSR) (G statistic, F statistic, and Ripley's K functions) in conjunction with the ArcInfo coverage in spatial-point pattern analysis (Kaluzny et al., 1998). Nearest-neighbor analysis may also be adapted for unsupervised image classification, as shown in PCI's K-Nearest Neighbor classifier.

Nearest-neighbor analysis may be easily applied to analyze settlement patterns, as illustrated by Lo's work (1984, 1986) in the North China Plain using Shuttle Imaging Radar-A (SIR-A) data (Figure 10.4), described in connection with quadrat analysis. For the study area of Baoding in the plain, it was found that the nearest-neighbor statistics [$E(r)$ in Equation 10.13] approached a value of 1, implying that the settlements could be described by a random distribution (Table 10.2). However, the nearest-neighbor statistic alone is insufficient to understand the process of settlement formation, and quadrat analysis should be conducted at the same time to shed more light on the process.

10.5 Trend Surface Analysis

Trend surface analysis has been mentioned in connection with spatial interpolation in Chapter 9 (see Section 9.5.2), and it is also a very useful data exploratory technique that will shed light on the spatial trend of the point data. Basically, this is the application of the general linear model to spatial data with the intention of discovering the regional and local trends. The technique was originally developed by geologists for oil prospecting (Davis, 1973). The trend revealed may be the underlying geological structure, topographic structure, broad trends of rainfall belts, or the pattern of housing and income surfaces, according to whether you are a geologist, a geomorphologist, a climatologist, or a population geographer, respectively. The general linear model takes the following form:

$$Z_n = b_0 + b_1X + b_2Y \quad \textbf{(10.17)}$$

where Z_n is the value at any point, X and Y are the rectangular coordinates, and b_0, b_1, and b_2 are the coefficients that describe the inclination of the plane or linear surface fitting the spatial variation of Z (Figure 9.17a). For a second-degree surface, or the linear plus quadratic surface, the following equation is used to fit the plane through the point data:

$$Z_n = c_0 + c_1X + c_2Y + c_3X^2 + c_4XY + c_5Y^2 \quad \textbf{(10.18)}$$

where all the c values are coefficients (Figure 9.17b). Thus the expansion of the trend surface to higher-order surfaces involves each variable raised to a higher level and the appropriate cross-product terms, such that the third-degree surface (or the linear plus quadratic plus cubic surface) is defined by the following equation:

$$Z_n = d_0 + d_1X + d_2Y + d_3X^2 + d_4XY + d_5Y^2 + d_6X^3 + d_7X^2Y + d_8XY^2 + d_9Y^3 \quad \textbf{(10.19)}$$

where all the d values are the coefficients describing the plane (Figure 9.17c). Higher-order surfaces of more complex form may be fitted, but it is increasingly difficult to comprehend the meaning of the plane fitted. It was observed that in practice, going beyond the cubic surface is rarely necessary because the quadratic and cubic surfaces provide good fit in most cases (Cole and King, 1968). Mapping the residuals from the fitted trend surface may shed useful light on the underlying spatial structure of the data. Trend surface analysis is one of the methods for spatial interpolation in the ArcView Spatial Analyst extension, thus making it fully integrated with GIS. It may also be loosely coupled to the GIS using S-PLUS, SAS, and SPSS programs. After the solution of the equations using these statistical analysis programs, the results may be imported back to the GIS for graphical display (scientific visualization).

10.6 Gravity Models

Gravity models are the most widely used types of spatial interaction models. Spatial interaction refers to any movement over space that results from a human process (Haynes and Fotheringham, 1984). Specifically, it refers to population migration, journey to work, information and commodity flows, and other phenomena involving the movement of people or goods or ideas from one place to another. The underlying concepts of the gravity models are borrowed from physics: Newton's law of gravity, which states that the force of attraction of two bodies is proportional to the product of their masses but inversely proportional to the squared distance separating them. Mathematically, it is shown as

$$F = \frac{m_i \times m_j}{d_{ij}^2} \tag{10.20}$$

where m_i and m_j are the two masses located at i and j respectively, and d_{ij} is the straight-line distance separating them. The analogy in human geography may be seen in the case of two cities with different population sizes, P_i and P_j (the mass). The two cities are separated by a distance, d_{ij}. The predicted interaction (T_{ij}) between the two cities may be expressed by the following formula:

$$T_{ij} = \frac{P_i P_j}{d_{ij}} \tag{10.21}$$

In other words, cities close to each other have greater interaction than cities further apart, and larger cities exert a greater influence than smaller ones.

The basic gravity model needs to be modified to the following form for application to spatial interaction analysis in the real world:

$$T_{ij} = k\frac{P_i^\lambda P_j^\alpha}{d_{ij}^\beta} \tag{10.22}$$

or alternatively,

$$T_{ij} = kP_i^\lambda P_j^\alpha d_{ij}^{-\beta} \tag{10.23}$$

In the real world, straight-line distances are not always the most appropriate distances to use. It is well known that the cost per mile of traveling may decrease with distance, as in air travel. Apart from the physical distances, there are other distances, such as time distance, economic distance, cognitive distance, and social distance (Gatrell, 1983). The effect of these different types of distance varies. In physics, the gravity law makes use of "squared" distance. The exponent of distance (β) is generally interpreted as the responsiveness of interaction to spatial separation. A larger exponent (β) for the distance implies an increased importance of friction of distance, which reduces the expected level of interaction between the two places. In countries where transportation is well developed, physical distance will have less effect in reducing interaction between the two places than in countries with poor transportation facilities. Therefore, the exponent term (β) is added to distance. Similarly, the sheer physical size of the population may not be the most relevant data to use in certain applications of the gravity model. For example, in shopping interaction, the average income level of the population at each location will be more relevant than the physical size. Therefore, exponents λ and α are added to P_i and P_j, respectively, in Equation 10.22. Finally, the constant, k, is added as a scalar to scale down large numbers used for P_i and P_j.

The gravity model may be further expanded and generalized so that it would include interactions among a set of centers in the system, not just between two centers. A second expansion is to include a set of variables other than just the population (P) variable at both the origin and destination. A third expansion is to include the impact of intervening opportunity and the agglomeration effect. The presence of an alternative destination (intervening opportunity) may reduce the interaction between the two centers. On the other hand, when opportunities cluster together, they generate an agglomeration effect and attract more flows from an origin, as seen in the case of shopping centers. Finally, the spatial structure of the origins and destinations matters. As a result, the gravity model is further generalized (Haynes and Fotheringham, 1984):

$$T_{ij} = f(V_i, W_j, S_{ij}) \tag{10.24}$$

where V_i represents a vector of origin attributes; W_j represents a vector of destination attributes; and S_{ij} represents a vector of separation attributes.

The gravity model in its various forms has been employed in migration study and market analysis. The following provides two examples to illustrate its use.

Example 1: Population migration analysis. The model given in Equation 10.22 may be applied to estimate the migration flows between two different places. This assumes the population size in each place is important. It is necessary to calibrate the model to determine the values for the parameters λ and α for the populations at the origin (v_i) and at the destination (w_j), respectively, and also what will be a meaningful measure of distance between the two places (d) and the proper power to raise ($-\beta$)? These parameters may be estimated only from an analysis of the history of previous migration flows between these two places. One should also note that these parameters may be different for the same place depending on whether it is the origin or the destination. Once these parameters are estimated, the number of people migrating from one place to the other is a matter of multiplying the population of the two places and the distance together (as modified by the appropriate parameter for each term).

Example 2: Determination of market area boundaries. The gravity model has been applied to retailing and marketing. The work of Reilly (1929) as modified by Converse (1949) becomes Reilly's law of retail gravitation. A city i's attractiveness for retailing (A) to individuals in city j is directly proportional to the population P at city i and inversely proportional to the square of the distance (d_{ij}) between city i and city j, as follows (compare Equation 10.22):

$$A_i = \frac{P_i}{d_{ij}^2} \tag{10.25}$$

Similarly, a city j with a population P_j is attractive for retailing to individuals in city i as follows:

$$A_j = \frac{P_j}{d_{ji}^2} \tag{10.26}$$

If there is an intermediate center x at distance d_{ix} and d_{jx} in proportion to the attractiveness of A_i and A_j, as computed in Equations 10.25 and 10.26, then

$$A_i = \frac{P_i}{d_{ix}^2} \tag{10.27}$$

and

$$A_j = \frac{P_j}{d_{jx}^2} \tag{10.28}$$

The point at which the attraction of each market center is equal is $A_i = A_j$. In other words,

$$\frac{P_i}{d_{ix}^2} = \frac{P_j}{d_{jx}^2} \tag{10.29}$$

Because $d_{ix} = d_{ij} - d_{jx}$, Equation 10.29 may be rewritten as

$$d_{ix} = \frac{d_{ij}}{1 + \sqrt{\frac{P_j}{P_i}}} \tag{10.30}$$

Putting back the parameters λ and α for the population at origin and destination respectively, and $-\beta$ for distance, a more general form of the equation is

$$d_{ix} = \frac{d_{ij}}{1 + \left(\frac{P_j^{\alpha}}{P_i^{\lambda}}\right)^{1/\beta}} \tag{10.31}$$

This equation will determine the location of the market boundary between i and j.

In Figure 10.6a, the market boundary around city C is to be determined. There are six other cities A, B, D, E, F, G, and H surrounding city C. The distances of these cities from each other as well as the population in each city are known. Using city C as the origin, Equation 10.31 may be applied to determine the market boundary between city C and A. This is then repeated for CB, CD, CE, CF, and CG. At the point where the market boundary is determined, a perpendicular line is drawn to the line connecting the origin with the destination. A market area for the city C is produced (Figure 10.6b).

The implementation of the gravity model in the GIS environment has the advantage that it can allow improved distance measures to be derived. As noted previously, there may be different types of distances, and the type of distance that reflects most realistically the situation should be used. Another advantage is the scientific visualization of the results, such as in the case of the market area boundary determination, using the graphical capabilities of GIS. A loose coupling approach is needed between the spatial interaction model and the GIS program. The ARC NETWORK is an add-on module to ArcInfo that allows the use of a set of commands to perform spatial interaction modeling (see Section 6.4.3). The Network Analyst extension to ArcView may also do the same.

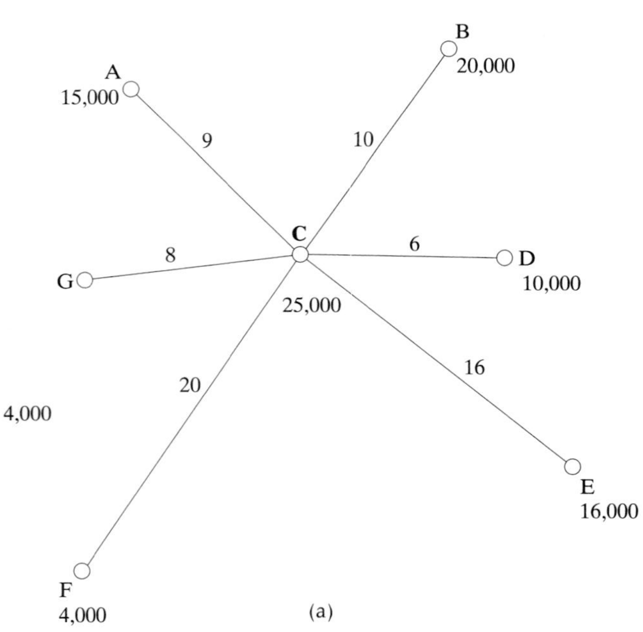

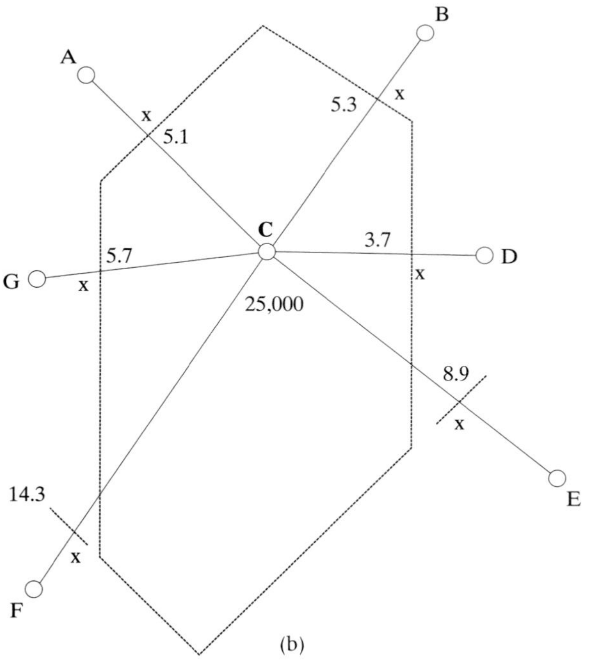

FIGURE 10.6
(a) City C and its linkages to other surrounding cities. *(Modified after Haynes and Fotheringham, 1984).* (b) Market area for city C. The numbers at the "x"s indicate the market boundaries of city C computed by applying Equation 10.28. *(Source: Haynes and Fotheringham, 1984)*

10.7 Network Analysis

Network analysis is closely related to spatial interaction modeling. A network is defined by Kansky (1963) as "a set of geographic locations interconnected in a system by a number of routes." A network refers to a system of lines topologically structured. Typical examples of networks are transportation lines and river systems. In analyzing networks, geographers' major concern is the characteristics of these networks, namely, how far the locations are from one another, whether the routes joining them are straight or curved, what commodity the network carries, whether the flow is continuous or intermittent, and so forth (Haggett and Chorley, 1969).

Networks may be reduced to topological graphs, which are arrays of points connected or not connected to one another by lines (Figure 10.7). This simplification facilitates the revelation of common topological structures of the networks. The following elements may be identified: nodes (vertices, v_1–v_7), links (edges, e_1–e_9), and regions (r_1–r_4). Connectivity matrices for these elements in binary form may be produced (Table 10.3). The number of edges (links) in the network (e), the number of vertices (nodes) in the network (v), and the number of isolated (i.e., nonconnecting) networks (subgraphs) (g) are employed to develop a series of topological measures to characterize the network structure (Haggett and Chorley, 1969; Kansky, 1963). It should be noted that an edge is defined by two nodes. There are two main groups of measures: (1) those based on gross characteristics and (2) those based on shortest-path characteristics. These measures allow a quantitative description of the network and a comparison of one network with another.

10.7.1 Topological Measures of Network Structure Based on Gross Characteristics

There are four useful measures in this group.

(1) cyclomatic number (μ) $= e - v + g$

The cyclomatic number measures the spatial structure of networks. Highly connected graphs have high cyclomatic numbers. In Figure 10.8, it is clearly seen that

Network

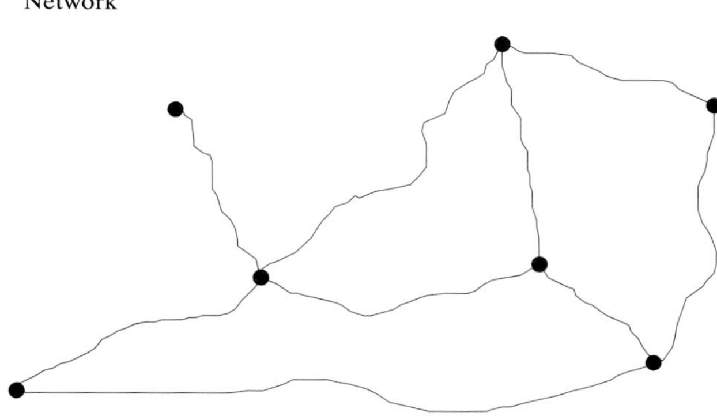

Graph

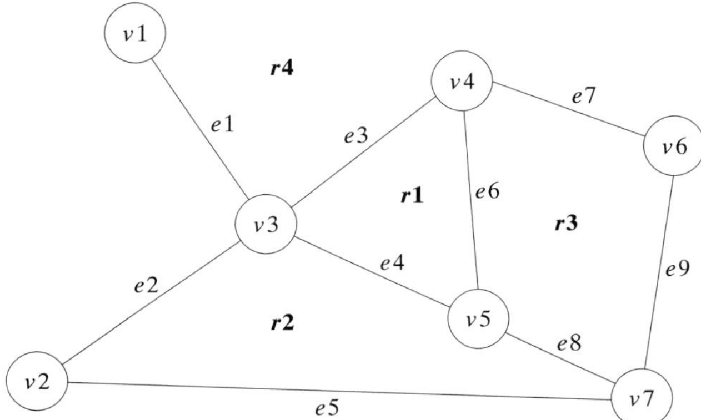

FIGURE 10.7
Reduction of a network to a graph. *(Source: Haggett and Chorley, 1969)*

Table 10.3
Connectivity Matrices Derived from the Network Graph in Figure 10.7

(a) *Vertices (v)*

	1	2	3	4	5	6	7
1	0	0	1	0	0	0	0
2	0	0	1	0	0	0	1
3	1	1	0	1	1	0	0
4	0	0	1	1	0	0	1
5	0	0	1	1	0	0	1
6	0	0	0	1	0	0	1
7	0	1	0	0	1	1	0

(b) *Regions (r)*

	1	2	3	4
1	0	1	1	1
2	1	0	1	1
3	1	1	0	1
4	1	1	1	0

(c) *Edges (e)*

	1	2	3	4	5	6	7	8	9
1	0	1	1	1	0	0	0	0	0
2	1	0	1	1	1	0	0	0	0
3	1	1	0	1	0	1	1	0	0
4	1	1	1	0	0	1	0	1	0
5	0	1	0	0	0	0	0	1	1
6	0	0	1	1	0	0	1	1	0
7	0	0	1	0	0	1	0	0	1
8	0	0	0	1	1	1	0	0	1
9	0	0	0	0	1	0	1	1	0

Key: 0 = not connected 1 = connected
Source: Haggett and Chorley, 1969

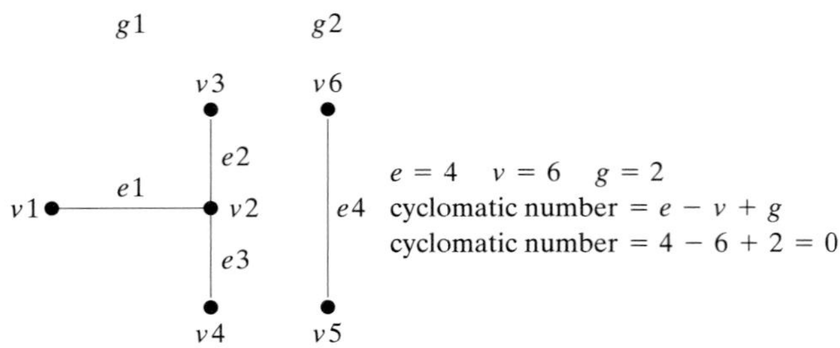

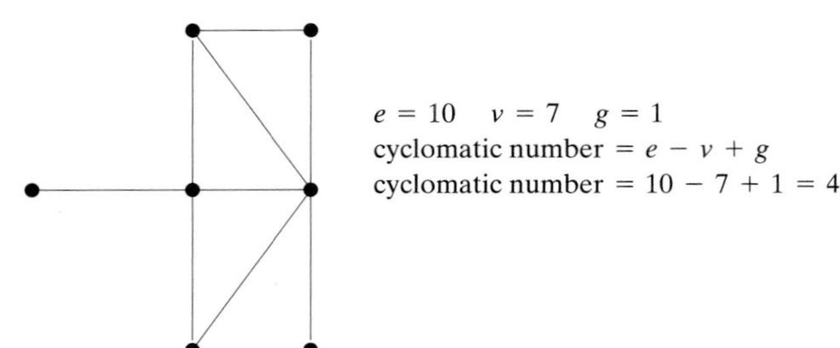

FIGURE 10.8
Computation of cyclomatic number of graphs. *(Source: Kansky, 1963)*

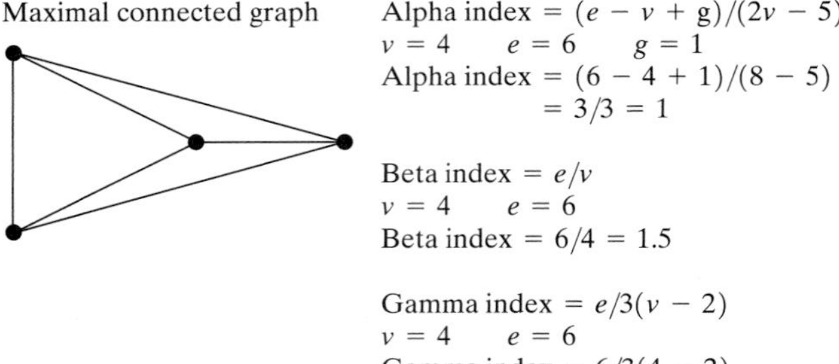

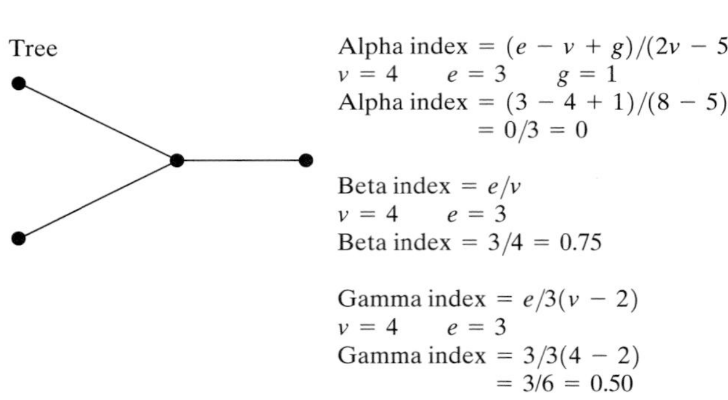

FIGURE 10.9
Computation of alpha, beta, and gamma indices. *(Source: Kansky, 1963)*

highly connected networks have higher cyclomatic numbers.

(2) alpha index $(\alpha) = [(e - v + g)/(2v - 5)]$
for planar graphs.
alpha index

$$(\alpha) = \{(e - v + g)/[(v(v - 1)/2) - (v - 1)]\}$$

for nonplanar graphs (i.e., three-dimensional).

Alpha index may be regarded as a ratio between the observed number of circuits and the number of maximum circuits in the networks. This index varies from zero to one. For completely interconnected networks (i.e., networks with the maximum number of edges), the alpha index is 1. By multiplying the alpha index by 100, the index value will vary from 0 to 100. It also measures the percentage of a maximum connectivity (Figure 10.9).

(3) beta index $(\beta) = e/v$

Beta index shows the relation between the edge and the vertex in the network. A more complicated network structure exhibits a higher beta index (Figure 10.9).

(4) gamma index $(\gamma) = e/3(v - 2)$ for planar graphs.

gamma index $(\gamma) = e/[v(v - 1)/2]$

for nonplanar graphs (i.e., three-dimensional).

The values of gamma index also vary from 0 to 1. Gamma index also measures the degree of connectivity of networks. Multiplying this index by 100 suggests the percentage connectivity. This index is independent of the number of vertices of a network. The value of 1 (or 100%) will be assigned to all completely connected networks whether they have 5 or 5000 vertices (Figure 10.9).

10.7.2 TOPOLOGICAL MEASURES OF NETWORK STRUCTURE BASED ON SHORTEST-PATH CHARACTERISTICS

There are three useful measures in this group:

(1) diameter (d) = maximum d_{ij}

where d_{ij} is the shortest path (in links) between *i*th and *j*th vertex (node). The diameter measures the maximum number of edges in the shortest path between each pair of vertices. In other words, it measures the longest shortest path in the network. For example, in one network it may take three edges to form the shortest path, while another network takes only two. The one with a diameter of 2 is better developed than the one with 3. One should note that two different graphs may exhibit the same diameter (Figure 10.10).

(2) dispersion index $= \sum_{i=1}^{v} \sum_{j=1}^{v} d_{ij}$

where *d* is the distance from a vertex *i* to a vertex *j*. This index, which measures the dispersion of the network, is the sum of a sum. First, the sum of all distances between a vertex *i* and all other vertices of the network

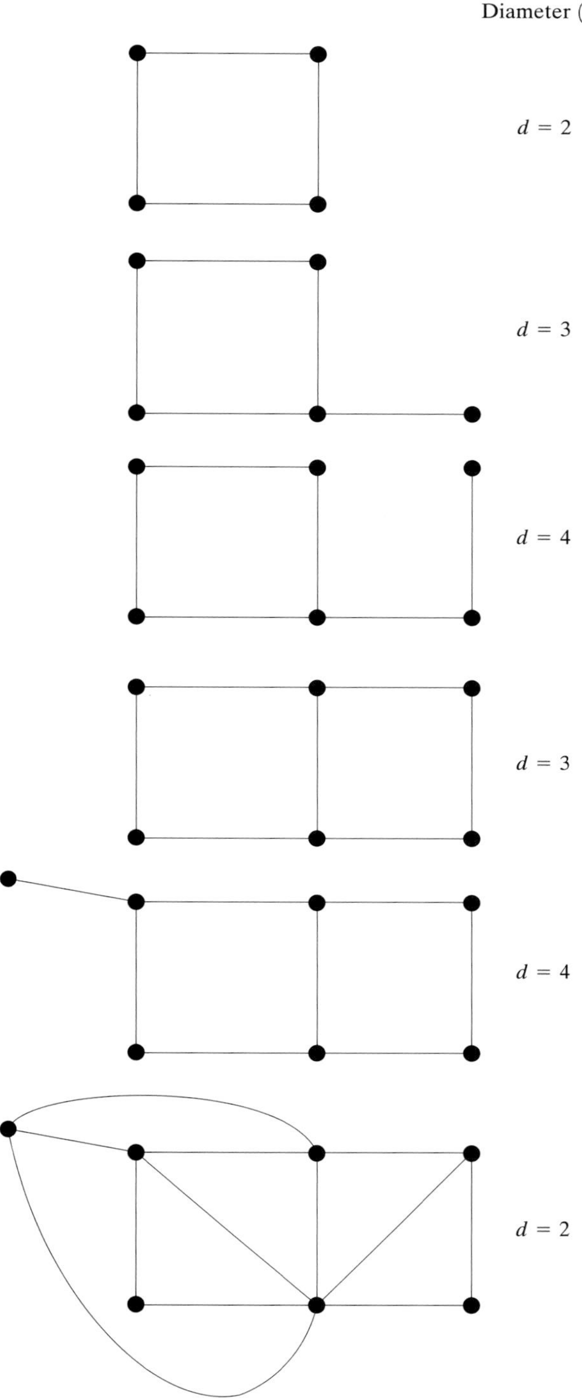

FIGURE 10.10
Diameter values of six different types of graphs. *(Source: Kansky, 1963)*

is determined. Then, all totals resulting from the first summation are added together. If the result of the first summation is large, the vertices of the network are far away from vertex i. The network is more dispersed. On the other hand, if the result of the first summation is small, the vertices are close to vertex i, and the network is more compact.

(3) accessibility index $= \sum_{i=1}^{v} d_{ij}$

This index with only one summation measures the accessibility of a vertex i to the network. Unlike the dispersion index, which measures the overall property of a network, the accessibility index measures the spatial relation between a given element of a structure and the remainder of the network.

10.7.3 SMEED'S INDEX

Closely associated with the shortest path and better connectivity is an index developed by Smeed (1968) that compares the efficiency of road networks. His main concern was congestion at peak hours of travel from the suburbs to the central business district (CBD). He made two assumptions: (1) The origins are equally distributed among the points at which the road leading into the CBD meets its boundary, and (2) the destinations are distributed either uniformly along the sides of the roads of the CBD or uniformly within the CBD (Haggett and Chorley, 1969). Assuming that the journey was made by the shortest route, Smeed measured the average lengths of journeys from the suburbs to the CBD for different types of street plans (Figure 10.11), each with the same area, and found that

$$d = 0.87\sqrt{A} \qquad \textbf{(10.32)}$$

with variation between $0.70\sqrt{A}$ and $1.70\sqrt{A}$, where d is the average length of journey (or average distance traveled along the road to the CBD from the suburb) and A is the area within which the journeys are measured.

From Equation 10.32, one can write Smeed's Index (SI) as follows:

$$SI = \frac{d}{\sqrt{A}} \qquad \textbf{(10.33)}$$

The smaller the value of SI, the more efficient is the road network. Hence in Figure 10.11, road plan (a) is the most efficient of all, while road plan (f) is the least efficient.

10.7.4 IMPLEMENTATION IN THE GIS ENVIRONMENT

For practical purposes, the shortest-path characteristics of the network have been implemented in vector GIS. A typical GIS can find the shortest route from one location to another based on real-world data (Figure 10.12). Addresses of locations have to be converted into point locations using geocoding. A digital highway map will provide the topological network structure to determine the routing. The ArcInfo GIS has a module called "network analysis," which provides a pathfinding program called "route" to model

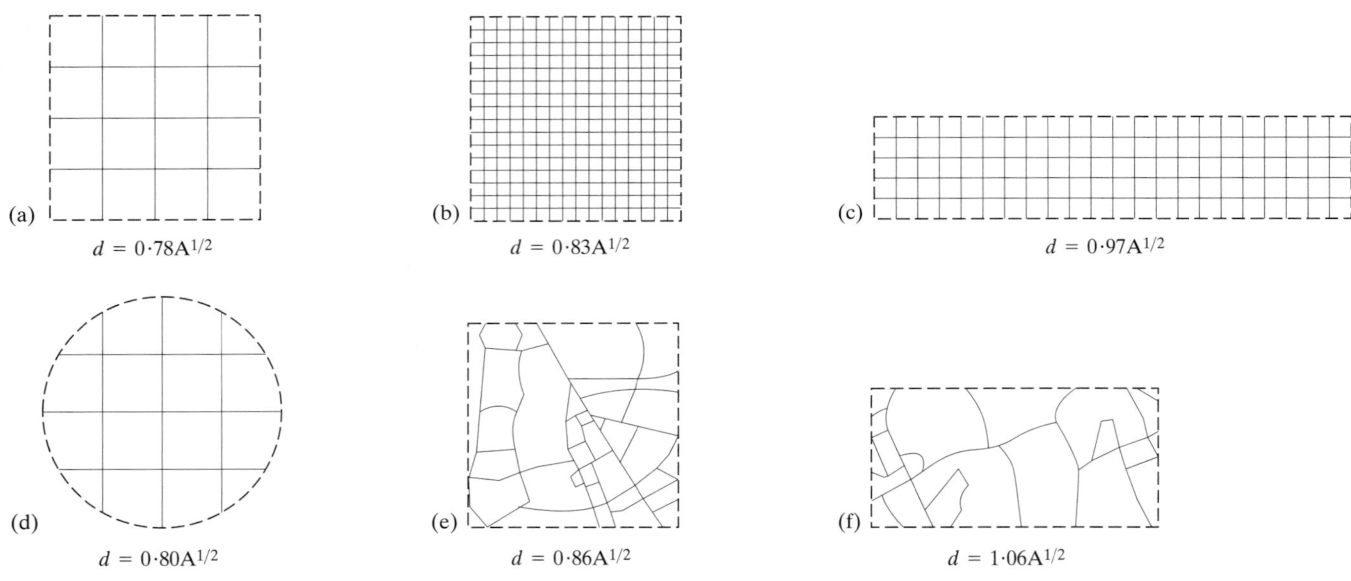

FIGURE 10.11
Average length of journey (d) on some imaginary and real networks. *(Source: Smeed, 1968; Haggett and Chorley, 1969)*

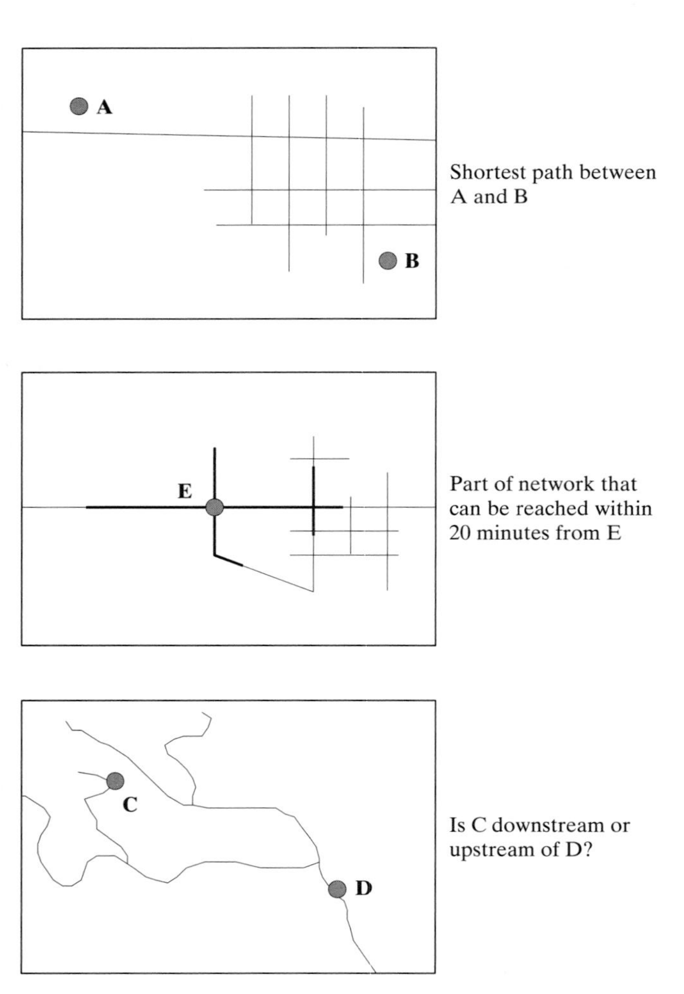

FIGURE 10.12
Implementation of network analysis in GIS.

the movement of resources between two or more points. The user may control the path of the route by specifying the origin, destination, and any stops or nodes the route must pass through. The program will find the optimal path between the points in the order specified by the user. Attributes of arcs and turns in the network represent *impedance*, which is the amount of resistance required to traverse the arc from one node to the other. The length of arcs is a good example of impedance: longer arcs having greater impedance than shorter ones. Another example of impedance is the number of turns required to move along an arc from one node to another. Intersections of roads with stop signs will slow down movement. By using impedance, the variable conditions of movement along the network are simulated. Thus, the optimal path is the path with the lowest impedance. This concept of the shortest distance to travel may be modified to the *best* route to travel, which is not necessarily the shortest. For example, if you are a tourist, you may be interested in traveling along the most scenic route.

Another function of network analysis implemented in vector GIS is to find the closest facility from a given location. An example is to find the closest store to your house. Alternatively, one may need to find the parts of the network that can be reached within a given travel time from a selected point (e.g., to find all the streets within a five-minute response time from a fire station). This type of network analysis requires the use of *allocation*, the process of assigning links in the network to the closest center (performed by the *Allocate* command in ArcInfo Network module). Allocation requires continuous evaluation of impedance along all paths from a specific location. This function really evaluates accessibility, which is the ease to get to a site, measurable in terms of travel time or distance. By specifying the travel time and the location of the site, the extent of the service area around a specific location in the network may be determined.

Finally, it is possible to determine service areas around any location on a network by applying the gravity model, notably Reilly's breakpoint equation (Equation 10.31). This requires a knowledge of the population size at various centers and the distances between these centers and the selected site on the network.

Before network analysis is possible, the impedance of the network needs to be set up. Vector GIS has the capability to set up "rules of the road," such as speed limits, one-way streets, prohibited turns, closed streets, streets to avoid, and overpasses and underpasses (i.e., nonplanar graphs). ArcView GIS through an extension called Network Analyst facilitates routing, allocation, geocoding, and creation of network without the need to write down the commands as in the case of ARC NETWORK module (Environmental Systems Research Institute, 1996a).

10.8 GIS MODELING

A model is a simplified representation of the real world and its processes. It summarizes data and makes general statements on the way phenomena exist and operate. GIS modeling is to use spatial analysis techniques to produce models of spatial phenomena. There are two main objectives in GIS modeling: to understand and to predict. For understanding, generalizations of spatial phenomena are made. Figure 10.13a is a detailed land-use map showing residential, commercial, industrial, and forest uses in a fictitious city. This map may be generalized to show simply two classes of land uses: urban and forest (Figure 10.13b). This is a simplification of the reality to achieve understanding. For prediction, an assumption is made that some more of the forest in the city will be depleted in 10 years' time (Fig. 10.13c). How will this change affect the rate of water runoff in the city or the local climate? This "what-if" scenario is prediction. The predictive capability of GIS modeling is particularly useful for a constantly changing world.

There are six basic steps to follow in GIS modeling: (1) stating the problem, (2) breaking down the problem into its components, (3) searching for data, (4) deciding on one or more suitable spatial analytical models to use, (5) deciding on a suitable GIS program (data model) to use, and (6) implementing the model in the GIS environment. These steps will be explained with reference to some actual examples of GIS modeling in the next section.

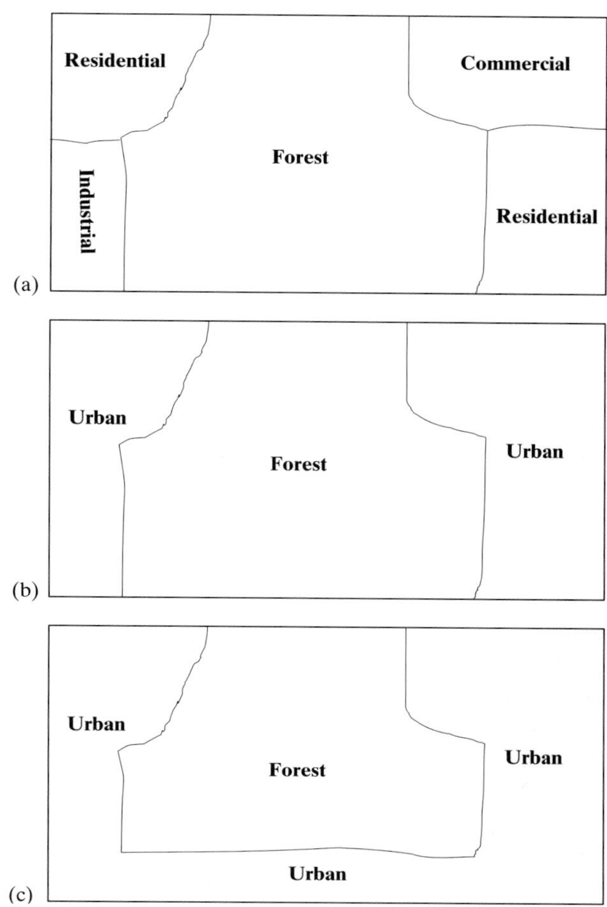

FIGURE 10.13
GIS modeling: (a) original detailed land-use map, (b) understanding: generalized land-use map, and (c) prediction: what if more forest becomes depleted?

10.8.1 GIS MODELING AND MANAGING AGRICULTURAL NON-POINT-SOURCE POLLUTION

An excellent example of GIS modeling is an application in which the agricultural non-point-source (AGNPS) pollution model is integrated with GIS (Engel et al., 1993). Non-point-source (NPS) pollution has attracted attention in recent years because of its high cost to the environment. NPS implies water pollution that originates from widely dispersed locations. For agricultural land use, NPS pollution includes surface and subsurface runoff from rainfall or irrigation that contains contaminants such as soil particles, nutrients, herbicides, and so forth. The water runoff induces soil erosion. Identification of non-point-source pollution problem areas is difficult because of the spatial, distributed nature of the processes involved. Once the problem areas are identified, alternative management practices designed to minimize the impact of agriculture and other activities on the environment can be evaluated. Therefore, the problem statement for the GIS modeling attempt is the prediction of areas in an agricultural watershed most susceptible to non-point-source pollution.

The AGNPS model was developed for the analysis of non-point-source pollution from agricultural fields. It is capable of incorporating the influences of the spatially variable parameters, such as topography, soils, and land use, to simulate conditions at all points within the watershed simultaneously (Young et al., 1989). It estimates the quality of surface water runoff and compares it to the expected quality of other land management strategies. It is a *distributed parameter model* as opposed to a *lumped parameter model*. A distributed parameter model is a mathematical model that attempts to account for spatial variability influences of its independent variables (parameters) by applying the governing equations to small elemental areas within which the parameters are assumed to be uniform. The outputs from one element become the inputs for adjacent elements. For such a model, it is computationally intense, but it accurately reflects spatial changes. By contrast, a lumped parameter model applies the parameters uniformly over the entire region being modeled, without considering spatial variability influences. Clearly, GIS is particularly suited for use in a distributed parameter model.

To apply the AGNPS model, the watershed is divided into a grid of square cells, typically about 100 m^2. It may be used for watersheds up to 20,000 hectares in size with cell sizes of 0.4 to 16 hectares. Within this spatial framework, runoff characteristics and transport processes of sediments and nutrients are simulated for each cell and routed to its outlet, thus allowing the runoff, erosion, and chemical movement at any point in the watershed to be examined. This breakdown to individual components represents the second stage of the GIS modeling process.

The AGNPS model makes use of a number of submodels to predict runoff, soil erosion, sediment transport, and nutrient movement. To predict runoff for each grid cell, the former Soil Conservation Service (SCS) curve number has to be applied to the following formula.

$$RO = (P - 0.2S)^2/(P + 0.8S)$$

where RO is the runoff volume in millimeters, P is the rainfall in millimeters, and S is a retention parameter in millimeters. The value of S is determined by

$$S = (25400/CN) - 254$$

where CN is the SCS curve number, calculated based on land use, hydrologic soil groups, hydrologic condition, presence of conservation practices and antecedent soil moisture.

Soil erosion is determined from a modified version of the Universal Soil Loss Equation (USLE) as follows.

$$A = 2.24\, KLSCP\, EI_{30} SSF$$

where A is the average soil loss for the time interval with the combined erosivity of rainfall and runoff (tons per acre), K is the soil erodibility factor relating to the cohesiveness of the soil type, L is the slope length factor, S is the slope steepness factor, C is the crop management factor, and P is the conservation practice factor, EI_{30} is the 30-minute energy intensity (ft^{-2}), and SSF is a factor to adjust for the convex or concave nature of slopes within the cell.

Sediment transport through the watershed was performed for five particle-size classes (clay, silt, small aggregates, sand, and large aggregates). The nutrient movement components of AGNPS are extracted from another model called CREAMS (Frere et al., 1980) and from a feedlot model (Young et al., 1982, 1989). CREAMS is a field scale model for predicting runoff, erosion, and chemical transport from agricultural management systems, applicable to field-sized areas. A total of 22 parameters for each cell was used for input (Table 10.4).

Table 10.4

AGNPS Input Data Required

Input Data Values (distributed parameter information) Required for Each Watershed Element
Cell number
Number of the cell into which it drains
SCS curve number
Average land slope (%)
Slope shape factor (uniform, convex, concave)
Average field slope length (feet)
Average channel slope (%)
Average channel side slope (%)
Mannings roughness coefficient for the channel
Soil erodibility factor (K) for USLE
Cropping factor (C) for USLE
Practice factor (P) for USLE
Surface condition constant (factor based on land use)
Aspect (one of 8 possible directions indicating the principal drainage direction from the cell)
Soil texture (sand, silt, clay, peat)
Fertilization level (zero, low, medium, high)
Incorporation factor (% fertilizer left in top 1 cm of soil)
Point-source indicator (indicates existence of a point-source input within a cell)
Gully source level (estimate of amount, in tons, of gulley erosion in a cell)
Chemical oxygen demand factor
Impoundment factor (indicating the presence of an impoundment terrace system within the cell)
Channel indicator (indicating existence of a defined channel within a cell)

Engel et al., 1993

Once the model is determined, the next stage in GIS modeling is to search for data. All the input parameters required for the AGNPS model may be extracted from eight GIS layers (maps): soils, elevation, land use, management practices, fertilizer nutrient inputs, type of farm machinery used for land preparation, channel slope, and slope length factor. Using Digital Elevation Model (DEM), such topographic information as slope, aspect, ridges, channels, watersheds, and overland flow paths may be extracted (Douglas, 1986; Jenson and Domingue, 1988). In extracting the required input data for each cell, various spatial analysis models have been used. Exploratory spatial data analysis within the watershed boundary has to be carried out.

In addition to extracting parameters from the eight GIS layers, the following five items of data relating to the whole watershed to be modeled have to be entered: (1) watershed identification/description, (2) precipitation (in inches), (3) erosion index (EI) for a storm/rainfall event, (4) the area of each cell (in acres), and (5) the outlet cell number.

The output from the AGNPS model generates 19 variables for each cell. These should be graphically displayed by the GIS as maps for scientific visualization (Table 10.5). Based on the visualization, evaluation of different management strategies to improve water quality may be carried out.

Table 10.5

AGNPS Output Data

AGNPS Output Values Items for Each Watershed Element
Runoff volume
Delivery ratios by particle size
Peak runoff rate
Sediment associated phosphorous mass
Sediment yield
Soluble phosphorus mass
Upland erosion
Soluble phosphorus concentration
Deposition
Fraction of runoff generated
Sediment generated
Enrichment ratios by particle size
Chemical oxygen demand concentration
Chemical oxygen demand mass
Sediment concentration
Sediment particle-size distribution
Soluble nitrogen mass
Soluble nitrogen concentration
Sediment-associated nitrogen mass

Source: Engel et al., 1993

Because of the use of grid cells in the watershed for the AGNPS model, a raster-based GIS is a clear choice. Indeed, the public-domain GIS software called GRASS (Geographical Resources Analysis Support System) developed by the U.S. Army (1987) has been used quite successfully for the development of an integrated AGNPS/GIS system (Engel et al., 1993). It has the advantage that the source code is available for GRASS to be integrated with the AGNPS model. However, the AGNPS model has also been successfully implemented in other raster-based GIS software packages, such as ERDAS Imagine or Idrisi for Windows, with varying degrees of integration. This completes the choice of GIS data model and implementation stages in the GIS modeling process.

10.8.2 IDENTIFICATION OF PRIORITY SITES FOR THE MANAGEMENT OF WHITE-TAILED DEER IN COSTA RICA USING GIS AND HABITAT SUITABILITY INDEX (HSI)

This application of GIS modeling relates to resources management using the concept of habitat suitability index (HSI) developed by the U.S. Fish and Wildlife Service (1991). Although the interest is in white-tailed deer (*Odocoileus virginianus*), the method may be applied to other types of animals or to endangered species. In Costa Rica, the evaluation of the white-tailed deer's potential habitat is important because this is a valuable game-ranching species, and a knowledge of the quantity and quality of available habitat is crucial (Lopez, 1998).

The study area is located at Bagaces, Guamacaste, Costa Rica, bounded by the Tempisque and Bebedero rivers, with a total area of 600.8 km^2. It consists of three wildlife zones: tropical dry forest, wet premontane forest, and tropical wet forest, with an average annual rainfall of 1624 mm.

Based on the habitat evaluation procedure (HEP) developed by the U.S. Fish and Wildlife Service, a model for the white-tailed deer's potential habitat was developed. From research, six variables were found to be significant in controlling the white-tailed deer's habitat, and for each variable, a habitat suitability index (HSI) value (from 0 to 1) was assigned. The six variables were:

1. *Food* (V_1). The deer prefer sites with higher plant diversity and biomass. Therefore, the richer the plants, the higher the HSI value.
2. *Slopes* (V_2). Deer prefer sites with slopes of less than 30%, or sites with a more regular topography.
3. *Distance Between Habitats* (V_3). The deer use pastures that are immediately adjacent to the forest. A greater distance presents more risk to the deer because of more exposure to predators.

4. *Water* (V_4). The distribution of water bodies influences the use of habitat by white-tailed deer in the dry season. The average radius of the home range of the deer for the dry season is about 750 meters. If the water bodies are located within the average home range, a high value is assigned.
5. *Vegetation Cover for Protection* (V_5). This refers to the horizontal visual obstruction of the site. The more horizontal vegetation cover present, the less visible is the deer and the safer it is from its predators.
6. *Human Factors* (V_6). The presence of small settlements and human activity could affect the deer's use of the habitat. Therefore, sites located far away from human presence score high values.

The following formula may now be used to calculate the final HSI values.

$$HSI = (2V_1 + V_2 + V_3 + 2V_4 + 2V_5 + V_6)/9$$

It should be noted that in this formula, a multiplier "2" indicates a larger weight assigned to that factor. As can be seen, food, water, and protection were considered to be more important and were given higher weights than other variables in computing the final HSI values.

To collect the data, some fieldwork was carried out. For vegetation mapping, sampling along secondary and tertiary paths commonly used by horses was carried out. Each path was mapped and subdivided into fifteen segments of 100 meters each. Sample sites were randomly selected from the segments available along each path. Once the sample site was located on the map, its location to the left or right of the path was randomly determined. At each sample site, five circular plots of 100 m^2 each, 20 meters apart, were established. About 0.01% of the total area was sampled, or approximately 60,000 m^2.

In each of the sampled plots, data relating to species of plants, bushes, and trees larger than 2 centimeters in DBH (diameter at breast height) were recorded. The horizontal visual obstruction was also measured by placing on the ground a target with a dimension of 1.2 m × 1.2 m with 25 black-and-white squares in the center of the plot. The intervisibility of the target at each cardinal direction was determined. Finally, the number of fresh feces left by the deer on each of the plots was counted.

The following maps were obtained: (1) an up-to-date land-use/cover map prepared from an interpretation of aerial photographs (1:35,000 scale), supplemented by Landsat TM images, and (2) a map of springs whose positions were fixed by using GPS. These maps were digitized and rasterized into pixels with a spatial resolution of 25 m × 25 m. The raster-based IDRISI program was chosen as the GIS software to implement the HSI model.

The GIS area and distance functions were used to compute areas and distances required for the six variables. An overlay (addition) function was used to calculate the final HSI. An HSI distribution map was created by Idrisi. The sample plots were superimposed on the map. The HSI values were then spatially correlated with the average numbers of fresh feces/plot/site. A correlation coefficient of 0.64 at the 0.01 level of significance was obtained, suggesting a significant but weak relationship. In other words, the HSI model could explain deer habitat.

By means of a reclassification function, seven categories of quality of habitat potential were obtained, as follows:

1. High: all favorable factors present
2. Medium (food): food as the limiting factor
3. Medium (cover): vegetation cover as the limiting factor
4. Medium (water): water as the limiting factor
5. Low (cover, food): both vegetation cover and food as limiting factors
6. Low (cover, water): both vegetation cover and water as limiting factors
7. Inappropriate: all limiting factors present

Based on this map, the area for each category of HSI values was computed using Idrisi. It was found that the majority of the study area was potentially optimal habitat (high category) for white-tailed deer, and about 50% of this category was located in the protected wildlife areas. The GIS methodology has succeeded in identifying and delimiting the habitat areas of white-tailed deer. The simple mathematical model is easily integrated with the built-in functions of the GIS.

10.8.3 OPTIMUM ROUTE FOR HAZARDOUS MATERIALS TRANSPORTATION

Network analysis is particularly useful for routing applications that require finding the best route between the origin and the destination. The best route may be the shortest, the safest, or the most scenic, depending on the purpose of travel. The following showcases a common application to determine the safest and cheapest route to transport hazardous materials on highways, as a case of application in Nevada (Yang and Sathisan, 1994).

In transporting hazardous materials, the best route is one that is both efficient and safe. Efficiency may be interpreted as minimizing transportation cost or travel time, while optimizing safety could mean minimizing accident rates on the selected route, minimizing population impacts or potential property damage cost along the route, or simply minimizing risks. Unfortunately, these are two conflicting conditions. An efficient route is normally the shortest route that may contain segments that are not safe, while a safe route is not always efficient and is often costly. The problem is therefore one of optimization or finding a route that strikes a balance between risk and cost.

To put these ideas in network analysis terms, finding the shortest route, or a path with minimum length from an origin to a destination, is achieved by identifying a path with the smallest accumulated impedance. If only risk factors are considered, a safe path should include a group of arcs (segments) with small variances, or, in other words, with risk values falling into a least-risk range.

Yang and Sathisan (1994) presented an approach in which GIS is used to solve the routing model for hazardous waste materials transportation. The model is designed to seek a route with minimum cost (*C*) and least-risk variance (*RR*). In most cases, the lower limit (*LR*) is equal to zero, so that $RR = UR$ (upper limit). Then, the objective is to minimize the upper-risk limit. The optimum solution may be found by decreasing RR value to an appropriate value. This is the routing optimization model.

An algorithm to solve the problem is presented as a flowchart in Figure 10.14. The left path of the flowchart finds a maximum-risk upper-limit value on a least-cost (*C*min) route without any risk constraints. This will find a least-cost route with risk values lower than or equal to *UR*. The right path establishes a lowest value for risk upper-limit based on the risk value among all links connecting to the origin and the destination (*UR*m). Search for a least-cost route by starting

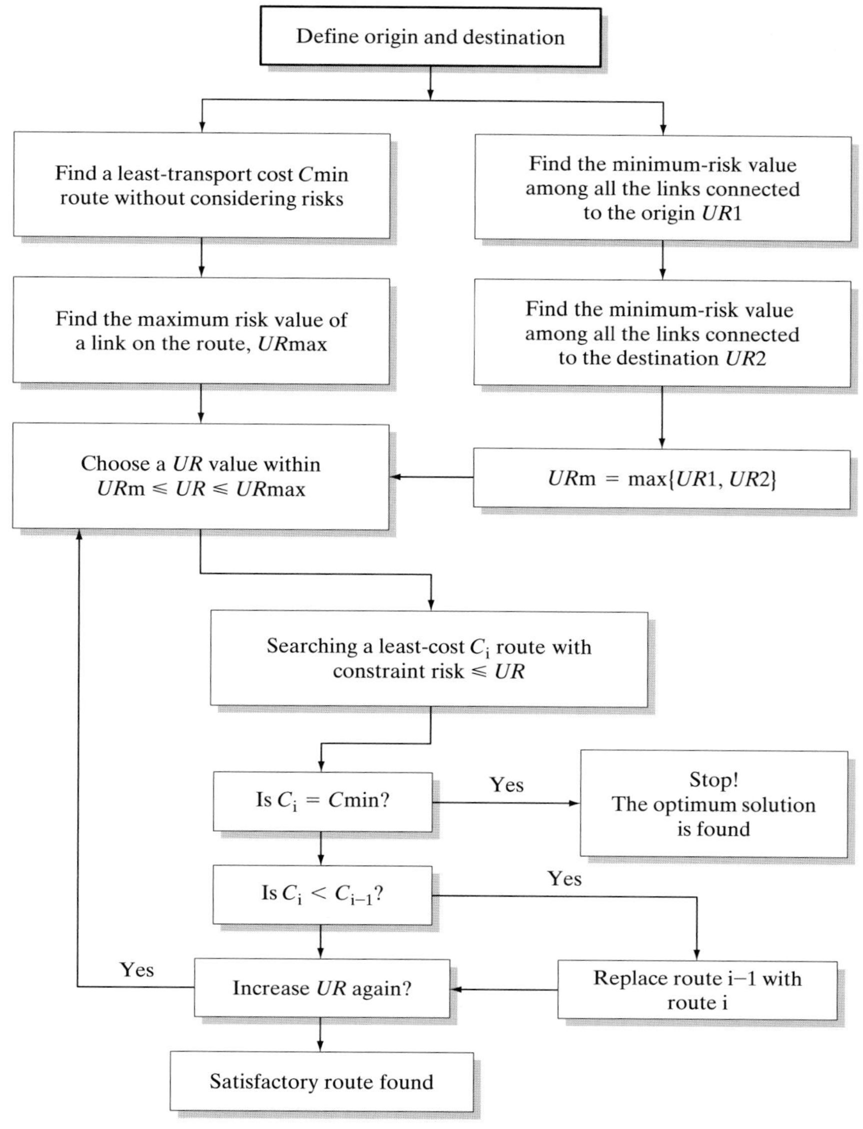

FIGURE 10.14
Algorithm for determining *UR* value. *(Source: Yang and Sathisan, 1994)*

from *URm*. If no route is found, increase the *UR* value until a route is found. If the cost of the route (*Ci*) is equal to *Cmin*, stop the search, and the optimum route is found. Usually, the cost of the route is greater than *Cmin*, a decision to increase the risk to decrease the cost has to be made. But how much risk should be increased? Eventually, a feasible solution may be found by choosing a *UR* value between *URm* and *URmax* (the maximum risk value of a link on the route).

This logic of determining the *UR* value may be implemented in the form of a routing system using Network Analysis of ArcInfo, a vector GIS. The network analysis functions used (for the workstation) are Pathfinding, Allocate, Tracing, and Spatial Interaction. Fortran and C programming are needed to provide the interface between the optimization model and ArcInfo. The routing system developed consists of three major subsystems: (1) static routing, (2) dynamic routing, and (3) emergency response (Figure 10.15).

Static routing refers to the route selection procedures when the locations and routing information are all known prior to operation. The basic data are travel distance, travel time, network characteristics, and risk effects. The user can choose the travel time, distance, and risk. In this model, there are two objectives according to what has been discussed in relation to Figure 10.14: (1) a function to minimize cost of a route and (2) a function to find a minimum upper risk limit on the route.

Dynamic routing subjects the routing system to imprecise or unknown inputs as well as greater time constraints in a dynamic setting. It has to make changes, mostly local reroutings, to accommodate real-time inputs for quicker solutions that may not be optimal. In other words, whatever route was determined under the static routing model will have to be modified on real-time settings assigned interactively by an operator. The real-time settings include setting barriers, blocking streets, and reassigning stops. New route selection processes will select a route for detouring the barriers and blocked streets, passing assigned stops.

Emergency response is to act quickly when unexpected incidents occur on the route selected. It involves the use of quickest-path algorithm. Time is the most important consideration. Before the operation of the route selection module, all the response centers should be located on the network. Using the Allocate module of ArcInfo, the network is divided into several areas or zones according to the response time for each response center. For emergency response, the location of the incident on the network is displayed, and the quickest path from all response centers to the incident location will be found by the system.

An example of the routing system for application to previously discussed situation in Nevada is shown in Figure 10.16. The basic impedance is set as

$$\text{impedance} = 0.7^* \text{ time} + 0.3^* \text{ distance}$$

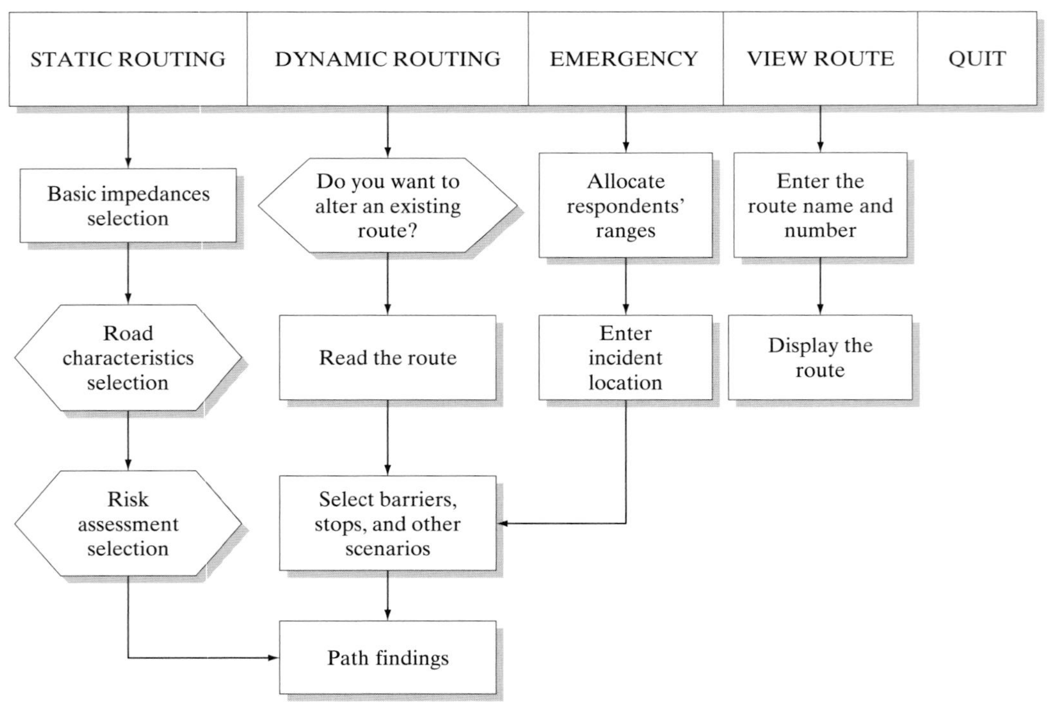

FIGURE 10.15
Flowchart showing an overview of the routing system design. *(Source: Yang and Sathisan, 1994)*

(a)

(b)

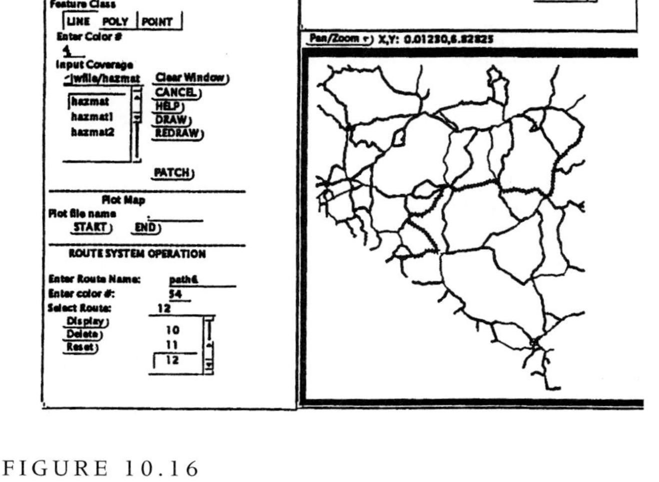

(c)

(d)

FIGURE 10.16
The routing system developed: (a) Route #10 and statistics with no risk constraints, (b) Route #11 and statistics with no risk constraints, (c) Route #12 and statistics with upper limit value = 9.0, and (d) a comparison of three routes with statistics. *(Source: Yang and Sathisan, 1994)*

The route selected as well as the route and network statistics are displayed by the computer. The statistics shown are the maximum risk, the minimum risk, and the mean risk for both the route and the network. In addition, the impedance is also shown for the route.

First, an optimal route (i.e., least impedance) is selected from a starting point to a destination without considering the risk constraints, and statistical analysis is carried out. Route #10 is such a route, showing the statistics (Figure 10.16a). The maximum risk upper limit (*UR*max) is found to be 44.085, and the route impedance is 127.14. After several trials, the minimum-risk upper limit (*UR*m) is found to be 5.5 for the least impedance route (#11) (Figure 10.16b). Therefore, a preferable upper-limit *UR* value may be chosen between 5.5 and 45. Finally, another route (#12) is found with *UR* = 9.0, *UR*max = 8.11, and impedance = 143.94 (Figure 10.16c). Both routes #11 and #12 are suitable choices (Figure 10.16d).

This application illustrates a different approach to integrating a spatial model with GIS. The optimization model may be solved by presenting different scenarios using different constraint settings. In this example, a balance between risk and cost in selecting a route for the transportation of hazardous materials has to be achieved. Unlike the previous two examples, this application deals with links and nodes, which form a network. The vector-based ArcInfo GIS is customized to provide a user-friendly graphical interface for a full integration.

10.8.4 DETERMINING THE BEST LOCATION AREAS OF A STORE

This is an example illustrating a typical application of GIS invoking gravity and spatial interaction models mentioned in Section 10.6, taken from ArcView Spatial Analyst manual (Environmental Systems Research Institute, 1996b). As the owner of a franchise of small neighborhood stores for upscale professionals in a city, you want to open another store. Where is the best location area for this store? This is your problem statement.

What conditions constitute "best"? Obviously, the store should not compete with your existing stores for customers, the trade area (the area around which you will attract customers) should have a high percentage of wealthy customers for upscale goods, and the population density should be high also.

To avoid competing with your existing stores for customers, it is necessary to know the spatial distribution of all your existing stores in the city. Based on a survey of the customers in your stores, you can determine that these customers will not travel more than 3 km to your stores. Your new store should therefore be located outside the 3-km belts from all your existing stores. This distance factor is the result of application of the spatial interaction model. A map showing the distances between all your existing stores needs to be produced.

More research needs to be conducted about the potential customer pools. What are the socioeconomic characteristics of the customers who will buy from your store? Perhaps the income level and age can provide good indicators, depending on the type of goods you want to sell. These types of data may be easily obtained from census statistics. Let's assume that your potential customers are in the age groups of 25–35 with an average annual income of $30,000 per capita. A map of the percentages of population possessing both these two characteristics within an enumeration unit (e.g., census tract) in the city should be produced.

Finally, population density is used to decide whether there are enough customers in the 3-km trade area around each location in the city. From your successful stores, you know that a minimum of 1000 people in the trade area is required to sustain a profitable business. A map of population density per 3-km trade area from all locations in the city is produced.

For each of the three maps produced, a suitability scale has to be constructed. There are different ways to construct a suitability scale, and for some phenomena, the scale is nonlinear. A good example is distance. Driving a distance of 10 km is not necessarily ten times as bad as driving 1 km. In this example, let's assume that for the distances between existing stores maps, five categories of values are produced with the following suitability rank scores, remembering that distances below 3 km (3000 m) are not suitable for locating the new store: (1) 0–3000 = 1, (2) 3000–4000 = 3, (3) 4000–5000 = 5, (4) 5000–6000 = 7, and (5) > 6000 = 10. The higher the rank score, the more suitable it is for the new store's location. For the map of the percentages of population possessing the right socioeconomic characteristics within the trade area, the higher the percentage, and the more desirable. An equal classification to ten classes in ascending order may be carried out. The higher the rank score, the better. Finally, for the population density within the trade area map, grouping into five classes with the following assigned rank scores is appropriate, recalling that 1000 is the minimum density to support a profitable business: (1) 0–1000 = 1, (2) 1000–1400 = 5, (3) 1400–1900 = 7, and (4) > 1900 = 10.

The best location areas for the new store are obtained by overlaying all three maps with the suitability rank scores together in the form of a union. Areas with the highest suitability scores will be the best location areas for the new store.

The GIS solution to the problem is best summarized in the form of a flowchart (Figure 10.17), which should always be used in developing a GIS model to solve spatial problems.

10.9 SUMMARY

In this chapter, spatial analysis and its use as a data exploratory tool in the GIS environment were introduced. The problems of integrating spatial analysis in the GIS environment were also examined. Three approaches to integration, namely, full integration, loose coupling, and close coupling, have been explained with reference to specific spatial analysis functions. Much research on integrating spatial analysis with GIS has resulted in the publication of at least four major volumes of papers at the time of writing (Fotheringham and Rogerson, 1994; Fischer et al., 1996; Longley and Batty, 1996; and Fischer

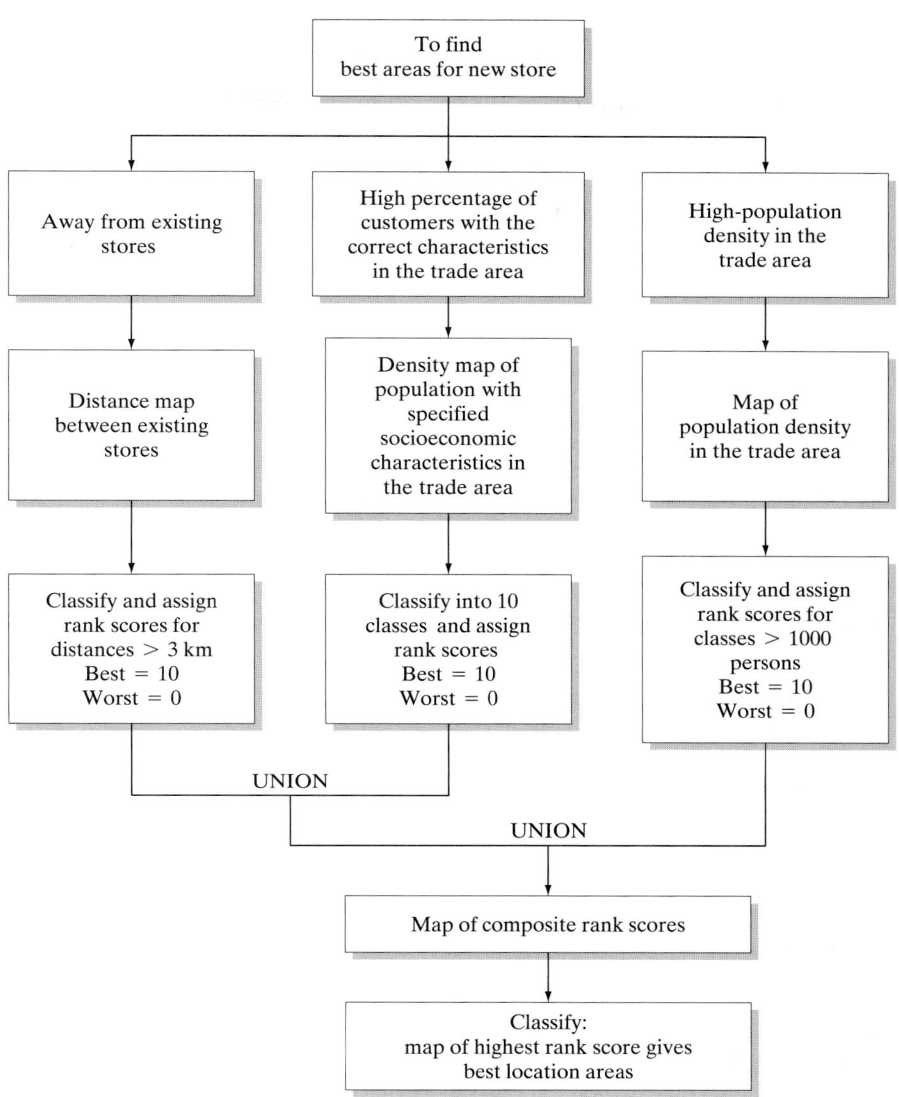

FIGURE 10.17
GIS modeling flowchart for locating the best areas for a store.

and Getis, 1997). However, full integration has not yet been achieved. Spatial analysis in the strictest sense of its definition, as an exploratory and explanatory tool of patterns of spatial distribution, is weak in the GIS toolbox. Fortunately, advances in computer technology and programming allow loose-to-close coupling of spatial analysis with GIS, notably with the use of spatial analysis programs developed by researchers, such as Ding and Fotheringham (1991) or by software vendors such as Insightful (formerly MathSoft) (with S+ SPATIALSTATS module, which can be closely coupled with ArcInfo and ArcView GIS programs).

Despite the weak link, all GIS practitioners should know the concepts behind some fundamental spatial statistical analysis techniques. Some useful statistical measures, such as mean, median, mode, standard deviation, skewness, and kurtosis, together with some useful spatial statistical measures, such as spatial autocorrelation, quadrat counts, nearest-neighbor analysis, trend surface analysis, gravity models, and network analysis, were explained. Attention is drawn in particular to the use of GIS to extract information about distances and the related concepts of accessibility and proximity. The gravity and spatial interaction models are particularly useful for GIS applications to solve real-world problems.

This chapter also explains the fundamentals of the GIS modeling process, as illustrated by four examples drawn from applications to natural resources management, routing, and retail location. This should help students learn to practice GIS in the real world. All students are encouraged to read more case studies of GIS applications in their own fields of interests from the journals and proceedings of annual conferences listed in the Appendix at the end of this book.

REFERENCES

Bailey, T. C. (1994) "A review of statistical spatial analysis in geographical information systems," in *Spatial Analysis and GIS*, by Fotheringham, S., and Rogerson, P. eds. pp. 13–44, London: Taylor and Francis.

Can, A. (1993) "Residential quality assessment alternative approaches using GIS," in Fischer M. M., and Nijkamp, P. eds. *Geographic Information Systems, Spatial Modelling, and Policy Evaluation*, Berlin: Springer–Verlag, pp. 199–212.

Christaller, W. (1966) *Die Zentralen Orte in Suddeutschland*, Jena: Gustav Fisher, Verlag, 1933; translated into English as *Central Places in Southern Germany*, by Baskin, C. W. Englewood Cliffs, N.J.: Prentice-Hall.

Clark, P. J., and Evans, F. C. (1954) "Distance to nearest neighbor as a measure of spatial relationships in populations," *Ecology*, Vol. 35, pp. 445–453.

Cole, J. P., and King, C. A. M. (1968) *Quantitative Geography: Techniques and Theories in Geography*, London: John Wiley and Sons Ltd.

Congalton, R. G. (1988) "Using spatial autocorrelation analysis to explore the errors in maps generated from remotely sensed data," *Photogrammetric Engineering and Remote Sensing*, Vol. 54, No. 5, pp. 587–592.

Converse, P. D. (1949) "New laws of retail gravitation," *Journal of Marketing*, Vol. 14, pp. 379–390.

Ding, Y., and Fotheringham, A. S. (1991) *The Integration of Spatial Analysis and GIS*, Working Paper, NCGIA, Department of Geography, State University of New York, Buffalo.

Davies, C. S., Holz, R. K., and Robertus, D. (1999) "A test of central place theory using Shuttle Imaging Radar (SIR-A) of China's North Central Plain," *Geocarto International*, Vol. 14, No. 1, pp. 13–22.

Davis, J. C. (1973) *Statistics and Data Analysis in Geology*, New York: John Wiley and Sons.

Douglas, D. H. (1986) "Experiments to locate ridges and channels to create a new type of digital elevation model," *Cartographica*, Vol. 23, No. 4, pp. 29–61.

Eastman, J. R. (1997) *User's Guide Idrisi for Windows Version 2.0*, Worcester, MA: Clark Labs for Cartographic Technology and Geographic Analysis, Clark University.

Eastman, J. R. (1999) *Idrisi32 Reference Guide*, Worcester, MA: Clark Labs, Clark University.

Engel, B. A., Srinivasan, R., and Rewerts, C. (1993) "A spatial decision support system for modeling and managing agricultural non-point-source pollution," in *Environmental Modeling with GIS*, by Goodchild, M. F., Parks, B. O., and Steyaert, L. T. eds. New York: Oxford University Press, pp. 231–237.

Environmental Systems Research Institute (1996a) *ArcView Network Analyst: Using the ArcView Network Analyst*, Redlands, CA, Environmental Systems Research Institute, Inc.

Environmental Systems Research Institute (1996b) *ArcView Spatial Analyst: Using the ArcView Spatial Analyst*, Redlands, California, Environmental Systems Research Institute, Inc.

Fischer, M. M. and Getis, A. eds. (1997) *Recent Developments in Spatial Analysis: Spatial Statistics, Behavioural Modelling, and Computational Intelligence*, Berlin: Springer–Verlag.

Fischer, M., Scholten, H. J., and Unwin, D. (1966) "Geographic information systems, spatial data analysis and spatial modelling: An introduction," in *Spatial Analytical Perspectives on GIS*, by Fischer, M., Scholten, H. J., and Unwin, D. eds. pp. 3–19, London: Taylor and Francis.

Fischer, M., Scholten, H. J., and Unwin, D. eds. (1996) *Spatial Analytical Perspectives on GIS*, London: Taylor and Francis.

Fotheringham, A. S., Brunsdon, C., and Charlton, M. (2000) *Quantitative Geography: Perspective on Spatial Data Analysis*, London: Sage Publications.

Fotheringham, S., and Rogerson ed. (1994) *Spatial Analysis and GIS*, London: Taylor and Francis.

Frere, M. H., Ross, J. D., and Lane, J. L. (1980) "The nutrient submodel," in *CREAMS, A Field Scale Model for Chemicals, Runoff, and Erosion from Agricultural Management Systems*, Washington, D.C.: U.S. Department of Agriculture, Conservation Research Report 26, pp. 65–85.

Gatrell, A. (1983) *Distance and Space: A Geographical Perspective*, Oxford: Clarendon Press.

Gatrell, A., and Loytonen, M. (1998) *GIS and Health*, London: Taylor and Francis.

Geary, R. C. (1968) "The contiguity ratio and statistical mapping," in *Spatial Analysis: A Reader in Statistical Geography*, by Berry, B. J. L., and Marble, D. F., pp. 461–478, Englewood Cliffs, NJ: Prentice–Hall.

Getis, A., and Boots, B. (1978) *Models of Spatial Processes: An Approach to the Study of Point, Line and Area Patterns*, Cambridge: Cambridge University Press.

Goodchild, M. F. (1986) *Spatial Autocorrelation, Concepts and Techniques in Modern Geography*, Norwich, England Geo Books.

Goodchild, M. (1992) "Integrating GIS and spatial data analysis: Problems and possibilities," *International Journal of Geographical Information Systems*, Vol. 6, No. 5, pp. 407–423.

Griffith, D. A. (1989) *Spatial Regression Analysis on the PC: Spatial Statistics Using Minitab*, Discussion Paper No. 1, Ann Arbor, MI: Institute of Mathematical Geography.

Griffith, D. A., and Amrhein, C. G. (1991) *Statistical Analysis for Geographers*, Englewood Cliffs, NJ: Prentice Hall.

Haggett, P., and Chorley, R. J. (1969) *Network Analysis in Geography*, London: Edward Arnold.

Haynes, K. E., and Fotheringham, A. S. (1984) *Gravity and Spatial Interaction Models*, Beverly Hills, CA: Sage Publications.

Jenson, S. K., and Domingue, J. O. (1988) "Extracting topographic structure from digital elevation data for geographic information system analysis," *Photogrammetric Engineering and Remote Sensing*, Vol. 54, No. 11, pp. 1593–1600.

Kaluzny, S. P., Vega, S. C., Cardose, T. P., and Shelly, A. A. (1998) *S+ SPATIALSTATS User's Manual for Windows and UNIX*, New York: Springer–Verlag.

Kansky, K. J. (1963) *Structure of Transportation Networks*, Chicago, IL: Department of Geography, the University of Chicago.

Lo, C. P. (1984) "Chinese settlement pattern analysis using Shuttle Imaging Radar-A data," *International Journal of Remote Sensing*, Vol. 5, No. 6, pp. 959–967.

Lo, C. P. (1986) "Settlement, population and land use analyses of the North China Plain using Shuttle Imaging RADAR-A Data," *Professional Geographer*, Vol. 38, No. 2, pp. 141–149.

Longley, P., and Batty, M. eds. (1996) *Spatial Analysis: Modelling in a GIS Environment*, Cambridge, England: GeoInformation International.

Lopez, W. S. (1998) "Application of the HEP methodology and use of GIS to identify priority sites for the management of white-tailed deer," in *GIS Methodologies for Developing Conservation Strategies, Tropical Forest Recovery and Wildlife Management in Costa Rica*, by Savitsky, B. G., and Lacher, Jr., T. E. eds. pp. 127–137, New York: Columbia University Press.

Mayer, J. D. (1983) "The role of spatial analysis and geographic data in the detection of disease clustering," *Social Science and Medicine*, Vol. 17, pp. 1213–1221.

Moran, P. A. P. (1948) "The interpretation of statistical maps," *Journal of the Royal Statistical Society*, Series B 10, pp. 243–251.

Odland, J. (1988) *Spatial Autocorrelation*, Newbury Park, CA: Sage Publications.

Openshaw, S. (1990) "Spatial analysis and geographical information systems: A review of progress and possibilities," in *Geographical Information Systems for Urban and Regional Planning*, by Scholten, H. J., and Stillwell, J. C. H. eds. pp. 153–163, Dordrecht: Kluwer Academic Publishers.

Reilly, W. J. (1929) *Methods for the Study of Retail Relationships*, University of Texas Bulletin, No. 2944, Austin: University of Texas.

Ripley, B. D. (1976) "The second-order analysis of stationary point processes," *Journal of Applied Probability*, Vol. 13, pp. 255–266.

Ripley, B. D. (1981) *Spatial Statistics*, New York: John Wiley and Sons.

Rogerson, P. A., and Fotheringham, A. S. (1994) "GIS and spatial analysis: Introduction and overview," in *Spatial Analysis and GIS*, by Fotheringham, S., and Rogerson, P. eds. pp. 1–10, London: Taylor and Francis.

Scholten, H. J., and LoCascio, A. (1997) "GIS application research: History, trends and developments," Geographic Information Research at the Millennium, GISDATA Final Conference, Le Bischenberg, France, 13–17 September, 1997, *GISDATA Newsletter* 1997, pp. 38–51.

Smeed, R. J. (1968) "Traffic studies and urban congestion," *Journal of Transport Economics and Policy*, Vol. 2, pp. 1–38.

Thomas, R. W. (1977) *An Introduction to Quadrat Analysis*, Norwich, England: University of East Anglia, Geo Abstracts Ltd.

U.S. Army (1987) *GRASS-GIS Software and Reference Manual*. Champaign, IL: U.S. Army Corps of Engineers, Construction Engineering Research Laboratory.

U.S. Fish and Wildlife Service (1991) *Habitat Evaluation Procedure (HEP)*, Washington, D.C.: Division of Ecological Services, Department of the Interior.

Yang, X., and Sathisan, S. K. (1994) "Development of a GIS-based routing model," Proceedings of the 1994 ESRI User Conference, pp. 1076–1091, Redlands, CA: Environmental Systems Research Institute.

Young, R. A., Onstad, C. A., Bosch, D. D., and Anderson, W. P. (1989) "AGNPS: A nonpoint-source pollution model for evaluating agricultural watersheds," *Journal of Soil and Water Conservation*, Vol. 44, No. 2, pp. 168–173.

Young, R. A., Otterby, M. A., and Roos, A. (1982) "A technique for evaluating feedlot pollution potential," *Journal of Soil and Water Conservation*, Vol. 37, No. 1, pp. 21–23.

CHAPTER

11

GIS Implementation and Project Management

11.1 Introduction

GIS implementation and project management can be learned most effectively by practice. There is no better way to acquire the skills of designing an information system and managing its implementation than to actually take part in real-world projects. This chapter explains the process of GIS implementation using the software engineering paradigm. By working through the different stages of the development of a simple land parcel information system as an example, students will be able to see how the concepts and techniques of software engineering are used in practice. We cover the basic knowledge to develop the skills that are needed in order to participate in, and contribute to, GIS implementation projects in the future.

The implementation of information systems is much more than the acquisition of hardware and software. It involves considerable commitment and effort in project planning, database design, application development, as well as systems installation and maintenance. As database technology advances, methods have also been developed to provide a structured approach to the implementation of information systems. Such an approach, which is usually referred to as *software engineering*, has become standard methodology in the software development industry. Software engineering allows systems to be implemented within the framework of business rules, information architecture, and the operational priorities of an organization (i.e., a business corporation, an academic or research institute, or a government agency).

11.2 Software Engineering as Applied to GIS

Software engineering has been defined by the Institute of Electrical and Electronic Engineers (IEEE, 1993) as (1) the application of a systematic, disciplined, quantifiable approach to the development, operation, and maintenance of software, that is, the application of engineering to software; (2) the study of approaches as in (1). Within the context of these definitions, software engineering is a method for developing high-quality computer applications. It encompasses both the *processes* by which software is designed and developed and the *technologies* that are used to populate the processes (Pressman, 1997). As the result of the phenomenal growth of information systems in the last couple of decades, there have been major advances in software engineering as a systems development paradigm.

11.2.1 The Nature and Scope of Software Engineering

In general terms, engineering is the analysis, design, construction, verification, and management of technical or social entities. As such, engineering may be perceived simply as a structured approach to problem solving that may be applied to many fields of science and technology. In software engineering, the problem is a *business function* of an organization that is in need of a change for higher productivity or efficiency. The entity of interest, on the other hand, is a software solution that is designed to address the problem. The work associated with "engineering" a particular software solution may be divided into three generic phases (Pressman, 1997):

1. *The problem definition phase* is focused on the questions of "what," for example, what processes are performed in the business function in question; what processes may be addressed by using appropriate software tools; what data are required as input; what data are available; what information is required by the user as output; what system behaviors may be expected; what design constraints exist; and what criteria are required to evaluate the design of the system.
2. *The development phase* has its focus on the questions of "how," such as how data are obtained and structured; how processes are implemented as a software architecture; how interfaces are designed; how the design can be translated into a software application; and how testing will be performed before the system can be signed off as operationally acceptable.
3. *The maintenance phase* attempts to deal with questions regarding "change," such as those changes associated with error fixing; design modifications and functional enhancements in response to changing user requirements; as well as software upgrades in response to advances in hardware and software technologies.

In information system development projects, these three generic phases of work are usually formalized as a *systems development life cycle* (SDLC). The term "life cycle" in this context carries two meanings (Figure 11.1): (1) the sequential steps performed during the entire "life span" of an information system development project, and (2) the "cyclical" nature of the development process. The term "life cycle" implies that the development of information systems is a continuous and recursive process. It stipulates modifications and enhancements to the system in response to changes in the user requirements and the technological environment almost immediately after its implementation.

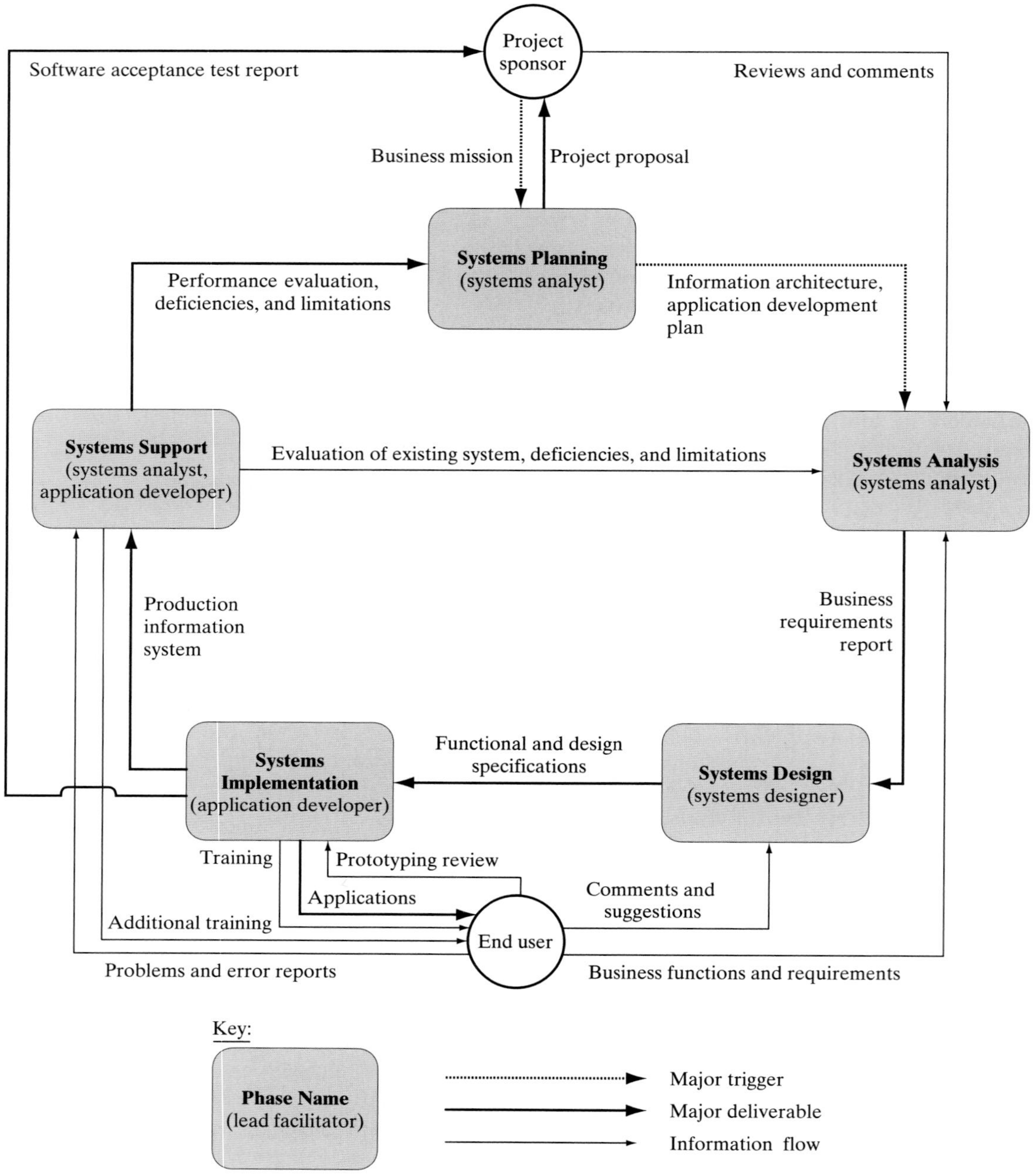

FIGURE 11.1
The process of information system development. This flow diagram illustrates the "cyclical" nature of the process of developing information systems and the concept of dividing the process into sequential phases of a systems development life cycle (SDLC). *(After Whitten et al., 1994, with modifications)*

The systems development life cycle (SDLC) is a top-down approach to systems development that may be depicted by a pyramid model (Figure 11.2). This model is layered to distinguish the different phases of work in the systems development process: planning, analysis, design, implementation, and support. The increasing volume in the lower layers implies that as the development process proceeds from one phase to the next, greater details are progressively introduced. These details are pertinent to the four components of information systems, that is, data, people, application, and technology (see Chapter 1). Generally speaking, a typical SDLC starts with relatively conceptual tasks and moves top-down through the phases to the technical installation and maintenance of the physical system. This signifies that the conceptual understanding of business goals is as important as the ability to use technology in the development of information systems.

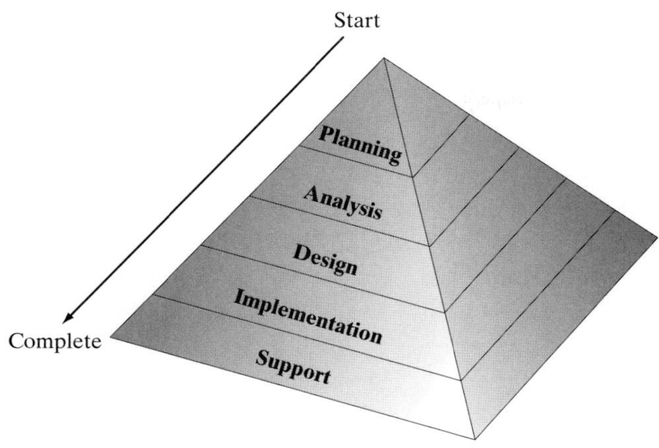

FIGURE 11.2
The pyramid model of systems development life cycle. The pyramid model of systems development is a top-down approach. As the systems development processes move from the top to the bottom of the pyramid, the focus is moving from business-oriented issues to technical details of building and maintaining the information system.

Conventionally, the emphasis of software engineering has been the development of better and more reusable codes (Marble, 1982). This was an *application-driven* approach based on the principles and methods of "structured programming." The objective is to improve the productivity of individual programmers and to develop modular computer programs that can be maintained more easily. Software engineering today is characterized by a *data-driven* approach. In this new approach, software is no longer viewed from a narrowly focused programming perspective, but from a more broadly based systems perspective. This means that the software engineering approach is now defined in terms of the entire software development process, rather than the individual software products. As a result, software engineering has also become known as *systems engineering*.

The new systems-oriented approach, as opposed to the conventional programming-oriented approach, of software engineering has led to the adoption of the philosophy of *total quality management* (TQM) in systems development. The objective of TQM is not only to assure the high quality of the software products being developed, but also to encourage the continuous improvement in the quality of the tools used to develop these products. In this regard, software engineering is a rapidly evolving technology, always striving to perfect its products as well as its own techniques. At the same time, the systems-oriented approach has also led to a changing role of the end user in the information system development process. As we will demonstrate later in this chapter, end user participation is a key feature of the software engineering methodology. Users contribute to the information system development process by having representatives taking part in every phase of the SDLC. Consequently, software engineering may be described as a people-oriented approach of systems development that aims to produce user-friendly software applications. Indeed, software engineering is not only about the use and development of technology, it is also very much concerned with the quality, management, and people aspects of technology as well.

The changing nature of software engineering means that its scope has been expanded considerably as well. The software engineering approach as we understand it today includes a variety of methods of developing high-quality information systems. It also embraces a myriad of project management activities that are carried out in support of the systems development life cycle, such as (Pressman, 1997)

- project planning, tracking, and control
- formal review of business cases and project proposals
- feasibility studies by simulation or prototyping
- data and software quality assurance and quality control (QA/QC)
- software evaluation and establishment of standards
- software acquisition, installation, and version control
- preparation and production of users' document and technical guides
- system performance evaluation and enhancement
- risk management (including contingency planning for data recovery and system failure)

11.2.2 SOFTWARE ENGINEERING IN GIS IMPLEMENTATION

The idea of using the software engineering approach to GIS is not new. In the early 1980s, for example, Marble (1982) was notably one of the strongest advocates of using software engineering methodology for the development of GIS. By adopting a structured approach to software development, he suggested, many of the problems that had plagued the earlier generations of GIS could be eliminated. These problems included the lack of flexibility in accommodating changed user needs, difficulty in adapting software to other hardware/operating environments, and difficulty in modifying and maintaining software. In a later technical paper, Marble et al. (1984) demonstrated how the principles of software engineering could be applied to the development of procedures for improving the productivity of map digitizing.

As GIS becomes increasingly integrated with mainstream information technology (IT), software engineering has gradually become the approach of choice for GIS application development. In response to this thrust toward the use of software engineering methodology in the computer industry, many GIS and database management system (DBMS) software vendors have made available compatible application development tools in their products. Conventionally, these tools have been provided in the form of fourth-generation macro or scripting languages. *Fourth-generation languages*, or 4GLs as they are commonly called, are *nonprocedural languages* that may be used for fast prototyping by using their built-in software tools for application generation, screen and menu design, and report formatting. More recently, the trend is toward the use of software components. These components, as we learned in Chapter 6, are low-level, reusable program controls that allow the programmer to embed geographic data-processing capabilities in application programs written in visual programming languages such as *Visual Basic* or *Visual* C++. With the aid of these application development tools, GIS software application developers may now take full advantage of new systems development techniques to deliver their products within a relatively short time frame and at a much lower development cost (see Section 11.6).

On the data side, it is now commonplace for large geographic databases to be designed using standard software engineering techniques such as conceptual, logical, and physical data modeling (see Section 11.5). These techniques have provided GIS database designers with an extremely rich set of tools that enables them to identify user requirements, structure data, and translate conceptual design into practical implementation in an orderly manner. By adopting the software engineering approach to geographic database design, users can be assured that the information needs of an organization are met, data integrity is maintained and data redundancy is kept to the minimum.

As an implementation paradigm for GIS, the software engineering approach has much more to offer. From the project management perspective, the structured approach can reduce the complexity in work scheduling, progress tracking, and budgetary control during the course of systems development. The practice of involving end users in the development process provides an excellent opportunity for them to learn GIS technology firsthand. It encourages users to actively participate in the systems design and evaluation process, thus allowing them to have more realistic expectations of what the system can and cannot do. Obviously, there are more than enough reasons to make it mandatory to use the software engineering methodology for all GIS projects.

11.3 GIS PROJECT PLANNING

As the first stage of the SDLC, the purpose of systems planning is to identify and prioritize those business areas of an organization that will benefit most from the use of information technology. During the systems planning process, formal business planning methods are applied to chart an information technology plan that supports the business goals of an organization. Since systems planning is concerned mainly with the high-level conceptual definition of the information needs of an organization, it is always assigned only to senior systems analysts with several years of systems development experience. The end product of systems planning is a *project proposal* that defines the objectives, scope, and architecture of the information system to be developed. This document is submitted to senior management of the organization to solicit and secure the necessary financial and human resource commitments to implement the proposed information system. It also forms the blueprint for the detailed technical design of the information system once the project has received the necessary stamp of approval.

To illustrate how the concepts and techniques of different phases of the SDLC may be applied in GIS implementation, we will use the example of setting up a simple land parcel information system in a typical city government. It is assumed that up to this point all the municipal business functions are still handled by conventional manual methods. The objective of the proposed land parcel information system is to automate the land title registration and property assessment functions of the city's administration.

11.3.1 PROJECT INITIATION

Traditionally, information systems have been developed as a reaction to one or more problems that prevented an organization from achieving its business goals. This is said to be the *reactive* approach to information system development. Information systems today are developed more commonly using a *proactive* approach. The concept of proactive information system development has grown out of modern business management philosophies that require managers to actively look for opportunities to improve an organization even in the absence of perceived problems. The implementation of GIS is a typical example of proactive information system development. In many cases, the use of GIS in an organization does not necessarily mean that there are problems to be fixed. Instead, it is often perceived as an opportunity to reengineer the way of doing business by taking advantage of geographic information (i.e., location, distribution, and spatial relationships) in business operation and decision making. In the case of our land parcel information system, for example, the initiation may be coming from the inability of city staff to handle the land title registration function manually. But it may also be perceived as an opportunity to change the old way of doing business and to integrate the land title registration function with related municipal business functions such as property tax assessment, engineering facility maintenance, land-use planning, and community development.

1.3.2 SYSTEMS PLANNING METHODOLOGY

The planning phase of the SDLC is typically divided into three major stages (Whitten et al., 1994): (1) studying the mission of the organization, (2) defining the information architecture of the proposed system, and (3) performing detailed business area analysis. These stages may be further broken down into different activities, each leading to a specific planning objective that will collectively define the purpose and scope of the proposed information system (Table 11.1).

Studying the Business Mission of the Organization

Information systems are developed to serve the specific needs of an organization, which are more commonly referred to as *user requirements* in systems analysis terminology. Therefore, the most logical way to begin systems planning is to study the business mission (i.e., goals and objectives) of the organization. The study of the mission usually starts with the formation of a systems planning team. This team normally consists of a systems analyst and several representatives from different

Table 11.1

Systems Planning: Major Stages of Work, Activities, Output, and Deliverables

Stages	*Activities*	*Outcome and Deliverables*
Study Business Mission	1. Form the systems planning team	Planning team and assignments of team members
	2. Define project and expectations	Organization chart and scope description
	3. Identify business performance measures	List of factors affecting business performance
	4. Develop a project plan	Project plan, Gantt chart, and network diagram
	5. Review and consolidate findings	Study report
Defining Information Architecture	1. Describe the organization at a general level	Enterprise business model
	2. Assess current business strategies	An association matrix (a chart tabulating the relationships among organization units, information needs, business functions, geographic locations, and performance measures)
	3. Assess current information services and strategies	Report summarizing technology and application assessments
	4. Identify and prioritize business areas	Proposed priority business areas
	5. Complete the proposed information architecture	Proposed information architecture (data, people, applications, and technology)
	6. Develop technology procurement procedure	Technology procurement schedule
	7. Review and consolidate findings	Consolidated information architecture report and technology procurement schedule
Performing Business Area Analysis	1. Form the business analysis team	Business analysis team
	2. Identify business area performance measures	Performance measures for specific business areas
	3. Model business areas	Expanded and refined business area models
	4. Assess current business area and information service performance	An association matrix and its interpretation
	5. Prioritize identified business processes in terms of corporate business objectives	List of prioritized business processes
	6. Plan application development strategy	Application development plan
	7. Review and refine application development plan	Refined application development plan

Source: Whitten et al., 1994, with modifications.

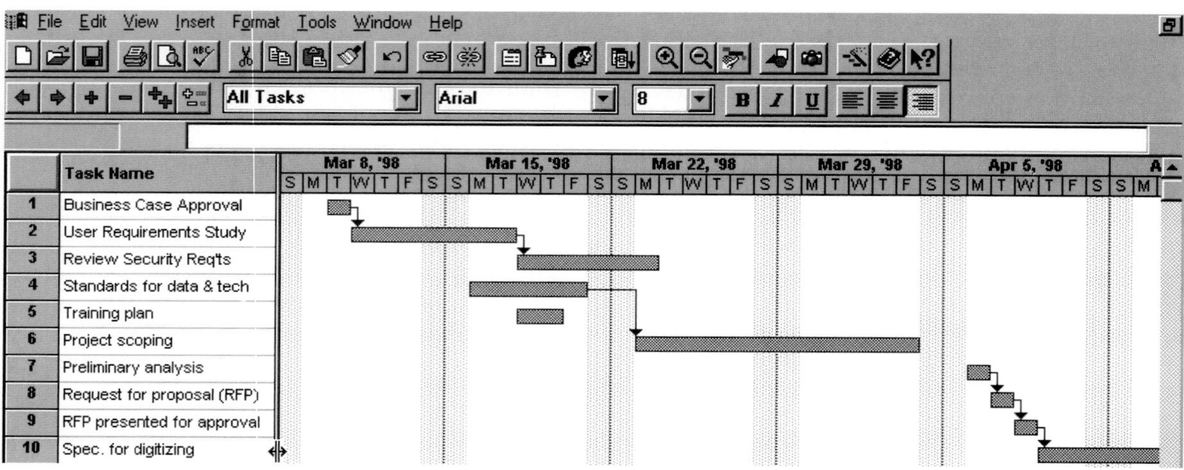

FIGURE 11.3
Example of a Gantt chart. The Gantt chart is probably the most commonly used project-planning and scheduling tool. It may be created using a variety of desktop applications, such as Microsoft Project.

business areas of the organization. In the case of large system development projects, external consultants are sometimes contracted to assist in the planning process as well. By using questionnaires, interviews, and workshops, the systems planning team studies the establishment, the management structure, and the goals of the organization. It also identifies the factors that affect the performance of the business. On the basis of the findings of these activities, a preliminary project plan, complete with a *Gantt chart* (Figure 11.3) and a *network diagram* (Figure 11.4), are drawn up. The study report and the preliminary project plan form the input to the two subsequent stages of work in systems planning.

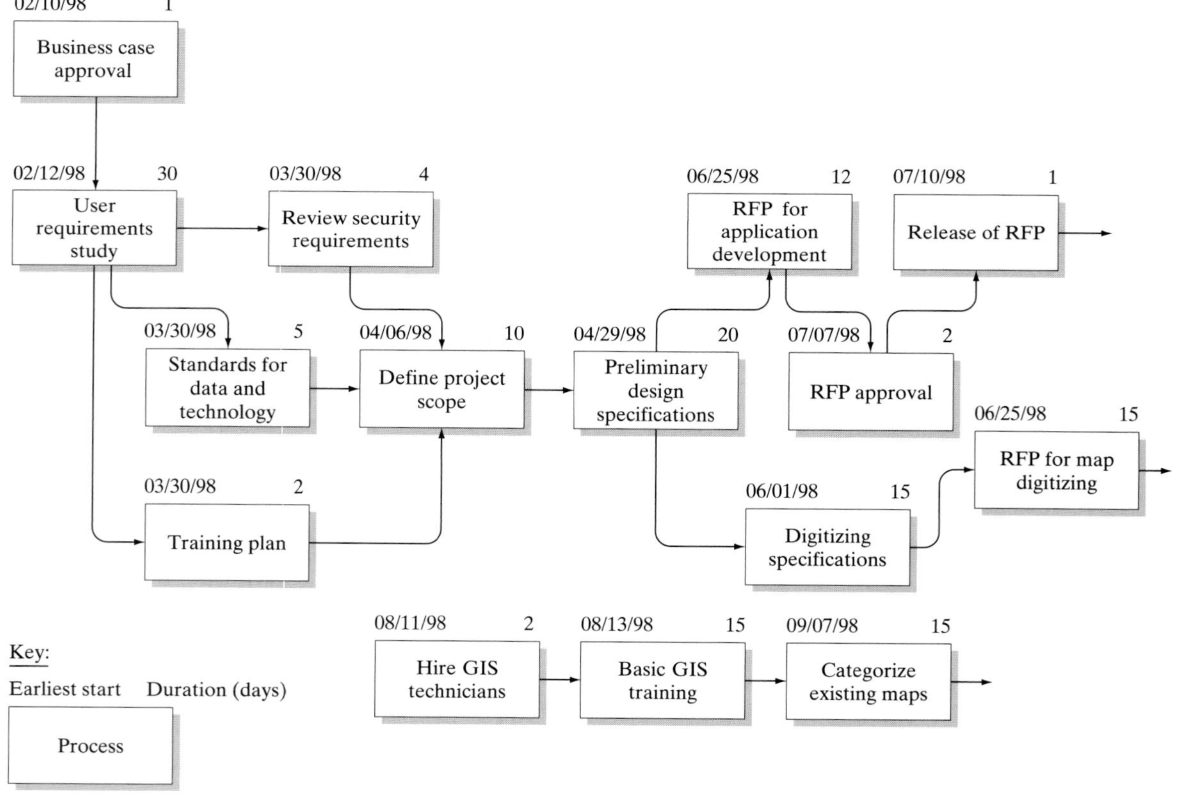

FIGURE 11.4
Example of a project network diagram. The network diagram provides a visual companion to the Gantt chart as a project-planning and scheduling tool.

Defining the Information Architecture of the Proposed System

By definition, an *information architecture* is a plan for using information technology and developing the information system necessary to support the business goals of an organization. An information architecture contains four building blocks corresponding to the four components of an information system (see Chapter 1):

1. *data architecture*, which defines the data resources required to support business operation and decision making
2. *people architecture*, which defines the human resources necessary to operate the proposed system
3. *application architecture*, which defines the critical business functions to be performed by the proposed system
4. *technology architecture*, which includes the hardware, software, and communication network infrastructure of the proposed system

The process of defining the information architecture starts with the development of an enterprise business model that describes the nature and characteristics of the business of the organization at a general level. Current business strategies and information requirements are then assessed against the opportunities of using information technology, as well as the risks of not using it. Finally, the results of the assessment are used to configure the four building blocks of the information architecture as noted above. At this point, it is also common to develop a schedule for procuring hardware, software, network, and other technical services pertaining to the project. This particular stage is always concluded by a review process that aims to refine and consolidate the findings of the different activities into one single *information architecture report*.

Performing Business Area Analysis

"Business area analysis" refers to a suite of systems planning techniques used to identify the requirements of individual business functions for the purpose of sharing databases and technologies. By mapping business information requirements to the proposed information architecture, useful information can be obtained for the design of a highly integrated information system capable of serving the diversified needs of various business areas of an organization. In the principles of enterprise computing and business administration, such an approach of building integrated information systems to enable business processes across departmental boundaries is referred to as *business process reengineering*. Since business area analysis is concerned with the study and redesign of fundamental business processes, it is usually necessary to expand the original systems planning team by including managerial staff from all the affected business areas. The results of business area analysis are used to identify those processes that can benefit most from the use of information technology. The results are also used to identify business areas where sharing of data and technologies are possible. The end product of business area analysis is an application development plan.

11.3.3 PROJECT PROPOSAL

On completion of these three stages in systems planning, the findings and materials are consolidated and compiled as a formal *project proposal* to be submitted for approval by the appropriate authority of an organization. A typical project proposal is a document, usually from 10 to 15 pages in length, that presents in a concise manner the objectives, feasibility, architecture, cost, and benefit of the proposed information system (Figure 11.5). It also contains general information about the systems analysis team, the project approval authority as well as the names of people to whom the document will be distributed for review or other purposes. Since not all readers are likely to be familiar with systems development terminology, it is always advisable to include a glossary of terms as an appendix.

Past experience has shown that many GIS projects failed not because of the lack of technology, but because of a variety of nontechnical factors, such as the lack of long-term corporate commitments, the fragmented approach to systems development, the lack of understanding of user requirements, and the pressure to complete the system within an unrealistically short time frame. The project proposal should aim to minimize the potential of repeating these pitfalls by convincing senior management to "institutionalize" the information system development project. This means that the system must be planned, designed, and implemented from a corporate and strategic perspective. In this regard, the acceptance of a project proposal not only signifies the milestone of completing the first phase of the system development life cycle, but also the commitment to a new vision of implementing information technology in an organization.

11.4 SYSTEMS ANALYSIS AND USER REQUIREMENTS STUDIES

If one of the major causes for the failure of GIS projects in the past was poor design because of the lack of understanding of the information needs of users, as suggested by Marble (1982), the importance of systems analysis can never be stressed enough. However, experience has shown that user requirements have been the most neglected part in information system development.

I. General

- Author(s)
- Contributor(s)
- Approving authority
- Distribution list

II. Executive summary

- General introduction
- Brief statements of problems, opportunities, constraints, and risks
- Brief statements of the objectives of the proposed information system
- Brief statements of the scope of the proposed information system

III. Background information pertaining to business case

- Mission of the organization
- Purpose of the proposed information system in relation to the mission
- Brief description of systems planning methodology and activities

IV. Observations on current business model

- Business performance analysis
- Data and information analysis
- Efficiency and productivity analysis
- Cost/benefit analysis
- Client/service requirements analysis

V. Detailed recommendations

- Objectives of proposed information system
- Information architecture
  - Data architecture
  - Human resources architecture
  - Application architecture
  - Hardware/software and network architecture
- List of constraints and risks
- Business case
  - Total estimated cost
  - Quantifiable monetary and nonmonetary returns
  - Payback period
- Development strategy and priority
- Project schedule, Gantt chart, and network diagram
- Technology procurement schedule

VI. Appendices

A. Systems planning team
B. Dates and summary of meetings and workshops
C. Summary of questionnaire surveys and performance measures
D. Samples of data and information products
E. Glossary of terms

FIGURE 11.5
Contents of a typical project proposal for information system development.

This is the case not only for GIS (Hadzilacos and Tryfona, 1996), but also for information system development in general (Batini et al., 1992). The objective of the systems analysis phase of the systems development life cycle is to ensure that user requirements are properly analyzed and identified before the project is allowed to proceed any further.

Contrary to the general belief that user requirements studies are relatively simple and straightforward tasks, such a study as a software engineering process is a very demanding undertaking, both intellectually and technically. It requires considerable commitments in terms of time and human resources from the project sponsor (which usually means senior management of an organization), the systems staff assigned to the project (i.e., the systems analysts), as well as representatives of end users of the proposed information system.

11.4.1 THE METHOD OF JOINT APPLICATION DEVELOPMENT

There are many ways to carry out a user requirements study: using questionnaires, conducting face-to-face interviews, or holding information-gathering workshops. For large-scale information system development projects, the method of *joint application development* (JAD) appears to be the most commonly used technique today. Originally developed by IBM, JAD is a highly structured workshop that brings together managers, users, and systems analysts to jointly define and specify user requirements. This particular method has many advantages over the conventional techniques of using questionnaire and face-to-face interviews (Whitten et al., 1994).

1. JAD tends to improve the working relationship between the end users and systems staff by working as a coherent group.
2. It provides an excellent educational opportunity for end users to improve their computer literacy; at the same time, it is an opportunity for systems staff to learn the nature and characteristics of the business that the system is designed to serve.
3. By having representatives from different business areas of an organization, JAD provides an excellent platform for conflict resolution in case of disagreements between different users or user groups.
4. JAD increases the productivity in systems development by consolidating multiple interviews into a series of structured workshops.
5. Systems are usually less costly to develop and maintain because JAD allows a good understanding of the user requirements to be correctly defined and prioritized at the beginning.
6. By participating in the development process, users generally have more realistic expectations of and greater confidence in the resulting information system.

Led by a facilitator who is usually an experienced systems analyst, a joint application development team is normally made up of five to seven members who are familiar with the organization. These members make up what is usually called the *core team* or the *focus group*. Other people may be co-opted to attend JAD sessions from time to time. These people include, for example, professional and technical staff who are familiar with the operation of the business but do not have the time to go to all the meetings; specialists in certain aspects of the business invited to provide expert opinion on a particular subject area; managers and supervisory staff who want to keep themselves informed of the progress of the project; and people outside the organization who have an interest in the project, such as clients and business partners.

The core team meets regularly over a period of time in a series of one- or two-day workshops. It is important to separate the workshops by several days so that members will have enough time to review the findings in the previous meeting and to prepare themselves for the next one. In the first meeting, the facilitator has to introduce the members, set the ground rules for the meetings, and explain the procedures to be followed. In the subsequent sessions, the facilitator starts by inviting members to review and refine the findings in the previous meeting. He or she then sets the targets for the session under way and guides them through the business analysis processes by asking them questions about the ways in which business is done. The facilitator may use a variety of techniques to elicit the opinions of the team members. A structured approach called *business function decomposition* has been most commonly used. This is a top-down approach by which individual business functions are identified and continuously decomposed into subfunctions. The process of decomposition may go on to several levels of subfunctions. It will stop only when a particular subfunction can be completed in one single activity or work step. As the process of business function decomposition progresses, the identified functions and subfunctions are uniquely referenced by means of a hierarchical numbering system as shown in Figure 11.6. For this reason, the resulting document is usually referred to as a *business function hierarchy* (BFH).

The BFH is a concise listing of business functions and subfunctions. It may contain many technical and business terms that are alien to the systems analysts and other people involved in the systems development processes. To avoid ambiguities and confusions that may arise when the BFH is used later in the systems design stage, an accompanying document called *business function definition* (BFD) is produced at the same time (Figure 11.7). As its name implies, the purpose of the business function definition is to explain in relative detail the meanings of individual business functions and subfunctions listed in the BFH. In addition to the explanation of the business functions, definitions

Business area: **Manage and produce parcel maps**	→	1. Compile a new parcel map 2. Maintain a parcel map 3. Archive a parcel map 4. Produce a parcel map

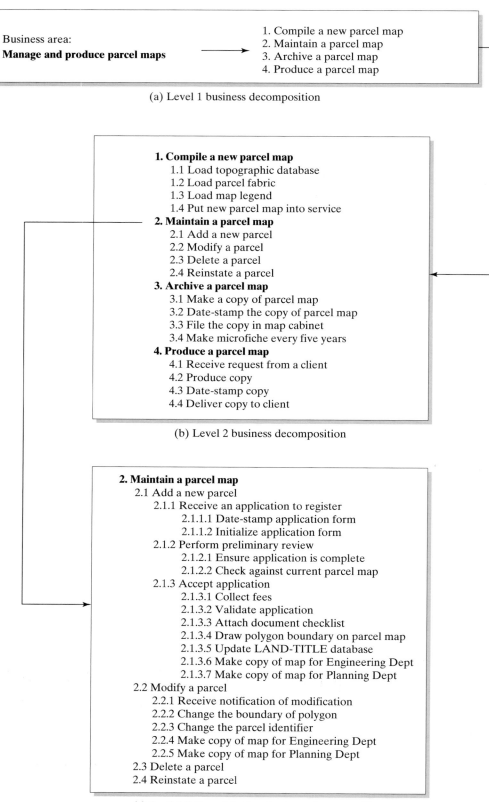

(a) Level 1 business decomposition

1. Compile a new parcel map
- 1.1 Load topographic database
- 1.2 Load parcel fabric
- 1.3 Load map legend
- 1.4 Put new parcel map into service

2. Maintain a parcel map
- 2.1 Add a new parcel
- 2.2 Modify a parcel
- 2.3 Delete a parcel
- 2.4 Reinstate a parcel

3. Archive a parcel map
- 3.1 Make a copy of parcel map
- 3.2 Date-stamp the copy of parcel map
- 3.3 File the copy in map cabinet
- 3.4 Make microfiche every five years

4. Produce a parcel map
- 4.1 Receive request from a client
- 4.2 Produce copy
- 4.3 Date-stamp copy
- 4.4 Deliver copy to client

(b) Level 2 business decomposition

2. Maintain a parcel map
- 2.1 Add a new parcel
  - 2.1.1 Receive an application to register
    - 2.1.1.1 Date-stamp application form
    - 2.1.1.2 Initialize application form
  - 2.1.2 Perform preliminary review
    - 2.1.2.1 Ensure application is complete
    - 2.1.2.2 Check against current parcel map
  - 2.1.3 Accept application
    - 2.1.3.1 Collect fees
    - 2.1.3.2 Validate application
    - 2.1.3.3 Attach document checklist
    - 2.1.3.4 Draw polygon boundary on parcel map
    - 2.1.3.5 Update LAND-TITLE database
    - 2.1.3.6 Make copy of map for Engineering Dept
    - 2.1.3.7 Make copy of map for Planning Dept
- 2.2 Modify a parcel
  - 2.2.1 Receive notification of modification
  - 2.2.2 Change the boundary of polygon
  - 2.2.3 Change the parcel identifier
  - 2.2.4 Make copy of map for Engineering Dept
  - 2.2.5 Make copy of map for Planning Dept
- 2.3 Delete a parcel
- 2.4 Reinstate a parcel

(c) Partial listing of further business decomposition

FIGURE 11.6
The creation of a business function hierarchy (BFH) by business function decomposition.

2. Maintain a parcel map
This function involves all processes for keeping the contents of the parcel map up to date.

2.1 Add a new parcel
A new parcel is added when a client files an application to register land title.

2.1.1 Receive an application to register
An application to register land title is the document (Form LT101) completed by a client; it can be submitted in person or by mail.

2.1.1.1 Date-stamp application form
Date-stamp the application form in the "Received stamp" box. It is important to circle the hour number.

2.1.1.2 Initialize application form
The receiving land title clerk must initialize the application form.

2.1.2 Perform preliminary review
This process involves reviewing new applications at the front counter with the client to ensure that there are no obvious errors such as
(a) incorrect or missing information on the application form
(b) no survey plan attached
(c) missing legal documents

2.1.2.1 Ensure application is complete
The land title clerk checks all boxes in the application form, as well as the declaration and the survey plan showing the location.

2.1.2.2 Check against current parcel map
The land title clerk retrieves the current parcel map of the area concern and checks to ensure that the parcel to be registered is not in conflict with existing parcels.

2.1.3 Accept application
If there are no obvious errors, the application is queued for processing. If there is an error, the application is rejected for corrections.

2.1.3.1 Collect fees
There is a fee for registering a parcel.

2.1.3.2 Validate application
The application form is validated by using the stamp of the cash register.

2.1.3.3 Attach document checklist
The document checklist is a prescribed form (LT102) that will be used to ensure that the correct sequence of registration procedures are followed in the land title registration process (see Appendix A).

2.1.3.4 Draw polygon boundary on parcel map
The parcel polygon is transcribed onto the parcel map; the parcel identifier is written at the visual center of the polygon.

2.1.3.5 Update LAND-TITLE database
Inputting of textual data to update the LAND-TITLE database is a separate process that is usually done after the parcel map is updated.

2.1.3.6 Make copy of map for Engineering Dept
Make a white print and send it to the Surveys and Plans Clerk in the Engineering Dept.

2.1.3.7 Make copy of map for Planning Dept
Make a photocopy of the parcel and surrounding area, not the full parcel map, and send it to Data Services Unit, Planning Dept.

FIGURE 11.7
Partial listing of business function definitions. Business function definitions explain the meaning of the functions identified in the business decomposition process. Note how the definitions are cross referenced to the business functions in the BFH in Figure 11.6 by using the same hierarchical numbering system.

may also include, for example, the frequency of performing the functions, the external events that trigger the functions, as well as the legislation and regulations that are associated with the functions. To facilitate the cross referencing between the BFH and the BFD, function definitions are referenced by identical hierarchical numbers as their corresponding functions and subfunctions.

The documentation of the BFH and the BFD is usually undertaken by a clerk, who is usually referred to as a *scribe*, specially assigned to the job. It is now commonplace for the scribe to do word processing using a computer with peripherals to project the BFH and BFD onto a screen while they are being compiled during a joint application development session. In this way, members may review the accuracy of the statements, refine the wording, and check the workflow by referring to the actual points in the two documents. At the end of a session, the facilitator rounds up by noting the issues that need further clarification. Members are then asked to research the issues according to their technical specialties and interests. They are required to report their findings in the next session. Members are also given a copy of the business function hierarchy and business function definition. They are expected to review the contents of these documents when they are back to their offices. From time to time, the scribe sends out the BFH and BFD documents to a selected group of people (e.g., client representatives and business partners) to solicit their comments. This is usually done when a certain milestone has been reached (e.g., after completing a major business function). Such an external review process helps the joint application development (JAD) core team validate its findings and refine the contents of the two documents.

11.4.2 SYSTEMS ANALYSIS AND THE IDENTIFICATION OF USER REQUIREMENTS

The focus of JAD is on business processes. Little is mentioned about data in JAD sessions. There is one major reason for the deliberate concentration on business processes: People usually think of their work in terms of the functions they perform, but not the data that they use. In order to obtain the business information requirements from the BFH, three further steps are required after its completion. The first involves the identification of those business functions and subfunctions in the hierarchy that require information to operate or to generate information as the result of their operation. This process of identifying information-driven business functions is usually done in the final JAD session when the business function hierarchy has been finalized. The business functions and subfunctions identified are then extracted from the BFH to form what is called a *function-information product table* (FIPT). This table contains two columns: One is for the business functions referenced by the same hierarchical number in the original business function hierarchy; the other column lists the information products associated with the business functions (Figure 11.8a).

The second step is to create an *information product data table* (IPDT) from the function-information product table (FIPT). This table is developed by identifying the data required in the information product, as well as the process that is required to obtain them (Figure 11.8b). This table is used as input for the third step, in which two further tables, called the *data specific table* (DST) and the *data attribute table* (DAT), are produced. The data specific table contains columns denoting the source, type, attribute, and accuracy of the individual data sets in the information product data table (Figure 11.8b). The DAT (Figure 11.8d), on the other hand, contains characteristics pertaining to attributes in the data specific table.

In practice, the process of identifying attributes and their characteristics for the DSTs and the DATs is usually carried out as part of conceptual data modeling in the design phase of the system development life cycle. These two tables are included here mainly for the purposes of giving a complete picture of how user requirements may be identified from the decomposition of business functions and the systematic construction of a series of tables. We will explain the techniques of determining attributes and their characteristics in detail when we talk about conceptual data modeling in Section 11.5.1.

11.4.3 GIS SOFTWARE EVALUATION AND SELECTION

In literature on systems development methodology, software and hardware evaluation is not normally considered an issue of systems analysis. We take exception in this book for a couple of reasons: (1) GIS applications have special software and hardware requirements that are distinct from computer applications in general; and (2) the most suitable time in the systems development life cycle to consider software and hardware requirements is at the end of the systems analysis phase when user requirements have been positively identified. This ensures the compatibility between the functionality of hardware/software and the requirements of the applications to be performed by the proposed information system.

Software evaluation, or *software bench marking*, is a highly subjective process by nature. In the context of software engineering, software bench marking is carried

Function	Information Product
1.1 Load topographic base	Topographic maps 1:1000 series
1.2 Load parcel fabric	Existing parcel maps

(a) Partial listing of function-information product table

Information Product	Data Required	Process
Topographic maps	Topo map layers: - streets - street names - water bodies	Determine extent of maps
Parcel maps	- street block boundaries - parcel boundaries - parcel identifiers	Determine extent of maps

(b) Partial listing of information product data table

Data Set	Source	Type	Attribute	Accuracy
Street segments	Engineering Dept	Line	Class-code	1:1000
Street names	Engineering Dept	Text	Name	
Water bodies	Engineering Dept	Line/polygon	Name	1:1000
Street blocks	Assessor's Dept	Line	Symbol-Code	1:1000
Parcels	Assessor's Dept	Polygon	Symbol-Code	1:1000
Parcel identifiers	Assessor's Dept	Text	Label	

(c) Partial listing of data specific table

Attribute	Type	Range	Domain	Description
Street segment class-code	N, 1	1–3	Street type	Street classification: 1 = street 2 = drive 3 = boulevard
Street names	C, 20		Street names	
Water body symbol code	N, 2	1–22	Carto, symbol set	Code for carto, line type
Water body names	C, 20		River/lake names	
Street block symbol code	N, 2	1–22	Carto, symbol set	Code for carto, line type
Parcel boundary symbol code	N, 2	1–22	Carto, symbol set	Code for carto, line type
Parcel identifier label	C, 15			Parcel identification no.

(d) Partial listing of data attribute table

FIGURE 11.8
The process of identifying business information requirements from BFH by the creation of a sequence of information product and data tables.

out using a suite of quantifiable measures against which functions of selected software packages may be assessed. The purpose is to identify a software package that best suits the requirements of the intended applications, but not the "best" software package in the market. With careful planning and adequate preparation, a relatively objective comparison between different software packages may be derived, and the most suitable package picked accordingly. A structured approach to GIS software bench marking may conducted by the following steps:

1. Using a top-down approach, develop an application requirements table listing the software capabilities required by the business functions identified in the systems analysis phase (Figure 11.9).

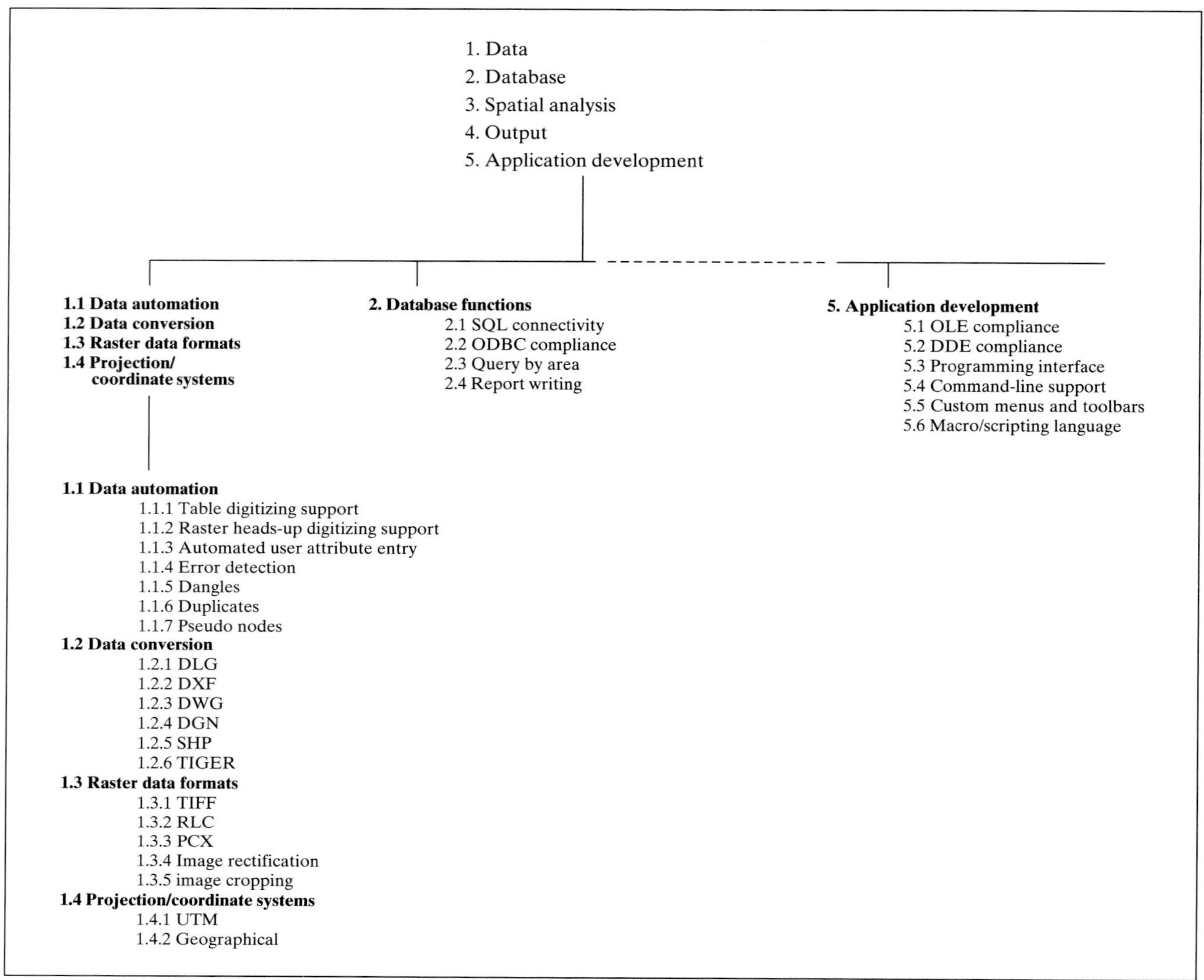

FIGURE 11.9
The top-down approach of developing GIS software requirements. The process starts by identifying the major functions (i.e., data, database, spatial analysis, output, and applications). These are then broken down to more detailed requirements as illustrated in the diagram.

2. On the basis of the application requirements table, translate the application requirements into technical specifications that indicate exactly what the software is expected to perform.
3. Classify individual items in the technical specifications into two categories: "mandatory" and "desirable" with respect to the business requirements established in the user requirements analysis (Figure 11.10).
4. Send out the technical specifications to software vendors and invite them to respond by indicating whether their software packages are capable of meeting the technical specifications.
5. In the invitation to respond, ask vendors to supply supplementary information such as the availability of a macro or scripting language for custom programming, the purchasing and maintenance costs of the software, conditions of license agreement, technical support and training available.
6. Evaluate the responses from software vendors, and select those packages that satisfy all the mandatory requirements for further evaluation.
7. Prepare a software bench-marking package containing test data that reflect the characteristics of the data used in the daily operation of business in the organization.
8. Form a software evaluation team made up of systems and non-systems staff; an external consultant may be hired to assist in the evaluation process if necessary.

1.1 Data automation

1.1.1 The software must be able to support table digitizing*.
1.1.2 The software must be able to support raster heads-up digitizing without the use of a third-party add-on software package*.
1.1.3 The software must be able to perform automated user attribute entry by reading both ASCII text and dbf files*.
1.1.4 The software must be able to detect digitizing errors automatically, using a user-defined tolerance*.
1.1.5 The software must be able to remove dangles automatically, using a user-defined tolerance*.
1.1.6 The software must be able to detect and remove duplicates automatically*.
1.1.7 The software should be able to remove pseudo nodes if required.

1.2 Data conversion

1.2.1 The software must be able to accept DLG data without the use of a third-party translator*.
1.2.2 The software must be able to accept DXF data without the use of a third-party translator*.
1.2.3 The software should be able to accept DWG data without the use of a third-party translator.
1.2.4 The software should be able to accept DLG data without the use of a third-party translator.
1.2.5 The software must be able to accept SHP data without the use of a third-party translator*.
1.2.6 The software must be able to accept TIGER data without the use of a third-party translator*.

1.3 Raster data formats

1.3.1 The software must be able to accept TIFF data without the use of a third-party translator*.
1.3.2 The software must be able to accept RLC data without the use of a third-party translator*.
1.3.3 The software should be able to accept PCX data without the use of a third-party translator.
1.3.4 The software must be able to rectify raster images*.
1.3.5 The software should be able to crop images using a user-defined window.

1.4 Projection/coordinate systems

1.4.1 The software must be able to transform data in UTM into geographical coordinates on the fly*.
1.4.2 The software must be able to transform data in geographical coordinates into UTM on the fly*.

Note: * denotes mandatory requirements

FIGURE 11.10
Technical specifications of software requirements. The software requirements in Figure 11.9 are translated into detailed specifications that are identified as "mandatory" and "desirable."

9. Invite the selected software vendors for a presentation of their products and a practical demonstration of all the functions that they claim their software has.
10. Evaluate the vendor's demonstration by assigning a score to individual functions according to the level of ease with which the function is performed; both mandatory and desirable functions are tested, but their weighting will be different; packages that fail to perform any one of the mandatory requirements are disqualified.
11. Evaluate the quality of the manuals, training materials, and technical support facilities (e.g., completeness of the manuals, types of training programs available, and the modes of technical support).
12. Consolidate the scores for different software vendors and, if required by law or corporate policy, check the references supplied by the vendor.
13. Select the software that has been awarded the highest score and can be acquired within an acceptable price range.

11.4.4 HARDWARE CONSIDERATIONS AND ACQUISITION

Just like software, hardware for a particular information project can be selected only after the completion of the systems analysis phase of the SDLC. Hardware selection usually centers around the selection of the appropriate hardware platform, which for GIS implementation often means the choice between workstations based on UNIX or the Windows operating systems (Maguire, 1998). As the technology stands, the differences between these two types of platforms are fast disappearing in terms of both cost and performance. In the GIS industry, this has made possible the migration of many UNIX-based software products to the Windows platform. The following is a checklist of factors that need to be considered when choosing a Windows-based GIS workstation (Graham, 1998). Similar factors may also be used for the selection of hardware for UNIX-based systems.

1. *Basic Hardware Configuration.* Based on the state of computer technology in early 2002, a typical

high-end Windows XP-based workstation was a 1.6 to 2.0 GHz Pentium 4 system, equipped with up to 2GB of memory, one or more 40 GB hard drives, CD-ROM drive, 17- to 19-inch monitor, removable 100 MB cartridges or other forms of back-up devices, graphics card, network cards, sound card, and speakers.

2. *Microprocessor (CPU) Speed.* Experience has shown that microprocessor speed doubles every 18 months, therefore it is advisable to purchase the fastest processor possible.
3. *Number of Microprocessors (CPUs).* The ability to use multiple microprocessors is software-dependent, which means that not every software package is able to take advantage of the improved power of a dual CPU system.
4. *Disk Controllers and Disk Type.* As GIS applications are data-intensive by nature, the use of the *small computer system interface* (SCSI) multithreaded input/output (I/O) technology is highly recommended so that the system can take full advantage of XP's multitasking capabilities.
5. *Random Access Memory (RAM).* Although many GIS software packages require only 32 MB RAM to run, it is advisable to install the maximum amount of RAM allowable by the system.
6. *Video and Monitors.* As a minimum, a 19-inch monitor is required, but a 21-inch one is always recommended.
7. *Printing and Plotting.* Make sure that the printer/plotter is able to generate information products required by all applications in terms of output quality, dimension, drawing media, and speed; it is also important to make sure that the printer/plotter is supported by the GIS software package selected. Check with the software vendor about the need to use add-on products (e.g., ArcPress for ArcInfo and ArcView).
8. *Digitizers Supported.* Make sure that the digitizer has the necessary driver for the GIS software package selected, and that its size, resolution, and other characteristics are compatible with the requirements of the applications; if digital data are to be acquired from government agencies or commercial data suppliers, or if scan-digitizing is to be provided by a service bureau, the purchase of a digitizer may not be necessary.
9. *Network and Remote Access.* Make sure that the system is configured with the necessary software for access to data residing in both Windows-based and UNIX-based servers. It is also essential to consider the estimated volume of data and their growth rate, the expected volume of data traffic, and the sharing of data between different servers when developing a network strategy.
10. *Systems Security.* Make sure that the system is capable of accepting hardware and software security mechanisms that will prevent unauthorized use of the applications.

It should be noted that while the capability of hardware is rapidly increasing, the price is decreasing. To capitalize the most from investments on hardware, it always makes economic sense to hold the purchase of hardware items until the last moment. This is probably the best strategy to ensure that the latest technology is acquired at the lowest possible cost in information system implementation.

11.5 Geographic Database Design Methodology

The design phase of the systems development life cycle (SDLC) is made up of two interrelated processes: database design and application software design. The input to both processes is the information obtained from systems analysis in the previous phase of the SDLC. The output will be a set of *functional and design specifications* of the proposed information system. Systems design is largely the work of systems analysts, but end users are also intimately involved in various aspects of the processes. In this section, we will explain database design by data modeling. The methods of application development will be studied separately in Section 11.6.

As noted in Chapter 3, data modeling is the process of defining real-world phenomena or geographic features of interest in terms of their characteristics and their relationships with one another. The objective is to turn *unstructured* real-world problems into *structured* descriptions and representations as an intermediate step in the development of software solutions. Conventionally, data modeling is carried out in three sequential stages: conceptual, logical, and physical modeling. When the data modeling process proceeds from the conceptual through the logical to the physical stage, the database becomes more rigorously defined, resulting in a series of progressively formalized descriptions and specifications of the database called conceptual, logical, and physical schemas.

11.5.1 Conceptual Data Modeling

As the starting point of database design, the purpose of conceptual data modeling is to define in broad and generic terms the scope and requirements of a database.

Conceptual data modeling is carried out independently of software and hardware that will be used to implement the database. It is high-level modeling in the sense that it describes the *contents*, rather than the *storage structure*, of the database in a natural way, using human languagelike expressions and diagrammatic representations known as *conceptual schemas*. It should be noted that despite the availability of automated systems design tools such as *computer-assisted software engineering* (CASE), conceptual data modeling has remained largely a manual process. The process of understanding and transforming user requirements into conceptual schemas is probably too complicated to be done effectively by machines. In practice, CASE tools are generally used only after the first stage of conceptualization.

There are many approaches to conceptual data modeling. The *entity–relationship* (E–R) model is probably one of the most widely used methods today. When it was first introduced, the E–R model included only the concepts of *entity*, *relationships*, and *attributes* (Chen, 1976). As the model became more widely used in later years, more sophisticated concepts such as *composite attributes* and *generalization hierarchies* were added as components of what has come to be known as the *enhanced* or *extended E–R* (EE–R) model (Batini et al., 1992). For the purpose of illustrating how this particular conceptual data modeling technique is used, we try to keep things simple by limiting our discussion to the original E–R concepts only.

In order to use the E–R model, it is necessary to understand the definitions of the basic concepts.

- *Entities* represent real-world objects (i.e., a person, a place, a thing, a concept, or an event of interest) about which data are collected and stored.
- *Entity types* are those that share the same characteristics and are sometimes also referred to as entity classes.
- *Relations* represent the natural association between two or more entities or entity types.
- *Attributes* represent elementary properties of entities and relationships.
- *Cardinalities of relationships* refer to the degree of relationships, which may be *one-to-one* (1 : 1); *one-to-many* (1 : m) or *many-to-many* (m : m).

Conventionally, conceptual data modeling using the E–R model involves four steps: (1) identifying entities, (2) identifying attributes, (3) determining relationships, and (4) drawing an E–R diagram.

Identification of Entities

The objective of this task is to identify those entities relevant to the business functions that the database is developed to serve. This can be done by one of the following two ways: (1) extracting key words from the business function hierarchy developed at the systems analysis phase of the SDLC (see Sections 11.4.1 and 11.4.2), and (2) sampling and analyzing existing data forms, a process that is usually referred to as *form analysis*.

To identify potential candidates of entities in a data form, examine the content of the form carefully and group the data into "information units" that satisfy the following conditions (Figure 11.11):

- Entities may be people, things, places, concepts, or events of interest.
- Entities have multiple occurrences (i.e., they occur hundreds or thousands of times).
- Entities must be relevant to the information objectives of the database (e.g., the name of the data collector in a data form is normally not what the end user will be interested in, and is therefore not considered as an entity).

To complete this task, it is necessary to give each identified entity a name that must be simple, meaningful, and business-oriented. As a rule, entity names are singular and the use of abbreviations or acronyms should be avoided. It is also necessary to define each entity (i.e., to write a concise description of each entity) and to keep a record of all the definitions in the project repository for later reference.

Identification of Attributes

Attributes are descriptions or characteristics of entities. They may be identified in the same way in which entities are identified by form analysis. Some data forms such as the one in Figure 11.11 are relatively simple, and attributes may be recognized fairly easily. Other data forms may be very complex, and the following structured approach by the process of elimination may be very useful in identifying relevant attributes (Figure 11.12):

1. Examine the form carefully, and circle each unique item.
2. Eliminate the following items by drawing a cross in the circle.
   - items that will not be stored in the system, i.e., items that are not relevant to the end users' information needs
   - items such as signatures and the date of data collection
   - constant information such as the address of the data collection agency and form identification numbers
3. Verify the remaining circled items as candidate attributes with the end users.

Municipality of Annytowne
Buildings and Lands Department

Ground Floor, East Tower, City Hall
1 City Centre Boulevard,
Annytowne, ZX, 90009-1234

Property Record Form

Form #: BLD-111-98rev

Date created: ______ Created by: ______

Particulars of owner: **OWNER**
Name: ______ Social Security No. ______
Address: Street # ______ Apart # ______ Street Name ______
City ______ ZIP ______
Telephone: Area code: ______ Telephone # ______ Ext. # ______

Particulars of property: **LAND TITLE**
Civic address: ______
Legal description:
Sale date: ______ Sale price: $ ______ Land price: $ ______
Deedbook page: ______ Deeded acreage: ______
Front footage: ______ Building type: ______ Year built: ______
School districts: Elementary ______ High ______
Voting district: ______ Census tract: ______ Fire district: ______
Soil type: ______ Flood zone: ______ Planning zoning: ______ **PARCEL**

Tax roll information (to be filled by Assessor's Department)
Tax roll #: ______ Tax rate: ______
Annual tax: ______ Tax map: ______ **TAX ROLL**

Duplication:
1) Planning Department Copied by: ______ Date: ______
2) Assessor's Department Copied by: ______ Date: ______
3) Works Department Copied by: ______ Date: ______

Data filing:
By: ______ Position: ______

Signature: ______ Date: ______

FIGURE 11.11
Identification of entities by form analysis. The first step in form analysis is to identify potential entities. Data items in the form are grouped into relevant "information units" with reference to users' requirements analysis. Information not of interest to the business requirements, for example, department name, department address, form identifier (form name and number), date of data entry, and signature and name of data collector, will be ignored. In this example, four "information units" have been identified: OWNER, LANDTITLE, PARCEL, and TAX ROLL.

4. Examine the names of the attributes carefully, and if required, change them to make them more understandable (e.g., from "File #" to "Registry File Number").
5. Map the attributes to the entities identified in the previous step.

Determination of Relationships

The relationships between entities may be established by drawing a matrix of the entities and giving a description of the association of the entities concerned. It should be noted that relationships are bidirectional in nature and they must be recorded as exemplified in

Municipality of Annytowne
Buildings and Lands Department

Ground Floor, East Tower, City Hall
1 City Centre Boulevard,
Annytowne, ZX, 90009-1234

Form #: BLD-111-98rev

Property Record Form

Date created: ______ Created by: ______

Particulars of Owner:
Name: ______ Social Security No. ______
Address: Street # ______ Apart # ______ Street Name ______
City ______ ZIP ______
Telephone: Area code: ______ Telephone # ______ Ext. # ______

Particulars of property:
Civic address: ______
Legal description:
Sale date: ______ Sale price: $ ______ Land price: $ ______
Deedbook page: ______ Deeded acreage: ______
Front footage: ______ Building type: ______ Year built: ______
School districts: Elementary ______ High ______
Voting district: ______ Census tract: ______ Fire district: ______
Soil type: ______ Flood zone: ______ Planning zoning: ______

Tax roll information (to be filled by Assessor's Department)
Tax roll #: ______ Tax rate: ______
Annual tax: ______ Tax map: ______

Duplication:
1) Planning Department Copied by: ______ Date: ______
2) Assessor's Department Copied by: ______ Date: ______
3) Works Department Copied by: ______ Date: ______

Data filing:
By: ______ Position: ______

Signature: ______ Date: ______

FIGURE 11.12
Identification of attributes by form analysis. Attributes in the data form are identified by the process of elimination. All names, dates, and addresses are circled as potential attributes first. They are then evaluated against their relevance to the business requirements of the intended applications. Those that are not regarded as relevant are eliminated by crossing them out as shown in the diagram. The remaining candidate attributes are then matched with the entities identified earlier.

Figure 11.13. It is helpful to make use of the matrix to determine the cardinalities of relationships between entities by figuring out how they are associated. The cardinality between landowner and parcel in Figure 11.13, for example, may be determined by the following ways of reasoning:

- An individual landowner must be associated with at least one parcel, but he or she may also be associated with many parcels (i.e., a landowner may own more than one parcel at the same time).
- A parcel has at least one owner, but it may be owned by more than one owner (i.e., a single

	Owner	Parcel	Land title	Tax roll
Owner		Owns 1:m		
Parcel	Is owned by 1:m		Is subject of 1:1	
Land title		Registers 1:1		Is used by 1:1
Tax roll			Uses 1:1	

FIGURE 11.13
Determining the relationships between entities and the cardinalities of relationships. The relationships between the entities may be explained as follows:
- Owner owns parcel (one owner can own many parcels).
- Parcel is owned by owner (one parcel can be owned by many owners).
- Land title registers land status of parcel (one land title for one parcel).
- Parcel status is subject of land title (one parcel has one land title).
- Tax roll uses land title information to calculate property tax (one tax roll item uses one land title).
- Land title information is used by tax roll (one land title is used by one tax roll item).

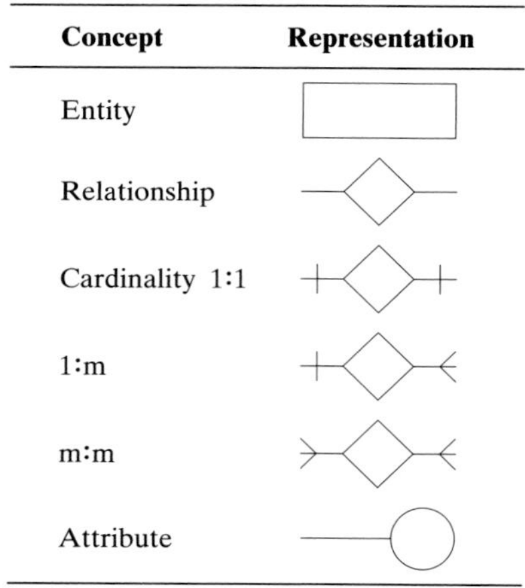

Concept
Representation
Entity
Relationship
Cardinality 1:1
1:m
m:m
Attribute

Example

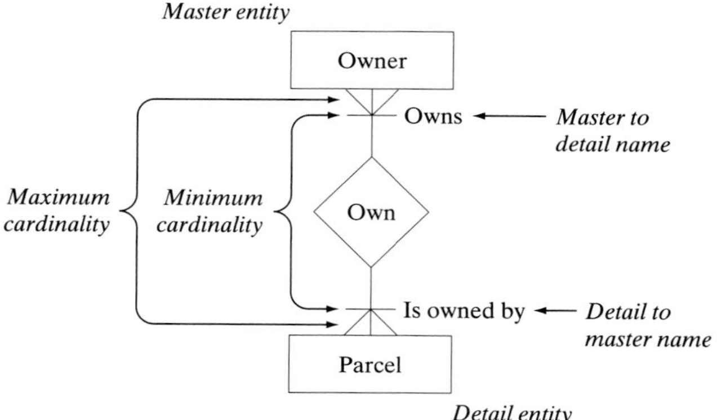

FIGURE 11.14
Symbols used to construct entity–relationship (E–R) diagrams.

parcel can be co-owned by more than one owner at the same time).

- Therefore, the minimum cardinality between the entities landowner and parcel is one-to-one (1 : 1), and the maximum cardinality is many-to-many (m : m). The cardinalities of relationships between other entities may also be determined in a similar manner.

Drawing an E–R Diagram

An E–R diagram is constructed using a set of symbols that represent the basic concepts of the E–R model graphically (Figure 11.14).

- An entity is represented by a rectangle.
- Attributes of an entity are represented by circles.
- A relationship is depicted by a diamond that is connected to one or more associated entities.
- The cardinality of relationship is depicted with either a numerical expression (e.g., 1 : m) or a graphical expression using the following representations.
    - A "crow's-foot" at the end of a line indicates many occurrences.
    - A bar on the line indicates a single occurrence.
    - An "O" on the line indicates zero occurrence.

In Figure 11.14, the two inner symbols show the "minimum" cardinality when the relationship is viewed from each direction; the outer symbols show the "maximum," also viewed from each direction. The entity at the end of the relationship that has a maximum cardinality of "one" is called the *master entity* for the relationship. The entity at the "many" end is known as the *detail entity*. If the maximum cardinalities are the same at each end of the relationship (i.e., both are "one" or "many"), the source becomes the master entity and the target becomes the detail entity.

In general literature on database design, discussion on the E–R model has always been limited to its use for modeling descriptive data. The E–R diagram may in fact be applied equally well to the conceptual modeling of spatial data. In Figure 11.15, for example, the E–R diagram is used to depict spatial entities and their relationships in the land parcel information system. Descriptive entities and their relationships in the same information system are represented in Figure 11.16

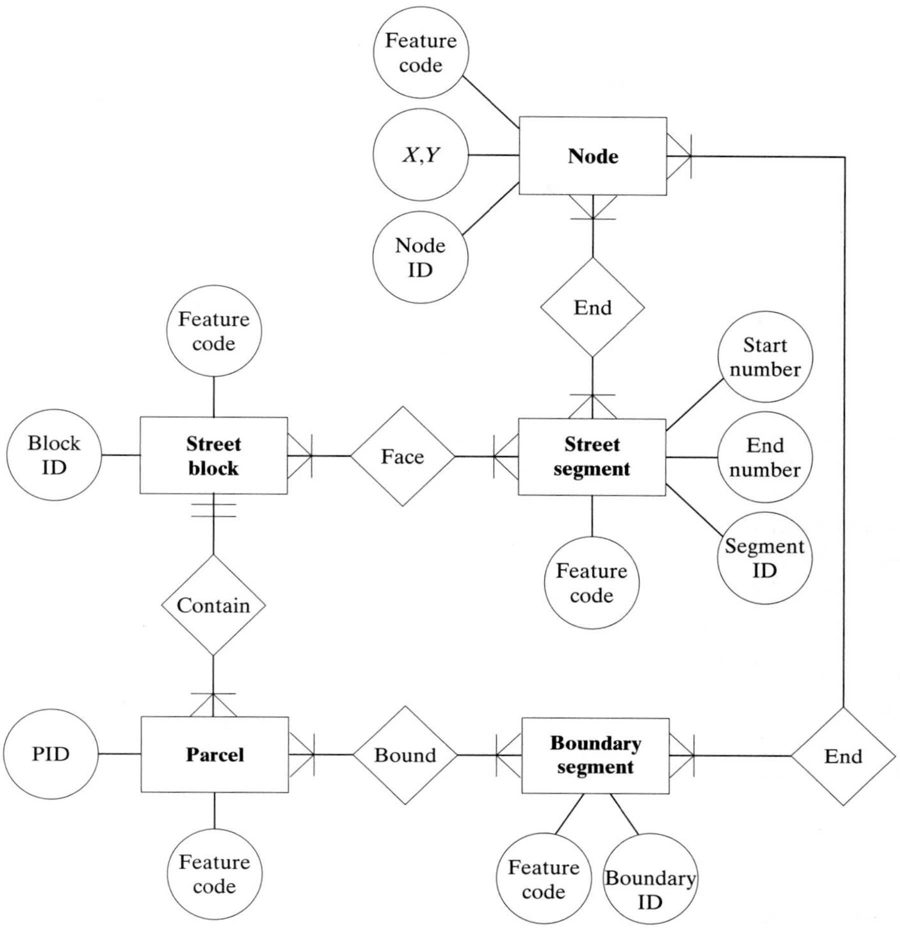

FIGURE 11.15
Modeling relationships between spatial entities by E–R diagram. This E–R diagram illustrates the relationships between polygon features (street blocks and parcels), line features (parcel boundary segments and street segments), and point features (nodes) in the spatial data database.

using the information obtained from Figures 11.11, 11.12, and 11.13.

11.5.2 LOGICAL DATA MODELING

Conceptual data modeling using the E–R model or other models is system-independent. In order to implement the database design as defined by the system-independent conceptual data model, it is necessary to map it to a system-specific model. Logical data modeling is the process by which the conceptual schema is consolidated, refined, and then converted to a system-specific logical schema. A logical schema is not yet a full implementation schema because it is expressed only in terms of database characteristics without any consideration of hardware requirements such as storage structures and data volumes. In logical data modeling, the proposed database is viewed in its entirety. The objective is to identify potential problems that may exist in the conceptual data model, such as

- irrelevant data that will not be used
- omitted or missing data
- inappropriate representation of entities
- lack of integration between various parts of the database

As logical data modeling is system-specific, the actual processes used to address these problems are dependent on the particular model of the database management system (DBMS) chosen for the information system. For the relational database model, which is the most commonly used database model for GIS

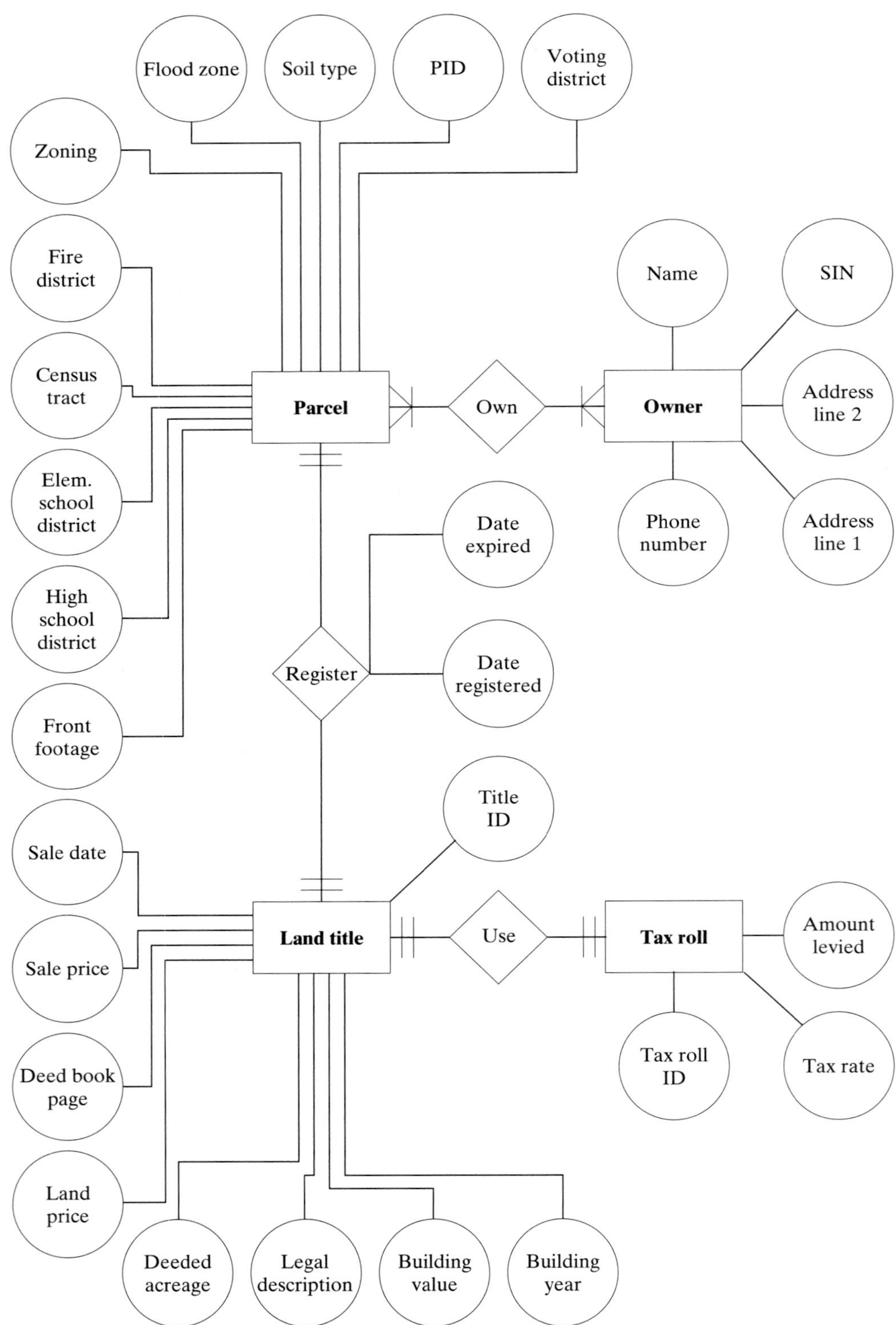

FIGURE 11.16
Modeling relationships between descriptive entities by E–R diagram. This E–R diagram illustrates the relationships between owner and parcel (owner owns parcel); land title and parcel (land title registers land status information of parcel), and land title and tax roll (tax roll uses land title information to calculate annual tax). Note that it is possible for a relationship to have attributes, for example, date-registered and date-expired for the "register" relationship indicate the dates of land title registration and expiry respectively.

today, the following processes are typically carried out: (1) mapping the conceptual schema to the logical schema, (2) identifying keys and foreign keys, and (3) normalizing the attribute tables. For geographic databases, it is necessary to additionally perform layer or coverage design according to the data structure of the selected software package to implement the GIS.

Mapping the Conceptual Schema to the Logical Schema

A conceptual schema such as the E–R diagram may be mapped into a system-dependent relational schema using either computer-assisted or manual methods. In large-scale database design, this process is usually done with the aid of CASE tools. Many of these automated database design tools accept an E–R diagram as input and allow the database designer to interactively develop the logical schema graphically. The resulting logical schema is then automatically converted into a relational database schema in the *data definition language* (DDL) of the selected relational DBMS. Mapping a conceptual design to a relational schema by a manual method is a relatively tedious task that is feasible only for small-scale projects. It may be carried out as a seven-step procedure as suggested by Elmasri and Navathe (1994). The resulting logical schema will be a suite of relations or tables for each relationship, entity and attribute in the E–R diagram. The LAND-PARCEL relational schema shown in Figure 11.17 has been developed from the E–R diagram in Figure 11.16 by following the method of Elmasri and Navathe (1994).

Identification of Keys and Foreign Keys

Tuples (i.e., occurrences of an entity) in a relation may be identified by means of an attribute called a *key*. If the attribute values allow users to identify individual tuples uniquely, the key is called a *primary key*. In the relations of the land parcel information system, for example, the parcel identification number (PID) may always be used as a primary key because it is unique to each parcel. In relational database design, the selection of an attribute to be used as a primary key is subject to what is called the *entity integrity constraint*. Such a constraint requires that no attribute with null values be permitted to serve as the primary key. This is because having null values for the primary key means that it is not possible to identify some tuples. If, for example, two or more tuples had null for their primary keys, it would be impossible to distinguish them from one another.

Occasionally, the use of a certain attribute alone in a relation cannot guarantee the unique identification of a tuple. Then it is necessary to adopt one of the following solutions: (1) use one or more additional attributes to form a concatenated or compound key, or (2) create an additional attribute to hold a unique identifier for each occurrence of an entity. In the second case, the attributes have no real business meaning other than the purpose of providing a unique identification for each

OWNER

Name	SIN	Add-line1	Add-line2	Phone-num

PARCEL

PID	Front-ft	Soil-type	Fire-dist	Elem-sch-dis	Hi-sch-dist	Cen-trt	Flood-zn	Vot-dist	Zoning

LANDTITLE

Title-ID	Parcel-ID	Sale-date	Land..................................Deeded-acre	Legal-desc	Bldg-yr

TAXROLL

TR-ID	Tx-rate	Tx-amt	LT-ID

REGISTER

PID	LT-ID	Date-reg	Date-exp

FIGURE 11.17
Logical schema developed from Figure 11.16. Explanatory notes:

- Each entity forms a relation with its identified attributes.
- Although abbreviations are not recommended for entity names, they are commonly used for attribute names in order to avoid long attribute names.
- The order or sequence of the attributes in a relation is unimportant.
- Different attribute names may be used in different relations to present the same attributes (e.g., SIN in OWNER relation and Owner-ID in PARCEL relation; PID in PARCEL relation and Parcel-ID in LANDTITLE relation; and Title-ID in LANDTITLE relation and LT-ID in TAXROLL relation).

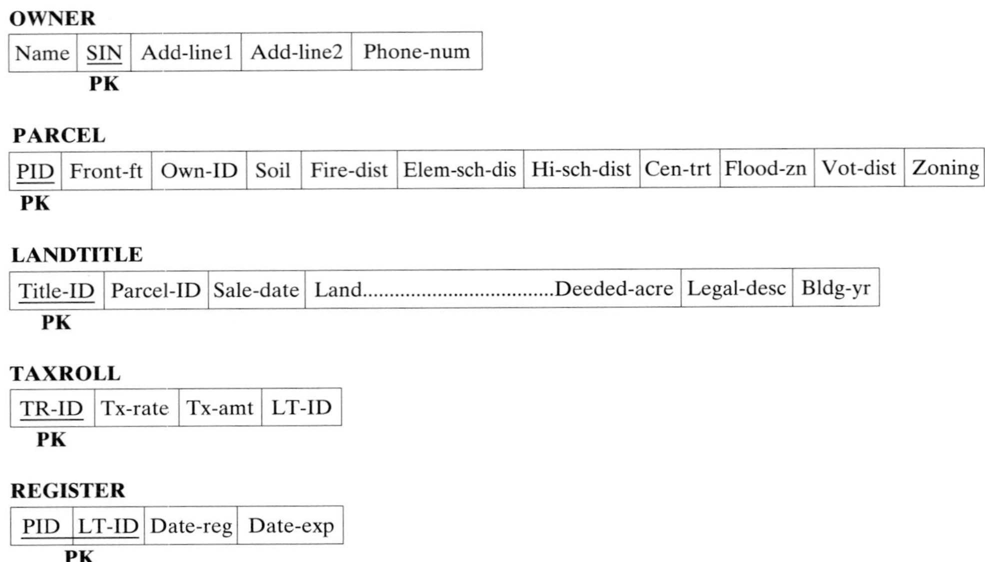

FIGURE 11.18
Identification of primary keys. Primary keys are underlined and marked by "PK" beneath. The primary key for the REGISTER relation is a composite key made up of PID from the PARCEL relation and Title-ID from the LANDTITLE relation.

tuple in a relation. Primary keys are conventionally identified in the logical schema by having their attribute names underlined, usually with an annotation "PK" written beneath them as well (Figure 11.18).

Data held in different relations may be linked to one another as long as they have at least one attribute in common. This leads to the concept of the *foreign key* which, by definition, is an attribute in a relation that is a primary key in another relation. The identical values of the primary and the foreign keys make it possible to logically link different tuples in different relations. Just like the primary keys, foreign keys are also indicated by an underscore in the logical schema. They are distinguished from the primary keys by a different annotation "FK" (Figure 11.19). The concept of the foreign key is used to enforce the *referential integrity constraints*

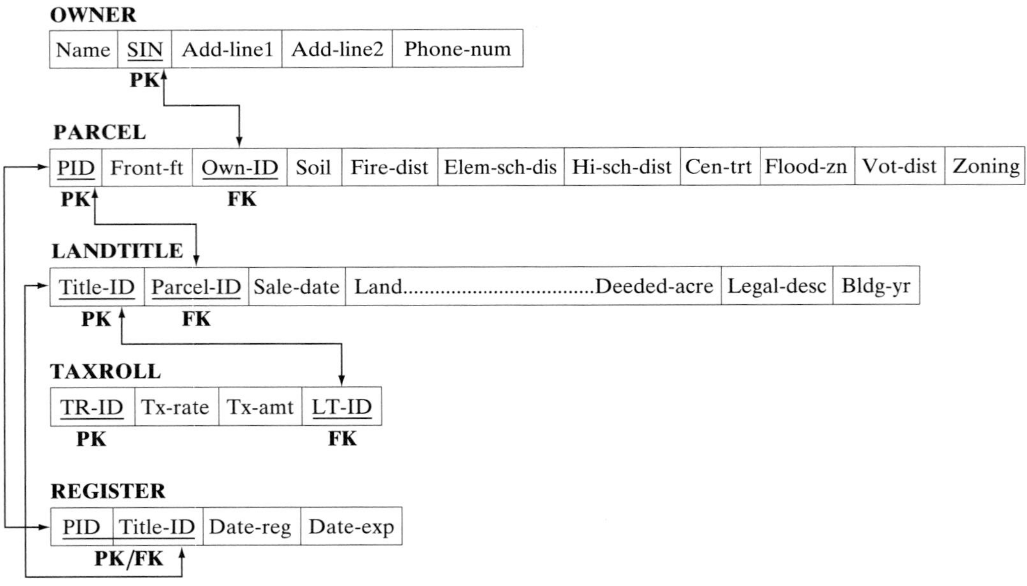

FIGURE 11.19
Identification of foreign keys and verification of referential integrity. Foreign keys are underlined and marked by "FK" beneath. Referential integrity is confirmed by joining primary keys and the foreign keys they refer to in other relations. The primary key for the REGISTER relation is also used as a foreign key.

between two relations in database design. The objective of this constraint is to maintain the consistency among tuples of the two relations by ensuring that a particular tuple in one relation that refers to another relation must refer to an existing tuple in that relation. In a database of multiple relations, there are usually many referential integrity constraints. These constraints are generally checked and displayed graphically by drawing a direct arc from each foreign key to the relation it references. For clarity, the arrowheads always point to the primary keys of the referenced relations as shown in Figure 11.19.

Normalization of Attribute Tables

Normalization of attribute tables is another process that is carried out to address the problem of database integrity at the logical data modeling stage. The objective is to make sure that all relations in the database conform to relational DBMS requirements by satisfying the following two conditions: (1) each attribute in a relation must be functionally dependent on the primary key, and (2) the storage of data must be minimized to maintain data integrity. The process of normalization is built on the concept of *normal form*. A relation is said to be in a particular normal form if it satisfies a certain prescribed set of conditions. When the concept of normalization was initially introduced (Codd, 1971), three normal forms were defined. Subsequently, more general definitions of these forms, which take into account all candidate keys other than the primary key, have been developed (Date, 1995). As a minimum, all relations in a relational database must satisfy the conditions of the first, second, and third normal forms as explained below. Higher normal forms are applied to advanced database development with more sophisticated integrity requirements.

A relation is said to have satisfied the condition for the *first normal form* (1NF) if it contains no repeating attributes (observations or measurements). If, for example, our land parcel information system contains a LANDUSE relation with the following attributes:

- PID: parcel identification
- LU90: land-use status of the parcel in 1990
- LU60: land-use status of the parcel in 1960
- AREA: area of the parcels
- LU-CODE: land-use classification code
- SYMBOL: cartographic symbol used to shade the parcel on a map

The LU60 attribute is regarded as a repeating observation because it provides land-use information about the same parcels, although at different points in time. This column is therefore not permitted and must be removed from the relation to make it satisfy the condition for the 1NF (Figure 11.20).

PID	LU90	LU60	AREA	LU-CODE	SYMBOL
00001	Commercial	Commercial	6000	3	shade-001
00002	Commercial	Residential	5800	3	shade-001
00003	Industrial	Open	8000	2	shade-002
00004	Residential	Residential	4500	1	shade-003
00005	Residential	Residential	6000	1	shade-003
00006	Commercial	Residential	9050	3	shade-001
00007	Industrial	Industrial	8500	2	shade-002
.....					
.....					
.....					
15601	Recreational	Recreational	10,000	7	shade-007
15602	Commercial	Residential	9000	3	shade-003
15603	Institutional	Institutional	7500	8	shade-008

(a) A flat file of land-use data in the years of 1990 and 1960

PID	LU90	AREA	LU-CODE	SYMBOL
00001	Commercial	6000	3	shade-001
00002	Commercial	5800	3	shade-001
00003	Industrial	8000	2	shade-002
00004	Residential	4500	1	shade-003
00005	Residential	6000	1	shade-003
00006	Commercial	9050	3	shade-001
00007	Industrial	8500	2	shade-002
.....				
.....				
.....				

(b) LANDUSE-90 relation of 1NF showing land-use data in the year of 1990

FIGURE 11.20
Formation of a relation of the first normal form (1NF). The table in (a) is a flat file. It is simply a collection of data records. This file is turned into a relation of the first normal form LANDUSE-90 in (b) by eliminating repeated observations in LU60. Data in the LU60 column are repeated observations because they represent the same kind of data (i.e., land use) at a different point in time.

To satisfy the *second normal form* (2NF), a relation must have satisfied the condition for the 1NF as well as the condition that all non-key attributes are functionally dependent on the primary key. In the new LANDUSE-90 relation in Figure 11.20, PID is the primary key. Of all the non-key attributes, only LU90 and AREA are functionally dependent on this key. The other two attributes LU-CODE and SYMBOL are functionally dependent on the attributes of LU90 rather than on the PID of the parcel. LU-CODE and SYMBOL should therefore be removed to make way for a new LANDUSE-90 relation of 2NF. These two attributes are used to form a new relation called LANDUSE-CODE as shown in Figure 11.21b.

Tables are said to be in the *third normal form* (3NF) if they satisfy the conditions for 1NF and 2NF, as well as the condition that there is no transitive dependency of attribute on the primary key. "Transitive" dependency in this context means *indirect* dependency. In the LANDUSE-CODE relation in Figure 11.21b, for example, the SYMBOL attribute is directly dependent

PID	LU90	AREA
00001	Commercial	6000
00002	Commercial	5800
00003	Industrial	8000
00004	Residential	4500
00005	Residential	6000
00006	Commercial	9050
00007	Industrial	8500
.....		
.....		
.....		

(a) LANDUSE-90 relation

LU90	LU-CODE	SYMBOL
Commercial	3	shade-001
Commercial	3	shade-001
Industrial	2	shade-002
Residential	1	shade-003
Residential	1	shade-003
Commercial	3	shade-001
Industrial	2	shade-002
.....		
.....		
.....		

(b) LANDUSE-CODE relation

FIGURE 11.21
Formation of a relation of second normal form (2NF). The LANDUSE-90 relation in (a) has been created from a relation of 1NF (old LANDUSE-90 in Figure 11.20b) by removing all non-key attributes that are not functionally dependent on the primary key. Both LANDUSE-90 and LANDUSE-CODE in this figure have satisfied the conditions of 2NF.

on LU-CODE, but only indirectly on LU90. Such a dependency of SYMBOL on LU90 is described as transitive. Its association with the primary key of this relation (i.e., LU90) is transitive through LU-CODE. As a result, the SYMBOL attribute must be removed from the LANDUSE-CODE relation to form two relations of the 3NF, namely, LANDUSE-CODE and SYMBOL relations, as shown in Figure 11.22b and c. In the LANDUSE-90 relation, since both LU90 and AREA are directly dependent on the primary key PID, this relation is regarded as having satisfied the condition for the 3NF.

At this point, we have created the following relations from the original relation: LANDUSE-90, LANDUSE-CODE, and SYMBOL. Of these relations, LANDUSE-90 contains the basic land parcel information. In the LANDUSE-CODE and SYMBOL relations, tuples with repeating values are removed to eliminate redundancy of data storage. These two new relations contain information that we can "look up" for occurrences in the LANDUSE-90 relation. For this reason, these two tables are sometimes referred to as *look-up tables*, as distinct from the regular relation LANDUSE-90.

The needs for normalization may now be explained by comparing the original relation with the three normalized relations. In database management, normalization is the major mechanism to enforce data integrity. If, for example, we wish to change the cartographic symbol for a certain land-use type, it is necessary to change the symbol code of the affected tuples one by one in the original LANDUSE relation. Any errors or omissions in the process will lead to the degradation of data integrity because there will be two or more symbol codes for the same land-use type. However, with the normalized relations, it takes only one single modification in the SYMBOL look-up table to effect all the necessary changes to the land parcels involved. The process of changing the symbol code in the normalized relations is not only much faster, it also

PID	LU90	AREA
00001	Commercial	6000
00002	Commercial	5800
00003	Industrial	8000
00004	Residential	4500
00005	Residential	6000
00006	Commercial	9050
00007	Industrial	8500
.....		
.....		
.....		

(a) LANDUSE-90 relation

LU90	LU-CODE
Residential	1
Industrial	2
Commercial	3
Transport	4
Public	5
Water bodies	6
Recreational	7
Institutional	8
Open	9
Others	10

(b) LANDUSE-CODE relation

LU-CODE	SYMBOL
1	shade-001
2	shade-002
3	shade-003
4	shade-004
5	shade-005
6	shade-006
7	shade-007
8	shade-008
9	shade-009
10	shade-010

(c) SYMBOL relation

FIGURE 11.22
Resulting relations of the third normal form (3NF) from the flat table in Figure 11.20. The new LANDUSE-CODE relation in (b) and the SYMBOL relation in (c) have been created by eliminating transitive dependencies of nonprimary attribute on the primary key from the old LANDUSE-CODE relation in Figure 11.21. It may be noted that after normalization, there has been considerable saving in data-storage requirements but practically no loss of information. To find the cartographic symbol for industrial land use in parcel 00003, for example, simply look up the LU-CODE for industrial land use in the LANDUSE-CODE relation and then the SYMBOL for LU-CODE #2 in the SYMBOL relation. This may be done relatively easily in relational databases by a standard process called relational join.

basically rules out the possibility of errors that would have otherwise occurred in the case of non-normalized relations.

Normalization also leads to considerable reduction of the storage requirements of the database. Assuming that the original LANDUSE flat file has 15,603 records, there is a total of 93,618 data items in this particular data file. (For definition of flat file, see Section 3.2.4 and Figure 3.12c). The total number of data items after normalization is equal to the sum of the numbers of data items in the three new relations (i.e., 46,809, 20, and 20, respectively), which in this case is 46,849. This represents a saving of almost 50% in the storage requirements of the data. Since databases are always made up of multiple relations that may contain tens of thousands of data items, the total saving in the data-storage requirements resulting from normalization is tremendous. Also, since database volume is directly related to the rate of data traffic and database access time, the reduction of storage requirements also has a very significant impact on systems performance during data processing.

Layer or Coverage Design

GIS based on the georelational data model organizes spatial data in layers or coverages (see Chapter 3). Layer or coverage design is therefore a significant aspect of logical data modeling for geographic databases based on this particular data model such as ArcInfo. The purpose of layer design is to organize spatial entities into logical layers that will optimize the physical implementation of geographic databases. As logical data modeling is system-specific, layer design must be carried out according to the data structure requirements of a specific GIS software package. The method explained here is based largely on the concepts and techniques of layer design pertaining to ArcInfo. However, the presentation of the materials is generic enough to make it useful for other georelational GIS software packages as well.

The primary objective of layer design is to group related spatial entities into logical layers. Whether one spatial entity is "logically related" to one another is determined by the following factors (Chambers, 1989):

1. *The Representation of the Entities by the Three Basic Graphical Elements.* Entities represented by points, lines, and polygons are organized separately in different layers. Mixing different basic graphical elements in a single layer is not permitted.
2. *Anticipated Use of the Data.* Basic graphical elements of the same type are organized into separate layers according to their function (e.g., roads and highways are normally placed in the same layer that is separate from a layer of rivers and streams).
3. *Scale and Resolution of Source Data.* The same data may be represented by different basic graphical elements, depending on the mapping scale (e.g., a river may be represented by a line at 1 : 250,000 scale but by a polygon at 1 : 5,000 scale).
4. *Data-to-Data Relationship.* Data having similar characteristics should be reflected in the design (e.g., administrative divisions with hierarchical association such as boundaries between countries, provinces/states, and regions/counties).
5. *Data-to-Function Relationship.* Data having similar functions should be reflected in the design (e.g., data that tend to be used for the same applications should be considered for inclusion in the same layer, but data that are maintained by different departments, have security restrictions, or need frequent updates should be isolated into separate layers).

For each of the layers identified, it is also necessary to design its feature attribute table and related look-up tables. The same logical data modeling techniques for descriptive data may be used for these tables. It is important to make sure that all feature attribute tables are properly normalized so that they may be integrated with and linked to the descriptive relations of the geographic database using the common identifiers as keys.

For workstation-based ArcInfo applications, another design element known as *map librarian design* is usually carried out as part of the layer design process. *Map librarian* is a special concept that is used in ArcInfo to hide data structure from the user. It is the way by which large spatial data sets are organized, so that they will appear to be a single seamless database. Other GIS software packages may use a similar mechanism but under different names (e.g., index map in GenaMap). Using the map librarian concept, geographic data are physically divided into "tiles" that are logically linked by the map librarian function of the GIS. As map librarian design directly affects database maintenance, data query, and general systems performance, it must be carefully exercised by taking into account the following factors (Chambers, 1989):

1. *Shape of Tiles.* ArcInfo accepts tiles of various shapes, but many systems allow only rectangular or square tiles.
2. *Tile Boundaries.* It is necessary to make sure that tile boundaries will remain stable over time, because it is technically very difficult to convert from one tile base to another once the map library has been set up.
3. *Tile Size.* Tile size has a direct impact on system performance. A tile size that is too large will result in the transfer, processing, and storage of data that may not be needed by an application. This is

because whenever any portion of the data in a tile is required by an application, the entire tile will be processed. On the contrary, a tile size that is too small will unnecessarily increase the frequency of data transfer between the database and the GIS. It will take several transfer processes to move small tiles around, which could have been done in a single process in a larger tile. Therefore, the determination of tile size is always a compromise between the intended applications of the data and their impact on database access during data processing.

11.5.3 PHYSICAL DATA MODELING

Unlike conceptual and logical data modeling, physical data modeling is low-level implementation modeling. As such, physical data modeling is concerned with defining specific storage structure and access paths to the database. While conceptual and logical data modeling focus on *what* a database is supposed to contain, physical data modeling attempts to specify *how* data are stored and *how* data flow between processes. Therefore, physical data modeling is both software- and hardware-dependent. The input to physical data modeling is the logical schema obtained in the previous stage of database design. The output is a *physical schema* that is also known variously as *data dictionary, characteristics of items,* and *physical database design specifications* (Figure 11.23).

In large-scale database development, physical data modeling may be complex. This is because for a given logical schema, there may be many physical design alternatives. It is not possible to make meaningful decisions in physical data modeling and performance analysis without knowing the queries, transactions, and applications that are expected to run on the database. This means that before an appropriate physical schema may be developed, it is necessary to analyze

PARCEL

The PARCEL relation contains data about nonlegal descriptions of individual land parcels not included in the LANDTITLE relation.

Table Definition

Name	Type	Size	Optional/ mandatory	Unique	Indexed	Key
PID	char	15	M	Y	Y	P
Front-ft	num	6,2	M	N	N	
Own-ID	char	15	M	Y	Y	F
Soil	char	2	O	N	N	
Fire-dist	char	2	M	N	N	
Elem-sch-dist	char	2	M	N	N	
Hi-sch-dist	char	2	M	N	N	
Cen-trt	char	5	M	N	N	
Flood-zn	char	3	M	N	N	
Vot-dist	char	3	M	N	N	
Zoning	char	2	M	N	N	

Name	Description
PID	Unique parcel identifier. This column value will be the primary key.
Front-ft	Front footage is the measurement of the street frontage of the parcel in meters.
Own-ID	Parcel owner's social security number
...	
...	
Vot-dist	Municipal voting district identified by abbreviation
Zoning	Zoning classification as designated by the city planning department

Initial volume: 30,000 Growth: 10%
Space: 5000 blocks, 15MBytes
Initial 1800 k

FIGURE 11.23
Example of a physical schema.

the applications, their expected frequencies of invocation and update operations, as well as any time and network constraints on their execution. As a result, physical data modeling is usually carried out with the aid of CASE tools. By using these tools, the designer can convert a logical schema into physical design specifications. Today, many CASE tools are capable of performing physical prototyping activities. They can generate program codes in various programming languages to represent data structure defined in the data dictionary and input/output screens for user interfaces. More advanced tools, known as *lower CASE tools* or *application generators,* can actually generate the programs for large portions of the final information system. There are now many utility routines in the market that can convert output specifications from the physical schema stored in CASE systems into specifications that may be input to application generators. The generators then produce the application programs automatically in the language of the chosen DBMS. When CASE tools are used for physical prototyping and code generation, the activities may no longer be considered simply as physical data modeling in database design. They are in fact a part of the application software design process already (see Section 11.6).

The process of physical data modeling for GIS is dependent on the software package and the hardware platform selected to build the information system. In ArcInfo, it is typically carried out as a two-step procedure (ESRI, 1995). The first step involves the determination of the characteristics of the identified attributes in the logical schema. These characteristics include item name, item width (i.e., storage space), output width, data type (character, integer, binary integer, numeric, binary floating point, and date), and the number of decimal places if the data type is numeric or binary floating point (Figure 11.24). In a sense, the resulting specifications, known as *characteristics of items,* are ArcInfo's equivalent of the physical schema.

The second step is to create the new feature attribute tables for each of the layers identified in the spatial database. This process, which uses the DEFINE command in the INFO module, is in fact a computer-assisted procedure that will automatically prompt the user to enter the values of an attribute item by item (Figure 11.25). The method of defining the physical database design in ArcInfo seems surprisingly simple and straightforward. This is because INFO is not a real relational DBMS, and feature attribute tables seldom have very sophisticated storage requirements. For large corporate information systems using ArcInfo in conjunction with a commercial relational DBMS such as Oracle, conventional physical data modeling techniques must be used to design the attribute database. The built-in facility of INFO should be used for creating the basic feature attribute tables only.

Parameter	Description	Examples
Item name	Any name (up to 16 alphanumeric characters, beginning with an alpha character)	LANDUSE-CODE
Item width	Number of spaces (for characters, integers, and number) or bytes (for binary integers and binary floating points used to store item values)	
Item output width	Number of spaces used to display item values	
Item type	Data type of the item, including:	
C	Character: any combination of alphanumeric characters	"J Smith", "200 Spruce St."
I	Integer number	200, 82, 33011
B	Binary integer (an integer stored in binary format)	200, 82, 33011
N	Number (a valid decimal number)	0.1, 12.5
F	Binary floating point (a decimal number stored in binary format)	123.456
D	Date (stored as 8 bytes; displayed as 8 or 10 spaces)	12/31/98, 12/31/1998
Number of	For N or F, the number of digits to the right of the decimal place	

FIGURE 11.24
Characteristics of items used in ArcInfo to define the physical structure of attribute tables.

11.6 GIS APPLICATION SOFTWARE DESIGN METHODOLOGY

As one of the two components of the systems design phase of the systems development life cycle (SDLC), application software design is usually carried out concurrently with its counterpart, that is, database design. While database design is concerned with the definition and structure of data, application software design has its focus on the use of data in terms of (1) processing rules, (2) the flow or navigation requirements of data, and (3) input/output forms, screens, and reports. Since the objectives and requirements of application software design are very different from those of database design, it is always carried out by systems staff with distinctly different technical expertise and experience in process analysis,

```
Arc: INFO
ENTER USER NAME> ARC
ENTER COMMAND> DEFINE PARCEL.DAT
ITEM NAME, WIDTH [,PUTPUT WIDTH], TYPE [, DECIMAL PLACES]
  1
ITEM NAME: PID
ITEM WIDTH: 15
ITEM OUTPUT WIDTH: 15
ITEM TYPE: C
  2
ITEM NAME: FRONT-FT
ITEM WIDTH: 6
ITEM OUTPUT WIDTH: 10
ITEM TYPE: N
DECIMAL PLACES: 2
  3
ITEM NAME: OWN-ID
ITEM WIDTH: 15
ITEM OUTPUT WIDTH: 15
ITEM TYPE: C
  4
...
...
...
...
```

FIGURE 11.25
Using the DEFINE command to interactively create a feature attribute table in ArcInfo.

application programming, and in GUI design. The input to systems design is the user requirements analysis documents obtained in the previous phase of the SDLC, and the output includes the application software design specifications of the information system.

Prior to the advent of X-Windows and Microsoft Windows technology that supports the application programming interface (API), computer applications were typically built using macro or scripting languages. An example of this approach to GIS application development is the Arc Macro Language (AML) of ArcInfo. The software engineering approach to application software design today is based heavily on the idea of reusable components (Anderson et al., 1996; McClure, 1997; Pressman, 1997; see also Chapter 6). In contrast to the concept of reusable software routines in conventional computer programming, reusable components in modern software engineering not only include reusable program codes, but also design specifications, the graphical user interface, the user documentation, and any other items associated with software (Kuhns, 1998). Application software design using component technology is made up of two sequential processes: (1) establishing design standards and (2) identifying and designing components. In the following two sections, we will address issues pertaining to these two software design processes with special reference to GIS.

11.6.1 APPLICATION SOFTWARE DESIGN STANDARDS

The objective of adopting standards in application software design is to maintain a consistent and logical style for all the programs written for an information system, no matter by whom they have been coded. The use of standards is one of the primary mechanisms for ensuring reusability of software components (McClure, 1997). It is also the best way to assure and control the quality of software products. There are many benefits of adopting standards in application software design, including:

- Individual programmers do not have to spend time developing a coding style and deciding how best to format the application programs, resulting in more time for the actual production of efficient and robust programs.
- Large-scale application software development may be carried out more easily by teamwork because standards provide a common language to facilitate communication and exchange of ideas among team members.
- Programs are easier to maintain and support because programmers are able to read and understand program codes written by others.
- Programs developed for different applications, by different development teams, and at different times may all be kept to a consistent coding style, which is important for long-term software maintenance, as well as to a consistent graphical interface, which is important to end users.
- Design specifications, codes, user interface, and documentation developed for one application may be easily reused for other applications, thus leading to considerable savings in both time and cost of software development.

The development of standards for application software design is no trivial task. There are now numerous resources from both software vendors and book publishers (e.g., Microsoft Corporation, 1992; Hix and Hartson, 1993) that may be used as a blueprint for developing corporate or project-specific design standards. However, such resources are both an asset and a burden to the application software design team. Although they have provided the standards design team the necessary conceptual framework and guidelines to start with its own design work, the amount of time and effort required to fully understand and use the materials presented in them may be tremendous. In addition, the process of developing design standards is further complicated by the need to satisfy the requirements of both end users and application developers while at the same time to be in compliance with industry standards. As a result, the task of developing software design standards usually takes weeks, if not months, to complete.

The types and number of design standards that must be developed depend on the nature and scale of individual systems development projects. In general, these standards may be grouped into four categories (Table 11.2):

1. *Coding or scripting standards* specify the format and style of program codes so that a program written by one programmer may be readily understood and used by another.
2. *Data-naming standards* specify the convention for naming data items to ensure that databases may be shared across organizational and application boundaries.
3. *Application and GUI standards* provide a consistent approach in designing and building software products from an end user's point of view, as well as from an application developer's view.
4. *Development and version management standards* are used to standardize the directory and file structure of the application and to control the revision of program codes in order to maintain the consistency and integrity of application software products.

11.6.2 IDENTIFICATION AND DESIGN OF COMPONENTS

In the context of software engineering, the design of components starts with the method of *process analysis* or *modeling*. The purpose of process analysis is to transform the data objects obtained in the data modeling phase into data flow information necessary to implement a business function. This is achieved by examining

Table 11.2
Categories of Software Application Design Standards

Category of Standards	*Contents*
Coding or Scripting	Templates for new programs Program header Syntax including comments Parameters and options Exit status User and global variable names Exit status Error handling and messages
Data Naming	Entity names Attribute names Data name suffixes Table names File names
Application and GUI	Window and menu-naming convention Windows and menu design and resources Display management Dialog design including DBMS and spatial queries Function keys Tool bars and icons Screen and hard-copy output forms Design documentation User documentation On-line help
Development and Version Management	Directory and file structure Revision control of codes Revision control of applications Access control for collaborative application development Component test and acceptance procedures

and identifying data objects pertaining to specific business functions in the business function hierarchy, as well as the processes that are performed to manipulate (i.e., add, modify, delete, or retrieve) these data objects. The different data objects and associated processes identified in this way are then packaged into different classes known as *candidate components*. The results of process analysis are represented by means of a *process flow diagram* that depicts the flow or navigation requirements of business functions in the system (Figure 11.26).

Once the candidate components have been identified, the component library is searched to determine if these components already exist. A component library is made up of three types of products:

1. *Off-the-shelf components* are existing software components that have been purchased from a third party (e.g., MapObjects from ESRI) or developed internally for a past project, fully validated, and ready for the current application.
2. *Full-experienced components* refer to existing specifications, designs, codes, or test data developed for past projects that are similar to the software to be built for the current project. They are so called because members of the current software development team have had experience in the application areas represented by these components. Generally speaking, it is relatively simple to turn a full-experienced component into a custom component that meets the functional requirements of the current project.
3. *Partial-experienced components* are similar to full-experienced components except that members of the current software development team have had only limited experience in the application areas represented by the components. It usually requires substantial modifications to turn a partial-experienced component into a custom component that meets the functional requirements of the current project.

Components that may be found in the library may be extracted and reused. If, however, a candidate component cannot be found in the library, it must be designed, built, and deposited in the component library (Figure 11.27). Creating a component is a relatively straightforward process in Visual Basic (Aitkin, 1997). It starts with a *form*, placing *controls* in it, and then writing codes to define *properties* and *methods* and to deal with *events* (Figure 11.28). In the context of Visual Basic programming, a "form" is the drawing board for visually designing an icon or a window; "controls" refers to the name and the shape of a screen object (e.g., a button or a window) that represents a component; "properties and methods" are border style, fill color, name, and foreground color of the control; and "events" are actions to be performed by the user (e.g., clicking a button and selecting an item from a list of options).

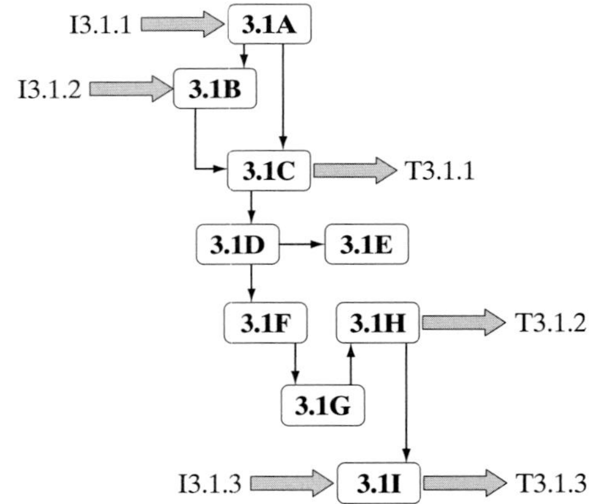

Process ID	Process Name
3.1A	Log new application to register
3.1B	Establish list of new applications
3.1C	Review application
3.1D	Digitize parcel boundary
3.1E	Print check plot
3.1F	Input attribute data
3.1G	Verify attribute data
3.1H	Update attribute database
3.1I	Print certificate of title registration

Initiation Events		Termination Events	
I3.1.1	Receive application to register	T3.1.1	Reject application due to incomplete information
I3.1.2	Process queued applications due to backlog	T3.1.2	Parcel is registered
I3.1.3	Print certificate of registration	T3.1.3	Certificate is sent to owner

FIGURE 11.26
Example of a process flow diagram. The process flow diagram depicts the functions to be performed by a particular program module. The process flow diagram is an application software design tool and will be used to develop the application programs (see Section 11.7.2 and Figure 11.29).

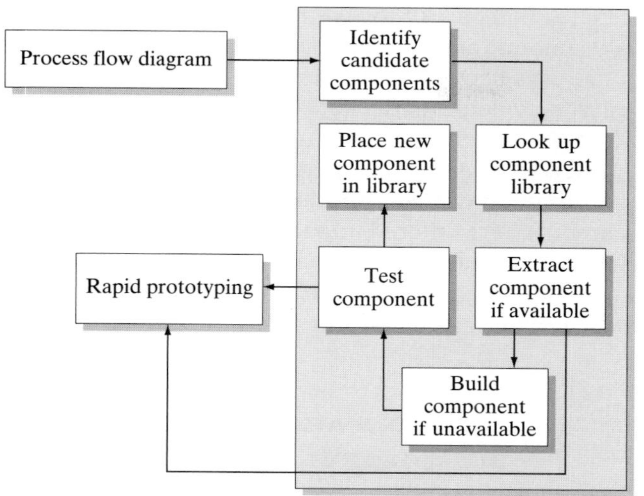

FIGURE 11.27
Using components in application software development.

All new components created must be subject to a stringent test and certification process as set out by the appropriate corporate or project-specific standards. They are then placed in the component library so that they can be accessed when required. The component library, or *component repository* as it is sometimes called, is an important part of the concept of reusable software components (McClure, 1997). The purpose of the library is to organize, store, and manage reusable components in a systematic manner. A component library is normally managed by a systems person specially assigned to the job. A search engine and component browser should be developed so that programmers can easily access components in the library. In addition, procedures and rules must also be in place to control the use and update of components. The objective is to ensure that the same version of component is used or upgraded across the board for all applications (see Section 11.8.3).

11.7 Systems Implementation and Technology Rollout

Systems implementation probably represents the most critical milestone of the systems development life cycle (SDLC). After all the efforts spent on the planning and design of the systems, it is now time to turn abstract concepts into an operational information system. As all the tasks to be completed at this particular stage are very labor-intensive and the number of people involved

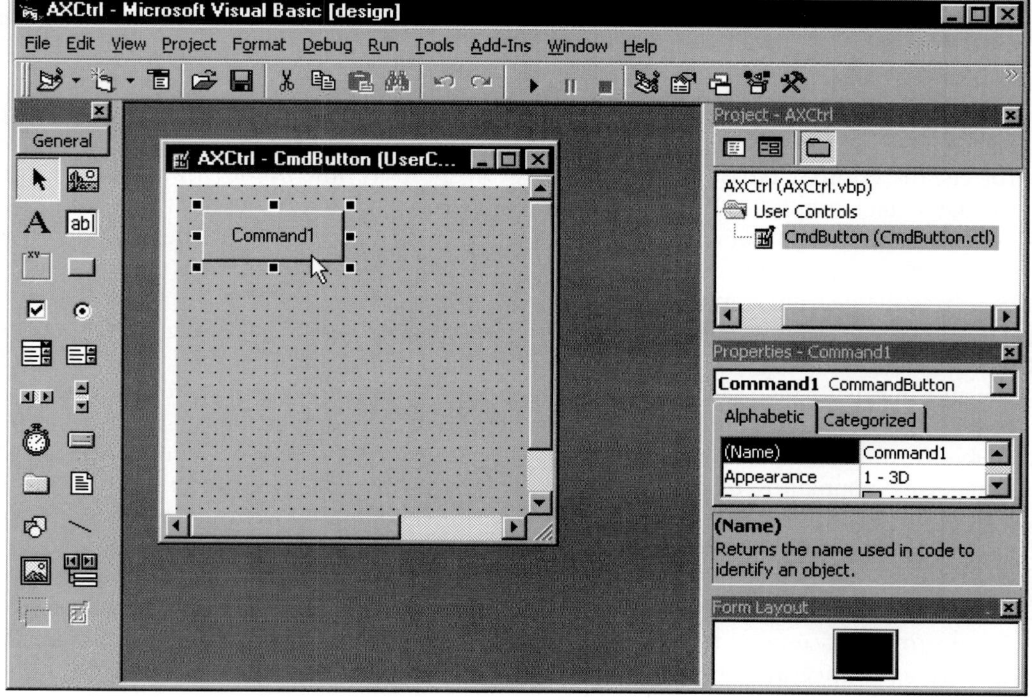

FIGURE 11.28
Creating an ActiveX control in Visual Basic. Visual Basic provides a very rich visual programming environment (collection of easy-to-use widgets such as buttons and multimedia players) for the interactive design, testing, and compilation of ActiveX components.

are numerous, systems implementation is also the most challenging phase in the SDLC from the project management perspective. It requires careful scheduling and coordination of a variety of tasks, including the acquisition of hardware and network, site preparation, database creation, application development, transition planning, and training. At the same time, it is critical to maintain effective communication among all parties involved in these tasks. Any error or oversight at this stage may quickly erode the achievements obtained in the previous phases of work, thus paving the way for the potential failure of the project.

Input to the systems implementation phase includes the functional and design specifications of the system, hardware and network architecture, as well as a technology rollout plan and a training plan. The tasks in systems implementation involve four major tasks: (1) populating the database, (2) constructing the application software, (3) testing the application software, and (4) rolling out the technology. The output will be an operational information system that is delivered in time and within the approved cost, as well as an end user community that has been adequately prepared to use the new technology.

11.7.1 POPULATING THE GEOGRAPHIC DATABASE

Implementing a GIS database design simply means the creation of both the necessary attribute relations and the graphical layers as set out by the design specifications. The amount and the complexity of the work involved are dependent on the scale and nature of individual projects. In some projects, the entire database may be purchased from third-party data suppliers. This is usually the case in business geographics applications where the data requirements are relatively standard and socioeconomic data are readily available from commercial, institutional, and government sources. At the other extreme, the whole database has to be built from scratch. In such cases, the collection and automation is an extremely costly and lengthy undertaking. However, as digital data are increasingly available from commercial data suppliers and government agencies, populating the database for many projects is now more concerned with converting or customizing existing data to corporate standards rather than with extensive data automation.

Regardless of the sources of the data, one area that has not always been properly attended to in database creation is the quality assurance and quality control (QA/QC) of the data. As the quality of data has a direct impact on the results of transactions and data analysis using the information system, the standards of QA/QC should never be compromised. Unfortunately, experience has shown that because of the pressure to complete the database in the shortest possible time and the large amount of resources involved, QA/QC of data has not always been given the priority that it deserves. In order to avoid potential data problems that may occur when the system is in operation, all data sets must be certified by a responsible person (e.g., QA/QC manager) to have met the corporate data quality standards before they are entered into the database. This rule applies to both packaged data acquired from third-party data suppliers and digital data converted by internal staff or external service bureaus.

11.7.2 APPLICATION SOFTWARE DEVELOPMENT STRATEGIES

Within the context of software engineering, numerous strategies have been proposed for application software development. These include, among others, the following methods (Pressman, 1997): (1) classic life cycle or waterfall model, (2) prototyping model, (3) rapid application development (RAD), (4) evolutionary software process model, and (5) fourth-generation techniques. Each one of these methods has its merits and limitations and may be more suitable to one development environment or application scenario than the others. To take advantage of the relative merits of individual methods while at the same time avoiding their drawbacks, it is now not unusual for application developers to adopt a hybrid approach. One such approach commonly used today is called the *rapid prototyping methodology*. This approach combines the concepts and techniques of prototyping and RAD to create an application environment that will enable the delivery of a working system within a relatively short time frame. It makes heavy use of reusable components to expedite the software construction process and to reduce the development cost.

The conventional prototyping paradigm of software development starts with a "quick design" that is based on an initial understanding of the data-processing requirements. The primary focus of the quick design is on those aspects of the software that will be most visible to the end users, that is, input and output forms, screens, and reports. If accepted by the end user, this quick design is turned into a prototype. The prototype is then evaluated by the end user and is refined in a series of iterations until it meets the operational needs of the end user. In the conventional application development process, the technique of prototyping is used basically as the means to identify and define the data-processing requirements of the software. Once the user is satisfied with the prototype, it is usually "thrown away," and a real application is built. In the rapid prototyping methodology, however, RAD tools and components are used to develop the prototype. This means that although the prototype begins as a mock-up, it is quickly turned into a real system when the development process progresses. Therefore, instead of a throw-away prototype, the end product of the rapid prototyping methodology is a fully operational application.

In the Microsoft Windows platform, there are two RAD environments: (1) *Dynamic Data Exchange* (DDE) or *Dynamic Link Libraries* (DDL), and (2) *ActiveX controls*, formerly known as *Object Linking and Embedding* (OLE). DDE is an interapplication communication architecture that allows application developers to build Windows-based GUIs using RAD tools. Application development in ArcView GIS is a typical example of the DDE environment (Cattran and Mackaness, 1998). ActiveX controls, on the other hand, represent an object-based software integration architecture which allows application developers to use RAD tools to seamlessly embed mapping and other components into Windows applications (Ngan and Dennerlein, 1997). Several GIS-oriented ActiveX controls packages are now available in the market (e.g., MapObjects from ESRI and MapX from MapInfo Corporation). Custom ActiveX controls may also be developed using Visual Basic as described in Section 11.6.2. The most commonly used RAD tools today include Visual Basic and Visual C++, Delphi, and PowerBuilder. These tools provide the necessary visual design environment for application software development by the rapid prototyping methodology.

Application software development is a relatively complicated process. Therefore it is essential to document carefully everything that needs to be done and that has been done. The starting point of documentation is a narrative description of the business processes to be manipulated by the application (Figure 11.29a). Several tables or *parameter lists* are then developed to assist the programmer in coding the application by following the standards established in the systems design phase (see Section 11.6.1) and by using resources available in the selected RAD tools. These tables include:

1. *action summary table* that lists the trigger, processes and results of the processes associated with and manipulated by the component (Figure 11.29b)
2. *program logic* table that depicts the sequence of actions to be performed by the system to execute the processes in the component (Figure 11.29c)
3. *Windows resources* table, that lists all the windows to be opened when the component is used (Figure 11.29d)
4. *text resources* table that lists all the text files (e.g., option list and error messages) that will be invoked when the component is used (Figure 11.29e)

When the prototype is refined and modified during the application development process, the contents of these tables should be updated accordingly to reflect all the changes that have been made. On completion of the application development process, all these tables must be properly filed in the project repository. These documents are indispensable sources of information for systems maintenance and upgrade in the future (see Section 11.8).

11.7.3 SOFTWARE TESTING METHODOLOGY

Software testing is a critical component of software engineering. No application software should be implemented in the production environment without first passing all the necessary tests. In practice, software testing is designed and carried out by a specialist software QA/QC team that works independently of the software development team. The purposes are to avoid potential conflict of interest and to ensure impartiality in the testing process. However, software testing should be ideally integrated as part of the application development process because test results always provide very useful direction to improve the design and construction of the software.

The software test plan, which sets out the objectives and procedures of all the software tests to be performed, is usually developed by the software QA/QC team when the application development process starts. This allows software tests to be carried out in phases as the application development proceeds from one stage to another. The first phase is called *unit test*. Unit testing focuses on the smallest unit of software construction: the module. Unit testing is usually carried out right after the source-level code has been developed, reviewed, and verified for correct syntax. Using the systems design specifications as a guide, important control paths, input/output parameters, and error-handling mechanism of the application are tested to uncover possible errors in individual modules. A unit test report listing the tasks tested, expected results, and actual results is prepared for each module that has been tested.

The second phase is called the *integration test*, which takes place once the unit test has been completed. Integration testing aims at uncovering errors associated with the interfacing between different modules. Different integration testing strategies have been proposed, but the method of *regression testing* appears to be one of the most commonly used in practice. This is an incremental approach of testing in the sense that the test is performed each time a new module is completed and added to the application. This helps to uncover immediately any error that may have been introduced to unit-tested modules by the addition of new modules. An integration test report detailing the test performed, the expected results, and actual results is prepared for each integration test performed.

The final phase is called *systems* or *acceptance test*. Systems testing is performed when the application

Process 3.1D Digitize parcel boundary

This process creates the boundary of a new parcel. The user can digitize by (a) key in the coordinates or (b) using the digitizing table.
Regardless of the method of digitizing, this process creates a single polygon of three or more vertices. The user can digitize more than one polygon in a single session.The boundary is to be in green dashed line type, on the PARCEL-BDRY layer. The label (i.e., PID) will be placed at the geometric centroid by default, but the user is allowed to drag the label to any point inside the polygon.

(a) Narrative description of process

Process ID: 3.1D	Digitize parcel boundary
Initiation event	N/A
Initiating function	Application has been approved (Process 3.1C)
User scenario	Choose (a) key in coordinate (b) use digitizing table; label PID
Processing description	Create parcel boundary in green dashed line type, on the PARCEL-BDRY, add PID
Termination event	N/A
Following process	3.1E Print check plot; 3.1F Input attribute data

(b) Action summary

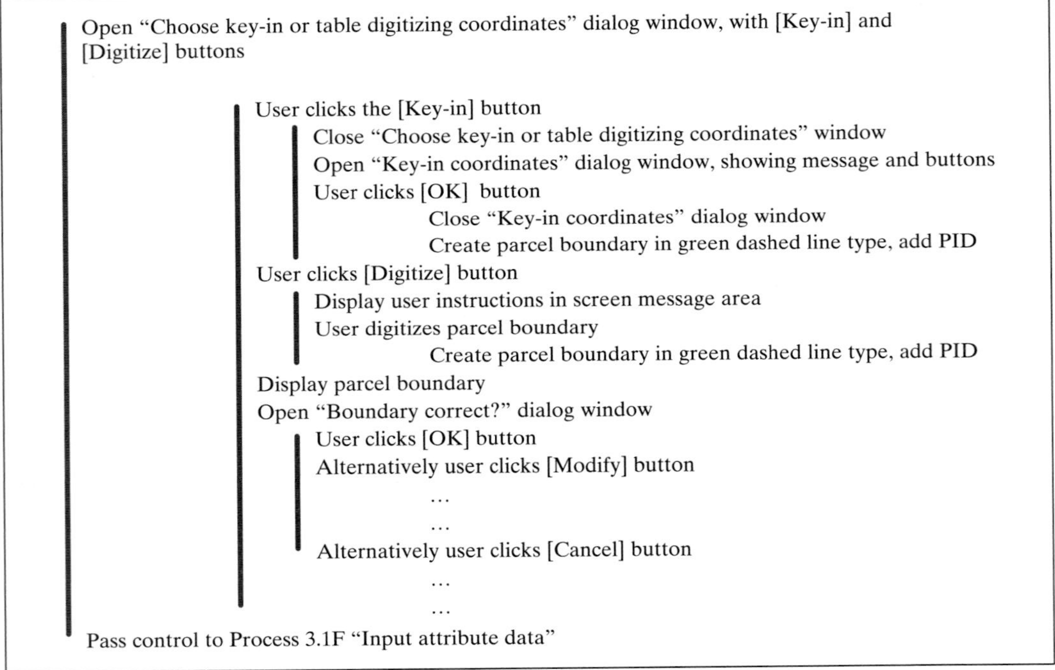
Open "Choose key-in or table digitizing coordinates" dialog window, with [Key-in] and [Digitize] buttons
User clicks the [Key-in] button
Close "Choose key-in or table digitizing coordinates" window
Open "Key-in coordinates" dialog window, showing message and buttons
User clicks [OK] button
Close "Key-in coordinates" dialog window
Create parcel boundary in green dashed line type, add PID
User clicks [Digitize] button
Display user instructions in screen message area
User digitizes parcel boundary
Create parcel boundary in green dashed line type, add PID
Display parcel boundary
Open "Boundary correct?" dialog window
User clicks [OK] button
Alternatively user clicks [Modify] button
...
...
Alternatively user clicks [Cancel] button
...
...
Pass control to Process 3.1F "Input attribute data"

(c) Program logic

FIGURE 11.29
Documentation for application software development. Tables and charts pertaining to process 3.1D (digitize parcel boundary) in Figure 11.26. Documentation to this level of detail seems excessive but is essential for large-scale application software development.

development process of the entire system is complete. In practice, systems testing is actually a series of different tests that aims to verify the functionality of the information system as a whole, as explained here.

- *Recovery testing* verifies the ability of the system to recover to the state before a systems failure.
- *Security testing* verifies the protection mechanism built into the system to protect it from unauthorized use.
- *Stress testing* examines the response of the system to abnormal situations that may break the system, such as excessive interrupts, increased input data rates, and maximum memory use.

Process ID	Window Name	Type	Interface object
3.1D1	Choose to key-in or table digitizing coordinates	Dialog	Static text (ids-choose) buttons [Key-in] [Digitize], [Cancel]
3.1D2	Key-in coordinates	Dialog	Static text (ids-key-in) buttons [OK] [Cancel]
3.1D3	Digitize coordinates	Dialog	Static text (ids-dig) buttons [OK] [Cancel]
3.1D4	Boundary correct?	Dialog	Static text (ids-bdryok) buttons [OK] [Modify] [Cancel]

(d) Window resources

TextID	Text
ids-choose	Choose "Key-in" to enter the boundary coordinates from the keyboard Choose "Digitize" to start table digitizing Choose "Cancel" to return to previous screen
ids-keyin	Enter the UTM coordinates for each boundary point in a clockwise direction
ids-dig	Digitize parcel boundary in clockwise direction, press [Enter] key to close to first point
ids-bdryok	Is the parcel boundary correct? Choose "OK" if this boundary is correct Choose "Modify" if you wish to modify it Choose "Cancel" to abort process and return to previous screen

(e) Text resources

FIGURE 11.29 *(continued)*

- *Performance testing* monitors the functioning of the system at run time with a view to evaluating whether the system has met functional and design specifications.

As system tests are normally carried out to determine whether the final system is functionally acceptable for implementation, they are also collectively called an *acceptance test*. When the acceptance test has been completed, the results are consolidated into a systems test report. This report is then signed by a responsible person (e.g., software test manager) and submitted to the project sponsor to signify the completion of the systems development stage of the SDLC.

11.7.4 PLANNING AND MANAGEMENT OF TECHNOLOGY ROLLOUT

The implementation of the information system in the end users' sites is usually called *technology rollout* because it is a process of delivering new technology from a development site to single or multiple production sites, often distributed over a large geographical area. The nature and complexity of the tasks at this stage of systems implementation are dependent on the characteristics (i.e., functions, size, and architecture) of the information system, the geographical distribution of the users' sites, and the number of potential end users. Technology rollout must always be carefully planned and managed to ensure a smooth transition from the old to the new working environment that results from the implementation of the new information system. In general, technology rollout involves the following tasks:

1. *Site preparation* includes the preparation of the space that will house the computers and peripherals; the acquisition, installation, and testing of hardware as well as the installation of communication networks.
2. *Training of end users* includes basic computer training and application-specific training according to end users' computer literacy and expected use of the new information system.

3. *Internal communication* aims to ensure that all users are aware of the technical support available, the error-reporting procedures, systems maintenance schedules, and information about software upgrades.
4. *External communication* has the role of notifying clients and business partners of the launch of the new information system and new information services, as well as any business practices that are introduced as a result.

11.8 Systems Maintenance and Technical Support

The systems maintenance phase of the systems development life cycle starts right after the implementation of the information system is completed. It will last as long as the information system is in operation. In practice, systems maintenance is always considered together with another process known as *technical support.* These two processes share a common goal, that is, to make sure that the system works, and the two are interdependent. In many cases, recurrent problems encountered in technical support trigger systems maintenance procedures. On the other hand, results of systems maintenance always have to be distributed to end users through the proper technical support channel. In the following sections, we will explain systems maintenance and describe how systems maintenance procedures may be built into the technical support infrastructure of an organization.

11.8.1 THE NATURE OF SYSTEMS MAINTENANCE

Systems maintenance may be formally defined as the technical and management activities carried out to sustain the ability of the information system to meet business objectives of an organization in a cost-effective manner. This implies that systems maintenance is concerned not only with fixing errors that may occur when the system is in use, it also includes all the refinements and enhancements to the system as the result of changing technology and user requirements. Depending on the nature and objective of the activities involved, systems maintenance may be classified into the following four categories (Pressman, 1997):

- *corrective maintenance,* which is carried out to correct errors
- *adaptive maintenance,* which results from external changes to which the system must respond, for example, changes in hardware technology or the operating system
- *perfective maintenance,* which represents all other changes to an operational information system that aim to make it more efficient and productive
- *preventive maintenance,* which is the "reengineering" of an information system for future use

It has been estimated that corrective maintenance usually accounts for about 20% of systems maintenance activities, and the other three categories share the remaining 80% (Pressman, 1997). This implies that what people usually regard as the major task of systems maintenance (i.e., fixing design and programming errors) actually represents only a relatively small part of the activities. Most of the efforts are in fact directed toward the improvement and enhancement of the functionality of the information systems. In the early 1980s, Boehm (1981) estimated that the cost of software maintenance was commonly over 60% of the total development cost. As computer systems are much more complex now than those in the 1970s and 1980s and technology is advancing at a far greater rate, the cost of systems maintenance may only be higher than ever before. Obviously, systems maintenance will significantly affect the operating cost of an information system. It follows that the development of an efficient mechanism for systems maintenance is absolutely essential for the implementation of information systems today from both the technical and economic perspectives.

11.8.2 PROCEDURE FOR CORRECTIVE SYSTEMS MAINTENANCE

An error occurs when an information system fails to function in the way it is expected to perform. Errors are a fact of life in all software development projects. Despite all the good intentions and preventive measures in the design, QA/QC, and testing processes, errors of one form or another will soon emerge after the information system has started operating. Some of these errors are fatal, which will basically stop the information system from working. Others may simply cause nuisance because the application concerned is not intuitive enough for use. Regardless of their nature, errors lead to user frustration, loss of productivity, disruption to customer service, and even the total functional breakdown of an organization. The objective of corrective systems maintenance is to develop the necessary procedures that will enable the systems support staff to minimize the impacts of errors by identifying and correcting them quickly.

Since errors are not avoidable, corrective systems maintenance should always be treated as a standard systems administration procedure. Preparation for corrective systems maintenance should be carried out well ahead of the date when the information system starts to operate. Usually, this involves the preparation of a service agreement with the internal systems development unit of the organization or, more commonly, a systems maintenance contract with an external software consultant. The contents of the service agreement or systems maintenance contract should state clearly several important items, including

- contact persons and error-reporting structure
- error-reporting procedure and standardized reporting form
- identification and criteria for the classification of errors
- priority of error correction and time frame for the delivery of corrections
- tracking of the progress of error correction
- responsibility for testing the corrections
- method of distributing the corrections

Corrective systems maintenance is a reactive process that is usually triggered by a problem encountered by an end user. It involves a chain of problem-reporting and correction-tracking responsibilities whenever an error occurs (Figure 11.30). In corrective maintenance, all errors must be logged properly in a standard problem-reporting form (Figure 11.31) so that the progress of correction can be tracked. When the correction is delivered, it will be subject to an acceptance test and then distributed to all users by updating their computers.

11.8.3 PROCEDURE FOR PROACTIVE SYSTEMS MAINTENANCE

Unlike corrective systems maintenance, which aims to fix software errors, the other three categories of maintenance activities are more oriented toward the improvement and enhancement of information system performance. These activities are described as proactive systems maintenance in the sense that they are initiated without the imminent threat of an existing application error that may stop the information system from functioning. In other words, proactive systems maintenance is more concerned with the long-term performance of an information system rather than with the immediate solution of an existing problem.

Decisions for proactive systems maintenance are usually dependent on three issues pertaining to the use of information technology (IT): (1) awareness of technology, (2) evaluation of technology, and (3) management of technology. Awareness of technology implies that in order to initiate any systems maintenance procedure, the user or systems staff must keep themselves constantly abreast of the availability of and trends in IT. This leads to the issue of evaluation of technology, which means that they must be able to assess the relevance of new technologies to existing and potential objectives of the information system. Evaluation of technology includes the feasibility of upgrading the existing software applications in terms of both cost and technical compatibility. Once it has been decided to start systems maintenance using new technologies, the issue then becomes one of managing technology.

Management of technology basically involves two major decisions. The first one is to choose the frequency of performing systems maintenance. A typical question that must be answered is: Should the applications of the system be upgraded for every new release of the parent software? The second decision is concerned with *version control* or *version management*. Version control is critical in today's client/server computing environment. The objective is to ensure the integrity of the applications in all client computers by using the same version of the application software. Whenever a systems maintenance procedure is executed, it must be done simultaneously to each and every computer within the system so that all affected applications will be changed across the board.

11.8.4 SETTING UP A TECHNICAL SUPPORT INFRASTRUCTURE

Technical support is indispensable for information system implementation. The objective of technical support is to provide a mechanism to help users solve any problem they may encounter when using the information system. For large-scale information systems,

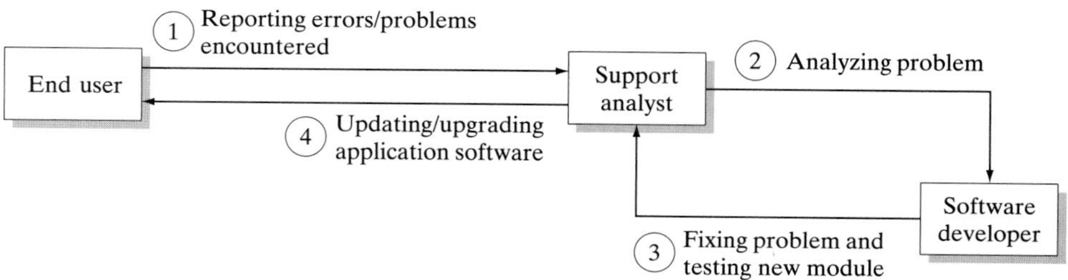

FIGURE 11.30
The process of error/problem reporting and fixing software support and maintenance.

PROBLEM REPORT FORM

Problem No.	Date:	Time:
	Reported by:	
Description of Problem	New Problem? [*y/n*] Data-related? [*y/n*] Priority [*Severe/Critical/Annoyance*] Name of system component/function where the problem occurs: Error message(s):	
Notify Contractor:	Date/Time:	
	By [*Fax/e-mail*]	
	Acknowledgment Date/Time:	
Problem Fixing Status	Clarify [ ]	Date/Time
	In Progress [ ]	Date/Time
	Fixed [ ]	Date/Time
	Distributed [ ]	Date/Time
Acceptance Test	Performed by [ ]	Date/Time
	Description of solution:	

Report(s) received from Contractor (MUST be filed with this Problem Report Form):

1. ______ Date/Time ______
2. ______ Date/Time ______
3. ______ Date/Time ______
4. ______ Date/Time ______
5. ______ Date/Time ______

FIGURE 11.31
Example of an error report form. Logging problems and keeping track of the progress of corrections are essential systems support procedures.

technical support should be properly instituted as an infrastructure. This means that it should be properly organized with permanent staff, standard procedures and well-defined client service protocols and responsibilities. Although information systems today have invariably been designed with an emphasis on user-friendly interface, many of them may still be too sophisticated for the average user. Without the proper technical support infrastructure, it may take a prohibitively long time to put the information system to productive use.

Just like systems maintenance, technical support may also assume a reactive or proactive approach. The method of the *help desk* is a typical example of reactive technical support. Using this method, technical support staff will respond to user requests or queries only when contacted. They remain hidden to users when no operational problems occur. In contrast, proactive technical support requires that support staff actively approach users to help them improve their productivity in using the information system. Proactive technical support activities include general and application-specific training, technical workshops, and newsletters. It involves input of and participation by both technical support staff and end users at large.

In practice, both the reactive and proactive approaches are required in the technical support infrastructure for an information system. The help desk function is particularly important at the initial stage of the implementation of large-scale information systems with a diversified user base. To properly set up a help desk function, it is necessary to make sure that all technical support persons are adequately trained in the use of the hardware, software, and data of the information system. It is also necessary to standardize the procedures to log and track all problems reported by users. As technologies and user requirements may change rapidly, it is always important for the technical support team to act proactively by working closely and continuously with the end users to monitor the performance of the information system. Such evaluation of systems performance should be carried out at regular intervals in order that opportunities for improvement and enhancement may be identified as early as possible. The technical support team should maintain a regular training program to ensure that all new users are properly trained. In addition, it should also keep the dialog with the end users open by means of a newsletter or other methods of communication.

A carefully planned technical support program not only provides the end users with timely and responsive feedbacks to their questions, it also represents one of the best sources of information that can be used to identify possible areas for improving and enhancing the functionality of existing information systems. Since the best way to uphold user satisfaction and productivity of an information system is to identify and correct any potential problems before they occur, it is always advisable for the technical support team to take a proactive approach. This will result in the tight coupling of technical support functions with systems maintenance functions, a more knowledgeable and enlightening workforce, as well as more committed users of modern information technology.

11.9 Summary

In the relatively short history of GIS, there have been many success stories. In the last couple of decades in particular, we have witnessed the phenomenal proliferation of GIS from the predominantly resource inventory arena to a wide spectrum of applications in digital cartography, facilities management, social research, and business planning. However, behind this apparently very promising picture, there have also been many failures. Although examples of failure have rarely been reported publicly, it is not too difficult to find systems that were unable to deliver as expected, went way out of budgetary control, or simply suffered a sudden demise because of the lack of committed support from senior management. Regardless of the nature of the failure of the systems, it appears that the root of most problems may be traced to poor systems design because of the lack of understanding of the user requirements, that is, what the information system is set up for.

The most critical factor for the successful implementation of GIS is the correct identification of user requirements at the beginning of a project. Advances in software engineering have now provided GIS users with an extremely rich set of tools to identify user requirements, design the information system conceptually, and to turn the design into practical implementation. Software engineering not only represents a structured approach to the use of technology, but also to the management of technology. In other words, it embraces the concepts and techniques of building a GIS, as well as the various project managerial activities that ensure the completion of a high-quality information system in an orderly manner. By using the software engineering approach, the systems designer is able to make sure that the functionality of the information system is properly aligned with the business objectives of an organization. This approach provides the systems staff with the conceptual framework and necessary software development tools to develop the databases and applications within a much shorter time frame. It also includes the necessary systems maintenance procedures to make sure that the information system continues to be compatible with the ever-changing technical environment and user requirements in the long run. In order to minimize the problems that have hindered the design and development of information systems in the conventional way, the software engineering approach is strongly recommended for all GIS implementation projects.

As a systems implementation paradigm, the software engineering approach may be largely described by the systems development life cycle. As noted, different variants of the life cycle have been proposed for the concepts and techniques of systems design. It should be understood that how the life cycle is divided into different phases is unimportant. What really matters is the structured approach to information system development and the idea that information system development is a continuous cyclical process. In the implementation of an information system, the end of one process always signifies the beginning of another. Even when an information system has been successfully put into operation, a new cycle of performance evaluation, redesign, and systems maintenance will immediately begin. This is the way by which the functionality of the information system is kept up to date with rapidly changing technology and the business objectives of an organization.

GIS implementation and project management may be learned more effectively by practice. There is no better way to acquire the skills of designing an information system and managing its implementation than to actually take part in real-world projects. In this chapter, we have presented an overview of the concepts and techniques pertaining to the full life cycle of a typical GIS implementation project. The depth and breadth of the treatment of the subject matter should be adequate to prepare students for participation in the design and implementation of large corporate GIS projects. However, it will probably take time to build up the necessary expertise to face the challenge and enjoy the fun of participating in or leading a successful GIS implementation.

REFERENCES

Aitken, P. (1997) *Visual Basic 5 Programming Explorer,* Scottsdale, AZ: The Coriolis Group, Inc.

Anderson, D., Ledbetter, M., and Tepovich, D. (1996) "Piecing together component-based GIS aplication development," *GIS World,* Vol. 9, No. 6, pp. 62–64.

Batini, C., Ceri, S., and Navathe, S. B. (1992) *Conceptual Database Design: An Entity-Relationship Approach,* Redwood City, CA: The Benjamin/Cummings Publishing Company.

Boehm, B. (1981) *Software Engineering Economics,* Englewood Cliffs, NJ: Prentice Hall.

Cattran, S., and Mackaness, W. (1998) "Rapid prototyping: A customization strategy," *Proceedings* (in CD-ROM), paper # 174, Arc/Info Users Annual Conference, Redlands, CA: Environmental Systems Research Institute, Inc.

Chambers, D. (1989) "Overview of GIS database design," reprint from *ARC News,* Vol. 11, No. 2.

Chen, P. P. (1976) "The Entity-Relationship Model: Toward a unified view of data," *ACM Transactions on Database Systems,* Vol. 1, No. 1, pp. 9–37.

Codd, E. F. (1971) "Normalized database structure: A brief tutorial," *Proceedings,* pp. 1–17, ACM-SIGFIDET Workshop on Data Description, Access and Control, New York: Association for Computing Machinery.

Date, C. J. (1995) *An Introduction to Database Systems,* 6th ed. Reading, MA: Addison-Wesley Publishing Company.

Elmasri, R., and Navathe, S. B. (1994) *Fundamentals of Database Systems,* Menlo Park, CA: Addison-Wesley Publishing Company.

ESRI (1995) *Understanding GIS: The Arc/Info Method,* 3rd ed. Redlands, CA: Environmental Systems Research Institute, Inc.

Graham, L. A. (1998) "NT vs UNIX: The battle is on!," *GIS World,* Vol. 11, No. 5, pp. 54–57.

Hadzilacos, T., and Tryfona, N. (1996) "Logical data modelling for geographical applications," *International Journal of Geographical Information Systems,* Vol. 10, No. 6, pp. 179–203.

Hix, D., and Hartson, R. (1993) *Developing User Interfaces,* New York: John Wiley & Sons.

IEEE (1993) *IEEE Standards Collection: Software Engineering, IEEE Standard 610.12-1990,* New York: IEEE (Institute of Electrical and Electronic Engineers).

Kuhns, R. D. (1998) "Strategies for designing and building reusable GIS application components," *Proceedings* (in CD-ROM), paper # 557, Arc/Info Users Annual Conference, Redlands, CA: Environmental Systems Research Institute, Inc.

Maguire, D. (1998) "Windows NT, UNIX and ESRI," *ARC News,* Vol. 20, No. 2, p. 5.

Marble, D. F. (1982) "On the application of software engineering methodology to the development of geographic information systems," in *The Design and Implementation of Computer-based Geographic Information Systems,* by Peuquet, D., and O'Callaghan, J. eds. pp. 102–111, Amherst, NY: IGU Commission on Geographical Data Sensing and Processing.

Marble, D. F., Lauzon, J. P., and McGranaghan, M. (1984) "Development of a conceptual model of the manual digitizing process," *Proceedings,* Vol. 1, pp. 146–171, International Symposium on Spatial Data Handling, Zurich, Switzerland.

McClure, C. (1997) *Software Reuse Techniques: Adding Reuse to the Systems Development Process,* Upper Saddle River, NJ: Prentice Hall.

Microsoft Corporation (1992) *The Windows Interface: An Application Design Guide,* Redmond, WA: Microsoft Press.

Ngan, S., and Dennerlein, J. (1997) "RAD and ESRI desktop solution," *Proceedings* (in CD-ROM), paper # 634, Arc/Info Users Annual Conference, Redlands, CA: Environmental Systems Research Institute, Inc.

Pressman, R. S., (1997) *Software Engineering: A Practitioner's Approach,* 4th ed., New York: The McGraw-Hill Companies, Inc.

Whitten, J. L, Bentley, L. D., and Barlow, V. M. (1994) *Systems Analysis and Design Methods,* 3rd ed., Burr Ridge, IL: Richard D. Irwin, Inc.

CHAPTER

12

GIS Issues and Prospects

12.1 Introduction

In the last several years, we have witnessed probably the most spectacular period of growth of GIS in its history. GIS is no longer simply a data analysis tool for scientific and academic research, but a core technology for information resource management and decision making in business and government. The proliferation of geographic information systems has led to numerous opportunities for hardware and software vendors, researchers, and practitioners in a wide range of fields in the humanities, science, and technology. At the same time, however, many of the old problems facing GIS have lingered on, and new problems arise as the user environment becomes increasingly sophisticated. These problems, if not properly addressed, will pose the major impediments to the healthy growth of GIS as one of the most important branches of information technology.

This concluding chapter provides an overview of the key issues facing GIS users today. We will also examine emerging concepts and techniques that are rapidly changing the nature of using geographic information. The objective is to provide a clear understanding of both the factors constraining the current state of using GIS and the trends of development that will shape the new generation of GIS in the future.

12.2 Issues of Implementing GIS

GIS in itself is a relatively sophisticated technology. Therefore, it is not surprising that despite the apparent success of GIS as one of the fastest-growing areas of the computer industry, many issues remain as barriers impeding its acceptance. The underlying causes of these issues are a combination of factors. These include, for example, the inherent complex nature of geographic information, the wide range and scopes of GIS applications, and the inability of many users and organizations to cope with the new working environment using GIS tools. There are various ways by which these issues may be examined. In the following discussion, we will make our observations with particular reference to the four components of GIS, namely, data, people, technology, and application (see Chapter 1).

12.2.1 Data Issues

In the early days of GIS, the major data issue was the absence or lack of digital geographic data. GIS users had to acquire their own data by digitizing paper maps. The cost and time incurred by digital conversion of existing map data was the major stranglehold to GIS implementation. In the last several years, however, the data problem has changed, as vast amounts of geographically referenced digital data became available. Conventionally, geographic data came mainly from government sources. However, they are increasingly obtainable from commercial data suppliers. Consequently, the problem of geographic data today is caused more by the extensive availability of digital geographic data rather than by the lack of them.

Access to Geographic Data

The increasing availability of digital geographic data today has apparently not increased the use of GIS significantly. This is because availability does not necessarily mean easier access to data. Very often, users are simply unaware of what data are in existence or where they are found. Even if they know where data may be available, they may have a hard time figuring out whether the data are suited for their applications. As GIS applications become more sophisticated, the need to access data from different sources increases. The integration of disparate data from different sources may be a complex task because these data are often incompatible not only in storage format but also in cartographic specifications such as scale, map projection, accuracy, and symbology. Although data translators are now more readily available than ever before, perfect conversion from one format to another is not always attainable (Lazar, 1998). GIS users sometimes find it easier to digitize their own data rather than going through the agony of finding, evaluating, and converting geographic data from existing sources.

In order to facilitate the sharing of digital geographic data, government agencies and the private GIS industry have joined forces to develop and promote the use of data standards, metadata, and what has now come to be known as national *geospatial data infrastructures* (see Chapter 1). Since the early 1980s, standards development organizations at both the international and national levels have been working very hard to develop a universal standard for geographic data (Thoen, 1998). This has been done in the hope that such a standard will be incorporated into commercial GIS software packages and used by both data producers and users alike. However, the need to address the diverse, and often conflicting, interests of a large number of users transcending various functional and political boundaries means that the resulting standards are often cumbersome and hard to implement. Up to now, geographic data sharing through format standardization has remained largely a goal rather than becoming a practical reality. It is still necessary for users to rely heavily on de facto data exchange standards such as AutoCAD DXF, ArcInfo E00, and ArcView shapefiles, among many others, when geographic data are exchanged between different systems. While the ideas behind standards, metadata, and data infrastructure are conceptually sound in the long run, they have not been able to address the short-term needs of many GIS users.

Data Ownership, Copyright, and Cost Recovery

Another major data issue is ownership, which is in turn closely related to the issues of copyright and cost recovery of using the data. Different jurisdictions have different approaches to data ownership. In the United States, data held by the federal government are regarded as public-domain properties. They are made widely available to the general public with the realization that an informed public is fundamental to the social and economic well-being of the country. The public's right to access government-held data is further reinforced by the fact that the public has paid, through taxes, for the collection of such data. As a result, the United States federal government has policies to limit the amount that can be charged for using public-domain data, including digital geographic data. Within the constraints of privacy and confidentiality laws, nearly all federally held geographic data in the United States are available to anyone, even for commercial reselling purposes, at nominal cost. The private sector is allowed to repackage public-domain data in the form of value-added information products and resell them commercially. The U.S. federal government in fact encourages these kinds of activities through policies that prohibit the copyrighting of data in the public domain.

The approach of the United States federal government, however, is an exception rather than a rule in other jurisdictions. Many state governments in the United States, for example, do not follow the federal information policy and have statutes that permit cost recovery through user fees (Masser, 1998). In countries such as the United Kingdom, government-held data are regarded as a valuable commodity and the provision of such data to the public is a chargeable service (Rhind, 1992). Government agencies are expected to recover a substantial part of the cost of data collection and processing through user fees. To this end, the use of government-held data is always subject to stringent copyright laws that aim at stopping illegal copying and reselling of the data.

Although such a policy appears to be relatively simple and straightforward, it is not always clear how it can be enforced or if it is even enforceable in practice. The attempt to recover cost, for example, is not easy to start with because data, which are not a tangible commodity, are hard to price. Full-cost recovery will definitely make the cost of using geographic data prohibitively high and, consequently, discourage more widespread use of the data. Additionally, while the copyright of the original data has been generally well-defined and understood, the copyright of the thematic data generated from the base data is far less explicit. Many jurisdictions require secondary data users to pay royalties to the agencies that produce the base data. This requirement increases the complication of enforcing copyright laws and the ultimate cost of using geographic data. The complication of the issues of ownership and copyright multiplies quickly as the data are continuously messaged and augmented. Obviously, it is difficult, if not impossible, for data-producing agencies to trace how their data are used and to determine and collect appropriate royalties accordingly. Although there are merits and justifications in this user-pay approach to the use of geographic data (Rhind, 1992), the complexity of the data ownership and copyright issues tends to discourage potential users from applying existing digital geographic data in their business and decision-making process, thus limiting the long-term growth of GIS technology as a whole.

Liability of Misusing Data

A problem that is directly related to data ownership is concerned with the legal liability of the suppliers of geographic data (Epstein and Roitman, 1990). In the past, this particular issue was concerned mainly with a relatively limited number of application areas such as cadastral surveying and engineering as well as hydrographic and aeronautical charting. As more geographic data are used in business decision making, it is increasingly possible for users to make flawed decisions as a result of the inherent uncertainties present in the data (see Chapter 4). In order to encourage the sharing of geographic data among data collectors and end users, it is essential to clearly define the limit of liability of the data suppliers when misuse of the data occurs. Legal liability has a particularly significant implication for the use of digital geographic data because many GIS practitioners are not members of professional bodies that offer malpractice indemnity (Rhind, 1992). In view of the absence of a mechanism for the certification of geographic data services, the need for people engaged in geographic data services to protect themselves against potential litigation is now more realistic and imminent than ever before.

Freedom of Information Legislation and the Issue of Privacy

Conventionally, the collection of geographic data has been largely a government mandate and responsibility. Many data sets are collected to satisfy legislative requirements (e.g., population censuses and cadastral surveys), and others are collected to serve specific areas of program delivery (e.g., forest resources, topographic mapping, and law and order). Freedom of information (FOI) legislation requires that data pertaining to the privacy of individuals be aggregated before they are released to the public. Defining privacy is certainly not an easy task and always represents a very delicate balancing act between the public's right to know and an individual's right to maintain privacy.

Privacy is normally not a critical issue when data are collected by government agencies, because people in general tend to be comfortable with government agencies regarding the handling of personal data. However, as the private sector is getting increasingly involved in the business of collecting and selling geographic data, the protection of privacy becomes a more obvious concern. In the United Kingdom, for example, Rhind (1992) noted that several private data suppliers commercially marketed data files containing names and addresses derived from the electoral roll. These files had been enhanced by inferring commercially relevant information from the census details for small areas to which individuals might be linked through their locations, thus exposing personal particulars that people may not be willing to reveal. Although most commercially supplied data files are intended primarily for marketing and business planning purposes, it is necessary for government authorities and data suppliers to avoid and prevent the abuse of personal data by enforcing freedom of information (FOI) legislation strictly.

Geometric Incompatibility

The problems of geometric incompatibility occur when digital geographic data captured from different map sheets or obtained from different sources fail to match. These may be due to a variety of causes. One of these results from the change of georeferencing standards. A notable example is the new North American Datum (NAD 83) that has been developed to replace the old georeference datum NAD 27 (see Chapter 2). As it is extremely time-consuming to change all data sets from the old to the new datum, problems of geometric incompatibility frequently occur during the transition period when geographic data referenced to both datums are in concurrent use.

Geometric incompatibility is also caused by using geographic data collected at different scales and map projections. National small-scale maps are usually based on the graticules of geographic coordinates. Each map sheet row is therefore ostensibly of a different size and the vertical sides of the map sheets are not parallel, resulting from the curvature of the line of longitudes. Medium- and large-scale maps produced by state, provincial, or local mapping authorities are commonly based on rectangular grids such as the Universal Transverse Mercator (UTM) coordinate system. To integrate data from different sources, it is necessary to convert data from one map projection or coordinate system to another. However, there is no guarantee that the resulting data sets will be geometrically compatible with one another because of rounding errors in the mathematical transformation of coordinates and cartographic generalization of topographic details.

Even when digital data georeferenced to the same datum and collected at the same scale are used, there are still mismatches between adjacent data files. This problem is particularly common in geographic data digitized from maps that are produced by different contractors, at different times, and in different map revision cycles. Reconciliation of mismatching features across map boundaries is no trivial task because the process can hardly be carried out by totally automated means. Human decisions are always required to determine the most logical way of shifting the feature on one side of the map boundary relative to its counterpart on the other, making data reconciliation an extremely time-consuming and costly undertaking.

Database Updating and Maintenance

It is important to understand the potential degradation of the quality of geographic data over time. Although data collection and digitizing are carried out using relatively stringent specifications as a rule, experience has shown that the same level of requirements is not always enforced when the digital databases are updated. For example, it is not uncommon for new graphical data to be visually fitted to the existing data. This has resulted in the degradation of the geometric accuracy of the contents of the database. It also tends to invalidate the data quality information attached to the metadata of the data sets concerned.

Another issue that has not always been given due attention is the proper maintenance of the digital database. Ideally, a spatial database is a faithful snapshot of the status of human activities and natural features that are found in a particular geographic area of interest at a specific point in time. When the characteristics of these human activities and natural features change, the database must be updated as well. Maintaining the database at current levels requires continuous monitoring of human activities and the natural environment in order to detect all the changes that have occurred. Change detection is a very time-consuming and costly task, and so is the process of updating the database to reflect all the changes that have been noted. As a result, it is impossible to keep a spatial database up to date. This always leads to serious uncertainties when the data are used for solving time-sensitive spatial problems.

Data Quality and Documentation Standards

The importance of information about data quality to the appropriate application of geographic data is now well understood by end users (see Chapter 4). However, many data suppliers today still treat the provision of data quality information as a luxury rather than an absolute necessity. This is due partly to the high cost incurred in the collection and maintenance of metadata information and partly to the difficulty of obtaining data quality information of existing data sets. As a result, many geographic data sets are being used today with incomplete or no data quality information. Even in the case of those data sets for which data quality information is provided, the lack of a standard for documenting such information makes it difficult for users to find the data and determine their suitability.

12.2.2 ISSUES PERTAINING TO PEOPLE

The root of the people issues in GIS stems from the diversity of users in terms of their respective technical backgrounds (e.g., different levels of proficiency in using computers), application requirements (e.g., from simple turnkey desktop data browsing to large-scale modeling of global change) and expectations (e.g., different degrees of understanding the limitations and possibilities of using GIS). Therefore, each of the three categories of users identified in Chapter 1 has posed relatively unique issues that must be properly addressed. For large-scale corporate implementation of GIS, training of end users always represents one of the most difficult tasks. This must usually be delivered within a relatively short time frame to a large number of users who may be distributed in many geographically dispersed locations. Senior management users require GIS user education rather than user training. GIS user education tends to emphasize the understanding of concepts and technology that distinguishes it from end user training characterized by the focus on the acquisition of skills. The objective of GIS user education for senior managers is to ensure that they have a good understanding of the technology and a realistic expectation of what the system can and cannot do. The importance of user training and education should never be underestimated. Experience has shown that many GIS projects failed not because of the lack of technology, but because of the lack of support from end users and senior managers.

The most critical people issue today, however, seems to be the chronic shortage of skilled GIS specialists who are responsible for making the systems and applications work. This class of users includes systems managers, database administrators, systems analysts, and application programmers. The human resources problem in the GIS industry has been identified for quite a long time (Antennuci, 1991). It has been caused partly by the emerging opportunities created by the spectacular growth of GIS technology and the concomitant construction of a wide variety of geographic databases and partly by the failure of educational institutions to produce enough graduates with the right skills (Kennedy, 1991; Marble, 1998). In a study on geomatics human resources conducted in Canada in the early 1990s, for example, less than half of the employers surveyed indicated that they were confident of the capability of the educational system to provide the numbers of graduates that would be needed to meet their expected requirements. The same study reported even fewer employers who were confident that graduates had the necessary skills to meet their expected requirements (Kennedy, 1991). It is hard to say how far the findings of this particular study in Canada reflect the general problem of GIS human resources in other countries. However, in view of the persistent shortage of skilled workers in the computer industry as a whole, the condition in the GIS sector can only be worse because of its specialized requirements.

In an article that addresses the particular issue of the ability of university GIS graduates to meet the demands of the industry, Marble (1998) was highly critical of the present approach of GIS teaching with special reference to geography departments in the United States. He argued that it is basically wrong to shift the focus of GIS teaching from the geography/computer science interface approach of the past to the present one that gives too much weight to the use of software packages. This has resulted in a generation of students who are capable of applying only a very small portion of the power of GIS technology. These students are poorly prepared for job opportunities in the industry and in academic research. In order to rectify the present situation, Marble (1998) proposed a new model of GIS education with a return to the strong fusing of geography and computer science as it was in the past. In this model, GIS teaching is stratified into six levels (Figure 12.1).

- *basic elements*, which include foundation knowledge in cartography, basic spatial analysis, computer programming, and data organization, as well as "thinking spatially"

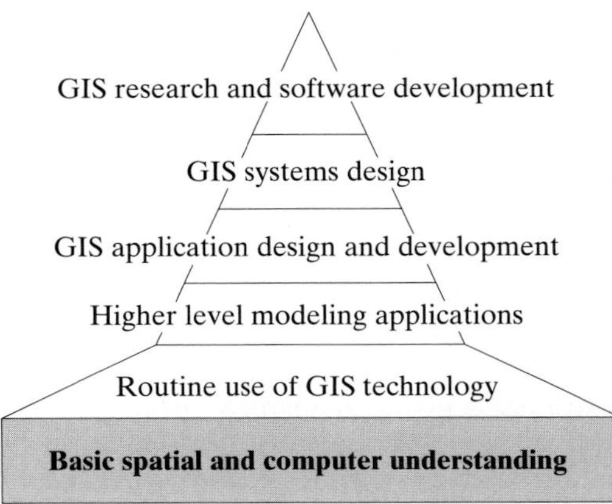

FIGURE 12.1
The pyramid model of GIS Education. *(Source: Marble, 1998)*

- *the first operational level*, which aims at developing the student's working knowledge of the full capabilities of GIS technology
- *the second operational level*, which represents a formal approach to spatial analysis, computer programming, and database management
- *the third operational level*, which addresses GIS application design and development using advanced techniques in computer programming, software engineering, and database systems, as well as advanced elements from geography and spatial analysis
- *the fourth operational level*, which places GIS technology in the new context of systems design, including database design and user interface design
- *the top of the pyramid*, which, with a concentration in research and the development of new software tools, will be open to only a relatively small group of individuals

A sound GIS education structure must be built on a rigorous yet useful first course that provides students with a solid foundation to enable them to make extensive use of GIS technology (Marble, 1998). To rebuild the upper tiers of the pyramid, it is necessary to reestablish the strong role of computer science education within GIS. At the same time, it is necessary to restructure GIS educational activities so that competence in computing may be fully integrated into the structure of these activities. This is achieved by the offering of advanced and highly technical courses in GIS that will meet the identified needs of individual students. It is noted that the evolution of this model of GIS education has already quietly begun as GIS has been increasingly offered as a degree specialty rather than as individual optional courses in the undergraduate curricula in universities. Opportunities for graduate studies have also increased tremendously both in the United States and elsewhere. However, as education is a long-term investment, only time can tell how these changes in GIS education will alleviate the persistent human resources problem in the GIS industry in the long run.

12.2.3 ISSUES PERTAINING TO TECHNOLOGY

From the technology perspective, the development of GIS is heavily dependent on the general trends of the hardware and software evolution in the computer industry. In recent years, growing awareness of the importance of geographic information in corporate information resource management has attracted many mainstream software companies to the GIS arena. There is a particularly strong interest among database software vendors to include spatial data-handling capabilities within the conventional database environments. Big information technology companies such as Oracle, Hewlett Packard, IBM, and Microsoft have now all had a presence in the GIS marketplace. This has led to several breakthroughs in database technology that make the integration of spatial and descriptive data a reality (Wilson, 1997). However, it has also created a substantial amount of hype and confusion among the ordinary users. For example, the entry of large technology companies into the GIS marketplace has engendered many takeovers, mergers, partnerships, and joint development pacts among them and conventional GIS software vendors. It has also given rise to intense marketing efforts with claims and counterclaims about the capabilities of products and the future directions of technology. Users are sometimes simply overwhelmed with excessive information when trying to configure their systems for implementation projects.

The technology issues of GIS today, therefore, have not very much to do with the lack or inadequacy of technology per se. Instead, they are concerned mainly with the ability of GIS users to evaluate and manage technology, as well as the capability of current technology to cope with the application requirements of GIS as explained next.

Evaluation of Technology

The evaluation of hardware and software technologies is a very subjective matter. It is a time-consuming and tedious process that requires considerable commitments from systems staff, end users, and hardware and software vendors. However, this is a task that simply cannot be ignored. It is most obvious that not all hardware and software products are created functionally equal. There are always unique features or strengths in a particular product that make it better suit a user's application requirements than others. The objective of technology evaluation is to identify as objectively as possible the best GIS by conducting a formal benchmarking exercise. Evaluation of technology involves both technical factors (e.g., hardware platform, operating system, network architecture, database size, modes of data processing, and conformance to standards) and nontechnical considerations (e.g., cost, size of client base, and user references). It is up to the GIS project manager, with input from the systems analyst and end users, to properly align the use of technology with identified application requirements as part of the systems development process (see Chapter 11).

Management of Technology

The objective of technology management is to make sure that once a particular hardware or software product has been chosen and implemented, its performance continues to meet design specifications. This is a crucial component of the systems maintenance stage of the systems development life cycle (see Chapter 11). There are two important considerations in managing technology: user training that aims to ensure that technology is used properly, and functional upgrades in response to changes in both the user requirements and technological environment. As computer technology is a very rapidly evolving technology, the operating procedures of any information system must include mechanisms to monitor technological changes, recommend systems upgrade at the appropriate time, and manage the transition from the old to the new systems environments (see Chapter 11). Failure to do so will lead to loss of productivity and shorten the useful life span of an information system.

Data-Related Technology Issues

There are two data-related problems that deserve our special attention when examining technology issues of GIS. One of these occurs because of the probable incompatibility between data collected by old data-acquisition technologies and those acquired by new ones. Many geographic data sets presently maintained by government agencies were digitized from existing paper maps compiled using old field surveying and cartographic techniques, often without metadata information. The quality of such data sets is inconsistent with the quality of new data sets obtained by means of modern positioning technologies such as GPS, digital photogrammetry, and total stations. Analysis using data sets of different qualities is problematic and may result in dubious interpretations and conclusions.

Another data-related problem occurs as the result of the current thrust toward the use of the distributed client/server architecture in information systems development. This has led to exponential growth in the transfer of geographic data over existing telecommunications networks. The problem is particularly serious when the data involved are in raster format. A USGS Digital Orthophoto Quadrangle (DOQ), for example, is 50MB in file size. It takes close to two hours to transfer such a file, using the *file transfer protocol* (FTP) via a standard copper phone line with a 56 kilobits per second (Kbps) modem. The cheapest alternative is an *Integrated Services Digital Network* (ISDN) line running on existing copper phone wires at 144 Kbps. The same 50MB DOQ file will take about 50 minutes to transfer, but the use of such a line incurs a monthly service cost that varies from place to place (Corbley, 1998). Faster carriers (e.g., T-1 lines, DS-3 lines, and cable modems) are now available, but these require higher costs as well. Before more affordable high-speed transmission systems are widely available, it appears that the data-transfer rate will remain the major barrier to the access and sharing of geographic data over existing telecommunications networks.

12.2.4 APPLICATIONS ISSUES

The large number and varieties of GIS applications currently in use have always been quoted as an indication of its maturity and success. However, if we examine more closely how GIS applications are developed and where they are used today, it will not be too difficult to discover that there is much to be desired. Of the many application issues that may be identified, the following are particularly worth noting: (1) the breadth and depth of the applications, (2) the approach to application development, and (3) the ability to integrate with other applications, especially spatial statistics, spatial modeling, and decision support.

The Breadth and Depth of Application Development

Until the late 1990s, GIS has been used mainly as a stand-alone application in natural resource inventories, surveying and mapping, land management, and facilities management. As a branch of information technology, it simply lacks the breadth and depth to be part of enterprise-wide applications in support of the business goals of organizations. In many GIS implementation projects, a large proportion of the resources has been devoted to structuring the system's architecture, stabilizing the technical environment and getting data into the database. These have often been done at the expense of the requirements of applications such as intuitive user interface design, integration with other types of computer-based data processing and more sophisticated use other than simple transaction processing. As GIS technology becomes mature and as digital geographic databases of various types are more readily accessible, a reorientation of resource allocation in GIS development has appeared in recent years. This has resulted in a new generation of applications that allows users to focus more on the jobs that they do (i.e., using the applications), rather than on the tools that they use (i.e., the information system).

The Approach to Application Development

Conventionally, GIS software packages are constructed using a toolbox approach. Users purchase the software packages and construct their applications using available commands and scripting languages as building blocks. As a result, each user or organization ends up with a unique proprietary application. It is not only necessary for the user or organization to bear the entire cost of application development, but also all the expenses incurred in the maintenance, enhancement, and upgrading of the applications in use. Obviously, such an approach to application development is expensive and inefficient. There is absolutely no need for each utility company, for example, to develop its own proprietary

GIS applications as they all perform essentially the same functions. Having recognized the problem in this approach to application development, GIS software vendors are moving quickly toward a standardized application development environment. Intergraph, for example, has created a Windows-based component architecture that forms the foundation of the company's engineering and design software. Based on a similar architecture, standardized task-specific applications may now be developed to address the needs of particular user groups (e.g., business analysis, data tracking, real estate, geological modeling, utilities, facilities management, and forest management) in ArcInfo, ArcView, MapInfo, and other GIS software packages.

Integration with Other Types of Applications

One of the major criticisms of GIS applications in the past was the lack of integration with other types of applications. This is mainly because of the special nature of geographic data, which require the use of software functions distinctly different from other types of applications. Also, old computer technology made integration of applications extremely difficult. Data used in or generated by one application could be moved between different applications only by transferring them in data files. At the same time, the interface between different applications was achievable only by very complex application programming. All these problems combined to make it practically impossible to integrate GIS functions with cognate techniques such as spatial statistics, spatial modeling, and decision support in a single data-processing environment.

The issue of integration with other applications is apparently coming to resolution with the advent of two emerging technologies. One of these is the breakthrough in database management technology that has made possible the incorporation of spatial data into the relational database management systems (DBMS), as exemplified by the *Spatial Database Engine* (SDE) of ESRI and *Oracle Spatial* (formerly known as *Spatial Data Cartridge* and *Spatial Data Object*) of Oracle. The other is component software technology (see Chapter 6). Built on today's key integration platform, Microsoft ActiveX using ActiveX controls (OCXs), component-based applications allow their objects to be plugged into a wide variety of development environments such as Visual Basic, Visual C++, Delphi, and PowerBuilder. Unlike their GIS toolbox predecessors, component software suites such as MapX (MapInfo Inc.) and MapObjects (ESRI) conform to programming standards of the Windows NT environment, thus allowing both GIS applications and data to be linked easily to their non-geographic counterparts. The removal of the barrier between GIS and other applications is in fact one of the principal factors for bringing geographic data processing into the mainstream of information technology (see Section 12.3.1). It also forms the foundation of GIS interoperability (see Section 12.3.2).

12.3 The Trend of GIS Development

If the current trend continues, GIS will soon enter a new era that is very different from what we have experienced so far. As a tool for information resource management, GIS is moving increasingly closer to the core of corporate information technology. Different methods and mechanisms are being developed at state/provincial, national, and international levels to make GIS interoperable not only in terms of data but in terms of processes. Advancements in new communication and information technologies, notably the Internet and multimedia technology, will totally change the ways by which geographic data are delivered, used, and visualized. In the discussion that follows, we will examine the prospects of GIS development with special reference to several emerging concepts and techniques that will most likely form the foundation of the next generation of GIS.

12.3.1 ENTERPRISE COMPUTING AND GIS

One of the most profound changes redefining geographic data processing is the rapid integration of GIS technology with mainstream information technology and its consequential incorporation into the enterprise computing environment. Many factors have contributed to moving GIS beyond its traditional role as a stand-alone application to become a cornerstone of corporate information technology in both the private and public sectors. Growing intensity in competition and economic conditions in the last few years have forced both private and public sector organizations to reengineer their ways of doing business. This has led to a completely new perception of what geographic data can do to serve critical enterprise business objectives and where they can fit into the overall information needs of an organization. In this regard, the integration of GIS with mainstream technology is not merely a convergence of technologies, it actually represents a fundamental change in the philosophy of using geographic data. Instead of focusing on the technology that uses geographic data, users have now turned their attention to what they use geographic information for.

The idea of coupling GIS with conventional enterprise computing applications has been around for many years. However, it has been practical to implement enterprise GIS applications only relatively recently with the advancements of information technology and systems implementation methodology in several key areas.

Database Management

Conventionally, GIS has been a graphics-based technology. The complexity of geographic data, together with the nature of transactions and the difficulty of tracking state transitions, have kept GIS separated from mainstream information technology. In the new generation of database management tools such as Spatial Database Engine and Oracle Spatial, spatial data may be encoded in tabular form and used to build graphics on the fly when the data are used. This creates a graphics-independent environment that allows the same data to be used in a GIS, computer-assisted design (CAD), multimedia production as well as any other enterprise computing functions such as work scheduling and inventory management in utilities and transportation companies.

Data Modeling Methodology

In the enterprise computing environment, the objective of data modeling is to identify corporate data requirements and relationships, as well as the responsibilities among different departments for the maintenance of corporate databases. Data modeling techniques have now been sufficiently well developed to absorb GIS concepts and requirements into the conventional systems development process, thus allowing GIS to be developed seamlessly within the conceptual framework and technical architecture of enterprise-wide information systems.

Application Development Environment

One of the major barriers that kept GIS a stand-alone application separate from mainstream information technology in the past was its toolbox paradigm of application development. With the adoption of the interoperable approach to application development as explained in Section 12.3.2, GIS components may now be integrated with other software components into one single application, resulting in a real-time interface between different software/hardware platforms. GIS software vendors have been quick to capture the opportunity by making their products compatible with this new application development environment. At the same time, they have also followed the general trend in the computer industry to develop off-the-shelf applications that can be used with only minimal customization from enterprise to enterprise (Wilson, 1996). An example of this new breed of GIS application is ArcFM from ESRI (1998a). This is a self-contained ArcInfo-based application specially designed for the enterprise computing needs of utility companies. It provides a user interface, utility-specific tools (e.g., network trace and automated mapping), and a *RuleBase Engine* (RBE) that encodes domain-specific business rules for electric, gas, water, and wastewater companies. Built on open architecture and technologies, ArcFM may be used as a stand-alone application that can be linked seamlessly to other applications in an enterprise-wide system.

Enhanced Capabilities of GIS for Network Analysis and Modeling

Conventionally, GIS has been developed mainly for applications in natural resources and land management. As data in these applications are fundamentally polygon-based, GIS software packages in general have well-developed polygon handling capabilities but are relatively weak in the manipulation of networks. The lack of network data-processing functionality is obviously a critical deficiency for organizations whose primary business is the management of facilities, such as utility, transportation and telecommunications companies, and engineering departments in local governments. The ability to integrate GIS and CAD technologies has greatly enhanced the power of conventional GIS to model and manage network-based geographic data, thus eliminating one of the major impediments of using GIS in the vast enterprise computing market dominated by organizations noted earlier.

Advances in Mobile Computing Technology

Outdoor or fieldwork accounts for a substantial part of the operation of major enterprise GIS users in utility, transportation, and telecommunications companies. As the creation, management, and maintenance of large-scale infrastructure networks require constant communication between head offices and field crews, effective communication is crucial for enterprise-wide field applications. Recent advancements in mobile computing technology (which is made up of wireless communications networks, Internet-based data-delivery protocol and field data-acquisition hardware and software), have basically removed the barrier of distance between office and field users of GIS in the enterprise environment. As a result, geographic data can now be transmitted over a mobile GIS in a myriad of applications, such as dispatch and crew tracking, remote data viewing and query, field data collection, as well as field-based engineering (Wilson, 1998).

To sum up, it is very obvious that the business environment, computing philosophy, and telecommunications technologies are all there to redefine and reposition GIS as a core technology in today's enterprise information resource management strategies. In many organizations, geographic data now play a central role in business operation and decision making, along with relational databases, document management, human resources, financial systems, and other program-specific enterprise applications (e.g., building permits in municipal administration, work management in utilities, and data tracking in goods delivery). A typical spatially-enabled

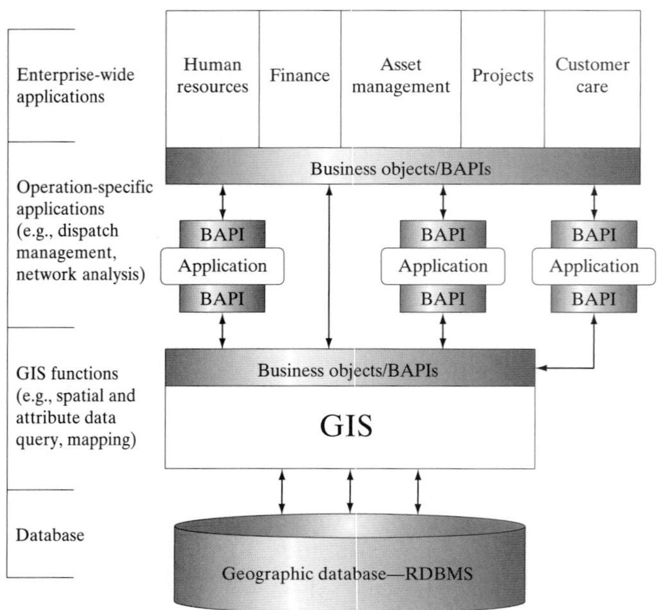

FIGURE 12.2
A multitier enterprise computing architecture. *(Source: Partially based on ESRI, 1998b)*

enterprise computing environment, as exemplified by the *Utility Solution Bundle* from ESRI (1998b), is a multiparty solution involving end users, GIS application developers, systems consultants and integrators, as well as hardware and software manufacturers. It has a multitiered architecture that is made up of geographic data, GIS applications, business application programming interfaces (BAPI), and a myriad of conventional enterprise applications (Figure 12.2). As the business application programming interfaces are defined in generic terms, each interface can be built as ready-made, drop-in options that require little or no customization. Additionally, since the architecture is in compliance with open industry standards, integration for all vendors may be accomplished relatively easily.

When compared with stand-alone GIS applications, using GIS in enterprise computing has many advantages. As data and applications are designed from a corporate perspective that transcends departmental boundaries, duplication of effort may be readily identified and avoided, thus making the use of information technology more cost-effective. Built on client/server and spatial data warehousing technologies, enterprise GIS may be used both as a powerful interface to numerous existing enterprise databases and as a data integrator linking across different business functions. At the same time, spatially enabled enterprise computing applications have greatly popularized GIS technology by bringing geographic information to numerous users in affordable desktop and laptop computing environments.

The next step is to bring geographic information to society in general. With the implementation of initiatives such as the *National Spatial Data Infrastructure* (NSDI) (see Section 12.3.3), all levels of government agencies in the United States are required to make public-domain data readily available to citizens. As geographic data make up a substantial part of public-domain data sets, it appears that a spatially enabled enterprise computing solution using GIS is the only mechanism that is capable of delivering this service cost-effectively.

12.3.2 SPATIAL DATA WAREHOUSES

The building of data warehouses is one of the cornerstones of the concept of enterprise computing. Inmon (1996) coined the term "data warehouse" to describe a central repository of all the data collected by the various information systems of an enterprise. To understand why a data warehouse is not just called a "super database," it is necessary to understand the history of computers and databases. In an organization, business data were conventionally stored by domain-specific databases designed to serve the needs of particular functional areas. Using a carefully designed and normalized database, which aims to improve systems performance by eliminating redundant database records, rapid and reliable changes to domain-specific databases may be made using a process called *on-line transaction processing* (OLTP). However, since domain-specific databases were often created using different hardware configurations and software standards, without a corporate computing standard and systems working environment, these databases tended to evolve in different directions according to the specific needs of the user departments. As a result, it was very difficult, if not impossible, to correlate and cross reference data in different databases in the same organization.

Data warehouses have been designed to unify disparate domain-specific databases without disrupting or changing them. A data warehouse is created by storing a clean copy of all the departmental databases in a central repository. In essence, this means that a data warehouse is created to supplement departmental databases, rather than to replace them, in order to meet the information needs of an organization. The departmental databases and the corporate data warehouses, therefore, serve different purposes. The main purposes of departmental databases are to capture and store business data, and to manage the transactions that changes the contents of the database. A data warehouse, on the other hand, uses a process called *on-line analytical processing* (OLAP) to determine the characteristics of both the data and the transactions. These characteristics include, for example, the relationships between different data sets, the trends of data use, and the behavior of the data users.

Many business and government organizations have now adopted the method of data warehousing as a key component in their spatial data management strategies, leading to the growth of spatial data warehouses. The function of a spatial data warehouse is not only to integrate disparate spatial data collections into a single repository, but also to translate formats between heterogeneous operating systems and data structure. Spatial data warehouses are designed to be frequently read, but not written to. While domain-specific databases in different departments are operated in support of their respective daily functions, spatial data warehouses provide the data and process for acquiring spatial knowledge for decision makers. In response to the needs of corporate users, GIS vendors produce software products for building, maintaining, and using spatial data warehouses. Therefore, spatial data warehousing is an integral component of GIS implementation in the enterprise computing environment.

12.3.3 INTEROPERABILITY AND OPEN GIS

Incompatibility of proprietary data used by different systems has been a major barrier to the sharing of data among GIS users. Conventionally, the solution has been to use data interchange standards (see Chapter 4). In this approach, source data in the proprietary format are translated into a "neutral" format, from which they are again translated into a second proprietary format. Such an approach has now been gradually taken over by the concept of interoperability of GIS. Interoperability is by no means unique to GIS but is a common problem with information systems in general (Wegner, 1996). When different organizations develop information systems to address their respective needs, they tend to use different hardware/software technologies and different data models. The problem of incompatibility will likely occur when these disparate systems are interfaced or connected. In the computer industry, "interoperability" usually refers to a bottom-up approach of integrating existing systems and applications that were not originally intended to be integrated but are later systematically combined to address problems that require the use of multiple DBMS and application programs (Litwin et al., 1990). This is exactly the context under which interoperability is used in GIS.

When applied to GIS, the objective of interoperability is not to develop a super data translator or common data model, but to establish a *data storage specification* that will enable different GIS applications to access disparate data sets residing in different hardware platforms by means of standard data-sharing protocols such as *Structured Query Language* (SQL) for interrogating relational databases and Microsoft *Open Database Connectivity* (ODBC) for connecting one system to another. Many strategies of interoperability go well beyond this basic level of data sharing by adopting a standard distributed computing model in systems development. The industry standards that make interoperability work in the distributed computing environment work include *Object Linking and Embedding/Component Object Model* (OLE/COM) and *Common Object Request Broker Architecture* (CORBA). OLE/COM is a general environment for interapplication communication for Windows-based desktop applications, whereas CORBA is its equivalent that allows distributed use of interoperable objects in UNIX-based applications. Using these application standards, different information systems are able to share data as well as processes in heterogeneous and distributed computing environments. Just as a journalist may invoke a GIS function when writing a report using a word processing package, a GIS user working on geographic data processing may perform statistical analysis, image processing, spreadsheet calculation, and other types of applications without switching between one application and another. This is true interoperability with a completely seamless synthesis of data and processes in a single working environment.

The *Open GIS Consortium Inc.* (OGC), a membership organization comprising government agencies, academic institutes, standards development organizations, as well as software and hardware manufacturers, is spearheading the promotion of interoperable GIS (Buehler and McKee, 1998). Its approach to interoperability is through technology development by consensus building, rather than by creating and enforcing standards. The consortium works closely with its members and makes use of published standards for the development of GIS technology. The end products are a series of Open GIS specifications that will enable and encourage the sharing of data, applications, and other systems resources between individual GIS as well as between GIS and other applications such as statistical analysis, numerical modeling, document management, image processing, and scientific visualization (Figure 12.3). OGC assures the interoperability of GIS products by testing and certifying the conformance of these products to OpenGIS specifications (McKee, 1998). OpenGIS specifications can be found in the Web site *http://www.opengis.org/techno/specs.htm*, which gives *Simple Features Specifications* for OLE/COM, CORBA, and SQL, as well as *Implementation Specifications* for Catalog Interface, Grid Coverages, Coordinate Transformation Services, Web Map Server Interfaces, and Geography Markup Language (GML) (Nebert, 1999). GIS products (i.e., data, application software including server software and interfaces) may be certified as being in conformance with one or more of the specifications in the form of a certification mark. The OpenGIS certification mark carried by a product indicates that this particular product is interoperable with other products

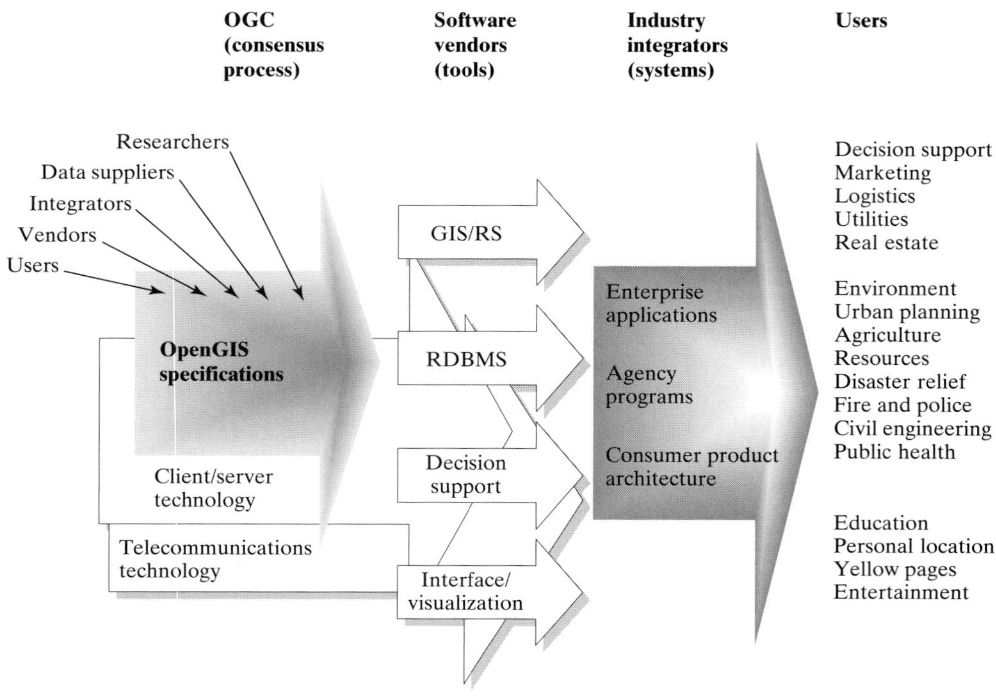

FIGURE 12.3
The OpenGIS model of interoperable GIS data processing. *(Source: Partially based on Buehler and McKee, 1998)*

carrying the same mark. OGC maintains a special page on its Web site *(http://www.opengis.org)* that lists products that have passed its conformance tests and cautions the public about false conformance claims.

With the growing acceptance of interoperability, GIS users are eventually able to see the possibility of an end to a decade-old problem caused by incompatible proprietary data formats and application environments. OpenGIS as proposed by OGC has many advantages to data suppliers, software/hardware vendors, and end users. It will bring greater flexibility to the development of application software, to the access of disparate geographic data set across a variety of hardware and software platforms, and to the integration of GIS and other forms of computer-based applications. GIS users must be aware that in spite of the large number of systems already at work, the problem of incompatibility between different systems and applications will not go away overnight. However, the popular acceptance of the concept as well as mechanisms of interoperability, such as the OpenGIS Specifications, will ensure that the next generation of GIS or geographic data-processing tools will be more cost-effective to implement and more user-friendly to operate. Interoperability between GIS and other forms of computer-based applications will also further strengthen the position of GIS as a core technology in the future enterprise computing environment.

12.3.4 NATIONAL SPATIAL DATA INFRASTRUCTURE

The spatial data infrastructure has its origin in the term "National Spatial Data Infrastructure" (NSDI) coined by the Mapping Science Committee (MSC) of the United States National Research Council (see Chapter 1). When the Federal Geographic Data Committee (FGDC) updated its NSDI strategic plan in 1997 following an extensive review and consultation process, four main goals were identified (FGDC, 1997; Tosta, 1997).

- building of relationships among organizations to support the continuing development of NSDI, i.e., the development and implementation of a process that allows stakeholders to define logical and complementary roles in support of the NSDI and the creation of a network of organizations linked through commitments to common interests within the context of the NSDI initiative
- planning for and coordinating production of a framework of basic digital geographic data from which other data can be derived or to which other data may be registered
- conceptualizing and implementing the National Geospatial Data Clearinghouse that aims to facilitate access to geographic data by developing

tools for easy exchange of data, application, and results, as well as by research and development of architectures and technologists that will enable data sharing
- fostering development of a variety of educational and training opportunities to increase the awareness and understanding of the vision, concepts, and benefits of NSDI, and to improve the collection, management, and use of geographic data

The United States NSDI was not the first initiative of its kind that aimed to foster the awareness of existing geographic data resources and to encourage harmonization in geographic data characteristics (Rhind, 1997). However, its impacts on the use of geographic data are probably more profound than any of its predecessors. Within a relatively short time after its implementation, NSDI has generated considerable interest in the United States and internationally. It has significantly increased the awareness about the value, use, and management of geographic data, particularly among federal agencies. It has also helped publicize the benefits of partnership among stakeholders of geographic data in the public, private, and academic sectors. What is more, it has triggered similar initiatives in many other countries (Masser, 1998), including national programs in Canada, Japan, the Netherlands and the United Kingdom, as well as international collaborations such as the *European Geographic Information Infrastructure* (EGII), the *Permanent Committee on GIS Infrastructure for Asia and the Pacific* (PCGIAP) and the *Global Geospatial Data Infrastructure* program (Coleman and McLaughlin, 1997).

It is interesting to note that the scopes and objectives of many of these national and international information infrastructure initiatives go well beyond the idea of facilitating access to and dissemination of geographic data. In many countries, the spatial data infrastructure is regarded as an integral part of national information infrastructure. As such, it is an essential means for political and social development, an important tool for sustainable land and resource management, as well as a core technology of the knowledge-based industry that is crucial to the economic well-being of the nation as a whole (ANZLIC, 1997; Industry Canada, 1998; NRC, 1997). From the perspective of the long-term development of GIS, raising the profile of geographic data to a high level of the agenda of national governments and international organizations is perhaps the most significant outcome of the concept of spatial data infrastructure. It provides a focus for the geographic information policies of different levels of government and ensures the continued political support from top-level decision makers in public offices.

12.3.5 THE INTERNET AND ITS IMPACTS ON GIS

Of the numerous technologies that have impacted on the development of GIS in its history, the Internet is definitely the most profound and far reaching. The Internet may be thought of as a configuration of international telecommunications networks linking computers in educational institutions, government agencies, military establishments, commercial organizations, and homes all over the world. In essence, it is a logical rather than a physical entity because it is not managed as a single system but as a cooperative system of computer networks based on the *Transmission Control Protocol/Internet Protocol* (TCP/IP). The Internet is made up of seven principal data communication facilities or application protocols.

- *file transfer protocol* (FTP), for exchanging files between computers
- *telnet*, a log-on procedure for accessing programs on remote computers
- *electronic mail* (E-mail), a mailing system for the delivery and exchange of messages among Internet users and between Internet users and users of networks external to the Internet
- *gopher*, a communication protocol for retrieving text-based or file-based data distributed in different computers across the Internet
- *newsgroup*, discussion groups that contribute information to communities of users in a particular topic area
- *World Wide Web* (WWW, or the Web), a graphical distribution hypermedia system that incorporates most aspects of these five services and delivers data files in multiple forms, including text, graphics, sound, and animation
- *wide area information searching* (WAIS), a distributed document retrieval system

Although all these protocols can be and have been used for GIS in one way or another, the World Wide Web is probably the most important tool from the perspective of distributing and visualizing geographic information. It provides not only access to a wealth of information all over the world, but also a standardized hypermedia format that allows users to navigate the Internet in a quick, intuitive, and consistent manner. The ability to provide an interactive interface regardless of computer platform allows the Web to subsume most of the functions of the other protocols, making it the de facto graphical interface of the Internet.

The *Map Viewer* developed by Xerox Corporation's Palo Alto Research Center (PARC) in June of 1993 was the first known experiment to deliver geographic data over the Internet (Putz, 1994) (Figure 12.4). This is a

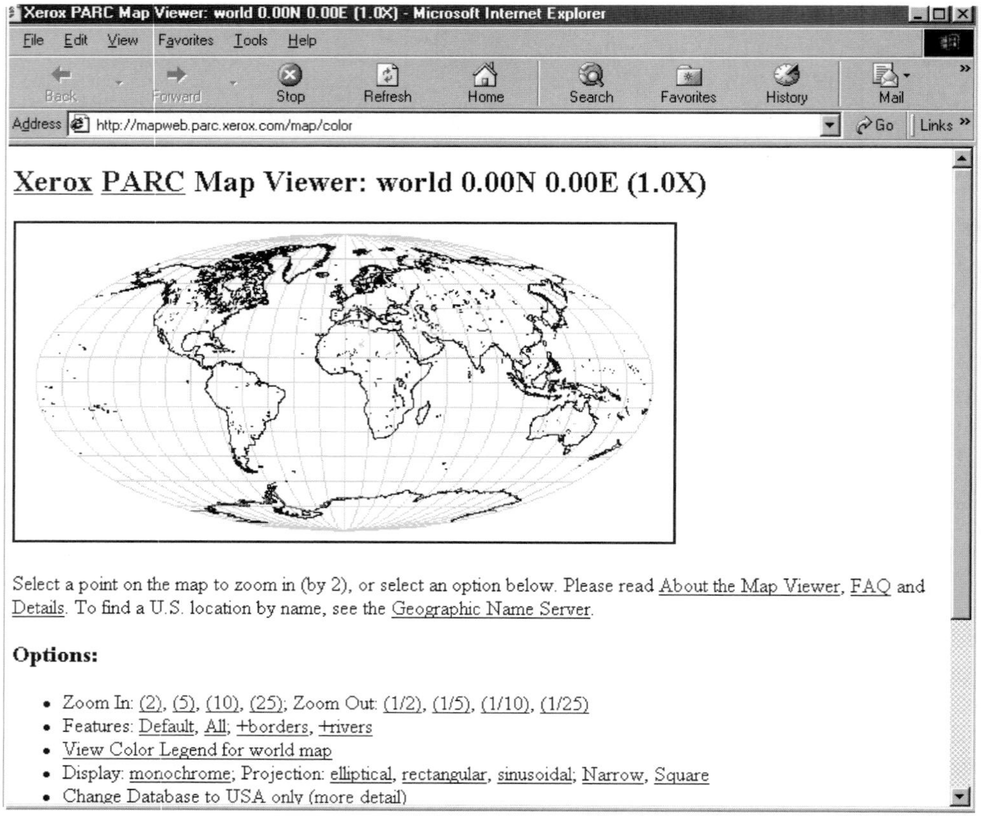

FIGURE 12.4
The Xerox Map Viewer. The Xerox Map Viewer was developed as an experiment to demonstrate the ability of transferring map data through the Internet *(Courtesy: Xerox Palo Alto Research Center)*.

simple (by today's standards) *HyperText Transfer Protocol* (HTTP) server that accepts requests for a world or U.S. map and returns it in a *HyperText Markup Language* (HTML) document. Users use links embedded in the HTML document to control the display options such as zoom-in and zoom-out, feature selection, and legends. The idea of using the Internet, or the World Wide Web to be exact, for serving mapping functions to a large number of users electronically was quickly accepted by GIS vendors and users alike. In a few years' time, the Internet has totally changed the ways that geographic data are accessed, disseminated, and used. A large variety of GIS applications may now be found on the Internet, and they are variably known by different names, such as *distributed geographic information* (DGI), *WebGIS*, *Internet-GIS*, and *Internet-enabled GIS* (Harder, 1998).

The need to manipulate georeferenced data makes the use of GIS on the Internet a more complex process than other forms of generic Internet applications. In order to perform GIS functions, an additional piece of software commonly referred to as a *GIS server* is required. The GIS server is an interface between the HTTP server and the geographic database. It has a twofold function that includes, first, interpreting the geographic component of incoming requests, and second, returning the results of geographic data retrieval and analysis in a form acceptable to the HTTP server (Figure 12.5). GIS servers associated with different GIS products are known by different names (Table 12.1). They are also constructed using different technologies that give them their respective characteristics in terms of several factors including, for example, the types of graphics (i.e., raster and/or vector) supported; the level of user/computer interactivity allowed; software requirements on the client computer; the amount and types of geographic analysis accessible from the Web; and restrictions on data and computer platforms (Limp, 1997).

Internet-enabled GIS applications are based on the client/server architecture, which consists of two programs that communicate across a computer network using a communication language called a *protocol*. For the Internet, the protocol is called HTTP. The most common client program is the *Web browser* (e.g., Netscape Navigator or Microsoft Internet Explorer). Using the Web browser as the interface, the user interacts with the server on a client machine. The Web browser is a special piece of software that allows the client to communicate with the server using the HTTP protocol. It accepts the user's requests and input data, and then interprets and displays the responses to the requests. Behind the screen, data-processing functions may be performed on either the client or the server computers (Peng, 1997). Client-side applications make

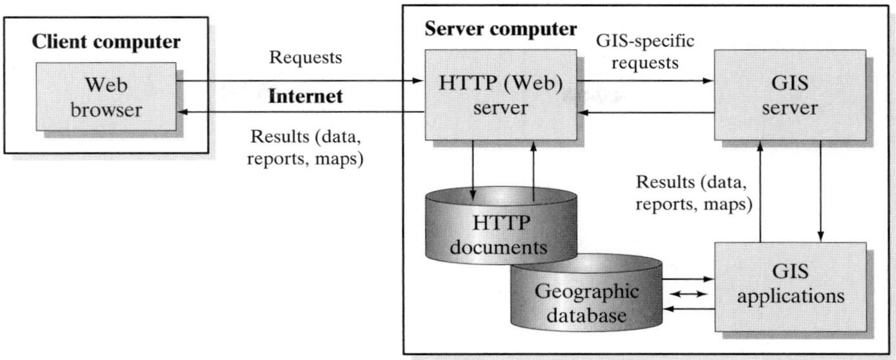

FIGURE 12.5
The three-tier architecture of Internet-enabled GIS applications. An Internet-enabled GIS is made up of three components: the client computer, the middleware (the server software), and the server computer. The function of the HTTP or Web server is to parse client requests and return information to the Web browser. Since the HTTP server is unable to understand GIS-specific commands, a GIS server is required to serve as an interface between the HTTP server and the GIS applications. The GIS server effectively extends the capabilities of the HTTP server by allowing it to respond to client requests that require the use of geographic data-processing functions of a GIS.

Table 12.1
GIS Server Software for Different GIS Packages

GIS Vendor/Software	*Name of Server Software*	*Remarks*
Autodesk AutoCAD Map	MapGuide	MapGuide is capable of distributing vector data, i.e., the resulting map is not a "dump" image. However, a plug-in is required by the browser to view the vector data.
ESRI ArcView	ArcView Internet Map Server (IMS)	ArcView IMS includes a setup wizard, a ready-to-use Java applet, called MapCafe, and several ArcView extensions to help user serve data over the Internet quickly.
ESRI MapObjects	MapObjects Internet Map Server (IMS)	MapObject IMS also includes MapCafe to enhance the browser functionality by allowing local zoom, scroll, and query capabilities.
Intergraph Modular GIS Environment (MGE), FRAMME and MGEDM	GeoMedia Web Map	GeoMedia is able to access Intergraph data as well as MicroStation design files; data are converted to the Computer Graphics Metafile (CGM) before serving them in the Internet. A plug-in is required to view the data.
MapInfo Corp. MapInfo	ProServer	User needs to know MapBasic, the programming language of MapInfo, in order to use ProServer.
MapInfo Corp. MapX	MapXsite and MapXtreme	These are tools (MapXsite for Windows NT 3.51 and MapXtreme for Windows NT 4 and Windows 95) for the construction of spatially enabled Web pages.

Compiled from Limp, 1997; and Culpepper, 1998

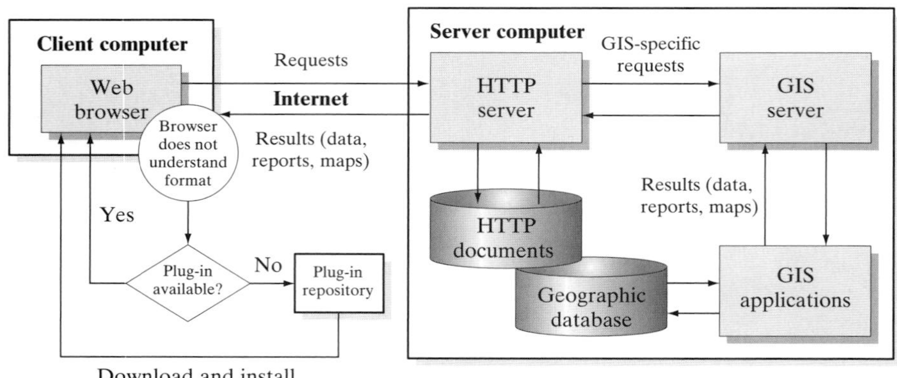

FIGURE 12.6
The method of using plug-ins by a Web browser. When the Web browser is unable to interpret the format of the incoming response to a request, it starts looking for the appropriate plug-in. If the plug-in is found, it is used to display the incoming data. Otherwise, it prompts the users to download the plug-in from a vendor site or from a third-party plug-in software repository.

use of *plug-ins* and *helper programs, GIS ActiveX controls* and *Java applets*, to overcome the limitations of the HTML in handling geographic data. These software tools initially reside on the server side, either on the server where geographic data are stored or on the server of third-party software vendors. Plug-ins and helper programs are computer programs that must be downloaded and installed before a GIS application can be processed (Figure 12.6). GIS ActiveX controls and Java applets work differently. GIS ActiveX controls are software components embedded in HTML programs. They are downloaded from the server to the client when the requested data and processes are transmitted. Java applets, which are small applications that run inside a Java-enabled Web browser, are also embedded in HTML programs. They are automatically downloaded from the server when the data and processes are requested, and are executed by the browser when the HTML programs in which they are referred are executed.

As the name implies, server-side applications depend on the server machine to perform geographic data processing and analysis. A server-side application is normally initiated by a user on the client computer by specifying the *Universal Resource Locator* (URL) of the application on the server machine. When the request for an application reaches the server, it is interpreted and directed to the GIS server where the application resides. The linkage between the Web and the GIS servers is established by *Common Gateway Interface* (CGI) scripts. These are small programs that enable the Web server to execute applications in the GIS server and return the results to the browser for display. This approach of interaction between the client and server computers is called the *client-pull* approach. Another approach is called the *server-push* approach, in which the server machine sends information to preselected client sites automatically at regular time intervals. This mode of server-side operations is most suitable for situations where client computers need updated information regularly. Examples of GIS applications that utilize server-push technology include the transmission of weather or traffic information from a central station or control room to customers who have subscribed to such services.

Numerous GIS applications may now be found on the Internet (Harder, 1998). These applications vary considerably in terms of objectives, functionality, and levels of user–computer interaction allowed. They may be broadly classified into different categories.

Static Maps or Snapshots

Static maps represent the simplest GIS application for delivering raster or vector geographic data across the Internet (Figure 12.7). Raster data may be displayed directly in the browser window, but vector data require a plug-in. Static maps are seldom distributed alone, but usually as a collection of maps or, more commonly, as an electronic atlas for a specific area, state, or country.

Dynamic Maps or Cartographic Animation

A dynamic map is a sequence of static maps that when displayed on the screen consecutively will create an impression of change and movement of the map features. Simple dynamic maps may be created using an animation software package that logically links successive static maps together (Figure 12.8). More sophisticated dynamic maps are always stored in a format designed for display movies such as QuickTime and MPEG. For GIS applications, dynamic maps are more a data visualization tool than a data distribution tool, as in the case of static maps. These maps are very useful for the presentation of geographic information that changes spatially over time, such as weather patterns, land-use change, and urban growth (for

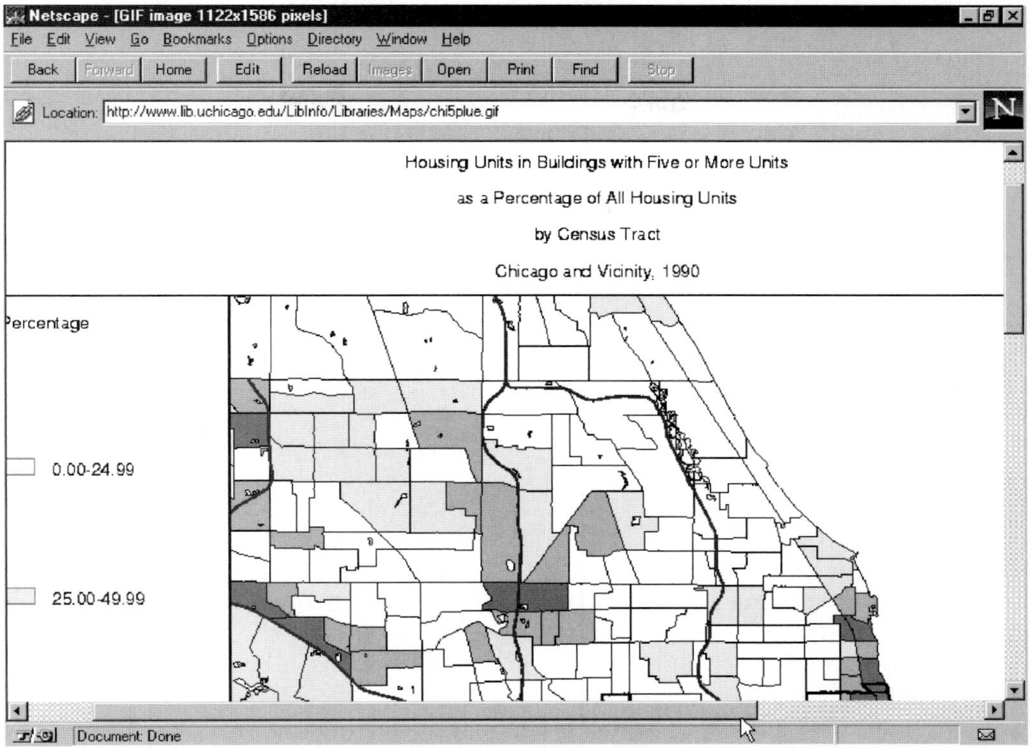

FIGURE 12.7
An example of a static map. Delivering static maps is the simplest GIS application of the Internet. The map can be in the form of raster image or vector objects *(Courtesy: University of Chicago Library)*.

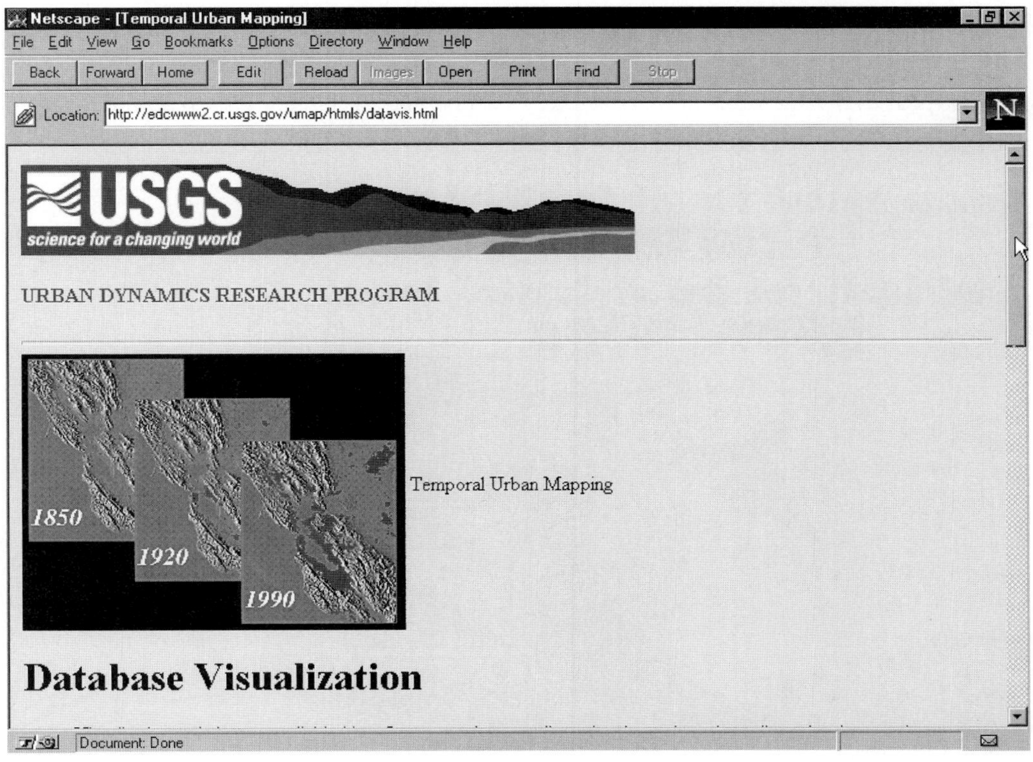

FIGURE 12.8
An example of a dynamic map. A dynamic map is an animation of spatiotemporal process created by displaying a sequence of static maps or images successively *(Courtesy: USGS)*.

an example of land-use change dynamic maps in Atlanta, see *http://svs.gsfc.nasa.gov/imagewall/AAAS/alanta.html*). They are also widely used in landscape visualization by the method of fly-bys (see Chapter 7, Section 7.4.2).

Spatial Data Catalogs

A spatial data catalog contains information about the geographic data holding of an organization (e.g., government agencies, academic or public map libraries, commercial data suppliers). It provides the means by which interested users may quickly search, preview, and retrieve required geographic data from the repository (Figure 12.9). A spatial data catalog, therefore, is more advanced than a collection of static maps in terms of the purpose, scope, and organization of the data.

Geographic Information Search Engine

A geographic information search engine is a special Web browser application for searching geographic information on the Internet. Instead of using keywords in ordinary search engines such AltaVista, Excite, Lycos, and Yahoo, geographic information search engines use geography or location information as the search criteria (Figures 12.10a and 12.10b). The location information may be in the form of a city name, a street address, or a clickable or image map.

Map Generators

Map generators allow the user to generate maps on the fly. The user enters specifications such as location, thematic layers, and symbology by means of a data input form (Figure 12.11). A CGI script at the Web server passes the request to a GIS server where the map is composed. The resulting map is then sent back to the client computer where it can be viewed using the native capabilities of the browser. The user may interact with the resulting map by zooming in or out and scrolling. This feature basically distinguishes a map generator from a geographic information search engine.

Real-Time Map Browsers

Real-time map browsers are high-end map generators in that they also generate maps on the fly (Figure 12.12). In fact, many real-time map browsers are created using the same software tools that are used to create map generators, such as MapObjects of ESRI and MapGuide of Autodesk. However, a real-time map browser is different from a map generator in a couple of ways: first, it is capable of accessing very large spatial databases, and second, it allows a higher degree of interaction between the user and the computer. The use of real-time browsers has made possible the implementation of interactive maps and real-time GIS.

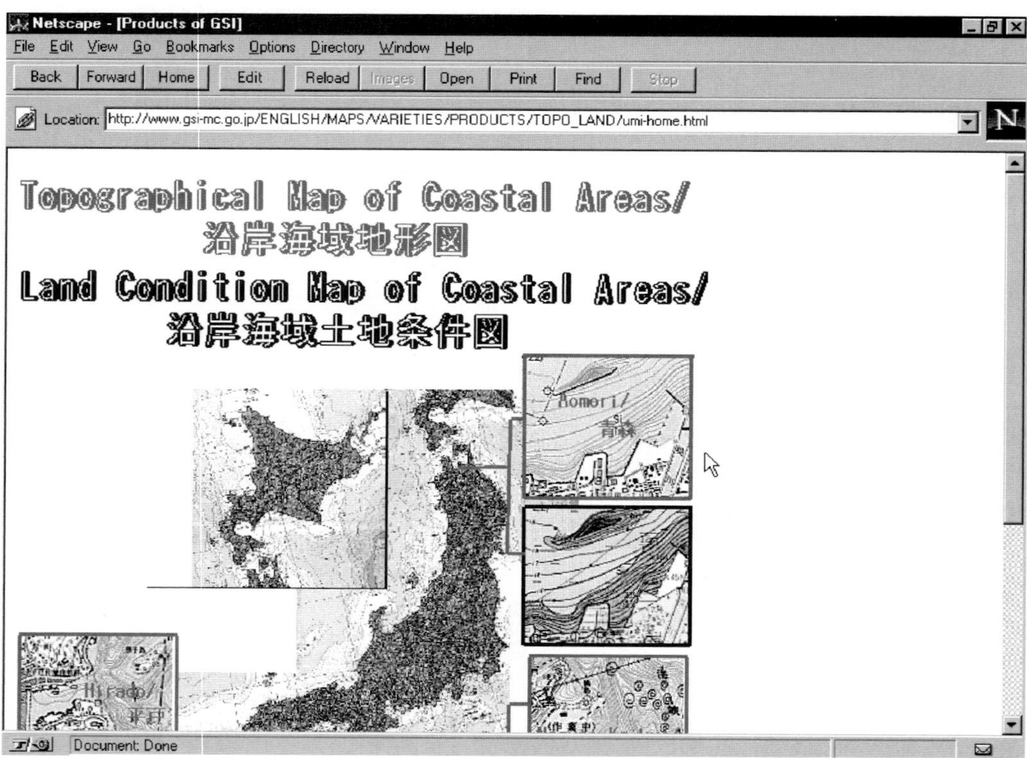

FIGURE 12.9
An example of a spatial data catalog. A spatial data catalog provides the mechanism for accessing an organization's spatial data holding, including maps, aerial photographs, and satellite images *(Courtesy: Geographical Survey Institute, Japan).*

(a)
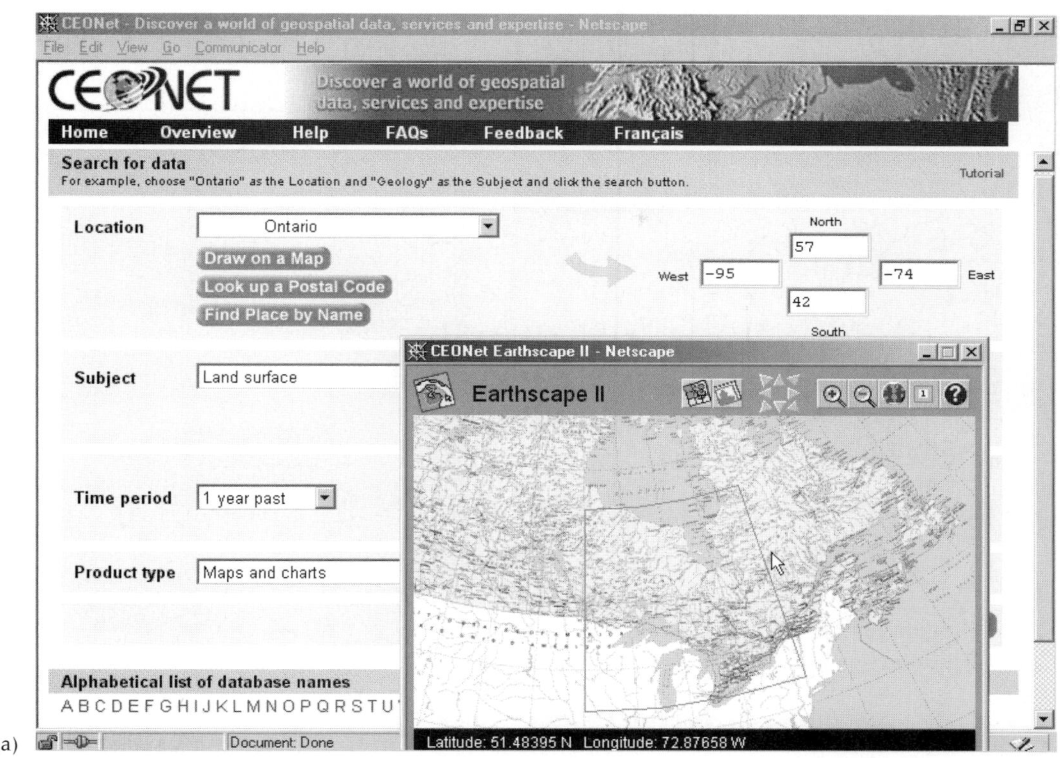

(b)
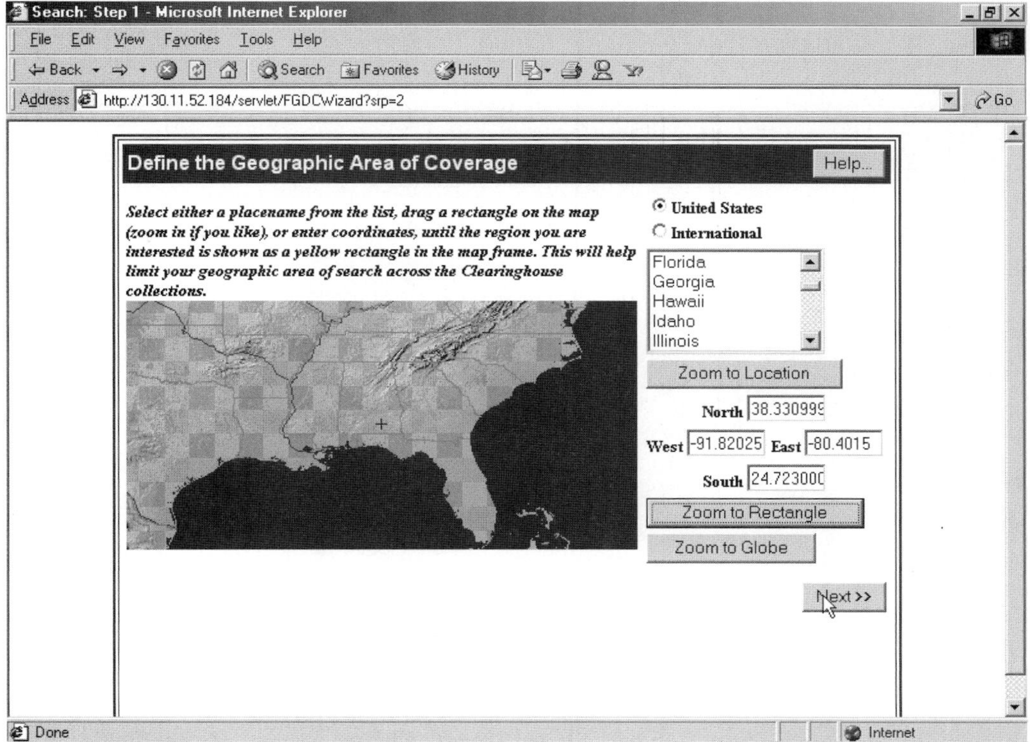

FIGURE 12.10
(a) An example of a geographic information search engine (This screen shot is based on information taken from the Natural Resources Canada Web site *http://ceonet.cgdi.gc.ca/cs/en/Header.html* ©2001 Her Majesty the Queen in Right of Canada with permission of Natural Resources Canada.) Unlike ordinary search engines, a geographic information search engine makes use of location (addresses, postal codes, geographical coordinates, or clickable maps) to find information on the Internet. (b) An example of a geographic information search engine (National Spatial Data Infrastructure Search Wizard, USGS) *(Courtesy: USGS)*.

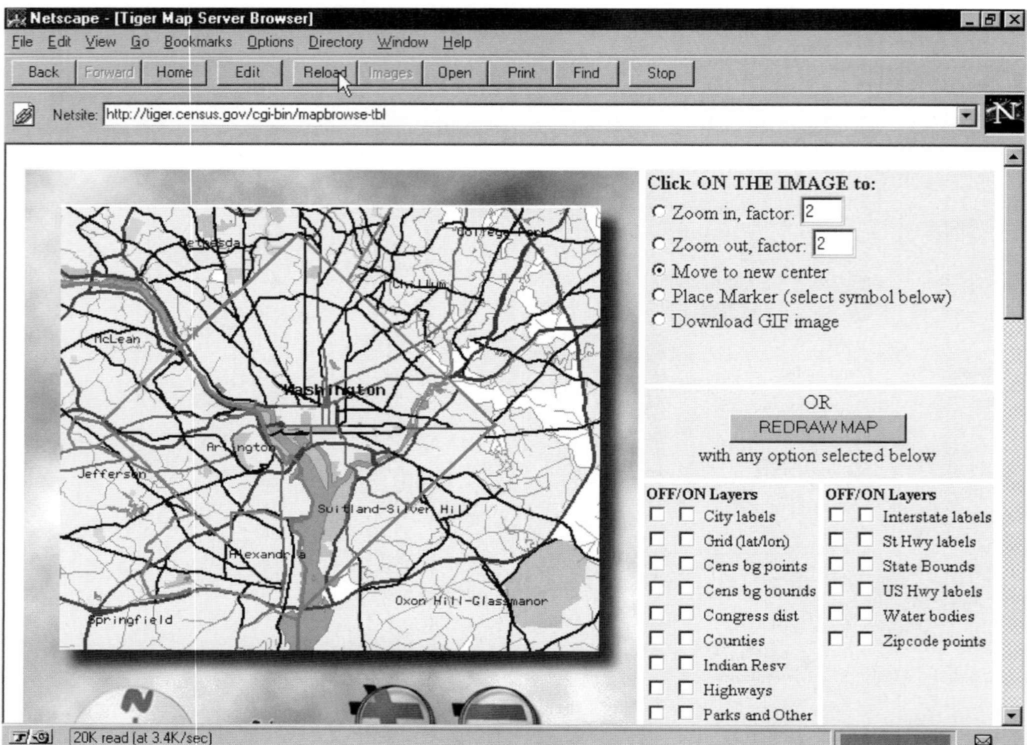

FIGURE 12.11
An example of a map generator (U.S. Bureau of the Census). A map generator allows the user to construct a map interactively by selecting or hiding different database layers *(Courtesy: US Bureau of Census)*.

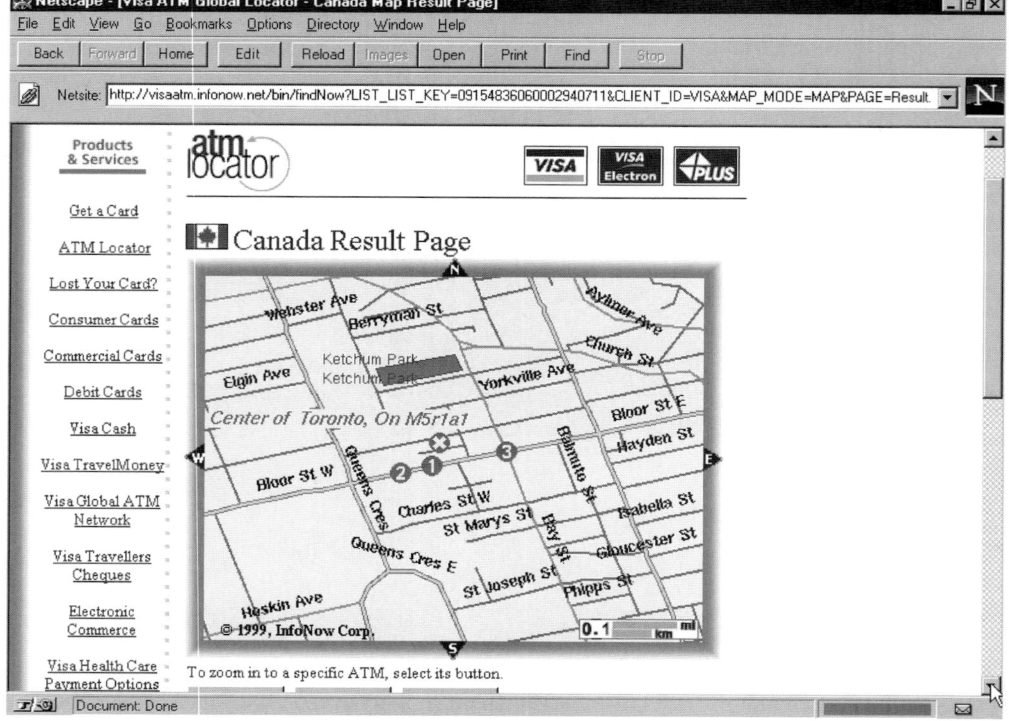

FIGURE 12.12
An example of a real-time map browser (ATM Locator of Visa Card). A real-time map browser allows the user to interactively browse data in a large geographic database *(Courtesy: Visa)*.

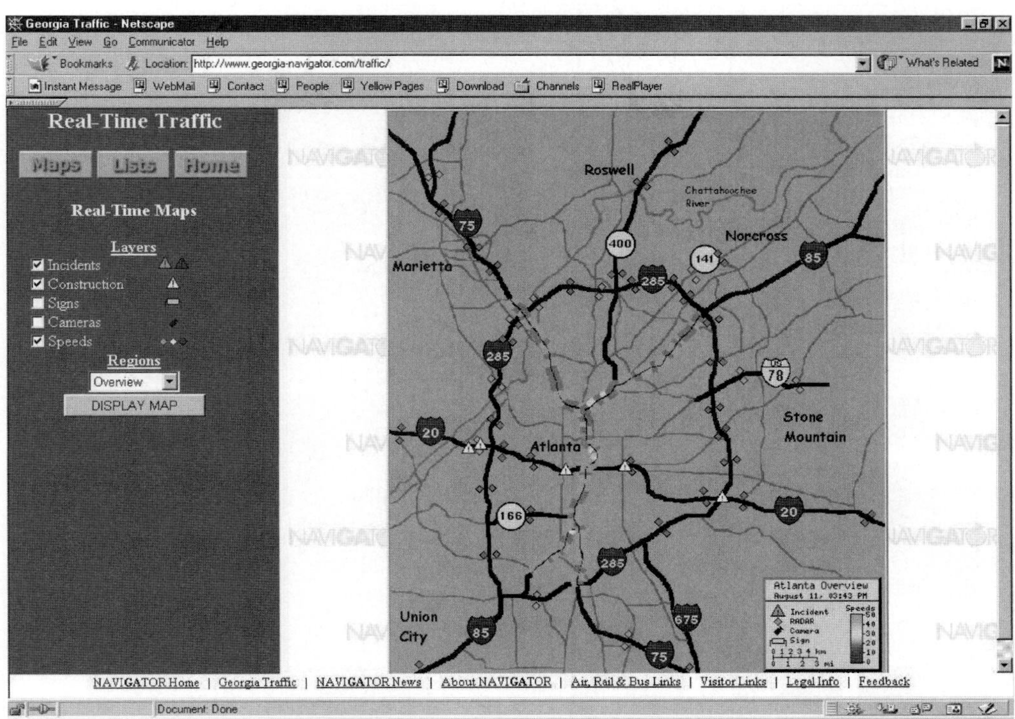

FIGURE 12.13
An example of a real-time map (Atlanta, GA, traffic management system). A real-time map is created by superimposing data captured by an on-line sensor onto a static base map; users are sometimes allowed to zoom in and out and scroll to different locations on the digital map *(Courtesy: Georgia Department of Transportation)*.

Real-Time Maps and Images

Real-time maps use on-line sensors to supply specific themes of geographic information (e.g., traffic information) to be displayed on a base map (Figure 12.13). Real-time images are captured by video cameras, usually referred to as Web cams, connected to a Web server (see Chapter 7). Web cams may be installed in the forms of terrestrial video cameras and satellite-based imaging systems.

Interface to a GIS

This category of applications represents the most advanced and sophisticated use of the Internet for the distribution and visualization of geographic information. An interface to GIS is in fact created by combining the capabilities of map generators and real-time maps, plus certain additional functions that will permit the user to interact with the geographic database by performing on-line query (Figure 12.14) and various forms of spatial overlay analysis typically found in a conventional GIS (Figure 12.15).

To sum up, the Internet is definitely the most powerful factor that shapes the development and use of GIS today. It has greatly popularized the use of geographic information by providing the mechanism for implementing enterprise GIS and the national spatial data infrastructure noted earlier. It is also one of the major driving forces behind the move toward GIS interoperability as well. In order for a technology to be successful, Kiernan and Black (1998) suggested that a "killer app" is necessary. Spreadsheets and word processors are typical examples of killer apps that have made desktop computing an unprecedented success in the history of computers. At present, the demands for GIS are much less than the demands for spreadsheets and word processors. However, with the advent of the Internet, the situation is changing rapidly. The Internet has already revolutionized the user base of GIS from a relatively small number of sophisticated specialists to an infinitely large audience. It will just be a matter of time before Internet-based applications will become the killer apps that will bring GIS to the same level of success that spreadsheets and word processors did to desktop computing.

12.4 Frontiers of GIS Research

Although GIS as a branch of information technology may be said to have reached a relatively high level of maturity, there is still much room for improvement in terms of science. The study of geographic information, whether it is from the "systems" or the "science" perspectives, has opened a new horizon of research

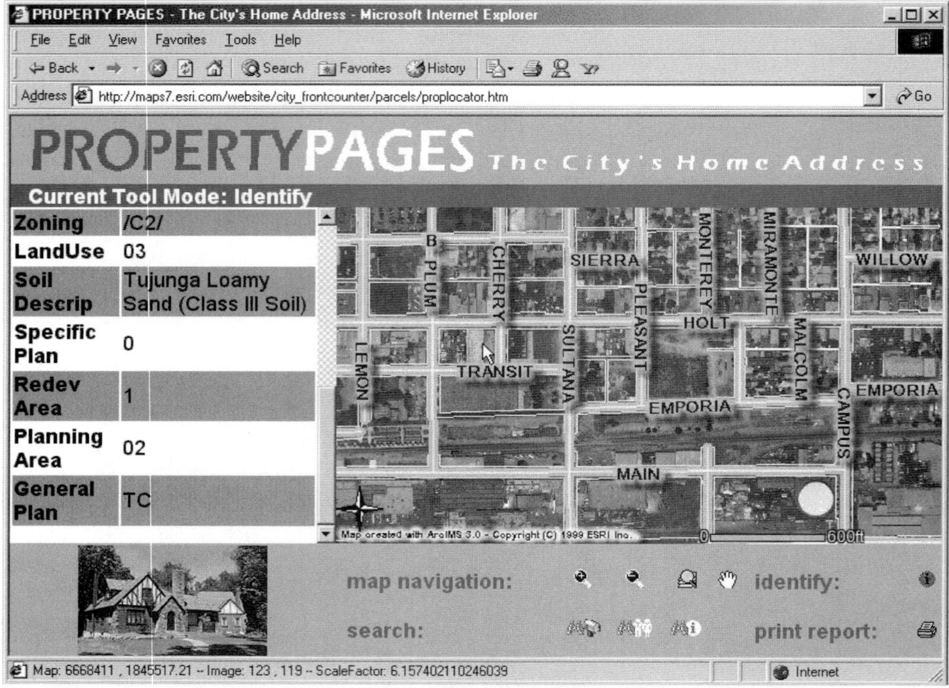

FIGURE 12.14
An example of an interface to GIS for interactive spatial database query. Interactive property information system of the city of Ontario, CA. This GIS allows users to interact with the city's property database and retrieve street address, ownership, zoning, and related data pertaining to individual properties by selecting them on an index map. It is also possible to locate properties spatially, using owner name, street address, and property identification number as the search criteria. (Graphic image supplied courtesy of ESRI. Image map courtesy of City of Ontario, CA).

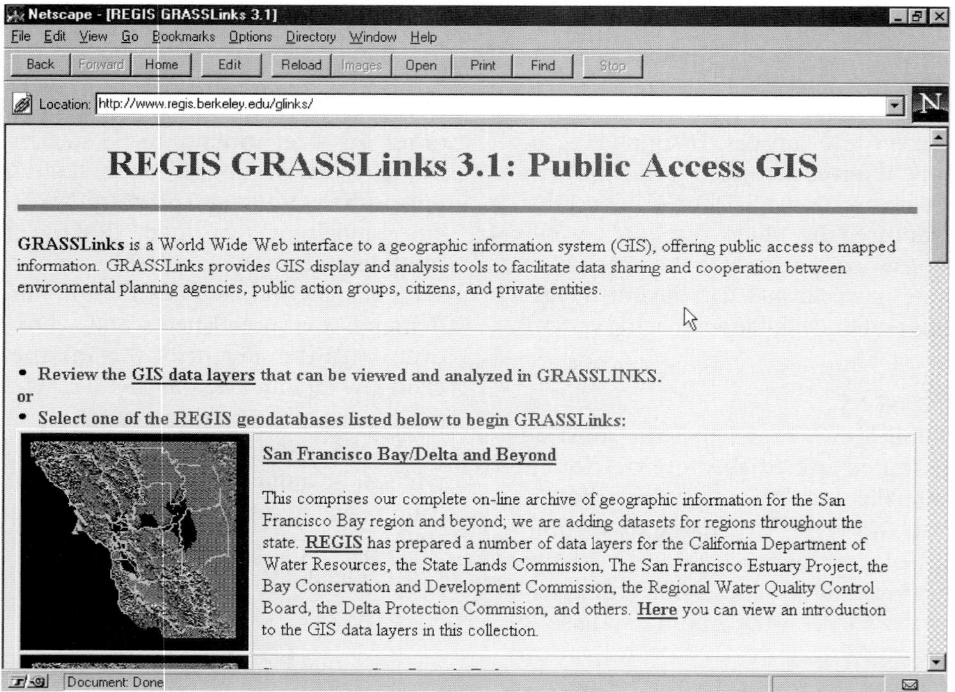

FIGURE 12.15
An example of an interface to GIS for spatial analysis. GRASSlinks, developed at the University of California at Berkeley as one of the components of the Research Program in Environmental Planning and Geographic Information Systems (REGIS), is an interactive Internet-enabled GIS. It is based on the GRASS GIS package and allows users to display and analyze geographic data in real time *(Courtesy: Dr. Robert Twiss)*.

opportunities in geography, geomatics, computer science, systems engineering, and many other disciplines. Some of the investigations have been carried out to address the issues that are now impeding the use of GIS as explained in Section 12.2. Others are undertaken with the aim of perfecting the emerging concepts and techniques that are now redefining GIS. In the following discussion, we will highlight the key areas of GIS research as we understand them today.

12.4.1 THE UCGIS RESEARCH PRIORITIES

The *University Consortium for Geographic Information Science* (UCGIS) is a nonprofit organization of academic and research institutions, government agencies, and hardware/software vendors dedicated to the advancement of the understanding of geographic processes and spatial relationships. One of the important goals of UCGIS is to develop a set of research priorities for geographic information science. After a meeting in June, 1996, and extensive consultation, UCGIS identified ten areas where priorities for research and scope for scientific agenda should be set (UCGIS, 1996). This UCGIS list was by no means exhaustive, but it did reflect the general view of the GIS research community in the United States at that time.

Spatial Data Acquisition and Integration

With the advent of new data-acquisition technologies such as GPS and new satellite sensors, spatial data may now be captured with ever-increasing accuracy, speed, and volume. Research in this area aims to develop methods to integrate data obtained by different technologies, with different accuracies, and in different formats.

Distributed Computing

Distributed computing is the predominant architecture for information systems development today. GIS is moving rapidly from the conventional monolithic mode of operation to this new paradigm of computing. Research on distributed computing from the GIS perspective is concerned with the impacts of both technical factors (e.g., network configuration and operation protocols) and nontechnical factors (e.g., economic and institutional implications) of distributed computing on geographic data processing.

Extensions to Geographic Representations

While it is true that, in general, GIS data models today are capable of representing complex associations among geographic features, they are not adequate for the representation of three-dimensional, temporal, and dynamic spatial phenomena. They are also obviously weak at handling the fuzziness and uncertainty that is inherent in geographic data. Geographic representations should receive a high research priority because they are a prerequisite for the development of databases and analytical applications.

Cognition of Geographic Information

This particular research area is concerned mainly with human–computer interaction, perception of geographic space, and the use of GIS for logical decision making. Cognitive research in GIS promises to improve our ability to turn geographic data into useful information, which will then be used to develop spatial knowledge and intelligence for geographic problem solving.

Interoperability of Geographic Information

The objective of interoperability, as explained in Section 12.3.3, is to develop mechanisms for cross-functional computing on multiple platforms. This is now achieved by sharing data using metadata, sharing files, as well as by sharing processes using standard software development paradigms and protocols. Further research is required to enhance GIS interoperability by improving the link between GIS applications and other applications such as image processing, document management, and dynamic modeling.

Scale

The term "scale" in the present context refers generally to the levels of detail adopted for data collection, representation, analysis, and communication. The transition from paper maps to digital databases has fundamentally changed the concept of scale in the use of geographic data. This has also led to considerable interest in the implications of multiple scale representation, variable resolution representation, and database generalization on geographic problem solving.

Spatial Analysis in a GIS Environment

At present, the link between GIS and spatial analysis in the physical and social sciences is relatively weak. It has not yet been able to take full advantage of well-developed algorithms in conventional geostatistics, spatial econometrics, and space–time modeling in GIS applications. In order to capitalize on the power of these established spatial analysis tools for the interpretation and understanding of very large and complex geographic databases, much work will be required to integrate these tools seamlessly into the GIS environment.

The Future of the Spatial Information Infrastructure

Despite the growing recognition of geographic data as one of the key components of the information infrastructure of modern society, several issues have not yet been satisfactorily resolved. For example, the problems of data ownership, copyrights, protection of privacy, and legal liability are all still in need of clarity. The goal of research on the future of the spatial information infrastructure is to help policymakers, data suppliers, and data users understand the interrelationships among factors such as government policies, geographic data resources, and the use of geographic information for the social and economic well-being of society.

Uncertainty in Geographic Data and GIS-Based Analyses
Although it is now well-known that errors occur at every phase of the geographic data life cycle (i.e., collection, database construction, processing, analysis, and presentation), our understanding of the characteristics of the resulting uncertainty of the quality of data is rather limited. In order to minimize the adverse consequences of using data of uncertain quality in geographic problem solving, intensive research is required to determine the nature of uncertainty, the methods of testing and quantifying it, as well as strategies of tracking and reporting it in GIS applications.

GIS and Society
Up to now, the growth of GIS has brought enormous benefits to society by sustaining an important sector of the computer industry and enhancing our geographic knowledge base to understand and manage Earth's resources. However, relatively little has been done to investigate the social impacts of the increasing use of geographic information in our daily lives. These impacts include, for example, the potential of abuse and misuse of geographic information that will lead to the invasion of privacy; the inequality of access to geographic information that may place some sectors of society in a disadvantaged position; and the effects of increased accessibility to geographic information on the relationships between the public and the government and the relationships among different social groups or communities. Although GIS in itself is largely used as a technology, it is the social implications resulting from the use of this particular branch of technology that will pose the most challenging objectives for the research community.

12.4.2 OTHER AREAS OF RESEARCH INTEREST

In addition to the UCGIS list of research priorities that represents a long-term goal of the GIS research community, there has been considerable interest in some specific areas pertaining to the more immediate needs of GIS today.

GIS Education and Training
The GIS research community has always had a keen interest in aspects pertaining to GIS education and training. In the past, the focus was mainly on the development of educational programs with an emphasis on *what* must be included in GIS curricula (Unwin, 1991). Recently, this emphasis seems to have extended to *how* GIS education and training should be delivered using multimedia and alternative modes of instruction such as distance learning (Cornelius and Heywood, 1994; Krygier, 1997). With a team of international experts, the National Center for Geographic Information and Analysis (NCGIA) is developing a new core curriculum in GIScience to be delivered over the Internet to assist teachers and students in teaching and learning the fundamental concepts of GIS *(URL:http://www.ncgia.ucsb.edu/giscc/)*. As GIS is a rapidly changing technology, there is a constant need for research into what students must learn in order to meet the demands of the job market and how practitioners in the current workforce can and should be kept abreast of the advancement of technology.

Evaluation and Use of Emerging Computing Technologies
The development of GIS has drawn heavily from concepts and techniques originated in various branches of computer technology. The research community should play a more active and significant role in the evaluation of emerging technologies with regard to their applicability in GIS. One important area of research is the use of parallel processing to solve complex geographic problems (e.g., network analysis, DEM interpolation, and modeling of global environmental changes) (Armstrong, 1994; Ding et al., 1992; Li, 1994). Another important area is the use of new software techniques and tools for geographic data processing and visualization (e.g., software component technology, multimedia, and application development methodology) (Dykes, 1995; Hartman, 1997; Hughes, 1996). On the database side, there is a strong need for investigations into the problem of database integration (Smith, 1994), as well as the possibility of extending the concepts and techniques of data warehouses and data. These will enable GIS to capitalize in the global enterprise computing market that is estimated to be U.S. $15 billion, growing 10–15% annually (ESRI, 1998c; MapInfo Corp., 1996).

Spatial Digital Libraries
The computer is rapidly changing the way libraries are organized and operated. The main focus of early efforts in computerization was on the automation of end-user services. Computer technology is now sufficiently advanced to support the design and deployment of large *digital libraries* that are not only capable of supporting conventional end-user functions, but also networked access to printed and nonprinted documents, as well as video and audio materials. In 1994, the National Science Foundation (NSF), in cooperation with the National Aeronautical and Space Agency (NASA) and the Advanced Research Projects Agency (ARPA), announced

the award of digital library research funding to six universities. Two of these experimental projects, the *Environmental Electronic Library* at the University of California, Berkeley, and *Alexandria Digital Library* at the University of California, Santa Barbara, have a particular focus on spatial data. Elsewhere in the United States and in Australia and Europe, the idea of a digital library has also been adopted by government agencies to deliver geographic data (i.e., maps and images) to the public (Von Rimscha, 1998). At this point, most of the research efforts have been directed toward the issue of access to information (i.e., metadata and information search methodology) (Smith, 1996). In order to turn digital libraries in different countries into a global network of geographic knowledge, it is necessary to carry out further research on legal issues (e.g., ownership and copyright), institutional structures (e.g., organization and management of the network), as well as social and economic implications (e.g., value of knowledge and increased productivity because of improved availability of information) pertaining to this new approach to accessing geographic data (Lopez, 1997; Wiederhold, 1995).

The Geography of the Information Economy

The information economy, or *knowledge-based* economy as it is sometimes called, embraces all human activities engaged in the collection, processing, and distribution of information. Broadly speaking, these include education, communications and telecommunications, the media, as well as the manufacture of information goods (i.e., hardware, software, and information products such as books, maps, digital data files, and movies), and the provision of information services (e.g., business and systems consulting and think tanks). Development of the information economy is rapidly changing the global economic landscape. Locations, sizes, borders, and time zones are all becoming increasingly irrelevant to the way business decisions are made and business operations are conducted. From the perspective of geographic research, such changes are of great intellectual interest in three important ways.

- the spatial distribution and interaction of the key elements of the information economy itself, which results in what has become known as the *geographies of cyberspace* (Dodge, 1998)
- the impacts of information technology, such as communications and telecommunications networks, on the ways people act and interact, both spatially and socially, which leads to regional economic restructuring (Castells, 1989) and urban transformation (Moss and Townshend, 2000)
- the importance of geographic information in business decision making in the context of the information economy, which leads to the development of new theories of location and spatial organization

As studies of the information economy involve a significant amount of data pertaining to spatial distribution and movements of people and goods, GIS provides the perfect tool for data collection and analysis. It helps identify the role of information in general, and geographic information in particular, in the global, national, and local economies. In this regard, the study of the geographic aspects of the information economy falls comfortably within the realm of GIS research and will remain one of the most challenging topic areas to GIS researchers.

12.5 Conclusions

The rapid development of GIS in the last several years has brought this particular branch of information technology to a new height in terms of functionality, applications, users, and data. GIS as we know it today is very different from what it was five to ten years ago. Advances in information technology in general, and telecommunications technology including the Internet in particular, have fundamentally changed the way GIS is used. At the same time, the growing awareness of the importance of geographic information has brought the issues pertaining to geographic data to the agenda of all levels of government. As a result, GIS is now perceived as an essential part of mainstream information technology, a major component of the information infrastructure of the modern society as well as a primary driving force behind many of the institutional, social, and economic changes that we witness today.

Despite its apparent success, however, GIS has a fair share of the problems arising from the growing use of digital information in society. The issues of data ownership, copyright, and liability have not yet been adequately addressed in most countries. The chronic problem of finding enough people with suitable expertise to meet the demands of the industry has largely remained. On the more technical side, the problems of interoperability have not been totally resolved yet and the GIS industry is just at an early stage in its attack on the decade-old problem of incompatibility among proprietary applications. As a result, although new data-acquisition technologies are now turning out vast amounts of digital geographic data at a speed never before experienced, it appears that society as a whole has not been able to capitalize fully on the increased availability of digital geographic data. Before the data problems are resolved and techniques for filtering through the vast volume of data to find meaningful geographic knowledge are better developed, today's society will continue to stay "data-rich and information-poor."

Several concepts and techniques have emerged that will hopefully turn the situation around very soon. One of these is the spatial data infrastructure. Although the approaches to implementing this particular concept vary in different countries, the common goal is to provide the necessary mechanisms to foster the awareness of the existence of geographic data resources and to disseminate such data in a cost-effective way. The timely advances of telecommunications technology have provided the necessary technical infrastructure to support national and international initiatives in spatial data infrastructure. As both the possibilities and requirements of data sharing grow, the focus of GIS development has quickly shifted from database creation to data integration. As a result, the next generation of GIS will be characterized by the ability to access a variety of data sources, serve the needs of a multidisciplinary user community, and work seamlessly with all kinds of computer-based applications in a single-system environment.

In the academic sector, the study of GIS has gradually evolved from a focus on the use of the technology to a very much broader scope of understanding, developing, and deploying the technology, leading to the formation of a distinct field of scientific specialization now referred to as geographic information science. The proliferation of GIS has not only made geographic information science one of the fastest growing fields of study in universities and colleges, it has also opened a new horizon of opportunities for basic and applied research on the concepts, techniques, institutions, applications, as well as the socioeconomic implications pertaining to the use of GIS. It is particularly interesting to note the rapid growth of GIS research from the perspective of the humanities and social sciences. This represents the increasing recognition of geographic information as a key indicator to many problems in human geography. It also signifies the growing understanding that as a technology, GIS cannot be developed in isolation from its social context.

The development of GIS owes much to the academe who formulated the basic concepts. As the underlying technologies and algorithms used to construct GIS become more sophisticated and as the socioeconomic impacts of using GIS become more widespread, academe is even more effective in shaping the future development of GIS. Universities and colleges are not only centers of learning GIS, but also centers of GIS innovation through participation in government initiatives and cooperative research with the hardware/software development industry. In this regard, students should be aware that the objective of studying GIS is not merely to learn to use one or more software packages. They should set their goals to include the understanding of the concepts, the values, and the vision of using GIS for the benefits of society. GIS empowers us to solve environmental problems of a changing world faced by humankind in the new millennium.

REFERENCES

Antenucci, J. (1991) "Management issues in implementing and utilizing GIS technology," *Proceedings*, Geographic Information Systems Seminar, pp. 6–14. Toronto, ON: Ontario Ministry of Natural Resources.

ANZLIC (Australia New Zealand Land Information Council) (1997) *ANZLIC Strategic Plan 1997–2000*, ANZLIC Secretariat, Belconnen, ACT, Australia.

Armstrong, M. P. (1994) "GIS and high performance computing," *Proceedings*, GIS/LIS '94, pp. 4–13, Phoenix, AZ.

Buehler, K., and McKee, L. eds. (1998) *The OpenGIS Guide*, 3rd ed. Wayland, MA: Open GIS Consortium, Inc.

Castells, M. (1989) *The Informational City: Information, Technology, Economic Restructuring and the Urban-Regional Processes*, Oxford: Basil Blackwell.

Coleman, D. J., and Mclaughlin, J. D. (1997) "Defining global geospatial data infrastructure (GGDI): Components, stakeholders and interfaces," Theme Paper #1, International Seminar of Global Geospatial Data Infrastructure, University of North Carolina, Chapel Hill, NC.

Corbley, K. (1998) "Open the flood gates! Data providers prepare to overcome access and delivery problems," *GIS World*, Vol. 11, No. 1, pp. 52–56.

Cornelius, S. C., and Heywood, D. I. (1994) "Professional development and multimedia: the development of a GIS 'Learning Station' for the InterGIS diploma program," *Proceedings*, GIS/LIS '94, pp. 181–190, Phoenix, AZ.

Culpepper, R. B. (1998) "Weave maps across the Web, 1998 edition," *GEOWorld*, Vol. 11, No. 11, pp. 46–52.

Ding, Y., Densham, P., and Armstrong, M. (1992) "Parallel processing for network analysis: Decomposing shortest path algorithms for MIMD computers," *Proceedings*, 5th International Symposium on Spatial Data Handling, pp. 682–691, Charleston, SC.

Dodge, M. (1998) "The geographies of cyberspace," *Abstracts of Papers presented at the 94th Annual Conference of the Association of American Geographers*, Boston, MA. (in CD form), p. 203.

Dykes, J. (1995) "Dynamic maps for spatial science: A unified approach to cartographic visualization," in *Innovations in GIS 3* by Parker, D. ed. pp. 177–187, London: Taylor and Francis.

Epstein, E. F., and Roitman, H. (1990) "Liability for information," in *Introductory Readings in GIS* by Peuquet, D. J., and Marble, D. F. eds. pp. 364–371, London: Taylor and Francis.

ESRI (1998a) *Arc Facility Manager (ArcFM)—A Powerful New Arc/Info-based Application for Utilities*, an ESRI White Paper, Redlands, CA: Environmental Systems Research Institute, Inc.

ESRI (1998b) *The Utility Solution Bundle ESRI, Hewlett Packard, Microsoft and Miner and Miner*, an ESRI White Paper, Redlands, CA: Environmental Systems Research Institute, Inc.

ESRI (1998c) *Spatial Data Warehousing*, an ESRI White Paper, Redlands, CA: Environmental Systems Research Institute, Inc.

FGDC (Federal Geographic Data Committee) (1997) *A Strategy for the National Spatial Data Infrastructure*, Reston, VA: Federal Geographic Data Committee.

Harder, C. (1998) *Serving Maps on the Internet*, Redlands, CA: Environmental Systems Research Institute, Inc.

Hartman, R. (1997), *Focus on GIS Component Software*, Santa Fe, NM: Onword Press.

Hughes, J. R. (1996) "Technology trends mark multimedia advancements," *GIS World*, Vol. 9, No. 11, pp. 40–43.

Industry Canada (1998) *Geomatics Technology Road Map, Special Report*, Ottawa, ON: Department of Industry.

Inmon, W. H. (1996) *Building the Data Warehouse*, 2nd ed., New York: John Wiley & Sons.

Kennedy, E. (1991) "Geomatics human resources planning studies," *Proceedings*, pp. 217–224, Geographic Information Systems Seminar, Toronto, ON: Ontario Ministry of Natural Resources.

Kiernan, B., and Black, J. (1998) "System performance: in search of the 'killer app'," *GIS World*, Vol. 11, No. 1, pp. 42–47.

Krygier, J. H. (1997) "Multimedia in geography education," *Journal of Geography in Higher Education*, Vol. 21, No. 1, pp. 17–39.

Lazar, B. (1998) "Break through spatial data translation obstacle," *GIS World*, Vol. 11, No. 6, pp. 48–51.

Li, B. (1994) "When to use parallel processing in GIS?," *Proceedings*, GIS/LIS '94, pp. 514–523, Phoenix, AZ.

Limp, W. F. (1997) "Weave maps across the Web," *GIS World*, Vol. 10, No. 9, pp. 46–55.

Litwin, W., Mark, L., and Roussopoulos, N. (1990) "Interoperability of multiple autonomous databases," *ACM Computing Surveys*, Vol. 22, No. 1, pp. 265–293.

Lopez, X. R. (1997) "The network as organization: Digital libraries for spatial information," *Proceedings*, UCGIS 1997 Annual Assembly and Summer Retreat **(http://www.spatial.maine.edu/ucgis/testproc/lopez/zlopex.html)**

MapInfo Corporation (1996) *MapInfo and the Data Warehouse*, a MapInfo White Paper, Troy, NY: MapInfo Corporation.

Marble, D. F. (1998) "Urgent need for GIS technical education: Rebuilding the top of the pyramid," *ARC News*, Vol. 20, No. 1, p. 1.

Masser, I. (1998) *Governments and Geographic Information*, London: Taylor and Francis.

McKee, L. (1998) "What does OpenGIS specification conformance mean?," *GIS World*, Vol. 11, No. 8, p. 38.

Moss, M. L., and Townshend, A. M. (2000) "How telecommunications systems are transforming urban spaces," in J. O. Wheeler, Y. Aoyama, and Warf, B. eds. *Cities in the Telecommunications Age: The Fracturing of Geographies*, New York and London: Routledge, pp. 31–41.

Nebert, D. (1999) "Interoperable spatial data catalogs," *Photogrammetric Engineering and Remote Sensing*, Vol. 65, No. 5, pp. 573–575.

NRC (National Research Council) (1997) *The Future of Spatial Data and Society*, Commission on Geoscience, Environment and Resources, Washington, DC: National Academy Press.

Peng, Z.-R. (1997) "An assessment of the development of Internet GIS," *Proceedings*, 1997 Users Conference, Paper No. 526, Redlands, CA: Environmental Systems Research Institute, Inc.

Putz, S. (1994) "Interactive information services using World Wide Web hypertext," *Computer Networks and ISDN Systems*, Vol. 27, No. 2, pp. 272–280.

Rhind, D. (1992) "Data access, charging and copyright and their implications for geographical information systems," *International Journal of Geographical Information Systems*, Vol. 6, No. 1, pp. 13–30.

Rhind, D. (1997) "Implementing a global geospatial data infrastructure (GGDI)," Theme Paper #2, International Seminar of Global Geospatial Data Infrastructure, University of North Carolina, Chapel Hill, NC.

Smith, T. R. (1994) "On the integration of database systems and computational support for high-level modelling of spatio-temporal phenomena," in *Innovations in GIS 1* by Worboys, M. F. ed. London: Taylor and Francis.

Smith, T. R. (1996) "The meta-information environment of digital libraries," *D-Lib Magazine*, July/August 1996 **(http://www.dlib.org/july96/new/07smith.html)**

Thoen, B. (1998) "The fairy tale of the ugly standard," *GIS World*, Vol. 11, No. 8, pp. 32–34.

Tosta, N. (1994) "Continuing evolution of the National Spatial Data Infrastructure," *Proceedings*, GIS/LIS '94, pp. 769–777, Phoenix, AZ.

UCGIS (1996) "Research priorities for geographic information science," *Cartography and Geographic Information Systems*, Vol. 23, No. 3.

Unwin, D. J. (1991) "The academic setting of GIS," *in Geographical Information Systems Volume 1: Principles* by Maguire, D. J., Goodchild, M. F., and Rhind, D. W. eds. pp. 81–90, Harlowm, UK: Longman Scientific and Technical.

Von Rimscha, S. (1998) "GeoServe project unlocks European geographic data," *GIS World*, Vol. 11, No. 5, pp. 58–60.

Wegner, P. (1996) "Interoperability," *ACM Computing Surveys*, Vol. 28, No. 1, pp. 285–287.

Wiederhold, G. (1995) "Digital libraries, value and productivity," *Communications of the ACM*, Vol. 38, No. 4, pp. 85–86.

Wilson, J. D. (1996) "Enterprisewide implementations transcends GIS boundaries," *GIS World*, Vol. 9, No. 10, pp. 42–46.

Wilson, J. D. (1997) "Technology partnerships spark the industry," *GIS World*, Vol. 10, No. 4, pp. 36–42.

Wilson, J. D. (1998) "GIS goes mobile," *GEOWorld*, Vol. 11, No. 12, pp. 54–57.

Appendix A

Internet Resources for GIS

A.1 Introduction

The Internet contains a tremendous amount of information resources pertaining to the study and use of GIS. These include sources of geographic data, scientific papers on the concepts, techniques, and applications of GIS, as well as information about hardware and software products. This appendix provides the locations of different types of GIS resources and explains the ways of obtaining them on-line. Students should go to the Web sites of selected examples of Internet-based GIS applications so that they can acquire firsthand experience of how the rapidly advancing Internet technology is changing the ways GIS is used today. Two useful guides on using Internet are Turlington (1999) and Newquist (2000). Useful hints on searching for geospatial data on the Web are given in an article by Thoen (2001).

Although all the Web sites quoted in this Appendix were tested when the information was compiled, it is expected that some of them will be changed or removed over time. New resources will also be constantly added to the Internet. Students are therefore encouraged to take advantage of the power of the search engines such as ***Google*** *(http://www.google.com)*, ***Alta Vista*** *(http://www.altavista.com)*, and ***Webtop*** *(http://www.webtop.com)* to update their knowledge base of GIS resources on the Internet from time to time.

A.2 Where to Start the Search?

The easiest way to jump-start the search for Internet resources on GIS is to make use of one of the existing indices created and posted on the Internet by individuals or organizations interested in the subject. These indices are usually organized into categories to facilitate access to different resources. There are many excellent indices that can serve as "jump stations" to GIS resources on the Internet, including, for example:

- **GIS.com**, a portal to GIS information on the World Wide Web created by ESRI, Inc. (Figure A.1)
  http://www.gis.com
- **GeoCommunity**
  http://search.geocomm.com
- **GISLinx**
  http://www.gislinx.com
- **University of Minnesota, Remote Sensing and GIS Resources**
  http://www.gis.umn.edu/rsgisinfo/rsgis.html
- **University of Nebraska-Lincoln, Center for Advanced Land Management Information Technologies (CALMIT)**
  http://www.calmit.unl.edu/calmit/gisrs.html

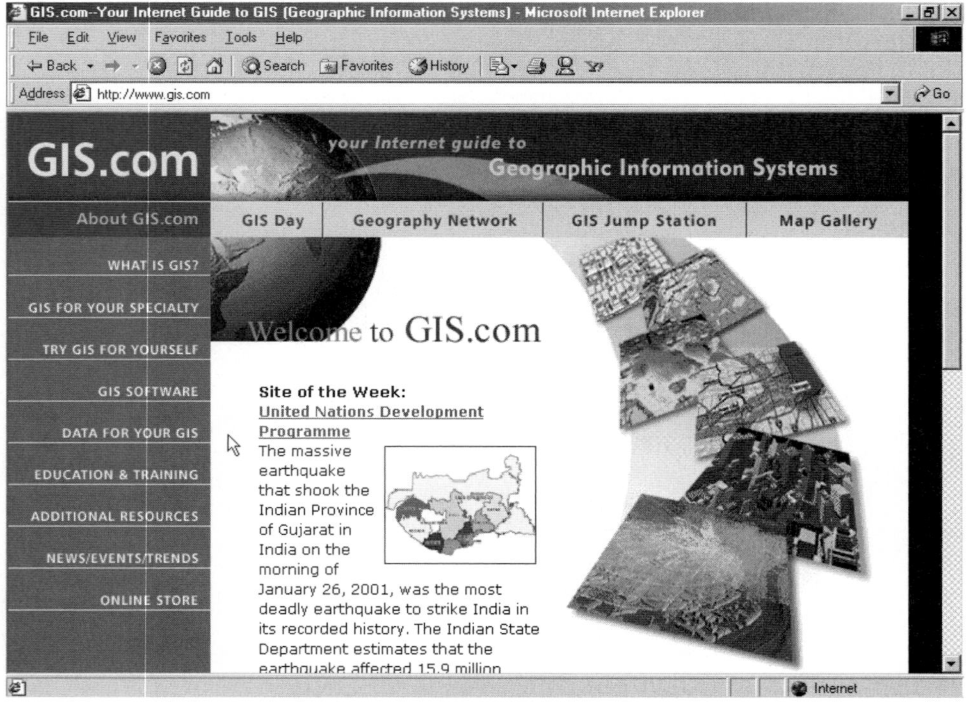

FIGURE A.1
GIS-specific portals, such as gis.com shown in the diagram, provide a convenient starting point to start the search for GIS resources on the Internet. *(Graphic image supplied courtesy of ESRI)*

- **US Bureau of the Census GIS Gateway**
  http://www.census.gov/ftp/pub/geo/www/gis_gateway.html

A.3 Pointers to Information Resources

As a result of the implementation of the National Spatial Data Infrastructure (NSDI), the U.S. federal government has taken full advantage of Internet technology by building a network of Web pages and on-line data sets open to the public. The Federal Geographic Data Committee (FGDC) is spearheading the effort to establish the National Geospatial Data Clearinghouse and is coordinating the federal government's effort to provide public access to its geographic data holdings. FGDC's home page provides information about its mandates, spatial data metadata standards, NSDI, as well as publications and other FGDC information. The URL of FGDC's home page is http://fgdc.er.usgs.gov.

Among the U.S. federal agencies, the USGS probably has the largest network of servers for geographic information. The USGS Web site at http://www.usgs.gov is the starting point for on-line information about the USGS, its programs, data products, and educational resources (Figure A.2). It also contains an extensive list of pointers to Internet geographic information resources not managed specifically by the USGS. Many other U.S. federal agencies have set up publicly accessible information resources on the Internet. Some of these include

- **NASA Earth Observation System (EOS)**
  http://eospso.gsfc.nasa.gov
  http://terra.nasa.gov
- **National Imagery and Mapping Agency (NIMA)**
  http://www.nima.mil
- **National Oceanic and Atmospheric Administration (NOAA), Environmental Satellite, Data and Information Service (ESDIS)**
  http://www.nesdis.noaa.gov
- **U.S. Bureau of Land Management**
  http://www.blm.gov/gis/nsdi.html
- **U.S. Bureau of the Census**
  http://www.census.gov
- **U.S. Department of Agriculture Natural Resources Conservation Service (NRCS)**
  http://www.ncg.nrcs.usda.gov/nsdi_node.html
- **U.S. Environmental Protection Agency, National GIS Program**
  http://www.epa.gov/ngispr
- **U.S. Fish and Wildlife Service**
  http://www.fws.gov

Information pertaining to geographic information activities in other countries may be found at the Web sites of the following organizations:

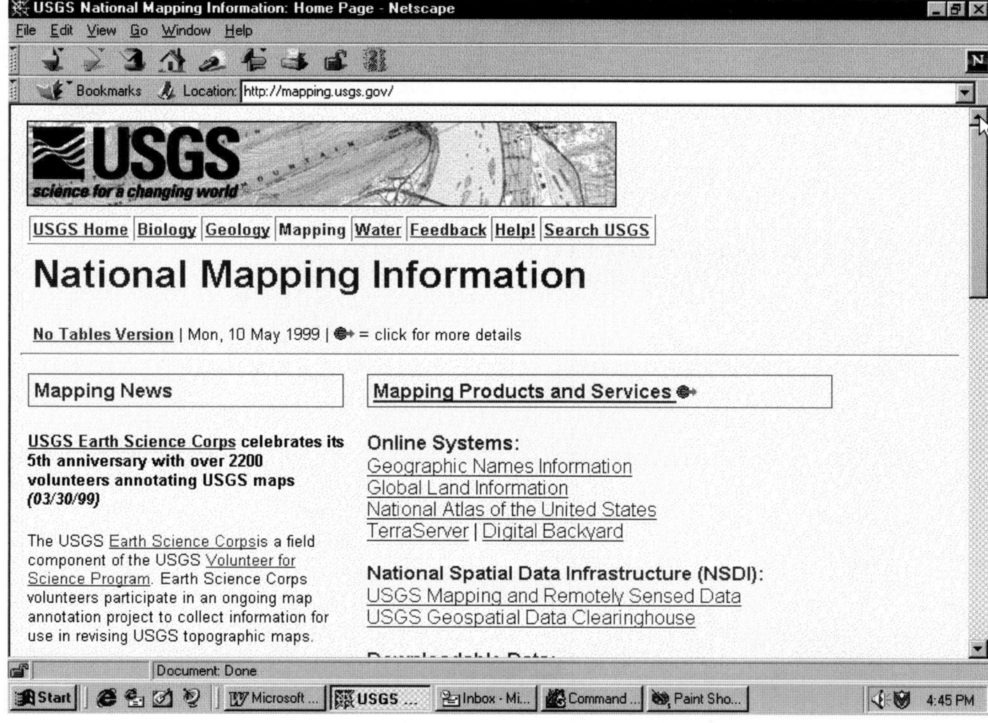

FIGURE A.2
The USGS Web site is one of the most important nodes of the National Spatial Data Infrastructure (NSDI) that serves as a gateway for obtaining information about GIS activities of U.S. federal government agencies. *(Courtesy: USGS)*

- **Australia Surveying and Land Information Group**
  http://www.auslig.gov.au
- **European Umbrella Organization for Geographic Information (EUROGI)**
  http://www.eurogi.org
- **Geomatics Canada** (Figure A.3)
  http://www.geocan.nrcan.gc.ca
- **Ordnance Survey, U.K.**
  http://www.ordsvy.gov.uk
- **United Nations Environment Program**
  http://www.grida.no

Very useful geographic information resources may also be obtained from the Web sites of professional organizations, learned societies, publishers of trade journals, universities, and colleges. Typical examples of such sites include

- **American Association of Geographers (AAG) GIS Specialty Group**
  http://www.cla.sc.edu/gis/aaggis.html
- **American Society of Photogrammetry and Remote Sensing**
  http://www.asprs.org
- **Association of Geographic Information (U.K.), Online Resources for GIS**
  http://www.agi.org.uk/pages/links.html
- **Geospatial Solutions (formerly GeoInfo Systems)**
  http://www.geospatial.online.com
- **GeoWorld**
  http://www.geoplace.com/gw
- **National Center for Geographic Information and Analysis (NCGIA)**
  http://www.ncgia.ucsb.edu
- **Open GIS Consortium**
  http://www.ogis.org
- **The Geographer's Craft, University of Texas at Austin**
  http://www.utexas.edu/depts/grg/gcraft/contents.html
- **University Consortium for Geographic Information Science (UCGIS)**
  http://www.ucgis.org
- **University of California, Berkeley, Research Program in Environmental Planning and GIS (REGIS)**
  http://www.ced.berkeley.edu
- **University of Edinburgh, Department of Geography GIS WWW Server**
  http://www.geo.ed.ac.uk/home/gishome.html
- **Urban and Regional Information Systems Association**
  http://www.urisa.org

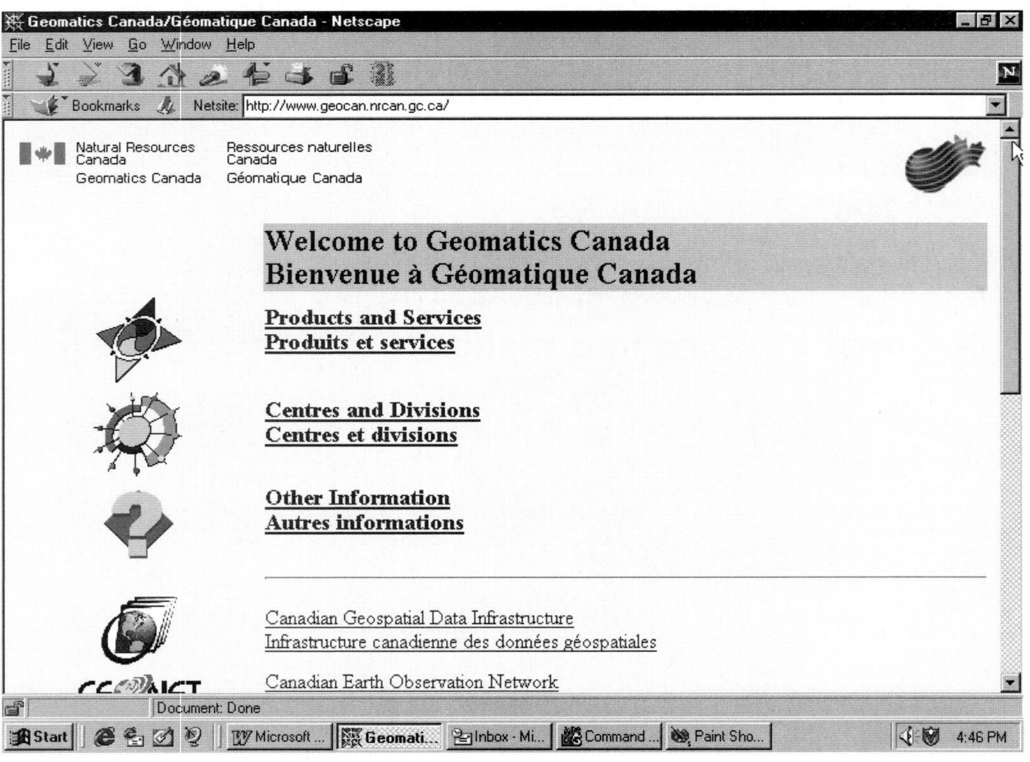

FIGURE A.3
The Web site of Geomatics Canada is the gateway for obtaining information about GIS activities of the federal government of Canada. (This screen shot is based on information taken from the Geomatics Canada Web site *http://www.geocan.nrcan.gc.ca/* ©2001. Her Majesty the Queen in Right of Canada with permission of Natural Resources Canada.)

A.4 Pointers to Data Resources

Sources of public-domain geographic data in the United States may be obtained from the Web sites of the federal agencies listed in Section A.3. Alternatively, a convenient starting point is the home page of NSDI at http://130.11.52.184/. In addition, there are many good collections of URLs for data resources posted by universities and research organizations, such as

- **The Alexandria Project, NCGIA**
  http://alexandria.sdc.ucsb.edu/
- **The Consortium for International Earth Science Information Network (CIESIN)**
  http:/infoserver.ciesin.org/
- **University of Arkansas, Center for Advanced Spatial Technologies (CAST)**
  http://www.cast.uark.edu/local/hunt/index/html
- **University of Iowa, GeoData Information Sources** (Figure A.4)
  http://www.cgrer.uiowa.edu/servers/servers_geodata.html

At the state level in the United States, URLs of geospatial data clearinghouses for individual states may be found at the Web site of the National States Geographic Information Council (NSGIC) at http://www.nsgic.org/links.htm. The GeoData Information Resources Web site also contains a list of state-level geographic data resources. A vast number of local-level data sets, some of which are free for online download, may also be obtained from commercial organizations. These include

- **ESRI Data Hound**
  http://nt1.esri.com/scripts/production/esri/marketing/datahound/main.cfm
- **ESRI Data Online** (Figure A.5)
  http://www.esri.com/data/online/index.html
- **Etak Inc.**
  http://www.etak.com/howard2
- **Geographic Data Technology**
  http://www.geographic.com
- **The GIS Data Depot**
  http://www.gisdatadepot.com/index.html
- **Land Info International**
  http://www.landinfo.com

In Canada, the Canadian Earth Observation Network (CEONET) provides a single point of access to geographic data on the Internet. This is the Canadian equivalent of the FGDC's geospatial data clearinghouse. The Web site of CEONET is at http://ceonet.ccrs.nrcan.gc.ca/cs/en/index.html. Resources and socioeconomic data may also be obtained from the National Atlas of Canada at

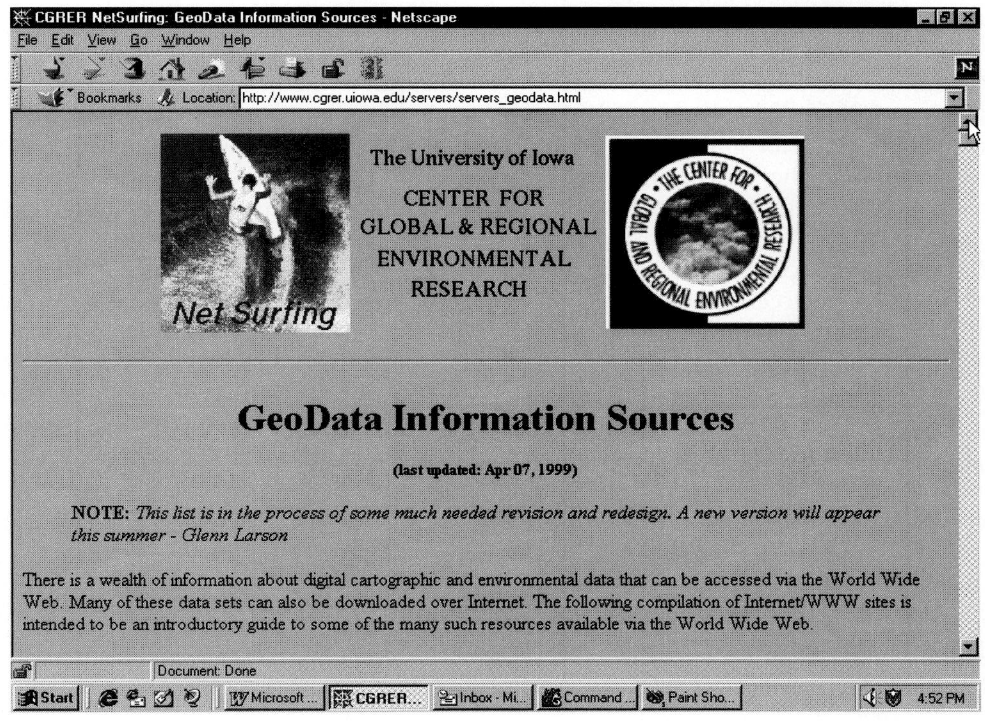

FIGURE A.4
The GeoData Information Resources at the Web site of the Center for Global and Regional Environmental Research, University of Iowa, provides a very comprehensive list of cartographic and environmental data resources that may be found on the Internet. The data resources cover the United States as well as other countries. *(Courtesy: Center for Global and Regional Environmental Research, University of Iowa)*

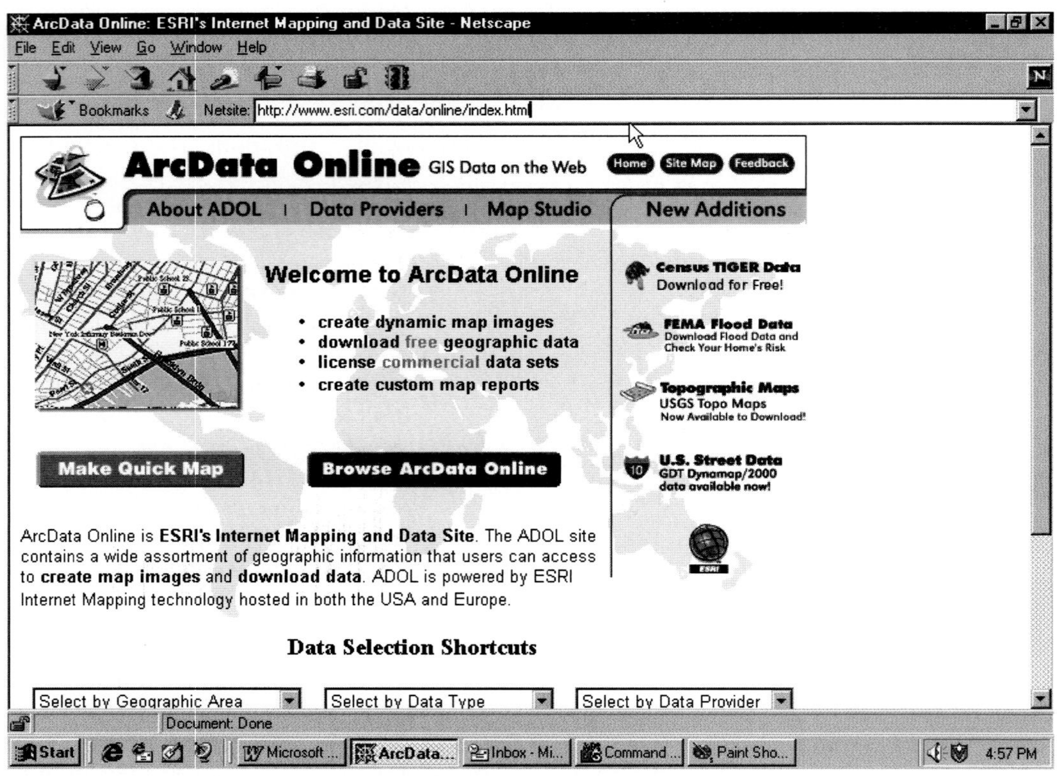

FIGURE A.5
ArcData Online, hosted by ESRI, is one of the commercial Web sites where digital geographic data are marketed. Many such Web sites offer sample data sets free of charge. *(Graphic image supplied courtesy of ESRI)*

http://www-nais.ccm.nrcan.gc.ca/english/home-english.html. Information pertaining to national spatial reference, standards, and networks for gravity and survey control may be obtained from the Web site of the Geodetic Survey of Geomatics Canada at http://www.geod.nrcan.gc.ca. In Australia, the Web site of AUSLIG at http://www.auslig.gov.au/products/digidat/digindex.htm provides a one-stop service to access digital geographic data of the country. Environmental data of Australia may also be accessed through Environment Australia Online at http://kaos. erin.gov.au/erin.html and the Australian Ministry of Environment's site at http://www.environment.gov.au/search/databases.html. Information about data sources of European Union (EU) countries may be found at the GI Data Description Web page of EUROGI at http://www.eurogi.org/directry/datadesc.html. Information pertaining to geographic data of the United Kingdom may be obtained from the Web site of the U.K. National Geospatial Data Framework at http://www.ngdf.org.uk.

For users who are looking for DEM data resources, Bruce Gittings' DEM Data Catalogue is probably the best starting point for the search. This catalog, which provides worldwide coverage, is located at http://www.geo.ed.ac.uk/home/ded.html. Other geographic data resources with an international coverage include

- **The Consortium for International Earth Science Information Network (CIESIN)**
  http://infoserver.ciesin.org
- **Digital Chart of the World (DCW)**
  http://ortelius.maproom.psu.edu/dcw
- **Microsoft TerraServer**
  http://www.TerraServer.microsoft.com
- **NASA Global Change Meteorological Data Center**
  http://gcmd.gsfc.nasa.gov
- **NASA Jet Propulsion Laboratory Shuttle Radar Topography Mission (SRTM)**
  http://www.jpl.nasa.gov/srtm/
- **USGS Earth Resources Observation System (EROS) Data Center, Distributed Active Archive Center**
  http://edcwww.cr.usgs.gov/landdaac
- **USGS Global Land Information**
  http://edcwww.cr.usgs.gov/webgis
- **USGS Landsat 7**
  http://LANDSAT7.USGS.GOV/

A.5 Pointers to Product Information

The best sources of information pertaining to GIS hardware and software on the Internet are the Web sites of individual vendors. The Internet has now become an indispensable marketing tool for manufacturers of hardware and software. GIS users may obtain up-to-date product information, demonstrations of applications, and very often evaluation versions of software programs from vendors' Web sites. The Internet also provides an important channel for user support and user-to-user communication (e.g., discussion groups) from which useful information about the performance, enhancement, and applications of the products may be obtained.

The major computer hardware suppliers for the GIS market include

- **Compaq**
  http://www.compaq.com
- **Dell Computer**
  http://www.dell.com
- **Hewlett Packard**
  http://www.hp.com/go/gis
- **IBM**
  http://www.ibm.com
- **Intergraph**
  http://www.intergraph.com/ics
- **Silicon Graphics**
  http://www.sgi.com/go/visual
- **Sun Microsystems**
  http://www.sun.com/products-n-solutions/

Information about data-acquisition products, including satellite data, photogrammetry, remote sensing, image scanning, table digitizing, and GPS, may be obtained from the Web sites of the following companies:

- **Altek**
  http://www.altek.com
- **ANAtech**
  http://www.anatech.scanners.com
- **Earth Resource Mapping**
  http://www.ermapper.com
- **EarthWatch**
  http://www.digitalglobe.com
- **ER Mapper**
  http://www.ermapper.com
- **ERDAS Inc.**
  http://www.erdas.com
- **GTCO CalComp, Inc.**
  http://wwwgtco.com
- **Intergraph**
  http://www.intergraph.com/ics
- **Laser-Scan**
  http://www.laser-scan.com
- **Leica**
  http://www.leica-geosystems.com
- **MicroImages Inc.**
  http://www.microimages.com
- **NovAtel Inc.**
  http://www.novatel.ca
- **Orbital Imaging Corp.**
  http://orbimage.com
- **PCI Geomatics**
  http://www.pci.on.ca
- **Radarsat**
  http://www.rsi.ca
- **Research Systems (ENVI)**
  http://www.rsinc.com
- **Sokkia Corporation**
  http://www.sokkia.com
- **Space Imaging (IKONOS)**
  http://www.spaceimaging.com
- **SPOT Image Corporation**
  http://www.spotimage.fr
  http://www.spot.com
- **Trimble Navigation** (Figure A.6)
  http://www.trimble.com

Information pertaining to GIS software products may be obtained from their respective developers. Major companies that produce GIS and related software products (e.g., CAD, OS, and DBMS) include

- **Autodesk**
  http://www.autodesk.com/gis
- **Bentley Systems Inc.**
  http://www.bentley.com
- **Blue Marble Geographics**
  http://www.bluemarblegeo.com
- **ESRI**
  http://www.esri.com
- **Insightful (S-Plus)**
  http://www.insightful.com
- **Intergraph** (Figure A.7)
  http://www.intergraph.com/software
- **MapInfo**
  http://www.mapinfo.com
- **Microsoft Corp.**
  http://www.microsoft.com/mappoint/default.htm
  http://www.microsoft.com/expedia
- **Oracle Corp.**
  http://www.oracle.com/products
- **Smallworld Systems Inc.**
  http://www.smallworld-us.com
- **Sun Microsystems**
  http://www.sun.com/products-n-solutions/
- **Universal Systems Limited (USL)**
  http://www.universal.com
- **VISION* Solution**
  http://www.gis.shl.com

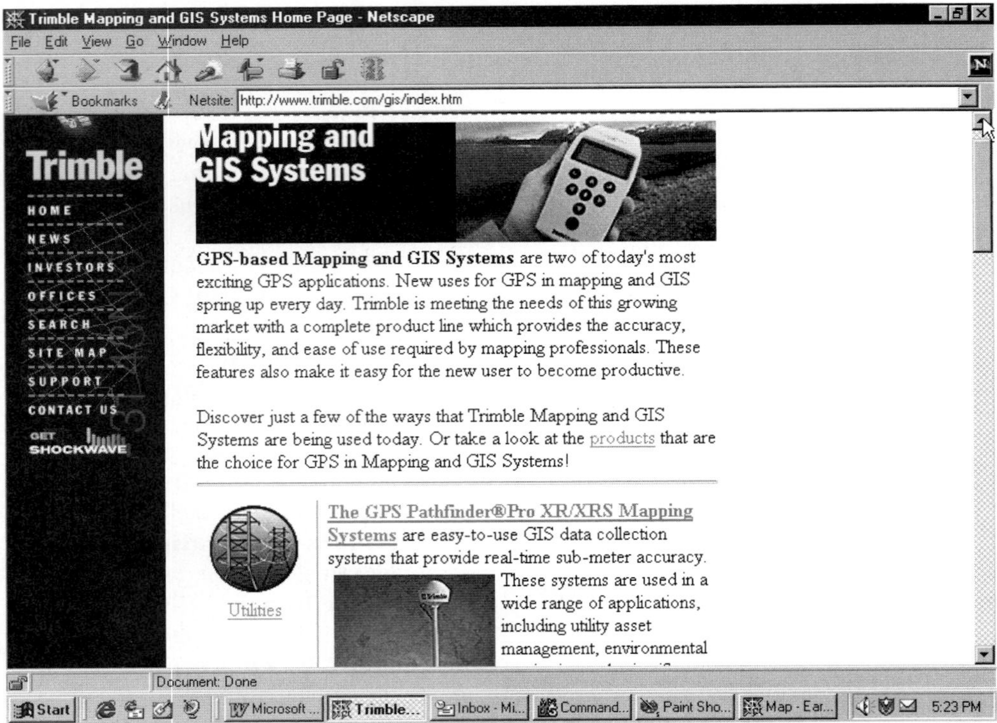

FIGURE A.6
The Internet is now the most convenient way to obtain up-to-date product information about GIS hardware and software, including equipment for data collection. These Web sites allow on-line retrieval of product specifications, descriptions, and illustrations. Some sites also take advantage of the multimedia nature of the Internet to provide live demonstrations of the application of the products. *(Courtesy: Trimble)*

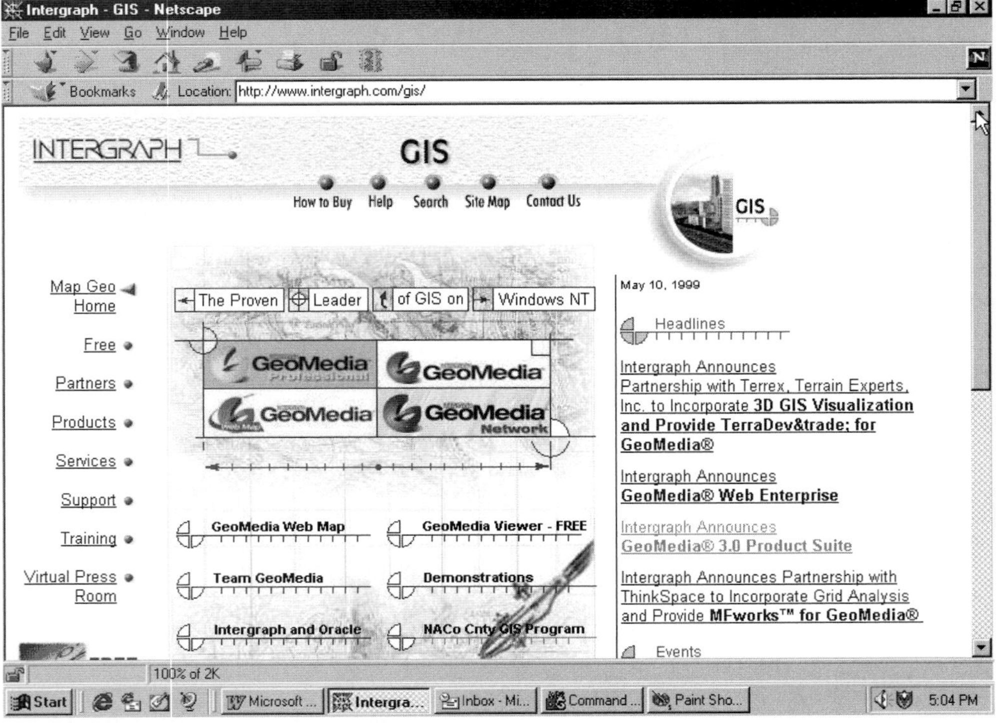

FIGURE A.7
The Internet allows GIS software producers to develop a one-stop shopping mechanism to provide users with release news, product specifications, application development partnerships, user support and training, as well as downloading of evaluation versions of software. *(Courtesy: Intergraph)*

A.6 Examples of Internet-Based GIS Applications

In Chapter 12, we identified eight categories of GIS applications that may be found on the Internet (see Section 12.3.4). The following list represents typical examples of applications in each of these categories.

(1) Static Maps and Snapshots

- Maps of China, University of Oregon
  http://darkwing.uoregon.edu/~felsing/cstuff/cmaps.html
- Forest Land Distribution Data for the U.S.
  http://www.epa.gov/docs/grd/forest_inventory

(2) Dynamic Maps and Cartographic Animation

- USGS Urban Dynamics Research program
  http://edcwww2.cr.usgs.gov/umap/htmls/datavis.html
- NASA EOS Project ATLANTA land-use change
  http://svs.gsfc.nasa.gov/imagewall/LandSat/atlanta_heat_background.html
- Normalized Difference Vegetation Index (NDVI) of Africa
  http://www.ceo.org/mtv-docs/ndvi_africa.html

(3) Spatial Data Catalogs

- USGS Global Land Information System Map Finder
  http://edcwww.cr.usgs.gov/Webglis/glisbin/finder_main.pl?dataset_name=MAPS_LARGE
- Digital Map of Ontario
  http://obms.pmo.mnr.gov.on.ca
- Georgia Environmental Resources Digital Data Atlas
  http://csat.gatech.edu/csat/statewide/statewide.html
- California Geographic Data Library
  http://www/gislab.teale.ca.gov/wwwgis/files_html/dataview.html

(4) Geographic Information Search Engines

- Big Book
  http://www.bigbook.com
- Map Machine (National Geographic)
  http://www/nationalgeographic.com/resources/ngo/maps
- The Aussie Traveller
  http://www.ozemail.com.au/~wilmap

(5) Map Generators

- U.S. Bureau of the Census
  http://tiger.census.gov/cgi-bin/mapbrowse-tbl
- Interactive Mapping at CAST, University of Arkansas
  http://www.cast.uark.edu/local/online_map/
- Cyber Route
  http://www.delome.com/cybermaps/route.asp

(6) Real-Time Map Browsers

- VISA Card ATML Locator
  http://www.visa.com/pd/atm/main.html
- Australian White Pages
  http://www.whitepages.com.au

(7) Real-Time Maps and Images

- Trailmaster, Arizona Department of Transportation
  http://www.azfms.com
- Southern California Real-time Traffic Maps
  http://www.maxwell.com/caltrans
- Puget Sound Traffic Conditions, Washington State Department of Transportation
  http://198.238.212.10/region/northwest/NWFLOW/
- Ontario Ministry of Transportation Webcams
  http://www.mto.gov.on.ca/english/traveller/compass/camera/cammain.htm
- University of Michigan Weather Underground
  http://cirrus.sprl.umich.edu/javaweather/

(8) Interface to GIS

This is probably the most exciting area of using the Internet for GIS applications. As the Internet is a very rapidly advancing field of technology, Internet-based GIS applications are expected to change rapidly as well. All GIS software producers have developed Web sites to showcase the capabilities of their Internet-based GIS solutions. These Web sites are the best places to look at the latest state of the technology. Students are advised to check the demonstration sites of the following companies and software products from time to time to keep themselves updated.

- Autodesk (MapGuide)
  http://www.mapguide.com/products/mapguide/customer.html
- ESRI (MapObjects Internet Map Server and ArcIMS) (Figure A.8)
  http://maps.esri.com
- Intergraph (GeoMedia)
  http://www.intergraph.com/software/geo_map/geo_tdrive.asp
- MapInfo (MapXsite and MapXtree)
  http://www.mapxsite.com/mapxsite/mapxsite_gallery.html
  http://www.mapxtree.com/mapxtree/mapxtree_gallery.html

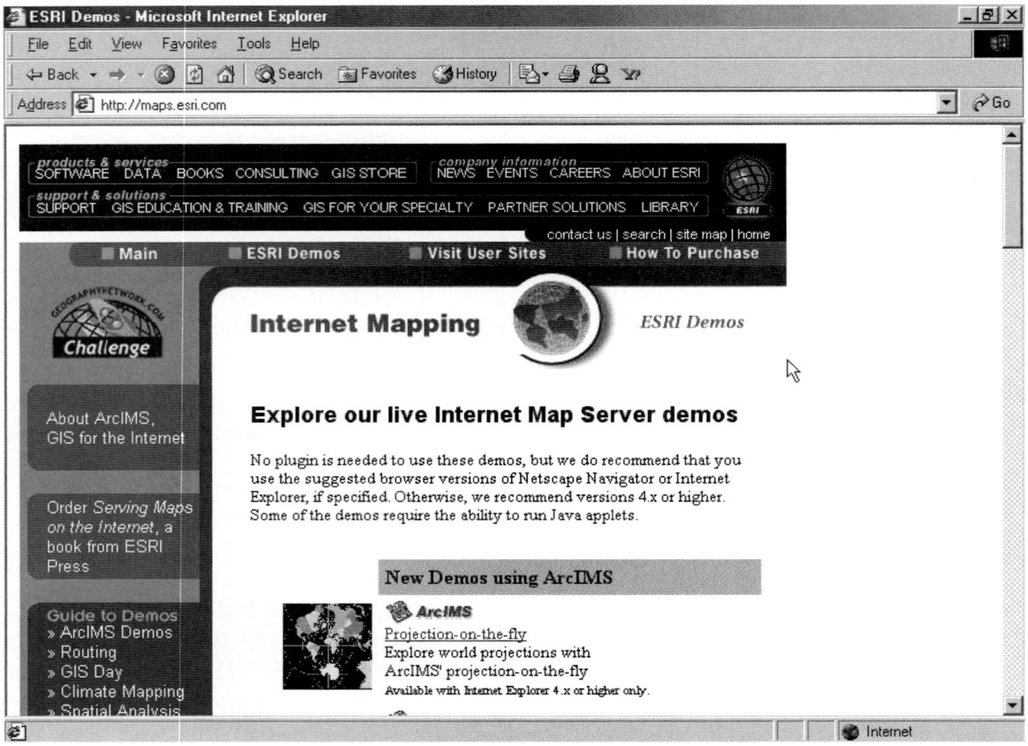

FIGURE A.8
One of the best ways to keep abreast of new developments and applications of Internet technology in GIS is to visit demonstration Web sites set up by software producers. *(Graphic image supplied courtesy of ESRI)*

A.7 Major GIS, Cartography, Remote Sensing, Photogrammetry, and Surveying Journals

American Cartographer
Cartographica
Cartographic Journal
Cartography
Cartography and Geographic Information Systems (official journal of ACSM)
Computers and Geosciences
Computers, Environment and Urban Systems
GeoCarto International
Geodesy, Mapping, and Photogrammetry
Geomatica (formerly, *The Canadian Surveyor*)
International Journal of Geographical Information Science
International Journal of Remote Sensing
ISPRS Journal of Photogrammetry and Remote Sensing (formerly, *Photogrammetria*, official journal of the International Society for Photogrammetry and Remote Sensing)
The ITC Journal
Journal of the Surveying and Mapping Division
Journal of Surveying Engineering
Photogrammetric Engineering and Remote Sensing (official journal of ASPRS)
Photogrammetric Record
Remote Sensing of Environment
Surveying and Land Information Systems
Survey Review (formerly, *Empire Survey Review*)
Transactions in GIS

REFERENCES

Newquist, H. P. (2000) *Yahoo! The Ultimate Desk Reference to the Web*, New York: HarperResource

Thoen, B. (2001) "Maximizing your search for geospatial data," *GeoWorld*, Vol. 14, No. 4, pp. 34–37.

Turlington, S. (1999) *The Unauthorized Guide to the Internet*, Indianapolis, IN: QUE.

Appendix

B

Glossary of GIS Terms

This glossary contains explanations of technical terms that have appeared in the chapters. Its purpose is to help students better understand concepts and techniques presented in this book by (1) defining concisely technical terms with which students may not be familiar and (2) cross referencing technical terms that are conceptually or semantically related to one another. It should be noted that all the definitions are made in the context of using GIS. The generic meanings of many of the technical terms may be different when they are used elsewhere.

Absolute location A point in geographic space that is measured with respect to the **origin** of a standard coordinate system (e.g., latitude/longitude geographical coordinates). See also **Relative location**.

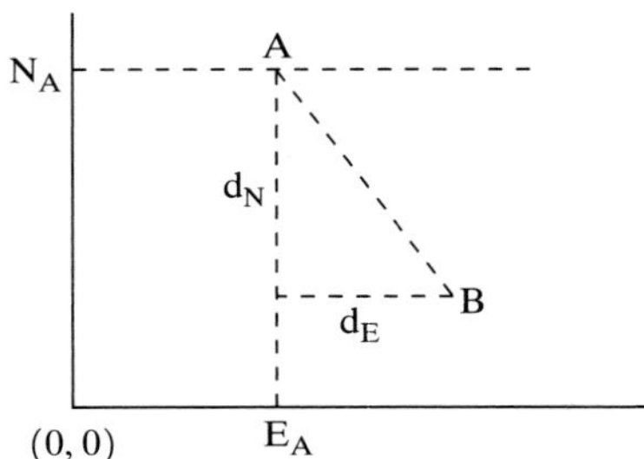

Absolute location of A = (N_A, E_A)
Relative location of B to A = (d_N, d_F)

Accuracy The degree to which a measurement approximates the true value of the object being measured. See also **Error**, **Precision**, and **Uncertainty**.

ActiveX control See **Object linking and embedding (OLE)**.

Address matching A geographic data-processing operation that relates street addresses to census blocks, census tracts, or other administrative units. See also **Geocoding** and **Quasi-spatial attribute**.

Ad hoc application An application designed and implemented for a specific or project-oriented purpose.

Adjacency A topological relationship that occurs when neighboring polygons share a common boundary. Adjacency is sometimes referred to as *contiguity*. There are two other topological relationships, namely, **containment** and **connectivity**. See also **Topology**.

Affine transformation A combination of linear transformations (rotation and scaling) followed by translation (shift of position). In geographic data processing, affine transformation is used to register maps of different scales and origins.

Annotation Descriptive text used to label features in a digital map. Annotations are used for the purpose of cartographic display only, not for spatial analysis.

Anonymous FTP A process that allows a user to download data files from a remote computer across the Internet without needing a password. Universities, government agencies, and business organizations around the world have made data and applications freely available to the public via the Internet. To transfer a file using anonymous FTP, the user signs in to the remote computer as "guest" or "anonymous" instead of using a user name assigned by the host computer. See also **FTP**.

API (Application Programming Interface) A suite of system calls or routines for application programmers to access services of the computer's operating system or other computers. An API allows different programs to work with one another. It is the foundation technology for **client/server computing**.

Arc A line represented by an ordered sequence of points usually referred to as *vertices*. The end points of an arc are called *nodes*. An arc may be generated by digitizing or by a mathematical equation. See also **Node** and **Vertex**.

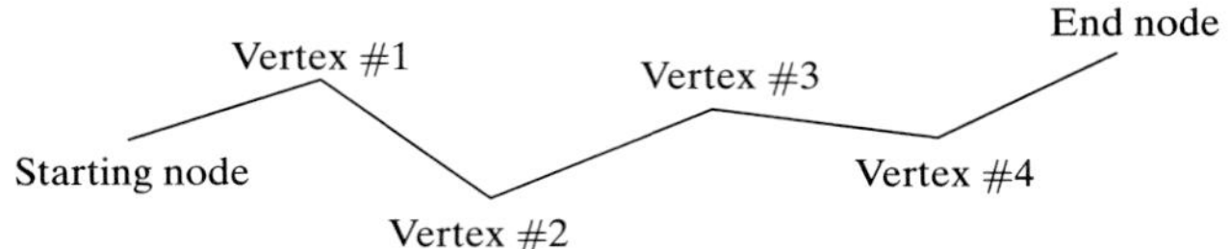

Arc/node topology Arcs represent linear and area features in a digital map. Arcs split where they intersect with one another. Arc/node topology is a data structure that makes use of arcs (the polygons to the left and right of an arc) and the nodes (the from-nodes and to-nodes of an arc) to define the relationships between neighboring polygons and connecting line features. See also **Adjacency** and **Connectivity**.

Arc second A unit of angular measurement that is used to denote the spacing of adjacent profiles in USGS DEMs. An arc second is approximately equivalent to 30 meters of linear measurement.

Area feature A closed figure that represents a geographic feature or phenomenon characterized by a certain homogeneous characteristic, also known as a **polygon**.

Array A collection of data items of the same size and type (though they may have different values). In mathematics, a one-dimensional array is called a *vector*; a two-dimensional array is called a *matrix*.

Artificial intelligence (AI) A branch of computer science that aims to simulate the thought process of the human brain, commonly through the use of software. AI is the fundamental technology for building

expert systems. See also **Decision support system**.

ASCII (American Standard Code for Information Interchange) A set of codes used to represent alphanumeric characters in computer data processing. Data files created using text editors (e.g., Windows Notepad) are often referred to as *ASCII files*.

Attribute (1) Description of a geographic feature in the form of numbers or characters, typically stored in a table and linked to the graphics of the feature by a common identifier. (2) A column of a relational table.

Azimuth A horizontal angle measured clockwise from the north to the direction of an observation. See also **Bearing**.

Base map In a GIS or digital mapping system, base map refers to those layers of map information (e.g., topography, roads, rivers, and lakes) used for georeferencing layers of thematic information (e.g., forest stands, mineral deposits, land use, and population distribution).

Basic graphical elements These are the points, lines, and areas in a geographic database.

Batch processing The method of data processing by submitting programs and data to the computer in one or more files rather than interactively from a terminal.

Bearing A bearing may be expressed as (1) a *whole-circle bearing* or (2) a *reduced* or *quadrantal* bearing. A whole-circle bearing of a line AB (α_{AB}) is the clockwise angle from 0 to 360 degrees at A between the direction to north and the direction to B. The reduced bearing is the angle lying between 0 and 90 degrees, between the direction to north or south and the direction of the line. In the diagram, the reduced bearing of the line AB is N α_{AB} E and the reduced bearing of line BA is S α_{BA} W, where α_{BA} is equal to α_{AB}.

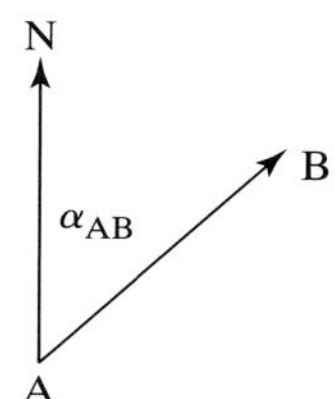

Bench mark (1) A point of known elevation above an accepted **datum**, established by surveyors for use as reference and control in mapping. (2) A standard test designed to evaluate the functionality of different software products by systematic comparisons and testing of their features.

Bit (binary digit) (1) A single number, with a value of either 1 or 0, in the binary number system. (2) The smallest unit of storage in computer memory, which can be switched to either the *on* (1) or *off* (0) states.

Bit map A form of computer graphics created by a pattern of tiny dots. A bit map requires a large amount of memory to store and becomes visually blurred (i.e., losing definition) when the image is enlarged. See also **Raster**.

Boolean expression An expression that contains logical (i.e., *true* or *false*) expressions and Boolean operators (i.e., *NOT, AND,* and *OR*), e.g., HEIGHT $>$ 50 AND DIAMETER $<$ 3.

Browser A computer program that allows a user to interact with the World Wide Web. It is also called a *Web browser*. See also **World Wide Web**.

Buffer (1) A zone of specified width surrounding a point, line, or area feature. (2) A temporary data-holding area in computer memory.

Buffering The process of generating a buffer zone from a point, line, or area feature by offsetting a user-defined distance from these features.

Business functional hierarchy A document that describes the breakdown of business functions or processes into increasingly detailed levels in systems design and analysis.

Byte A group of consecutive bits treated as a unit of computer storage. In current computer implementation, eight bits constitute a byte. See also **Bit** and **Word**.

Cable modem A network connection technology that is capable of transmitting data at a speed of 10 Mbit/s (megabits per second).

Cadastre A public land register or survey that defines boundaries of public or private land for the purpose of recording ownership and taxation.

Cartesian coordinate system A coordinate system consisting of intersecting straight lines called axes, in which the lines intersect at a common origin. Usually it is a 2-D surface in which an *x, y* coordinate defines each point location on the surface. The *x*-coordinate refers to the horizontal distance, and the *y*-coordinate to vertical distance. Coordinates may be either positive or negative, depending on their relative position from the origin. In a three-dimensional space, the system may also include a *z*-coordinate, representing height or depth. The relative measurements of distance, direction, and area are constant throughout the surface of the systems.

CASE tools (Computer-Aided Software Engineering tools) Defined programming rules for applying engineering principles and techniques in systems design and development.

Categorical data Data measured or observed by descriptive terms (e.g., very fertile, fertile, moderately fertile, and infertile), rather than by numerical values.

Central meridian A line running north and south, at the center of a graticule, along which all points have the same longitude. See also **Meridian**.

Centroid The geometric center of a polygon, computed mathematically from the locations of all the vertices defining the polygon. Sometimes centroids are estimated visually. Centroids determined in this way are usually referred to as *visual centroids*.

CGI (Common Gateway Interface) A method for running computer programs, called *CGI scripts*, on the **Web server** based on input from a Web **browser**. CGI scripts enable a user to interact with the World Wide Web: to search for an item in a database, to fill in data entry forms, to select several items from a list, and to get a customized reply in return.

Classification The process of placing items into groups or categories, according to a predefined set of rules. For example, maps and map projections may be classified according to their use, scale, properties, or appearance. See also **Classification scheme**.

Classification scheme A table showing the number of categories in a classification, their names, and respective characteristics that distinguish one category from another. See also **Classification**.

Client/server computing A method of separating the functions (i.e., applications and processes) and sharing the resources (i.e., programs and data) of computers in a network environment. A client is a computer that requires the use of resources, and a server is a computer that provides the resources. Depending on the nature and requirements of individual applications, data-processing functions may be put either in the client or the server, or both.

COM (Component Object Model) The underlying object architecture developed by Microsoft that allows objects written by different software developers in different programming languages to interact with one another. See also **Object linking and embedding**.

Complex object Spatial features that cannot be represented by simple geometry, e.g., the coastline of the United States.

Composite attributes Groups of attributes that have an affinity in meaning or usage, e.g., ADDRESS is a composite attribute because it denotes a group of attributes including STREET_NUMBER, STREET, CITY, STATE, ZIP_CODE, and COUNTRY.

Compression A variety of techniques used to reduce the physical size of data files.

Conceptual database design The initial stage of database design that defines user requirements globally. Conceptual database design is both software- and hardware-independent, i.e., software and hardware are not considered at this stage of database design.

Conformal A map projection property that preserves the local shape of features on maps.

Conic projections A class of map projection involving the projection of part of the globe onto a cone-shaped surface.

Connectivity A topological relationship that occurs when arcs are connected using shared nodes. Connectivity is used for network analysis in geographic data processing. See also **Topology**.

Containment A topological relationship that occurs when one polygon is located within another polygon, e.g., a lake on an island or an island in a lake. See also **Topology**.

Coordinate geometry The method used to construct graphics mathematically in engineering design. It is usually referred to as *COGO*.

Coordinate system A system used to register and measure horizontal and vertical distances on a map.

CORBA (Common Object Request Broker Architecture) A general and open **industry standard** for working with distributed objects in the UNIX environment. Developed by the Object Management Group (a nonprofit consortium made up of large computer manufacturers such as Sun, IBM, Apple, and Hewlett Packard and users), CORBA allows the interconnection of objects and applications regardless of the computer language that provides or uses the object, the hardware architecture, and the geographical locations of the computers involved.

Coverage (1) A digital map that forms the basic unit of vector data storage in ArcInfo. A coverage stores map data in the three basic forms of graphic elements, i.e., points, lines, polygons. (2) A single theme or layer of data (e.g., soils, land cover, roads, built-up areas) in a geographic database.

Cursor A hand-controlled computer hardware device consisting of a graphic pointer or cross hairs and keys, used to point to locations on a computer terminal screen. In digitizing, a cursor is used to trace and capture the location of analog features in *x*, *y* coordinates.

Cylindrical projections A class of map projection that projects part of the globe onto a cylinder-shaped surface.

Data automation The sequence of processes for converting a paper map into digital data, including table digitizing and/or **screen digitizing**, **topology** building, and storing the digital data in a selected **data structure** or **format**.

Database A collection of data organized in a systematic manner so that they can be accessed on demand. See also **DBMS**.

Database administrator The person who is responsible for the day-to-day operation of a database (e.g., managing user accounts, allocating computing resources, database security, and backup). See also **DBMS**.

Data definition language (DDL) The language used in database management systems (DBMS) to set up databases. It allows the database administrator to define the components of the database, using commands such as CREATE, DROP, ALTER, and REVOKE. See also **Data manipulation language** and **DBMS**.

Data dictionary (1) A comprehensive collection of the design specifications used to define and describe all the components of a system. Also referred to as *design dictionary* and *physical data model*. (2) The part of the database that contains information about the files, records, and attributes pertaining to the data in the database.

Data exchange standard Specifications for a "neutral" data format developed for exchanging data in different **proprietary data formats**. The use of a common data exchange standard eliminates the need to develop data translators between individual proprietary formats. The source data in one proprietary format may always be translated first to the data exchange standard and then to the proprietary format of the system where the data are used. Also referred to as *data transfer standard*.

Data manipulation language (DML) The language used in database management systems (DBMS) to access and retrieve data. It allows database users to access and manipulate the database using commands such as SELECT, INSERT, UPDATE, DELETE, and ROLLBACK. See also **Data definition language** and **DBMS**.

Data mining The procedure that makes use of the power of computers to sift through vast data stores in order to unveil important information (e.g., patterns, trends, and relationships) for business or management decision making.

Data model A logical way of organizing and representing data in an information system.

Data structure The logical and physical way of organizing digital data in a database.

Data transfer standard See **Data exchange standard**.

Data translator A program that converts data from one proprietary format to another. See also **Data exchange standard**.

Data warehouse An active intelligent store of data that is designed to manage data from multiple sources and distribute them where needed for business operation and decision support.

Datum A point, line, or surface selected as the origin or reference for measurements. A datum may be mathematically defined (e.g., Earth's ellipsoid) or established by field observations (e.g., the mean sea level). See also **Ellipsoid** and **Mean sea level**.

DBMS (Database Management System) A collection of software tools, procedures, and rules for the creation, management, and use of databases. See also **Database** and **Database administrator**.

DDE (Dynamic Data Exchange) Microsoft's underlying technology to provide and manipulate dynamic objects between applications, e.g., a word processing application may instruct, via DDE, a spreadsheet application to update spreadsheet objects embedded in a document it is working on.

Decision support system (DSS) A computer application that analyzes business data and presents it so that users can make business decisions more easily. DSS is an "informational" application as distinguished from the "operational" applications that collect data in computer data processing. A DSS can present information graphically and may include an **expert system** or **artificial intelligence** in its configuration.

De facto standard See **Industry standard**.

DEM (digital elevation model) A data file containing an array of elevation values. See also **Digital terrain model** and **TIN**.

Desktop mapping The use of PCs to generate maps for viewing geographic data and performing simple spatial data analysis. Desktop mapping normally uses *canned data* (i.e., data obtained commercially from data suppliers) and *turnkey mapping applications* (i.e., off-the-shelf applications that do not require extensive training to use).

Desktop metaphor The use of physical analogy or icons to help PC users interact with computer applications.

Device-independent The ability of a software product to run on computer systems of different hardware configuration.

DIGEST (Digital Geographic Information Exchange Standard) A data transfer standard used by NATO (North Atlantic Treaty Organization) for the exchange of geographic data among the military establishments of member countries.

Digital Chart of the World (DCW) A digital cartographic database of the world, developed by the U.S. Defense Mapping Agency [DMA, which is now part of National Imagery and Mapping Agency (NIMA)] with the cooperation of Australia, Canada, and the United Kingdom to support the display and analysis of geographic data. It is generally based on the DMA's 1 : 1,000,000 scale Operational Navigation Chart (ONC) base map series. The DCW is in vector format and is suitable for GIS applications.

Digital Orthophoto Quad (DOQ) A U.S. national mapping program that aims to cover the lower 48 states at a 1-m ground resolution using monochrome air photographs in digital format with a 1 : 12,000 equivalent ground extent.

Digital photogrammetry The photogrammetric method of using scanned images, rather than conventional diapositives or paper-based aerial photographs, for measurement and map production purposes. Also commonly referred to as **soft-copy photogrammetry**.

Digital terrain model (DTM) A topographic surface or computer representation of terrain stored in a digital data file as a set of three-dimensional (*x*, *y*, *z*) coordinates. The image may be displayed on a computer monitor or portrayed on a map. See also **DEM** and **TIN**.

Digitizer An electronic device for capturing digital data from paper maps. Small-sized digitizers are usually referred to as *digitizing tablets*, and large-sized ones are called *digitizing tables*.

Digitizing The process of converting paper maps into digital files. Digitizing may be done by table digitizing or scanning followed by **screen digitizing**.

DIME (Dual Independent Map Encoding) A concept developed by the U.S. Bureau of the Census to permit the semiautomatic editing of databases that describe the urban street network and statistical units such as street blocks and census tracts. See also **GBF** (Geographic Base File).

Dissolve The process of automatically eliminating the boundary between polygons that have identical attribute data. This process is usually carried out to remove the **neat lines** of digital maps after several maps are joined. See also **Mapjoin**.

Distortion The deviation of the shape, scale, or position of geographic features from their true shapes, scales, or positions as a result of data acquisition or data processing.

DLL (Dynamic Link Library) An implementation of Microsoft's **COM** (Component Object Model) architecture that allows components objects to communicate with one another, e.g., using a statistics component to generate a table of statistics from data displayed in a histogram produced by graphing component.

DLG (Digital Line Graph) (1) A digital data format used by the USGS for large-scale digital topographic maps. (2) A digital data file standard published by the USGS for exchanging digital cartographic data files.

Download The retrieval of data files from a remote computer over a network.

DOQ See **Digital Orthophoto Quad.**

Double precision The method of storing numerical values in computers using twice the regular word length. In geographic data processing, coordinates are usually stored and manipulated in double precision to minimize the effect of rounding errors.

DRG (digital raster graphics) A scanned image of USGS topographic maps. The purpose of DTG is to provide a backdrop on which other digital data (e.g., vector data from **DLG**) can be overlaid.

DS-3 A North American standard denoting communication lines in a network that is capable of transmitting data at a rate of 44.73 Mbit/s (megabits per second). See also **cable modem** and **T-1**.

DXF (Drawing eXchange Format) The digital file format for exchanging data between AutoCAD and other graphics data-processing software packages. It has now become the industry standard for graphics file exchange between CAD and GIS applications.

E00 The data transfer format for ArcInfo.

Edge matching The process of matching corresponding features across map boundaries. This is a prerequisite for using multiple digital maps in geographic data processing and analysis.

Editing The process of modifying and updating graphics and attribute data.

Elevation The vertical distance or height above a **datum**.

Ellipsoid The "imaginary" or "mathematical" surface of Earth used by surveyors for the computation of geodetic and astronomic coordinates. See also **Geodesy**, **Geoid**, and **Spheroid**.

End node The beginning and ending coordinates of an **arc**.

Enterprise computing The organization of computer resources (e.g., databases, document management) in business organizations to serve both corporate information needs (e.g., decision support, human resources management, and accounting) and program-specific or departmental information needs (e.g., client orders processing, goods delivery tracking, and issuing of permits and licenses).

Entity A term widely used in the database community to mean any distinguishable geographic feature or phenomenon that may be represented individually in a database. See also **Feature** and **Object**.

Entity–relationship diagram (E–R diagram) A diagram showing the attributes of and relationships between entities identified in **entity–relationship modeling**.

Entity–relationship modeling (E–R modeling) An approach of data modeling that aims to identify the entities of interest and their relationships in database design. The major tool for E–R modeling is the **entity–relationship diagram**.

Equal area A property of map project that preserves the sizes of map features.

E–R diagram See **Entity–relationship diagram**.

Ergonomics The science of designing equipment for a comfortable and safe working environment. Also known as *human factors* or *human engineering*.

Error The deviation, variation, or discrepancy of a measurement or observation from the true or accepted value of an object. See also **Accuracy** and **Uncertainty**.

Expert system A computer system that is programmed to mimic the procedure and decision that human "experts" make. See also **Decision support system**.

FAQ (Frequently Asked Questions) Questions and answers, usually posted on network newsgroups or printed in users' manuals, that serve to save new users the trouble of asking old questions again and again.

Feature A geographic object or phenomenon that may be discretely identified or measured in a spatial data collection. See also **Entity** and **Object**.

Feature attribute tables Relational tables used to store attribute data associated with graphics data in a geographic database. Feature attribute tables also contain system-generated parameters (for internal reference by the computer) and user-assigned feature identifiers (for external reference by users). See also **Feature identifier**.

Feature identifier (FID) A user-assigned number for each **basic graphical element** in a geographic database, used to relate graphical data to attribute data in **feature attribute tables**. See also **Tag**.

Field-based engineering Engineering work carried out in the outdoor environment, e.g., construction engineering, highway engineering, surveying engineering, and electrical engineering concerned with cross-country power transmission.

File A collection of logically organized data in a storage device. A file is identified by a *file name* and has a *file size*.

File header The initial part of a data file that contains information about the content of the file.

File server A computer whose sole function is to store data and make them available to applications in computers in a network. See also **Client/server computing**.

FIPS (Federal Information Processing Standards) Specifications and rules, maintained by the U.S. National Institute of Standards and Technology, that govern a wide range of information technologies used in the U.S. federal government, including hardware, software, storage media, coding practice and programming languages, data transfer, network and communication, documentation, and security.

Flat file A collection of data stored in a tabulated form. The rows of a flat file contain measurements or observations pertaining to individual objects, and the columns contain measurements or observations for different attributes of the objects.

Fly-by A method of animation in computer graphics that generates an oblique moving view of the landscape as if the viewer were flying in an airplane over the area. Also often referred to as *fly-through*.

Fonts A set of characters in a specific design (e.g., typeface and size).

Format The organization or structure of a digital data file.

Fourth-generation languages (4GL) These are programming languages having a syntax closer to human languages than the conventional high-level computer languages such as FORTRAN, C, and C++. Most 4GLs are used to access databases. A typical 4GL command is like this: FIND ALL RECORDS WHERE CITY_NAME IS "TORONTO."

Freeware Software applications that are made freely available to anyone who wants to use it.

FTP (File Transfer Protocol) One of the protocols of the Internet that provides a standardized way of transferring files between computers in a distributed network.

Fuzzy tolerance The distance within which individual discrete points are snapped to form a single point during geographic data processing.

Gateway An entry point to a network of server computers associated with a particular project or organization.

GBF (Geographic Base File) A data file encoded in the **DIME** (Dual Independent Map Encoding) format, used by the U.S. Bureau of the Census in the 1980 census. It has now been replaced by **TIGER**.

Generalization The process of simplifying the form or shape of map features, usually carried out when the map is changed from a large scale to a small scale.

Generalization hierarchy A concept in **E–R modeling** that defines a subset relationship between the elements of two or more classes, e.g., the class PERSON is a generalization of the class MAN and WOMAN.

Geocoding The process of determining the coordinates of a point or an attribute (e.g., an address) so that it can be located/displayed graphically.

Geodesy The scientific discipline that is concerned with the study of the size and shape of Earth and the measurement of its gravitational and magnetic fields.

Geographic coordinates The geographical referencing system using latitudes and longitudes.

Geoid Geoid is the "imaginary" surface of Earth that is obtained by projecting the highly irregular surface of Earth onto a more regular surface formed by extending the average level of the oceans under the continents. This is a concept used in **geodesy** and map projections. See also **Ellipsoid** and **Spheroid**.

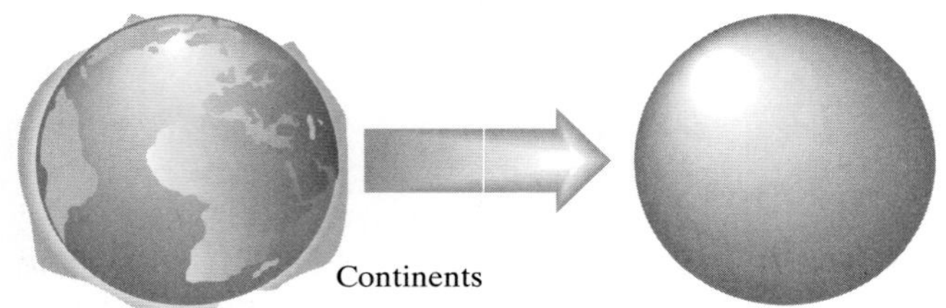

Georeferencing The process of registering a geographic data set to an accepted coordinate system. See also **Geocoding**.

GPS (Global Positioning System) A satellite-based navigation system, funded by the U.S. Air Force, that is used to determine accurate geodetic positions on Earth's surface. Specially designed GPS receivers, when positioned at a point on Earth, can measure the distance from that point to 3 or more of the 24 orbiting satellites. The coordinates of the point are determined through the geometric calculations of triangulation.

Graphics device interface (GDI) A piece of software whose function is to translate commands from the computer to a plotter or printer for the generation of graphical products.

Graticule The grid of latitudes and longitudes drawn on a map or a globe.

GRS 80 (Geodetic Reference System of 1980) A standard defining the size and shape of Earth as adopted by the International Union of Geodesy and Geophysics in 1979.

GUI (graphical user interface) The method of interaction between users and computers using windows, menus, icons, and pointers.

Help desk A support structure, including people (analysts and programmers), procedures (for logging problems), and protocols (e.g., the time frame within which all logged problems must be answered), that is established to assist end users in the use of information resources, including application, data, and hardware.

Helper program A computer program developed to enhance the capability of Web **browsers**. For example, as Web browsers do not support video files, a helper program is required to display video files on the Internet. Helper programs are product-specific and are installed separately. They are invoked when the browser detects file formats that need their service. See also **Plug-ins**.

Hierarchical data model One of the classical data models that is based on layers of data sets and subsets organized as parents-children.

Hill shading A method of portraying topography on a map by shading different locations with tonal differences (i.e., different levels of gray) according to illumination by an assumed light source from the northwest, thus creating a pictorial representation of relief.

Histogram A graph showing the relative values of attributes that are represented as bars.

HPGL (Hewlett Packard Graphics Language) An **industry standard** for defining and generating vector graphics in geographic data processing.

HTML (Hypertext markup language) A collection of codes and rules for the creation of documents used in the **World Wide Web**.

HTTP (hypertext transmission protocol) The language by which computers in the **World Wide Web** communicate with one another.

Hypermap A map-based document that incorporates **hypertext** technology to allow it to seamlessly reference other documents and retrieve information from them.

Hypertext The method of linking different documents in HTML by hot links or hyperlinks. It allows the logical association of documents distributed at different locations in a computer network.

Identity The topological overlay of a line or polygon coverage with another polygon coverage that computes the geometric intersection of the two coverages. The resulting coverage preserves all the input features plus those portions of the polygon coverage that overlap the input coverage. See also **Intersect** and **Union**.

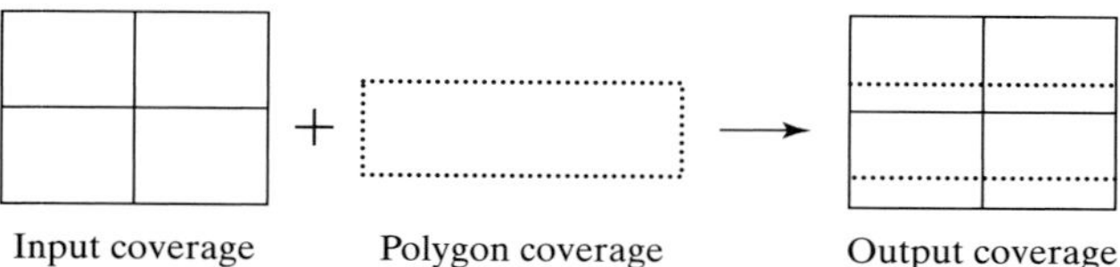

Image map (1) A map created from aerial photographs or satellite images by assembling them spatially (i.e., making a mosaic), correcting them geometrically (i.e., georeferencing and scaling), and adding cartographic symbols and place names, making it usable just like an ordinary line map. (2) A digital map obtained by scanning a paper map. (3) A screen display, not necessarily in the form of a "map," that allows a user to interact with the computer by clicking different parts or locations on the display. Such an image map is also commonly referred to as a *clickable map*.

Import The process of getting data from an external source into a particular GIS.

Industry standard Specifications and procedures adopted by common use, rather than established by standard development bodies or regulated by government legislature. Industry standards are sometimes referred to as *de facto standards*.

Information economy Economic activities engaged in the collection, processing, and distribution of information, including the manufacturing of hardware, software, and information products (e.g., books, maps, digital data files) as well as the provision of consulting services. Also commonly referred to as *knowledge-based economy*.

Integrated Services Digital Network (ISDN) A digital telephony technology that supports the high-speed transfer of voice and data over telephone lines.

Internet A logical supersystem of local, regional, national, and international communication networks. Computers on the Internet can access the resources of one another by using common communication protocols such as the **World Wide Web** and **FTP**.

Interoperability The cross-platform use of computer applications, i.e., the ability to use applications in different computer hardware environments and operating systems without any need to change the programs and retrain the users.

Intersect The topological overlay of two coverages that preserves a feature falling within the spatial extent common to both coverages.

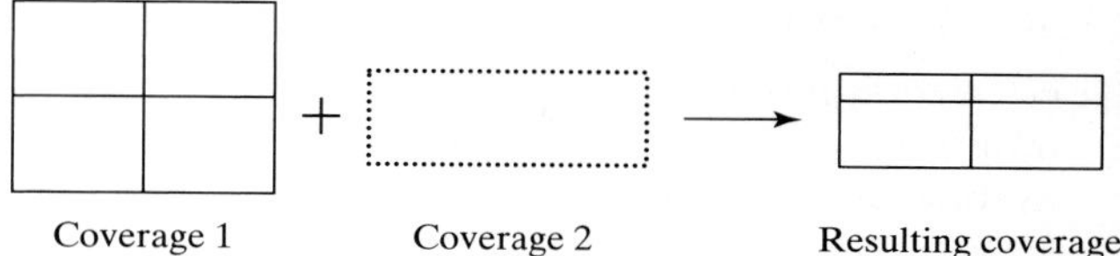

Interval data Data whose values may be set on a scale of measurement indicating the distance or time between differing values. This scale of measurement is usually depicted and explained in the map legend. A rainfall map indicating the number of millimeters per year in different locations would illustrate interval data. See also **Level of measurement**, **Nominal data**, **Ordinal data**, and **Ratio data**.

ISO 9000 A set of five universal standards, sponsored by the International Organization for Standards (ISO) and accepted by about 90 countries as national standards, for quality assurance in business organizations. The purpose of these standards is to help organizations ensure that their products and services meet the prescribed quality. The most comprehensive of the standards is ISO 9001, which applies to industries involved in the design, development, manufacturing, installation, and servicing of goods and services. Organizations have to register with ISO and pass the necessary evaluations to be certified as ISO 9000–compliant. Governments in many jurisdictions now require business organizations to have this certification before they are allowed to bid for publicly funded contracts, including GIS-related contracts.

Isohyet A line on a map or weather chart connecting locations with the same rainfall.

Isotherm A line on a map or weather chart joining locations with the same temperature.

Java applets Small computer programs written in the Java language, commonly used to support Internet-based applications.

Job control language (JCL) Commands that appeared on the first several cards in a stack of punch cards submitted for batch processing in the old time-sharing computing environment. The function of JCL was to identify the user and to provide instructions to the computer regarding resources required for data processing (e.g., the name of the compiler to be used and the device names of the card reader and output writer).

Joint application development (JAD) A highly structured workshop that brings together users, managers, and information systems specialists to jointly define and specify user requirements, technical options, and external designs (e.g., input and output formats as well as screen and menu design). See also **User requirements analysis**.

Key A unique feature identifier that serves as a common thread to relate records in one relational table to those in other tables.

Killer app A computer application that is indispensable to users because it provides a way of doing a commonly used and essential task. In the desktop environment, word processing and spreadsheets are killer apps.

Kriging A mathematical interpolation method based on the use of a generalized least-squares algorithm. It was first developed for mining applications but is now commonly used for three-dimensional geographic data processing.

Label A text string that describes a geographic object in a digital map, e.g., elevation value of a contour line. Unlike **annotation**, labels are used for display, query, and analysis purposes. In some systems, labels are also known as **tags**.

Label placement The process of adding labels to a digital map. This may be done manually or by using an automated label placement program. Ordinary label placement programs simply locate labels relative to their associated graphic objects according to some predefined rules. More advanced programs use the principles and technique of artificial intelligence to resolve conflicts caused by overlapping labels during the label placement process.

Latitude The angle, measured from Earth's geometric center, between the Equator and a point on or above Earth's surface. Latitudes are also referred to as *parallels*.

Layer A subset of a geographic database containing data associated with a particular theme. See also **Coverage**.

Least squares A statistical method of fitting a mathematical model. It is based on computing the minimum sum of the squared deviation between measured data and their estimates (i.e., the mean value).

Level of measurement The degree of subjectivity that is applied when performing measurements or observations in data collection. There are four levels of measurements, namely: **nominal**, **ordinal**, **interval**, and **ratio**.

Lineage A document that describes the characteristics of the source materials from which a specific data set has been derived. See also **Metadata**.

Line feature A geographic feature represented on a map by connecting an ordered sequence of points.

Logical database design One of the phases of database design that is carried out after the completion of **conceptual database design**. The objective of logical database design is to refine the finding of the conceptual design. The output is a software-specific but hardware-independent design specification to be used as input to the next phase of database design called physical database design.

Longitude The angle, measured on the plane of the Equator, between the position of a point on or above Earth's surface and the reference meridian (i.e., 0°) that passes through Greenwich, England. Longitudes range from 180 degrees west (−180) to 180 degrees east (+180). Latitudes are often referred to as **meridians**.

Macro (1) A computer programming language, usually proprietary to a specific software vendor, used for developing customized application modules to extend the functionality of the software. (2) A program written in a macro language.

Map algebra A suite of arithmetical operations (i.e., addition, subtraction, multiplication, and division) and their combinations for manipulating and analyzing raster data.

Mapjoin The process of physically joining digital map files into a larger map or a **seamless geographic database**.

Map projection The principles and techniques of transforming and representing three-dimensional features on Earth's surface on a two-dimensional flat map sheet. This process is accomplished by the use of geometry or, more commonly, by mathematical formulas. Map projection can be best visualized by imagining a lightbulb placed at the center of a transparent globe and having its lines of longitude and latitude cast upon either a flat sheet of paper or a sheet of paper rolled into a cylinder or cone placed over the globe.

Matrix A table of numerical values with a given number of rows and columns. See also **Array**, **Flat file**, and **Relational Table**.

Mean sea level (MSL) The numerical value representing the level of the sea surface at a particular location, obtained by making repeated measurements over an extended period of time in order to take into account the effects of tides and seasonal changes. The mean sea level is used as the datum for elevation measurements in surveying and mapping.

Menu A component of a graphical user interface **(GUI)** that provides the users with a list of options to choose from during computer data processing.

Meridian See **Longitude**.

Metadata Data about data or descriptions of the data in a data file, including date of collection, sources, map projection, scale, quality, format, and custodian.

Model A representation of the characteristics of real-world phenomena or objects and their relationships. A model may be conceptual, analogous, descriptive, graphical, or mathematical. See also **Modeling**.

Modeling (1) The process of creating a model. (2) A stage in scientific investigation involving the development of abstraction, theories, and mathematical formulas to simulate real-world processes or phenomena.

Mosaic A composite of aerial photographs or satellite images.

Motif A graphical user interface standard used on UNIX workstations.

Multimedia The simultaneous use of a variety of media for the storage and presentation of documents, including sound, graphics, animation, and hypertext.

NAD 27 (North American Datum of 1927) The datum adopted for earlier national surveying and mapping programs by Canada, the United States, and Mexico. It was based on the Clarke 1866 reference ellipsoid.

NAD 83 (North American Datum of 1983) The new datum that replaces NAD 27 for national surveying and mapping programs in Canada, the United States, and Mexico. NAD 83 is redefined from NAD 27 by using a new Earth-centered reference **GRS 80**. The modification of existing NAD 27 coordinates to NAD 83 coordinates involves transformations, which range from simple shifts through scaling and rotations to distortion modeling and local refinements. The magnitude of the coordinate adjustment ranges from almost 0 to approximately 250 meters, depending on the location.

National Spatial Data Clearinghouse A World Wide Web resource, sponsored by the U.S. Federal Geospatial Data Committee (FGDC), that contains field-level descriptions of digital spatial data in the public domain and other organizations. This descriptive information, known as **metadata**, is collected in a standard format to facilitate query and presentation across multiple participating sites.

Neat line The neat line of a map refers to its frame or border. In geographic data processing, the neat line serves to define the extent of the digital data. It is not used for data analysis.

Network (1) A configuration of two or more computers connected so that they can share functions (i.e., applications and processes) and resources (i.e., data). (2) A system of linear features, e.g., roads, in a geographic database.

Newsgroup A discussion forum on the Internet that allows many-to-many discussions on a specific topic.

Node (1) The end points of an arc in a digital geographic database. (2) The points created after the arcs in a coverage have been broken down at their intersections in topology building. (3) A computer in a communication network.

Node snap The process of forcing the end point of a newly digitized line to close at an existing node automatically when the end point is within a small distance (called a snap distance or tolerance) of the node.

Nominal data Simple data that provide only qualitative differences. There is no ranking, and data are classified without hierarchy. Examples of maps containing this class of data would show the location of different types of soil, vegetation, or employment.

Normalization In relational database design, the process of organizing data to minimize duplication and ensure integrity. Normalization usually involves dividing a database into two or more tables and defining relationships among the tables. The objective is to isolate data so that additions, deletions, and modifications of a field may be made in just one table and then propagated through the rest of the database via the defined relationships.

Object (1) A primitive element that combines properties of procedure and data in object-oriented programming. (2) A geographic feature or phenomenon characterized by a set of attributes in a relational database. (3) A digital representation of a geographic feature or phenomenon in cartography.

Object linking and embedding (OLE) The specifications for object technology developed and used by Microsoft in all its operating system, development tools, and application software packages. Based on the underlying **COM** (Component Object Model), OLE is the foundation for component software. OLE makes it easy to create compound documents consisting of multiple sources of information from different applications, e.g., including a PowerPoint presentation in a Word document. OLE has now been renamed **ActiveX control**.

Object orientation (1) A software development of programming approach that uses one of the number of object-oriented languages (e.g., Ada 95, C++, Eiffel, Smalltalk). (2) A new view of software engineering that encompasses the use of object-oriented methodology not only for programming, but also for requirements analysis, systems design, software testing, and databases. Such a broadened view of object orientation is often referred to by the term *object technologies*.

OCX (OLE Custom Control) See **Object linking and embedding**.

ODBC See **Open Database Connectivity**.

Odessey An early generation GIS developed at the Harvard Graduate School of Design. It was based on the principle of **arc/node topology**.

Open Database Connectivity (ODBC) An **API** standard for Microsoft Windows that supports cross-platform database access.

Open GIS The concept of interoperable GIS that allows the cross-platform use of geographic data and geographic data-processing applications by the adoption of common specifications and standards. See also **Interoperability**.

Open GIS Consortium A not-for-profit membership organization established by the computer industry and GIS user community to promote the idea of **interoperability** of GIS and to coordinate the use of common specifications and standards to achieve such a goal.

OpenGL A three-dimensional graphics language developed by Silicon Graphics. There are two implementations: (1) Microsoft OpenGL, developed to improve the performance of hardware that supports OpenGL standard, and (2) Cosmo OpenGL, a software-only implementation designed for computers that do not have a graphics accelerator.

OpenLook A graphical user interface standard used in UNIX workstations.

Ordinal data Quantitative data that provide the map user with information about differences in rank or hierarchy but not about their class or amount. For example, a map showing populated places classified as either city, town, or village would depict ordinal data. See also **Level of measurement**, **Nominal data**, **Interval data**, and **Ratio data**.

Origin The reference point in a coordinate system at which both the easting and northing (i.e., x and y) values are equal to zero.

Orthophoto map A map generated from assembling and cartographically enhancing aerial photographs that have been geometrically corrected for the effect of relief and photographic tilts.

Overlay A process in geographic data processing that involves registering and combining layers of thematic data to form a new composite layer.

Pan Moving around in the display on a computer. It is also referred to as scroll and is used in conjunction with **zoom** in data browsing.

Parallel A line of constant **latitude**. See also **Longitude** and **Meridian**.

Parallel processing The coordinated processing of a computer program by multiple processors that work on different parts of the program, using their own operating system and memory resources. With the aid of a messaging interface, 200 or more processors can work on the same application. The setup for parallel processing is typically complicated and requires careful thought about how to partition a common database among the processors and how to assign different parts of the application to different processors.

Parcel A tract of land demarcated and registered according to laws governing land administration and property registration.

Physical data modeling The final stage of database design that takes place after logical data modeling. The outcome is the functional and design specification called a systems-specific data dictionary for database creation.

Pixel A term derived from "picture element," which is the smallest unit of resolution in raster data storage and display.

Plug-ins Small computer programs developed to enhance the capability of Web browsers. For example, browsers do not support only raster graphics in certain **bit-map** formats, and plug-ins are required in order to display vector data. Plug-ins are product-specific and are installed separately. They are loaded automatically when the browser program starts. See also **Helper program**.

Point feature A spatial object recorded in a geographic database as a point. A point feature is represented by a pair of (x, y) coordinates, and it may carry one or more attributes (e.g., the depth, water quality index, and ownership of a well).

Point mode of digitizing The method of manual map digitizing that picks up the location of geographic features point by point. This is achieved by pressing the cursor button at each point the operator wants to capture the coordinates. See also **Stream mode of digitizing**.

Polar coordinate The expression of location in geographic space in terms of an angle and a distance. The point B is determined by angle α and distance d.

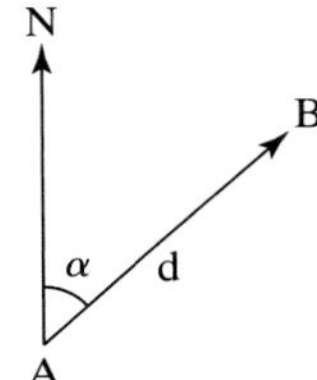

Polygon An object in a map formed by an arc closing to itself or by a multiple of arcs joining together. Polygons are also referred to as areas or **area features**.

Precision (1) The number of digits used to store a numerical value in computers. (2) The number of digits that a measuring device can use to record the numerical values of an observation.

Prime meridian The meridian, or line of longitude, adopted by international agreement in 1884 to be the 0° meridian from which all longitudes worldwide would be calculated. Also called the *Greenwich* or *international meridian*.

Process modeling A systems design technique for organizing and documenting a system's processes including inputs, outputs, and data stores. The most commonly used processing modeling tool is a *data flow diagram*.

Proprietary data format A data format that has been copyrighted by a systems developer. People or organizations wishing to use this format have to obtain permission from and usually pay a royalty to the copyright holder.

Pubic domain Data or applications placed in repositories that can be accessed and used by the general public with no conditions of copyright and patent attached, and usually at no cost other than that incurred in the storage medium.

QA/QC (Quality assurance/quality control) The rules and procedures exercised to ensure that the end products from a development project or contract meet design specifications.

Quad tree A hierarchical data structure based on the recursive decomposition of a given geographic space into square cells.

Quasi-spatial attribute A description of geography, such as a place name and a street address, that is not explicitly related to a georeferencing system and, as a result, cannot be displayed graphically and used for spatial analysis. A quasi-spatial attribute must be address-matched or geocoded before it can be used for cartographic display and geographic analysis. See also **Geocoding** and **Address matching**.

Query An input to the computer that aims to retrieve data from a database. This is done with the aid of a **database management system** (DBMS).

RAD (rapid application development) A sequential software development approach that emphasizes an extremely short development cycle. This is achieved by using **CASE tools** to assist in business and data modeling and **fourth-generation languages** to build applications through the method of **rapid prototyping**.

RAM (random access memory) The part of a computer's memory that may be accessed more rapidly than the regular memory.

Random Without any designed or discernible structure of sequence.

Rapid prototyping An approach of systems development that is characterized by the use of prototyping tools in **fourth-generation languages** to quickly create and test input and output designs, screen menus, and dialogs as well as simple procedures around a computerized **data dictionary**, not a DBMS. Once the designs have been approved, codes may be generated quickly to build the system in the appropriate technical environment.

Raster A data structure that is based on the use of grid cells. See also **Tessellation** and **Vector**.

Ratio data Quantitative data that show the relationship between two values that represent similar information, such as distance. The relationship is expressed as a fraction that is determined by the number of times one value contains the other. For example, the scale of a map is usually expressed as ratio data (1 : 50,000, or one unit of distance on a map represents 50,000 of the same units on the ground). See also **Level of measurement**, **Nominal data**, **Interval data**, and **Ordinal data**.

Reclassification The processes and techniques of recording attributes in one classification scheme into another classification scheme. Since it is possible only to reclassify from classification scheme with a larger number of categories to another one with a smaller number of categories, reclassification is in effect a generalization process for attribute data. See also **Resampling**.

Record A collection of attribute values pertaining to a geographic object. In a **flat file** or a **relational table**, records make up the rows and individual attributes make up the columns. In relational database terminology, a record is formally called a **tuple**.

Rectification The processes and techniques of eliminating the effects of relief and tilts in aerial photographs.

Registration The process of fitting thematic data layers to the base map. See also **Georeferencing**.

Relate A process in relational databases that logically merges two or more databases through the use of keys.

Relational DBMS A database management system based on the relational data model.

Relational Table A table in a relational database is formally called a relation.

Relative location A position that is measured or described with respect to another location, not the origin of the coordinate system. See also **Absolute location**.

Representative fraction The ratio between the measurement on a map and the measurement on the ground, expressed in the form of $1:x$ where x can be, for example, 1000, 1,000,000, 50,000, or 63,360.

Resampling A technique in digital image processing that interpolates pixels in the source data to new locations of transformed pixels, usually coinciding with a georeferenced grid.

Resolution The minimum distance between two objects that can be measured or distinguished by a data capturing or imaging device (e.g., table digitizer and image scanner). See also **Temporal resolution**.

RGB (red, green, and blue) The system of specifying colors by their red, green, and blue saturations.

Riparian buffer A buffer zone that is created along the bank of a waterway or the coastline of a lake.

Rubber sheeting The method of registering one layer of digital data to another by using a set of common points, rather than by coordinates of control points.

Run-length encoding A method of data compression by replacing repeated sequences of values with a single value and a multiplier, e.g., 333333333333 444422222 becomes (12)3(4)4(5)2 when stored as a run-length code.

Sample A subset of the population selected for measurement in a survey.

Sampling (1) The process of determining or selecting features to be used as samples in measurement. (2) The method of survey by measuring samples rather than the whole population.

Scale (1) The ratio between measurements on a map and measurements on the ground, expressed as a representative fraction or a statement. (2) A representative fraction, a statement, or a horizontal bar symbol that is displayed on a map.

Scanning The method of converting hard-copy documents into the digital form using an electronic data capture device known as a scanner. The output from scanning is a dump image, and the process of scanning itself represents only the initial stage of geographic data automation. This means that after a map has been scanned, the resulting image must be georeferenced before it can be used for measurement. It also requires extensive work in order to turn the raster image to vector data.

Schema The "description" of a database within the framework of a data model. This means that a schema provides only the metadata of a particular database; it does not contain any contents of the database populated from the data model.

Scientific visualization (SciVis) The principles and techniques of using advanced computer graphics for the analysis and interpretation of a large volume of numerical data. Also referred to as *visualization for scientific computing* (ViSC).

Screen digitizing The process of acquiring vector graphics by tracing the raster image displayed on the computer screen. This may be done manually or with the aid of computer software. Also commonly referred to as *heads-up* digitizing. See also **Scanning** and **Vectorization**.

Script A synonym for **macro** of a batch file. It is a list of commands that can be executed without user interaction. A *script language* is a simple programming language used to write scripts.

SCSI (Small Computer System Interface) Pronounced as "scuzzy," SCSI is a type of controller card that allows peripheral devices to be hooked up one after the other so that the CPU can address several devices through one controller card.

Seamless geographic database The method of organizing geographic data in such a way that the data physically stored as individual files, known as **tiles**, can be retrieved as if they were in one single file or database.

Search agent A computer program that is capable of using multiple **search engines** to find information on the World Wide Web.

Search engine A computer program that is designed for finding information in the World Wide Web using directories or keywords.

Selective availability (SA) The intentional degradation of the satellite signals in the **Global Positioning System** (GPS) by the U.S. Department of Defense. The purpose is to control the availability of high-precision positioning data for national security purposes.

Sensitivity analysis An analysis that is carried out repeatedly, each time varying the values of different input data sets. The objective is to investigate how changes in the input data will affect the result of the analysis.

Separates (Map Separates) Film positives or negatives used in map production. A USGS topographic map is printed using six separates, one for each color (black, blue, green, red, brown, and purple).

Server A computer that is dedicated to serve the data and application needs of client computers in a network environment.

Shapefiles The file format for graphical data used by ArcView GIS.

Shareware Computer applications that are marketed commercially. They are downloadable from either the developer's Web site or shareware repositories in the World Wide Web and can be used for a predefined trial period. If a user decides to continue using an application after the trial period, a license fee is payable to the developer.

Single precision The storage of numerical values in the computer using the default precision that stores up to 7 significant digits.

Skew Distortion from the true symmetrical form. (1) In remote sensing, skew denotes distortion of a digital image caused by Earth's rotation relative to the movement of the satellite. (2) In table digitizing, skew refers to the lack of orthogonality (i.e., intersecting at exactly 90°) between the *x* and *y* axes.

Sliver polygon An extremely small polygon, usually of elongated shapes, that results from errors in data capture and overlay analysis when identical linear objects fail to register. It is also known as a **spurious polygon**.

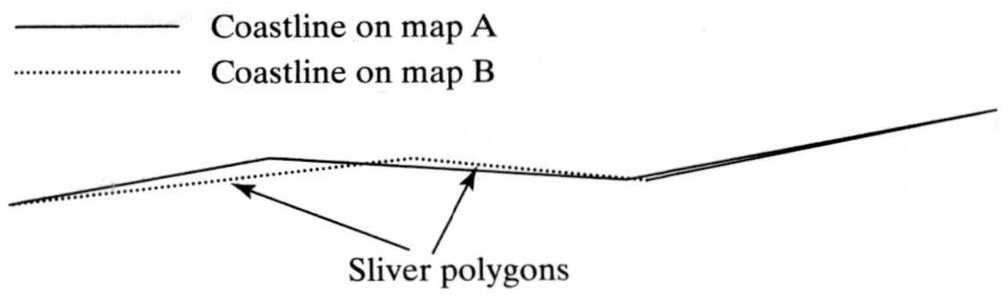

Snapping The process of forcing two or more nodes within a given distance between one another to become a single node.

Soft-copy photogrammetry See **Digital photogrammetry**.

Spaghetti digitizing The method of map digitizing that does not require the operator to follow a particular sequence in order to work. The operator digitizes map features in any order he or she deems suitable. The totally unorganized data will be properly structured later automatically at the topology building phase of work.

Spatial autocorrelation The degree of relationship that exists between two or more spatial data variables (e.g., amount of organic matter in soil, gradient, suitability for agriculture). When one variable changes in one direction (e.g., increases of soil organic matter), another variable also changes in the same direction (e.g., increase in suitability for agriculture), the two variables are said to be positively autocorrelated. If however the changes are in different directions (e.g., an increase in gradient leads to a decrease in suitability for agriculture), the two variables are negatively autocorrelated.

Specifications (specs) A set of detailed instructions used to describe the layout, content, and procedures involved in the production of a map or the creation of a digital database. Specifications may be in text form and contain graphical representations of map symbology, along with measurements, colors, and screen percentages.

Spheroid The standard globe that is obtained by projecting the ellipsoid onto a regular surface closer to the shape of a sphere. The spheroid is used as the model for the determination of common spherical coordinates of latitudes and longitudes. There are now many spheroids in use. ArcInfo, for example, supports over twenty spheroids in its map projection and coordinate transformation function.

SPOT (System Proprietaire pour l'Observation de la Terre) A French satellite system that is capable of capturing stereo images with a 10- to 20-m resolution.

Spurious polygon See **Sliver polygon**.

SQL (structured query language) The standard database query language for relational database systems. See also **Data definition language** and **Data manipulation language**.

Standard deviation A normalized measure of the magnitude of deviation from the mean within a set of measurements.

State plane coordinates A plane coordinate system used in surveying and mapping in the lower 48 states of the United States. Each states has its own central meridian for its coordinate system, which can be the Transverse Mercator or Lambert Conform Conic Projections.

Stream mode of digitizing The method of capturing point coordinates semiautomatically. Once the operator has pressed the cursor button to start digitizing, he or she needs only to move the cursor cross hair over the line feature and the cursor will capture point coordinates at a very small distance. Stream mode of digitizing is most suited for digitizing long line features such as coastlines. However, the number of points captured is always excessively large, and line **generalization** is usually required as a postdigitizing process.

Surface fitting The generation of a mathematical surface to pass through, or close to, a set of existing elevation points.

Systems development life cycle (SDLC) The stages of work involved in the building and implementation of information systems, including design, development, testing, implementation, and maintenance.

T-1 A North American standard that denotes network communication lines capable of transmitting data at a rate of 1.54 Mbit/s (megabits per second). See also **cable modem** and **DS-3**.

Tag A term that is commonly used as a synonym of **label**. See also **Feature identifier**.

Tagging The process of linking attribute data to their associated graphical elements, e.g., giving contour lines their respective numerical elevation values after digitizing. See also **Label placement**.

TCP/IP (transmission control protocol/Internet protocol) Computers in a network need not run on the same operating system, but they do need to communicate with one another using a common language or set of rules called protocol. TCP/IP is the protocol used by the Internet to facilitate the communication among computers connected to the network.

Temporal resolution The time interval between measurements in data collection.

Tessellation The recursive decomposition of a given geographic space into regular (e.g., square cells) or irregular (e.g., irregular triangles) units. **Quad tree** is another common form of tessellation in geographic data processing.

Ticks Control points used for map registration in table digitizing.

TIGER (Topologically Integrated Geographic Encoding and Referencing) A standard digital geographic data format currently used by the U.S. Bureau of the Census for street-level mapping of the United States. See also **GBF**.

Tile A digital storage unit that is the equivalent of a paper map, i.e., digital map files are stored by tiles.

Time stamp A date in a database, collected at the same time when its associated data are collected. Time stamp is important for database transactions which are time- or date sensitive, e.g., the closing date of real estate transaction and the date when a building permit is issued.

TIN (triangulated irregular network) A vector-based data structure for storing terrain information in digital terrain modeling. In a TIN data model, each sample point has an x, y coordinate and a height, or z value. All the points are connected by edges to form a network of nonoverlapping triangles that collectively represent the terrain surface. TIN is also referred to as *irregular triangular mesh*. See also **DEM**.

Topology The spatial relationships of adjacency, connectivity, and containment between geographic features.

Total quality management (TQM) A new management-led and customer-oriented approach to quality assurance in business and government organizations. Unlike conventional **QA/QC** practices that are mainly concerned with product quality, TQM emphasizes performance assessment and improvement on all aspects of the manufacturing of goods as well as the provision of services pertaining to the delivery, use, and servicing of goods. See also **ISO 9000**.

Transverse A class of map projection in which the axis of the map sheet is aligned in a pole-to-pole position rather than along the Equator.

Triangulation (1) The process of subdividing a two-dimensional geographic space into triangular-shaped units in digital terrain modeling. (2) The geodetic surveying process that establishes a network of horizontal control points for mapping purposes. See also **Tessellation** and **TIN**.

Tuple A row in a relational table. See also **Record**.

Uncertainty The lack of confidence in the fitness of use of a particular data set, usually resulting from the presence of an unknown amount of **errors** in the data.

Undershoot A topological error that occurs when an arc that is supposed to intersect with another arc fails to reach that arc. See also **Overshoot**.

Union A topological overlay of two polygon coverages that preserves features that fall within the spatial extent of either input data sets, i.e., all features from both coverages are retained. See also **Intersect** and **Identity**.

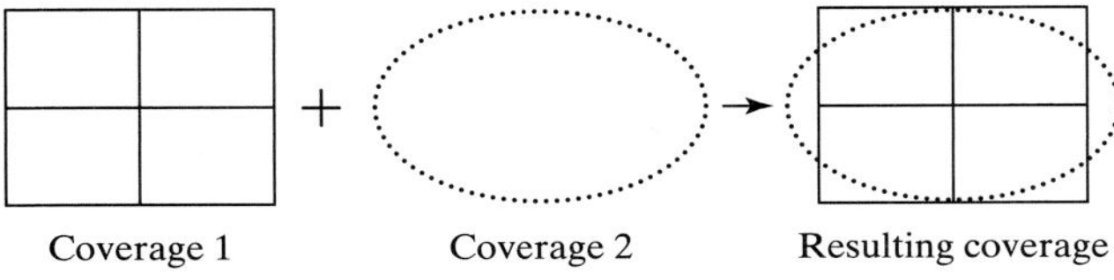

URL (Universal Resource Locator) An address or pointer to a specific bit of information on the Internet.

User requirements analysis Formal investigation of the business information needs of users in an organization, conducted at the conceptual data modeling stage in the systems design process. See also **Joint application development**.

UTM (Universal Transverse Mercator) A coordinate system based on the Transverse Mercator Projection. The UTM grid extends north–south from 84° N to 80° S latitude. It is divided at the 180° meridian eastward into 60 6-degree zones. The UTM has been widely used for military and civilian topographic mapping.

Vector (1) A quantity that has magnitude and direction. (2) A form of computer graphics in which objects are represented as points, lines, and polygons. (3) A coordinate-based spatial data structure in which the data are represented as vectors. See also **Raster** and **Arc/node topology**.

Vectorization The process of obtaining vector graphics from raster images, usually by **screen digitizing**.

Version control The systems management procedure that is carried out to ensure that all copies of a particular software application in use in different computers are of the same release or version.

Vertex Intermediate coordinated points along an arc. See also **Arc** and **Node**.

Viewshed analysis A method of spatial analysis that aims to determine locations visible from a particular point of observation called the *vista point*. The end product is a *viewshed map* depicting all areas that can be seen from one or more vista points.

Virtual reality (VR) A computer-generated three-dimensional imaginary world, also referred to as *virtual environment* and *virtual world*. Special VR software allows the user to enter this imaginary world and interact with the images it contains. Different levels of VR exist, from *partial* to total *immersion*.

Visualization See **Scientific visualization**.

VRML (Virtual Reality Modeling Language) A scene description computer language used to create three-dimensional virtual reality images called *virtual worlds* or *virtual environments*. See also **Virtual reality**.

Web server A server computer dedicated to providing services to client computers in the World Wide Web. See also **Client/server computing**, **Server**, and **World Wide Web**.

Weed tolerance Weeding is the process of reducing the number of vertices defining an **arc** while preserving its general shape. Weed tolerance is the minimum allowable distance between any two vertices along an arc. Weed tolerance is used as a parameter in two processes: (1) In weeding, any vertex within the weed tolerance from the preceding vertex will be removed; (2) in digitizing, any input point within the weed tolerance from the last input point will be disregarded.

Window (1) Portion of the computer screen through which a user controls or performs an interactive application. (2) Portion of a digital map selected for display on the computer screen.

Wizard A utility within an application that helps a user perform a certain task, e.g., a "letter wizard" within a word processing application is designed to lead the user through the steps of writing different types of correspondence.

Word A group of **bytes** treated as a unit of storage in computer memory. The size of a word varies from one computer to another, depending on the CPU. For computers with a 16-bit CPU, a word is 16 bits (i.e., 2 bytes). On large mainframes, a word can be as large as 64 bits (i.e., 8 bytes).

Workspace A directory or folder in which all data files associated with a particular project are stored.

Workstation A single-user, stand-alone micro or mini computer with both high processing power and large storage capacity.

World Wide Web One of the protocols of the Internet that uses a graphical user interface known as a browser to access information resources in a distributed network of computers.

WYSIWYG The acronym standing for "What you see is what you get," pronounced as "wizzy wig." WYSIWYG means that the printout from the computer matches exactly what is displayed on the screen.

X-Windows A public-domain user graphical interface for the UNIX operating system. It is developed and supported by the Massachusetts Institute of Technology (MIT) and is the standard for most UNIX-based shareware on the Internet.

Zoom The process of increasing (zoom in) or decreasing (zoom out) the size of objects displayed on the computer screen. In geographic data viewing, zoom is always used in conjunction with the process of **pan** or scroll.

Photo and Figure Credits

Chapter 2

Figure 2.7: Reproduced from Maling, D. H. (1973), *Coordinate Systems and Map Projections,* Figure 22, p. 33, with permission from Butterworth-Heinemann. Figures 2.10, 2.12, 2.21, and 2.24: Reproduced from Snyder, J. P. (1987), *Map Projections—a Working Manual,* with permission from United States Geological Survey (USGS). Figure 2.15: Reproduced from Seeber, G. (1993), *Satellite Geodesy: Foundations, Methods, and Applications,* with permission from Walter de Gruyter GmbH & Co. KG Publishers. Figure 2.18: Reproduced from Schwarz, C. R. (ed) (1989) *North American Datum of 1983,* with permission from National Oceanic and Atmospheric Administration (NOAA). Figure 2.19, modified from Zilkoski, D. B. (1992), *They Are Now One and the Same,* with permission from National Geodetic Survey (NGS), NOAA. Figure 2.25: Reproduced from Clarke, K. C. (2001), *Getting Started with Geographic Information Systems,* with permission from Dr. Clarke and Prentice Hall. Figures 2.35 and 2.36: Reproduced with permission from United States Census Bureau. Plate 2.1: Courtesy of Leica Geosystems. Plate 2.5: © CNES 2002/ Courtesy SPOT Image Corporation. Plate 2.6: RADARSAT-1 imagery © Canadian Space Agency 1996. Plate 2.7: Reproduced with permission from Trimble Navigation Ltd.

Chapter 4

Figure 4.4: Reproduced with permission from the author (C. P. Lo). Figure 4.9: Reproduced from Berry, B. J. L. and Marble, D. F. (1968), *Spatial Analysis: A Reader in Statistical Geography,* with permission from Prentice Hall.

Chapter 5

Figure 5.26: Reproduced from Fellmann, J., Getis, A., and Getis, J. (1995), *Human Geography: Landscapes of Human Activities,* with permission from McGraw-Hill Companies. Figure 5.35: Reproduced with permission from Martin D. Adamiker. Figure 5.37: Reproduced with permission from the authors (C. P. Lo and B. Faber). Figure 5.39: Reproduced with permission from the authors (J. Gao and C. P. Lo).

Chapter 7

Figure 7.5: Reproduced with permission from ESRI. ESRI Graphical User Interface is the intellectual property of ESRI and is used herein with permission. Copyright © ESRI. All rights reserved. Figure 7.8: Reproduced with permission from Professor Michael Batty, University of College London. Figure 7.9: Reproduced with permission from Professor G. B. Hall, University of Waterloo. Figure 7.11: Reproduced with permission from Webscape. Figure 7.20: Photograph supplied by Aristomat, and reproduced with permission. Figure 7.25: Based on information taken from the National Atlas of Canada website http://atlas.gc.ca © 2001. Reproduced with permission from Natural Resources of Canada. Figure 7.26: Reproduced with permission from the USGS from the National Atlas of the United States website (http://nationalatlas.gov/). Figure 7.27: Reproduced with permission from the Ontario Ministry of Transportation, Canada. Figure 7.29: Reproduced with permission from Pearson Education Inc. Figure 7.30: Reproduced with permission from the Department of Geography, University of California, Santa Barbara.

Chapter 8

Figures 8.1, 8.3, and 8.6: Reproduced from Curran, P. J. (1985), *Principles of Remote Sensing,* with permission from Pearson Education Inc. Figures 8.2, 8.4, 8.7, and 8.8: Reproduced from Lo, C. P. (1986) *Applied Remote Sensing,* with the permission of the author. Figure 8.9: Reproduced with permission from McGraw-Hill Companies. Figures 8.10, 8.12, 8.15, 8.17, 8.22, 8.23, 8.24, and 8.25: Reproduced with permission from American Society for Photogrammetry and Remote Sensing. Figure 8.13: Reproduced with permission from Blackwell Publishing. Figure 8.18: Reproduced from Lillesand, T. M. and Kiefer, R. W. (2000), *Remote Sensing and Image Interpretation,* with permission from John Wiley and Sons, Inc. Figure 8.19: Reproduced from Jensen, J. R. (1996), *Introductory Digital Image Processing,* with permission from Prentice Hall. Plate 8.1: Reproduced with permission from Space Imaging. Plate 8.2: Reproduced with permission from NASA. Plates 8.3, 8.4, and 8.5: Reproduced with permission from NASA Jet Propulsion Laboratory, Caltech for the Shuttle Radar Topography Mission (SRTM).

Chapter 9

Figures 9.4 and 9.5: Reproduced with permission from the United States Geological Survey (USGS). Figure 9.11: Courtesy of Leica Geosystems. Figure 9.13: Courtesy of Intergraph Corporation. Figure 9.17: Reproduced from Burrough, P. A. (1986), *Principles of Geographical Information Systems for Land Resources Assessment*, by permission of Oxford University Press. Figure 9.23: Reproduced with permission from USGS. Figure 9.26: Reproduced with permission from ESRI. ESRI Graphical User Interface is the intellectual property of ESRI and is used herein with permission. Copyright © ESRI. All rights reserved. Figure 9.28: Reproduced with permission from 3D Nature, LLC. Figure 9.29: Reproduced with permission from Martin D. Adamiker. Figure 9.30: Reproduced with permission from Professor Ron Eastman, Clark University. Figure 9.31: Reproduced with permission from 3D Nature, LLC. Figure 9.32: Reproduced with permission from Poitra Visual Communications. Figure 9.33: Reproduced with permission from Stage 22 Imaging.

Chapter 10

Figure 10.1: Reproduced with permission from Springer-Verlag GmbH & Co. KG. Figure 10.2: Reproduced from Cole, J. P. and King, C. A. M. (1968), *Quantitative Geography: Techniques and Theories in Geography*, with permission from John Wiley and Sons Ltd. Figures 10.3 and Table 10.1: Reproduced from Getis, A. and Boots, B. N. (1978), *Models and Spatial Processes: An Approach to the Study of Point, Line and Area Patterns*, with the permission of Cambridge University Press and Professor A. Getis. Figure 10.6: Reproduced from Haynes, K. E. and Fotheringham, A. S. (1984), *Gravity and Spatial Interaction Models*, with the permission of Sage Publications. Figures 10.7 and 10.11 and Table 10.3: Reproduced from Haggett, P. and Chorley, R. J. (1969), *Network Analysis in Geography*, with the permission of Arnold. Figures 10.14, 10.15, and 10.16: Reproduced from Yang, X. and Sathisan, S. K. (1994), "Development of a GIS-based routing model," with the permission of the authors.

Chapter 12

Figure 12.4: Reproduced with permission from Xerox Palo Alto Research Center. Figure 12.7: Reproduced with permission from University of Chicago Library. Figure 12.8: Reproduced with permission from USGS. Figure 12.9: Reproduced with permission from Nishio Satoko, Geographical Survey Institute, Japan. Figure 12.10 (a): Reproduced based on information from the Natural Resources Canada website http://ceonet.cgdi.gc.ca/cs/en/Header.html © 2001 Her Majesty the Queen in Right of Canada with permission of Natural Resources Canada. Figure 12.10 (b): Reproduced with permission from USGS. Figure 12.11: Reproduced with permission from United States Census Bureau. Figure 12.13: Reproduced with permission from Georgia Department of Transportation. Figure 12.14: ESRI Graphical User Interface is the intellectual property of ESRI and is used herein with permission. Copyright © ESRI. All rights reserved. Image map courtesy of City of Ontario, California. Figure 12.15: Reproduced with the permission of Dr. Robert Twiss, REGIS, University of California at Berkeley.

Appendix A

Figure A.1: ESRI Graphical User Interface is the intellectual property of ESRI and is used herein with permission. Copyright © ESRI. All rights reserved. Figure A.2: Reproduced with permission from USGS. Figure A.3: Reproduced based on information from Geomatics Canada website http://www.geocan.nrcan.gc.ca/ © 2001 Her Majesty the Queen in Right of Canada with permission of Natural Resources Canada. Figure A.4: Reproduced based on information of Center for Global and Regional Environmental Research, University of Iowa website http://www.cgrer.uiowa.edu/servers/servers_geodata.html. Figure A.5: ESRI Graphical User Interface is the intellectual property of ESRI and is used herein with permission. Copyright © ESRI. All rights reserved. Figure A.6: Reproduced with permission from Timble. Figure A.7: Reproduced with permission from Intergraph. Figure A.8: ESRI Graphical User Interface is the intellectual property of ESRI and is used herein with permission. Copyright © ESRI. All rights reserved.

INDEX